Siege Warfare and Military Organization in the Successor States (400-800 AD)

History of Warfare

Editors

Kelly DeVries

Loyola University Maryland

John France

University of Wales, Swansea

Michael S. Neiberg

United States Army War College, Pennsylvania

Frederick Schneid

High Point University, North Carolina

VOLUME 91

The titles published in this series are listed at brill.com/hw

Siege Warfare and Military Organization in the Successor States (400-800 AD)

Byzantium, the West and Islam

By

Leif Inge Ree Petersen

BRILL

LEIDEN · BOSTON
2013

Cover illustration: Monumental gate leading to the Church of St. John: Ephesus, Selçuk, Turkey. Constructed with spolia from the Artemision, probably in the reign of Constans II (641-669).
©Photograph by Özel Hansen.

Library of Congress Cataloging-in-Publication Data

Petersen, Leif Inge Ree.
 Siege warfare and military organization in the successor states (400-800 AD) : Byzantium, the West and Islam / by Leif Inge Ree Petersen.
  pages cm. -- (History of warfare ; volume 91)
 Includes bibliographical references and index.
 ISBN 978-90-04-25199-1 (hardback : alk. paper) -- ISBN 978-90-04-25446-6 (e-book) 1. Siege warfare--History--To 1500. 2. Sieges--History--To 1500. 3. Military art and science--History--Medieval, 500-1500. 4. Byzantine Empire--History, Military. 5. Islamic Empire--History, Military. 6. Europe--History, Military--To 1500. I. Title.

 UG443.P47 2013
 355.4'4095609021--dc23

2013019487

This publication has been typeset in the multilingual "Brill" typeface. With over 5,100 characters covering Latin, IPA, Greek, and Cyrillic, this typeface is especially suitable for use in the humanities. For more information, please see www.brill.com/brill-typeface.

ISSN 1385-7827
ISBN 978-90-04-25199-1 (hardback)
ISBN 978-90-04-25446-6 (e-book)

Copyright 2013 by Koninklijke Brill NV, Leiden, The Netherlands.
Koninklijke Brill NV incorporates the imprints Brill, Global Oriental, Hotei Publishing, IDC Publishers and Martinus Nijhoff Publishers.

All rights reserved. No part of this publication may be reproduced, translated, stored in a retrieval system, or transmitted in any form or by any means, electronic, mechanical, photocopying, recording or otherwise, without prior written permission from the publisher.

Authorization to photocopy items for internal or personal use is granted by Koninklijke Brill NV provided that the appropriate fees are paid directly to The Copyright Clearance Center, 222 Rosewood Drive, Suite 910, Danvers, MA 01923, USA.
Fees are subject to change.

This book is printed on acid-free paper.

Printed by Printforce, the Netherlands

για τη Βασίλισσά μου

Bizans güzelime

CONTENTS

Preface and acknowledgements xvii
Conventions of transcription, translations, references and resources   xxi
List of Maps... xxv

PART ONE
MILITARY ORGANIZATION AND SIEGE WARFARE

Introduction 1
  0.1 Historiography 2
      0.1.1 Exceptionalism, Eastern and Western 2
      0.1.2 The (Even More) Exceptional Rise of Islam 6
  0.2 Methodological and Theoretical Approaches 10
      0.2.1 Thick and Thin Descriptions 10
      0.2.2 Co-evolution and Continuity 14
      0.2.3 Technological Diffusion: The Cultural and Institutional Foundations 16
      0.2.4 Construction of Identity and the Diffusion of Knowledge and Technology 18
  0.3 Sources and Limitations 19
      0.3.1 Limitations 19
      0.3.2 General Observations 21
      0.3.3 Greek Sources 23
      0.3.4 Syriac Sources 25
      0.3.5 Arabic and Other Eastern Sources.................. 27
      0.3.6 Latin Sources 27
  0.4 Structure of the Argument 29

Chapter One An Age of Transition: From the Fall of the Roman West to the Early Middle Ages..................................... 34
  1.1 From Late Roman to "Barbarian" Poliorcetics 34
      1.1.1 Late Roman Siege Warfare......................... 35
      1.1.2 The Thin Description: Visigoths and Romans, 376-474 39
      1.1.3 The Thick Description: Huns and Romans, 441-452   46

1.2 From Emergency Measures to New Institutions 49
    1.2.1 The Regular Army in the 5th Century 49
    1.2.2 New Ways of Recruiting Troops 53
    1.2.3 The Military Following (*obsequium*) in East Roman Warfare 56
    1.2.4 The Rise of Private Military Forces in the West 63
    1.2.5 The Origins of Medieval Military Obligations: *Munera Publica* 67
1.3 Where Did All the Romans Go? The Military Implications of Ethnogenesis... 74
    1.3.1 Roman Influences beyond the Frontier.............. 75
    1.3.2 Civil Wars by Proxy and the Involution of the Frontier 78
    1.3.3 The Legions on the Rhine Become Franks 84
    1.3.4 The Last Roman Civil Wars in the West, 496-511..... 90
1.4 Conclusion: From Emergency Measures to Medieval Institutions... 92

Chapter Two   East Rome to Byzantium: Survival and Renewal of Military Institutions 94
2.1 Continuity and Change in East Roman Warfare and Society, 450-800.. 94
    2.1.1 The Strategic Situation of the East Roman Empire: A Brief Overview................................... 95
    2.1.2 The East Roman Army in the 5th and 6th Centuries.. 97
    2.1.3 The "Two Hundred Years' Reform," or before the Thematic System 103
    2.1.4 From Late Roman client management towards a Byzantine Commonwealth 111
2.2 Organization of Siege Warfare I: The Army 115
    2.2.1 Specialist Skills among the Regular Troops........... 115
    2.2.2 Military Engineers 116
    2.2.3 New Developments from the Late 6th Century....... 119
2.3 The Many Faces of East Roman Siege Warfare: The Example of the Anastasian War 123
    2.3.1 The Background to the Anastasian War and the East around 500 124
    2.3.2 Abject Surrender: *Theodosiopolis and *Martyropolis 502 125
    2.3.3 Fierce but Flawed Resistance: *Amida (502-3) 126

        2.3.4 Multiple Approaches: *Constantina-Tella 502-03 130
        2.3.5 Complex Operations against Country and City:
              *Edessa 502-03 131
        2.3.6 Complex Operations and the Fog of War: *Amida
              503-04 133
   2.4 Organization of Siege Warfare II: The Militarization of
       Society 135
        2.4.1 The Construction of Dara, 505-06 135
        2.4.2 Civilian Cooperation in the 6th Century 139
        2.4.3 Use of Civilians in the 7th and 8th Centuries 143
   2.5 Conclusion 147

Chapter Three   The Successor States in the West: Ostrogoths,
Visigoths, and Lombards 149
   3.1 The Ostrogoths, 493-554 149
        3.1.1 Ostrogothic Ethnogenesis 150
        3.1.2 Strategic Situation 152
        3.1.3 Military Organization 153
        3.1.4 Logistics: Adminstration, Labor and Supplies 157
        3.1.5 Ostrogothic Siege Warfare 162
   3.2 The Visigoths in Spain, 508-711 164
        3.2.1 Strategic Situation of the Visigothic Kingdom of
              Toledo 165
        3.2.2 Visigothic Military Organization 166
        3.2.3 Visigothic Siege Warfare 173
   3.3 The Lombards 176
        3.3.1 Ethnogenesis on the Middle Danube 176
        3.3.2 The Lombards in Italy 179
        3.3.3 Lombard Military Organization 183
        3.3.4 Lombard Siege Warfare 188

Chapter Four   The Last Legions on the Rhine: Siege Warfare in the
Frankish Kingdoms 192
   4.1 Frankish Warfare and Military Organization in the 6th
       Century 192
        4.1.1 The Problem of Gregory of Tours 193
        4.1.2 The Establishment of Frankish Burgundy and Wars
              with the Visigoths, 534-89 196
        4.1.3 Austrasian Interventions in Italy, 538-90 200

        4.1.4 Frankish Civil Wars 206
        4.1.5 Military Organization and Siege Logistics in the 6th
              Century 211
  4.2 The 7th Century: Ascendancy of Military Followings and
      Proprietal Warfare 224
        4.2.1 Fredegar, the *Liber Historiae Francorum*, and the
              United Frankish Kingdom 224
        4.2.2 Bishops, Magnates and Monasteries 229
        4.2.3 Late Merovingian Siege Warfare 232
  4.3 The Carolingian Ascendancy in the 8th Century 234
        4.3.1 The Debate Revisited 235
        4.3.2 Size, Composition, and Distribution of Carolingian
              Military Forces 238
        4.3.3 Objectives and Means: Charles Martel and Pippin the
              Short 245
        4.3.4 Organization and Supplies: Charlemagne and Louis   250
  4.4 Conclusion 254

Chapter Five   The Anatomy of a Siege: Tactics and Technology.... 256
  5.1 Siege Strategy and Tactics: Basic Definitions 256
        5.1.1 The Blockade 259
        5.1.2 The Storm 264
  5.2 Siege Tactics 267
        5.2.1 The Basic Approach: Archery and Ladders 268
        5.2.2 Artillery 272
        5.2.3 Siegeworks: Camps and Encircling Fortifications 278
        5.2.4 Siegeworks: Firing Platforms—Mounds and Towers   281
        5.2.5 Wallbreaking: Machines 283
        5.2.6 Wallbreaking: Engineering 286
  5.3 Defensive Responses 289
        5.3.1 Technological Responses 289
        5.3.2 Sorties 290
        5.3.3 Relieving Armies 293
  5.4 Conclusion: Towards a Thick Description of Sieges 295

Chapter Six   The Anatomy of a Siege: Economy, Society and
Culture 299
  6.1 The Topographies of a Siege 299
        6.1.1 Defensive Topography: Infrastructure and Fortifica-
              tions 300

6.1.2 The Topography of Settlement: City and Country in the "Dark Ages" 307
6.1.3 Cultural Topographies: Morale and Ritual under Siege Conditions 316
6.2 The Urban Community at War 326
6.2.1 The Politics of a Siege: Loyalty and Dissension 327
6.2.2 Societies at War: Garrisons and Civilians 336
6.2.3 Specialists at War 343
6.3 Ending the Siege 347
6.3.1 Consequences of Survival 348
6.3.2 Consequences of Fall 350
6.4 Conclusion: Deconstructing, or Reconstructing, Thin Sources 357

Chapter Seven   Appropriation of Military Infrastructure and Knowledge 360
7.1 The Hunnic, Persian and Visigothic Templates 361
7.1.1 Client Integration and State Formation: The Visigoths 361
7.1.2 Inter-state Transfers: The Sassanids 363
7.1.3 Conquest Appropriation: The Huns under Attila 365
7.2 The Balkans, 530-825: From Client Assimilation to Conquest Appropriation and Back 369
7.2.1 Huns as Clients: Utigurs, Kotrigurs, Sabirs and Bulgars in the 6th Century 369
7.2.2 Slavs and Appropriation 371
7.2.3 Avars and Appropriation 378
7.2.4 The Bulgars, 680-825 383
7.2.5 On Northwestern Peripheries: Western Slavs, Saxons and Danes 387
7.3 The Arabs and Islam: Appropriating and Domesticating the Late Antique System 389
7.3.1 Background and Early Events 390
7.3.2 The Sources of Expertise 392
7.4 Conclusion: From Appropriation to Domestication 405

Chapter Eight   Diffusion of the Traction Trebuchet 406
8.1 State of the Question: Historiography and Technical Aspects 406
8.1.1 Historiography of the Traction Trebuchet 406
8.1.2 Technical Aspects of the Traction Trebuchet 409

CONTENTS

    8.2 The Philological Evidence 410
        8.2.1 Generic, Classicizing and Uncertain Terms 411
        8.2.2 *Manganon* and Its Derivatives 413
        8.2.3 Descriptive and Functional Terms 417
    8.3 The Diffusion of the Traction Trebuchet: The Historical Context 419
        8.3.1 The Early Introduction of the Traction Trebuchet: Diffusion or Independent Invention?................ 419
        8.3.2 The Wider Diffusion of the Traction Trebuchet within the Former Roman World 422
    8.4 Epilogue... 425

Appendix One   Reconstructing the Arab invasion of Palestine and Syria from contemporary sources and the importance of Arab siege warfare 430

Appendix Two   'Iyad ibn-Ghanm's invasion of Armenia in 640 and the Arab capacity for storming cities without heavy siege engines   434

Appendix Three   Arab grand strategy, 663-669: 'Abd al-Rahman ibn Khalid's invasion, Saporios' revolt, and the battle for Anatolia   439

PART TWO
CORPUS OBSIDIONUM (CATALOG OF SIEGES)

Conventions Adopted 457
    The 5th Century... 460
    The 6th Century... 484
    The 7th Century... 613
    The 8th Century... 693
    The Early 9th Century 754

Bibliography 765
    Abbreviations of Series, Reference and Collective Works........ 765
    Sources... 768
    Secondary Literature 772
Index Obsidionum 789
General Index 811

PREFACE AND ACKNOWLEDGEMENTS

In 2004, a year before embarking on the project that became this book, I joined my prospective advisor on a research trip to the Byzantine rite monastery Grottaferrata, an hour from Rome. It has a very large collection of Byzantine palimpsests, manuscripts whose original text is mostly illegible to the naked eye. However, the chemical residue of medieval ink still remains in the parchment, and can be examined in ultraviolet or infrared light. The library had acquired the technological apparatus that allowed us to see a fragment of the Chronicle of Symeon the Logothete that had not been read in centuries. As we (mostly Staffan) were making out the individual letters on a computer screen, the monastery librarian, the ancient and most venerable-looking Padre Marco passed by. We had seen him several times, but he never said a word until now, as he stopped and peered over our shoulders at the Greek letters in fluorescent purple on the screen, and said in Greek, *ipomoní, ipomoní* (patience, patience). Although I made light of it then, his brief statement has reverberated in the back of my mind ever since.

I owe a considerable debt to my teachers at the University of Minnesota, especially Bernard Bachrach, who inspired my early research interests and taught me to get work *done* under pressure, an essential element of *ipomoní* as I have since discovered. At the Department of History and Classical Studies at NTNU, where I am now happily ensconced thanks to a very generous grant from the Norwegian Research Council, I have been helped and encouraged by many of my teachers and colleagues throughout my career (which has also involved a brief, but treasured foray into teacher education at HiST). Since these debts have been accumulated over so many years and in so many ways, it would be impossible to mention some and leave out others, but I must thank my former advisor, Staffan Wahlgren, who taught me Greek thirteen years ago, Tore Iversen, who first introduced me to late antique and early medieval history, and my students, who have learnt the Roman battle cry *Deus adiuta Romanis* with great enthusiasm. At the Faculty of Humanities as well as the NTNU and HiST libraries I owe a great "thank you" to the whole staff; as well as to the IT experts at NTNU, who have rescued me many a time from many a difficulty, ranging from silly mistakes pushing buttons with unreasonable consequences, to catastrophic meltdowns on the part of both laptop and operator.

I would further like to thank the members of my committee, Judith Herrin, Jan Retsö and Marek Thue Kretschmer, who provided excellent suggestions for revising the dissertation from which this book derives. I must also thank Claudia Sode and John France for valuable feedback, Sara Elin Roberts for effective copy-editing, and the anonymous reader for suggesting crucial improvements to the manuscript. Brill has expedited the process with remarkable efficiency; for which my gratitude goes to Kelly DeVries, Julian Deahl, and not least Marcella Mulder who has been a most patient editor whenever I found myself outpaced by events; at the last moment I must also thank Tessel Jonquière for seeing the book swiftly and safely through production, and finally Erik Goosmann for producing excellent maps. I remain in debt to Albrecht Berger, Noel Lenski, David Bachrach and Wadad al-Qadi for sending me offprints of in-proof or hard-to-get articles; Michael Featherstone, Charlotte Roueché, Judith Josephson and Marek Jankowiak for excellent responses to my queries; the scholars at the Institut für Byzantinistik und Neogräzistik in Vienna, especially the director Johannes Koder for welcoming me and letting me try out some of my ideas on his students; Ernst Gamillscheg for a delightful example of scholarly generosity, and Mesrob Krikorian, the Armenian Archbishop of Vienna, who assisted by taking me through crucial parts of the Armenian text of Sebeos. The editors of the *Oxford Encyclopedia of Medieval Warfare and Military Technology* gave me an opportunity to publish unused materials on Scandinavian warfare from "mark 1" of my original dissertation. I would particularly like to thank the editor-in-chief, Clifford J. Rogers, for inviting me to the conference *Medieval Frontiers at War* in Cáceres in November 2010, the host Manuel Rojas Gabriel, whose kindness and friendship has been a source of constant encouragement, and all the other scholars, especially Stephen Morillo, for (even more) encouragement and positive feedback. Others are probably not aware of the impact the have had upon me: after I heard James Howard-Johnston's mesmerizing talk at the Late Antique Archaeology conference at Oxford in March 2007, my original dissertation project was completely derailed. Peter Sarris provided a wonderful forum to present my ideas at his seminar and has given much treasured, indeed indelible, feedback. John Haldon deserves a special mention not only for several of the points mentioned above, but also for encouraging this project many years ago, far more than he knows.

On research trips I have met a number of kind and generous villagers, museum curators, archaeologists and historians from Palermo and Bari to Adana and Amorion. I would like to single out Cemal Pulak for having his students guide me through the excavations of Byzantine ships at the Harbor of Theodosius in Constantinople, and Eric Ivison for giving me the tour of the Amorion excavations in the summer of 2007 and suggesting that I might want to focus on a single aspect of warfare. The same thanks go to Roger Scott, who encouraged my ideas but rightly suggested I stop at Charlemagne. As a result, I decided to focus on siege warfare as a case for analyzing military organization in a comparative perspective in the late and post-Roman world—not necessarily because I regard it as the "most important" form of warfare, but because it can be so effectively compared across a range of societies and not least be related to current fruitful debates on society, economy and culture in late antiquity and the early middle ages.

Those who have shown so much *ipomoní* to me also deserve my gratitude: my friends, who still remember me, and my family, from whom I have been far too absent. For many in my closest circle, these last few of years have unfortunately been truly harrowing times. I would therefore like to thank my parents for all their love, concern and support through it all; my father has on top of everything else provided the best role model an aspiring historian could have. While I am sorry that they both had to relive the struggles of their own youth through me, they have made the whole process more bearable because they understood so well.

My final word of gratitude goes to my very own *vasilissa*, Özel, the only true Byzantine that I know. We have battled together against life's hardships since the earliest stages of this project. But between all the difficulties, Özel and I have also had many glorious moments: together we have gazed upon the dome and mosaics of Hagia Sophia, climbed Vesuvius and circled Etna, braved the mighty Taurus and looked down on the Cilician gates, mused over the ruins at the *Forum*, marveled at the golden domes of Norman churches in Sicily, poached figs in the heat of the *Mezzogiorno*, witnessed sunrise from high above the volcanic Cappadocian landscape and sunset over the Euphrates, wandered through the groves around the Areopagos in the shade of the Acropolis, explored the little villages, churches and cathedrals of Italy, cruised the blue Aegean and its islands and coasts in pursuit of Arab and Byzantine fleets, and ascended innumerable rocks and mountains to inspect forts and castles in the sweltering heat all over the Mediterranean. No scholar could ever hope to have a better translator, secretary, photographer and driver, who can navigate through microscopic Italian village streets, rush-hour traffic in downtown Istanbul, and every impassable road from Greece to Anatolia with consummate skill. Indeed, no magnate, bishop or general ever had such a *solacium* or *therapeia* to support and protect them; no army ever had a better guide on the march through the wilderness; no soldier ever had a steadier comrade to cover his right side in battle; nor has any man ever had a better companion. This is dedicated to her.

Çeşme, Chios, Athens and Trondheim
Easter 2013

CONVENTIONS OF TRANSCRIPTION, TRANSLATIONS, REFERENCES AND RESOURCES

Technical terminology and quotes are transcribed according to the usual conventions in each specialist field. Lengthy Syriac quotes are given in *Estrangela* (a few occasions, e.g. at *Caesarea 640, in *Serto*) while the most important terms and phrases are transcribed. Arabic quotes are given in full, mostly in the original form but always with accompanying transcription, which follows the system in *EI²*, with the usual substitution of "q" for "ḳ" and "j" for "dj" etc. For languages of which I have no knowledge, such as Ge'ez, Persian and Armenian, I use transcriptions as they occur in the standard translations or secondary literature, occasionally made consistent with modern transcription practices (e.g. "ā" for "â"). For translation of quotes, I follow existing standard translations to English (or German), which are listed along the original sources in the bibliography; the translations are only referred to in the apparatus when multiple sources are found in an anthology, such as Palmer et al. 1993. In order to ensure consistency in transcription or emphasize certain points, they have been slightly modified according to need. Since I have consulted most of the sources in the original, I have felt free to modify translations somewhat without alerting the reader. This should not present any problem, as translations are never modified unless accompanied by the original language. Significant differences of interpretation are always pointed out and discussed as they occur. If no translation is listed in the bibliography, the translation is my own.

As a rule of thumb, names of people and places are transcribed according to the conventions in each field, but with the following principles: English or other conventional forms are always preferred (e.g. John for Johannes, Florence for Firenze or Florentia, Thessalonica for Thessaloniki or Thessalonike), while Latin, Greek, Syriac or Arabic forms are used according to source language. Locations in the former West Roman Empire are rendered in their modern, local form, while those in the East in their classical form; normally in Latin along the Danube and in the upper Balkans, Greek in the rest of the East. Significant alternative forms from Syriac, Armenian, Arabic, or Coptic are used as they occur in the sources but are cross-referenced in the index and given in *Corpus Obsidionum*, either in parenthesis or in the relevant source quotations. Within these parameters, utility rather than abstract consistency has been the goal.

Uppercase East(ern) and West(ern) consistently refer to the area comprised by the two halves of the Roman Empire, even after the 7th century; lowercase east(ern) and west(ern) simply refer to geographic direction. As long as the West Roman Empire and all of her immediate successors existed, I rather consistently use "East Roman;" after the end of the Western Empire, I vacillate between (East) Roman and Byzantine, although there is no conscientious difference in meaning, except that "Byzantine" becomes slightly more prevalent towards the end of the period under consideration. Similarly, for early Islam I vary between terms based on convention: while "Arab" before c. 630 always refers to the (mostly) Christian client tribes in northern Arabia and the extended Syrian Desert, I later specify "Christian Arabs" as opposed to Muslims or Muslim Arabs. Unless otherwise specified, after 630, "Arab" is used synonymously with "Muslim," although by the end of our period, c. 800, "Muslims" began to include a large number of ethnic identities.

This study builds on a large number of siege anecdotes assembled found in sources from (nearly) the whole of the former Roman Empire ranging across four centuries. In order to reduce repetition, save space and keep the number of references and cross-references in the footnotes manageable, most quotes, references and historiographical or source critical discussions are to be found in the second part of the book, the *Corpus Obsidionum*, or catalog of sieges, abbreviated CO. The references to this in the main text and footnotes of the first part of the book (as well as cross references within the *Corpus*) are marked with an asterisk and year. Thus *Constantinople (717f) will refer the reader to a summary and discussion of the Arab siege of Constantinople, in the chronologically arranged *Corpus*, at 717-18, with references to modern literature and quotes from a number of sources. Conversely, siege entries refer to the main text whenever a siege is discussed at some length there, so that important issues can easily be followed up. Finally, for a few crucial points of fact or terminology a footnote specifies the exact reference if the *Corpus* entry is very extensive. Other conventions are described in the introduction to the *Corpus*.

The scope in time, geography and themes required some knowledge of at least four rather distinct fields with potential for delving into further subspecialties: late Roman/late Antique, early Byzantine, early medieval Western, and early Islamic history, archaeology, epigraphy, papyrology, art history, sigillography and numismatics. While I have read as widely as possible, I must confess increasing ignorance in the order of specializations listed—epigraphy has only been used on a few occasions, while for more remote fields, such as papyrology, I have relied on the work of specialists. I also realized early on that I would have to limit myself to works of synthesis and a few selected sites in order to grapple with developments of which it is only possibly to acquire knowledge through archaeological research, but I intend to return to some specific cases where physical remains might be checked against narrative evidence in some detail. Due to the enormous amount of studies potentially of interest to this topic, then, I have tried to limit the footnotes and bibliography to the most relevant, and, in most cases, recent, scholarly literature, and only attempted some completeness in the field of early medieval military history, which is most pertinent to the problem at hand. The readers will have to decide for themselves whether my efforts in crafting this study have been worthwhile. For all its possible faults, it certainly has been for me.

> Μέγα κακὸν εὖ οἶδ' ὅτι ὁ πόλεμος καὶ πέρα κακῶν
> I know well that war is a great evil, indeed the worst of evils...
> Author of a Byzantine military manual, c. 800 AD

LIST OF MAPS

I	West Roman Empire, c. 400: cities, regions, provinces and clients	xxii
II	The Frankish World	xxiii
III	The Central and Eastern Mediterranean to around 600	xxiv
IV	Italy	xxv
V	The Balkans	xxvi
VI	The Middle East	xxvii

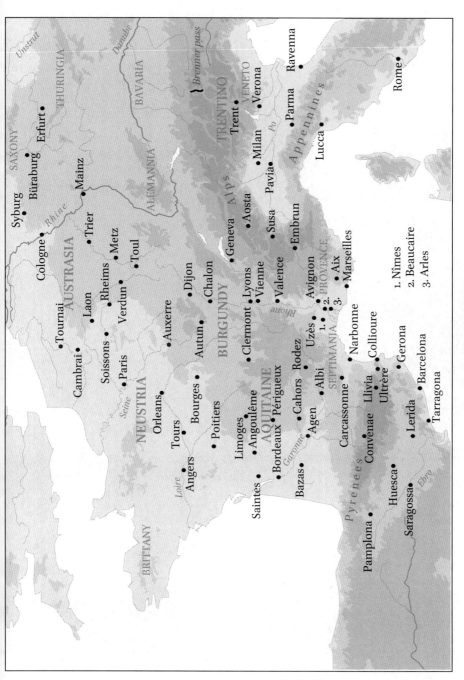

Map II: The Frankish World

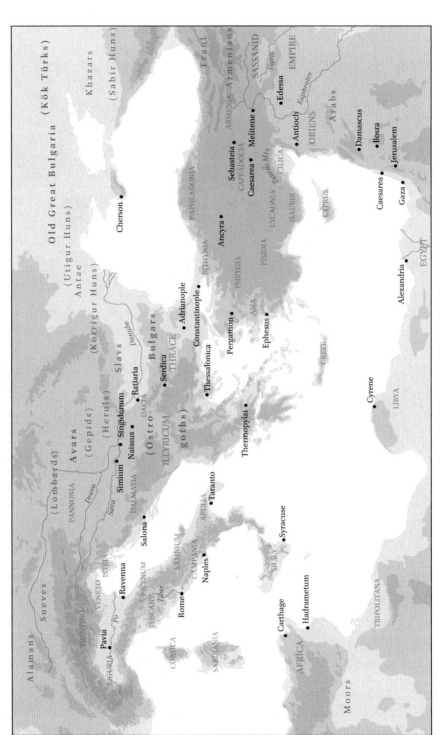

Map III: The Central and Eastern Mediterranean to around 600

MAPS   XXV

Map IV: Italy

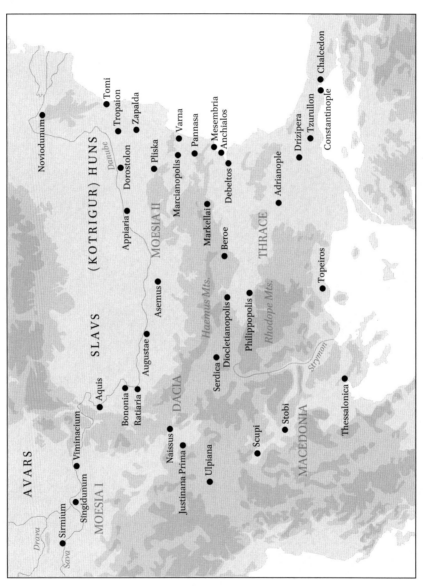

Map V: The Balkans

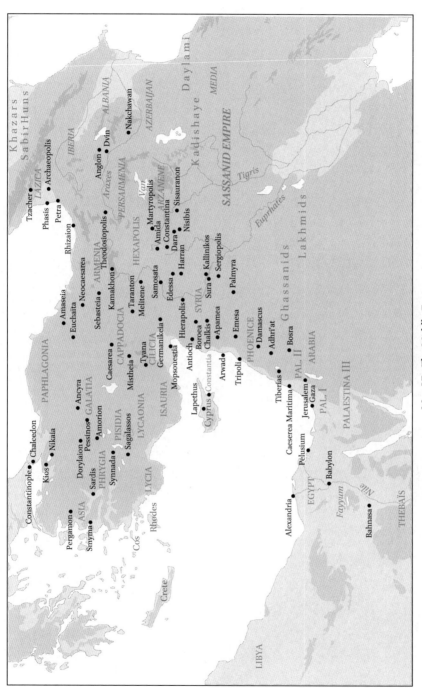

Map VI: The Middle East

PART ONE

MILITARY ORGANIZATION AND SIEGE WARFARE

INTRODUCTION

In a recent review of Guy Halsall's *Warfare and Society in the Barbarian West, 450-900*, which is ostensibly meant to be a new standard in the field, Bryan Ward-Perkins approves of the author's findings, summarizing of early medieval warfare that:

> It remains very difficult to imagine a seventh- or eighth-century army, except as a hairy and ill-equipped horde, or as a Beowulfian band of heroes, and almost impossible to envisage what such an army did when faced with an obstacle such as a walled town[.][1]

Ward-Perkins is being willfully—not to say wonderfully—provocative. He is well aware of the substantial and growing body of scholarship on early medieval warfare that has rendered clichés of hairy barbarians obsolete.[2] But quips aside, he highlights a serious gap in existing historiography. While many scholars recognize the similarity (and to varying extents continuity) between Roman and high medieval siege warfare, this has never been demonstrated in any detail. Hence, a notable number of scholars favor a minimalist interpretation of warfare in which the siege is accorded little importance, especially in the West, arguing that post-Roman society lacked the necessary demographic, economic, organizational and even cultural prerequisites. Continuity of military institutions is unquestioned in the case of the East Roman, or Byzantine Empire, but Byzantine siege warfare has received only sporadic attention, and it has been even more marginal in explaining the Islamic expansion.[3]

This study seeks to examine the organization and practice of siege warfare among the major successor states of the former Roman Empire: The

---

[1] Ward-Perkins 2006, reviewing Halsall 2003. Quote at 524.
[2] Older preoccupations, such as Germanic essentialism, the feudal paradigm, and focus on battlefield history, are found in varying degrees in the standard works of Delbrück 1920/90, Oman 1924, Lot 1946, Verbruggen 1954/97, Ganshof 1968a, Beeler 1971 and Contamine 1984. Their contributions increased in sophistication when dealing with the Carolingians and are still useful on certain issues, but much of their conceptual framework has long been abandoned. Nevertheless, similar clichés still appear, couched in more fashionable anthropological or sociological terms, as will be seen presently.
[3] For Roman and high medieval siege warfare, see *MCuS*, Marsden 1969-71, Kern 1999, Rihll 2007, Rogers 1992; for observations on continuity, see e.g. Bradbury 1992, Bachrach 1994, and Morillo 1994:136ff. For a discussion of existing historiography on late Roman and early medieval siege warfare, see below.

East Roman Empire, the Western successors (Ostrogoths, Visigoths, Lombards and Franks), and conquest states that were established later, or on the fringes of Roman territory, most importantly the early Islamic Caliphate. The point of departure is the united Roman Empire shortly before 400, when it still had a fairly homogenous military infrastructure, to around 800, when the three major and distinct civilizations were well established and the exponential increase of source materials allows for a more nuanced image.[4] While a large undertaking, the geographic and chronological scope avoids significant fault lines in scholarship, as it treats the problematic transition from late antiquity to the early middle ages as a whole, and also comprises societies that all too often are studied in splendid isolation.

0.1 *Historiography*

The relative paucity of sources with reliable, detailed information on warfare in general in this timeframe, compared with Roman and high medieval periods, has made it difficult to study. Existing research is therefore episodic, mostly found in works with a different focus, and often determined by the scholar's view of society as a whole.[5]

0.1.1 *Exceptionalism, Eastern and Western*

To some, the question of siege warfare in the West before the 9th century is nigh irrelevant. They believe that early medieval society lacked the capacity for organizing large-scale military campaigns, necessary to conduct siege warfare, which, apart from naval warfare, has always been one of the most resource-intensive forms of war.[6] In this view, which may be termed

---

[4] While a dedicated study of siege warfare is still a *desideratum* for the 9th-11th centuries, the basic framework for large-scale warfare in an increasingly urbanized and diversified economy is less in doubt. For a distinct Western military history, see e.g. the works by France, Gillmor, B. Bachrach, D. Bachrach and Bowlus listed in the bibliography. For perspectives on military organization in a wider economic and political setting in West and East Frankia, see Nelson 1986, 1992, 1995; Goldberg 2006; D. Bachrach 2012; and Hill 1988 on the role of cities in warfare. For Byzantium and Islam, see below. However, this still leaves a significant gap of well over four centuries, from about 400 to 800 AD. Purton 2009 has a comprehensive survey, but generally follows existing scholarship.

[5] Bradbury 1992: 1-19 is symptomatic: he argues briefly for continuity from Roman siege warfare and interrelationships between successor states, but spends most of his energies on the high and late medieval West. Even Bachrach 1972 hardly identifies any instances of siege warfare among 7th-c. Franks, but see chapter 4.2 below.

[6] On siege warfare in general, see n. 3 above. On naval warfare, see Ahrweiler 1966, Fahmy 1966, Haywood 1999, Rose 2002, Pryor and Jeffreys 2006. Thompson 1958 regards the

minimalist, the objective of warfare was to reinforce socio-political bonds through the redistribution of booty, and since most cities were nearly abandoned (or in ruins) anyway, they were simply not worth the effort of a costly siege.[7] Anything that indicates complex organizational capabilities is played down as atypical or merely lifted from ancient sources,[8] and siege narratives have consequently been ignored or given a minimalistic or even a ritual explanation.[9] That cities and fortifications had any significant role in early medieval warfare is thus hotly contested; so is indeed the very existence of urban life in some areas of post-Roman Europe.[10] Ward-Perkins himself has recently tried to revive the narrative of cataclysmic decline in *The Fall of Rome and the End of Civilization*. While Halsall is somewhat less pessimistic,[11] he has argued at length for a post-Roman society in the core Frankish territories around the Rhine valley (specifically Metz) that differed profoundly from the classical period, despite the obviously Roman origins of most Franks.[12] Due to the importance of the Franks in Western military history, this has implications for his interpretation of warfare as well. Based on this, Halsall argues that early medieval society had fundamentally different priorities that are better explained through anthropological models. This leaves distribution of plunder, webs of gift-giving, competitive display, a peculiar early medieval mentality, and the process of ethnogenesis as the driving forces of early medieval warfare.[13] Other scholars believe that this is insufficient to explain a post-Roman society that was still infused with surviving elements of Roman local administration, infrastructure, ideology and economy, and point out exten-

---

lack of logistical capacity as the most important impediment to "Germanic" siege warfare. Cf. Amt 2002 for an illuminating account from a well-documented pre-modern siege, at Bedford in 1224, which required resources, craftsmen and specialists to be brought in from most of eastern and southern England, and forced the English to abandon a projected expedition to Poitou.

[7] The classical minimalist statement is found in Reuter 1985 and 1990, but see further chapter 4.3.1 below.

[8] See Abels and Morillo 2005 for a recent critique with a response by Bachrach 2007.

[9] See Halsall 2003 and chapter 4.1.1 below for a discussion of his ritualistic interpretations of Merovingian sieges.

[10] For the most recent (and for many areas, largely negative) synthesis of late antique urbanism into the early 7th century, see Liebeschuetz 2001. A full discussion of urban models and siege warfare is found in chapter 6 below.

[11] Ward-Perkins 2005; see Halsall 2008 for a critical review.

[12] Halsall 1995. His 2007 survey establishes the Roman origins of the Franks and is in general more positive on continuities in the Mediterranean basin, but maintains northern Gaul experienced extensive decline. For a discussion of the Roman origins of the Franks and the implications this has for military history, see chapter 1.3 below.

[13] Halsall 2003, but I follow his 2007 approach to ethnogenesis and late Roman politics.

sive evidence for administrative sophistication in the Merovingian and Carolingian periods.[14] They thus prefer to see a high degree of continuity from Roman practices, with early medieval warfare continuing on a similar scale, mode of organization and practice; these may hence be termed maximalists.[15]

The extreme poles of this debate betray startling and apparently irreconcilable differences. While there are many possible nuances in between, most arguments are dominated by an isolationist paradigm: warfare in the West is compartmentalized or treated as essentially different from developments in Byzantine and Islamic societies. Furthermore, varying schemes of periodization mean that scholars of early medieval military history treat their field of study as a *tabula rasa*: while the early to mid 7th century is a standard ending point in late Roman studies, scholars of the early medieval West still tend to treat the Merovingians as a fundamental break with the 5th century, leaving the Carolingians a world apart.[16]

In contrast, recent trends in scholarship on late antiquity demonstrate that in many fields, understanding post-Roman society requires a full evaluation of several societies against a late Roman background, and it is increasingly common to trace developments from the later Empire well into the Carolingian age.[17] Many collective studies and a few comparative ones have thus appeared on the West and Byzantium, or Byzantium and Islam, but few consider all societies due to different scholarly traditions and source languages.[18] There are also substantial debates on military history,

---

[14] Werner 1968 provides a strong argument based on documentary evidence that it is largely ignored or sidestepped by minimalists. Other notable maximalists include B. Bachrach, D. Bachrach and Bowlus. For maximalism in administrative history relevant to military affairs, see e.g. Durliat 1990 and Goffart 2008.

[15] For a brief summary of the two positions, see *EncMedWar* 3: 188-91. Stephen Morillo pointed out to me (personal communication, Dec. 2010) that there need not be a conflict between minimalism and acceptance of continuity from Roman times, but in practice this is often the case, especially in siege warfare, one of the most "Roman" arts of war.

[16] E.g., Lee 2007 treats late antique warfare only as long as the Roman Empire lasts in the West, but extends his treatment of the East to the reign of Herakleios (610-41). For the West, contrast the handful of studies on Merovingian warfare to the much larger amount of research on the Carolingians in chapter 4 below; see e.g. Collins 2010 for his review of McKitterick 2008, calling attention to her disinterest in the Merovingian background as a sign of "chronological chauvinism" (p. 359), as well as a disinterest in neighboring societies typical among many Carolingian specialists: among military historians, we can mention Verbruggen, Ganshof, France etc.

[17] The classical study is of course Brown 1971, which has engendered a vast field of its own represented by journals such as *Journal of Late Antiquity* and *Antiquité Tardive*. See the bibliography in e.g. Wickham 2005.

[18] A representative collection covering all societies is edited by Howard-Johnston and

such as organization, recruitment, military ritual, payment and tactics, where historians of late antiquity are steadily breaking new ground and demonstrating continuity or adaptation of Roman practices as well as interrelationships between Rome's successors.[19] Comparative research on early medieval military history is still rare, however, and only touches on specific points without systematic overall treatment. Bachrach has compared Byzantine and Frankish military lands in a late Roman perspective, and occasionally brought relevant Byzantine and Islamic parallels and connections to attention in his research, but his publications generally focus on the West.[20] Others simply avoid the issue or actively try to dismiss even the possibility of using such comparisons. Halsall mostly ignores neighboring Byzantine and Islamic societies, despite longstanding contacts and conflicts in Spain and Italy throughout most of his period.[21] Otherwise, comparisons remain superficial and are mostly made for rhetorical purposes, such as casting Byzantium and the Islamic world as bureaucratic, oriental others, essentially superior to and different from the backward, rural West.[22]

For present purposes, the problem is compounded by the fact that late Roman siege warfare lacks a comprehensive synthesis,[23] while research on Byzantine and Islamic siege warfare before 800 AD is still rudimentary. The Byzantines possessed both the institutional and technological skill for complex siege warfare throughout the period studied here, yet siege warfare in the Byzantine Empire has received scant attention, especially after

---

Hayward 1999. See further the series *Late Antique Archaeology* (*LAA*) and *Transformation of the Roman World* (*TRW*) with studies on Byzantium and the West and the *Byzantine and Early Islamic Near East* (*BEINE*) series on Byzantium and Islam.

[19] See especially McCormick 1986, Kazanski et al. 1993 and the studies published in *BEINE* 3.

[20] Bachrach 2001b and works listed in the bibliography.

[21] Halsall 2003.

[22] Collins 1998: 174 believes that Byzantine and Islamic bureaucracies were inherently more stable and efficient than the Carolingian "administrative organization" which was unable to keep up a large and complex military establishment. In a rather circular process, Whittow 1993: 181-96 uses Western parallels on the minimalist side (e.g. ignoring Werner 1968) to establish a grand total of the 10th-century Byzantine army of around 30,000, in the process conflating particular field armies with total army strength. Neither engages critically with the historiography they rely on, far less use any pertinent sources.

[23] Nicasie 1998, the most comprehensive survey of the 4th-centruy Roman army, lacks a systematic treatment of siege warfare. Brief sections on the topic are found in Elton 1996: 257-64 and Southern and Dixon 1996 (chapter 8). Individual sieges have received more attention; see individual entries in *Corpus Obsidionum* below (*passim* for references). For brief surveys of the current status of Byzantine warfare and military technology, see Haldon 2008, 2010 and further chapters 2 and 8 below.

the 6th century.[24] Existing research on the army in general and certain aspects of military technology nevertheless provide a very good starting point for an analysis of Byzantine siege warfare. Yet the focus on formal aspects of the army has reinforced the impression that Byzantium was essentially different from the West, and may explain some of the reluctance on the part of "westerners" to engage in comparisons. There is however good reason to believe that the informal and illicit arrangements in the East were far closer to Western realities than has hitherto been recognized, especially until c. 600. Once we take into account private military forces, the logistical role of the late antique estate economy, the effect of ethnicity on military organization and practice (or *vice versa*), and the continuation of late Roman labor obligations through widespread use of civilians in military organization and local defense, the differences between East and West in organizational capabilities and military practices are dramatically lessened. This will obviously have an impact on how Western sources are interpreted, and this approach will be followed here.

0.1.2 *The (Even More) Exceptional Rise of Islam*

The similarities and close connection between East and West was broken up in the 7th century by the Avar, Slav, Persian and Arab invasions. The complexity of the period means that not only is comparative work between East and West increasingly rare, but several new, even more different actors are introduced. While the Avars and Persians are known for their poliorcetic abilities, the Arab conquests are not generally explained in similar terms. According to a systematic survey of sieges in the early conquest phase (634-56) by D.R. Hill, based largely on Arabic sources, most sieges were resolved by surrender.[25] This implies that the Islamic invasions were somehow different from typical late antique wars, but this deceptively simple categorization often camouflages the fact that surrender could take place days, weeks or even months into a hard-fought siege, and that the frequency and reasons for surrender differed little from the Roman-Persian wars in the previous centuries right up to c. 630. Similarly, Lilie's work on the Arab raids and invasions of Anatolia in the following century and a half recognizes Arab strategic goals, but following Arabic (and highly problem-

---

[24] E.g., only six pages are devoted to siege warfare in the survey by Haldon 1999: 183-89, but see the note above.

[25] Hill 1971. Donner 1981, also based on Arabic sources, reaches similar conclusions. See Appendix I-II below for a different interpretation of the conquest of Palestine and early Arab raids.

atic Greek) sources, he inescapably emphasizes the *razzia*-like quality of the fighting.[26] While it is recognized that the Byzantines responded with an extensive program of fort-building and urban relocation,[27] Kennedy argues that: "Siege warfare played a comparatively small part in early Islamic military activity" and: "Until the third [Hijri]/ninth centuries, there are few mentions of sappers or specialists in artillery in Muslim armies (though that does not mean that they did not exist)."[28] Arabists have indeed long recognized that Muslim military engineers were preeminent during the Islamic "Golden Age" from the 9th century onwards, but the focus on Arabic sources has to a large extent obscured the origins of their technological skills. Indeed, Robinson pointed out that we can learn more about siege warfare from two pages of the Greek historian Theophylact Simocatta than from 200 pages of Tabari, and Syriac sources would be even more informative. The existing literature thus portrays the development of technology in the Caliphate in general and siegecraft in particular as a gradual evolution that only came to maturity in the Golden Age, based on the increasingly detailed information found in the Arabic sources towards the turn of the millennium.[29]

In the 1970s, glaring differences between the perspectives on 7th century society offered by the rising field of late antiquity, on the one hand, and traditional Islamic studies, on the other, were becoming increasingly obvious. This led to an extensive revisionist movement, spearheaded by Patricia Crone and Michael Cook. They pointed out that most of the Islamic traditions on the 7th century were in fact codified in the 9th century. While this has long been known, scholars had believed they could identify reliable layers of older materials, and by internal source criticism prune out unreliable narratives and legendary anecdotes. In stark contrast, Crone and Cook argued that in the two to three centuries between events and codification, most anecdotes and narrative complexes had been so pulverized and reshuffled many times over according to later power struggles and theological battles, that they could have little bearing on the period they purportedly described.[30]

---

[26] Lilie 1976. For a similar approach see Haldon and Kennedy 1980; for a different interpretation, Appendix III.
[27] For the transitional period of urbanism in Anatolia see Brandes 1989 and Niewöhner 2007; see also chapter 6.1.1.
[28] Kennedy 2001: 183, 185.
[29] Robinson 2000: 28; cf. the literature surveyed in chapter 8, the main focus of which has been artillery technology.
[30] Crone and Cook 1977.

While Crone and Cook's critique of the whole early Islamic historical tradition may be taken too far, they inspired a new field of studies that has showed that it is virtually impossible to rely exclusively on the Arabic sources for the first Islamic centuries.[31] Consequently more critical approaches have been applied, which have confirmed that in most cases, it is impossible to demonstrate continuity in Islamic historiography earlier than the 720s. However, this does not mean that all is false: while whole narrative complexes can be shown to be spurious or only marginally related to events that are well known from contemporary sources, in other cases there may be startling correspondence between comparatively late Islamic narratives and e.g. contemporary Syriac chronicles. It has in practice become a requirement to use non-Islamic sources in *conjunction* with traditional Islamic sources, while Islamicists have begun to critique and use traditional sources in new ways. The work is still ongoing, as early Islamic historiography is vast and the innumerable anecdotes have to be evaluated one by one. Many of them have indeed been rehabilitated, as it were, while good parameters are being developed for evaluating them on a more systematic scale. Interestingly, battles, the traditional staple of Islamic conquest narratives (and modern histories) have been shown largely to derive from a limited set of literary *topoi*, while sieges show less evidence of this, and more likely derive from genuine, local traditions.[32]

Islamic jurists assembled a mass of information in the form of siege narratives in order to explain local variations in taxation, since the canonical texts and legal literature imposed tax levels according to whether a city was captured by treaty ($ṣulḥ^{an}$) or by force ($'anwat^{an}$). The precise legal categories used for taxation purposes were the result of centuries of scholarly debate and systematization. Thanks to papyrus finds from Egypt dating to the first century after the conquest, it has been shown that most of the legal categories employed by 9th- and 10th-century jurists simply did

---

[31] For an example of excessive revisionism, see Nevo and Koren 2003. Although their analysis of proto-Islamic theology based on contemporary Arabic inscriptions is interesting, it is undermined by their theory of a deliberate East Roman turnover of the eastern provinces to Arab federates, which is demonstrably nonsense; cf. chapter 7.3, the campaigns analyzed in Appendix I-III below and next note.

[32] For correspondence between Syriac and Arabic sources, see Robinson 2004; for source critical studies evaluating Islamic against non-Islamic sources see Conrad 1992, Hoyland 1997, Retsö 2003; for internal criticism of traditional Islamic scholarship, Donner 1981; Noth's 1994 critique of early Islamic narratives does not find as systematic use of *topoi* in siege narratives as in other types of narrative. The most important recent contribution is Howard-Johnston 2010, who vindicates much traditional Islamic source material by systematically and meticulously comparing it with non-Islamic sources contemporary with events.

not exist at the time of conquest, or had a different meaning. Instead, it has been found that taxation indeed originated with local arrangements, but these were far more complex and very different from the jurists' schematization. Since Hill's analysis of the early Islamic conquests hinges on these precise legal categories, it has more bearing on how legal scholars utilized their own history than 7th-century events. However, when the narratives themselves are read without this later imposition, *qua* narratives, and with attention to episodic detail (below), the sieges of the Arab conquests become more easily comparable to Late Roman and Persian practices. Hence the narratives so laboriously collected have been used here.[33]

The result of these advances is also an awareness of how small the first generation elites were, and that much of what has been termed "Islamic" often involved the pre-conquest majority well into the 9th century. Thus, recent focus on e.g. the client class and the slow rate of conversion has made it clear that many aspects of early Islamic civilization continued and adapted earlier Roman and Persian institutions and practices.[34] This conclusion is borne out by investigations into early Islamic technology and craftsmanship, which was dominated by clients and *dhimmis* (non-Muslim subjects) well into the second century of Islam.[35] Papyri and Christian Arabic sources prove that Islamic naval warfare was largely organized by clients while the fleets were built and manned by Christians.[36] While Arabic sources often refer to Persians in the early Islamic armies, they only occasionally mention Persian military engineers, who are usually credited with introducing technological skills to the Arabs. However, Syriac (and some other Christian) sources reveal that the majority of military engineers and craftsmen used by land-based early Islamic armies were in fact Syrian Christians who were pressed into service, apparently *en masse* as Coptic craftsmen were in Egypt. Hence, early Islamic warfare to a large extent was based on and preserved East Roman and Persian military organization.[37]

---

[33] See e.g. Frantz-Murphy 1991 and 2007 on a comparison between later legal abstractions and contemporary 7th-century papyrus documents.

[34] Clients, Arabic *mawlā* (pl. *mawālī*), were converts but had second-rate status as affiliates of Arab tribes until the Abbasid revolution. Many of these insights are collected in the volumes of *FCIW*. See further the collections edited by Sijpesteijn and Sundelin 2004, Seijpesteijn 2007, and *NCHI* 1.

[35] See esp. Judah 1989 and the other articles collected in *FCIW* 12 (ed. Morony 2003) for use of Islamic sources; Morony 1984 for a combination of Syriac and Arabic sources on the social history of early Islamic Iraq.

[36] See chapter 7.3.

[37] Chapters 2 and 7.3 below. A sharpened version of this argument will be published in

0.2 Methodological and Theoretical Approaches

Once it is established that Eastern and Western societies shared common ground, it becomes essential to use new insights from late antiquity to explain phenomena that appear obscure in Western and Islamic sources, and identify which sources may be used for interpretative purposes. Peter Sarris has used the detailed papyrus material from Egypt to build up a coherent image of estate management in the 6th century, identifying the bipartite estate, a combination of demesne land directly administered by the lord and dependant villages owing rents or labor, as the predominant mode of organization among great landowners. While this was believed to be an innovation in 8th-century Francia, Sarris has applied his model to scantier evidence in other provinces to discover a far earlier and more widespread distribution.[38] Walter Goffart has similarly used the wealth of late Roman legal and administrative material to demonstrate how the various successor states adopted and adapted what were essentially Roman techniques of administration to extract labor, military service and wealth from their subjects.[39] Similar approaches, including anthropological models, are used throughout chapters 1-4.

0.2.1 Thick and Thin Descriptions

Since this study is based on the analysis of surviving narratives, anthropological tools are very useful in addition to the approaches of Sarris and Goffart. Clifford Geertz introduced the terms "thick" and "thin" descriptions to critically analyze (and less relevant here, produce) ethnographic descriptions.[40] The basic premise is that apparently straightforward, "thin" descriptions can fundamentally distort our understanding if we lack the necessary context, especially if the cultural differences are great: observers need background information, or a "thick description" to make sense of a certain behavior. The premise may be explained with a vignette used by the philosopher Gilbert Ryle, who inspired Geertz' model, in his lecture "What is *le Penseur* doing?"

> Two boys fairly swiftly contract the eyelids of their right eyes. In the first boy this is only an involuntary twitch; but the other is winking conspirato-

---

the proceedings of the conference "Medieval Frontiers at War" held at Cáceres in November 2010 (Petersen, forthcoming).

[38] Sarris 2004, 2006, but note the different interpretation in Wickham 2005, 2009.
[39] Goffart 1972, 1980, 1982.
[40] Geertz 1973.

rially to an accomplice. At the lowest or the thinnest level of description the two contractions of the eyelids may be exactly alike. From a cinematograph-film of the two faces there might be no telling which contraction, if either, was a wink, or which, if either, were a mere twitch. Yet there remains the immense but unphotographable difference between a twitch and a wink. For to wink is to try to signal to someone in particular, without the cognisance of others, a definite message according to an already understood code. It has very complex success-*versus*-failure conditions. The wink is a failure if its intended recipient does not see it; or sees it but does not know or forgets the code; or misconstrues it; or disobeys or disbelieves it; or if any one else spots it. A mere twitch, on the other hand, is neither a failure nor a success; it has no intended recipient; it is not meant to be unwitnessed by anybody; it carries no message. It may be a symptom but it is not a signal. The winker could not *not* know that he was winking; but the victim of the twitch might be quite unaware of his twitch. The winker can tell what he was trying to do; the twitcher will deny that he was trying to do anything.

Our equivalents of the "cinematograph-film" are the narrative sources. "Thin" descriptions from our period are often anecdotal and incidental in nature: they may be *episodic*, i.e. focus on certain episodes but leave out the larger framework, or *schematic*, as they telescope complex sets of events down to the level of a simple phrase or sets of phrases. Neither type allows us to understand the complexity of events, nor the complex of institutions that underlie events, which scholars then attempt to reconstruct from other sources. This problem is all too typical for most of the sources available in this period: for much of the 5th, 7th and early 8th centuries sources are extremely brief, while others tend to focus on specific episodes for reasons of genre and rhetoric. The methodological problem has been how to contextualize these "thin" notices.

When looking at early medieval sources, especially as obscure as those dealing with warfare in general and sieges in particular, the meaning often hinges on a single word. Verbs like *obsidebant* present the type of problem typical in the historiographical debate noted above, and are typical for schematic sources that only present one or two "nuggets" of information. While all can agree on the basic lexical meaning, "they were besieging," the semantics and pragmatics are more difficult: was it a blockade or a storm, or did sieges actually involve machinery, engineering, fighting on the walls? The answer of course lies in attempting to use other sources with the same or similar terminology that go into further detail, and extrapolate what was normal practice when people *obsidebant*. Similarly, the deceptively simple [*de*]*vastavit*, "he devastated" may in fact be very complicated. The word should, according to a simple lexical reading, mean no more than a raid that caused an unspecified, but noteworthy amount of destruction (the twitch). However, whenever we can identify the context in which a "thin" devastation takes place from other sources or incidental information in the same source, it surprisingly often involves the following surrender of a city, or later on, further military conflict that may end up in a formal siege. The tendency has been to take such a phrase at face value, but an evaluation of alternative sources and analogous situations demonstrates that raiding was often deliberately planned as a prelude to siege warfare (the conspiratory wink). Since raiding could have a devastating impact on the livelihood of virtually everybody within miles, the raid and the ultimate outcome may very well be the only aspect of a conflict that is recorded, especially if a source has been excerpted or epitomized. However, what often drops out in the process are the months of fighting, engineering, negotiation, sally, attempted storms, and so on; only schematic information, such as the place name and perhaps an action verb, remains.

Two concrete examples may illustrate my point. It is difficult to demonstrate Frankish besieging skills around 500 in detail from the few notices found in Gregory of Tours, our main source. Hence scholars' conceptual framework (Germanism, ghost city, continuity) "fill in" the thick description. Indeed, historians focus on the Frankish victory at the battle of Vouillé in 507 to explain their conquest of Visigothic Aquitaine, and rapidly pass over the ensuing campaign, which Gregory describes more or less as a march through a series of cities merely to establish Frankish authority. In most cases, nothing is mentioned other than the Franks passing through, although *Toulouse was apparently burnt, lots of booty captured, and the walls of *Angoulême spontaneously collapsed (in the fashion of Jericho, no doubt).[41] It is clear that the surviving two-line description of the campaign does not do justice to events. Even if we do allow ourselves to imagine Ward-Perkins' "hairy barbarians" roaming around the countryside at this early date, standing below city walls eating cattle by the head, bellowing at the locals and scaring them to hand over tribute, while the Visigoths flee head over heels across the Pyrenees, all the specific details, such as the routes of march, methods of pillaging, negotiations, hauling of booty, ransoming of prisoners, flight of the Visigoths and their Roman supporters

---

[41] See the entries in *Corpus Obsidionum* [hence cited as CO; entries marked in the main text with * and date] following the *Vouillé campaign 507 and the detailed analysis in chapter 1.3.4 below. Halsall 2003: 217 n. 9 (at 293) takes Gregory's account of *Angoulême quite literally, using it as evidence for the poor state of the walls due to lack of maintenance in the post-Roman period.

could all merit detailed description by a contemporary. Gregory is simply a very bad source and cannot be used to study the nature of warfare in this case. However, a large amount of information survives on the siege of the former Roman capital of Gaul, *Arles, in 508. Not even mentioned by Gregory, it is attested in a series of contemporary or near-contemporary sources composed by people involved in events.[42] The bishop Caesarius of Arles was present during the siege, and his *vita*, composed by his local supporters shortly after he died in 542, describe the depredations of the Franks and their Burgundian allies, who demolished extramural buildings to provide materials for their siegeworks and blockaded the city and river below. A Gothic garrison fought alongside the civilian population; the latter held particular stretches of wall according to existing social organization (the Jews are singled out for holding one such stretch, which they were accused of being about to betray), and were expected to keep the enemy off by throwing missiles. This they did successfully, as there was a wide no man's land outside that was patrolled by guards who ventured outside the walls. To complicate matters, the bishop's political allegiance was dubious, as he was accused of colluding with the enemy, and while under house arrest, it transpires that the garrison of professional soldiers was so large that it had to be billeted throughout his palace—one of them even had the temerity to lie on the bishop's bed, but was of course struck down by God for this. After the siege was raised with the assistance of an Ostrogothic relieving army, the bishop proceeded to ransom large numbers of prisoners, who were part of his household for years afterwards. The Ostrogothic administrative documents composed by Cassiodorus largely corroborate this image: the garrison was held strictly accountable for their conduct, it had in fact cooperated well with the civilians, who were commended for their bravery and loyalty to the Ostrogoths, and the Ostrogothic state remitted taxes to the citizens due to the cost of wartime damages, especially to the city walls that needed to be repaired after the siege. Clearly some barbarians were not yet too hairy and still knew what to do when faced with such an obstacle.

The additional sources are of course episodic—they relate whatever the author thought relevant for specific purposes, but put together, they present a composite image that seems far more convincing (and understandable) than the simple phrase king X captured/raided/marched to city Y of Gregory, and warn us that for the most part, we cannot understand what

---

[42] For references, quotes and discussion, see the relevant entry in CO and cf. the preceding note.

is going on when the source is too brief. Instead, we must rely on long string of incidental details, told to justify or glorify the saint Caesarius and the administrative records of Cassiodorus to find revealing sidelights on social structures, military procedures, problems of garrisoning and internal political conflict is a type of conflict. The image gleaned from these episodic sources very different from the schematic one offered by Gregory, but closely reflects common issues in siege warfare in strictly contemporary East Roman sources on the Anastasian war. What they leave out of technical details, such as the use of engines, methods of fighting, or supporting infrastructure, may again be pieced together from other cases where these are mentioned incidentally, and held up against the common framework identified in the first four chapters.

At the other end of the geographical and chronological spectrum, some two and a half centuries later, an Abbasid army besieged the East Roman border fort of *Kamakhon (766). Both traditional sources used for the event, the Greek Theophanes and the Arabic Tabari, only briefly state that there was a siege. If taken with the scores of similar events recorded in the same sources, we could conclude that blockades of or surprise attacks on fortifications were a common feature during the Islamic summer raids against Roman territory (as is the conclusion of Lilie, Haldon and Kennedy), but little could prepare us for the mass of details found in the Syriac chronicle composed by a monk at Zuqnin, who lived near the events and spoke with many of the participants. This is full of details on the siege itself, the engines used, people involved, problems of logistics, morale, and the great complexity of operations around the siege itself, including raids to shield the besieging force, maneuvers, marching in reinforcements, and battles with counter-raiders. Any of the features might have been emphasized by a later chronicler who only had limited space at his disposal.[43] Both these anecdotes warn us that there is always very much more to the brief allusions and mentions found in the standard chronicles and histories.

0.2.2 Co-evolution and Continuity

Cultures and states that interact militarily affect each other profoundly, especially when contact draws out over centuries.[44] Steppe nomads, such

---

[43] See chapter 5.4 below.
[44] For the concept of coevolution, see W.R. Thompson 2006, refining the theory of paradigmatic armies proposed by Lynn 2006.

as the Sarmatians, Alans and Huns had a profound effect on the societies with which they were in contact, from Rome (later Byzantium) to Persia and China.[45] Roman and Byzantine cavalry tactics, equipment, and even clothing were heavily influenced by the steppes. The Roman Empire fought innumerable wars with Persia, to the extent that their military practices became very similar; by the age of the Sassanid dynasty (224-651 AD), there is little to distinguish their technological and organizational capacities.[46] A similar mutual adaptation is observable between Byzantium and the Arabs—while the Arabs soon adopted Byzantine-style border fortifications, the Byzantines adapted to Arab raiding tactics. Mutual influences between Byzantium, the Islamic world and the Christian West are also known for the crusading period, ranging from equipment, tactics, and organization to cultural perceptions.[47] The dynamic behind it is a simple Darwinian rationale: Any culture that fails to appropriate an enemy's advantageous military practices (or adapt countermeasures to them somehow) will by necessity suffer from heavy military defeats, which either lead to adaptation in order to counter the challenge, or conquest if not. The theoretical concept is called "co-evolution," meaning that societies in military contact will often develop at roughly the same pace.[48] Influences can be particularly strong in cases of outright conquest, since the conquerors tend to absorb functioning elements into their own structures. In contrast to the Darwinian rationale, despite the breakup or end of polity, we can very often observe continuity of complex military systems. This is because the failure is primarily political, and the "conquest" actually involves a large number of those belonging to the old regime going over to the new. This explains why significant features of the Sassanid Persian military system survived in Iran under local elites during Umayyad (661-750) and Abbasid (750-945) rule, forming part of the larger Islamic war machine. Persian dynasties re-emerged with the breakup of the Abbasid caliphate as leaders of state-bearing, Persian-led armies in Iraq, Iran and Afghanistan under the Buyids, Saffarids, and Ghaznavids in the 10th and 11th centuries.[49] Many

---

[45] Bivar 1972 and Nickel 2002 for brief introductions. The literature on this field is vast; see e.g. *CMAA*.

[46] Börm 2007: 171 regards the two as "ebenbürtig." See further below, chapter 2 *passim* and 7.1.2.

[47] Baladhuri's great legalistic composition on the status of conquered territories also include details on a number of fortifications set up in the 8th and 9th centuries against the Byzantines. For perspectives on Roman-Arabic and Crusading warfare, see Kennedy and Haldon 1980; McGeer 1995; Haldon 1999; France 1994; Smail 1995.

[48] W.R. Thompson 2006.

[49] Bosworth 1963, 1966, 1968, 1978; Madelung 1969.

of the regions or ethnic groups that supported these dynasties had been prominent in Sassanid military organization. Similar survivals of complex Chinese systems have been observed from the Han to the Tang dynasties,[50] while the early Roman Republic systematically adopted Greek, Hellenistic, Italic, Celtic and Carthaginian practices.[51]

0.2.3 *Technological Diffusion: The Cultural and Institutional Foundations*

The theory of diffusion of technology developed by Everett Rogers is based on the idea of social carriers: technological development and diffusion is not necessarily propelled by its inherent utility, it needs someone with a certain social and/or political prestige to introduce it.[52] We therefore have examples of complex inventions and technologies that have become fundamental to the development of the modern world, but first appeared in recognizable form in antiquity or the middle ages, when they failed to gain widespread use. To mention only two famous cases, we have the invention of steam power, which was only used for tricks and gadgets in antiquity, and the *tarsh*, an early Islamic printing technology that was only used for printing incantations and spells on amulets. In both cases, the cultural and economic environment was not conducive towards wider application—labor-saving devices were not an issue in the ancient slave economy, while high society in the Islamic world frowned upon the status and activities of *tarsh*-printers, and were more than happy with the large class of educated scribes writing on paper, newly introduced by the Abbasid Caliphs—who must be regarded as social carriers *par excellence*—for use in their chanceries.[53]

Due to the inherent pressures of military competition, as we saw, this is less of an issue with military technology. Political and military leaders were rarely restrained by cultural factors when adapting new practices and technology. Furthermore, the focus on social carriers disregards the institutional foundation for knowledge preservation and dissemination.

---

[50] Graff 2002.

[51] Any good introduction to the Roman Republic will include a survey of Roman absorption of and adaptation to enemy fighting styles. For example Scullard 1982: 345-48 has a brief survey, while the more popularizing Warry 1980 and Connolly 1981 have excellent illustrations of these developments as well as of Rome's enemies.

[52] Rogers 2003; Edqvist and Edqvist 1979. For theoretical approaches relevant to technology and its institutions in late antiquity, see the essays *LAA* 4.

[53] For the steam engine in antiquity, see Landels 2000: 28ff; he is however critical of the current social explanation, p. 227 (with references to other views). For the *tarsh*, see the classic study by Bulliet 1987.

Particular sets of practices and technologies often come in packages, requiring complex institutions to maintain them. Too often, only *practitioners* (i.e., those who actually practiced the required trade) were introduced, but without regard for the system that produced them. Thus even when social carriers facilitated the successful integration of a large number of practitioners, either as prisoners, conscripts, slaves, clients or allies, their ventures might fail within a generation or so. The reason is that their knowledge has been produced over generations by self-perpetuating socioeconomic systems, such as the ancient/late antique city, guilds, military units, aristocratic households, and so on. If people are taken out of a self-perpetuating institutional context and no viable alternative has been created, they are unlikely to transfer complex knowledge for more than another generation. This was the case with many of the great nomadic empires, such as the Huns and Avars—they tried to conquer Roman territory and thus gain control over their institutions, but often resorted to deporting people *en masse* as virtual slaves. The "Germanic" invaders, in contrast, systematically sought integration into Roman society through formal recognition as officers and soldiers in order to gain access to the Roman logistical system.

The alternative was to slowly adopt and develop institutions of their own. Successful conquerors, such as the early Romans, tried to preserve and harness the abilities of conquered peoples by granting them political rights and rewards more or less according to their acceptance of Roman dominance. The Arabs similarly solved this problem by creating a conscript/client system where people retained their old cultural attachments and social organization when they were conscripted into labor service; others were semi-integrated into the social fabric of the conquerors as clients, almost equals to their masters but still maintaining important features of their pre-Islamic identity. In both cases, this created social tensions in the long run that were only resolved by conflict and integration; in the former case, the Social Wars of the late Republic followed by a rapid extension of citizenship; in the latter, by the Abbasid revolution and subsequent full-scale islamization of society.

0.2.4 Construction of Identity and the Diffusion of Knowledge and Technology

This brings us to the final theoretical approach, construction of identity.[54] This is another anthropological approach with much promise for the early middle ages. Ethnic identity, in particular, was extremely fluid in this period, and both "Arabs" and "Germanic peoples" were socio-political constructions caused by massive upheavals. Many scholars used to treat ethnicity as an essential characteristic, obscuring fundamental continuities in people and practices; a few scholars still hold similar ideas.[55] However, Roman identity, against which various barbarian, tribal ethnicities have been contrasted, both in ancient historiography and modern scholarship, was essentially a *political* identity.[56] In the late empire, a Roman was whoever was loyal to the Roman cause, no matter his ethnic, geographic or linguistic background. This meant that disloyalty put "Romans" beyond the pale, while "barbarians" from beyond the borders were commended for spending their whole lives in service dedicated to the Roman state. The problem was particularly evident during civil wars, when groups and individuals had to choose whom to support. Perfectly good "Romans" by culture, self-identification, or previous political alignment were, by virtue of temporary political divisions within the Roman body politic, permanently assigned to one or another barbarian group, who, acting as good and faithful clients, actually had to support the side that provided patronage. Due to the failure of the central Roman authority in the 5th century, many thus reassigned retained their new identities, although they might not have differed at all from friends, relatives, neighbors and peers who made other political choices.[57]

---

[54] See the classic studies edited by Barth 1969; this is used by Heather 1991 and Amory 1997 on the early medieval period, especially the various Gothic groups, but with somewhat diverging conclusions. See also chapter 1.3.2 and 3.

[55] See e.g. the classic treatment of "Germanic" warfare by Thompson 1958. Ward-Perkins 2005 bases much of his argument on the "otherness" of the invaders.

[56] See the collection edited by Mitchell and Greatrex 2000, especially the editors' own contributions; further on Roman identity, Chrysos 1988.

[57] The Balkans military culture in the 4th, 5th and 6th centuries is a case in point, as the Visigoths, Ostrogoths and Lombards were formed from groups that could imperceptibly have been integrated into the Roman state given different political circumstances. Instead, due to opportunities, choices, and contingent developments mostly caused by Roman elites, these client groups "flaked off" as independent units, who had to identify themselves in opposition to the Roman state. With them, they brought a large number of Roman provincials. See chapter 3 for this.

The result became the search for a usable past, often directly modeled on the ancient ethnography, which was celebrated and used to distinguish one's own group from others.[58] While labels assigned by ancient authors have long since ossified, the contents they delineate may constantly change.[59] What often happens is that a group label based on political allegiance creates the defining signposts (diacritic markers) of a new group. Although the "signposts" may be similar to those familiar from classical ethnography, the "content" they delineate has changed radically.[60] While ancient tribal labels were a common and important ethnographical *topos*, there is no reason to assume continuity of social praxis and organization: they were hybrids of people with diverse origins, some of whom were born outside of the empire, but all of which operated solidly within a world shaped by Rome and inherited her knowledge regardless of what labels were applied to them.[61]

0.3 Sources and Limitations

For those familiar with the fields of late Roman, Byzantine, and early medieval Western history, the sources used here should be quite well known and have been critically analyzed in the historiography cited. What follows, therefore, is a brief survey of the narrative sources used in light of the historiographical and methodological issues noted above, and the self-imposed limitations of this study. For documentary and other sources, see the literature referred to *passim*.

0.3.1 Limitations

Due to the large geographical and chronological span of this study, some limitations have been imposed on sources consulted. The main focus is, firstly, on the identification and analysis of siege narratives, which have been catalogued almost exhaustively throughout the period for the East

---

[58] This is the approach of e.g. Amory 1997 and Halsall 2007. For modern parallels observed in "real time," see Keesing 1989 and Friedman 1992. A useful theoretical introduction is Anderson 1983, whose approach is based on the emergence of modern nationalism. A similarly oriented theoretical work would be illuminating for this period.

[59] See especially Barth 1969a, b and c, who introduced the phrase "diacritic markers" for ethnicity.

[60] Halsall 2007.

[61] For this view, see Goffart 2006; *contra*, Heather 2009, but see full discussion in chapter 1.3. below.

Romans (Byzantines), Visigoths, Ostrogoths, Lombards and Franks. In addition, significant attention has been paid to the West Romans, Huns, Avars, Slavs and Bulgars. The Arabs receive so much coverage that they merit mention in the title, but due to the inherent source critical problems I have had to limit my use of Arabic and other oriental sources, and grouped the analysis of Arab siegecraft with the other conquest groups (chapter 7), where they receive the lion's share of attention, while source critical problems that affect the interpretation of warfare have been dealt with separately in a series of appendices.[62] As a result, in excess of 500 sieges (including some dubious cases) have been included in the *Corpus Obsidionum*, or catalog of sieges. As a rule of thumb, I avoid overburdening the source references with parallel passages from minor chronicles if they add little or nothing to the discussion. Unless the contrast between thick and thin is particularly illuminating, the relevant standard literature cited should provide a guide to other sources that mention the same events.

The break-up of the Roman Empire in the West was, needless to say, an extremely complex affair, and 5th-century narrative sources reflect this. A vast mass of fragmentary classicizing histories, disparate chronicles and brief annals cover a wide range of groups involved. Most energy has therefore been put into establishing the survival of Late Roman structures and the emergence of siegecraft among those groups that have been mentioned above (and followed up in chapters 3-4 and 7). This means the exclusion of certain prominent groups, unless they directly impinge on the main narrative. Those who invaded the West in 406 (Sueves, Vandals, Alans) are mostly left aside, except where they are involved with any of the primary actors, such as the Visigoths or the East Roman conquest of Africa. More could also have been done on the Huns in the 5th century and the Papacy in the 7th-8th centuries, but enough material has been presented here to demonstrate the validity of my basic arguments.[63] The Anglo-Saxons have been completely excluded, as has most of the Islamic conquest of the Maghreb and Spain, since that would involve late Arabic chronicles with their own source critical problems.

Finally, I have been selective in using Arabic sources in general. While I argue that Hill's study has become somewhat obsolete, his survey of siege narratives in Arabic sources show that the vast majority of siege narratives come from Baladhuri's *Conquest of the Nations* and Tabari's *History of the*

---

[62] See below and chapters 0.3.4 and 0.3.5 for more detailed explanation.

[63] For the Huns, see chapter 1.1.2; for the Papacy, see chapter 3.3 (on the Lombards) *passim*.

Prophets and Kings. The former is most detailed, and has therefore been used here for the 7th century. The latter has fewer details on siege warfare, although an immense number of campaigns are mentioned in passing. In order to avoid filling up the *Corpus Obsidionum* with scores of one-line siege references, I therefore only use Tabari sparingly for insights on technology transfer or other details unknown from other sources. The same applies to the late Christian chronicles composed in the East, such as Michael the Syrian, the *Chronicle of 1234*, and Agapius (see below), who pose many of the same problems. In most cases, the reader can consult Hill and Lilie's works for the 7th and 8th centuries, respectively. My own contributions to the source-critical problems are set forth in the appendices (I-III).

Within these limitations, most major and a great number of minor narrative sources from the period 400-800 have been examined, as well as other sources that provide some information on sieges.[64] This includes some hagiography, legal and administrative records, and miscellaneous sources that have proven valuable. The next step, analyzing the siege narratives in their socio-economic and political context has required an even more extensive span. Here the focus shifts towards legal materials and administrative records, as far as they survive. I have chosen to exclude the vast majority of poetry, panegyric, homilies and epistolography due to the obscure and difficult style typical of late antiquity. While these sources sometime provide information on warfare, the source critical problems are sometimes insurmountable (or returns diminishing, since much can be known from existing narrative sources), so I have limited myself to the most relevant letters and poems of Sidonius Apollinaris, the *Epistulae Austrasiacae*, the *Codex Carolinus*, and other incidental references in poetry.[65]

0.3.2 *General Observations*

The strong historiographic tradition in ancient culture (Herodotus, Thucydides, Livy, Tacitus) came to a halt in the Latin-speaking West around 400 AD. This genre provides the most important sources for siege warfare, since the classicizing authors were obliged to include at least a handful of detailed military descriptions in their works. However, in the West, the last

---

[64] There are some exceptions in the 5th and 8th centuries, while on a few occasion 4th and 9th-century sources have been consulted when they throw further light on certain issues.
[65] References to these sources can be found in the secondary literature referred to for each entry. An important omission, the letters of Gregory the Great, has been ably treated by Brown 1984 and Marcus 1997.

great Latin historiographer of the classical tradition was Ammianus Marcellinus, who wrote a Roman history up to the battle of Adrianople in 378. A few classicizing historians writing around the turn of the 5th century are preserved in fragments in Gregory of Tours, but by around 450 AD, the patrons and authors of such literature, the leisured senatorial aristocracy, began to take on new functions as militarized landholders or ecclesiastics with different literary priorities.

While the new military elite in the West gradually moved away from leisured culture towards military and administrative functions, the ecclesiastics who wrote most of the preserved sources from the 6th century onward were equally involved in power politics even to the extent of leading military forces of their own. Bishops in the East Roman Empire helped organize urban defense, but they were not as directly responsible for carrying out military operations as in the West. In extreme circumstances, clerics were in charge of organizing defenses, as did Pope Gregory the Great (590-604) at Rome, when the East Roman government was unable to assist the city against the Lombards. Thus, we find eastern ecclesiastics, from Mesopotamia to Rome, who in great detail spell out their involvement in defending their flocks and describing miracles which also contain much incidental military detail. In the Visigothic and Frankish territories, however, bishops were, or were becoming, men of war, who increasingly used their forces on foreign expeditions and in internal power struggles. This was clearly a breach of canon law, and fighting bishops were denounced as an embarrassment to their office. Nevertheless, from Caesarius of Arles and Gregory of Tours onwards, the role of bishops in politics and warfare is frequently alluded to through their writings and biographies, but the contradiction in their behavior means that this is very much toned down, and that is one of the primary reasons that Western sources seem to provide fewer details on warfare in the early middle ages.[66]

In the East, the source situation is better during the 5th and 6th centuries due to the survival of the ancient historiographic tradition written in Greek, often written by a participant in the events. The prime examples are the fragments of Priscus and Olympiodorus from the 5th century and the complete Procopius and Agathias in the 6th. However, as the 6th cen-

---

[66] It is only in the 9th century we again see the ecclesiastics' in warfare through their own eyes. We have abbots like Lupus of Ferrières, whose private letter collection has survived, and Hincmar, bishop of Rheims, who was unapologetic about the church's military service but disliked direct military participation by clerics and tried to limit the proportions of ecclesiastic resources that were expropriated for military purposes by the Carolingian kings.

tury wore on, the sources became ever more arcane, and by the time of Theophylact, who wrote about the reign of Maurice (582-602) in the 630s, the military information is ever more difficult to use due to high-flowing rhetoric and distance from the events. After that date, the Greek chronicles become nearly as thin as the Latin sources. Traditionally Arab sources have been used to fill much of the gap, but they present immense problems of their own, as we have seen. Even if treated critically, they still provide a distorted image of the late antique social fabric. In contrast, Syriac and Armenian sources open up completely new perspectives that are far more informative.

0.3.3 *Greek Sources*

The Greek sources are by far the fullest for the 6th century, but the whole period is well represented. Since a number of specialized studies dealing with narrower timeframes or regions have appeared in recent years,[67] we only need to note some of the most important difficulties with these sources. The sources used here may be divided into four basic categories: *Narrative histories*, which may be subdivided as follows: *classicizing authors*, such as Priscus, Procopius, Agathias, Menander, Theophylact; *ecclesiastical historians*, such as Theodoret, Evagrius and John of Epiphania. These traditions ended rather abruptly around 600, so that afterwards, our view is mostly dominated by the great *chronicles*: *Chronicon Paschale*, Theophanes (see discussion of the Syriac tradition below), and Nikephoros, although aiming to write classicizing history, in practice his work was a stylistic retouching of now lost chronicle sources. The vast majority of these compositions have a universal nature, attempting to cover the history of the Empire or Church (or both) at large, leaving little room for local concerns. We can thus get to grips with how groups of cities and provinces interacted with the central government in times of war, but often, we are left in the dark about what happened at a specific location. City and regional chronicles would compensate greatly for this, but few are preserved in Greek, although this tradition is reflected in some of the Syriac histories and chronicles. For *hagiography*, the most important text is the *Miracula sancti Demetrii*, a contemporary collection of miracles ranging from the

---

[67] See e.g. Rohrbacher 2002 (4th-5th centuries), Blockley 1981-83 (5th century), Greatrex 1998 (early 6th-century eastern frontier), Curta 2002 and Whitby 1988 (Balkans), Whitby 1992 and Howard-Johnston 2006 (late 6th-early 7th century), Brubaker and Haldon 2001 (8th-9th centuries), Lillie 1993 and Howard-Johnston 2010. Furthermore, the standard translations of the various works have full and detailed commentaries.

late 6th to mid/late 7th centuries detailing the miraculous interventions of Thessalonica's patron saint Demetrios against famine, fire, plague, and most importantly, invading Slavs and Avars. *Normative sources*: the preservation of so many 6th-century authors means that little recourse has been made to legal and didactic sources. However, Justinian's *Corpus Juris Civilis* and military manuals such as the *Strategikon* are essential for certain points, while *De cerimoniis*, although a 10th-century compilation, provides some insights into earlier rituals and military organization. In addition, there are other *literary texts* that to a greater or lesser degree provide information on military affairs (panegyrics, poetry, homilies, religious commentaries, oddities such as astrological texts and the *Parastaseis syntomoi chronikai*). Only a few of these have been used in the pursuit of specific philological points or other difficulties.

In the histories and chronicles from Procopius to Theophanes, we find that the majority of siege accounts are in fact one-liners or short paragraphs with only a few details. Even when containing interesting information, they often prove to be highly episodic in nature. While the modern historian may lament this compressed writing style, it is quite understandable. If Procopius had provided a full description of every siege in his works, there would be room for little else. Since very brief, episodic accounts form the bulk of siege descriptions, even in late antique historiography, it has been easy to downplay the scale and importance of sieges. Fortunately, classicizing historians were bound by convention to write very detailed set-piece siege accounts on a few, selected occasions. Thus, Procopius' detailed description of the siege of *Rome (537f) has been fundamental to understanding defensive siege warfare in late antiquity (and therefore very awkward for minimalists). However, in another challenge to writing military history based on these sources, several scholars have criticized the set-piece siege convention as a literary *topos*, arguing that most of the events are only literary inventions based on famous sieges in classical sources. Thus, Priscus' detailed description of the siege of *Naissus (442) was dismissed by Thompson for the use of Thucydidean vocabulary and perceived similarity to Thucydides' account of the siege of Plataea and to Dexippus on the Gothic invasions in the 3rd century. However, Thompson's objections have been rejected in detail by several specialist studies, and Priscus' account is now generally accepted by modern historians. As we shall see, this has profound implications for assessing the state of technological knowledge and military practice in the mid-5th century.[68] While Procopius on some occasions was similarly inspired by the style and presentation of older historians (notably Thucydides), his military accounts, and especially details of technology and military practices, have mostly been impervious to criticism of *mimesis*. Averil Cameron criticized Agathias' description of the siege of *Onoguris (555) because the physical efforts involved appeared exaggerated. In light of the great number of similar accounts assembled here, as well as archaeological evidence for the enormous efforts that went into fortification, this argument cannot stand. Furthermore, when we examine the Syriac tradition, it becomes evident that none of the Greek authors' descriptions can be rejected as mere *topoi*, since they sometimes provide independent confirmation of the complexity of the same event, or a number of similar events that confirm the general accuracy of Greek sources.

0.3.4 Syriac Sources

The dual nature of the Greek tradition—either full, classicizing history with very strong limitations of genre, or brief chronicles with even more limitations of space—means that much of what we would like to know (e.g. army organization, interaction with local populations, the contributions of various social groups, details of morale and so on) is only mentioned in passing. For these aspects, the Syriac sources prove invaluable. Most were written by locals in an area frequently affected by warfare, and this concern is reflected in their compositions. Pseudo-Joshua wrote a detailed chronicle that covered northern Mesopotamia and Syria in the last decades of the 5th and the first decade of the 6th century, providing an immense amount of detail on social and economic conditions, relations between the government, the army and society at large, and a very detailed exposition of the Anastastian war between Persia and East Rome in 502-06.[69] On a number of points, he is supplemented by Pseudo-Zacharias of Mytilene's *Ecclesiastical History*, which however is fragmentary and less relevant for military history as it proceeds through the 6th century.[70] John

---

[68] Thompson 1945, refuted by Blockley 1972 and others. See the discussion in chapter 1.1.3 and the relevant entry in CO. Similar objections are dealt with as they occur.

[69] Translated to English by Trombley and Watt 2005 with extensive introduction. On several occasions, however, I have preferred Watt's fine rendering from 1882, but I have used both carefully in conjunction with the Syriac text, for which see the bibliography. There is also a German translation with introduction by Luther 1997.

[70] The old translation by Hamilton and Brooks 1899 is now superseded by the 2011 translation by Phenix and Horn edited by Greatrex.

of Ephesus wrote an ecclesiastical history that covers much of the late 6th century, which is of immense value for the wars under Justin II, Tiberius and Maurice until 594, when he died.[71] Parallel to Greek historiography, the Syriac tradition of connected narrative history was replaced by briefer chronicles in the 7th century that nevertheless provide invaluable details on the Arab conquests not found elsewhere, and especially how the local population responded to the conquest. The most significant are the *Chronicle of 637*, the *Chronicle of 640*, the *Maronite Chronicle*, and the *Chronicle of 819*.[72] The *Chronicle of 775*, also called the *Chronicle of Zuqnin*, briefly covers the 7th century in a similar manner, but becomes extremely detailed as the 8th century progresses, and shows many of the same social groups involved in warfare under the Arabs as under the Romans over 250 years earlier.[73]

Finally, there is a cluster of sources that derive from what has been called the *Syriac Common Source* (CS), but now established as the chronicle compiled by Theophilos of Edessa around 750. Fragments of this survive in four versions. Two of them are the late Syriac compilations in Michael the Syrian (d. 1199) and the *Chronicle of 1234*. The relevant sections from the 7th and 8th centuries have passed through another Syriac intermediary, the lost 9th-century chronicle written by Dionysios of Tell-Mahré. The third version of the *Common Source* is the Christian Arabic chronicle of Agapius of Manbidj, which has only been consulted on a handful occasions for information not preserved elsewhere. The Greek Chronicle of Theophanes derives in part from the Common Source for much of the 7th and 8th centuries, and has been used throughout.[74]

---

[71] Whitby 1988 does an excellent job at relating John to other sources, but Payne Smith's translation from 1860 desperately needs an update.

[72] The short chronicles have been translated by Palmer 1993, with further analysis in Hoyland 1997.

[73] Introduced and translated by Harrak 1999; extensively used in Petersen, forthcoming.

[74] The first part of Dionysios from 582 to 717 has been reconstructed and translated by Palmer 1993, whose work I have quoted as such and used extensively for the 7th century and events up to the siege of *Constantinople (717). A complete translation of all derivations from Theophilos with an important introduction has recently been published by Hoyland 2011. The original of Agapius, Michael and *Chr. 1234* have been consulted where necessary. Parallel passages after 718 clearly deriving from Theophilos are normally only noted with reference to Hoyland. See also his very important introduction, which supports some of Speck's arguments on the complex textual history of CS.

0.3.5 Arabic and Other Eastern Sources

In Arabic, we have noted the most important sources used here, namely Baladhuri and Tabari. The latter is quite massive, with a translation running in 40 volumes. In addition come the Christian Arabic sources. Agapius has been mentioned above. There is also the history of Sāwīrūs ibn Muqaffa', a 9th-century Coptic bishop, who wrote a collection of biographies of the bishops and patriarchs. For this and Egyptian papyrus evidence in Greek as well as Arabic, I rely on Sijpesteijn, Trombley, Savvides and al-Qadi's work on Egyptian craftsmen.[75] Other traditions are preserved in various languages. The Coptic (represented by the Ge'ez translation from the 16th century) Chronicle of the 7th-century bishop John of Nikiu is invaluable for the Roman civil wars at the beginning of the 7th century, but has a lacuna before recommencing in the middle of the Arab invasions, for which it provides many valuable insights. Georgian sources also provide evidence for the 7th century; here I have used the translation of Antiochus Strategius on the Persian siege of *Jerusalem (614). The most important tradition in the East besides Syriac, however, is the Armenian. Since I cannot use the original version, I have only used selectively the history attributed to Sebeos, which has recently received an excellent translation with detailed commentary, and covers the last Persian wars and the rise of Islam.[76]

0.3.6 Latin Sources

The main source for much of the 6th century, Gregory of Tours, did not like war and thought warfare and politics were ridiculous (or at least, that was how he portrayed them).[77] But he himself was often involved in the events of his day, and had to provide some military details, if only to prove a certain moral point. In some respects, his style resembles Evagrius, who was similarly fond of manipulating sources to prove a moral point or emphasize the divine at work in the world.[78] Unfortunately, we do not have the array of original sources against which to check Gregory's manipulation as we do with Evagrius. This is further complicated by his status as a powerful actor in Merovingian politics; as noted at 0.3.2 above, Western bishops were

---

[75] Trombley 2004, 2007; *EI*² s.v. Miṣr (Savvides), discussed in chapter 7.3.

[76] Thomson and Howard-Johnston 1999.

[77] Goffart 1988 argues that Gregory in fact wrote Christian satire; for a full implication of this, see the discussion in chapter 4.1.1. Thorpe's translation has to be used with much care; see chapter 4.1.4 for examples of this.

[78] See Whitby's introduction to his 2000 translation of Evagrius and his 1988 analysis of all relevant sources.

great magnates with military obligations that went far beyond defense. The popes were less apologetic, since they were mostly involved in defending their flocks. Therefore we have much good information in papal letters, as well as the papal biographies found in the *Liber pontificalis*, which often commemorates defensive works organized by various popes.[79] These biographies often provide evidence of late antique and contemporary Byzantine modes of organization being used for urban defense in the West.[80] However, hagiographies are very informative, although they betray the same sensitivity to and reticence against the involvement of their saintly protagonists. For example, the *Vita St. Genovefae*, *Life of Caesarius of Arles* and *Lives of the Visigothic Fathers* are useful for the earliest period. Other literary sources include Paul the Deacon, who wrote a history of the Lombards from the beginnings to 744, but bases much of his 6th-century information on Gregory,[81] and a host of minor 5th-7th century chronicles.[82] The most significant literary text is the *Historia Wambae*, which provides valuable information on late Visigothic warfare alongside the substantial Visigothic legal corpus.[83]

Fredegar and his continuators are the main source for the 7th-century Franks, and continue in the same vein as Gregory, but only much more compressed.[84] Only ten brief anecdotal chapters (less than nine printed pages, mostly lifted from the *Liber Historiae Francorum*, composed in 727) cover the period 657-721. The historical sources for the later Merovingian period are thin indeed, and consequently there has been a shift towards using hagiography, although some administrative and diplomatic sources survive to provide context.[85] The situation improves somewhat with the rise of the Carolingians, as they sponsored a number of chronicles on their military successes from Charles Martel to Charlemagne (714-814). Here the *Royal Frankish Annals* have been used alongside the latter chapters of Fredegar's continuation, as well as the early part of the Astronomer's *Life of Louis the Pious*, when he was sub-king of Aquitaine under his father Charlemagne.[86] Most valuable are Carolingian legal and administrative records that survive in increasing numbers towards 800.

---

[79] Gregory the Great's large collection is used only indirectly here, cf. 0.3.2 above.
[80] See the translations with introduction by Davis 2000, 2007.
[81] See Goffart 1988 for a discussion of Paul's methods.
[82] Most of those relevant to Gaul are translated in Murray 2000, those relevant to Spain in Wolf 1999.
[83] Martínez Pizarro 2005.
[84] For the effects of this compression, see chapter 4.2.1.
[85] For a discussion, see Fouracre 1990, Gerberding 1987.
[86] For a survey of the source situation in this period, see McKitterick 1989, 2004.

0.4 Structure of the Argument

Chapters 1-4 show how the organization of siege warfare worked and how it fit into late antique and early medieval societies. The general Roman background is laid out in the first chapter. By progressing from the East Roman/Byzantine Empire, we establish a "thick" description from which to evaluate Ostrogothic, Visigothic, Lombard and Frankish warfare. Chapters 5-6 establish the common ground shared between the post-Roman societies in tactics, technology, and socio-political life before, during and after the siege. It clearly emerges that all the societies treated here shared fundamental similarities in culture, mentality, and technological ability that soon leveled differences. Chapters 7-8 and Appendix I-III demonstrate how technological knowledge diffused within this society and how it was spread to external conquering societies that did not share the common Roman background. All these sections are supported by an extensive *Corpus Obsidionum* (analytical collection of sieges) that list all sieges found in the sources for the period under consideration.

Recently it has been argued that late and post-Roman magnates acquired and exercised power through enormous, newly organized bipartite estates that produced vast income and posed serious challenges to all central authorities from Egypt to Gaul in the 5th and 6th centuries. Such landowners effectively co-opted large parts of the Roman administration and army in the West, and came close to doing so in the East as well. Magnates raised private armies through administrative techniques adopted or usurped from the Roman state and applied to their vast estates. These efforts were directed by disgruntled generals, who as they opted out of the imperial system began to be called kings. While some of their soldiers were deserters from regular units, others came from federate armies. Both were products of Roman military frontier culture. These soldiers took on barbarian identities in opposition to the central government without changing their military practices, which still included advanced siege warfare. Logistical and technological support was provided through the decentralization and militarization of the late Roman government's practice of demanding labor corvées for public works and military logistics. This was only reinforced by the magnates' age-old ability to draw on their own rural and urban dependents. Thus, while the infrastructure of the army appears to have evaporated in the West well before the deposition of Romulus Augustulus in 476, it had actually begun to morph in to the estate-based framework familiar from the Carolingian Empire as early as around

400 AD, well aided by Roman emergency measures. The institutional foundations for this transition can be examined in some detail in the 5th century and into the 6th (chapter 1).

The long-term effects on Eastern (Byzantine) and Western ("Germanic") successor states only becomes apparent when examining the long transitory period of the 6th through the 8th centuries, with major divergences only occurring after around 600 AD. The East Roman government struggled to cope with the centrifugal forces of semi-independent armies in the Balkans as well as powerful, militarized magnates throughout the empire. Emperors like Anastasios and Justinian were able to harness these forces to a certain extent by directing them outwards. It was only when external pressure from Avars, Slavs, Persians and Arabs caused the destruction or loss of the estate economy by the 7th century that magnate power declined beyond repair. What remained into the middle Byzantine period was a civilian population that had become accustomed to arranging their own defenses and cooperating with the regular army (chapter 2).

In the West this development was partly masked by the so-called barbarian invasions, a process dominated by formerly Roman armies that opted out of imperial ideology and collaborated with local magnates to establish the successor states. Three of the Balkan field armies are peculiar for their development into peoples when they established territorial states in opposition to central authorities. Much recent research has begun to accept these "barbarians" more as products of Roman military frontier culture that began to intrude noticeably on civilian metropolitan culture around 400. The problem of Visigothic and Ostrogothic origins has obscured how similar these armies were in composition, organization and outlook to the armies that remained loyal to Rome. The Lombards are a case in point: while they were late arrivals, they became militarily indistinguishable from their East Roman contemporaries due to their formative years on the fringes of the Balkans military culture long after central authority had collapsed. All of these armies (rather than peoples) continued to fight as their contemporary East Roman neighbors as long as they existed, organizing their campaigns, sieges, and defense in much the same manner (chapter 3).

The Roman army on the Rhine followed a similar pattern, outliving the emperors and producing, by around 500 AD, a state that eventually came to define Frankish identity. Due to the fairly low degree of urbanization of the northwest ever since the principate, Frankish warfare is sometimes made to appear different from that of other periods and places. However, a thorough comparison with their neighbors and competitors shows how the Franks continued the 5th-century innovations and made them the basis of their military organization, and differences between East and West were still modest until c. 600. From an estate-based economy and military system, they were able to raise vast armies that conquered large swaths of Western Europe through consummate mastery of siege warfare (chapter 4).

Sieges could be fought in many ways, ranging from loose blockades to fierce storms. While a basic definition of sieges as either blockades or storms is found in Vegetius' *De re militari*, there was in fact a continuum between the two, far better described by the 4th-century BC Greek military author Aeneas Tacticus, who furthermore demonstrates that sieges also included harassing tactics, threats and political pressure. Often the sources indicate only one aspect of a complex process by mentioning a raid, *or* blockade, *or* storm, *or* surrender. If taken at face value, we are utterly confused as to what exactly took place. But in actual fact most sieges shifted between several of these alternatives. This is clear from the more comprehensive Eastern sources, but few Western chroniclers could be bothered to write down all stages comprehensively. Being aware of this, we can more easily understand that events were often far more complex than they might appear at first glance. Any siege can be placed on a continuum of possibilities that was determined by a range of varying conditions, not by any inherent ability associated with ethnicity, area, or period. When the various sources are read against the common late antique framework spelt out in the preceding chapters, it becomes clear that the ability to organize armies, supply them, equip them with advanced military technology and employ them in the field was startlingly similar across the board, from Aquitaine to Armenia. While there was some significant technological change, this occurred in a framework where pragmatic adaptation to unforeseen challenges was the norm (chapter 5).

Sieges were characterized by remarkably similar demographic, economic, social and cultural contexts throughout the former Roman Empire. Very often, the defense of a city had to be organized by the local population in cooperation with magnate retinues, regular garrison forces, and other government representatives. On the positive side, this mobilized all classes of the population, from goatherds to bishops, into concerted action to protect their city. The negative side was often divisiveness. In foreign invasions no less than civil wars, each group (or members within a group) could have diverging motives to resist, depending on personal networks or risk of losing property and position. Where we have sufficient sources, it is often possible to establish the existence of devastating fault lines *within* a given city. Another major feature in siege warfare was Christian beliefs and practices. In order to protect their cities, urban populations and their spiritual leaders devised a whole range of public rituals designed to reinforce morale and ensure divine support. This was accomplished through a range of means, including miracle stories, saints' cults, public processions, fasts, blessing of the walls, and so on. Christian (and later Muslim) armies would also perform similar rituals to ensure successful attack. The necessity of divine support and urban cohesion was paramount, since the consequences of loss were horrific, even by 20th-century standards (chapter 6).

Urban craftsmen complimented by those drafted in from magnates' estates provided much of the framework for necessary repairs and operation of engines. This facilitated the spread of technology despite the decline of regular army institutions. These features were quickly adopted by foreign invaders, from Huns to Arabs. However, the invaders differed greatly in how they were able to maintain these skills over time. The least successful means in the long term was to deport skilled craftsmen, treat them as prisoners, and exploit their skills until they deserted or died. Others tried to preserve as much as possible of the Roman urban fabric, allowing the institutions that supported technological knowledge to thrive within a client framework. This worked well for the Arabs, who were able to use former Roman craftsmen on a massive scale for naval and siege warfare; the effective preservation of late Roman institutions proved instrumental in establishing the first Islamic state (chapter 7; Appendix I-III).

The traction trebuchet was the most important technological innovation during this period and replaced classical torsion artillery. It had a phenomenal rate of fire and accuracy, outshooting the more complex Roman artillery pieces. Its origins have long been debated, but here it is shown how it arrived in the eastern Mediterranean area no later than 500 AD, a century earlier than previously thought. Its spread throughout most of the Mediterranean within the 6th century provides a valuable case study of how the decentralized network of urban and rural craftsmen throughout the former Roman Empire could transmit advanced technology (chapter 8).

Part Two, *Corpus Obsidionum* collates and analyzes all siege narratives found in the sources for the period c. 410-814. This includes a very large number of brief siege descriptions as well as some dubious cases that might seem superfluous or excessive. However, even brief references provide valuable information on vocabulary, frequency, context and stylistic choices made by different sources. Furthermore, the definition adopted here makes it clear that it is not always straightforward to dismiss a military event as a siege, while a closer inspection of certain events normally taken as sieges actually proves that they have been misinterpreted. A few selected sieges are used in the main text to illustrate preceding and following developments (c. 350-825), but are not included in the *Corpus*. This body of information is used actively in the text of the dissertation by marking the siege with an asterisk followed by the date, if not absolutely clear from the context. Thus *Rome (537f) refers to the first siege of Rome by the Ostrogoths against the East Roman (Byzantine) garrison, which began in 537 and lasted into the next year. The sources, relevant quotes and technical terms in translation and the original language, as well as references to modern scholarly treatment will be found there. This will lighten the burden upon the text by avoiding much repetition and too great a focus on incidental details.

CHAPTER ONE

AN AGE OF TRANSITION FROM THE FALL OF THE ROMAN WEST TO THE EARLY MIDDLE AGES

During the 5th century, Roman siege abilities passed to the successor kingdoms through a variety of means. Client armies drifted into and out of Roman service and control, especially during civil wars. Invaders appropriated the personnel and infrastructure of Rome. Roman elites, often in collaboration with invaders or usurpers, took over military administration, including the recruitment, pay and maintenance of troops, based on vast estates that had long been obliged to shoulder similar burdens under the Roman state. Much of the infrastructure needed for warfare in general and sieges in particular also passed under magnate control, as their military retinues and urban or rural dependents were recruited from, or acquired skills that had been monopolized by, the Roman army. Otherwise the necessary resources were provided through the militarization of traditional civic burdens and voluntary civic munificence, such as the obligation to repair infrastructure and competitive monumental construction. While all successor kingdoms utilized these administrative techniques to a greater or lesser extent, at least some of the new ruling élites originated as Roman field armies taking on new identities, a process that partly occurred in the East as well. Successor fighting methods remained the same. Not only did they keep the traditional Roman knack for breaking down or defending walls; sieges became more prevalent in the 5th century than they had been in the 4th.

1.1 *From Late Roman to "Barbarian" Poliorcetics*

The terrifying realities of siege warfare in late antiquity have been well attested by the finds at Dura-Europos in Syria, where the Sassanid Persians besieged the Romans in 256 AD. The city fell and the population was massacred or deported, leaving behind a ghost town with no-one to clear the debris. This has allowed modern archaeologists to explore the spectacular remains of the siege, which include evidence of complex engineering skills on both sides.¹ A Persian siege ramp led up to the breastwork, internal Roman terraces supported the wall against battering and artillery, Persian saps undermined a corner tower and a long stretch of the adjoining wall, while Roman counter-mines intercepted the Persian siegeworks. One of these counter-mines was used in an attempt to undermine the Persian ramp. Another was directed against the Persian sappers undermining the tower. Evidence of what happened when the Romans intercepted the Persian sappers is dramatic in the extreme: when the Romans broke into the Persian sap, a brutal under-ground struggle ensued, resulting in the deaths of at least one Persian soldier and several Romans when the mines collapsed.² The Persians subsequently blocked off their mine with stones to prevent further Roman penetration. The fighting over ground was no less dramatic: a great number of artillery projectiles have been found around the walls, indicating fierce artillery duels as the Persians tried to overpower the Roman defenders on the ramparts in order to cover the Persian siegeworks or before attempting to storm the walls. These remains are clear evidence of Persian as well as Roman siege skills.³ Yet no literary descriptions of the siege are extant.

1.1.1 Late Roman Siege Warfare

Roman siege practices and their supporting infrastructure in the following century and a half are well known from the works of Ammianus and Vegetius in particular, and have received some measure of attention in modern scholarship, although conclusions are mixed.⁴ The engineering skills so

---

¹ Excavations have taken place since 1920 with extensive, but incomplete reports published in the following decades. There is still vigorous publication, including on military affairs. Here I follow the survey of the siege by Leriche 1993.
² S. James 2011 has proposed that the Persians used poisonous gas to kill the Romans: when the Persian sappers heard the Romans approaching, they prepared a trap of sulphur and bitumen that was ignited by a sole Persian soldier just when the Romans were about to break through. The effect would have been instant, as the gas rose quickly into the higher Roman tunnel, killing the Romans in seconds. The Persian soldier probably lingered too long to make sure that the mix ignited and was killed himself.
³ Further on Persian siege skills, see chapter 7.1.2 below (as well as chapters 2, 5 and 6 *passim*). For a survey of Roman siege warfare during the Republic and Principate, see e.g. Kern 1999: 251-351.
⁴ See most recently Elton 1996: 257-263 for a positive assessment on Late Roman methods of siege warfare in general; Rihll 2007: 251f is positive and argues for continued innovation by 4th century military engineers. Similarly Nicasie 1998 *passim*, but his focus lies elsewhere. Southern and Dixon 1996: 148 argue for an increase in the incidence of sieges and technological advances, but their treatment is mostly based on well-known anecdotes from Ammianus, Procopius and Vegetius. Marsden 1969: 195-98 is more dismissive of late Roman siege skills.

well in evidence at Dura-Europos remained common. The late Roman army deliberately sought recruits with skills in carpentry, masonry, smithing and other relevant crafts to fill the ranks to the exclusion of "fishermen, fowlers, pastrycooks, weavers" and textile workers. In addition to these skilled soldiers, specially trained engineering experts were attached to many units or specific border fortifications, and most legions were capable of building and repairing military infrastructure.[5]

There is however some dispute over how artillery knowledge was maintained after the reforms of Diocletian and Constantine. During the Principate, all legions had been equipped with artillery of various sorts that added up to a vast baggage train whenever they went on campaign,[6] and had an infrastructure of highly trained experts who were recruited for additional service (*evocati*) in order to pass their knowledge on to the next generation of troops.[7] Marsden identified a shift of artillery organization sometime in the late 3rd or early 4th century, when he believed the smaller Diocletianic and Constantinian legions lost their artillery capacity. Instead, designated units of *ballistarii*, ballista-operators/constructors, were set up along or near the frontier with the particular task of providing artillery to the now less competent infantry units.[8] However, Marsden paid little attention to the connection between artillery knowledge and supporting skills in engineering in late antiquity, which were necessary for the repair of fortifications and the construction of siegeworks. These two branches of military engineering were closely interrelated. During the 3rd and 4th centuries there were enormous advances in Roman fortification techniques as evidenced by huge numbers of complex fortifications identified all over the Roman Empire in the period, while *ballistarii* began to oversee fortification work.[9] These fortifications were uniformly adapted for more extensive defensive artillery use than previously, so that bastions provided interlacing fields of fire and mutual support with different types of artillery and missile weapons.[10]

---

[5] Varying utility of trades: Vegetius 1.7. The practice was common throughout Roman history: for Republic and Principate, see Adams 2007: 224ff and Marsden 1969: 182-85; for the Byzantine period, see chapter 2 below.

[6] Roth 1999: 83f; Marsden 1969: 190.

[7] That is, after a regular tour of duty of 20 years. See Marsden 1969: 191-95.

[8] Marsden 1969: 195-98.

[9] Their role had been subordinate to the legion *architectus* during the principate. For the new role of the *ballistarii*, see chapter 2.2.2 below. Petrikovitz 1971 discusses late Roman fortifications on the Rhine and Johnson 1983 the West more generally. There are also a mass of specialist studies, listed by site in e.g. Brühl 1975, 1990, and *TIB*.

[10] Marsden 1969: 116-63 has extensive discussion with rich illustrations of classical

The conclusion that artillery skill declined is based on Ammianus' description of the Persian siege of Amida in 359, which he barely escaped just as the city fell.[11] There were two Gallic legions present at the siege, recruited by the usurper Magnentius in his bid for power in the West (350-353), but after his downfall these legions had been transferred to the East.[12] According to Ammianus, "they were not helpful in operating siege engines or constructing fortifications."[13] Marsden took this to prove that Roman legions by that date were less competent than their predecessors.[14] Yet from the context of the full narrative, it is clear that their inability was exceptional and that Ammianus was not being quite straightforward about their treatment. They seem to have been treated more like a penal battalion than a regular legion. There was no lack of expert artillerymen in the city, as it had a fully equipped artillery workshop,[15] and the other five legions present apparently had no problem helping to operate the variety and great number of artillery pieces available to great effect.[16] Secondly, since the Magnentian legions had recently taken part in the wrong side of a civil war, they had probably not been entrusted with a siege train so shortly afterwards, as usurpers could hole themselves up in fortified cites (for instance, shortly after, rebels were besieged at Aquileia).[17] Instead they had

---

fortifications, but as the examples in the previous note show, the possibilities for artillery use became even *more* advanced. There was a very active intellectual interest in military technology at the time. In addition to Vegetius (see 5.1), an anonymous author, possibly commissioned by Valens, described fantastic contraptions in *De rebus bellicis* alongside practicable weapons such as spikes (*triboli*). It was better on economic policy than technology: Lenski 2002: 299ff and *passim*.

[11] Amm. Marc. 18.7.1-19.2.2.

[12] For a brief overview of the context, see Hunt 1998: 14-22.

[13] Amm. Marc. 19.5.2, "cum neque in machinis neque in operum constructione iuvarent."

[14] A similar sentiment is found as recently as Blockley 1998: 412, who commendably argues that infantry legions were still efficient in the late 4th century, but in the process has the Magnentian legions discard the walls and sally out as a result of "frustration" over their lack of defensive siege skills (the incident is discussed in the following).

[15] Unknown from the *Notitia Dignitatum*, but according to Amm. Marc. 18.9.1 it was placed there when the Caesar Constantius fortified the city, "establishing there an armory of mural artillery" (*locatoque ibi conditorio muralium tormentum*).

[16] See Amm. Marc. 18.9.3 for a survey of the legions defending the city; 19.1.7 for impeccable ballista aim; 19.2.7 for massed artillery and archery used by the defenders. For a similar argument, see Nicasie 1998: 65f.

[17] Ammianus called the Magnentian legions "untrustworthy and troublesome" (*fallaces et turbidos*, 19.1.3), and they were hence sent to a region where they had no previous ties of patronage and could thus cause less trouble. Civil war sieges: the fierce civil wars that destroyed the Tetrarchy and resulted in the rise of Constantine ended with the siege of Byzantium in 324 (Lenski 2006: 75f); the siege of Aquileia in 361, where Julian had rebel troops led by a group of senators besieged, was extremely hard fought with a range of artil-

the unenviable job of performing dangerous sallies with very high losses, the only way they could vindicate their recent treason.[18] Thus interpreted, the description of Ammianus actually demonstrates an *increasing* skill in artillery use among the legions. Not only were most regular legions fully capable of operating siege engines; in addition there were now at least seven legions of *ballistarii* who had this as their primary task. These dedicated units could bolster siege defense and attack at selected sites while providing an institutionalized body of engineers with a high degree of specialization. They could train legions and garrison artillerymen at need (or perhaps dispatch them from their ranks), as Roman fortification technology had advanced significantly and required many more skilled defenders. In effect, the reforms of Diocletian and Constantine increased rather than decreased the poliorcetic capacities of the Roman army, a rather more logical conclusion as the enemies of Rome became ever more skilled in siege warfare.

To support the Roman army's logistical and technological needs, there was a significant infrastructure of imperial *fabricae*, workshops for arms and other necessities, in many cities in the empire, especially along or near the frontiers. These had staffs of specialist craftsmen, *fabricenses*, who were regarded as members of the military.[19] While most of these provided arms and armor for individual troops, some specialized in artillery. The artillery *fabricae* were not the only providers of engines: similar institutions have been identified in the narrative sources, which provide sporadic evidence that add to the workshops of the *ballistarii* units known from *Notitia Dignitatum*, which only mentions artillery *fabricae* in Gaul, at Augustodunum (Autun) and near Trier. It is clear from narrative evidence that facilities with similar capabilities were far more common.[20] Amida clearly had its

---

lery and engines. When assault failed, Julian's troops set up a blockade, which was resolved by negotiations and the execution of the offending senators (Amm. Marc. 21.11.2-12.25). Cf. the early 5th-century examples at 1.1.2 below.

[18] They were apparently so desperate to vindicate themselves that they incurred a casualty rate of 20% dead in *one* sally. It paid off, as the authorities put up statues of their commanders at Edessa to commemorate their bravery (Amm. Marc. 19.6.3-12).

[19] For a general overview, see S. James 1988.

[20] Jones 1964: 834ff; Elton 1996: 116f; Lee 2007: 89-94. Rihll 2007: 46f argues that the *ballistarii* operated small hand-held *ballistae* (i.e., early crossbow-like weapons), but this must have formed only part of their arsenal. S. James 1988 believes that much of this competence, especially along the Danube and the East, was maintained either by legionary workshops that remained in use through the reforms, or by urban craftsmen who continued traditions from the Hellenistic era; cf. Justinian's laws on the *fabricenses* (discussed by S. James and in chapter 2.2.2 below).

arsenal of engines. The same was the case at Adrianople (see below), while East Roman evidence abounds with garrison artillery outside the arsenal cities mentioned in the *Notitia Dignitatum* and 4th-century narrative evidence.[21] Needing to bolster his authority after a recent revolt in the Balkans, Valens proposed to invade Gothic territory north of the Danube in 366-67. In preparation Valens instituted a program of fort-building and renovation, equipping installations with *ballistae*.[22] Finally, the extensive Danube and Rhine flotillas were armed with artillery.[23] It should thus be clear that at the end of the 4th century, Roman skills in fortification building, siege engineering and urban defense showed no apparent signs of decline, while the institutional framework in some respects seem to have become better organized. In contrast, due to Visigothic and Hunnic involvement in Roman affairs, siege warfare became even *more* prevalent in the early 5th century; only the narrative evidence is poorer.

1.1.2 *The Thin Description: Visigoths and Romans, 376-474*

A large number of Goths asked the Roman government to accept that they be settled south of the Danube in 376, but revolted against corrupt officials. The ensuing war was an important stage in the formation of the Visigoths, although the name is inappropriate at this early date. The fighting also illustrates challenges faced by barbarian invaders when attempting to assault fortifications, as well as Roman defensive practices. The Goths tried to storm Adrianople (their second attempt) in the aftermath of the famous battle in 378. After the Romans received and rejected Gothic demands for submission, Ammianus relates how the Romans made standard defensive preparations:

> ...the rest of the day and the whole night were spent in preparing defensive works. For the gates were blocked from within with huge rocks, the unsafe parts of the walls were strengthened, artillery was placed in suitable places for hurling missiles or rocks in all directions, and a supply of water that was sufficient was stored nearby [to relieve thirst of defenders].[24]

---

[21] See chapter 2.2 below.
[22] Lenski 2002: 127.
[23] See Lenski 2002: 136 for the Danube flotilla to the early 5th century, when the fleet was officially supposed to number 225 vessels, and Curta 2001: 184f for a brief survey of the 5th-6th-century situation.
[24] Amm. Marc. 31.15.6: "... parandis operibus diei residuum et nox omnis absumpta. Nam intresecus silicibus magnis obstrusae sunt portae, et moenium intuta firmata, et ad emittenda undique tela vel saxa, tormenta per locos aptata sunt habiles, aggestaque prope sufficiens aqua."

The city was yet another base for Roman arms manufacture and thus well prepared for the challenge.[25] Taking part in the defense were regular Roman garrison troops, *fabricenses*, provincials and elite soldiers from the field army.[26] In the face of a formidable barrage of artillery, javelins, sling stones, column drums and other objects bearing down on them, the Visigoths persisted and managed to kill a good number of defenders while setting scaling ladders up against the walls. It also seems that they adapted their tactics to this terrifying barrage, as they "no longer fought in order" (which would make them an easy target for artillery and concentrated missile fire), "but rushed forward in detached groups", which Ammianus claims was "a sign of extreme discouragement" but rather appears to be a sensible adaptation to overwhelming fire from above.[27]

Visigothic reactions to the realities of siege warfare have been exaggerated in modern scholarship, and have helped perpetuate an image of senseless barbarians bumbling about city walls.[28] Their approach was rather conditioned by their lack of a logistical infrastructure to support them. Instead of persisting in attacking the walls without logistical backing and with a relieving Roman army on the way, they followed the sensible course of giving up their first attempt on Adrianople (in 377) and turned to ravaging the countryside. The tactics they employed in their second attempt (in 378) could have found success against a less well fortified city, as numerous examples from the following centuries show (cf. 5.2.1). Their strategic sense was not bad either—if they had taken Adrianople, they would have controlled one of the great logistical centers of the Roman

---

[25] Amm. Marc. 31.6.2-4, where the *fabricenses* and the population were rounded up and armed by the local *duumvir* (chief magistrate) to punish the Goths for pillaging the *duumvir*'s nearby villa. The ensuing Gothic victory led to the first siege of Adrianople. Fritigern advised not to storm the city, but to blockade it and plunder the countryside. Incidentally, Julian's troops had similarly abandoned their attempt at storming Aquileia in 361, turning instead to a blockade of the city; cf. above.

[26] Amm. Marc. 31.15.10, "At cum armatis provinciales et palatini" fought the Goths, presumably still alongside the *fabricenses* who had survived the debacle the previous year, since artillery was still available and the provincials were clearly armed. The *palatini* were probably the elite *comitatenses* discussed by Jones 1964: 125f, 608f rather than civilian officials (as in the Loeb translation), who bore the same name.

[27] Amm. Marc. 31.15.12-15.

[28] Even Wolfram's highly regarded standard work contains some curious assertions not supported by the sources. Wolfram 1988: 129 claims that the Goths were terrified of a sudden downpour of rain at Adrianople, thinking "the heavens were collapsing." No such sentiment is found in the relevant passage in Ammianus (31.15.5); instead, the Goths simply returned to their fortified camp to compose a letter demanding the surrender of the city.

Empire, containing arms factories, surrounded by rich farmland, and straddling the Thracian road network only a few days from Constantinople.[29]

Theodosius eventually managed to force a deal and settled them in the Balkans in 382 until his death in 395.[30] During this period, these Goths formed an integral part of the Roman army, but were supplemented by a host of other Gothic and barbarian groups. They provided recruits for regular units to the extent that the field armies in the East were dominated by Goths around 400.[31] The intervening years may mask a substantial change in the nature of the Goths who were settled in 382 through integration with Romans and other barbarians in Roman service. Their activities were eminently normal: Gothic contingents garrisoned border fortifications alongside Roman regulars along the Danube.[32] Their leaders sought a secure and recognized position within the Roman system. The title *magister militum* would give their commander formal control over military administration and infrastructure in their area of settlement, and protect against haphazard resettlement as *dediticii* in small scattered groups. During two Roman civil wars, the Goths provided large contingents that took massive losses in battle, which largely explains the ensuing revolts and desertions. They were then used as pawns in court politics that accidentally left the Goths marginalized and in revolt during the years 395-97.[33] The relationship of this group to those who defeated the Romans at Adrianople in 378 is actually constructed by late antique authors and has been taken for granted since. It is only after the rise of Alaric (called their "king" in much later sources) as commander of Roman units designated "Vesi" around 394 that we can begin to trace their activities and call them

---

[29] This is where Licinius fought his last great field battle against Constantine before he was besieged at Byzantium. See *RE* s.v. Adrianopolis for the strategic importance of the city.

[30] See the discussion of the treaty in Halsall 2007: 180-85, who points out that there is little evidence that the arrangements were substantially different from those of other barbarian groups settled on Roman soil.

[31] For the situation around 400, see Liebeshuetz 1992. While he argues that the Goths were purged when the *magister militum* Gainas (see *PLRE* 2 s.v.) was deposed, the temporary eclipse of "Gothic" troops should probably be viewed in light of the discussion on identities in chapter 1.3 below: political circumstances may have made it inconvenient to emphasize "Gothicness" at certain times, while it was more accepted or even favorable at others.

[32] Lenski 2002: 117 for the 360s; for the later period, see Heather 2005: 211f. Heather is probably wrong in arguing that all or most Goths in the Balkans were (the descendants of) those who took part in the battle of Adrianople and were subsequently settled by Theodosius.

[33] Heather 1991: 157-92; Wolfram 1988: 142f; for a more recent interpretation, see Halsall 2007: 186-95.

"Visigoths" in order to distinguish them from the other Gothic groups in the area.[34]

When the Visigoths were resettled in Macedonia in 397 after two years in revolt, Alaric probably received the long desired title of *magister militum*, and during the years 397-400 and 402-405, the Visigoths were again part of the regular army, first serving in the Eastern army, then probably joining the Western army under Stilicho. Part of the arrangement attested in 397 (but probably much older) was equipment from Roman arsenals—thus if they did not already look Roman by the time they crossed the Danube, they certainly did by around 400.[35] During the years 400-02 and 408-10 the mechanisms of court politics pushed them into the cold again. The Visigoths did not wish to break with the imperial authorities, and on both occasions invaded Italy to put pressure on the government to grant them provincial commands and a secure place within the Roman army's logistical and remunerative system. The first invasion was successful, leading to settlement in Pannonia, contested between the two halves of the empire but temporarily brought under Western control.[36] However, with the fall of Stilicho in 408, they were again out in the cold, joined by Roman troops of barbarian origins who had been purged at the same time. The second attempt to pressure imperial authorities failed with tragic consequences when they entered and plundered *Rome in 410, after being refused settlement in Istria and Venetia.[37] By then, the Visigothic threat was eclipsed by the Vandal and Alan invasion of 406 and a Roman civil war in the West that successfully brought them into the fold for the next generation.

The Visigoths survived because they were rarely were a threat to the legitimate emperor, as opposed to the string of Roman generals who sought the purple in the years 407-413.[38] Alaric's successor, Athaulf (r. 410-15), managed to maneuver the Visigoths into Roman political culture on the highest levels during the internal Roman struggles—in fact, the Visigoths were an integral part of the conflicts and civil wars at the time, and at first they

---

[34] For this perspective, see Halsall 2007: 193f, who points out that the term is first attested in *Notitia Dignitatum*. For a discussion of Alaric's kingship, which was a later feature of his career, ibid.: 202-06. *Contra*, Heather 1991.

[35] Heather 1991: 205; Wolfram 1988: 131-50; Halsall 2007: 200ff. See also below, chapter 1.3.1.

[36] Although very little is known of the details. They seem to have captured Aquileia and other cities in northern Italy, for which see Wolfram 1988: 151 and the following note.

[37] Halsall 2007: 212-17; Wolfram 1988: 155-59; further the loss of troops, ibid.: 152; on the purge, Nixon 1992: 66.

[38] For the events and context, see Drinkwater 1998 and Kulikowski 2000.

supported the usurper Jovinus, whom they betrayed to Honorius before attempting to set up a legitimately Roman regime of their own.[39] When the Visigoths gave up and surrendered to the future emperor Constantius (III) in 416, they provided valuable service for the Romans by destroying part of the Vandals and Alans in Spain, before they were finally settled in the Garonne valley in Aquitaine in 418. This turned out to be their final settlement and lasted until 507, well after the fall of the West Roman government, although that was probably not the intention of anyone concerned.[40] Their integration into Roman society was such that it is impossible to identify them archaeologically despite their presence in the region for nearly a century.[41] With only the name "Goth" (and perhaps the royal title of their leadership) to distinguish them from their Roman protagonists, there are only a few place names to provide some indication as to their distribution.[42] Even their Arianism was a holdover from the Eastern imperial regime of the mid-4th century.

Visigothic poliorcetic skills improved between 378 and their final settlement in Gaul in 418 through service with Roman units and access to Roman military infrastructure.[43] While they may have kept peace with walls in the 370s (although, as we have seen, this is debatable), this was certainly not the case in the early 5th century. The fighting between imperial usurpers Constantine III, his successors, and legitimist generals supporting Honorius most often ended in sieges, e.g. *Arles (411). Likewise, Visigothic operations were directed against the control of the fortified cities of southern Gaul.[44] The list is extensive, but consists only of brief references in chronicles and fragmentary historians. *Valence was stormed in 411, although possibly by another Gothic unit in Roman service;[45] *Narbonne and *Toulouse were stormed by Athaulf's Goths in 413;[46] at *Marseilles king Athaulf himself was wounded in the fighting and withdrew; during the siege of *Bazas, the Goths were joined by *bacaudae* (local rebels) and Alans, but

---

[39] Wolfram 1988: 150-171.
[40] Halsall 2007: 231ff.
[41] This is universally recognized; see e.g. E. James 1977, Heather 1996, and Todd 1998: 463f.
[42] Wolfram 1988: 228-31.
[43] Thompson 1958 ignores evidence of "Germanic" siege warfare between 378 and 536, which, of course, includes the whole period under discussion here.
[44] For a detailed survey of the civil wars, see Drinkwater 1998.
[45] In service of Honorius, capturing the usurper Jovinus. *Gallic Chronicle* of 452 s.a. 411 (Murray 2000: 81).
[46] Held by Honorius' generals, this was done in frustration over Honorius' failure to keep his promises.

the Alans went over to the citizens, prompting a Gothic withdrawal. They had come a long way from their "fumbling" around city walls in 378—if that indeed had ever been the case. By now, they were certainly able to storm heavily fortified Roman cities manned by the regular army. The Visigoths continued to rely on siege warfare throughout the 5th century. After their final settlement, they possessed the logistical wherewithal to settle down for year-long sieges in order to starve out their enemies. In the course of the wars in the 420s and 430s, they besieged the provincial capital of Gaul, *Arles, in 425 (and 430, and 453, but were several times turned back by Aëtius). *Narbonne was also besieged on several occasions, significantly for a whole year in 436-7, but again the Visigoths were turned back by Aëtius' second in command, Litorius, and were (possibly) besieged in turn at *Toulouse in 439.[47]

Throughout the first period of their settlement, c. 418-465, the Goths, though preoccupied with their own self-preservation, remained in effect Roman clients that bought into Roman imperial ideology. The problem, of course, was which contender to the purple they should fight for and which local aristocratic faction to rely on whenever Roman civil wars erupted, which became painfully apparent at *Bazas in 413, when the federate Visigoths and Alans chose to support two different Roman factions. Instead of regarding them as foreign invaders, it is more profitable to see them as active participants in the greater game of imperial infighting where aristocratic faction played an immense role. They became more independent after a treaty in 439 granted kingship, but this may still be in the mode of client recognition.[48] They still fought for the Romans against the Sueves in Spain and the Huns in Gaul, and were also used against the Alans holding *Orléans in 453. They even supported Avitus, from the Gallic senatorial aristocracy, as Roman emperor (r. 455-456), but after his demise they became embroiled with his successor Majorian (r. 456-61) and the latter's general Aegidius, whose power base was in northern Gaul.[49] Again the conflict revolved on sieges; the Goths managed to hole up Aegidius in *Arles in 458, but were defeated by a relief army under Majorian. After Majorian's death in 461 the Visigoths supported his successor, Libius Severus (461-65), which meant more fighting against Aegidius: they

---

[47] It is unclear whether this was a siege or a battle; see the discussion in CO.

[48] Halsall 2007 argues that this, rather than the settlement of 418, is a significant turning point in Visigothic history.

[49] Wolfram 1988: 173-81 for the basic narrative. For the factionalism and aristocratic machinations, see e.g. Mathisen 1979. For a larger collection of perspectives on the chaotic situation, see the essays in *5th-C Gaul*.

received *Narbonne in 461 to keep it in loyalist hands; along the Loire, *Castrum Cainonense was besieged by Aegidius in 463, and in the same year, many parties were involved in the fighting over *Angers, which was the objective of a loyalist army; presumably the Franks mentioned were allies or subordinates of Aegidius.[50]

The Visigothic kingdom took final form under Euric (466-484), who achieved real independence and expanded the Visigothic kingdom to the Loire and Rhône in Gaul as well as large parts of Spain. *Pamplona, *Saragossa and *Tarragona were all captured in 472, while *Arles and *Marseilles fell in 473. The Visigothic siege of *Clermont 474, which was defended by Sidonius Apollinaris and his supporters, was conducted by a Gothic army large enough to blockade the city and clearly resulted in the destruction of parts of the wall (the citizens observed the relief army *e semirutis murorum aggeribus*). Although Sidonius' account does not refer to the methods employed, in a *carmen* to his friend Pontius Leontius in the 460s extolling the latter's *burgus* (fortified residence), he outlined possible threats that it would surely withstand:

> Those walls no engine (*machina*), no battering-ram (*aries*), no high-piled mound (*strues*) or near-built ramp (*agger*), no catapult (*catapulta*) hurling the hissing stones, no tortoise-roof (*testudo*), no mantlet (*vinea*), no wheel rushing onwards with ladders (*scalis*) already in position shall ever have power to shake.[51]

It might be tempting to dismiss this detailed catalog of siege technology as mere classicizing flattery, but Sidonius' description of the destroyed walls of *Clermont would also have to be dismissed, as well as what else we know of Visigothic military organization. Destroying a defended wall required substantial skills in engineering and a logistical ability that would reflect Visigothic absorption of Roman institutions and personnel. Indeed, by the 470s, the Visigoths had a system of border guards patrolling the Loire, an Atlantic navy that could reach to the Rhine estuary, and a significant number of Gallic aristocrats and erstwhile Roman army commanders in their service.[52] They also had extensive experience in siege warfare for over a century, both as attackers and defenders, and the nature of siege warfare in the mid-5th century is independently confirmed by a number of sources, both Greek and Latin, describing the Hunnic wars.

---

[50] See further chapter 1.3.3 below.
[51] Sid. Ap. *Carm.* 22.121-25; see also Wickham 2005: 171 and Mathisen 1993: 55 for context.
[52] Wolfram 1988: 217ff.

1.1.3 *The Thick Description: Huns and Romans, 441-452*

Most of the 5th-century siege descriptions mentioned above are so brief that it is difficult to conclude anything about their nature, except that they frequently involved logistically demanding blockades, while some were settled by storms. The general context of Visigothic activities indicates that they operated as a regular Roman army at the time, certainly from the 380s onward. The problem, of course, is establishing the normal *modus operandi* for contemporary Roman armies. The inherent problem is demonstrating continuity: while Ammianus provides excellent evidence of the situation in the 4th century, and Vegetius largely corroborates this image a generation or so later, there are obviously methodological problems in applying these sources to a situation where it is precisely continuity that is under debate.

Sidonius apparently confirms survival of advanced siege technology, but only alludes to it in actual use. The most detailed 5th-century siege description relevant to the situation in Europe is found in one of the fragments of the Greek historian Priscus. He was a high-ranking diplomat with good sources of information, and described the Hun capture of *Naissus in the Balkans in 442. The anecdote provides direct evidence against the assumption that barbarians were inherently incapable of adapting to complex (Roman) styles of warfare. We can demonstrate that Hunnic besieging skills were largely derived from Roman captives taken from active service on the Danube frontier; hence it is also evidence that the Romans maintained the necessary infrastructure to the mid-5th century, when the Visigoths and other barbarian groups were heavily involved in Roman civil wars or outright conquest. Some scholars have therefore sought to eliminate the embarrassing level of detail by discrediting this description as a literary invention. Recent research has however refuted the arguments, so that the general consensus today is that Priscus' description is accurate.[53] The Huns acquired these skills through a combination of historical experience from Central Asia and Persia, service as Roman auxiliaries, and the capture of a great number of Roman prisoners.[54]

During their invasion of the Balkans in 442, the Huns bridged the river that separated them from *Naissus—which in and of itself was a signifi-

---

[53] Thompson argued forcefully for this in an article published in 1945; hence his 1948 book on the Huns is virtually devoid of any siege warfare. His assertions have been dismissed by Blockley 1972 and Tausend 1985/6. See CO for further discussion and full references.

[54] See chapter 7.1.3 for a discussion of this.

cant feat of engineering[55]—and brought across siege engines, which included towers, described by Priscus as [a framework of] "beams mounted on wheels" (δοκοὺς ἐπὶ τροχῶν κειμένας)

> ...upon which men stood who shot across at the defenders on the ramparts. At the other end of the beams [towers] stood men who pushed the wheels with their feet and propelled the machines wherever they were needed, so that one could shoot successfully through the openings made in the screens. In order that the men on the beams should fight in safety, they were sheltered by screens woven from willow covered with rawhide and leather to protect them against other missiles and whatever fire darts might be shot at them.[56]

These towers provided dense cover fire that scared the defenders off the ramparts, allowing rams (κριοί) to be brought up close to the walls. The Romans used the time-tested method of dropping large boulders onto them,[57] and clearly had artillery capable of firing missiles, incendiary or otherwise, against the approaching Hun towers. This was partly successful but the Huns had prepared far too many rams to stop them all, and the towers were making sections of the wall untenable.[58] The Huns could then exploit the breaches made by the rams and set up ladders where the defenders had been driven off.[59]

The use of rams is attested as a particular Hunnic skill by an independent source. Gregory of Tours, based on the *Vita Aniani*, bishop of *Orléans, informs us that in 451,

> Attila the King of the Huns marched forward from Metz and ravaged a great number of other cities in Gaul. He came to Orléans and did all he could do to capture it by launching a fierce assault with his battering-rams ... [The bishop and population prays for help, which miraculously arrives] ... The walls were already rocking under the shock of the battering-rams and were

---

[55] For the significance of bridge-building in relation to military engineering, see chapter 6.1.1 on infrastructure and 7.2.3 on Avar appropriation of Roman skills.

[56] Priscus fr. 6.2.

[57] Vegetius 4.23; see also the discussion of the siege of Aquileia in 361, chapter 1.1.1 above, where the defenders employed similar tactics against the sappers and engines that approached the wall.

[58] Priscus fr. 6.2: "From the walls the defenders tumbled down wagon-sized boulders which they had prepared for the purpose when the machines were brought up to the circuit. Some they crushed together with the men working them, but they could not hold out against the great number of machines.

[59] Priscus fr. 6.2; note further that the methods employed against *Naissus 442 are consistent with Theodoret's description of the siege of *Theodosiopolis 421.

about to collapse when Aëtius arrived, and with him Theoderic, the King of the Goths, and his son Thorismund.[60]

The exact course of the Hunnic invasions remain unresolved, but these are not the only cases where the Huns showed a formidable capacity for besieging and capturing well-fortified Roman cities.[61] Their most successful conflicts with the Roman Empire between 440 and 452 were sieges. *Viminacium and other cities and forts were taken in the Balkans in 441, before the capture of *Naissus, *Singidunum and *Sirmium in 442.[62] Another invasion resulted in the capture of *Ratiaria in 447. In Gaul, *Metz was attacked and possibly stormed in 451, but with no traces in the archaeological record. *Trier was another center of artillery production and erstwhile Roman capital with defenses that would match those of the middle Danube; it too was attacked at that time. When the Huns invaded Italy in 452, *Aquileia was held by a Roman garrison that bravely withstood a fierce storm involving wall-breaking machines supported by artillery. *Milan and *Pavia were similarly assaulted and defended.

The East Roman government thought it necessary to expend immense efforts on rebuilding the Theodosian Walls of Constantinople in just two months after an earthquake in January of 447. The Huns were raiding the Balkans (cf. *Ratiaria 447), so restoring the original single-course wall was of course expeditious in any case. What emerged, however, was not only a restoration of the old course of wall, but the famous and formidable double wall in addition to outworks (*proteikhisma*) and a deep moat, which withstood virtually every siege for the next millennium.[63] The episodic narratives from *Naissus (442), *Orléans (451) and *Aquileia (452) provide us with enough evidence to understand the nature of the Hunnic threat and the context in which the Visigoths fought. It also shows that regular Roman garrisons and auxiliary forces not only survived, but were still geared towards defending cities in much the same manner as in the 4th century.

---

[60] See CO for further discussion and quotes. The alternative account found in Jordanes has Aëtius already in the city and emphasizes Roman countermeasures, such as constructing additional earthworks, but this is probably confused with a later siege of the city discussed at *Orléans 453.

[61] Maenchen-Helfen 1973: 108-29 contains a detailed discussion of the Balkans campaigns in 441/2 and 447. He never finished the section on the war in Gaul, but proceeds (129ff) with a discussion of the war in Italy. Thompson 1948/1996 is less detailed, as he disregarded much of the narrative evidence for the sieges but at the same time emphasizes how the Huns captured fortress cities with arms-producing facilities.

[62] Tausend 1985/6 has a detailed survey of the siege methods used against Balkan and Italian cities.

[63] See Foss and Winfield 1986 and *EncMedWar* s.v. Constantinople (Walls).

1.2 From Emergency Measures to New Institutions

Taking institutionalized poliorcetic knowledge to c. 425 presents no substantial problem, since most of the regular army and its infrastructure was still operational and can readily be identified thanks to the *Notitia Dignitatum*. By the mid-5th century, however, several regions, especially Britannia and North Africa, had slipped permanently outside central control, but the examples above show the traditional infrastructure still in operation on the Danube (see also *Noviodunum 437), in Italy and Gaul into the 440s and 450s, providing an environment in which later clients were still acculturated to Roman styles of warfare. The eastern frontier is treated in the next chapter.[64] Demonstrating a functioning military infrastructure in the West in the following decades is something else, as central authority collapsed in the 460s and 470s. Clear references to a functioning army after 425 are few; to military infrastructure and institutions even more scarce. However, as the narrative evidence above shows, the two probably can not be separated: the survival of Roman or Romanized troops will in most cases mean the survival of supporting institutions. Even if these declined beyond repair by the end of the 5th century, there were several factors that facilitated the continuation of siege warfare. New forms of "privatized" administration (such as the great bipartite estates; for definition see below) and recruitment (private retinues) gained in importance, drew directly on existing systems and personnel, and sustained Roman-style warfare far beyond the lifespan of the Roman state. At the same time, substantial regional commands survived the collapse of central authority. Most of these in turn were either absorbed by or morphed into "barbarian" peoples: ethnogenesis produced "barbarians" from the direct descendants of Roman provincials.

1.2.1 The Regular Army in the 5th Century

By the end of the 4th century, the Roman army was still a formidable force comprising some 4-500,000 men in regular units paid, equipped and commanded by central authorities. There existed a distinction between *comitatenses*, originally field troops in the retinue (*comitatus*) of the tetrarchic emperors, and *limitanei*, a term that evolved for garrisons on the frontier (also called *ripenses*—"those on the riverbanks," *burgarii*, *castrenses*—

---

[64] Although examples are few in the 5th century; see *Theodosiopolis 421.

"those in fortifications" and the like).[65] Elaborate theories of grand strategy, with defense in depth, based on increasingly barbarized mobile strike units of cavalry used to support inferior border troops of mere peasant militias, have been thoroughly demolished.[66] Isaac demonstrated that the chief differences between *comitatenses* and *limitanei* were administrative rather than qualitative.[67] Scholars now recognize that the two very often fought together, were often transferred between categories (*limitanei* could e.g. become [*pseudo-*]*comitatenses*), and had in almost every respect similar conditions of service, property holding, recruitment, equipment and tactics. However, Whitby demonstrated that in the Balkans, *limitanei* had more cavalry than *comitatenses*, belying the notion of a cavalry-dominated mobile strike force. The simple reason was that border troops had to patrol and police long stretches of frontier.[68]

Non-Romans certainly served on a significant scale, either recruited into regular units, which as a result often carried an "ethnic" unit name, or as *foederati*. Originally a term for client units from beyond the border, serving for limited periods based on a treaty (*foedus*), by the early 5th century some federate units had become partly or wholly integrated into the formal structure of the army. Serving permanently in Roman provinces, they eventually drew on Roman recruits to uphold numbers, changing the nature of manpower and meaning of the term. Olympiodoros confirms that this was the case already in the early 5th century; by the early 6th-century East *foideratoi* were regular units by another name.[69] In the West, however, such processes of integration came to an unnatural halt in the 5th century due to civil wars and the subsequent involution of the frontiers. This meant that units of Romans or thoroughly Romanized barbarians ended up as political groups in opposition to central authorities at various times, while the conflation of Roman and barbarian identities led to non-Roman labels derived from the original unit or federate designation (see further 1.3).

Elton argued that the West Roman army remained efficient during the first quarter of the 5th century.[70] Although named units are extremely dif-

---

[65] The literature on the late Roman army is vast, but for its basic organization, see especially Grosse 1920, Jones 1964 chapter 17, Hoffmann 1969, Southern and Dixon 1996, Elton 1996, Nicasie 1998, Lee 2007.

[66] Notably Luttwak 1976, based largely on outdated views and selective use of evidence.

[67] Especially Isaac 1988, but see also Jones 1964 and the relevant sections in the most recent standards referred to above.

[68] Whitby 2007.

[69] Elton 1996: 91-94.

[70] Elton 1996 finds no major flaws or signs of disintegration in the Roman army in

ficult to identify securely after that date, there is still evidence for regular and federate Roman units that phase out of central control and either form the basis for many of the independent Roman warlords in the period, or become absorbed by one of the successor kingdoms.[71] Occasional notices in the 420s and 430s do however attest to the existence of Roman units in the West,[72] and Valentinian III (425-55) still legislated actively for the return of deserters to their units in 440 and the recruitment of citizen soldiers from the dependents of senators and landholders as late as 443. There were substantial garrison and field forces in Italy during the Vandal scare of 440, but not enough to protect all cities under threat in the peninsula.[73] In northern Italy, a Roman garrison still defended *Aquileia against Attila in 452. In southern Gaul, Roman troops acclaimed Avitus at Arles in 455.[74]

Recognizably Roman troops may however have been most prevalent in northern Gaul. The scarcity of Roman troops under Aëtius at the battle of the Catalaunian plains in 451 has been taken by many to prove that the Roman army in northern Gaul had declined beyond repair.[75] However, Jordanes' famous list of federate participants on the Roman side reflects a 5th-century literary fad of listing outlandish and anachronistic barbarian tribes—the more the merrier—and probably camouflages a number of still active Roman units. Furthermore, it is often forgotten that the Romans needed to garrison a large number of *urbes* (well over 60 great cities were within range of the Huns), suburban fortifications and border forts in the face of an overwhelming invasion by an enemy skilled at siege warfare. Assembling a field army from these garrison forces was theoretically possible, but it would strip civilians from desperately needed protection as well as deprive the Romans from any form of strategic reserves or secure bases in case of failure. The fate of the Danube cities in 441-42 is instructive.

Europe before 425. The problems were, on the whole, political, not ethnic or institutional. Cf. the general argument of Halsall 2007.

[71] MacGeorge 2002 and just below for warlords; see chapter 1.3 below for assimilation into "barbarian" ethnicities.

[72] Some pertinent examples: *Chr. Gall.* 452, 98-100 (a. 425); Hyd. a. 430; where soldiers still play a significant political role in Gaul and Italy.

[73] *NVal* 6.1 and 6.2, respectively. The latter provision was followed up shortly afterwards by a demand for cash payments from a different category of Senators (inactive) as well as counts and court officials. They were apparently exempt from the recent demand for recruits, but now required to pay money instead as an exceptional levy to cover emergency costs (*NVal* 6.3). Hence Valentinian granted permission for citizens to arm; see discussion at 1.2.5 below.

[74] MacGeorge 2002: 153; see Mathisen 1979 for the political context.

[75] Wood 1998: 534f takes the evidence of Jordanes literally.

Garrisons had been reduced in order to provide troops for the campaign against the Vandals in North Africa, but the Huns were able to exploit the situation to a devastating effect (see 1.1.3). Indeed, following recent archaeological advances made on Roman border fortifications on the Rhine, a strong case can be made that the frontier was still manned and defended until c. 450.[76] I would therefore argue that by far the heaviest concentration of Roman troops in the West after 450 was in northern Gaul, as is evident from developments in the 450s to 470s, when a large number of Roman warlords were active in the region, but now shifting their troops towards the Loire due to conflicts with central Roman authorities and their Visigothic allies. Their troops were organized and equipped in Roman fashion, still clearly recognizable to their Eastern colleagues well into the 6th century. Although the context is somewhat muddled,[77] Procopius' description is well worth quoting:

> Now other Roman soldiers, also, had been stationed at the frontiers of Gaul to serve as guards. And these soldiers, having no means of returning to Rome, and at the same time being unwilling to yield to their enemy who were Arians [i.e. Visigoths], gave themselves, together with their military standards and the land which they had long been guarding for the Romans, to the Arborychi [Armoricans, i.e. independent Roman provincials] and Germans [Franks and Alamans]; and they handed down to their offspring all the customs of their fathers, which were thus preserved, and this people has held them in sufficient reverence to guard them *even up to my time*. For even at the present day they are clearly recognized as belonging to the legions to which they were assigned when they served in ancient times, and they always carry their own standards when they enter battle, and always follow the customs of their fathers. And they preserve the dress of the Romans in every particular, even as regards their shoes.[78]

Remaining troops in *Noricum ripense* (along the upper Danube) held on until the 480s, but were finally ordered back to Italy by Odovacer (who famously deposed the last western emperor in 476).[79] A little to the south, *Noricum mediterraneum* continued to be a part of Odovacer's and Theoderic's kingdoms requiring protection, though little is known of this. To the

---

[76] Thus Wickham 2005: 102f and n. 117 for references.

[77] For context, see e.g. Bachrach 1975. While Halsall 2007 argues for the survival of Roman units and allies in northern Gaul and also for a militarily significant Loire frontier in the late 5th century, I believe, with Wickham 2005: 178-84 and *passim*, he somewhat exaggerates economic and administrative decline after 400, cf. ch. 1.3.3-4.

[78] Proc. 5.12.53f. My emphasis.

[79] Halsall 2007: 287. For an explanation of the curious circumstances, see below chaper 1.3.1.

west, however, Rhaetian *limitanei* continued to guard Italy's northern borders under the Ostrogoths.[80] Romans appear to have been the basis of Roman officers Nepos' and Marcellinus' power in Dalmatia, though Marcellinus hired great numbers of Scyths (most likely Pannonian Goths) to join his Italian expedition in 450;[81] Nepos was still plotting his return to the Western throne at his death in 480. There are two very peculiar reports of Roman troops that moved over large distances in search of new employers. The Briton Riothamus came to Gaul around 470 and was allied with the last emperors against the Visigoths, but after a complete defeat at Visigothic hands ended up in Burgundy.[82] A Western aristocrat, Titus, was invited to the Eastern Empire around 468 with his personal followers due to his military reputation.[83] Thus, Western commanders and soldiers were still prized by the East Roman government.

1.2.2 New Ways of Recruiting Troops

It is common wisdom that barbarian federate armies replaced the salaried, professional units in the West, and that these armies eventually formed their own kingdoms as the Roman state atrophied.[84] Visigoths (Aquitaine, 418) and Burgundians (upper Rhône, 443) formed the largest blocks of effective federate troops in the 5th century, although Alans were also significant in e.g. Armorica and southern Gaul. They were settled by Roman authorities according to administrative procedures long debated by scholars. In Spain and Africa, in contrast, the invading Sueves and Vandals helped themselves, but may still have utilized similar, available administrative techniques in order to ensure orderly government and supply. The question is, basically, which of the many sophisticated tools available in the Roman bureaucratic toolbox were used, and implicitly, which of them best explains the evolution of early medieval military and political institutions. The Romans clearly had many options, ranging from land via cash payments to compulsory service. They had often settled defeated or surrendered peoples (*dediticii*), some as military colonists (*laeti*), on tax exempt land in return for providing recruits to the army. Land was also

---

[80] Wolfram 1988: 301.
[81] MacGeorge 2002: 62f. This was the environment in which the Ostrogoths arose, cf. chapter 3.1.1 below.
[82] Ibid.: 74, 242f and *PLRE* 2 s.v. Riothamus (p. 945).
[83] *Life of St. Daniel the Stylite*, 60ff. Needless to say, the emperor Leo was frustrated after Titus went to get a blessing from St. Daniel and decided to take up asceticism himself.
[84] For a much more sophisticated update, see Halsall 2003 and 2007 with references to older debates and literature.

important for regular troops (*limitanei* as well as *comitatenses*) who were stationed in the same district for many years and invested in lands and houses. Due to their military services, they too were exempt from public burdens normally required of landowners.[85]

Due to the importance of landowning to military organization in the early middle ages, a recurring theory has been that property was expropriated from imperial estates and other landowners and distributed to individual soldiers. Another old suggestion is that procedures for temporary billeting of troops and officials on the march, *hospitalitas*, became a permanent solution. Goffart has the most elegant and comprehensive solution, combining elements of these theories with the immense administrative sophistication of the late Roman state: soldiers were not given land as such, but the right to collect tax income from specific tracts of land; possession remained with the original owners, taxes were still assessed by the government, but tax payments were made directly to troops who had received official documentation from central authorities detailing properties liable and amounts owed. Again, some have suggested a model in veteran settlement, while others rents from land (i.e. *emphyteusis* rather than *possessio*), as compensation for service.[86] No absolute position shall be taken now; clearly a variety of adaptations were available depending on period and region. While the debate between various modes of settlement has received an enormous amount of attention the past generation, the evidence is amenable to many interpretations and anyway has little bearing on the military practices examined here.[87]

---

[85] Jones 1964: 620; Lee 2007: 81f for settlements; Lee 2007: 60 for exemption from public burdens.

[86] I owe the quite ingenious suggestion of *emphyteusis* to Peter Heather, who mentioned it to me during a discussion at the "Medieval Frontiers at War" conference at Cáceres, November 2010.

[87] The main statement is by Goffart 1980, where he laid out his argument for the division of tax receipts among barbarian troops settled on Roman soil. A summary of the following debates (often highly critical) is found in Halsall 2007: 422-54, who also discusses the problems of landowning and military service in general. Wickham 2005 has a good case for land distribution, certainly for Vandal Africa and extends the argument to Ostrogothic Italy. Sarris 2011 also argues for landed troops, adducing both his reinterpretation of late antique estate structure (1.2.3 below) as well as 7th-century Byzantine parallels; I similarly argue for parallels between Byzantine and Western military landholding in the following centuries, but without accepting a "thematic" system based on land in the 7th century (see 2.2.3). Goffart 2006 responds to criticism and adduces new evidence; particularly important is his insistence that terms for land are *not* straightforward, but complex terms in late Roman administration. Some notable alternatives have been suggested for e.g. the Visigoths by Sivan 1987 (veteran settlement) and Jiménez Garnica 1999 (a variety of gradual measures),

In the areas of the West under consideration here, salaried troops on the Roman (or rather, Goffartian) model may have been typical in Ostrogothic Italy (and the first phase of Visigothic settlement in Aquitaine), while self-supporting landowning soldiers were more common in Gaul and Visigothic Hispania from around 500. If indeed large numbers of barbarians only collected taxes and rents, as Goffart suggested, they soon invested their wealth in acquiring more land. Thus all soldiers with any significant means would become landowners in their own right, no matter the original arrangements, and could seek exemptions and immunities from taxes and obligations on newly acquired properties in return for continued military service. Thus, during the whole period under examination, military obligations were held for a combination of historical, economic, family or ethnic reasons, or personal inclination.[88] No matter the views adopted, most of these arrangements stemmed from 4th-century imperial practices or 5th-century improvisation based on the imperial administrative apparatus. The importance, status and obligations of landowning soldiers were complex and very sensitive to personal, economic and agricultural changes, as well as the burdens of campaigning. As a consequence, such troops receive much attention in early medieval legislation, giving the impression that a large proportion of early medieval armies were raised from among free, landowning men who owed individual military service. However, the cost of military service for a qualified individual landholder was very high and constantly had to be regulated by royal decree (hence the laws). Finally, siege warfare required institutionalized knowledge and logistical support above the level of individual landowners/tax-collectors/rentiers.[89]

As will be argued presently, large-scale landholders, even if at first disadvantaged by the burden of supporting troops, were soon able to turn the situation around, and maintained their own military followings that became far more important than individually landed troops. Not only were they able to negotiate directly with the emperor or king and his officers (or

---

but Goffart's model still works well for the Ostrogoths (cf. discussion in Halsall 2007: 443ff). See also *TRW* 1.

[88] Both Bachrach (e.g. in Kazanski (ed.) 1993, with many other pertinent contributions) and Halsall 2003 emphasize the importance of such troops, though from thoroughly different perspectives. The most famous example is of course from the early Frankish *Pactus Legis Salicae*, which legislated against women inheriting *specific* types of land. Anderson 1995 demonstrates that this peculiarity in fact derived from the Romans settling Franks in Gaul in order to provide recruits—since only men could serve, only they could hold land associated with conditions of military service. All other land was unproblematic, but the meaning was soon forgotten. See 1.3.3 and 4.1.5 below.

[89] See in general the discussions in chapters 5-8 below.

rather, *were* his officers), their economic and logistical capabilities would dwarf those of individual troops settled on or near their domains (1.2.5). By the sheer gravitational pull of their wealth, magnates were continually able to attract Roman soldiers into their orbit and make them part of their military followings; alternately they could take possession of their land in return for shouldering their military burdens. They must in fact be credited with keeping Roman soldiers active into the early middle ages in Gaul and possibly Hispania as well.

1.2.3 *The Military Following* (obsequium) *in East Roman Warfare*

Research on the private military following (*obsequium*)[90] has focused on its origin as private bodyguards for officers and magnates recruited from various barbarian peoples, whether the institution was inspired by Germanic custom, and how they related to the Roman state. There has however been little focus on their use as military forces compared to the regular army. In the 6th-century East, we see a fully developed model in which the private forces were heavily drawn from and often supplemented the regular army in all types of warfare. This model can be traced back to common late Roman developments and may thus be used to shed light on the West in the 5th century, showing in detail how privately raised forces gradually took over the functions of a centrally controlled army while maintaining continuity in personnel and practices.[91] The spread of militar-

---

[90] Current English renderings (e.g. war-band, retinue, following) tend to be value-laden and imprecise, but I nevertheless use the latter two. When speaking of the institution in a more abstract sense, I attempt to stay with the Latin term *obsequium*, which is in itself problematic (meaning anyone who was obliged to follow a lord, or less feudally, a patron), but occurs in the early legislation on military followers (see below). Another viable alternative is *buccellarii*, which occurs in both Greek and Latin, cf. below.

[91] Schmitt 1994 provides a good overview of the evidence and previous debates, the imperial legislation, and division into free and slave troops, but argues for Germanic origins of this institution in the late 4th century, before going on to point out that the necessary mechanisms are actually attested in Roman evidence decades earlier. Liebeschuetz 1993 similarly argues for a close connection between Roman commanders' recruitment of federate barbarian troops and the rise of private retainers, but Whittaker 1993, 1994 points out that Roman landlords played an important part in establishing this practice in the Roman West. Liebeschuetz 2007 (wrongly) criticizes Whittaker's views (cf. 1.2.4 below). For the Roman East, see Gascou 1976, who argues that the 6th-century *buccellarii* known from Egyptian papyri were a reflection of the public obligations imposed on the vast Egyptian estates, but Sarris 2006 demonstrates that this was in fact a case of illegal appropriation. Private retinues were nevertheless applied in state service, as will be argued below. For the 5th-century background in the East, see Lenski 2009 on armed slaves; for later developments, see Haldon 1984 and Whitby 1995.

ily significant private troops (as opposed to bailiffs, bodyguards and aristocrats' thugs) supported by estate income developed simultaneously with the rise of the bipartite estate and its spread from East to West, and may be connected.[92]

Based on a thorough analysis of Egyptian papyri, Peter Sarris has recently demonstrated how East Roman magnates in the 6th century appropriated public prerogatives on a large scale. Amongst other things, they maintained considerable private forces of *buccellarii* dispersed in small units over many tenements (*epoikia*) of their lord's domains.[93] They cannot be dismissed as thugs and barbarians, as a significant proportion of these forces were illegally drawn from regular military units, enticed into the temporary or permanent service of an aristocratic household. This was possible because garrison troops were chronically underemployed in times of peace. They often had months to spare every year during which they could do private business or find employment in a magnate's retinue.[94] Their organization generally reflects that of the regular army, albeit with a few peculiarities. In 6th-century Greek literary sources, there is normally a distinction between higher-ranking *doryphoroi* (δορυφόροι, "spear-bearers" or bodyguards, i.e. someone with high personal trust with the patron), and *hypaspistai* (ὑπασπισταί, shield-bearers), denoting regular troops, although other ranks are encountered; the Latin *buccellarii* is used frequently as a blanket term in Greek administrative records.

While the papyri provide details of individuals and administrative procedures in Egypt, there is good narrative and legal evidence that the practice prevailed elsewhere in the East Roman Empire in the 5th and 6th centuries. Belisarius had the most spectacular following of seven thousand heavy cavalry, which was used faithfully in service of the Roman Empire. He was reputed to take good care of soldiers and their equipment, supplied from his own household (ἐκ τῆς οἰκίας), and "each of them could claim to stand first in the line of battle and to challenge the best of the enemy."[95]

---

[92] For the significance and spread of the bipartite manor, see Sarris, 2004, and the end of this section.

[93] Sarris 2006: 162-75.

[94] Sarris 2006: 166. *Stategikon* 1.7 (14) limits the annual winter leave to three months, but implies that it could be more in time of peace when soldiers were quartered in their home province. See further Palme 2007: 26off for more Egyptian evidence. While Egypt may have been a special case within the empire, and scholars have rated Egyptian troops as inferior, Palme warns against underestimating the defensive needs of Egypt and the efficiency of the units stationed in the province.

[95] Proc. 7.1.8-15; 18-21. See also *PLRE* 3 s.v. Belisarius.

Other prominent generals also had their own retinues that took part in combat, usually numbering from 300 to 1000 men.[96] There was a high degree of interchangeability of personnel between regular units and private retainers. The practice of drafting skilled regular soldiers into generals' *buccellarii* was widespread and apparently more acceptable on expeditionary campaigns. Likewise, skilled and experienced *buccellarii* (the *doryphoroi*) could be commissioned as high-ranking officers in the regular army, while larger units were used both for garrison and expeditionary service. In terms of military practices and areas of use, there was little to distinguish the *buccellarii* from the regular army. Both categories of troops were frequently mobilized for expeditionary campaigns and regularly served together from Italy to the Caucasus and were frequently involved in sieges.[97]

The existence of private troops depended on a particular set of socioeconomic conditions that prevailed in the late and post-Roman Mediterranean. Just before the period under consideration here, agricultural organization varied between independent villages (largely in the East), while vast *latifundia* cultivated by bound tenants (*coloni*) and centered on opulent *villae* dominated in much of the West, especially Italy, North Africa and parts of Gaul and Spain. Lesser *villae rusticae* and individual, scattered farmsteads were otherwise the norm.[98] In contrast, many of the estates (Latin *domus*, Greek *oikoi*, οἶκοι, lit. "houses") that supported private forces in the East were organized in the form of bipartite manors, consisting of demesne lands for the immediate needs of the landowner (but often distributed over a wide area), which were surrounded by dependent villages, whose inhabitants paid rents off their own lands and further provided labor service on the landowner's demesne. This organizational form, once believed to be a late Merovingian or early Carolingian innovation, originated among the Imperial service aristocracy in the 4th-century East and became widespread in the Mediterranean world by the early 5th century.[99]

---

[96] For further examples, see Jones 1964: 666f and Schmitt 1994, who also discusses the attempts of the authorities to control private troops. However, in estimating numbers, he tends to disregard the practice that even *buccellarii* were detached from their unit for special assignments. Hence the numbers quoted are in fact minimums.

[97] E.g. *Rimini II 538, *Osimo 539, *Fiesole 539, *Rome 549f, *Petra 550, *Phasis 556.

[98] Lewit 1991, Dyson 2003, *TRW* 9; Wickham 2009 extends this model well into the Carolingian period.

[99] For the classic view, see e.g. Verhulst 2003, Wickham 2005 and 2009; for the early dating, Sarris 2004: 303-11; for the relevance for the Visigoths, see the following section and chapter 3.2.2; for the Franks chapter 4.1.5. The Ostrogoths seem mostly to have avoided the intrusion of magnate landowners into military organization until late in Justinian's war,

Although large-scale Roman *latifundia* could produce on an industrial scale for export, the bipartite manor was much more efficient, as the system was organized for closely supervised production and specialization in different regions, villages and production centers. Tenants were grouped into villages, and could thus be better controlled than tenants holding scattered farmsteads, but they also kept a significant proportion of their own production. Finally, the demesne centers were more tightly and rationally administered.[100] The practicalities of maintaining private forces were modeled on the regular army's peacetime hierarchy and distribution in very small units of a handful of men in each location. This required an administration that could keep track of and draw on a wide range of resources, including cash income and a variety of supplies that could be moved around between production centers. Again the model was the imperial supply system established for the army (see further 1.2.5). The organization of such estates was conducive to maintaining troops scattered in small groups across a large number of villages and production centers belonging to the main *oikos*.[101]

Although the practice was at times opposed by the central authorities, they often had to accept realities and instead found ways to exploit these illicit contracts. As long as the private forces did not cause political scandal and were used for local defense, policing and estate administration, the central authorities tacitly accepted a certain loss of control.[102] Instead, the emperors focused on how to mobilize their resources in whatever way they could. Anastasios used magnate power to his advantage while trying to limit its centrifugal tendencies. He appointed the great Egyptian magnate Apion (whose family's *buccellarii* are so prominent in the 6th-century papyri) to organize the supplies of the combined armies that faced Persians in Mesopotamia in 502-03. However, he put bishops in direct charge of organizing urban defense in many Mesopotamian cities. He thereby avoided entrusting secular magnates, who had similar organizational capabili-

---

but bipartite estate organization has nevertheless been attested in 5th century Italy (and later) by Sarris. During the 540s, both Romans and Ostrogoths had landowners organize their dependants as fighting forces, but in Italy this never evolved beyond stopgap measures. See further Noyé 2007.

[100] For the most recent and comprehensive treatment, see Sarris 2006.

[101] Schmitt 1994: 167 & n. 172 provides good examples of how troops were distributed in very small units.

[102] See Lenski 2009: 145 for how a landowner used his private retainers against the governor of Syria in order to bully his way into public office in 444. Theodosius had to degrade his aristocratic rank as punishment.

ties, with too much control of military administration on the volatile eastern frontier.[103]

Yet the centrifugal tendencies remained strong, partly due to local needs, and by the mid-6th century, private military initiatives were a regular feature in the East Roman Empire. The Syrian countryside near the Persian frontier provides ample evidence for magnate and ecclesiastic sponsorship of local defenses.[104] On at least one occasion in 528, Justinian, who otherwise legislated energetically against private forces, ordered landholders to move to the East with their retinues in order to defend against a Persian invasion, and in 540, the defense of *Antioch was organized by Germanus, Justinian's nephew, who arrived with a small retinue of 300 men.[105] While this was famously unsuccessful, the *Miracula sancti Demetrii* show how the battle-hardened *douloi* (δοῦλοι, servants) and *therapeia* (θεραπεία, retinue) of local landlords and officials were regarded as integral to urban defenses in *Thessalonica (586) along with regular garrison troops and the citizen "militia" (see 1.2.5 below for the term).

Magnates had a powerful gravitational pull on the administrative and military apparatus, especially if they were civilian or military officials in charge of taxes and supplies. Agathias provides a particularly egregious example how Justin, a general related to the emperor, let one of his trusted retainers (from the context, clearly a *doryphoros*, although Agathias is loath to apply such an honorable title to him) exploit normal forced purchase and requisitioning practices while on campaign in Lazica. He presented himself as being out to buy supplies, but since he required forced purchases of livestock essential to the population (or even nonexistent in the area), they were compelled to pay for immunity. In addition to money, he also requisitioned foodstuffs that were sold off abroad or used to supply Justin's large retinue instead of the regular army.[106] There was also a risk that particularly powerful officers could actually take control over large parts of the regular army in their province. In Egypt, magnates already exercised considerable social control over regular military forces due to complex networks of patronage and their essential role in collecting and redistributing taxes and supplies. A famous case in point is the wealthy Aristomakhos in late 6th-century Egypt.[107] He was probably the military

---

[103] See chapter 2.4.1-2 on this.
[104] For privately funded defenses, see Trombley 1997.
[105] See Whitby 1995: 117 for discussion of events in 528 and Sarris 2006 for Justinian's legislation.
[106] Agathias 4.21f.
[107] *PLRE* 3 s.v. Aristomachus 2 (118f).

commander (*dux et augustalis*) of the province of Thebaïs, but had in addition a vast military following based on his personal income. While he used his powers to defend Egypt against severe raids from Nubians and Moors, he went far beyond accepted bounds by confiscating estates and commanding regular units from outside his province. At first he was tried for misconduct, but the emperor Maurice (582-602) probably needed his support and tried to appease him with honors in Constantinople and formalized his administration of the imperial estates, making him *curator domus Augustae*. Eventually his case was so serious that he was banished for plotting against the emperor and presumably the regular command structure was put back into place.

Even Justinian had immense difficulties in controlling his magnates. One obvious reason to fear their power was their direct control of military forces in distant provinces, a threat that became very real in the late 6th century as the example of Aristomakhos shows. Hence he spent much legislative energy on banning private forces and arms production.[108] A further measure was to formalize the episcopal oversight of urban defenses that Anastasios had instituted.[109] Since Justinian could not suppress all the great military magnates, he instead harnessed their forces by employing a large number of generals' followings in his wars of expansion. Based on narrative evidence from Procopius, such private soldiers accounted for a significant percentage of expeditionary troops during the reign of Justinian.[110] This may go some way in explaining why Justinian seems to have starved his expeditionary armies of supplies and reinforcements at critical junctures. In a climate of real logistical and economic stress, potentially dangerous rivals were effectively kept in check. Otherwise, the very largest followings could be broken up. Falling out of favor after his first tour of duty in Italy ended in 540, Belisarius saw his estates confiscated and his *doryphoroi* and *hypaspistai* dispersed among magnates whose loyalty was less

---

[108] For Justinian's struggles with his aristocrats, see Sarris 2006. Much of his social legislation might seem intolerant (against homosexuals, heretics, pagans etc.) and hence widely noted today, but it was used selectively against particularly wealthy individuals who may have been difficult to get at otherwise, cf. Proc. *Secret History* 11.30.

[109] See further the discussion in chapter 2.4.1 below.

[110] It seems that at least 10 % of the forces in Italy were always private retainers, but sometimes considerably more. Procopius never states the number of Belisarius' retainers on campaign in the West. Most scholars assume that they would have numbered no more than a thousand or so men, but Procopius refers to the Goths' awe as "one house" (i.e. that of Belisarius) destroyed the whole kingdom of the Ostrogoths.

in doubt. Alternatively, they found service in the regular army.[111] A final option was simply to pay private retainers to leave their lords, which Justinian resorted to during a recruitment drive in the Balkans around 550. His nephew Germanus (with imperial permission) simply paid cash bounties from his personal resources in order to attract volunteers from the *doryphoroi* and *hypaspistai* of local dignitaries.[112] Since their retinues were technically illegal, there was little they could do to protest. The practice continued in the East in the latter half of the 6th century, when several large-scale recruitment drives in the Balkans resulted in regular units that were named *buccellarii*, presumably from their origins. The middle Byzantine *thema* of the *boukellarioi* probably originated from the last remnants of private forces that were recruited under Maurice and Herakleios. Although dating is problematic, I would suggest that they were gradually recruited as their employers lost the ability to support them.[113]

The great independent estates that supported private troops disappeared from the East Roman Empire with the crisis of Avar, Slav and Persian invasions in the early 7th century.[114] Until then, however, large estates had important tax-collecting and army-supplying functions, both East and West, which were developed by Roman authorities before 400 AD and continued through the 5th and 6th centuries (see also 1.2.5). These practices eventually devolved on landowners in the West, where the most important adaptation was the widespread conversion of direct taxes to services. These were then merged with traditional obligations of labor and organization required by the Roman state.[115] However, the development is obscure and

---

[111] Proc. *Secret History* 4.13 for the distribution of *buccellarii*; see further Schmitt 1994: 171f.

[112] *PLRE* 2 s.v. Germanus 4; Proc. 7.39.16f. Dewing translates the problematic ἄρχοντες as "commanders," but the term might as well mean civilian officials and other high-ranking dignitaries.

[113] See Haldon 1984: 101f for recruitment under Maurice and *passim* for the history of these units and the establishment of the theme of the *boukellarioi*.

[114] According to Haldon 2004: 207-21, there were still some substantial estates in the 8th century, especially in Paphlagonia, furthest from the Arab raids, but the evidence seems to show little continuity in secular elites, and at any rate the estate economy was no longer on the scale that it could support a large proportion of the empire's soldiers.

[115] Durliat 1990 states the maximalist position, arguing that a precise fiscal and redistributive regime based on taxable income from the countryside survived and was administered by each *civitas* well into Carolingian times according to detailed land records. Goffart 1972 and 1982 rather identifies the underlying Roman mechanisms to show how one tier of a two-tiered tax system survived in the medieval *seigneurie* and other manifestations of labor obligations to early medieval kings, as the upper, more visible tier of taxes in cash was abandoned. The process is traced in more detail in chapters 2 and 3 *passim*, 4.1.5, and 4.2 and 4.3 *passim*.

many researchers have assumed that there was a lack of continuity in estate management from the 5th to the 8th century, while others have been unable to stomach the idea of a Roman-descended tax system operating in the Frankish kingdoms. Yet this form of economic organization became the norm in the West and a cornerstone of Carolingian political and economic life.[116] It is also clear that both the fiscal (above) and military (below) aspects of this system can be demonstrated in many Western provinces as early as around 400 AD, well before the collapse of imperial authority.

1.2.4 The Rise of Private Military Forces in the West

The realities of magnate power over local garrisons that are so evident in the 6th-century East also apply to the 5th-century empire as a whole. As early as 409, a significant force of "slaves and servants" was raised by two aristocratic brothers in northern Spain to resist the regular troops of the usurpers in Gaul.[117] While their forces were ultimately defeated, they had proven effective in the short term, and the financial ability to raise a large military force through private means was formidable. Olympiodorus informs us that around 400 AD, the great "Roman households received an income of four thousand pounds of gold per year from their properties (ἀπὸ τῶν κτημάτων αὐτῶν), not including grain, wine and other produce, which, if sold, would have amounted to one-third of the income in gold. The income of the households at Rome of the second class was one thousand or fifteen hundred pounds of gold."[118] The sums are astounding. A great senator had in theory enough income to equip, supply and pay the annual salaries of 48,000 heavy infantry.[119] It may be tempting to associate the decline of conspicuous consumption (games, inscriptions, opulent *villae*) and urban munificence after 400 AD with a shift towards investment in military rather than civic competition. Indeed, civic munificence in the West experienced significant decline almost exactly proportional to this

---

[116] Most recently, see Verhulst 2003.
[117] Olympiodorus fr. 13.2.
[118] Olympiodorus fr. 41.2.
[119] At the rate of 72 solidi to the pound (Jones 1964: 107), 4000 pounds make 288,000 solidi. A regular infantryman in the late 4th century cost about 6 solidi per year to pay and equip, including donatives and discharge bounties (Elton 1996: 123). Treadgold 1995 operates with a much higher pay of 20 nomismata, which however he has derived from later Byzantine numbers and thus less reliable here. At any rate, Roman troops were often paid in kind, so if their new masters provided their military equipment they could live comfortably on his estates.

development, while levies of recruits for the army were imposed on the great landowners (see below).[120]

The economic ability to raise troops began to be applied at a time of great turbulence. Many Gallic and Hispanic landowners had been involved in the civil wars of the early 5th century, and were hence subject to persecution from the other side (cf. the brothers discussed above). At the same time, as we saw at Amida and Aquileia, the treatment of military units as well as senators who had participated in revolt could be unforgiving (ch. 1.1.1). Units from all over Britain, Gaul and Hispania had supported Constantine and the other usurpers, but by the time the dust had settled, Britain was outside Roman control, while the Sueves, Vandals and Alans occupied large parts of Spain. Unable to go home and probably not too eager to face the victors' justice, many presumably deserted and sought service with magnates wanting to secure protection they could rely on. For example, the Roman officer Edobich, after his failure to relieve *Arles in 411 in support of the usurper Jovinus, fled to the estates of a perceived friend, Ecdicius, who however beheaded Edobich and forwarded the evidence to the emperor Honorius. Despite this violent protestation of loyalty, no reward was forthcoming; we may surmise that Ecdicius was suspected of involvement in the revolt. Indeed, Paulinus of Pella's complained that his estate alone lacked a Gothic guest during the later stages of the civil wars.[121] The implication is that billeted troops would be able to protect the property and position of the landowners who supported them. While this incident involved Visigoths who were formally settled shortly afterwards, a novel of Valentinian from 440 throws light on the fate of Roman soldiers in similar situations. Such laws are normally interpreted in light of the famous 4th-century legislation that was enacted in order to prevent landowners from withholding their due quotas of recruits from the army.[122] While similar to earlier legislation in emphasizing non-mili-

---

[120] Thus also Wickham 2005, but I would suggest that the relationship is perhaps even more direct.

[121] Paulinus of Pella, *Eucharisticon* 285: "[domus...] hospite tunc etiam Gothico quae sola careret."

[122] See in general *CTh* 7.18. for the laws on deserters and their harborers, and discussion in Elton 1996: 152ff; Valentinian's law begins much in the same vein as it states that "no person shall misuse an occasion of public loss and suppose that he should receive profit from fiscal expenditures, and that for the purpose of cultivating his land or of engaging in business, or on the pretext of any service whatever, he should detain a person who has been marked once for all with a military title[.]"

AN AGE OF TRANSITION 65

tary pursuits of withheld recruits, it is followed up by an interesting provision:

> ...We have ordered by this edictal law that if any person should suppose that he should harbor on a rustic or urban landed estate any recruit or *also a man of previous military service who had deserted his own service units and standards*, he shall both restore the man whom he has concealed and shall be compelled to pay as a fine three other men suitable for military service.[123]

In effect landowners were not only retaining *potential* recruits to keep as agricultural labor, but were also maintaining, on their estates, experienced Roman soldiers from regular units which were still in existence at the time. We should hardly expect that they were bribed out of their units in order to cultivate cabbage: the practices that were denounced conform remarkably well to the Egyptian situation and what else is known of the issue in the East.[124]

Valentinian's constitution also helps to explain the gradual demise of professional Roman citizen units in the second quarter of the 5th century as outlined (at 1.2.1) above. Constant infighting blurred loyalties, while the lack of fiscal ability on the part of the state, especially after the loss of Africa to the Vandals,[125] meant that many troops and aristocrats became disillusioned with Roman ideology and looked to their own interests. While some troops certainly had joined "barbarian" invaders and settled along with them, a great number found an intermediary in the Roman aristocracy, especially in southern Gaul. What exactly became of Edobich's and his colleagues' troops is uncertain, but a notorious example a generation later was the papal vicar, bishop Hilary of Arles, who not only deposed neighboring bishops at will, but to the exasperation of pope Leo the Great had acquired quite uncanonical expertise to assist him:

> A band of soldiers (*militarus manus*), as we have learned, follows the priest [Hilary] through the provinces and helps him who relies upon their armed support (*armati praesidii*) in turbulently invading churches, which have lost their own priests.[126]

---

[123] *NVal* 6.1.1. ...hac nos...edictali lege iussisse, quisque de tironibus aliquem nec non prioris militiae virum proprios numeros et signa desertentem rustico urbanove praedio crediderit occultandum, et ipsum restituat quem celavit et tres alios aptos militiae poenae nomine cogatur inferre... My emphasis.
[124] Cf. Lenski 2009 and chapter 1.2.3 above.
[125] For the fiscal significance of this loss, see Heather 2005.
[126] Leo *Ep.* 10.6.

The emperor Valentinian was equally shocked, and when he issued his novel in 445 in an attempt to redress the situation, he described how the bishop went about ousting his opponents:

> ...he gathered to himself an armed band, and he either encircled the enclosures of the walls by a siege, in the manner of an enemy, or he opened them by an attack, and he who was to preach peace led his band to the abode of peace through wars.[127]

Clearly this "band" had some significant military skills and would fit the type of troops described by Valentinian a few years earlier as entering magnate service; if they had been of barbarian origin this would surely have been noted. Gallic aristocrats such as Sidonius Apollinaris and his close friend Ecdicius (descendant of the namesake in 411?) were the last independent Romans to organize forces against the Visigoths. Sidonius on several occasions alludes to the retinues and military activities of his peers in the third quarter of the 5th century, and for a period hosted a substantial Burgundian force on his estates, although he seems not to have been too happy about their taste in songs, culinary habits and grooming products.[128] He himself led the defense of *Clermont in 474, while Ecdicius severely defeated the Visigoths besiegers in a series of battles that led up to his famous charge on the Gothic camp. Sidonius only alluded to the origins of these troops when he praised Ecdicius for "raising what was practically a public force from your private resources, and with little help from our magnates."[129] While the expression "public force" (*publici exercitus*) warns us that we may be dealing with the remnants or descendants of regular Roman troops, the reference to other magnates also indicates that the Gallic aristocracy would normally be expected to contribute from their resources.

Imperial legislation was divided in the matter, reflecting different challenges in the two halves of the empire. Valentinian in 440 and Majorian in c. 459 permitted civilians to carry arms, although it is uncertain whether the former meant retinues or militias, and only the heading of the latter law survives. In contrast, authorities tried to limit the spread of private

---

[127] *NVal* 17.1.1: ...manum sibi contrahebat armatam et claustra murorum in hostilem morem vel obsidione cingebat vel adgressione reserebat et ad sedem quietis pacem praedicaturus per bella ducebat.

[128] See e.g. Sid. Ap. *Ep.* 1.6, 3.9 and *Clermont 474 for retinues; *Carm.* 12 for Burgundian guests.

[129] Sid. Ap. *Ep.* 3.3.7: "... collegisse te privatis viribus publici exercitus speciem parvis extrinsecus maiorum opibus adiutum..."

troops in the East in 468, when Leo upheld the *Lex Julia*, banning armed slaves and retainers.[130] The Visigothic Code of Euric (r. 466-484) from 475 established (or rather, legalized already existing) procedures for the recruitment of private soldiers by landowners.[131] A fundamental element of Visigothic military organization in the latter half of the 5th century was the followings of erstwhile Roman army officers and landowners.[132] Hence, it was important for the Visigothic leadership to ensure their support by legalizing the Roman aristocratic power base. In the bargain the Visigoths acquired the direct descendants of regular Roman troops who had fallen outside the control of the central government. Incidentally, these forces probably formed a significant proportion of the "Gothic" force defeated by Ecdicius outside *Clermont. Sidonius agonized over the role of his friend Calminius in the besieging army, while he himself was exiled to the fortress of Livia after the Visigoths had finally taken over Clermont. He was performing compulsory public service (*per officii imaginem vel...necessitatem*; *officia*) within the walls of the fort, strongly implying some sort of military function, presumably with a retinue.[133]

1.2.5 *The Origins of Medieval Military Obligations:* Munera Publica

The loss of Roman Africa in 439 had severe repercussions for the ability of the Roman government to pay and supply its troops. Losses in tax revenue, both from imperial estates and through the regular levy, were immense, and further compounded by the loss of African grain that supplied the city of Rome. However, one should not exaggerate consequences or underestimate the inherent flexibility of the Roman state in covering its fiscal and material needs. Often ignored in discussions about late Roman finances is the vast complex of public obligations, *munera*, which complemented

---

[130] *CJ* 9.12.10.
[131] *Codicis Euricani fragmenta*, MGH LL 310 (p. 18f): "Si quis buccellario arma dederit vel aliquid donaverit, si in patroni sui manserit obsequio, aput ipsum quae sunt donata permaneant. Si vero alium sibi patronum elegerit, habeat licentiam, cui se voluerit commendare; quoniam ingenuus homo non potest prohiberi, quia in sua potestate consistit; sed reddat omnia patrono, quem deseruit. Similis et de circa filios patroni vel buccellarii forma servetur; ut, si ipsi quidem eis obsequi voluerint, donata possideant; si vero patroni filios vel nepotes crediderint relinquendos, reddant universa, quae parentibus eorum a patrono donata sunt." See ibid. 311 for the *saio*, who was more tightly bound to the service of his lord.
[132] Wolfram 1988: 217ff; chapter 1.1.2 above.
[133] Sid. Ap. *Ep.* 5.12 for friend in Visigothic besieging army; 9.3.2 and 8.3.2 for service at Livia.

taxes in money or in kind.[134] No-one has to my knowledge attempted to calculate how much public obligations were worth compared to regular budget items (army, bureaucracy, court, public munificence). Clearly these represented immense values in the maintenance of all public infrastructure (roads, bridges, aqueducts, temples and churches, city walls, fortifications, public post), and supplemented regular taxes in kind through forced purchases (anything from horses via foodstuffs to uniform items). They could also be extended in time of emergency to compensate for any eventuality, ensured regular supply to urban populations and vital stretches of frontier (especially the upper and middle Danube), and most importantly, were levied upon the whole population from great landowners to *coloni* or bound tenants through an administration of local notables, whose service again was a public obligation.

The *munera* (Greek *leitourgiai*) were regular annual obligations, levied parallel with taxation. For the higher classes, decurions to senators, they involved performing routine administrative work and providing public munificence, while physical labor and services were imposed on peasants and craftsmen. *Munera* were levied on the basis of personal status (*munera personae*) and landed wealth (*munera patrimonii* or *possessionis*). In addition to those listed above, common *munera* recognized by Roman jurists and dating back to the Principate or late Republic also included such personal obligations as defense (of *coloniae*) and landowners' obligation to provide transportation.[135] Extraordinary levies (*munera extraordinaria* or *superindicta*) were exactions above those calculated by the pretorian prefecture for any one indiction (year of the tax cycle), and eventually had to be approved by the emperor himself.[136] *Munera sordida*, although ostensibly denoting physical labor unworthy of high status, became a legal term for those obligations that exceeded labor and services already conscripted for the indiction, and does not therefore equal the vast range levied regularly every year, although exemptions from *munera sordida* provide the most comprehensive lists of common labor services.[137] Particularly inter-

---

[134] Wickham 2005 is concerned with demonstrating how the Roman tax system broke down; Jones 1964 only devotes just over half a page of concentrated discussion with incidental mentions elsewhere.

[135] For legal definitions, see Neesen 1981, especially 205-15, and further Horstkotte 1996. Drecoll 1997 has the most comprehensive recent discussion of the vast range of services attested in Egyptian papyri with parallel evidence in epigraphy and literary sources elsewhere in the late Roman Empire.

[136] See e.g. *CTh* 11.16.7, 8, 11.

[137] For this point, see Jones 1964: 452 and n. 100 (at p. 1189).

esting for our purposes are those obligations relevant to warfare: materials for construction and metalworking (lime, wood, coal); transport obligations (any public building project, public post, logistical need of garrisons); building and maintenance of infrastructure, particularly fortifications; administration and organization of the above.

Conscripted labor was organized in several manners. In all cities of some size, craftsman and merchant guilds were responsible for assigning members to perform relevant duties in any given indiction. This covered most state needs for specialized and technical labor. Otherwise, the responsibility fell upon landowners (*possessores*) according to wealth, just as tax payments. Furthermore, in order to spare poor farmers busy in the fields, services were to be levied on great landowners first. They routinely organized their dependent laborers and artisans (*operarum atque artificum diversorum*), providing draft animals and wagons for logistical needs; in regions where villages were common, this was delegated to a village headman.[138]

The distinction between the legal term *munera sordida* and similar labor obligations within an indiction's regular exactions is important, because many immunities were only granted for the former. The principle was that service in the state (military, bureaucracy, municipality) brought immunity from ordinary and/or extraordinary exactions in recognition of services rendered or dignities achieved, and could not be inherited nor extended to family members, even spouses. Furthermore, remission of taxes normally occurred only in cases of long-term delinquency either caused by corruption or emergency.[139] However, the type of emergency that engendered a fall in taxable income would often require more energetic exploitation of public labor, such as natural disasters or war, to repair infrastructure or supply hungry urban populations and garrisons. *Munera* were exacted by the local administration with such regularity that even imperial estates (*res privata*), whose tenants were exempt from extraordinary obligations by virtue of providing the emperor with his personal revenue, were continually badgered for *munera extraordinaria* or *sordida* by local authorities, as a long string of laws throughout the 4th century demonstrate.[140] It was only in the mid-6th century that Justinian legislated for

---

[138] For rich first, see e.g. *CTh* 11.16.4; obligations revealed by the immunities granted in *CTh* 11.16.15 (384) and 11.16.18 (390).

[139] For legal exceptions, see the detailed table in Neesen 1981: 216-23 and examples quoted in previous note.

[140] *CTh* 11.16.1, 2, 5, 9, 12, 13, 17, covering the period from 312 to 390.

a procedure of subtracting extraordinary exactions from regular taxes and *munera*; the Theodosian Code shows that while remissions and exemptions were fairly frequent, they were by no means automatic whenever extraordinary levies were dictated by state needs.[141]

The importance of public obligations, then, can hardly be overstated. We must belabor the point that landowners, *possessores*, controlled the physical means (wagons, animals, stocks of materials or ownership of resources in question) and labor (tenants, rural artisans, urban clients), and had the competence to administer them whenever they were up for duty. This means that landowners near frontiers and garrisons during the Principate must have been closely and routinely involved in most of the army's daily needs.[142] This involvement certainly became ever more routine and intimate as barbarian invasions and civil wars from the 3rd century affected most regions, not only those near the frontiers. By the turn of the 5th century, as we have seen, landowners increasingly participated in military affairs as they patronized troops, both regular and federate; between the 440s and 470s, most remaining West Roman units outside of northern Gaul, Italy and parts of the Danube and Dalmatian frontiers drifted into magnate service.

The transition to private military followings was facilitated by notable adaptations to the system of obligations just as the empire lost control of Africa. Within a year of permitting the public to bear arms in defense against the Vandals in 440, Valentinian abolished all immunities from taxes and public duties, whether from tenants of the imperial domain, members of the church, senators of illustrious rank, or anyone else who would normally have been exempt. Buried deep within a rhetorical question of how it was contrary to reason to use the term *munera sordida* for public services, it lists obligations to be shouldered by the wealthiest first, regardless of privilege:

> ...the building and repair of military roads, *the manufacture of arms*, the restoration of walls, the provision of the annona, and the rest of the public works through which we achieve the splendor of public defense...[143]

---

[141] *NJust* 128 & 130; see further e.g. Brubaker and Haldon 2011: 478.

[142] For early examples of civilians conscripted to provide labor and logistical services to the Roman army, see Roth 1999: 91-115.

[143] *NVal* 10.3 (441): ...instructio et reparatio itinerum pontiumque, instauratto militarium viarum; armorum fabricatio; murorum refectio; apparatus annonae; reliqua opera, per quae ad splendorem defensionis publicae pervenitur. My emphasis.

Most importantly, the production of arms was now levied as a public obligation, whereas it had traditionally been the sole preserve of the state, at least officially. The change might not be as dramatic as it first seems. Supplies and logistics to arms factories had always been a public obligation, requiring close coordination between the factories and those charged with supplying them with raw materials and transportation. In addition, the vast majority of non-weapon items used by the army were produced by civilian craftsmen outside of the *fabrica* system. Jones calculated that only one sixth of clothing to army and court was provided by state factories; the rest was requisitioned through forced purchases or levies in kind from civilian craftsmen.[144] Finally, most commentators have noted that *fabricae* seem not to have been distributed in such a manner as to cover all needs of the army in all regions of the empire, so it has long been assumed that garrisons had their own workshops, as we have shown was the case with artillery. It should come as no surprise if local commanders commissioned such work from local craftsmen, perhaps trained within *fabricae* or garrison workshops to produce standardized items, but now in semi-private employ. The new levy of arms is in fact official recognition of such a situation: arms production was probably not meant to be created *ex nihilo*, rather those with appropriate organizational and technical skill (*possessores*, guilds, civilian craftsmen and workshops) now provided arms from their own facilities or expanded official production by serving periodically in garrison workshops and public factories as a *munus publicum*. A similar situation existed in the East and must have arisen at about the same time. As we shall see in the next chapter, Justinian ordered that craftsmen knowledgeable in arms production report for service in *fabricae* and all illegally produced arms be confiscated.

Constructing walls was a more straightforward obligation. The Roman army traditionally provided labor for the forts it constructed along the frontier and routes of march, although supplementary civilian labor could be conscripted or recruited for pay during large-scale building activity, such as Hadrian's Wall. A normal procedure was to divide up stretches called *pedaturae* ("footages") for which each involved unit, whether military or civilian, would be responsible. A measure of *pedatura* also became the basis for assigning maintenance work and defensive duties: each unit was to patrol and defend its specific stretch.[145] The massive increase in

---

[144] Jones 1964: 434.
[145] Vegetius 3.8 for this procedure while building marching camps; Fulford 2006 for civilian teams participating at specific *pedaturae* on Hadrian's Wall.

fortification all over the empire in the 3rd and 4th centuries required civilian conscription on a larger scale than previously. The same rules and practices applied as for other public construction, such as roads, monuments, religious buildings and aqueducts, but fewer immunities were granted for building walls, and all exemptions were cancelled at Rome in 440 and everywhere else the next year.[146] The method of construction was similar as for military units. Each *possessor* or corporate unit received their particular stretch to construct, maintain, and defend in time of emergency. For example, in return for maintenance, those who owned land on which the Theodosian Walls had been built were allowed to use towers for storage and other private purposes.[147]

Roman defensive practices did not entail a militia as such, nor did Valentinian's law of June 440, allowing provincials to bear arms to defend against the Vandals, establish one.[148] Rather, the personal obligation (*munus personae*) to defend one's *colonia* at a time when they were islands of Roman control among conquered populations was gradually extended as *ad hoc* solutions to changing political, social and fiscal circumstances. Ideally, the Roman army was to provide security, but during the civil wars and disturbed circumstances of the late empire, any community might be exposed to danger. In a law published three months earlier, Valentinian was in fact quite explicit that guild members in Rome, who already guarded gates and walls routinely, were not subject to military service in any other capacity.[149] Perhaps the vague wording in the novel of June 440 actually masks a temporary concession allowing magnates to use private military followings. Such a practice required legal sanction that could be modified in different circumstances, whereas defense of one's city was an ancient customary obligation that was implemented by administrators and military commanders as need arose, and eventually became foundations for early medieval military organization. When the narrative evidence for the following centuries is examined in detail (see 6.2.2), it becomes clear that the general citizen body rarely had any formal military training, and only occasionally ventured outside the city walls. It was nevertheless a universal

---

[146] Immunities not applicable to wall-building and transport obligations to Illyricum in 412, *CTh* 11.17. 4 and 15.1.49; cancelled at Rome, *NVal* 5.1.3; for general abolition, see the novel quoted above (*NVal* 10.3).

[147] *CTh* 15.1.51 (413).

[148] *NVal* 9. Relevant treatments of the problem of urban militias, with references to the debates and further literature can be found in Alan Cameron 1976, Haldon 1995 and Bachrach 2001.

[149] *NVal* 5.1.2.

practice that whenever a siege threatened, *all* citizens participated in defenses, organized and armed *ad hoc* whenever crisis arose, just as had been in theory and practice in the 5th century and before.[150]

There is no doubt that *munera* continued to be performed until the end of the Western Empire and were absorbed by the Western successors. Sidonius relates in detail how he used the public post to travel from Clermont to Rome in 468, and in another letter shortly after vividly portrays the contrast between how the corrupt Seratonus and honorable Evanthius went about organizing maintenance of public infrastructure: the latter led his laborers by personal example, "like the pilot-fish" leading "the lumbering whale;" the former used chain-gangs and intimidation. Nonetheless, roads were cleared and taxes levied, and the very same class of ex-Roman officials soon served the Visigoths if they did not already do so.[151]

By the turn of the 6th century, then, it is clear that great magnates possessed formidable organizational capabilities cultivated over centuries by the Roman state to be directed towards military purposes. Such organizational skills continue unabated, even if no central authorities or municipal curias directed efforts, because magnates applied them in their own interests. St. Genovefa was able to protect *Paris (490) from famine by supplying the city from her estates, using barges to transport foodstuffs on the Seine. She could also draw on her dependent urban and rural craftsmen to organize the construction of a large new basilica. This included the industrial-scale production of tiles and lime, both of which were typical *munera*.[152] At around the same time, Caesarius of Arles organized the building of a large nunnery. The workforce was probably formed from his dependents, over whom he had extremely tight control.[153] Like his predecessor, he soon acquired a large military following, this time from Burgundian and Frank-

---

[150] Military commanders continued the practice whenever the population was deemed reliable, such as Belisarius at *Rome 537f. In contemporary military manuals civilian participation is assumed: Vegetius 4.7 obliquely refers to civilian men taking part in the defense, since he recommends removing "those unfit for fighting by reason of age or sex" in order to preserve supplies; the need to evacuate non-combatants and utilize fit men was clearly formulated in *Strategikon* 10.3.32-35; the anonymous 8th-century *De re strategica* (13 *passim*) refers to *politai* (citizens) as well as soldiers taking an active part in defending against enemy siege engines. This is further discussed in chapter 2.4.2.

[151] Sid. Ap. *Ep.* 1.5 (public post); 5.13 (repairs and taxes). See further Stroheker 1948 and Mathisen 1993.

[152] Daly 1994: 624-29. See also the *Vita Genovefae virginis parisiensis* cc. 17-21 for the miraculous discovery of industrial-scale lime kilns; the collection of craftsmen (*carpentariis collectis*, c. 21) was apparently less problematic.

[153] See *Arles (508) for the monastery, which was torn down for the siege, as well as *Vita Caesarii* 1.19, 21.

ish captives whom he ransomed after their armies were defeated at the siege of *Arles in 508. The industrial facilities of estates that survived the fall of the West Roman Empire or were reorganized according to this model in the following centuries were perfectly able to take over military production, as recognized by imperial legislation. While this will be reprised in chapters 3.2 and 4, it is worth pointing out some concrete examples similar to and contemporary with East Roman developments mentioned above. In northern Gaul in the early 6th century, legal evidence indicates that craftsmen with military skills were available on the estates of the militarized elite.[154] Later that century, both Gregory of Tours and the Patrician Mummolus had carpenters, *lignarii*, in their households.[155] The building and organizational skills of Merovingian and Visigothic bishops were directly transferable to military uses. There is also concrete evidence of military applications. At Trier, one of the longest-living centers of self-conscious *Romanitas* and Roman military organization on the Rhine, the bishop Nicetius of Trier (525/6-post 561) had in his retinue operators of the *ballistae* that guarded his private fortified residence. This capacity obviously must have originated in the artillery *fabrica* at Trier, which was held by a Roman officer as late as 477 and would have passed under local magnate control.[156]

1.3 Where Did All the Romans Go? The Military Implications of Ethnogenesis

Wherever Roman institutions survived or were adapted to new realities, it allowed the survival of the logistical, organizational and technological in-

---

[154] See chapter 4.1.5 on this.
[155] For Gregory's carpenter, see GT 5.49 (a *faber lignarius* named Modestus); for Mummolus' large-framed carpenter, see GT 7.41. This skill was even practiced by nobles; Leo, the bishop of Tours in 526 was a skilled *faber lignarius* as well (GT 3.17 and 10.31). Their military potential can be inferred from the sieges of *Avignon (583), which was held by Mummolus' retainers through advanced engineering, and *Convenae (585), the most complex siege attested by Gregory (and where his troops should have been). In the West, most skilled craftsmen came under magnate control or patronage, while in the East the traditional structures of independent urban craftsman continued.
[156] *Ballistae*: Venantius Fortunatus, *Carmina* III.12.35: "illic est etiam gemino ballista volatu." The classic statement of continuity is of course Bachrach 1972: 4, but see also Wallace-Hadrill 1962: 159; Jones 1964: 834ff and associated maps; especially note *ballista* production at Trier. On Nicetius, see Weidemann I: 221f; survival of *romanitas* at Trier, see Ewig 1979: 33-50; for the origins of the Franks, see chapter 1.3.3 below; for further discussion of the significance of Nicetius' *ballista*, see chapters 4.1.4, 5.2.2 and 8.2.1.

frastructure necessary for siege warfare to continue. Thus, for the Roman upper classes, it was possible to (ab)use networks of patronage to gain control over public institutions. For the Visigoths, this was achieved by integrating into the Roman army, its supply system, the administration and elite culture, and *then* co-opting the magnates. For the Huns, this was achieved through the rather rough and temporary appropriation of the necessary personnel (although not without connections and high-level support), and maybe the control of some Roman forts and cities on the middle Danube.[157] However, few of the other invaders seem to have had as long time as the Visigoths to adapt on Roman soil before the empire disintegrated, so a natural assumption has been that invasions by necessity destroyed much infrastructure, or the sheer influx of fighting men to the Roman army from outside the traditional Roman structures diluted or replaced the Roman style of war completely. There is only one problem: there was no "Germanic" (or any other barbarian, for that matter) way of war in the 5th-century. Roman influence from the first through the fifth century in effect erased or fundamentally mutated any traces of ancient "Germania," especially in the field of war.[158]

1.3.1 *Roman Influences beyond the Frontier*

For most barbarians, the Roman Empire was an opportunity for trade, military service, and political support for their tribe or faction. Centuries of client management in the form of deliberate economic, cultural and political intervention meant that Rome had considerable control beyond her borders. Chiefdoms and tribes clustered along the northern border where they developed a close symbiosis with the empire. It was practically impossible for the puny chiefdoms within marching distance of the Rhine and Danube to have anything resembling independent policies vis-à-vis the Romans for extended periods, except during civil wars when the system temporarily broke down or was manipulated by different parties inside the empire. They were there simply because Rome had use for them and stimulated their existence. The consequences of non-compliance were severe. The empire revoked trade rights, gifts and subsidies, which deprived local chieftains of prestige and the ability to award their followers.

---

[157] See the discussion in 7.1.3.
[158] For examples of older views, see Thompson 1958. Marsden 1969 (among many other scholars) implies that the Roman army of the fourth century was unable to handle complicated siege engines because it became less intelligent after being barbarized. For the transformation of Germanic society, see e.g. Todd 1992 and Hedeager 1992.

Alternately, it replaced ruling dynasties, reordered hierarchies of tribes and chieftains, or even forced whole tribes to relocate outside of the favored belt near the frontier, replacing them with others. Finally, direct military intervention through punitive raids or encouraging other tribes to attack could end with offending tribal rulers being deposed or even thrown to the beasts in provincial amphitheaters, and their populations subdued, enslaved, or forced to ask for settlement on Roman soil as *dediticii*.[159]

For Rome, the advantages were clear. It added another layer of defense to the legions on the Rhine and Danube, so that troublesome newcomers could be dealt with by clients before they ever saw a Roman city. It seems that the Romans even helped construct and man defensive infrastructure well beyond the Danube, or commissioned trusted groups to take over abandoned installations.[160] The petty chiefdoms also provided recruits to the army or temporary mercenary contingents on a large scale, while successful barbarian commanders became thoroughly Romanized and began to play an increasing role in Roman politics. Not all of their troops were local "tribesmen." Many came from as far away as Scandinavia (see below), while non-Germanic groups were prominent along the whole middle and lower Danube. By the end of the 4th century, continuous movement of people from the north and east into the frontier region had profound influences on the ethnic composition of the "tribes" that often had venerable old names. As a result, many of them no longer had the biological, linguistic or cultural coherence normally associated with tribal society, and only very vague connections with their ethnographic predecessors: they were defined by the control of a ruling dynasty who achieved dominance through their contacts with Rome. Some chiefdoms were grouped under regional identities, which were largely Roman creations but did not reflect any political coherence. Thus the Franks, Alamans, and Goths were regional designations that became common in the 3rd century, perhaps coinciding with Roman patronage and recruitment networks that were responsible for funneling newcomers into the Roman Empire. Lesser groups had old names that can be traced over the centuries, but scholars have recently begun to argue that the ruling classes of the petty border

---

[159] For most of this and the following paragraph, see Whittaker 1994, Heather 2001, Goffart 2006, and Halsall 2007. Heather 2009 in contrast presents a view of Roman-barbarian relations as highly antagonistic.

[160] For the former (Constantinian dykes in Dacia), see Whittaker 1994: 174-78; for the latter (the Alaman takeover of Agri Decumantes), see the recent argument by Halsall 2007.

polities were in fact familiar with the Roman ethnographic tradition and knew the value of an old, prestigious, and, especially, awe-inspiring name.[161]

While the creation of a highly Romanized border culture has attracted most attention, Roman influences completely reshaped Germanic societies much further away, which, before the Roman era, were comparatively primitive. The social structure of "Germania" became more stratified, agriculture and economy more complex, and beyond direct Roman influence, the rise of political structures of some significance. Scandinavia may provide an illuminating example of how deep influences ran.[162] The extensive finds of Roman-produced or Roman-inspired arms and armor (and recently bodies) at Illerup in Denmark belonged to an army of at least 1,000 men, possibly "Norwegians" going to or returning from Roman service. Defeated by some local enemy, their equipment was ritually destroyed and deposited in a lake that since dried out. The weapons and evidence of hierarchical structure within the army shows that Scandinavians served the Romans *en masse* as mercenaries as early as the second century AD.[163] Scandinavians at home organized and equipped their troops in Roman fashion, bringing specialist craftsmen on campaign to maintain equipment and even field surgeons to attend to wounded troops.[164]

While some stayed in the empire,[165] many returned with wealth, prestige, and political power. Scandinavia saw the rise of petty kingdoms that thrived on sending recruits, receiving payment, and trading with the Ro-

---

[161] See especially the relevant contributions in Mitchell and Greatrex 2000.

[162] Scandinavian historians tend to be preoccupied with a Germanizing framework, but this often arises from lack of familiarity with Mediterranean late antique history in general and recent scholarly advances in particular. See e.g. Petersen 2010a and 2010b; for collection of studies aiming at integration into a larger framework, see *MASS*.

[163] For a recent, lavishly illustrated survey of the finds, see Ilkjær 2000. The archaeological evidence has been published in 14 volumes since 1990, ed. by Ilkjær et al. While estimates of the Illerup finds have been based on partial equipment finds, in 2009 a mass grave was found nearby, probably containing the victims of the battle who were killed or sacrificed. As far as I am aware, the results have not yet been formally published, but some details have been made available to the media, e.g. at http://jp.dk/indland/aar/article1750118.ece and http://www.berlingske.dk/danmark/jernalderhaer-rasler-med-knoglerne. The "Norwegian" army that was defeated probably counted at least 1,000 men based on estimates of equipment and preliminary assessment of the mass grave.

[164] For repair and engineering capabilities, see Dobat 2008; for medical equipment, Frölich 2009.

[165] E.g. the usurper Constantine III's *magister militum* Nebiogastes (*PLRE* 2 s.v.), possibly *Nevjogastiz in proto-Norse, meaning something like "Knuckle-guest." Based on runic evidence, the name form –*gast* was common in Scandinavia in the period 300-400 AD. I am working on an article on the possible Scandinavian origins of several commanders in the West Roman army during this period.

man world. Archaeological excavations demonstrate that Scandinavian warfare in the 2nd to 7th centuries was in fact deeply influenced by Roman patterns. The petty kingdoms fought brutal wars that involved up to thousands of troops. Their conflicts revolved around stone fortifications, which were ubiquitous and seem to have been taken by storm and burnt on many occasions during this period. The nature of military organization is clear from clusters of great boat-houses that have been found on the Norwegian coast long before the Viking era, though the evidence for Danish and Swedish naval warfare is just as strong. The Norwegian boat-houses can be associated with estates that have been identified in place-names or archaeologically, in a pattern that resembles late Roman and Merovingian developments. To defend against such extensive fleets, complex, large-scale naval defenses were erected in many Scandinavian ports. Vast grave-mounds required the labor of hundreds of men over many months. Tactics evolved in tandem with organization. On Gotland as late as the 7th century, warriors were equipped with a panoply closely resembling that of East Roman and Sassanian heavy cavalry. Although we cannot conclude from this that the Götar were proficient cavalrymen, their equipment does prove that Scandinavian military practice was influenced by their Mediterranean neighbors.[166] Indeed, that most "Germanic" of tactics, the *svínfylking* (swine [head] formation, i.e. a wedge similar to a boar's head), an indispensible element in any textbook on the Vikings (as "Germanic" warriors *par excellence*), is simply the Roman *cuneus*, which in fact was already known as the *caput porci(-num)* in classical times.[167] These are fundamental changes that took place in a region where no Roman (that we know of) set his foot. There is no *a priori* reason to assume that Germanic-speaking peoples who lived on the border for centuries and afterwards settled inside the empire would have been any less influenced or able to adapt.

1.3.2 *Civil Wars by Proxy and the Involution of the Frontier*

The empire's immense gravitational pull over centuries destroyed and reshaped whatever tribal and ethnic structures existed in the age of Caesar or Tacitus, and left far less room for independent political maneuvering than was available beyond, say, the Elbe, resulting in small, often ephemeral polities under favored or successful dynasties, which I would argue

---

[166] See Petersen 2010a for an overview of developments and 2010b for the historiography with an extensive bibliography.

[167] Rance 2007: 364 for discussion and notes; see also the critique of "German" tactics by Nicasie 1999.

was the driving force behind barbarian ethnogenesis.[168] The resulting multiplicity of petty chiefdoms along the northern border was of course difficult to manage, and required constant attention. Before c. 400, barbarian groups might exploit Roman political weakness during civil wars to raid and plunder, perhaps to compensate for lack of patronage. This seems to have been the occasion for many of the large-scale barbarian invasions from the 3rd century onwards, as clients were especially interested in acquiring captives with skills necessary for prestige and military competition.[169] Otherwise it might be more appropriate to speak of brigandage and banditry.

Due to the profound integration of many of these petty chiefdoms into Roman political structures, we cannot discount the possibility that many "barbarian" invasions were in fact extensions of Roman civil strife by proxy. The famous raiding activities of Alamans and Franks in 350s, defeated with much effort by the Caesar Julian and well described by Ammianus Marcellinus, may in fact be the continued fallout of Magnentius' revolt, with clients egged on by surviving Roman partisans or aggrieved by broken promises. Thus interpreted, Julian's policy of punitive expeditions across the Rhine closely reflected events on the Danube in the following decade. There, the usurper Procopius revolted against Valens, relying firstly on his family ties to the house of Constantine to attract Roman supporters and military units, but secondly, he could exploit the loyalty of Gothic clients across the Danube to bolster his cause. Constantine had imposed a treaty on the Goths in 332 which entailed a combination of cross-border trade and subsidies in return for military service. This led to a remarkably peaceful frontier up to the 360s, as Gothic contingents served on Roman expeditions on several occasions. Procopius, claiming to be a legitimate emperor of the house of Constantine, similarly required a Gothic contingent to support his cause. While the Goths obliged, the troops had not reached him before he died and his revolt had been defeated in 365. As a result, Valens took 3,000 Goths captive and press-ganged them into the Roman army where they served in Roman border forts on the Danube. Lenski points out that it did not help to produce in their defense documentation of a formal request of *auxilium* from what they thought was the legitimate government; Valens' authority had been challenged, and in the subsequent years,

---

[168] See *TRW* 2 for theoretical problems and different views of ethnogenesis, as this process is known, a very difficult concept used here in only a very general fashion; further, *TRW* 13 for case studies of particular peoples.

[169] Lenski 2008.

he instituted a policy of punitive expeditions, the strengthening of border fortifications noted above (1.1.1), and a withdrawal of subsidies and trade across the border.[170]

Such instances of client involvement in civil wars began to multiply around 400. Kulikowski only hints at some cynical Roman involvement in the great Rhine crossing in 406 (or 405, as he dates it). This gave the usurper Constantine III (407-11) an excellent excuse for a revolt, but nevertheless he seems to have had his back free to pursue imperial ambitions while the barbarians were under control in northern Gaul for two years or so without much attested activity before 409.[171] While the problem of the Rhine crossing and its fallout is too complex to unravel here, we do have some quite explicit examples of how clients became drawn into the civil war shortly afterwards. The hapless Edobich, who was of Frankish origins, had risen within the ranks of the army to become *magister militum* under Constantine III, and was sent across the Rhine to gather Frankish and Alamannic client troops in order to relieve *Arles (411), besieged by troops loyal to Honorius. Similarly, when Constantine's revolt collapsed and Jovinus revolted in 411, he drew in part on barbarian clients, but for some contemporary observers, this was merely another Roman army rising up and the presence of client troops go unnoticed. This should warn us not to lend too much weight to any one barbarian appearance, or assume they were operating on their own initiative.[172] Edward James eloquently identified the problem:

> Events in Gaul in the first decade and a half of the [5th] century are confusing in the extreme; various Roman usurpers used different groups of barbarians against legitimate authorities, and those fighting for the legitimate Emperors also used barbarian support. At times it must have been difficult as for the barbarians involved to have known whether they were fighting 'for' or 'against' the Empire as it was for modern historians.[173]

The extension of civil wars by client proxies, as well as the settlement of some of these groups by legitimate or illegitimate Roman officials, greatly complicates our view of Roman civil wars in the last century of the Western empire, and all too often draws attention from the willed exploitation of

---

[170] Lenski 2002: 136f.
[171] Kulikowski 2000.
[172] See Scharf 1993 for references and discussion, although he conjectures a rather complicated build-up of the revolt in order to explain the lack of reference to barbarians in some accounts.
[173] E. James 1988: 54.

these groups in internal political strife to the "barbarianness" of some of the participants. Client troops would in most circumstances have returned home or been absorbed into the regular army if political circumstances had evolved differently. The view of who was barbarian was complicated by the fact that the frontier environment on the Roman side of the border had for centuries been a hodgepodge of Roman provincials, army units, veterans, and their dependents; settlements of barbarian military colonists from various periods of Roman history; recent individual barbarian volunteers who were serving in the army; and whole units of mercenaries and allies from beyond the border.[174] This environment was particularly heterogeneous in the Balkans, and dubbed "Balkans military culture" by Amory, but operated along the whole frontier. It was a powerful force in the formative years of the Visigoths, and the reason why they are essentially invisible in the archaeology.[175] We must conclude that as long as this system remained in operation, intermittently through much of the 5th century in the West and throughout the 6th century in the East, any "barbarians" that were exposed to the effects of Roman patronage, diplomacy, warfare and military service were profoundly formed by this experience; those that settled on the Roman side of the border, were effectively Romanized.

The frontier culture began to intrude on metropolitan Roman society in the first half of the 5th century, not because of any inherent barbarian strength or independent objectives, but because they were useful for Roman actors at one point or another. The Vandal, Sueve and Alan invasion of 406 was a dramatic instance of frontier groups entering the empire. The civil wars of 407-13 and subsequent internal conflicts led to the settlement of other groups on a large scale, notably the Visigoths and Burgundians. The Romans were mostly able to reestablish their authority on the old frontiers as soon as they could muster armies to make a show of force. Furthermore, Roman generals continued to use client groups in support of their cause; thus Aëtius drew support from and settled a number of groups, but the demise of Roman patrons could leave such groups stranded to fend for themselves between multiple Roman factions, adding to those already marginalized by political circumstances.

Two major factors that made a West Roman revival increasingly difficult were both indirectly caused by Roman actors. Firstly, Vandals and Alans,

---

[174] Halsall 2007.
[175] Amory 1997; cf. Heather 1996: 166-78 and 2009 for a different perspective. Although he correctly criticizes Amory on a number of points, Heather seems to disregard or underestimate the formative power of Roman policies.

even if they had been pawns in shady dealings during and immediately after the Rhine crossing, were left in opposition to the legitimate government of Honorius and later Valentinian III by the outcome of the civil wars in 413. Left to their own devices at a very early date, their invasion of Africa might again have been spurred by internal Roman conflict: the general Bonifatius invited them over when he fell out of favor, but the net result was the permanent loss of North Africa to the West Roman government in 439, with catastrophic financial results visible in legislation from the 440s. The rise of the Huns as a great power at the same time further complicated the situation. Originally patronized by Aëtius as "his" client group, and often used to further his ambitions, they were very different from other clients in that they had the political traditions of empire building from Central Asia. In the 440s and early 450s, the Huns developed a realistic alternative to Roman patronage, which meant that barbarians even close to the Roman Empire could begin to contemplate independent policy on a modest scale. This still only meant a choice in patrons: Aëtius and Attila supported different contenders for one of the Frankish petty kingdoms.[176] The situation would have been similar for most clients on the northern Roman border in the 440s and early 450s.

Even after the collapse of the Hunnic Empire, many barbarians still wanted the same things as before: employment in the Roman army, settlement inside the empire, or subsidies outside of it. The loss of North African revenue after 439 meant that this was increasingly difficult to achieve for central authorities, leaving initiative to local actors, while the involution of frontiers brought alternative power structures and identity shifts. West Roman authority fragmented intermittently during civil wars, creating new frontiers on every occasion, but such divisions only became permanent after c. 461 with the demise of Majorian. Even then only the upper Danube frontier collapsed; both Visigoths and Burgundians fought what were essential Roman civil wars into the 460s and 470s respectively, while the trans-rhenan Franks and Alamans remained comparatively docile and were incorporated into the Frankish kingdom as clients (see 1.3.3). Odovacer, the general and subsequently king of Italy, was himself a recent arrival from the Hunnic breakup seeking employment in the Roman army. Ousting the Burgundians from their position as a loyalist imperial army in 474, he withdrew the last Roman garrisons from the Danube to Italy, probably in order to keep infrastructure and personnel out of the hands of the invad-

---

[176] See Priscus fr. 20.3 and discussion in chapter 7.1.3.

ing Rugi whom he could no longer control.[177] The nearby Heruls still provided recruits for the army in Italy until demoted by the Ostrogoths; they subsequently moved to the East and found service there for a generation.[178] The collapse of the Roman Danube frontier produced no significant political structures, however, as the Franks, Ostrogoths and East Romans reestablished a client system. Independent political entities beyond the Roman frontier only arose in the 6th century in the East; in the West, this occurred as late as the 7th century under Frankish stimulation.

While some of the barbarian invaders and settlers formed kingdoms in the West, it is often forgotten that the majority of "barbarian" and "tribal" identities simply disappeared or were subsumed by much larger political and socio-economic entities. For instance, there were a great number of distinct Gothic groups, of which only two formed lasting kingdoms.[179] A similar process happened to provincial Romans, who were absorbed into new identities. Roman military units often used distinct names that were associated with particular warlike qualities; however, having an ethnic designator took on political meaning in the 5th century.[180] As Roman identity hinged upon loyalty to the Roman state, disloyalty was held as grounds for losing one's Roman identity.[181] Salvian pointed out that many Romans were "forced" to become barbarians by Roman authorities, presumably because they ended up on the wrong side of an internal conflict; indeed, the Theodosian Code effectively deprived those who had colluded with barbarians of their civil rights.[182] Usurpers and their supporters were thus little better than barbarians, according to a centralist Roman view, but the political fragmentation of the 5th century left very few within the boundaries of Roman identity. After the deposition of the last Roman emperor in 476, there was no other possible focal point for Roman loyalty. In most

---

[177] MacGeorge 2002: 284-93 for his career up to 476; afterwards, see *PLRE* 2 s.v. Odovacer and Halsall 2007: 278ff.

[178] The Ostrogoths, in turn, seem to have reestablished a client system, but possibly somewhat short of the Danube. See chapter 3.1.1 below for context and Sarantis 2011 for the Heruls.

[179] See Goffart 2006 for a survey of all the peoples that disappeared without political issue.

[180] Halsall 2007: 55ff has an extremely important discussion of the use of "barbarian chic" among Roman units and how this could be used as symbols of political belonging and ethnicity.

[181] In general, see the essays in Mitchell and Greatrex 2000, and further Chrysos 1996 for the related problem of late Roman citizenship; see chapters 2.1.2 and 7.2.2 for further discussion.

[182] Salvian *De gub. Dei* 5.5f; cf. provisions in *CTh* 9 *passim*.

cases, a group detached from Roman loyalty only when they were forced to do so. The Visigoths bet on the wrong contenders intermittently, but nevertheless tried to remain in Roman service until the late 460s, the "Franks" seem to have opted out at the same time, while the Burgundians were left stranded by the events of the 470s.

1.3.3 *The Legions on the Rhine Become Franks*

In most scholarship since Gregory of Tours, use of the term "Frank" conflates three distinct groups: trans-rhenan clients; Franks settled south of the Rhine by the Romans; and everybody else who adopted Frankish identity before and after the creation of the Merovingian Frankish kingdom. As a result, modern scholars recreate a Frankish migration or an invasion through a connect-the-dots methodology, where each mention of a Frank or someone with a Germanic-sounding name is taken as another stage in an imagined Frankish expansion. However, the evidence is flimsy in the extreme: a few snippets of the last classicizing West Roman historians quoted out of context by Gregory of Tours in search of the origins of Frankish kingship; an extremely vague allusion to Frankish expansion in Sidonius Apollinaris' panegyric to the emperor Avitus (455-56);[183] some of Salvian of Marseille's even more vague invective, which fails to identify the offending "barbarians"—who may not even have been that barbaric;[184] and the clever fitting of items of uncertain provenance into early Frankish history by the author of the 8th-century *Liber Historiae Francorum*, who in the process connected the Franks with the Trojans (independently of Fredegar, who did likewise).

Contemporary chronicles are almost as vague, but amongst the notices preserved by Gregory survives a brief note on the Frankish capture of *Trier c. 413. Either ignored or misinterpreted, the context is the end of the usurpation of Jovinus (411-13), which in light of the discussion above would

---

[183] Sid. Ap. Carmina 7.372f: Francus Germanum primum Belgamque Secundam spernebat.

[184] Salvian, *de gub. Dei* 6.8, 15 refers to the destruction of Mainz, Cologne and Trier; the latter no less than four times, but apparently Trier still had a large enough elite and population to request imperial aid to put on public spectacles. No barbarians are specifically mentioned in these instances, however; the blame for the alleged destruction of Cologne is assigned to the Franks on the basis of Salvian's first letter, where he refers to a young relative who had been taken captive by someone near Cologne, but later released, and whose mother remained in the city in the employment of barbarian women. Which barbarians are never mentioned, but their tastes must have been refined since they desired noble Roman women to wait on them. The whole context could just as well be interpreted as the fallout of various revolts and civil wars as barbarian invasions; see also below and next note.

indicate normal client involvement in a civil war, most likely by Frankish groups loyal to the regime of Honorius. The only late source that adds anything to the picture is Fredegar, who claims that a certain senator Lucius used the Franks to take control of *Trier (456) in opposition to the Avitus.[185] Although the notice of a capture of *Trier is placed correctly between notices that belong to the usurpation of Jovinus and the activities of a general Castinus in Gaul c. 420, it is however problematic, since Fredegar explicitly connects an event that must have occurred around 413 with Lucius' opposition to Avitus a generation later. Still, the notice cannot be dismissed out of hand. A senator calling in auxiliaries is instructive not only of Roman internal politics during a civil war, but also the predicament of barbarians with treaty obligations. Sidonius' oblique reference to "the Frank" taking control of *Germania Prima* and *Belgica Secunda* during the reign of Avitus provides corroboration for precisely such an event. He places the blame at barbarian feet, but for reasons of politics and meter fails to mention the fall of Trier, which lies between the two mentioned provinces, in *Belgica Prima*. Sidonius thus deliberately masks the (embarrassing) role of Roman opposition to his emperor behind a Frankish veil, but Fredegar, having obtained the notice of Roman involvement in another, unknown source, simply misplaced it into the framework left behind by Gregory of Tours.

As Frankish "invasions" are deconstructed one by one, it becomes clear that the Frankish client chiefdoms across the Rhine had no part in the breakdown of the Rhine frontier; in fact, they fought *for* the Romans throughout the 5th century. For instance, during the barbarian breakthrough of the Rhine in 406 and the siege of *Arles in 411, the Franks followed their obligations, in the first case defending the border fiercely, although in the latter case they were on the wrong (Roman) side in a civil war. The first priority of the régime of Honorius in Gaul after the end of the civil wars of 407-13 was to suppress *bacaudae*, i.e. Roman rebels, in Armorica in 417. Any Frankish support or involvement was eminently likely due to their obligations to local Roman patrons. This may have prompted the expedition of Castinus in order to pacify Franks, known from a brief notice that Gregory of Tours clipped from the historian Renatus Profuturus Friderigus and later picked up by Fredegar. Halsall takes this to mean that Rome had lost control of northern Gaul in the early 5th century, but we have no way of knowing whether this was in fact a police action across the

---

[185] Fredegar 3.7; the evidence from the *Liber Historiae Francorum* and Salvian is discussed further in the context of *Trier (413 and 456).

Rhine, as had been common in the last decades of the 4th century and attested in the fragments of Sulpicius Severus preserved by Gregory.[186] It may even have been a rather brute demonstration of authority to terrify a client group that had supported the wrong side in a civil war, as Valens had done against the Goths after he had defeated the revolt of Procopius. Thus most Frankish "invasions" across the Rhine were in fact clients trying their best to fulfill their treaty obligations to one or another party inside the empire who had some claim to their loyalty.

In this view, the Franks across the Rhine did not invade to create a barbarian kingdom on Roman soil. What has caused confusion is the fact that Frankish ethnicity became the basis of the kingdom that arose in northern Gaul in the last decade or so of the Western Empire. To a certain extent, the ethnic makeup of the northernmost provinces explains this, but not in the sense of a Frankish infiltration or usurpation of Roman power. In addition to those Germanic-speakers who already dwelt on the west bank of the lower Rhine at the time of the conquest, further groups were settled there by the Roman Empire to provide recruits for the army since the 1st century AD settlement of the *Batavi*, right through to the *Salii* in the 4th century.[187] As this process continued over the centuries, the concentration of "Franks" in the Roman army in northern Gaul was very high, probably many (tens of) thousands serving by the early 5th century. The effect of this environment was somewhat similar to the Balkans culture in the East, but here integration into Roman material culture and ideology was even more rapid and complete. As we saw, the area had long been a melting pot of Romanized peoples. While groups descending from Germanic-speakers ("Franks" etc.) predominated, Halsall has effectively demolished the idea that one can identify archaeologically any sign of a Frankish or any other trans-rhenan culture inside Roman borders, showing that influences overwhelmingly ran the other way and that "evidence" of Germanic culture actually reflects evolving Roman military fashion.[188] Those Franks and other "Germans" we know from written sources, mostly high-ranking army officers, were often of mixed ethnic descent, belonged to families Romanized generations before and intermarried into the Roman military

---

[186] GT 2.9.
[187] E. James 1988: 34-52 and Anderson 1995. For a recent reinterpretation, see Halsall 2007 (index s.v. "Franks").
[188] Schmauder 2003 bases his argument on the presence of "Germanic" artifacts, but the resulting image is extremely diffuse; Halsall 2007: 152-61 summarizes the evidence to quite devastating effect.

aristocracy,[189] commissioned Latin inscriptions,[190] composed panegyrics,[191] and were generally regarded by contemporary Roman senatorial snobs as torch-bearers of Roman civilization in the north into the 470s.[192]

Although northern Gaul may have been imperfectly controlled by central authorities for much of the early 5th century, the client system was maintained by local military officers and continued to function across the Rhine, which of course meant that the clients were mobilized for both defense and participation in Roman civil wars. The failure to recognize this has puzzled most historians, for whom a "migrating tribes" framework has been all-encompassing.[193] The few sources we have of conflicts normally associated with a "Frankish" takeover of northern Gaul have been impossible to shape into any coherent form, simply because they were either typical client management operations, as we have seen, or literary fictions. For instance, the "Battle of Vicus Helena," held up as an early Roman defeat of a Frankish invading enemy, was actually a case of some troops breaking up a rowdy wedding party.[194] The very late *Liber Historiae Francorum* has been used as evidence of the Frankish conquest of *Cologne and *Trier c. 456, but the latter has been lifted from Gregory of Tours describing an event in 413, although a "Roman" explanation for an actual event at that date cannot be ruled out, as the evidence from Fredegar shows.[195] Thus, we

---

[189] Demandt 1970 and 2007.

[190] E.g. Hnaudifridus at Hadrian's wall and the Burgundian prince Hariulf at Trier, both serving in the regular army; see Halsall 2007: 160 for references.

[191] Clover 1971.

[192] For example, Arbogast, the last known Roman officer at Trier (c. 477), was regarded as a beacon of Roman culture on the Rhine by his contemporary Sidonius Apollinaris. In fact, he was of Frankish descent. See *PLRE* 2 s.v. Arbogastes. St. Genovefa of Paris, a rich aristocratic lady, was probably half Frankish but this seems to have affected neither her social status nor her *romanitas*, for which see Daly 1994: 627f.

[193] An excellent example is MacGeorge 2002, who assesses the highly disparate sources in light of recent discussions, but does not reach the conclusions suggested here. Her comment on the possible kingship of the Roman general Aegidius (d. 465) is highly revealing: "I have a persistent feeling, however, that the story has some greater significance, and that the relationship between Aegidius and Childeric's Franks was an unusual one (reflecting perhaps the complex nature of Roman/barbarian relations in this period)." By classifying the Franks in the Roman army as barbarians, she forces the discussion into the Roman-Barbarian framework so familiar from the Visigoths and other, genuine, migratory Germanic groups. Halsall 2007 is more on the mark, but does not acknowledge the consequences of his own conclusions on continuity in his own work on warfare, e.g. Halsall 2003.

[194] Firmly established by James 1998: 57; however, even Halsall 2007: 249, who favors the Roman origins of the Franks, follows the traditional view.

[195] Gerberding 1987: 37ff for the tendency of *LHF*'s author to enhance material (sometimes producing very intelligent conjectures to link events unrelated in his sources; cf. *Trier and *Cologne 456) from Gregory of Tours.

cannot look to a Frankish "invasion" to explain the rise of the Frankish kingdom under the Merovingians. They had been there all the time as Roman citizens and soldiers.[196] The kingdom was rather the creation of a group of Roman officers and soldiers that had run out of options, pushed out in the cold by Roman civil wars.

While Franks (or more accurately, Germanic-speakers from the tribes that were eventually designated as Franks) had been settled in northern Gaul by the Romans for centuries, the full-scale adoption of a Frankish identity by the Romans (including thoroughly Romanized "Franks") only occurred after it received political expression through the development of a Frankish kingdom. As with several other ethnic designations in the Roman army, Frankishness began to take on political significance sometime after 461.[197] This was when the Roman general of Gaul, Aegidius, broke with central authorities due to the deposition of his friend and ally, the emperor Majorian (457-61). The conflation of "Roman" and "Frankish" identity by external observers began shortly after. The chronicler Hydatius refers to a war in Armorica in 463 between the Visigoths under Frederic (the brother of king Theoderic II, and hence an ally of the new emperor Libius Severus 461-65) and the independent Roman general Aegidius. A minor Gallic chronicle, referring to the same event, states: "Frederic, the brother of King Theuderic, was killed on the Loire fighting with the Franks."[198] These "Franks" were evidently the troops of Aegidius (i.e. as regular Roman troops as one could expect in the area at the time). However, on the model of the Visigoths and other barbarian kingdoms, external observers were beginning to regard the inhabitants of northern Gaul as something "other" than Roman, since they were no longer associated with central Roman control. For such observers, the (remote) Frankish origin of many of the troops was conflated with unit designations related to Frankish ethnicity, making the choice of ethnic designator obvious.[199] This is demonstrated

---

[196] Anderson 1995; Frankish units in the *Notitia Dignitatum* are numerous and were stationed from Gaul to Egypt; cf. E. James 1988: 39. Presumably many more carried no obvious ethnic designator. See also Goffart 2005; Böhme 2009.

[197] It may have begun with a shift in terminology: it was high fashion for skilled authors to throw about barbarian names, such as Sidonius Apollinaris' catalogue of tribes in Majorian's army in 460; apart from the obvious anachronisms, the list is actually dominated by provincials (Pannonians, Dacians) or known participants in and products of the Balkans military culture, such as the Sarmatians, Getans, Visigoths, Ostrogoths etc. See MacGeorge 2002: 205 for this example and Mathisen 1993: 39ff for further discussion of "catalogues of peoples."

[198] Hydatius s.a. 463; *Gallic Chronicle* of 511 s.a. 463; tr. Murray 2000: 99.

[199] Halsall 2007 chapter 14.

by the fact that those not involved in the immediate power struggles (or found a different emphasis politically useful), still regarded northern Gaul as "Roman" territory around 500. In fact, British observers called northern Gaul *"Romania"* well *after* Clovis had effectively established the Merovingian kingdom.[200] Indeed, it can be argued that Anastasios regarded Clovis as a legitimate Roman magistrate and military officer who had the right to be conferred the title *patricius* and hold a public triumph (see below).

The political situation in northern Gaul was very complex after the Hun collapse and the toppling of Majorian. Aegidius led an independent Roman province in opposition to the new regime in Italy and was probably the first to take the title king in order to legitimize his position. His "kingdom" apparently fragmented upon his death around 465. Subordinates, some of whom at least were engaged in a struggle for succession, are attested in Trier (Arbogast), Angers (Paul), and Soissons (Syagrius, probably the son of Aegidius) into the 470s and 480s. Another of Aegidius' local successors was a high-ranking Roman officer named Childeric (d. 481) in control of *Belgica secunda*. In the same vein, the postulated "Ripuarian" tribe of Franks centered on Cologne did not result from an external invasion. They were simply the local Roman troops on the river frontier, as their name implies Roman military organization, *ripenses* or *riparii*, and has no connection whatsoever with a putative "Ripuarian" Frankish tribe, which is not in fact attested until centuries later.[201] In sum, these commanders reflect the survival and re-focusing of Roman military organization to hold new borders. The Rhine became much less important after the dissolution of the Hun threat while the Loire now became the focus of defensive efforts.

No longer having imperial appointments to look to as a legitimizing factors, these Roman officers sought an alternative in a royal title on the model of other barbarian kingdoms.[202] That some of the contenders bore Germanic as opposed to Roman names is irrelevant, since that had been accepted by high Roman society for centuries. It has however confused historians ever since Gregory of Tours. Thanks to the discovery of Child-

---

[200] Wickham 2009: 150.
[201] Springer 1998 refines older views such as Zöllner 1970, who recognized the time-lag but still insisted on a *rheinfränkisch* polity. See further Scharf 1999 for a discussion of *ripari* in a 5th-century context.
[202] While the Visigoths had had a king since the late 4th century at least, the other mighty Roman figure in the West, Odovacer, *Magister Militum* and the one who deposed the last western emperor in 476, took a royal title that very year. This may have prompted a "Roman" king in Gaul. See MacGeorge 2002: 287-93.

eric's grave, however, there is no doubt that he regarded himself as a Roman officer, and would have been indistinguishable from his compatriots and competitors for power among the Roman officer class in northern Gaul. Childeric's son, Clovis (r. 481-511), inherited his father's province, and then absorbed most territories in Gaul outside of Visigoth and Burgundian control in the 480s. He defeated the Roman "kingdom" of his competitor Syagrius at Soissons in 485, and the kingdom centered on Cologne was taken shortly after.[203] Clovis thus expanded *northwards* from the Seine-Loire basin to small *civitas*-based principalities south of or on the Rhine. Other minor polities were mopped up in the following years, and the client system across the Rhine (Thuringians, Alamans, trans-Rhenan Franks) was reestablished and incorporated into the new kingdom. Clovis' subsequent conquest of Visigothic Aquitaine, full of Roman aristocrats and senators with their own military followings, only strengthened the Roman element in later Frankish armies.[204]

The final conclusion must be that a very large part of the early "Frankish" army in fact consisted of Romans who exchanged one identity with another. This early process of ethnogenesis from fundamentally Roman elements and led by a Roman officer is clouded behind the selections made by Gregory of Tours from older authors and the anti-barbarian bias in contemporary sources. Thus, Clovis' acclamation as *Consul aut Augustus* at Tours with East Roman approval in 508 was a legitimizing act not only in the eyes of his newly conquered Roman subjects in Aquitaine, but would perhaps have carried even more weight among his own troops of largely Roman stock.[205] So would Fredegar's and *LHF*'s myths of the Trojan origins of the Franks—they were, in a sense, true.[206]

1.3.4 *The Last Roman Civil Wars in the West, 496-511*

When the Franks under Clovis with their Burgundian allies faced the Visigoths in 496-508, there is little to separate this conflict from the Roman civil wars fought a century earlier. The conflict was punctuated by a *Burgundian civil war that led to Frankish intervention; it was decided in 500

---

[203] For this process, see E. James 1988: 72-85.

[204] Thus adding further dimensions to the conclusions reached by Bachrach 1972.

[205] Gregory of Tours, 2.38. McCormick 1986: 335ff identifies the particular form of ceremony recorded by Gregory as typical of a provincial Roman commander returning to his base after a successful campaign. We then have a successful Roman general following traditional forms. Se further Daly 1994 on the Romanness of Clovis.

[206] See James 1988 and Wood 1995 on the putatively Trojan origins of the Franks in 6th-7th-century literature.

by the siege of *Vienne. Although the Franks supported the losing side, peace was patched up in 502. When the war against the Visigoths was renewed in 507, Clovis exploited a political weakness in the Visigothic defensive system. Instead of invading deep into Visigothic territory, as he had attempted against *Saintes in 496 or *Bordeaux in 498, he appears to have devastated the lands of the Gallo-Roman nobility along the Loire frontier, using a large army to conduct persistent raids. Normally the safe course of action for the defenders would be to wait out the raiders, either allowing them to run out of supplies or drawing them further inside home territory, and ambush them while divided or cut off any besieging army. However, the Gallo-Romans demanded immediate action, as it was their lands that were under threat and the Franks were unlikely to go away soon, since they had short supply routes. King Alaric could not risk their defection to the Franks; his political and ecclesiastic reforms of the previous year clearly demonstrate that he was anxious to maintain support.[207] Thus it came about that the Visigoths were drawn into a premature battle and lost at Vouillé in 507.[208]

With the main field army of the Visigoths defeated, the Franks rapidly took advantage of the situation and pressed on to their capital *Toulouse, which was sacked. Little noted by most military historians, who emphasize the importance of the battle, is the fact that most of the cities and forts of central and northern Aquitaine were still garrisoned by Visigoth troops and had to be systematically reduced in the winter following the battle (507-8). The list includes *Albi, *Rodez, *Clermont and *Bordeaux. Thus the Franks engaged in sieges only when there was no major field army that could relieve a besieged fort or city, corresponding to Roman strategy in Mesopo-

---

[207] Wolfram 1988: 196f.
[208] See Heather 1996: 214f; E. James 1988: 86f; Wolfram 1988: 193; Thompson 1969. James points out that Gregory in typical fashion has compressed the events, and that the lead-up to Vouillé was far more complex. Wolfram's contention that the Goths fought as cavalry against Frankish infantry cannot be supported by the sources; Gregory only says (2.37, tr. Thorpe): "Some of the soldiers engaged hurled their javelins from a distance, others fought hand to hand. The Goths fled, as they were prone to do, and Clovis was the victor, for God was on his side." The original text is even less specific: "... et confligentibus his eminus, resistant comminus illi." "and one side fought from a distance, the other fought at close quarters" (my translation). Two Goths rushed up to Clovis and stabbed him with their spears, but he was saved by his leather corselet and his fast horse. From this description it is clear that we know little of the tactics; the implication might be that the Franks used missile weapons and the Visigoths tried to close; however it may also be taken to mean mixed tactics of missiles and heavy infantry on both sides. Wolfram argued that the Visigoths were horsemen and the Franks infantry, but apart from Clovis' scare, no horses are mentioned.

tamia in the Anastasian war of 502-6 (cf. 2.3). This process still took a whole year. Gregory of Tours' description of the fall of *Angoulême in 508, recalling the miraculous fall of Jericho, leads us to believe that Clovis undermined the city walls, one of the most common methods of siege at the time. Shortly after, Franco-Burgundian armies unsuccessfully tried to conquer Visigothic Provence and Septimania. Although the Burgundians had some success capturing *Narbonne, they were soon forced to withdraw due to an Ostrogothic intervention in 508. The Visigothic cities along the Mediterranean coast were heavily assaulted by the Franks. Defended by Ostrogothic troops and despite intense efforts and great damage, they survived the onslaught. *Arles in particular suffered a hard-fought siege lasting nearly a year in 508. Caesarius' *Vita* and Cassiodorus' letters give vivid testimony to the difficulties experienced by the Ostrogothic expeditionary forces in Gaul. They had to maintain good relations with the local landowners and population while fighting a brutal war which wrought devastation on the Gothic cities. In order to support their efforts, the Ostrogothic authorities granted tax remissions, not only to compensate for Frankish destruction, but also to offset the cost of the maintaining defenses and garrisons that had been shouldered by the local *possessores*.[209] The Franks under Clovis did not succeed in penetrating to the Mediterranean coast for another generation, only after turning on the Burgundians, their erstwhile allies, and with the active assistance of the East Romans' *reconquista*.

1.4 Conclusion: From Emergency Measures to Medieval Institutions

Roman siege warfare, supported by the traditional structures of the Roman army, can be demonstrated to the mid-5th century in the West. Beyond this, military infrastructure ostensibly disappeared in the West but survived in the East. However, the regular army in the East was complemented by private military forces, supported by income from estate complexes that also had important administrative and logistical capabilities that were either used by the state or usurped by magnates who exploited positions in government service. A similar institutional framework has been ignored or misunderstood in the West since developments there have been studied in isolation, but the apparent un-bureaucratic and highly personalized

---

[209] For this, see chapter 3. A defensive strategy based on fortifications was still viable into the 8th century. The Visigoth system appears in fact to have been so effective that it was briefly maintained by the Arabs in the 720s.

mode of government among early medieval states was in effect created by the last generations of Roman bureaucrats themselves. Regular units coexisted with private retinues into the 6th century, further diminishing the difference in institutional basis for warfare between East and West. Frontier acculturation meant that clients were not only equipped and fought as Roman troops, but they also participated in Roman civil wars; ethnogenesis produced new peoples out of predominantly Roman or Romanized provincial populations; while the extension and militarization of Roman public obligations meant that early medieval states were able to draw on wide range of non-monetary resources, including labor and military service. Thus, in the first wars of the 6th century in the West, there were at least a dozen sieges involving Franks, Visigoths, Burgundians and Ostrogoths, all of which originated as Roman client armies or Romans that took on new identities. There are strong indications that these sieges were every bit as violent and logistically demanding as those a century before. The destruction they caused required extensive repairs organized by local landowners and communities, while the costs were covered by remission of taxes. Such practices eventually became entirely regularized by the successor kingdoms, producing a model of military administration that supported fairly large armies and advanced military technology that could absorb new technology and adapt to challenges from the East Roman Empire as well as the challenge posed by internal demographic and economic developments.

CHAPTER TWO

EAST ROME TO BYZANTIUM: SURVIVAL AND RENEWAL OF MILITARY INSTITUTIONS

The East Roman (or Byzantine) Empire experienced massive upheavals in the course of these centuries. The army of the 5th and 6th centuries was in every respect a continuation of the late Roman army. Although some temporary institutional changes have been noted, other elements persisted, undergoing only gradual evolution into the 9th century and beyond. As the united empire in the 4th century, the East still had a profound gravitational pull on client states, even after their ostensible superiority was shattered by the Arab invasions in the 7th century. The Byzantine Empire was able to reestablish client relations or otherwise influence states and peoples from Italy through the Balkans to the eastern frontier. Furthermore, the rapid Arab conquest preserved significant elements of Byzantine military organization under new rulers. Despite massively constrained economic and demographic resources in the late 7th and 8th centuries, the Byzantines profoundly shaped how warfare was waged throughout the Mediterranean. This has a bearing on our analysis of Western successors, as Gothic, Lombard and Frankish military organization in the 5th and 6th centuries was formed not only by surviving West Roman structures, but significantly, also adopted East Roman innovations in administration, logistics, tactics and technology that responded to many of the same tensions and problems as affected the Western parts of the former empire.

2.1 *Continuity and Change in East Roman Warfare and Society, 450-800*

The East Roman Empire survived the dramatic upheavals of the 7th century, but the cost was a disruption of the antique social fabric, especially the characteristic urban civilization that produced spectacular monuments and the leisured upper classes that produced great works of literature. Despite this apparent decline, many aspects of late antique society, including military administration, knowledge and practice, survived in recognized form into the middle Byzantine period. At the same time, Byzantium's enemies became ever more resourceful, assimilating administration and technology that had once made Rome superior. In response to massive challenges on all fronts, the militarization of the Byzantine state and administration that had begun in the 6th century extended down to virtually every settlement in the remaining empire by the 7th century, and in large measure explains this survival.

2.1.1 *The Strategic Situation of the East Roman Empire: A Brief Overview*

In the Balkans, the Romans faced a reasonably stable border after the decline of the Hunnic menace in the 450s until the 560s. During this period, the Ostrogoths took control of Dalmatia, Gepids and Lombards settled in Pannonia along the middle Danube, and Slavic and Hunnic groups along the lower Danube occasionally raided across the river. While each of these groups could cause disturbance, none of them provided a direct threat to Roman control and could often be played off against each other. In fact, the Roman Empire was a far greater danger to these groups than vice versa: Rome destroyed the Gepids and Huns by diplomacy and contained the Slavs to a large extent. The only potential threats to Rome in the West were the Ostrogothic and Vandalic successor states, but these had found a *modus vivendi* with the empire until Justinian's project of reconquest destroyed them (533-554). Roman control in the Balkans and Italy was disrupted by the establishment of the Avars in Pannonia in 568, pushing the Lombards into Italy and the Slavs across the Danube. The Avars themselves launched intermittent large-scale expeditions anywhere from Dalmatia to Thrace until their failure before *Constantinople in 626, using subject peoples under their leadership.[1] Although the Romans failed to restore full control of the Balkans after that date, the remaining Thracian and coastal provinces were reasonably safe despite the establishment of the Bulgar khaganate on the lower Danube in 681, which was only able to directly challenge Roman hegemony after 800.[2] After c. 600, due to pressures in the Balkans and the East, the Romans were unable to send sufficient reinforcements to Italy, so that the situation deteriorated, but very slowly, for the troops still loyal to the empire.[3] The Roman province in

---

[1] The degree of Avar control of expeditions against Thessalonica is disputed. The basic works are Whitby 1988, Pohl 1988, and Lemerle's commentary to the *Miracula St. Demetrii*. For further discussion, see chapter 7.2.2-3 below and the relevant entries on *Thessalonica: 586, 604, 615, and 618.

[2] The basic study is Beševliev 1981. The Bulgars are discussed more fully in chapter 7.2.4 below.

[3] The Byzantine-Lombard wars are discussed in the context of Lombard military organization (chapter 3.3 below). Brown 1984 is the standard introduction to Byzantine Italy.

Spain, established in 554, was probably possible due to a power vacuum in the region, as the Visigoths were still expanding and did not have full control of the peninsula.[4] Once entrenched, the Roman presence was extremely tenuous in the face of a consolidated Visigothic kingdom in the late 6th century, and only gave way in the first quarter of the 7th due to relentless pressure and no hope of reinforcements during civil wars and invasions by the Avars and Persians. While Ravenna and Rome were eventually lost to the Lombards, or effectively independent by 751, other outposts remained into the 9th century and formed the basis for a revived Byzantine Italy.

The Persian Empire under the Sassanid dynasty continued to pose the main threat until 630. Due to the intensity of conflict and the combination of Greek, Armenian and Syriac historiographic traditions, Roman defenses are best attested in the East. Roman-Sassanid relations in the 5th century had been reasonably stable, with a few episodes of tension and two brief, localized wars that were quickly settled by negotiations, as neither power desired full-scale war.[5] The Anastasian war of 502-506 over the great fortified cities of Mesopotamia, which is analyzed in detail below, clearly shows why.[6] It was extremely costly, destructive, inhumane, and ultimately indecisive, but set a pattern that repeated itself many times over in the following 120 years. Roman-Sassanid wars evolved into fierce contests for years on end, raging (sometimes by proxy) from the Caucasus to Yemen and from Chalcedon to Ctesiphon.[7] The last and most destructive of these wars, that of 603-628, ended with the restoration of the *status quo* for a few years. The Arab-led Islamic invasions began in 633. Having recently established hegemony over the various tribes of the Arabian Peninsula, the early Muslims exploited the vacuum created by the collapse of Persia into civil war. By the mid-640s, Syria, Egypt and Mesopotamia were lost after over a decade

---

See Christie 1995 for the Lombard perspective; Marcus 1997 and Noble 1984 for the gradual rise of the Pope by Byzantine default.

[4] See Vallejo Girvéz 1999 on Maurice's reorganization of the province around 590 and chapter 3.2 below.

[5] See most recently the survey of Persian-Roman relations in late antiquity by Dignas and Winter 2007; studies covering more specific periods include Greatrex 1998 on the early 6th century; Isaac 1990 on the Roman army in the east from the beginnings to the 6th century and 1995 into the late 6th and early 7th centuries; Rubin 1986 covers political relations in the 5th century, while Howard-Johnston 1995 is a thorough comparison of "The Two Great Powers" into the 6th century. The Persian Empire's military organization is discussed in chapter 7.1.2 below.

[6] See chapter 2.4 below; for further political setting and discussion, see Greatrex 1998.

[7] There is as of yet no comprehensive studies of these wars, but see the works cited above. Greatrex and Lieu 2002 have edited a collection of translated sources.

of hard fighting; by c. 700, the rest of North Africa had been conquered, although fighting continued over Cilicia, Armenia and (by proxy) the Caucasus throughout the 8th century.[8]

The Roman (or, conventionally from this date, Byzantine) Empire from the mid-7th century survived these pressures thanks to a thorough reorganization and militarization of society. Central administration and the military are those areas with the most clearly recognized continuity.[9] Settlement and the economy went through dramatic changes. As the Byzantines were mostly on the defensive, they had to rely heavily on fortifications from which to control territory and provide safe bases for the army and refuges for the population. Hence, much effort went into fortifying settlements or moving whole cities to defensible sites, as well as providing them with adequate means of defense in conjunction with the army. The urban civilization of late Antiquity was transformed and most monumental centers were abandoned for smaller, more defensible sites. The new fortified towns (*polismata, kastra*) used the same methods of defense as in the 5th and 6th centuries, so that the decline in urban civilization by no means meant a decline in the importance of siege warfare. Indeed, Niewöhner recently pointed out that the characteristics of late antique border fortifications were used for urban settlement in Anatolia in the 7th and 8th centuries.[10]

2.1.2 *The East Roman Army in the 5th and 6th Centuries*

In virtually every respect, the army of the 6th century was the direct descendant of the late Roman Army, with essentially the same organization, distribution, recruitment patterns, fiscal backing, technological and tactical capacities, and of course limitations and difficulties.[11] For much of the

---

[8] See Howard-Johnston 2010 for a recent interpretation of the last Persian war and subsequent Islamic invasions.

[9] E.g. Haldon 1979, 1984, 1990, 1999 shows in detail how various administrative and military institutional forms and practices survived and adapted in the transitory phase.

[10] Niewöhner 2007.

[11] For basic surveys of most of the early and middle Byzantine period, see Haldon 1999 and Treadgold 1995. While they disagree on a number of significant points, both emphasize continuity from late antiquity to middle Byzantium, and are in reasonable agreement for the period to the early 7th century. The early transition from the united Roman Empire of the 4th century to the Eastern Empire of the 6th century is discussed in Southern and Dixon 1996, but both they and Grosse 1920 pay little attention to the crucial 5th century. The best general survey of this period is still Jones 1964: 607-86, although Lee 2007 has a fresh analysis of the social history of warfare. Ravegnani has written two good, but partly overlapping monographs (1998, 2004) on the army under Justinian. The recent proliferation of

5th and early 6th centuries, the East Roman Empire had to deal with small-scale raids and internal policing, suppressing banditry or revolts. Such was daily life along most of the African,[12] Egyptian,[13] Syrian,[14] and Danube[15] frontiers, as well as internally in Isauria, the Caucasus, Syria and Egypt.[16] This, along with logistical and political considerations, determined the continuation of late Roman strategic dispositions, which entailed widespread distribution of troops in small detachments throughout the provinces under Roman control.[17] Most of the time, soldiers in small groups of a few score to a few hundred were stationed in forts or billeted in fortified cities along with urban populations. When such postings were permanent (more often than not), troops struck deep local roots, establishing families, buying property, and engaging in business.[18] They were still available for service throughout the Empire, though were often deeply concerned with the protection of their homes, properties and families if threats were rumored while on campaign. The advantage of such a policy was evident in defensive warfare against minor enemies, bandits and localized raids, which was the daily business of the army but of which we know comparatively little. The difficulties of organizing defenses against large-scale invasions (especially if occurring on different fronts at the same time) are better known. These difficulties were compounded by the normal distribution of troops, as it often took months or even years to plan and execute an adequate response to a determined invader.

Research on the East Roman army has generally focused on its formal organization, which is now fairly well understood. The logistical system went through many reforms, but the ability to supply the army continued much as before. In arms production, private craftsmanship may have been more common in the production of military equipment. As in the 4th (and frequently ignored 5th) century, the 6th-century army had an administra-

---

handbooks relating to the period means that there are also many chapter-length surveys with valuable observations that are addressed here when relevant.

[12] Diehl 1896; Pringle 1981.
[13] The basic study is Maspero 1912, but see now also Palme 2007.
[14] For the eastern frontier, see Honigmann 1935 and Isaac 1990.
[15] For the Balkans, see Whitby 1988 and 2007, Curta 2001, and *TLAD*.
[16] For internal security, see the works cited in the preceding notes *passim*.
[17] See chapter 1.2.1 above. The basic document is of course the *Notitia Dignitatum*; according to the secondary literature in the preceding notes, the situation was much the same in 500 and beyond; see further Haldon 1984.
[18] The garrison pattern is obvious in *Notitia Dignitatum, Oriens* and discussed in chapter 6.2.2. See e.g. Jones 1964: 662f for some examples of small-scale business; Whitby 1995: 110-116 for a discussion of landholding.

tive distinction between mobile field troops (the institutional descendants of the *comitatenses*) and frontier troops, or *limitanei*. The former were largely garrisoned around Constantinople, its extended Balkans and Anatolian hinterlands, and in regional commands near or on the frontier under the supreme command of *magistri militum* (masters of soldiers): firstly, *Praesentales I* and *II*, i.e. in the "presence" of the emperor, then *per Thracias, per Orientem* (later a separate command was established *per Armeniam*), and *per Illyricum*. Troops near the frontiers (*limitanei*) sorted under provincial *duces*, who commanded both categories within their provinces.[19] There was little practical difference between the two categories in recruitment, social status, property, composition, equipment, or skills, and the originally administrative difference may have disappeared completely by c. 600.[20] Both categories were garrisoned in small units in cities throughout the empire, took long to assemble in an emergency, and often performed the same tasks, including siege warfare. There was a certain reluctance on the part of the *limitanei* to join expeditions far from their provinces when their homes were threatened, but the same problem affected those field units that were normally posted near the frontier. Thus, Illyrian, Thracian and Anatolian troops tended to dominate Justinian's expeditionary armies in the West. Even field troops who were on active service for years on end sometimes had families and properties in the East that were important for their morale and loyalty to the empire.

Informal institutions, i.e. those described in chapter 1.2 outside the regular units, were important for the army. The extensive use of private troops, *buccellarii*, was institutionalized to some extent by imperial legislation, before being integrated into the regular army by 600 AD. Barbarians from the client system continued to have a moderately significant role in military recruitment, but the preponderance of barbarians should not be exaggerated.[21] Since the trans-danubian client system took some time to reestablish after the Hunnic collapse, and as several barbarian groups now controlled significant Roman populations, urban centers, and infrastruc-

---

[19] Haldon 1999: 67ff with references and next note.
[20] Cf. discussion in chapter 1.2.1 with notes above; Isaac 1988 is fundamental for our understanding of the 6th century as well. See further Petersen (in press) for the new consensus on the merging to a single category.
[21] See chapter 1.3.2 above for the client system in the 4th and 5th centuries. Teall 1965 exaggerates the increase in barbarian numbers after the Justinianic plague; cf. the discussion by Whitby 1995: 103-110 who makes it clear that the use of barbarians was fairly limited compared with regular troops. I would in fact argue that private soldiers (*buccellarii*) were far more important, cf. chapter 1.2.3 above.

ture along the Danube, they could be somewhat more difficult to manage than before. As in the 4th and early 5th-century empire, we must posit three types of "barbarian" troops: those provided from beyond the border as ready-made mercenary units for limited time periods; barbarian volunteers or military colonists who served in the regular army; and erstwhile Romans who adopted an ethnic unit designator, or was perceived as such by an observer. With the collapse of the Western Empire, the western edges of the client system (especially along the middle Danube, in Dalmatia and parts of Illyricum) were extremely hazy, and eventually became an area of contention between various groups formerly within the West Roman or Hunnic orbit seeking new patronage and the Romans attempting to reestablish control. Thus, several Hunnic and Gothic groups contended for integration into the East Roman army in the 5th century.[22] The Heruls, ousted from their position as suppliers of troops to the West Roman army (including that of Odovacer) in the 490s, moved to the East, where they formed a loyal client kingdom that provided both mercenary contingents and recruits for the regular army. The Lombards did likewise in the mid-6th century. Even after the Avars destroyed the balance and the Lombards moved into Italy, older models prevailed as several Lombard dukes served the Byzantines efficiently on a large scale.

All of these groups formed part of the larger "Balkans military culture," and may explain how federate troops, formerly barbarian mercenaries, by the 6th century had become regular units recruited from both citizens and barbarians:[23] client armies that had gradually been assimilated to Roman fighting styles, culture and economy over generations only required official sanction to transit from *de facto* to *de jure* Roman soldiers. Conversely, Romans on the fringes of Eastern authority assimilated into the ethnicity of whichever group was ruling them, further making distinctions between "Roman" and "barbarian" armies and groups difficult.[24] Although Heather claims that the Ostrogoths were a fairly homogenous block that disappeared from the East when Theoderic led them to Italy in 489, Balkan Goths remained so prominent in Illyrian and Thracian armies that soldiers from those armies in Syria in the early 6th century were generally called

---

[22] See chapter 7.2.1 below for the former, 3.1.1 below on the latter.

[23] In the early 6th century they should be regarded as a component of the regular army and fought like it; see 1.2.1 and cf. the discussion in Amory 1997: 281f.

[24] For instance, the *Strategikon* was skeptical of Romans who had gone over to the Slavs, apparently in significant numbers; see 7.2.2 for further discussion and references.

"Goths" by the local population.[25] This did not mean that all individuals were of Gothic origins; only that, for a time, the association between "Gothic" and "soldier from Thrace or Illyria" was very strong due to the fearsome and warlike reputation of "Goths" in Roman ethnographic imagination, combined with their significant role as recruits at the time. While local patriotism was certainly a feature of ancient society, quite a few provincial identities had rather weak ethnic connotations.[26] Instead, political and military solidarity played a large role in the construction and solidification of a particular identity. What made the distinction permanent was whether a particular military unit remained faithful to the Emperor or not.[27] In this interpretation, the Ostrogothic army that followed Theoderic into Italy consisted of a conglomerate of soldiers of varying degrees of attachment to the Roman army and empire. One Gothic group had long been settled in the Balkans and attached to the Roman army, while another had recently passed out of Hunnic hegemony. The mix also included Balkan provincials from different ethnic groups, perhaps even Roman ex-servicemen, as Theoderic and other groups known (or assumed) to be Goths are simply called tyrants or rebels by Syriac sources, e.g. Zachariah Rhetor.

In contrast, others chose to enter or remain in Roman service. Vitalian had a career remarkably similar to that of Theoderic, including a full-scale revolt and projected siege of Constantinople, but remained within the Roman imperial orbit to become *magister militum* of the central field armies. He was nevertheless regarded as a Goth by some, while his troops are variously labeled as Romans, Goths, Huns, Bulgars and Scythians, all of which are probably accurate. Indeed, many individuals have wildly diverging ethnic designations depending on the proximity and sympathies of the author. Mundo, a freelance Balkan commander variously described as belonging to different ethnicities, was eventually integrated into the East

---

[25] Thus, Syriac sources simply call Illyrian soldiers in the East "Goths," although their personal names can be either Christian or Germanic; it appears that they are in fact Illyrians called by a generic term no matter what their origins were. Many were originally recruited in Illyricum but were later posted in Palestine, Syria and Mesopotamia. See JS 93f, especially Trombley and Watt 2005: 111f with n. 119 for a case in 505/6; Pseudo-Dionysius of Tell-Mahre *Chronicle* part III, s.a. 837 (Witakowski 1996: 25, 32) for a case in AD 525/6. Interestingly, on both occasions the term Goth seems to be used in a derogatory fashion about violent and abusive troops. Contra, see Heather 2007, although he discounts all evidence of Goths in Greek sources as exceptions, relies on the fragmentary John of Antioch's silence about Goths in the Balkans after Theoderic, and ignores the Syriac evidence.
[26] See in general the studies edited by Mitchell and Greatrex 2000.
[27] Greatrex 2000.

Roman army and fell on an expedition against the Ostrogoths.[28] In the 570s, when other groups were perceived as dominant due to recent Roman recruitment drives in Italy, Syrian bishops could threaten the Persians that the Emperor would soon arrive with "60,000 Lombards."[29] They were certainly aware of the distinction between Lombard federates or mercenaries and regular Illyrian troops, but there was a strong tendency towards applying the most warlike name to any army.

The conflation of various ethnic groups happened among regular Anatolian soldiers too. Procopius designated eastern units as either Armenian or Isaurian. While the former were mostly tied up by the Persian wars, the latter are prominent in Justinian's war of reconquest. As first bandits or rebels and later soldiers from the southern Anatolian highland province of Isauria, the Isaurians had acquired a particularly fierce reputation in the East in the 4th and 5th centuries, similar to the Goths in the Balkans from Adrianople to Theoderic. With the rise of the Isaurian Zeno to the throne in 474, there was a "flurry of literary activity" concerning Isaurian origins and their wars, and Isaurian identity was deliberately cultivated by the Roman state. This designation was subsequently extended: "Isaurian" was used of men who were of Cilician and Lycaonian origins, while the geographic (and hence, presumably, ethnic) term was also applied to areas normally included in Cappadocia, Pamphylia or Galatia; even Antioch was referred to as an "Isaurian" city by Marcellinus Comes.[30] While naval crews were recruited on the coast and eastern Anatolians were of Armenian or Tzanian origin, most recruits from the interior of Anatolia seem to have gone by the name "Isaurian," absorbing the more feeble ethnic feeling (and contemporary perceptions) of other provincials.[31] Procopius, then, consistently referred to Anatolian soldiers as "Isaurians," presumably in the after-

---

[28] See the entry for Mundo in Amory 1997: 387ff; more examples are found in Amory chapter 8. On Vitalian's army, see Malalas 16.16; *PLRE* 2 s.v. Vitalianus, who is called a Goth by the Syriac Pseudo-Zachariah.

[29] J. Eph. 6.13.

[30] For a discussion of Isaurian identity, the government's role, and quotes, see Elton 2000.

[31] For Cilician and Ionian sailors, see Proc. 3.11.14; for ethnicity in the Roman army, see Mitchell 2000; see further Elton 2000 for examples of groups of Roman soldiers and Isaurians proper lumped together simply as "Isaurians" by contemporary observers. "Isaurian" is often used as a generic term for troops recruited in Anatolia, which in other sources are called Phrygians, Lycaonians, Cappadocians. Cf. Greatrex' discussion (2000: 201 and n. 24) of the low quality of Lycaonian troops, who were apparently regularly confused with the Isaurians. Procopius states that poorly performing troops at the battle of Kallinikos were Isaurians, but corrects this misconception afterwards.

glow of their glory under Zeno, but soldiers from the Balkans were consistently called Illyrians or Thracians. It was hardly politically correct at the time to call to attention the close relations between the Ostrogoths and Roman soldiers who shared the same origins; indeed, one of the most prominent Thracian commanders was Bessas, an ethnic Goth who used his linguistic skills to negotiate with Ostrogothic garrisons during the war in Italy. In contrast, there was no problem in "ethnifying" the Isaurians as loyal citizens and soldiers who were supporters of legitimate Roman emperors.

2.1.3 *The "Two Hundred Years' Reform," or before the Thematic System*[32]

The 6th century structures continued into the reign of Herakleios (610-41). Although one should assume that civil war and the Persian conquest of Syria (from 614) and Egypt (from 619) caused significant disruption, Herakleios won the war in part because he was able to improvise supplies and pay, most significantly by forced loans from the church. His greatest achievement was keeping most of the army together, motivated, trained and in operation until victory in 628. Several units, listed by *Notitia Dignitatum* as being stationed in Syria two centuries earlier, are again attested operating against the Arabs in Syria and Palestine in the 630s, reposted in territories that had been under Persian occupation. They must either have been withdrawn to safe territories (Anatolia, Cyprus, North Africa) or reconstituted around a core of veterans. In reorganizing the lost territories after the Persians had withdrawn in 630, one would expect significant reforms in provincial and military administration. However, Herakleios' reforms clearly aimed at re-establishing previous arrangements with only modest changes. For instance, the defense of *Palaestina* in 634 was led by a *dux* as one would expect based on older arrangements.[33]

---

[32] The phrase is the title of Lilie's 1984 survey of the literature on the evolution of the middle Byzantine military provinces, called "themes" (from Greek *thema*, pl. *themata*). See also *idem* 1995.

[33] See the relevant sections in Haldon 1979, 1984: 164-82, and especially his 1993b article on continuity to 640, but note the somewhat diverging, but more detailed reconstruction by Schmitt 2001. Herakleios' provincial organization has led some to argue he instituted an early version of the middle Byzantine themes (military provinces) in the East at this date. For example, Shahîd 1994 and 2002 postulates a thematic reform under Herakleios, but Haldon 1995 has convincingly demonstrated that not only did Herakleios retain the traditional military structures, but the modifications he made also became the basic template for the early Islamic military organization. For further discussion of Herakleios' policies, see Kaegi 2003.

Defeated by the Arabs, Roman troops were withdrawn from the eastern provinces and finally settled in Anatolia between the 640s and 680s. The Greek names of later middle Byzantine armies are attested from the 7th century, and clearly descended directly from the late Roman field armies. Thus the army division of *Opsikion* (Latin *Obsequium*), renamed thus under Herakleios, was the direct institutional descendant of the two amalgamated *praesental* field armies stationed in Constantinople and its extended Balkans and Anatolian hinterlands, settled in northwest Anatolia; *Thakesion* was the army under the *magister militum per Thracias* withdrawn to western Anatolia after serving in Egypt as late as the 640s; the *Anatolikon* derived from the army under the *magister militum per Orientem* (in Greek, *Anatole*), and finally, *Armeniakon* from the army under the *magister militum per Armeniam*. The naval forces, represented by the *Karabisianoi*, may in part be based on the administrative framework of the Justinianic *quaestura exercitus*, a special district comprising Aegean islands and part of the Anatolian coastline, used to supply Danube garrisons in the 6th century, although this is one of the more vexed questions in the transition from late Roman to Byzantine military organization. By the early 8th century these armies were firmly associated with, and beginning to give name to, the districts in which they were settled. Further subdivisions, for political, strategic and logistical reasons, demonstrate that late Roman structures persisted into the 8th century and beyond; units of Justinianic, Theodosian and even tetrarchic origins are in some cases attested at various points into the 10th century and formed the basis for a number of smaller new themes, administrative divisions within themes, and *kleisourai*, or lesser frontier provinces commanding important passes.[34]

How these troops were supplied, equipped, and trained, and not least how they fought, have since been subject to lively debates, particularly as the origins of the thematic system familiar from the middle Byzantine period has been dated to this period. 10th-century legal evidence indicates that Byzantine soldiers were often landholders exempt from taxes, obligated to provide their own equipment according to how wealthy they were, alternatively supported by their local community who were required to assist troops who lacked means. Hence the predominant view was long that Herakleios or one of his immediate successors had supplied his troops with land from which to supply and equip themselves, in effect creating

---

[34] For the most recent treatment with extensive literature, see Brubaker and Haldon 2011, chapter 11.

an army of farmer-soldiers who defended their own lands in guerilla-style warfare.[35]

In light of recent research, it is clear that the idea of "military lands" and the romantic notion of "farmer-soldiers" as a basis for middle Byzantine military organization must be rejected. Haldon points out that there is no evidence for legally recognized "military lands" until the 10th century, and that this was only the institutionalization of a very long and slow development of troops investing capital in land and gaining tax exemptions in return for using their wealth to support military equipment and long-term campaigning.[36] Haldon has further demonstrated that even Constans II (641-69), to whom some attribute such a reform, maintained the basic structures as they existed under Maurice and Herakleios. Zuckerman has since shown conclusively that the word *"thema,"* long thought indicative of a specific settlement on land, was actually not used in Byzantine administration until the reforms of Nikephoros I (802-11). Contemporary documentary evidence from lead seals shows that Byzantine military districts in the 8th century were called *strategiai* or *strategides*, i.e. "regions of a *strategos* (general of a field army)," rather than themes.[37]

Troops were supplied by centralized production supplemented by provincial workshops and corvées that clearly derived from late Roman practices. For example, horses were supplied both by imperial stud farms and levies on taxpayers.[38] Government-run factories and arsenals seem to have still played their part, but centralized production became less important relative to local initiative under regional commanders after the mid- 7th century, but again based on 6th-century precedents that allowed for decentralized production. Similarly, instead of using land grants to supply soldiers, the fiscal administration of the late Roman Empire was wholly directed towards military needs, and replaced exchanges in cash (tax, payment, allowances for equipment) with exchanges in kind as far as necessary in the 7th and 8th centuries.[39] While the central administration survived and to a large extent maintained tax records as before under a modified praetorian prefecture, the specifics of redistributing goods and equipment to the army fell to what had been a fairly small section of Ro-

---

[35] The classic opinion is found in e.g. Ostrogorsky 1969, who has found a champion in Treadgold 1995.

[36] For a comprehensive survey of literature and evidence, see Haldon 1993a.

[37] For continuity into the reign of Constans, see Haldon 1993b; for terminology, see Zuckerman 2005, whose argument is integrated by Brubaker and Haldon 2011.

[38] Haldon 1984: 318-23.

[39] Brandes and Haldon 2000.

man fiscal administration, namely the *apotheke*, a network of state customs and warehouses. The officials employed by the *apotheke*, the *kommerkiarioi*, are well attested by a large number of lead seals that give precise information about their organization and geographic distribution over time. Originating as state officials in warehouses responsible for high-value luxury goods such as silk, the staff of the *apotheke* clearly had the administrative and logistical competence required to collect, store and transport taxes in kind on a massive scale. Furthermore, the *kommerkiarioi* were apparently highly trusted state officials who were accustomed to dealing with very large sums. Its logistical potential seems to have been first developed with the loss of Egypt to the Persians in 619, when Constantinople had to be supplied from North Africa and Sicily. Not only were these needs perpetuated by the Islamic conquests, but the relocation of substantial field armies to semi-permanent settlements, multiple-year expeditions to e.g. Italy and the Caucasus, and mobilization for defensive counter-attacks (e.g. following *Constantinople 670f) drove these new developments.[40]

While field armies were probably very much on the move in the 640s through the early 670s, more permanent postings gradually developed by around 680, with some further modifications until c. 720 as frontier territories were slowly lost to the Arabs. It seems that for both strategic and logistical reasons, the new situation required a more decentralized settlement pattern than in the 5th and 6th century, when individual units had mostly lived together in their designated garrison city. Scattered settlement facilitated the acquisition of property by soldiers, who only later received tax exemptions in return for continued service, but again exempting soldiers from taxes and *munera* was a typical late Roman arrangement. Because of their property and salary (whether in cash or in kind), soldiers became a new local provincial elite in the 8th century, but despite local ties and dispersed settlement on farmsteads, they were still regarded by the government as standing forces and treated as such. They remained attached to their units and could be called out for long-distance campaigns: even in the 8th and early 9th centuries, Roman troops were to be found on expedition to the Danube delta, Italy, Cherson and deep into the western frontier of the Caliphate.[41]

---

[40] Brubaker and Haldon 2011 chapter 10.
[41] For settlement patterns, see Haldon 1984: 217 and Brandes and Haldon 2000; for expeditions to Bulgaria, see chapter 7.2.4, for Italy, 3.3 *passim*; otherwise, see *Cherson 710 and 711, *Germanikeia 745 and *Melitene 750.

Identification of subdivisions within armies and their institutional history is difficult due to a tendency since the 6th century to replace formal unit designations (often descended from ancient legion names, filtered through tetrarchic reforms) with a simple regional designation; it is only in the 9th century that evidence becomes more abundant. Thus most units were simply called the *arithmos* (or Latin, *numerus*; see below) of so-and-so city or district, a practice continued beyond our period. Nevertheless, several tetrarchic unit names survive into middle Byzantine military organization, as we have seen, and the basic logic of tactical and administrative subdivisions were informed by late Roman practices. Due to the survival of more documentary material, we can follow the evolution of the Byzantine garrisons in Italy in some detail. While Brown rejected the older notion that Italian evidence could throw light on the evolution of the themes, we can now turn this premise on its head and use it to amplify reconstructions provided for the more poorly attested East. In Italy, Roman *numeri* (i.e. *arithmoi*) and *banda* attested in the 7th and 8th centuries clearly originated in the East, presumably having arrived with any one of the expeditions from the 530s through the 590s and perhaps later; a Bulgar division that found service with the Lombards was probably part of Constans' *Italian expedition in 663. Brown enumerates several units that must originate on the eastern frontier: a *numerus Armenorum*, a unit of *Persoarmeniaci*, another of *Persoiustiniani*; from the Balkans a *numerus Sermisianus* (Sirmium), and a *numerus Dacorum* (surrounding province on the lower Danube). While most of these units were probably raised in the 6th century, others had a pedigree that went back to *Notitia Dignitatum*, such as the *primi Theodosiani* and *felices Theodosiaci*. Others again took name after the locality in which they had been settled, so we have a *bandus Mediolanensium*, *bandus Veronensium* and *bandus Ravenna*. Some of these unit names even survived as designations for urban corporate bodies in Lombard and Carolingian Italy.[42]

While large-scale reinforcements to Italy from the East were increasingly rare after the 660s, commanders and smaller units regularly transferred from the East and intervened until the early 8th century, if only to suppress insurrections or enforce imperial policies. Units could relocate after the loss of territory to the Lombards, but Roman garrison forces in Italy were mostly settled in surviving territories by c. 600 and gradually became an élite of middling landowners. They used their status and re-

---

[42] Brown 1984: 89ff.

sources to acquire property from the mid-6th century as the old senatorial aristocracy atrophied. Although Italian units developed strong local ties and identified increasingly with local concerns by the early 8th century, military and fiscal organization was still integrated into the rest of the empire and followed Eastern practices. Most importantly, these forces maintained unit cohesion and military and effectiveness well into the 8th century despite consisting of what was in effect a landowning gentry (a term used by Whittow for Byzantine soldiers in Anatolia). Indeed, Brown notes successful military action by units from the last Roman enclave around Ravenna as late as 739.[43]

The mid-8th to 9th century saw the establishment of the *tagmata* in and around Constantinople, which eventually became a central field army that was to act as a strategic reserve, much the same way as the praesental field armies had in the 6th and its descendant *Obsequium* (*Opsikion*) in the late 7th and early 8th century. The debate has focused on the size and function of the tagmatic troops. Treadgold has argued for a large, professional force that formed the core of Byzantine armies as the remaining thematic forces in effect were part-time militias. Haldon, in contrast, has argued for a force of much more limited size that initially had a predominantly political function of supporting the reigning emperor against challengers. While Haldon, too, accepts a more militia-like thematic army, during the first decades of this arrangement, there appears to have been fairly little qualitative difference between the *tagmata* and the *themata*. Irene, for instance, could replace tagmatic forces *en masse* with thematic troops from Anatolia when the former proved disloyal in 787.[44] A functional and qualitative difference did arise at a later date, but in our period, the thematic and tagmatic forces were qualitatively interchangeable and had similar skills. This applies particularly to defensive siege warfare, as we shall explore below.

Surprisingly, the extreme pressures the empire underwent in the 7th and 8th centuries did not result in the predominantly cautious, guerilla-style warfare advised in 10th-century military manuals. Lilie has shown quite conclusively that the Byzantine army in the 8th century willingly engaged in conventional warfare with twenty battles attested in just over a century against the Arabs alone, but it follows that sieges outnumbered

---

[43] For the Italian evidence, see Brown 1984; 31f, 82 and passim; for identification of Byzantine soldiers in Anatolia as a landed gentry see Whittow 1996: 173.

[44] See the various works by Haldon, especially 1984, and on some accounts *contra*, Treadgold 1995.

battles by a factor of five to one, or more. The source situation is more difficult in the mid to late 7th century, but if large-scale expeditions in Armenia, Caucasian Albania, Media, Italy, Africa and the Balkans are included, conventional warfare based on sieges (and by definition, exposing armies to battle) was in fact the norm, even during most of the 7th century. It seems that the Byzantine army, then, was quite battle-ready despite serious defeats. Elements of guerilla warfare, harassment, and delaying tactics were of course always present, and the Byzantine army never recklessly committed to battle. While an "official" guerilla strategy seems to have been formulated c. 780 in the face of a resurgent Abbasid caliphate, battle and siege were still fundamental components of warfare.[45]

Training and maintenance of skill within army units was similar from the late Roman army to 10th century and beyond. The regular army and private troops spent much time on practice and drills. Firstly, all soldiers began as recruits (*tirones*) for a period of several years. It is clear from late 6th-century evidence that two years of service was not enough to prepare new recruits for battle. All military manuals recommend drills on both individual and unit level, minimally enforced by the regular muster of all Roman troops required by law and custom. For garrisoned forces, muster was apparently a daily event whenever troops were not on winter leave or detached for other assignments.[46] In the *Miracles of Anastasios the Persian*, more complex battlefield drills combining several units were held according to ancient custom in Palestine in the early 630s, in connection with the spring muster.[47] Units garrisoned together seem to have exercised much more regularly. A bizarre story of how neighborhood districts, named after different gates, formed up outside the city walls of Ravenna and attacked each other every Sunday is reported by Agnellus as taking place in around 700. Due to its regular nature and the identification of Roman unit names stationed along defined sections of the wall when Ravenna was threatened by attack shortly after, it must be a garbled recollection of how urban garrisons formed up in their respective units and practiced battle formations against each other, perhaps involving the local civilian militia.[48] Finally,

---

[45] Lilie 1976: 93 n. 92 and *passim*.

[46] For 6th-century practices and legal provisions, see Ravegnani 1995: 53-72; for the turn of the 7th century, see Whitby 1995; for relevant military manuals, see e.g. Vegetius, *Strategikon* and *Peri strategias*, *passim*

[47] Anastasios the Persian, *Miracles Anciens*, III.8, in Flusin 1992.

[48] For lining up every Sunday and infighting at a particular occasion, see Agnellus cc. 125-28; on Roman units, see Brown 1984:95ff for references and discussion. In his account, Agnellus gives the impression that participants were regularly killed or injured although

whole armies were trained for battlefield-scale maneuver whenever assembled for campaign, not only at the beginning of an expedition, but even using quiet periods such as winter quarters during extended campaigns. A form of exercise underestimated in modern research, hunting, was not only recommended in military manuals but practiced on a large scale when East Roman units were preparing for or during breaks in extended campaigns.[49]

Although an effective Roman naval service seems evident from the seaborne expeditions against the Western successors in the 6th century, Zuckerman argues that an actual navy was a 7th-century innovation, established under Constans in response to Islamic naval raids from 649. Analyzing Constans' poorly understood activities in Sicily in the 660s, Zuckerman argues not only for extensive preparations for a navy based on Italian, Sicilian and North African resources, but also suggests that methods of finance (by way of a poll-tax inspired by the early Islamic *jizya*) and recruitment of naval crews by conscription were inspired by the Caliphate. Cosentino, however, maintains that a naval establishment did indeed exist continuously from the 6th-century establishment under Justinian, and argues that Constans' tax reforms were only an extension of existing Roman forms of taxation. Brubaker and Haldon are similarly skeptical of influence from Caliphate without completely dismissing the possibility. An effective naval capability does seem evident from the temporary reconquest of *Alexandria from 646 and subsequent expeditions. Certainly a naval province in the mold of other commands (later *strategides*) was being developed in the last decades of the 7th century, and proved instrumental in keeping far-flung enclaves supplied and manned while threatening the Caliphate's tenacious control from Syria to Carthage. While naval battles were

---

he elaborates on a specific instance. If there is some reflection of reality, we may surmise that sometimes mishaps occurred during exercise, or inter-unit or neighborhood rivalry made tempers flare. Acrimonious intra-urban fighting certainly was a feature of late antique society, but there is reason to believe he was reading events of his days, after 800, when the Roman unit designations had developed into names for neighborhood militias, back to their origins when Roman units still were cohesive, distinct groups in Italian society. This certainly was the case in the other instance when the local commander assigned sections of the walls of Ravenna to eleven named units as well as the *familia* of the clergy.

[49] *Strategikon* 12 (D) gives detailed instruction for how to organize hunts on a large scale for the purposes of morale-building and unit exercise. The actual practice is confirmed by Joshua the Stylite (JS) 90, who relates an anecdote relevant to training: "... Pharazman set out from Edessa, and went down and dwelt at Amid ... [where] he ... used to make great hunts after the wild beasts, especially the wild boars, which had become numerous there after the country was laid waste. He used to catch more than forty of these in one day; and as a proof of his skill he even sent some of them to Edessa, both alive and dead."

extremely rare, fleets were fundamental to both Roman and early Islamic warfare, and both empires operated fleets across the Mediterranean. In addition, wars against the Bulgars, civil wars and diplomacy (or client policing) show Roman fleets operating on most Black Sea coasts. Behind these fleets obviously lay an extensive infrastructure of ports, dockyards, and arsenals, as well as the administration of materials, supplies, craftsmen and sailors. It is clear that the Arabs exploited existing Roman structures and practices in order to build, crew and supply their fleets, but also expanded traditional obligations on a massive scale (see chapter 7.3); a similar policy is evident under Constans, but the common origins of infrastructure and exactions means that it is difficult to discover who took the initiative first.[50]

2.1.4 From Late Roman client management towards a Byzantine Commonwealth

Neither Constans II nor Constantine IV (662/69-85) were great innovators, then, but fought to hold together the empire of Maurice and Herakleios by using tried and tested means adapted to new circumstances. This is further borne out by their policies of trying to maintain client states beyond the frontier (wherever they might be at a given point of time), recruiting grounds in the Caucasus and a sufficient hinterland against Arab invasions by campaigning in the Balkans and Italy.[51] Similar policies are clearly in evidence for Justinian II and the emperors of his interregnum (685-711), who could exploit Arab civil wars to foment unrest in Syria and attempted to maintain authority from North Africa to present-day Azerbaijan. It was only consolidation after the end of the second Arab civil war in 692 that allowed the Arabs to concentrate virtually all their resources on attacking a diminished Byzantium, and force Leo III (717-41) to shift to a more defensive strategy. Client management remained in many respects similar to late Roman practices well into 8th century. Clearly Byzantium still envisioned itself as a great power, and although the empire struggled to enforce its will with limited resources as compared to the 5th and 6th centuries, the gravitational pull of Roman prestige, wealth and military might was still significant. Again there were close precedents: Herakleios in the 620s found an ally against Persia in the great Kök Türk khaganate that had held sway

---

[50] Zuckerman 2005; Cosentino 2008; Brubaker and Haldon 2011; for previous discussions, general context and narrative, see Fahmy 1966, Ahrweiler 1966 and Pryor and Jeffreys 2006.
[51] See the relevant entries in CO in the 650s and 660s, as well as Appendix III.

over all Central Asia since the 6th century and so terrified the Avars; subsequently, he went about reestablishing or even extending Roman control over Christian client rulers in the Caucasus.[52]

Despite the collapse of Roman Danube frontier due to Avar and Slav pressure, the gravitational pull of the empire again became evident with the collapse of the Avars. Between c. 630 and 660, Slavs in the Balkans were becoming increasingly assimilated to Roman socio-economic and cultural patterns, a natural effect of large numbers of Roman captives, interlopers, or stranded communities scattered throughout the Balkans, along with substantial client policing that seems to have become efficient once Avar power declined. During the same timeframe, Kubrat ruled over of Old Great Bulgaria, a Turkic khaghanate established with support from Herakleios north of the Crimean Peninsula, in the wake of the collapse of the Kök Türk khaganate. Kubrat (and possibly a successor) seems to have remained a loyal ally until the 660s. Although we have little information about political details, the partly assimilated Balkan Slavs and Bulgar khaganate were in effect sections of a vast client belt stretching from the Balkans to the Caucasus, where Christian rulers in Lazica, Iberia, Armenia and sometimes even Albania were under Roman sway. It held firm despite the first Islamic conquests, as no disturbances are reported in the Balkans or north of the Black Sea during these decades. However, this delicate balance was wrecked by the rise of the Khazars, another Turkic group north of the Caucasus, whose rise can be dated to the 660s and whose reach was not much inferior to the western branch of the Kök Türk khaganate a century before. New Bulgar rulers emerged, presumably heirs of Kubrat, forced out of Old Bulgaria by the Khazars. Of five Bulgar groups, three remained within the Byzantine orbit. Alzeco and his followers probably found employment on Constans' *Italian expedition in 663, but went over to the Lombards. Kouber found Avar employment and was set to rule over a large group of Roman captives from the vicinity of Sirmium, deported either in the 580s or 610s. Sixty years later, depending on when they were captured, Kouber was causing trouble for *Thessalonica in 682 under the guise of seeking Roman patronage.[53]

---

[52] For the strategic priorities and objectives of 7th-century emperors, see now Howard-Johnston 2010.

[53] For the general context and Caucasus, see Howard-Johnston 2010; for the Slavs, see chapter 7.2.2 below; for the Bulgars, see Sophoulis 2012 and chapter 7.2.4 below; for the Khazars, see Zuckerman 2007.

Asparuch moved a group to just north of the Danube estuary before 680, but possibly years earlier. Constantine's expedition in 680 must have been meant to re-establish imperial control up to the Danube frontier and project Roman authority onto perceived client groups. However, when it went horribly wrong, the Bulgars were free to cross the river. Although the Bulgars came to occupy nominally imperial territory and pulled many Slav tribes out of the Roman orbit, a client relationship of sorts was nevertheless reestablished with the Bulgars by c. 700, benefiting Byzantium during the Arab siege of *Constantinople in 717-18. Byzantine interventions against Bulgaria under Constantine V were made possible by the turmoil in the Caliphate during and after the third Arab civil war (744-751). Extending direct Byzantine control far into the Balkans, this aggressive policy not only forced significant Bulgarian concessions, but also shook the Bulgar ruling elite deeply, and under Nikephoros I nearly stamped out the khaganate. Ironically, Sophoulis argues, the Byzantine aggression forced the Bulgar aristocracy to close ranks if it was to survive, and led to a formidable resurgence under Krum in the early 9th century.[54]

The extremes of East Roman power even in the mid-7th century stretched from North Africa and the city of Rome to present-day Iranian Azerbaijan. In the far west, Lombards came under severe pressure in 660. While Paul the Deacon claims that Constans was defeated by the Lombards, it is perhaps more likely that his assassination in 669 and the subsequent revolt in Sicily that gave them a respite. Constans clearly had the authority and prestige to impose his will on the pope in Rome and revamp Roman military and fiscal administration on a massive scale for six years—hardly likely after a severe defeat—before he fell victim to a court conspiracy. Even in Armenia and Caucasian Albania, there was real and extended competition with Caliphate until c. 700. In addition to the direct conflict with the Caliphate, relations were complicated by Khazars, who had their own ambitions and could not easily be controlled. Nevertheless there emerged a strategic alliance which forced the Arabs to expend enormous energy on the eastern Caucasus rather than the approaches to Byzantium.[55]

As with late Roman client management, the elements used to control and manipulate clients were subsidies, titles, trade privileges, police expeditions and use of other clients. In many cases we are poorly informed of

---

[54] Sophoulis 2012; see also chapter 7.2.4 below.
[55] See Howard-Johnston 2010 for context and Zuckerman 2005 for the reforms of Constans II.

specifics, but the benefits varied according to circumstances, most importantly relative power. Christian Caucasian rulers had their own military followings and had long served both Roman and Persian armies. While their full potential in the form of a grand counter-offensive against the Caliphate was never realized, the very threat must have been menacing to the early Umayyads and must be credited with helping the empire survive in the second half of the 7th century. But just as important was to keep them outside of the Umayyad orbit, as they on occasion provided valuable manpower, expertise and not least political advice which could be turned against Byzantium. To a large extent the situation was neutralized by the Khazars, who by the early 8th century had made Roman client politics in parts of the Caucasus irrelevant while at the same time severely challenging the Caliphate and providing a new staunch ally. Slav assimilation to East Roman socio-economic, cultural and even political forms seems to have been quite advanced, as Slav agriculturalists had settled down, their engineers were beginning to fully master late antique siegecraft, and their chieftains had access to the court in Constantinople. Slavs could therefore provide substantial troops to East Roman armies, which however is only revealed by large-scale desertion on two occasions in c. 665 and the 680s. In the first instance, at least, they may have been involved in a larger revolt against Constantinople under Saporios. As Brubaker and Haldon point out, the demographic and political realities in the Balkans were quite complex, and there were at times a very fine line between a Slav or any other semi-Roman war-lord, on the one hand, and Byzantine military commanders and officials, on the other. A case in point is the Bulgar aristocracy, who not only relished their Roman titles, but even advertised their apparent client status as *arkhontes* or minor imperial officials in Greek inscriptions. Despite the initial confrontation c. 680, relations were either rather peaceful, supportive of the empire in general, or in service of an imperial contender. As we have seen, trouble within the Caliphate complicated affairs. While Byzantium antagonized the Bulgars and failed to reestablish any large client blocks in the Caucasus—this was only achieved later in the 9th century—Christian Syrians and Armenians were systematically deported from the Caliphate's frontier region, where they had been in Umayyad service, and deported to Thrace in order to bolster expansion there (see p. 403). By 800 or so, a compact, but still multi-cultural and well-organized empire had emerged; its most important institution, making the transition possible, was the army. Nevertheless, the organization of siege warfare built on and greatly extended civilian participation from late Roman precedents, as we shall see.

2.2 Organization of Siege Warfare I: The Army

The East Roman army had a corps of engineers and specialist soldiers who maintained Roman siegecraft well into the middle Byzantine period. Soldiers with craftsman skills were found in many units, while engineering specialists were either employed in urban arsenals or attached to a field unit. In both cases, their responsibilities overlapped, as all specialists could be called out for offensive operations, while field units often assisted in urban defense.

2.2.1 Specialist Skills among the Regular Troops

Byzantine skills in siege warfare cannot be separated from the general technological ability of the army. As in the previous period, the 6th-century Byzantine government systematically sought to recruit troops of a particular social or ethnic background with craft skills, which typically involved working with masonry and construction, useful for basic field engineering, such as entrenching and fortifications. Isaurian troops (or, as argued above, Anatolian highlanders) were often associated with much of the heavy manual labor that was typical of siege warfare, especially during the Italian campaigns. Thus at *Naples (536), Isaurians were responsible for opening up an aqueduct that allowed the Romans to penetrate and eventually overwhelm the city. At *Osimo (539), "five Isaurians who were skilled in masonry" (a typical Procopian circumlocution for "masons") were sent by Belisarius to destroy an extramural cistern that the Goths were using as their water supply. It was located right under the wall, requiring a large Roman operation against the Goths manning the walls in order to provide cover to the Isaurian masons. However, the construction was too solid to be destroyed easily, and Belisarius settled on poisoning the water supply instead. Isaurians were also used for constructing emergency field fortifications. During the siege of *Rome (537f), newly arrived Isaurian troops dug a deep trench in order to protect the harbor at Ostia.[56] At *Rimini (538), they had to stealthily extend the Roman moat during the night in the face of a Gothic siege tower that threatened to overtop the wall on the following morning.

---

[56] Procopius 6.7.1-2; Elton 2000.

While the Isaurians appear to have been particularly adept at such work, often the sources specify only that engineering works were done by the "whole army," such as when Belisarius retook *Rome in 546. Totila, when he abandoned the city just weeks before, had destroyed sections of the walls to make it indefensible, but Belisarius was able to perform emergency repairs in only 25 days. A lack of skilled military engineers (*tekhnitai*) meant that more complex work, such as the building and fitting of gates, could not be finished before the Goths tried to retake the city, so the Romans prevailed through a combination of overwhelming archery fire and caltrops to cover the still open gates. Belisarius' army was a scratch force that he had been allowed to raise in the Balkans after returning to favor in 544. Since it consisted of fresh troops, it may not yet have been completely set up as standard Roman units with designated military engineers. The army's competence in emergency repairs does however indicate that some of these troops were, despite their lack of military experience (as Belisarius complained), recruited from a background in craftsmanship.

2.2.2 *Military Engineers*

As in the 4th century, the Roman army continued to employ specialist soldiers with more advanced skills in engineering and siege warfare than the regular recruit. This class of troops was responsible for siege engines, ladders, field engineering, fortification work, and many other tasks, and they were simply designated *tekhnitai* (τεχνῖται) by Procopius, the generic Greek word for (specialist) craftsman. This term obscures the division of labor and hierarchy within the engineering corps, as well as the distinction from civilian craftsmen, whom Procopius also called *tekhnitai*, e.g. at *Carthage (533). We thus have to rely on context, incidental notes elsewhere in Procopius, and additional information from other sources in order elucidate who actually participated in siege warfare. During the Roman siege of *Naples (536), apparently distinct from the Isaurians skilled in masonry, there were *tekhnitai* responsible for constructing siege ladders. At first they miscalculated the height of the walls and produced ladders that were too short to reach the parapet, so that the troops massing were unable to ascend. It was only by some frantic lashing of ladders together two and two that the Romans were able to commence the storm. The performance of these regular engineers improved during the Italian campaign. At the first Gothic siege of *Rome (537f), the military engineers operated a large array of complex machinery and directed necessary repairs and engineering, such as modifying the shape of the merlons on the breastwork so that the soldiers were better protected against enfilading fire from besiegers who managed to approach the walls. The most spectacular results were accomplished by those who operated the artillery, who terrified the Goths and played a large role in keeping them and their siege engines at a safe distance (see 3.1.5). If indeed these *tekhnitai* were distinct from the Isaurians, they belonged to the regular units, *katalogoi* and *foideratoi*, recruited for the Italian campaign among the field army garrisoned in the Balkans, and were thus closely associated with their permanent garrison units as one would expect of defensive artillerymen.[57]

For the organization of artillerymen and military engineers we must turn to legislation and inscriptions before examining the remaining narrative sources. In the Novels of Justinian, we find provision for urban *ballistarii* who were responsible for the construction, maintenance, and storage of arms in public arsenals, although distinct from other arms producers:

> We also desire that those who are called *ballistarii,* and whom We have stationed in different cities, ordering (or "organizing") them and those who know how to manufacture weapons, that they shall only repair and place in good condition those belonging to the government, which are deposited in the public arsenals of each town. Where any workmen have manufactured arms they must surrender them to the *ballistarii,* to be placed with those belonging to the public, but they must by no means sell them to anyone else. The *ballistarii* shall, at the risk of the municipal magistrates of the cities to whom they are subject, observe what We have decreed, and the responsibility for this, as well as for the preservation of the public arsenals, shall attach to these magistrates; and where any of the workmen called *deputati,* or armorers, have been detected in selling weapons, the local magistrates shall subject them to punishment; shall deprive the purchasers of these weapons without refunding the price paid for them; and shall claim them for the benefit of the public.[58]

---

[57] Procopius 5.5.2-3; see Hannestad 1960 in general on the composition of Byzantine troops in Italy.

[58] *NJust* 85.2f: "Hoc autem observare volumus et eos qui in ordinibus balistariorum sunt, quos per diversas statuimus civitates, ordinantes eos et arma facere scientes, ut et ipsi sola publica arma in armamentis publicis uniuscuiusque civitatis recondita corrigant atque renovent. Si qui vero novum aliquid armorum instruxerint, hoc similiter et ipsi inter publica arma praestent et nulli alteri penitus vendant. Hoc quoque custodiant qui in balistariis deputati sunt periculo patrum civitatum, sub quibus et ipsos balistarios statuimus et publicorum armamentorum diligentiam atque custodiam constituimus: quatenus si qui visi fuerint aut deputati aut fabricensii vendentes arma, provideant per loca iudices eos suppliciis subdere, insuper et arma ab his qui ea comparant sine pretio auferre et publico vindicare."

The *ballistarii* were thus enrolled and paid as regular troops (*deputati* or καταλεγόμενοι, i.e. enrolled in the κατάλογοι), but placed under the immediate authority of municipal magistrates (*patres civitatum*, literally "city fathers"), who were responsible for the proper conduct of arms production and prevention of arms sales.⁵⁹ The *ballistarii*, in turn, were in charge of the public arsenals (*publica armamenta*, δημόσιαι ὁπλοθῆκαι), were senior to regular state arms producers, *fabricenses*, and responsible for collecting weapons produced by armorers in (illegal) private employment and keeping them in storage.⁶⁰

The name of the *ballistarii* (Greek *ballist(r)arioi*, βαλλιστ(ρ)άριοι) is derived from one of the most fearsome defensive artillery piece of the age, the *ballist(r)a* (see chapter 5.2.2), which obviously indicates responsibility for artillery as well as supervising arms production and storage. In addition to their eponymous engines, the *ballistarii* were certainly responsible for personal arms and without doubt for all types of artillery and siege engines within their jurisdiction. However, artillerymen had a much wider range of responsibilities, as they also supervised military construction work. When walls were constructed or repaired, those who actually defended them had an important role in their design and execution. This inscription found in the Crimea, dated to 476, attests how the emperor Zeno distributed tax money to those responsible for repairing the walls:

> [...] I am speaking of the vicarate of the most devoted *ballista* artillerymen (τοῦ ἐνταῦθα βικαράτου τῶν καθωσιωμένων βαλλιστραρίων)—through whom we have rebuilt the walls (δι' ὧν ἀνανεοῦντε[ς] τα τίχη [*sic*]) for the safety of this same city and express thanks in setting up this inscription in perpetual remembrance of our reign. And this tower was rebuilt in the year 512 (of the era of the Chersonese) in the fourteenth year of the indiction, with the most exalted count Diogenes accomplishing (the work).⁶¹

It is uncertain exactly how the corps of *ballistarii* was organized, but there appears to have been a specific hierarchy. A chief artilleryman, *arkhibalistarios*, is attested at Philippopolis (Arabia), sometime after it was founded by Philip the Arab in 244, while simple *ballistarioi* are attested at other sites.⁶² The role of senior *ballistarioi* appears to have been important in planning and executing urban defense and supervising workshops or ar-

---

⁵⁹ *NJust* 85.3.
⁶⁰ For the *fabricae* and *fabricenses*, see chapter 1.1.1.
⁶¹ *CIG* IV no. 8621, translated by Trombley and Watt 2000: xlviii (for which see reference).
⁶² Thus *archibali(starios)*, αρχιβαλι(σταριος) SEG 7: 989. *ball(istariou)*, βαλλ(ιστariου) MAMA 3.93; at http://epigraphy.packhum.org/inscriptions/.

senals. Junior artillerymen, such as Procopius' *tekhnitai* (the generic term, as typical in Procopius; the technical term would probably be *ballistarioi*) at *Rome (537f), were responsible for setting up, maintaining and operating individual machines, which included both *ballistrai, onagroi*, and other types of devices. During offensive warfare, *tekhnitai* were responsible for offensive siege engines, such as rams, as was the case at *Petra (550). There seems to have been no difference in the artillery and engineering capabilities of mobile field armies and garrison forces (*limitanei*) based on information from the late 6th century.

2.2.3 *New Developments from the Late 6th Century*

The East Roman state preserved a number of features typical of late antique military organization into the 9th century and beyond, but modified the terminology to fit new developments. With the rise of a new type of siege engine, the traction trebuchet, a new type of senior artillery officer emerged, the *manganarios*. Originally this was an operator of *mangana*, a catch-all term for construction machinery and other mechanical devices based on levers, pulleys, counterweights and the like.[63] The term took on military significance when *manganon* and its derivatives became used for machines with military application in the 5th-6th centuries. By around 600 it was the regular term for trebuchets, while *manganarios* began to supersede *ballistarios* as a term for artillerymen.[64] The famous Bousas was abducted by the Avars from his post as *manganarios* at the fortified border town (φρούριον) *Appiaria (586) on the Danube sometime in the 580s.[65] Since he was able to teach the Avars how to use siege engines, he must have had the responsibility for constructing as well as operating them. He was stationed there permanently as a regular garrison soldier (στρατιώτης) living there with his wife, and was thus a good fit for the model of garrison artillery experts proposed above.[66] At about the same time, at *Thessa-

---

[63] These are operators of *mangana* or *manganika*, varieties of a traction trebuchet. See chapter 8.2.2 for the rise of this term.
[64] This is discussed in chapter 8.2.2 below.
[65] The story is found in Theophylact Simocatta 2.16.1-11 and Theophanes 258.21-259.5.
[66] The exact term for his position is somewhat uncertain, as the oldest source, Theophylact, studiously avoided contemporary vocabulary, while Theophanes might be drawing on the language of an original source (postulated by Whitby, rejected by Mango and Scott), or simply inferring from the context and applying the terminology of his own day. Theophanes (or a later scribe) applied the colloquial form, μαγγανάριν (*manganarin*) for μαγγανάριον (*manganarion*). His profession is translated by MS 381 as "engineer." See further discussion in the entry in CO; the story was evidently misplaced by Theophylact and thus impossible to date accurately.

lonica (586), a military *manganarios* is attested as directing defensive artillery fire against Avar trebuchets.⁶⁷

The function of the the office *manganarios/ballistrarios* in the field is known from the Strategikon, which in c. 600 used the term *ballistrarios*, perhaps a conservative usage (or still the technically "correct" term), for the officer responsible for field catapults called *ballistrai*. In this instance, the word may have meant traction trebuchet (see chapter 8.2.1). The *ballistrarios* was accompanied by carpenters (λεπτουργοί) and smiths (χαλκεῖς) who were his subordinates.⁶⁸ However, the *Strategikon* also states that *tekhnitai* were responsible for setting up both gates and *manganika* on the walls when preparing for siege defense.⁶⁹ It is unclear whether the term includes *ballistrarioi*, the *tekhnitai* operated under their direction, or were themselves mostly civilians. Probably all combinations were possible depending on the circumstances. From the overlap between conservative and colloquial usage, it is clear that the term *manganarios* had the same meaning as *ballistrarios*, and its usage became increasingly common in the 7th century, but military engineers were also complemented by civilian craftsmen (see 2.4 below). At *Thessalonica (618), the Slavs and Avars are portrayed as attacking with a wide array of forces, including *manganarioi*, and at an even later siege of the city (*Thessalonica 662), a particularly adept Slav *manganarios* was commissioned by his rulers to build a fearsome siege tower with artillery in several stories, but was miraculously prevented from accomplishing this by being divinely afflicted with insanity.⁷⁰

Within the borders of the empire, the source situation makes it a bit more difficult to ascertain how the formal organization of siege warfare survived into 7th and 8th centuries. We can certainly postulate the presence of trained artillerymen and arsenals whenever we find mentions of artillery in use, although this conclusion is complicated by the fact that civilians were playing an increasing part in siege warfare, as a detailed examination of sieges will show. For now, we can note that *Constantinople was heavily guarded with artillery, a fact attested several times through the 7th and 8th centuries.⁷¹ Specialist manpower would have been provided by its central arsenals, but it is clear that civilians, both craftsmen

---

⁶⁷ See at *Mir. St. Dem.* 206.
⁶⁸ *Strategikon* 12 (B) 6.
⁶⁹ *Strategikon* 10.4.7f.
⁷⁰ See at *Mir. St. Dem.* 262.
⁷¹ See CO 626, 663, 715. Furthermore, a bowyer (*toxopoios*) healed for a painful testicular hernia in *MirStArt* 29 indicates the presence of military workshops in Constantinople in the mid-7th century.

and ordinary citizens, were heavily involved in defense (see chapter 2.4.2). Elsewhere, the distinction between civilian and military craftsmen is unclear: At *SYLWS (664), the Arabs forced a Paphlagonian master carpenter, ܢܓܪܐ ܚܕ ܪܒܐ (*naggārā ḥad rabbā*, actually a calque of Greek ἀρχιτέκτων) to build a huge traction trebuchet for them.[72] He guided them through the process under compulsion, but deliberately let the Arabs misconstruct it so that they were unable to operate it correctly, firing shorter and shorter of their mark. The Byzantine artillerymen, in contrast, had impeccable aim, smashing the framework and killing the operators of the Arab trebuchet with a single round (after mocking them for their poor handling of their artillery). While the Paphlagonian master carpenter may have been a civilian craftsman who had learnt military skills,[73] at least some of the defenders, designated in the chronicle as *rhōmāyē* (representatives, i.e. soldiers, of the Roman state), were regular garrison artillerymen in the 6th-century mould.

The fearsome defensive skills of Byzantine artillerymen are in fact well attested. At *Thessalonica (586) the Byzantine *petroboloi* smashed Avar trebuchets, while the Avars could barely hit the walls. At *Constantinople (626), the Byzantine *manganika* again outshot those of the Avars, who were thus never able to attempt a storm of the wall. Barely a generation later, *Constantinople (663) still had skillful operators of *ballist(r)ai*, though now probably under another name: they were nonetheless able to keep an Arab raiding party at a safe distance by firing accurate ballista bolts, and thus save a failed Roman sally from being massacred under the walls. One of the regions of incessant siege warfare, the Taurus Mountains, is poorly attested in narrative sources. Thanks to the extensive description of the Zuqnin chronicler, we know that the Roman troops stationed in *Kamakhon (766) were formidable artillerymen as well as inventive engineers, and successfully outshot the artillery of the Abbasid forces, who ultimately had to abandon their assault.

The Justinianic model based on mobilizing garrisoned engineers for offensive campaigns also prevailed. At the Roman siege of the heavily fortified Persian frontier bulwark of *Nisibis (573), the Roman generals organized a number of engineers to direct a very complex siege. While regular troops were probably responsible for the field fortifications,[74] specialist

---

[72] See Appendix III in addition to the note in CO.
[73] See discussion in chapter 2.4. *passim* below.
[74] Field fortifications: Syriac ܩܠܩܘܡܐ *qalqūmē*, via the form χαλκώματα, a corruption of χαρακώματα, Greek for ditch and embankment crowned by a wooden palisade.

engineers constructed large towers and other machines.[75] We know from the description of the Persian siege of *Dara (573) that the Persians had brought Roman engines, including trebuchets, left behind around *Nisibis.[76] Herakleios' grandfather, when campaigning against the Persians, delegated the responsibility for artillery fire to his lieutenant general (ὑποστρατηγός), who proceeded to bombard an unnamed *Persian fort (587) without stop for several days.

The organization of offensive siege warfare after the Persian Wars was fairly rare and has to be inferred through incidental information in the following century and a half. Herakleios had led complex sieges on his campaigns in the Caucasus,[77] and the Roman army that was implementing the peace conditions in Syria in 630 had the wherewithal to enforce it. When the Persian garrison at *Edessa (630) along with its Jewish population refused to surrender, the Romans bombarded them into submission with *manganīqē*, trebuchets. We can only surmise that these were operated by *manganarioi*. Constans II (641-69) used regular army formations in his invasion of Italy in the 660s, which, as we saw at the beginning of the chapter, were still organized on traditional lines. Since garrison artillerymen were still common in the 7th century, Constans' siege train would have been provided by the traditional organization. Its capacity was certainly significant, as Constans successfully besieged *Lucera (633) and several other unnamed cities. It failed at *Acerenza (633) for unknown reasons, but at *Benevento (663) a Lombard relief army from Pavia saved the Beneventine Lombards; trebuchets were significant in the Roman assault.[78] The initial performance of the Roman expedition against *Carthage (698) was successful against a formidably walled city and its suburban fortifications. Although no details are given, a fleet-born siege train was raised only a few years later by Justinian II (685-95; 705-11) for an expedition against rebels at *Cherson (711). The fleet carried "a ram, trebuchets, and every [type of] *helepolis* for siege warfare."[79]

---

[75] Engineers: Syriac ܡܟܢܝܟܘ *myknykw* (pl.) = Greek μηχανικοί. See 2.4. below for a discussion of the various classes of craftsmen and their military or civilian affiliation at various times.

[76] Both "machines" and *manganōn*. The original rendering is found in J.Eph. *textus* 287.28: ܡܟܢܝܡܛ ... ܡܢܓܢܘܢ *myknymṭ'* ... *mngnwn* (both in pl.). Brooks in his *versio* (218.5) renders them thus in Latin and Greek: machinamenta / μηχανήματα ... ballistas / μαγγανόν. For a discussion of these terms, see chapter 8.2.2.

[77] Reported in the Armenian historian Movses Dasxunertsi, not in CO, but discussed in e.g. Dennis 2000.

[78] See the discussion of this in chapter 8.2.3.

[79] πρὸς καστρομαχίαν κριόν, μαγγανικά τε καὶ πᾶσαν ἑλέπολιν; see the relevant entry in CO for details.

Although poorly attested in detail, large fleet-supported invasions, consisting of thematic troops, were regularly sent against the Bulgars and their many fortifications in the reign of Constantine V (741-75), while the Byzantines could launch successful siege operations against heavily fortified Arab border cities such as *Germanikeia (745), *Melitene and *Theodosioupolis (both 750), and *Germanikeia again in 778.[80] By the early 9th century, information is again more abundant. Military engineers were enrolled in the army and stationed at important border forts, such as *Serdica (809), where the Bulgarians had massacred a large number of troops. When survivors asked the emperor for respite and economic compensation, they were rebuffed, so they defected to the Bulgarians: "among them the *spatharios* Eumathios, an expert in engines."[81] A similar incident occurred a few years later; an experienced engineer (πάνυ ἔμπειρον μηχανικῆς) of Arab origins stationed at Adrianople defected in time to allow the Bulgarians to prepare a full siege train against *Mesembria (812), which contained even more valuable military arsenals with stores of Greek fire and siphons for using it. The position of the *manganarios* preserved the competence of the *ballistrarios* in supervising military arsenals and remained a feature of the regular army throughout the 9th and 10th centuries, when their activities are far better attested in military manuals, laws and administrative documents.[82]

2.3 *The Many Faces of East Roman Siege Warfare: The Example of the Anastasian War*

In less than four years of war, from 502 to 506 (often called the Anastasian War), the Romans and Persians fought each other to a stalemate after mobilizing enormous armies and committing numerous atrocities against each other's populations.[83] Neighboring cities with similar population size,

---

[80] See chapter 7.2.4 for a discussion of Bulgar siege capacities. For a full survey of the very extensive tit-for-tat warfare across the Taurus during this period, see Lilie 1976, Haldon and Kennedy 1980.

[81] ἐν οἷς ἦν καὶ Εὐμάθιος ὁ σπαθάριος μηχανικῆς ἔμπειρος; see the relevant entry in CO for details.

[82] *Mekhanourgoi*, presumably with the same meaning as, or working in support of, *manganarioi* (cf. the the usage of *De obsidione toleranda*), built machines for Thomas the Slav in the 820s; the precise term is found in e.g. in Leo, *Tacticae Constitutiones* 15.35; *De Obsidione Toleranda* p. 47, alongside *mēkhanopoioi*; and several of the documents preserved in *De Cerimoniis* pertaining to Leo's operations against Crete (see Haldon 2000).

[83] For the historical context, see Greatrex 1998 and Luther's 1997 historical commentary to Joshua the Stylite.

social structure, ethnic composition, culture, religion, resident garrison, fortification types and so on suffered radically different fates, as they reacted to threats with anything from abject submission to heroic resistance. Any number of contingencies could rapidly affect the course of events. It is therefore futile to point out any one particular form of siege warfare as prevalent at a particular time or place based on insufficient source materials. As the following, well attested events demonstrate, the means and objectives of a siege could rapidly shift from one mode to another. This war therefore provides a framework for understanding how sieges could be conducted in particular circumstances (chapters 5-6). It also gives us the necessary insight into *informal* modes of military organization that were paralleled in the Western successor states, especially the involvement of civilians, craftsmen and clerics in the organization and practice of siege warfare (ch. 2.4; cf. 3 and 4).

2.3.1 *The Background to the Anastasian War and the East around 500*

The Anastasian war was the first major war to affect the East Roman Empire since the days of Attila, and the first large-scale armed conflict between Persia and Rome after two brief and inconclusive wars in 421-22 and 441-42.[84] The long period of relative peace meant that when the first major war in nearly three generations broke out, defenses took some time to organize. However, these problems should not be exaggerated. Apparent lack of preparation can be ascribed to the Romans owning up to their treaty obligations with the Persians. These often stipulated a limit to the degree of military activity near the border. Thus it appears that quite a few cities right on the border, such as Martyropolis, only had token garrisons, if any at all.

It took far less than a full-scale Persian invasion to begin preparations for a siege, however; internal revolts or nomadic raids could also provoke a mobilization. This happened in 484 at *Edessa, where the rebel generals Illus[85] and Leontius[86] wanted to install a garrison of 500 cavalry, deeming this sufficient for controlling the city. The citizens, loyal to the emperor Zeno (474-91), responded by closing the gates and "guarded the wall after

---

[84] See Rubin 1986 for an overview of the background for these conflicts and the pattern they set for the 6th century.

[85] Former *magister militum* of the East. See *PLRE* 2, s.v. Illus 1.

[86] He was current *magister militum* of Thrace, but joined Illus when sent against him by the emperor Zeno, and was subsequently crowned emperor by the rebel forces. See *PLRE* 2, s.v. Leontius 17.

the fashion of war." The usurpers, lacking a safe base beyond the Euphrates, had to stay in Syria to face the remaining army of Zeno. However, after a crushing defeat near Antioch, they and the remnants of their army fled to a fort named *Papyrius (484-88) in Isauria, where they were holed up for four years before being betrayed and beheaded. Even two of the most experienced and high-ranking officers of the Roman army with the armies of Oriens and Thrace under their control were reluctant to besiege an important fortified border city such as Edessa. This clearly indicates a high degree of preparedness and motivation of the Edessenes, as well as the logistical difficulties of undertaking a siege,[87] especially in a rapidly changing political situation. The situation was similar on the Persian side of the border. While Persia's Arab federates, the Lakhmids, fell out with the Sassanids and raided Mesopotamia, the Kadishaye (mountaineers of uncertain ethnicity),[88] rose in revolt and besieged *Nisibis for some time. The situation was only restored when both groups were convinced to join Kawad on his invasion of Roman territory in 502.[89] The various reactions of the Roman cities to this crisis demonstrate the full range of possible Roman responses to a determined invader, from outright submission to fierce resistance.

2.3.2 *Abject Surrender: *Theodosiopolis and *Martyropolis 502*

The first cities on the Persian march, *Theodosiopolis in Roman Armenia and *Martyropolis in Northern Mesopotamia, surrendered, apparently without a fight. The walls of both cities were dilapidated or too weak to withstand a full-scale siege. While the commander of Theodosiopolis may have acted treacherously, the leader and citizens of Martyropolis went to great lengths in their submission, offering the last two years of public taxes. The local leaders retained their positions and the citizens were hence recognized by the Persians as their subjects.[90] The sheer scale of the Persian invasion was unprecedented and there were few Roman troops in the region to resist such a massive army. In fact, none of the contemporary sources mention the presence of regular Roman troops at all, though there were some Armenians in Theodosiopolis.[91] There certainly were small,

---

[87] They had dismissed a large number of troops who simply went home on regular leave before facing Zeno's army.
[88] Trombley and Watt 2000: 19f n. 96; possibly Kurds or some other Persian nomadic group.
[89] For the context, see Luther 1997: 96-152 and Greatrex 1998.
[90] Greatrex 1998: 79ff.
[91] Discussion of the appointment and status of the commander with references in

permanent garrisons of a hundred to a few hundred men in most of the cities of the East, but these were meant to police the local countryside, not face the full might of the Persian army with thousands of Arab and Hun allies.[92]

2.3.3 Fierce but Flawed Resistance: *Amida (502-3)

With the army unprepared for a major war, most of the immediate responsibility for defending the cities fell on the inhabitants of the affected cities. While commanders and citizens of the first two cities chose not to fight (2.3.2 above), Amida endured a siege for 97 days in the autumn and winter of 502-3 while the Persians sent out raiders to tie up Roman forces further west. The city probably only had a very small garrison, but due to its status as a key border fortress since the 4th century, its population appears to have been highly motivated and able to come up with and organize effective countermeasures to Persian engineering. The Persians tried to take the city by storm, constructing a large ramp, called a "mule," in order to overtop the wall, while attacking with rams at various points. The Romans responded by increasing the height of the wall, but the Persians brought up a large ram, presumably up the ramp, "and after they had struck the wall violently, the part newly built became loosened, because it had not yet settled, and fell." However, it appears that the regular course stood firm against ram attacks at other points, protected by bundles of reeds taken from mattresses and lowered by chains according to Zachariah. Alternatively, the defenders used large beams set obliquely according to Procopius; these absorbed the blows of the rams. Dewing's translation reads far too much into the phrase, giving the impression of a machine that chopped off rams' heads. Although such devices existed, they are not attested in this instance.[93]

Trombley and Watt 2000: 50f n. 244. It seems that the relatively recent acquisition of Armenia was not fully integrated into the imperial military structure.

[92] Cf. above for the general context, but especially Luther 1997: 177ff. For the situation in the 5th century, see *Notitia Dignitatum* (Oriens 36, Amidae). In Amida there were *Equites scutarii Illyriciani* and *Equites ducatores Illyriciani*. Cavalry was more useful chasing off raids than defending against set-piece sieges, and thus reflected the priorities of the Roman government in the East during the 5th century. Since raiding and banditry were the greatest problems in the region, there is every reason to assume that such forces remained in the city.

[93] Note the sentence: Ἀμιδηνοὶ μὲν τὴν ἐμβολὴν ἀεὶ δοκοῖς τισιν ἐγκαρσίαις ἀνέστελλον. The key phrase is *embole*, which Dewey translated as ram's heads (as it is used in Thucydides—here perhaps expectations of mimesis has distorted the modern interpretation), but can also simply mean "attack" or "strike (of a missile)." *Anastello* simply means repulse;

In the meantime, the Romans had undermined the mule, propping up the structure with wood. When the Persians were ready to storm the city by laying large beams from the mule to the wall, the Romans responded by throwing hides soaked in oil on the beams, making them so slippery that the Persians were unable to cross. While this went on, the Roman sappers fired the props that held up the mule and destroyed it with great Persian losses. The Persians had to begin all over again, throwing in beams and sandbags in order to repair their collapsed mound. The Persians protected their workers by stretching across the mound thick, moistened cotton cloths folded many times over. However, again the Romans came up with a counter-device, a large throwing-machine, called *ṭubbāḥā* (ܛܒܚܐ) translated as "the Crusher" (Wright) or "the Striker" (Trombley and Watt), able to destroy the Persian protective covers that were impervious to arrows, slings and lighter artillery. This may in fact have been a traction trebuchet, considering the power, range and contrast with other weapons that were in common use at the time.[94]

In the end it was not Persian ingenuity, but Roman overconfidence that brought the city down. They crudely taunted the Persians, having prostitutes display their private parts from the city walls, and when the Persians demanded a token ransom for the city, the local leaders in a fit of hubris responded by demanding that the Persians pay for damage done and produce taken from the fields around the city. As they successfully warded off all attacks and the Persians were discussing withdrawal, the citizens of Amida began to take their guard duties lightly. One night the monks guarding a particular tower, known as the *tripyrgion*, fell asleep from drinking wine in the cold, and others had left the walls to take shelter in their houses. A group of Persians, discovering this by accident, snuck up the walls with a single ladder, and proceeded to kill the monks, taking control of the

---

dokos is a large beam (originally bearing beam in a house); *egkarsios* means athwart, oblique. Thus "the *Amidenes* repulsed the strike (of the rams) by means of oblique beams." One explanation is that Procopius garbled a slightly more detailed account which he might not have understood completely (cf. some types of trebuchet, with a large wooden frame of "oblique beams" for a pulling crew), or conflated what are two separate elements in Zachariah: beams used by the Persians to cross from the mound to the wall, and bundles of rushes dropped by the defenders to dampen the blows of the rams. It could be understood literally, that the Amidenes simply set up some very large beams obliquely (something like this, with the wall to the left, the beams represented as a backslash, and the ram protruding from the shed to the right: |||\ -^) from the top of the wall to the ground. This would absorb much of the force of the ram. Most likely, however, Procopius has misunderstood something. See chapters 5.2.5 and 5.3.1 for a discussion of rams and countermeasures against them.

[94] See chapter 8.3.1 for context and likely identification.

tower and a section of the walls, while the Persians began a general assault with ladders. Desperate fighting continued on the walls, and Kawad himself approached the base of the ladders, shooting his bow, spurring on his men by personal example as well as by dire threats, ordering death for anyone who descended unwounded. It took several days before the Persians had cleared all the towers one by one and could descend into the city itself, where they began to massacre the population. According to Joshua, 80,000 people were killed in the process, while some survivors were taken out of the city and stabbed, stoned or drowned, probably in retaliation for great Persian losses (claimed to be 50,000).[95] Although the massacre was clearly horrific, enough of the population remained to appoint Roman governors to administer the city, and the remaining population figures prominently later on in Joshua's account.

While the broad outline of the siege is fairly well known and has received some attention, little is said of how the major engineering projects were organized. The siege displays some interesting continuities in organization and practice all the way back to the 4th century, with many elements recognizable from the sieges of Dura-Europos and the fall of Amida in 359 (see 1.1.1). Formidable in their engineering capacities, we can assume that the Persians in 502 had skilled labor available in the form of their infantry levies (*paygān*), conscripted troops who doubled as medium infantry and sappers, but further clues as to their organization must be found elsewhere.[96] With the Amidenes, however, we are on firmer ground. A modest garrison was reinstalled after the city was recovered from the Persians, though we hear nothing explicit of the artillery arsenal that Ammianus mentions (see 1.1.1). Yet artillery was an important component in the defense of the city, and the machine used by the Amidenes was apparently of a very innovative and destructive sort. Furthermore, the countermeasures taken demonstrate a very high competence in siege defense in general, indicating the presence of trained specialists. There is a final clue in Zacharia's description of the harrowing aftermath. Kawad first ritually degraded the leaders of the city, having one of them carry a sow on his shoulders.

> But at last the great men, *and all the chief craftsmen*, were bound and brought together, and set apart as the king's captives; and they were sent to his country with the military escort which brought them down. But influential

---

[95] See the CO entry for *Amida, but especially JS 50, 53.
[96] Persian military organization is fully discussed in chapter 7.1.2, but see chapters 5 and 6 *passim* for many relevant examples.

men of the king's army drew near and said to him, "Our kinsmen and brethren were killed in battle by the inhabitants of the city," and they asked him that one-tenth of the men should be given to them for the exaction of vengeance. And they brought them together and counted them, and gave to them in proportion from the men; and they put them to death, killing them in all sorts of ways.[97]

The chief craftsmen, in Syriac ܪܒܝ ܐܘܡܢܐ (*rabbay ūmānē*),[98] are noteworthy. The Persians naturally wanted the leading men as bargaining chips or for ransom, but when it came to craftsmen, the Persians had a long tradition of deporting economically or militarily significant populations from captured Roman cities in order to enhance their own wealth and power.[99] These particular captives were held apart as the king's own property and were to be sent to Persia under special guard. In these circumstances, even after such a horrific massacre, the demand that some of them be given over to the troops to be killed in revenge indicates that Persians regarded them as particularly responsible for Persian losses. The construction and operation of siege engines and counter-engineering would certainly be involved in causing massive Persian losses. It therefore appears that Amida had a significant, militarily competent class of craftsmen that provided the backbone of much of the resistance, and that their chief craftsmen were singled out by the Shah to be protected for their skills, just as Persian notables desired vengeance for the death of their relatives. These men were probably the head engineers of a reconstituted military arsenal, along with civilian master craftsmen who were responsible for supporting them.

Although the fall of Amida was a devastating blow which terrified the population east of the Euphrates, the Romans managed to shore up their position by moving reinforcements, some all the way from the Balkans, eastwards. These troops may have helped to prevent the mass exodus of the population by force, but the presence of such a large army (said to number 52,000 in addition to the regular garrison forces in the east) meant that morale was stiffened and the Romans soon ready to organize a counterattack.[100] It was less than successful due to lack of Roman cohesion, as no Roman army this big had been assembled since Julian's failed expedition against Persia in 363, and soon the Persians regained the initiative. While they had only probed the defenses of Byzantine Mesopotamia in the

---

[97] Zachariah 7.4. My emphasis.
[98] Emended by Brooks from a corrupt passage, though the corruption is slight. See *textus* 22.1 and *versio* 20.16 with apparatus.
[99] See chapter 7.1.2.
[100] For this, see chapter 2.3.6 below.

summer of 502, they launched full-scale attacks on several cities in 503, but all of them failed.

2.3.4 Multiple Approaches: *Constantina-Tella 502-03

While the Persians were surrounding Amida in October-November 502, they sent out raiders into the surrounding countryside. The first party affected the *territorium* of Constantina (also called Tella) to the west, as the raiders spread out to plunder the surrounding villages. Constantina was the seat of the commander (*dux*) of Mesopotamia, who had substantial troops at his disposal. He was further reinforced by the *dux* of Armenia I and II, stationed at Melitene, who may have prevented a Persian probe into Anatolia, and then shadowed the Persian army south from Theodosiopolis. With these substantial troops, perhaps meant to relieve Amida, the Romans set out to clear the surrounding villages of Constantina of raiders. Arab and Hun auxiliaries had been sent from the main Persian army at Amida, but were also supported by some Persian troops. This operation was successful, but a report of 500 raiders in a nearby village drew the Romans out on another sortie, which chased them off for some distance to the east. The Romans were drawn into a Persian ambush as night fell, and the infantry, abandoned by the cavalry, was cut down by Persian, Hun and Arab cavalry. After this, the raiders continued on to Harran and Edessa (see below). In the meantime, the remaining garrison forces of Mesopotamia withdrew to their bases, while the dux of Armenia headed back north to recapture Theodosiopolis.

It was only the next campaigning season, in June 503, that the Persians attempted to invest Constantina with a full-scale siege, having defeated the Roman counter-attack. Joshua relates an interesting anecdote where some of the Jewish defenders decided to betray the city as the Persians camped around it. They had received the responsibility for defending a specific tower near their synagogue, and those behind the betrayal used the opportunity to excavate a tunnel from the synagogue, which was near the tower they defended, under the wall towards the Persian lines. However, their intentions were revealed by a Roman officer in Persian captivity who was allowed to approach the walls to ask for a pair of trousers. When the plot was revealed to the defenders, the Jewish population was subject to a fierce pogrom which was barely stopped by the bishop and *comes* (count) of the city. Joshua continues to relate how the Romans maintained morale during the rest of the siege, in markedly more appropriate forms than the Amidenes:

They guarded the city carefully by night and by day, and the holy [bishop] Bar-hadad himself used to go round and visit them and pray for them and bless them, commending their care and encouraging them, and sprinkling holy water on them and on the wall of the city. He also carried with him on his rounds the Eucharist, in order to let them receive the mystery at their stations, lest for this reason any one of them should quit his post and come down from the wall. He also went out boldly to the king of the Persians and spoke with him and appeased him.[101]

The Persian raids in late 502 had clearly had a dual purpose: tie up Roman forces massing to the west, so that they were unable to relieve Amida, and prepare for a coming siege of Constantina and other Mesopotamian cities, by depriving the citizens of their harvests and livestock. If successfully executed, the defenders should, by the next summer, be short on supplies, demoralized, and ready to surrender rapidly. However, the strategy backfired. Constantina lay in one of the most barren regions in northern Mesopotamia, and the raids had left little for the Persian army to forage on, while the Roman forces from Armenia had left the area and thus relieved pressure on supplies for their own part.[102] The Persians were instead appeased by the bishop, and moved on towards Edessa before the Romans could reassemble. The Persian field army was still superior to local Roman forces, but found that the Edessenes were well prepared after nearly a year of threats.

2.3.5 Complex Operations against Country and City: *Edessa 502-03

The measures taken at Edessa from the late autumn of 502 until late summer 503 show that the authorities and citizens took every threat very seriously, and began preparations for siege as soon as enemy troops were in the vicinity. While the Persians were surrounding Amida and a raiding party was approaching Constantina, the Edessenes, Harranites, and local villagers had gone out to the vintage. They probably believed that the Roman troops massed in Constantina, just to the north-east, beyond a mountain, would keep them safe, and anyway they had to bring in the harvest—if not, they would have serious problems surviving the winter, let alone a siege. The raiders who had been repulsed from Constantina defeated the Romans at Tell-Bashmai, far to the north-east of Edessa, when the vintage was going on. The Arab auxiliaries of the Persians then made a long detour and arrived surprisingly at Harran, which was *south* of Edessa.

---

[101] JS 58.
[102] See CO for further references to this; especially Greatrex' comments.

This caught many in the open, and Joshua reports that 18,500 people were taken captive at the time, many killed, and much livestock and booty taken.

> Because of these things Edessa was closed and guarded, and ditches were dug, and the wall was repaired; and the gates of the city were stopped up with blocks of stone, because they were decayed. They were going to put new ones, and to make bars for the sluices of the river, lest any one should enter thereby; but they could not find iron enough for the work, and an order was issued that every [aristocratic?] house in Edessa should furnish ten pounds of iron.[103]

These were preliminary preparations, since they clearly took the longest and required much labor and resources, from compulsory labor to the levy on household iron. Preparing the ground around the city before a set-piece siege was common practice, but extremely destructive to one's own property and had to be put off until the final moment.[104] When it finally became clear in the summer of 503 that the Persians were approaching,

> ...the Edessenes pulled down all the convents and inns that were close to the wall, and burned the village of Kephar Selem, also called Negbath. They cut down all the hedges of the gardens and parks that were around, and felled the trees which were in them. They brought in the bones of all the martyrs (from the churches) which were around the city; *and set up engines on the wall,* and tied coverings of haircloth over the battlements.[105]

The weapons and protective coverings were set up only three days before the arrival of the vanguard of the Persian army, but plans and preparations must clearly have been ready in advance, especially with regards to the engines. Engine parts must have been kept in storage and maintained in workable condition.

When the Persian vanguard arrived, they at first attempted to negotiate, but demanded an extortionate ransom as well as annual tribute. When this failed, the vanguard set off to raid the vicinity (and presumably reconnoiter for possible Roman reinforcements) before Kawad arrived from Constantina with a huge army. His army surrounded the city in a wide arc from the south gates along the eastern wall to the north gate, with pickets posted on the hills to the west. The population, including young boys, went out

---

[103] JS 52.
[104] For a survey of the topography of the walls, buildings and surroundings at Edessa (and hence the problems faced by the citizens), see Luther 1997: 153-59.
[105] JS 59; translated as "weaponry" by Trombley and Watt 2005: 75; in Chabot I. 286.21 we find ܫܘܪܐ ܙܝܢܐ ܘܐܣܩ *w-aseq zaynā l-šūrā* (*aseq* means to raise, place above, thus "and they set"; *zaynā* is indeed weapons, but can also be used of engines, which is what must be inferred here from the context).

with slings and forced the Persians to move their camp away from the walls, originally at bowshot range, to a nearby village.¹⁰⁶ Women helped carry water to those fighting outside. Villagers who had sought refuge in Edessa later proved courageous in fighting off a full-scale assault outside the walls with slings, despite charges from Arab, Hun and Persian cavalry. These villagers were publicly rewarded for bravery by the commander Areobindus in the cathedral, while Kawad departed.¹⁰⁷

Another extremely important aspect of Edessa's morale, according to Joshua, was their strong faith in Christ's protection of the city and its ritual expression through services on the walls. Clearly, many stories circulated concerning Christ's direct intervention to protect the city, including striking an arrogant Arab chief on the Persian side for his blasphemous intention to take the city. Sorties were another important element in keeping off a powerful enemy. This was a normal strategy for defending against minor raids and practiced by the Romans in the Balkans as well as in Syria.¹⁰⁸ Civilian participation in defense of the city emerges very clearly from the narrative, although there had been little precedence for this in a very long time.

2.3.6 Complex Operations and the Fog of War: *Amida 503-04

The Roman authorities did not garrison cities east of Euphrates with enough professional troops to bear the full brunt of a possible siege until after the fall of Amida. Edessa survived through a combination of skill, bravery and luck (or divine protection). When Amida fell, most of the population east of the Euphrates was preparing for a mass exodus to safe Roman territories, but were instructed by authorities to stay put as a huge army began assembling in the spring. The Roman counter-offensive and the effort to regain Amida began as soon as the armies were ready. While the Romans won many victories in the field, defeating several Persian armies and raiding deep into Persian territory, the Roman siege of Amida drew out over a long period of time. The Persian garrison made skilled defenders and the lack of coordination between Roman armies and com-

---

¹⁰⁶ Chabot I. 288.28f: ܘܟܠܗܘܢ ܐܚܝܕܝ ܩܠܥܐ ܗܢܘ ܐܘܟܝܬ ܛܠܝܐ ܫܒܪܐ ܕܥܡܗܘܢ ܡܢ ܩܪܝܬܐ. The expression *ṭelāyē šabrē* could conceivably have been used of armed servants, but the expression "innocent" hardly squares with this; *ṭelāyē* children; collectively, lads, young people, servants; *šabrē*: small, but also innocent; note also *qel'ē*: slings, so the expression *šādīn bqel'ē* means slingers. Cf. *Constantinople 663 for *ṭelāyē* guarding an Arab camp.
¹⁰⁷ JS 60, 62f.
¹⁰⁸ See further chapter 5.3.2.

134　　　　　　　　　　　　CHAPTER TWO

manders also played a great role. Romans had to contend with Persian reinforcements and allies, thus limiting their ability to concentrate on the siege, which had to be intermittently abandoned when other parts of the army suffered defeats and hurriedly withdrew.[109]

When the Romans had mustered enough troops to attempt a storm, they tried at first to recapture the city building three iron-reinforced wooden towers for overpowering the walls, but had to burn them: a Roman field army operating on its own had suffered a defeat, and the besiegers had to abandon the siege temporarily in order to pursue the unchecked Persian army.[110] The Romans tried several methods when they were in a position to attack again. Shortly afterwards, they set up an ambush, sending a flock of sheep past the city walls. The Persians sallied out, but 400 of them were caught, and when their officer failed to deliver on a promise to hand over the city, he was impaled in view of the city walls and the Romans went off to winter camp. The Persians had left a garrison of 3,000 that supplied itself through the winter by trading with the locals. An immediate reaction came from the *dux* of Melitene, who attacked and massacred anyone found bringing supplies to the Persians and established a loose blockade by pitching camp in the vicinity of the city. This also prevented Persian reinforcements from bringing supplies. From then on, the Persians were on starving rations and the remaining civilians in the city were treated abysmally. When time came for a renewed siege season, the Romans:

> ...again encamped against Amida, and Patricius sent and collected unto him artisans (ܐܘܡܢܐ *ūmānē*) from other cities and many of the villagers (ܩܘܪܝܝܐ *qūryāyē*), and bade them dig in the ground and make a mine beneath the wall, that it might be weakened and fall.[111]

When the mine was completed, the Romans propped up the wall with wood and fired it, but the wall only partially collapsed, so that it was impossible to storm. Instead, they continued the mine into the city, but this was discovered by a woman who in her excitement cried out, inadvertently warning the Persians who killed the first man to emerge. The Persians blocked off the mine with stones, dug ditches inside the walls and filled them with water in order to prevent new attempts.

---

[109] Cf. Greatrex 1998: 94.
[110] JS 56.
[111] JS 66. For the Syriac text, see Chabot's edition, I. 293.16-20: ܘܫܪܐ ܥܠ ܐܡܕ ܘܫܕܪ ܦܛܪܝܩ ܟܢܫ ܠܘܬܗ ܐܘܡܢܐ ܡܢ ܡܕܝܢܬܐ ܐܚܪܢܝܬܐ ܘܩܘܪܝܐ ܣܓܝܐܐ ܘܦܩܕ ܐܢܘܢ ܕܢܚܦܪܘܢ ܒܐܪܥܐ ܘܢܥܒܕܘܢ ܫܘܩܐ ܬܚܝܬ ܫܘܪܐ ܕܢܡܚܠ ܘܢܦܠ.

In the following months, there were many inconclusive skirmishes, and Joshua relates many of the anecdotes in great detail. But eventually, the Romans settled on starving out the Persians, so the siege continued into the winter of 504-05. By then most of the soldiers appear to have had enough of the whole affair, and instead of enduring camp conditions during an exceptionally cold winter, many simply went home with booty assembled on other assignments. Others went to live in nearby cities. Shortly afterwards, the Romans and Persians concluded a peace that restored the status quo.

2.4 *Organization of Siege Warfare II: The Militarization of Society*

In addition to the great variety of siege options (the Romans tried to retake Amida with siege towers, surprise and massacre of supply lines, ambush of the garrison, persuasion, terror, mines, archery duels and starvation, but settled on negotiation), the Anastasian war also set new patterns in how siegeworks were organized and by whom cities were defended or attacked. Humble civilians played an important role at *Edessa, while the chief craftsmen of *Amida suffered heavy retaliation for their effective defense of their city. Craftsmen and village laborers were conscripted for offensive as well as defensive siege warfare, a marked extension of civilian involvement from the late Roman period. Civilian aptitude for such work began to affect how large-scale public military projects were organized, and this remained a feature of East Roman and Byzantine siege warfare for centuries to come.

2.4.1 *The Construction of Dara, 505-06*

The new pattern emerges clearly with the repairs undertaken after the war, and the construction of Dara in 505-6, which was to anchor the Roman defenses in Mesopotamia and provide a safe base for Roman troops. After most field armies were withdrawn, the governor Eulogius received a grant from the emperor of 200 pounds of gold for reconstruction at Edessa. The instructions specified the circuit wall, aqueducts, bath, governor's residence and other buildings. Also a smaller amount of 20 pounds was allotted by the emperor to the bishop, who was to help renew the wall.[112] Further repairs were undertaken at the *kastron* of Batnan d-Serug, where broken

---

[112] JS 87.

walls were rebuilt by the same governor.[113] The costs of the damage caused during the invasions and whatever had to be contributed by the local landowners was offset by a tax remission granted to all of Mesopotamia, in some contrast to 5th-century policies of extracting taxes and services.[114]

An even more ambitious project was undertaken at a small village called Dara, on the south side of the Tur Abdin mountain range, overlooking the Persian fortifications in upper Mesopotamia. There, the Roman authorities decided to build a heavily fortified military base that would close a recognized gap in the defenses that allowed the Persians to move deep into Roman Mesopotamia, while it would provide a safe base for Roman troops operating in the region. Its construction required deliberate planning of strategy as well as considerable organization of labor and resources. Joshua the Stylite gives eloquent testimony to the strategic considerations that lay behind the construction:

> The generals of the Roman army informed the emperor that the troops suffered great harm from their not having any (fortified) town situated on the border. For whenever the Romans went forth from Tella or Amida to go about on expeditions among the Arabs, they were in constant fear, whenever they halted, of the treachery of enemies; and if it happened that they fell in with a larger force than their own, and thought of turning back, they had to endure great fatigue, because there was no town near them in which they could find shelter. For this reason the emperor gave orders that a wall should be built for the village of Dara, which is situated on the frontier.[115]

Joshua is rather brief on the organization of the works,[116] for which we must turn to Zachariah,[117] who attributes Thomas, bishop of Amida, with a fundamental role in the planning and organization of the works. Zachariah also emphasized the need for a logistical base in his presentation of the deliberations at the court of Anastasios. The generals, who had to de-

---

[113] Note that *kastra* played a modest role in JS's narrative, but were probably far more important in operations.

[114] JS 92: "After Pharazman went down to Amida, the *dux* Romanus came in his place, and settled at Edessa with his troops, and bestowed large alms upon the poor. The emperor added in this year to all his former good deeds, and sent a remission of the tax to the whole of Mesopotamia, whereat all the landed proprietors rejoiced and praised the emperor." Cf. the policies outlined in 1.2.5.

[115] JS 90. Further on the construction process, see also Zanini 2007: 385ff for a brief survey of the construction of Dara in a larger context.

[116] JS 90, "They selected workmen from all Syria (for this task), and they went down thither and were building it; and the Persians were sallying forth from Nisibis and forcing them to stop."

[117] Zachariah 7.6. References to the Syriac text in the following are drawn from Brooks' edition (and Latin tr.).

fend their poor showing in the recent war, said that on the one hand, it was hard to fight against a king and his huge army sent by God to punish the Romans for their sins, and on the other,

> "that it was no easy matter for them in his absence also to subdue Nisibis, *because they had no engines ready*,[118] nor any refuge in which to rest. For the fortresses were far away and were too small to receive the army, and neither the supply of water in them nor the vegetables were sufficient. And they begged of him that a city should be built by his command beside the mountain, as a refuge for the army in which they might rest, *and for the preparation of weapons.*[119]

Thus, a major desideratum was the establishment of a safe base for arms production and storage, especially for the heavy machinery required to capture the nearby Persian fortress city of Nisibis. Furthermore, it would help stop raids and invasions by Arabs and Persians. After deliberating on the location, Anastasios "sent a message to Thomas, the bishop of Amida, *and he despatched engineers who drew up a plan, and this holy Thomas brought it up with him to the king.*"[120]

Anastasios proceeded to buy the site for the treasury, freeing serfs and granting them property and he also established a fund for constructing a church. Thomas was then given the responsibility for organizing the construction and a guarantee of continued imperial funding. While the (unspecified amount of) gold paid for the village covered the first stages of the project, the wording implies that Thomas was to shoulder the costs of construction and send the bill on to Constantinople. All labor was to be paid (as opposed to mobilization through public burdens, cf. 1.2.5), since the fortification had to be finished while Kawad was busy with his enemies. Thomas clearly recognized that the laborers would be much more motivated if they profited (ܝܬܪܝܢ *yātrīn*) from the project, and strictly ensured that laborers promptly received their pay. The task was daunting, since

---

[118] *Textus* II.35.12f ܕܠܐ ܗܘܐ ܕܠܝܬ ܠܗܘܢ ܡܐܟܢܣ ܡܛܝܒܬܐ *b-hay d-layt hwā l-hōn mēkanas mṭibātā*. The apparatus reads ܡܐܟܢܐ as an alternative for MKNS, which is obviously the Greek μηχανάς. Thus a safe base for engines necessary for offensive operations.

[119] *Textus* II.35.18 ܘܠܛܘܝܒܐ ܕܙܝܢܐ (*wa-l-ṭūyābā d-zaynā*). From the above, it should be clear that *zaynā* means weapons in a generic sense that also includes engines.

[120] *Textus* II.35.21ff ܘܫܕܪ ܗܘ ܡܛܟܢܝܩܘܣ ܘܥܒܕ ܣܩܪܝܦܘܣ ܘܐܣܩ ܠܡܠܟܐ ܥܡܗ ܗܘ ܗܢܐ ܩܕܝܫܐ ܬܐܘܡܐ *w-šaddar MYKNYQWS w-'bad SQRYPWS w-aseq l-malkā 'ammeh hū hānā qaddīšā tāwmā*. Syriac can be notorious for its lack of clear agent with verbs, but from the context it is clear that it should be understood that "*Anastasios* sent (*šaddar*) the engineer (μηχανικός), and he (i.e. *the engineer*) made the ground-plan (σκάριφος) and sent it with the holy man Thomas to the king." Cf. Brooks' *versio* II.24.15f: "et μηχανικὸν misit et σκάριφος fecit, et secum ad regem attulit hic sanctus Thomas."

large amounts of material had to be assembled before construction could get underway. For the first stage, then, Thomas recruited "craftsmen and workers and laborers."[121] While these workers were responsible for transportation, hauling, carrying, woodwork for scaffolding, cranes and so on, Thomas sent stonecutters and masons to quarry the necessary stone.[122]

When construction could begin, Thomas used his own clerical staff, "Cyrus 'Adon and Eutychian the presbyters, and Paphnout and Sergius and John the deacons, and others from the clergy of Amida" "as overseers and commissaries" over the fortification works.[123] He personally visited the site to observe progress. Since it was publicly announced that the project provided reliable pay for honest work, day laborers and craftsmen[124] soon

---

[121] The vocabulary of the passage is a bit difficult. For the original, see Brooks' *textus* II.36.13ff: ܟܕ ܝܬܪܝܢ ܘܡܒܪܟܝܢ ܐܘܡܢܐ ܘܥܒܕܐ ܘܦܠܚܐ ܘܡܬܒܥܝܢ ܒܟܘܢܫܐ ܕܗܘܠܐ ܕܒܗ. *kad yātrīn wa-mbarrakīn ūmānē w-'ābdē w-palāḥē w-matba'īn b-kūnāšā d-hūlā d-bāh*, translated by Brooks in his *versio* II.24.27-30: "ut ita urbs statim conderetur, artificibus et fabricatoribus et agricolis quibus ad ὕλην in ea colligendam opus erat lucrum et munera adquirentibus." *ūmānē* is straightforward, meaning professional craftsmen and artisans as could be organized in guilds (cf. the *rabbay ūmānē* at *Amida 502f & 2.3.3 above). *'ābdē* is more difficult. Easily confused with "servants" (*'abdē*; note short vowel which is not distinguished in Syriac orthography), here it appears to mean paid laborers or workmen. Brooks translates with "fabricatoribus" and notes in the apparatus of his versio (II.24 n.11) that "Vox ܥܒܕܐ hunc sensum (lexicis ignotum) hic habere videtur, non 'servis' significare; cf. p. 25, l. 7." It is possible that they were craftsmen who were in the service of a magnate as opposed to organized in a guild. However, since it is a substantivized participle (the participle form in PS for "makers" is *'ābdē*, often in cstr. pl. *'ābday*), it might be a calque of *fabricenses*. If so, it shows that Amida, where they came from, still had operational *fabricae*.

A similar problem arises with PLḤ, which has two meanings; the first, *pālḥā*, has a special plural for groups more than ten, *pālaḥwātā*, meaning "servant, attendant, worshipper; worker, artisan"; also "soldier". The other form, *palāḥā*, the one used here, appears to have regular plural and fits well here: "laborer, husbandman, vine-dresser", i.e. [agricultural] laborer. The villagers who were emancipated and settled here by Thomas must be included in this group.

[122] *Textus* II.36.15. ܘܫܕܪ ܦܣܘܠܐ ܣܓܝܐܐ ܘܐܪܓܘܒܠܐ *w-šaddar pāsūlē saggī'ē w-argūblē*. Brooks *versio* II.24.30f: "Et lapicidas multos et ἐργολάβους misit." *pāsūlē* means "stone-cutter, stone-mason;" *argūblē*, "stone-mason;" the Greek version is given in LS as "contractor." The verb "sent" (*šaddar*) implies that they were under Thomas' authority, but the relationship is unclear; cf. Western bishops in e.g. chapter 4 *passim*.

[123] *Textus* II.36.19f. ܐܝܬ ܗܘܘ ܬܡܢ ܩܝܘܡܐ ܘܫܪܝܪܐ ܥܠ ܥܒܕܐ *īt-hwaw tammān qāyūmē w-šarrīrē 'al 'ābdē*. *Versio* II.25.3f: "erant ibi operum praepositi et curatores." *qāyūmē*, "manager, superintendent"; *šarrīrē*, "confident servant, commissioner, prefect." In order to clarify Brooks, it should be "*fabricatorum*/*fabricensium* praepositi et curatores," i.e. managers and commissioners over the *laborers*."

[124] When Zachariah discusses pay, he tells us that everyone received 4 *keratin* a day (double pay if they brought a donkey), whether they were craftsmen or "makers of all things." See *versio* II.36.25. ܠܐܘܡܢܐ ܘܠܥܒܕܐ ܕܟܠ ܨܒܘ *l-ūmānē wa-l-'ābdē d-koll ṣabū*[*n*]. *Versio* II.25.7: "artificibus et omnium rerum fabricatoribus." Together with Brooks II.37.1 ܦܥܠܐ *pā'lā*, "hired laborer, day laborer" and II. 37.5. ܦܥܠܘܬܐ ܘܐܘܡܢܐ *pā'lwātā w-ūmānē*, these

began to assemble from the whole region. Bishops had wide experience in organizing and overseeing extensive building projects.[125] This organizational ability could clearly be transferred to constructing fortifications and other public works. The practice became regularized by Justinian, who specified that the bishop alongside a committee of three "reputable landowners" oversee the maintenance of public buildings such as aqueducts, walls and towers.[126]

This marks a clear shift. Before around 500, the army in conjunction with local and government officials organized defenses. Thus, the agent responsible for restoring the walls at Crimea was a government officer (*comes* or count), while in other attested cases, the regular army set up defenses, sometimes in conjunction with local officials. After about 500, however, bishops were beginning to play a more prominent role in paying, organizing and supervising the construction of fortifications and urban defense, a development reflected in other aspects of society. Trombley and Watt note that this apparently deliberate administrative reform began in the East at the end of the Anastasian war, when several bishops were commissioned to construct fortifications. The new responsibility arose as the result of a host of civic duties, such as maintaining civilian infrastructure, being transferred to bishops, and has been taken as a sign of the dissolution of the traditional decurionate, which had previously been responsible. Although regularized by Justinian in 530, this was probably the greatest involvement bishops had in East Roman military and civil organization.[127] Here we shall focus on the military aspects.

2.4.2 Civilian Cooperation in the 6th Century

The late Roman administrative system relied heavily on various public duties to be performed by any corporate group, either as a regular *munus* (obligation), or as an extraordinary levy prompted by some emergency. In the latter case, taxes, from the reign of Justinian, were normally remitted in return for the expenses. Whoever possessed skills or resources required

---

expressions show the three basic categories of [specialized] laborer, contract craftsmen and day laborers.

[125] Trombley and Watt 2005: 106 n. 498.
[126] *CJ* 1.4.26, issued in 530.
[127] Trombley and Watt 2000: xlviii-xlix for further references. See further Rapp 2004 for context, while Liebeschuetz 2001 notes that this legislation was actually the high point for episcopal involvement in the East. As will be seen in chapter 4, especially 4.2.2, the episcopate in the West, especially Francia, which began on a similar trajectory but ultimately developed very differently.

by the state or army could be drafted in or levied according to need in return for exemptions from fiscal dues.[128] The regular procedure for maintaining a city wall was to assign a certain number of feet to a corporate group (e.g. a guild of craftsmen, the local Jews, or a religious house) or a rich individual (e.g. a *possessor* or *negotiator*). This was called the *pedatura*, and was also the section for which the said group or individual (i.e. his retinue) were responsible for defending in times of crisis.[129] Garrison troops were normally assigned to gates, towers and machines, while refugees were distributed according to need.[130] The *Strategikon* clearly expresses this policy: "If there are civilians in the city, it is necessary to mix them also together with the soldiers on the *pedatourai* of the wall."[131] At Constantinople and a number of other cities, the city walls were so extensive that civilians must have taken part; thus we find in the *De Ceremoniis* mention of *manganarioi* of the Blues and Greens taking part in imperial processions.[132] Since the circus factions had a regular place in urban defense, it is possible that these *manganarioi* were in fact trained operators of siege engines. It is likely that other groups with recognized stations at Constantinople had their own artillerymen attached to them as a public *munus*, analogous to and perhaps in cooperation with landowners who disposed of storage facilities in the towers of the Theodosian Walls in return for maintenance.[133]

In addition to these regular obligations, labor could be hired or conscripted for construction work as well as sieges (cf. 1.2.5). At the construc-

---

[128] See 1.2.5 for the Justinianic legislation; for later examples into the middle Byzantine period, cf. the texts edited by Haldon 1990b.

[129] Trombley and Watt 2005: xlvii; 72 n. 343 for references to epigraphic evidence and how the Jews of *Constantina 502 could use their assigned *pedatura* to dig a tunnel from their synagogue out to the Persians without being discovered, since it was "their" sector. See also chapter 1.2.5 for the late Roman background and 2.3.4 for details and context.

[130] Thus *Rome 537f, where it seems that some of the civilians enrolled by Belisarius came from outside the walled area of the city. The Blues and Greens are described by Malalas 14 (p. 351.8) seated in their *pedatoura* during public functions at the Hippodrome in Constantinople, but it is unclear whether this term merely reflects seating arrangements or may also refer to their assigned divisions of wall.

[131] *Strategikon* 10.3.32ff: Εἰ δὲ δῆμός ἐστιν ἐν τῇ πόλει, δέον κἀκείνους συμμίξαι ἐν ταῖς τοῦ τείχους πεδατούραις τοῖς στρατιώταις.

[132] The Blues and Greens were assigned their own *pedatourai* and seem to have had *manganarioi* (attested as receiving imperial donatives *De Cerimoniis* 1.23f) to man them; I believe that they were distinct from the civilian *manganarioi* who operated the Hippodrome gates, whose titles were clearly related to the ancient usage as civil engineer; cf. chapter 8.2.2.

[133] See the obligation to keep the walls in order when lower stories were used for private purposes at 1.2.5.

tion of Dara and siege of *Amida (503), the Romans could mobilize workers and craftsmen from all over Mesopotamia. Throughout the 6th century, the combination of military expertise and civilian participation was common. In addition to the very detailed information from the Syriac sources, we have a few instances mentioned by Procopius: when the Romans had taken *Carthage in 533, they prepared for a possible Vandal siege by assembling local *tekhnitai* who repaired the walls at a speed that astonished the captured Vandal king Gelimer shortly afterwards. At *Antioch 540, Justinian sent his nephew Germanus with a small following to Antioch, where he oversaw the defenses along with the "architects of the public buildings" (τοῖς τῶν οἰκοδομιῶν ἀρχιτέκτοσι) before the arrival of the Persians. At *Dara (540), the Persians were excavating a mine to reach inside the double walls. The Romans discovered this, and frantically dug a counter-trench in the interval between the walls under direction of the city engineer, "Theodorus, a man learned in the science called mechanics (ἐπὶ σοφίᾳ καλουμένῃ μηχανικῇ)."[134] In straightforward Greek, he would simply be called a *mēkhanikos* (μηχανικός).

These anecdotes demonstrate the division of expertise in late antique society: *mēkhanikoi* were academically trained architects with a solid theoretical foundation, such as Anthemios and Isidoros, the builders of Hagia Sophia. *Arkhitektones*, on the other hand, were master builders (literally "chief carpenters"), often private contractors employed with their teams of subordinate workers, or employed by the municipality.[135] The difference lay in the level of theoretical schooling, but both groups supervised construction projects, and unless involved in extremely complex projects requiring a mastery of geometric theory (such as the dome of Hagia Sophia), the day-to-day difference between the two were probably small. It is possible that some of these were in fact enrolled into the military,[136] which meant they were in the *katalogoi*, or rolls, and received pay. Otherwise, their organization and daily activities differed little from un-enrolled

---

[134] See CO and Proc. 2.13.26 for this particular quote.

[135] For a discussion of the different classes of late antique and Byzantine craftsmen, the fundamental discussion is Downey 1946. See also Ousterhout 1999: 43f and Cuomo 1997: 134ff, who essentially confirm Downey's basic survey of the Greek terminology for late antique craftsmen, but Ousterhout extends the image to the middle and late Byzantine periods, while Cuomo has a valuable discussion of epigraphic evidence, that demonstrate how prevalent the *arkhitektōn* was in late antique society. This may perhaps be due to the conflation in meaning between *arkhitektōn* and *mēkhanikos*. See further Zanini 2007 for the most recent discussion of their status and evolution.

[136] Such as the weapons producers noted in chapter 1.1.1 above.

craftsmen who could also be used for military purposes (cf. Justinian's law on weapons producers). *Tekhnitai*, when not applied to military engineers, were specialized craftsmen in a relevant civilian trade such as carpenters or masons.

Endemic warfare in the East meant that urban craftsmen had become accustomed to cooperating with the military during the defense of cities, while intermittent periods of peace would have allowed military engineers to engage in civilian trades. Both groups would have been similarly organized as civilian contractors with teams of workers. Cooperation between civilians and military is attested elsewhere as the result of constant threats from the Persians and their Arab and Hun allies. Trombley has shown from epigraphic evidence that many of the new 6th-century fortifications in Syria were built on local initiative by a combination of military commanders with local ties, landowners and bishops.[137] This presupposes a significant infrastructure of civilian expertise in defensive construction beyond the army specialists.

This cooperation is confirmed by Syriac sources, which provide further evidence on the terms for highly educated specialists and their social status. During a Roman siege of the Persian border city of *Nisibis 573, we are told by the Greek ecclesiastical historian Evagrius, quite sarcastically, that the Roman general marched against the city "having a few rustic laborers and herdsmen, whom he had pressed into his service from among the provincials" (ἔχων καί τινας σκαπανέας καὶ βοηλάτας ἐκ τῶν συντελῶν ἀφῃρημένους). John of Ephesus provides a more coherent image. While not as close to the lower social groups as the Joshua and Zachariah, he tells us that the Romans

> laid siege to Nisibis, the frontier town and bulwark of Mesopotamia, and then in possession of the Persians. And having strongly invested it, and constructed round it a palisade, he commenced, with the aid of the skilful *mekhanikoi* whom he had brought with him, to erect more scientific works, consisting of lofty towers and strong covered approaches.[138]

---

[137] Trombley 1997.
[138] See CO for references. Payne-Smith's translation leaves something to be desired, cf. *textus* 278.14-19:

ܘܡܢ ܕܫܪܝ ܥܠ ܗܕܐ ܡܕܝܢܬܐ ܚܕܬܐ ܕܝ ܒܝܬ ܢܗܪ̈ܝܢ ܘܬܚܘܡܐ ܕܪ̈ܗܘܡܝܐ ܕܦܪ̈ܣܝܐ ܘܒܚܙܝܩܘܬܐ ܐܚܕܗ ܘܩܡ ܗܘܐ ܠܗ ܘܐܪܝܡ ܚܣܝܢܐ܆ ܫܪܝ ܕܝܢ ܥܡ ܡܟܢܣܐ ܚܟܝ̈ܡܐ ܕܡܝܬܝ ܗܘܐ ܥܡܗ ܐܦ ܚܘܪܩܝ̈ܐ ܚܟܝ̈ܡܐ ܕܡܓܕ̈ܠܐ ܪ̈ܡܐ ܘܡܣܩ̈ܢܐ ܬܩܝ̈ܦܬܐ܀

Brooks' Latin rendering is more accurate, cf. *versio* 210.29-211.3: "Quam cum fortiter expugnasset ut eam expugnaret χαλκώματα contra eam circa aedificavit; et quoniam mechanicos

Together, these two narratives provide a more nuanced image: this is not merely a band of farmers pressed into service (although some probably were), but many of the civilians would have been skilled craftsmen under *mēkhanikoi*, as during the Roman siege of *Amida (503; see also 2.3.7 above). Skilled craftsmen could work well under the direction of these experts, although this term poses some difficulty. Syriac generally rendered Greek technical terminology quite accurately and with the same meaning. Hence it could be used here to refer to a theoretically schooled expert. Unfortunately, it is not specified whether they belonged to the military, as the garrison engineers noted above (2.2). There is however reason to believe that the difference was decreasing between master craftsmen (*arkhitektones*) and specialist architects (*mēkhanikoi*). Indeed, the latter term fell out of use in the 7th century, and middle Byzantine evidence shows a strong tendency to conflate terms for various craftsmen.[139] This is confirmed by the Syriac evidence on the Avar siege of *Sirmium (579ff), which refers to the master craftsmen or architects (*myknyqw* {pl.}, i.e. μηχανικοί) sent by the Romans to the Avars in order to construct baths. They were accompanied by builders (*banāyē*), thus reflecting a two-tiered structure of specialists or master builders along with their teams of craftsmen. They could build civilian buildings as well as military infrastructure, and considering the convergence of terminology between *Nisibis/*Dara (573) and *Sirmium (579ff) were probably also involved in transferring siege technology to the Avars.[140]

2.4.3 Use of Civilians in the 7th and 8th Centuries

Although it had long been the practice in emergencies, civilian participation was expressed as official military policy around 600. The Roman army of course maintained the capability to construct and defend fortified sites and conduct sieges with its own infrastructure, especially on enemy territory, but as we have seen, the *Strategikon* also expected soldiers to enlist the help of civilians during urban defense.[141] Indeed, the *Miracula sancti Demetrii* shows the urban population actively engaged in a range of defen-

---

[apparatus: μηχανικός] etiam secum habebat, machinamenta [apparatus: μηχανήματα] turrium altarum et πύργων validorum contra eam erexit..."

The *mēkhanēmata* include trebuchets, here called *manganōn*, as revealed later in the text; cf. 2.2.3 above.

[139] Ousterhout 1999: 44.
[140] See further in chapter 7.2.3 on this.
[141] *Strategikon* 10.3 and chapter 2.4.2 above.

sive activities. The inhabitants of *Thessalonica (τοὺς τῆς πόλεως) constructed complex defensive works and engines (including trebuchets) in 615, and civilian catapult crews (ἐν τῇ ἔνδον τῶν πολιτῶν πετραρέᾳ) operated trebuchets, again at *Thessalonica, in 618. As mentioned above, we have the civilian master carpenter (rabb ḥad naggārā) at *SYLWS in 664 with his household. His professional description seems to have taken over the range of arkhitekton (of which it is a calque), mēkhanikos, and tekhnites, although the simple naggārā also seems to have had a wide range of meaning. Kallinikos from Heliopolis, called arkhitekton in the Greek translation of the Syriac Common Source, is simply known as naggārā in the parallel Syriac versions. Sailors added to this expertise when coastal cities were under threat; ships' crews were instrumental in the defense of *Phasis in 556, *Thessalonica in 618, *Constantinople in 626, and again at *Thessalonica in 662.

We thus have a pattern of cooperation between parallel structures: the formal structures of the army on the one hand, and civilian specialists and workers conscripted according to regular procedures or used as emergency labor on the other. This practice derived from late Roman precedents, continued throughout our period, and must have been far more extensive than the few instances collected here. The Arabs were able to base a formidable siege capacity on the forced recruitment of Syrian and Egyptian craftsmen, who provided for their logistical and engineering needs.[142] This was so extensive that in the mid-8th century the Byzantines began a massive program of moving (the very same) Syrians (and Armenians, whose independent military traditions were maintained even under Arab hegemony) to Thrace, where they were settled to provide defenses against Bulgars and Slavs.

From Greek literary sources, we have very little information to go on between the late 7th century and c. 900. However, Anastasios II is credited with organizing extensive repairs in 715 when it was known that the Arabs were preparing for their great siege; many other 7th-8th century repairs are attested in inscriptions.[143] We know that the Byzantines could recruit skilled laborers, who continue to display significant specialization, for large-scale public building projects. A notable example is when the aqueduct of Valens was repaired in 766/67 when other sources of water failed:

---

[142] See chapter 7.3 and below.
[143] See *Constantinople 715, dealing with Anastasios' preparations for *Constantinople 717f. Foss and Winfield 1986: 53 present a brief survey of repairs, which were increasingly common in the 8th and early 9th centuries.

"There ensued a drought, such that even dew did not fall from heaven and water entirely disappeared from the city. Cisterns and baths were put out of commission; even those springs that in former times had gushed continuously now failed." A very good survey of specialist craftsmen available at the time is subsequently given by Theophanes:

> On seeing this, the emperor set about restoring Valentinian's aqueduct, which had functioned until Herakleios and had been destroyed by the Avars. He collected artisans from different places and brought from Asia and Pontos 1,000 masons and 200 plasterers, from Hellas and the islands 500 clayworkers, and from Thrace itself 5,000 labourers and 200 brickmakers. He set taskmasters over them including one of the patricians. When the work had thus been completed, water flowed into the City.[144]

Such organizational abilities could obviously be directly transferred to military projects, as in the 4th through 6th centuries, and were probably the basis for the repairs of the city walls at Constantinople.

Continued civilian involvement may explain some apparent anomalies in later Byzantine military organization. The Arab siege of *Tyana in 708-9 is sometimes discussed due to the strange appearance of a relieving army that may be taken as a sort of peasant militia.[145] Theophanes states that the army was accompanied by a "rustic crowd" (the commander arrived μετὰ στρατοῦ καὶ γεωργικοῦ λαοῦ). At first this crowd seems rather strange; Theophanes claims they were sent to relieve Tyana, which had undergone a harrowing 9-month siege where the Arabs had partly broken down the walls with engines and reduced the inhabitants to despair. However, the Arabs were themselves on the verge of abandoning the siege due to lack of

---

[144] AM 6258, AD 765/6 (MS 607f, deBoor 439ff).

... τοῦτο ἰδὼν ὁ βασιλεὺς ἤρξατο ἀνακαινίζειν τὸν Οὐαλεντινιανὸν ἀγωγὸν μέχρι Ἡρακλείου χρηματίσαντα, καὶ ὑπὸ τῶν Ἀράβων [sic] καταστραφέντα, ἐπιλεξάμενος δὲ ἐκ διαφόρων τόπων τεχνίτας, ἤγαγεν ἀπὸ μὲν Ἀσίας καὶ Πόντου οἰκοδόμους χιλίους καὶ χριστὰς διακοσίους· ἀπὸ δὲ τῆς Ἑλλάδος καὶ τῶν νησίων ὀστρακαρίους διακοσίους· ἐξ αὐτῆς δὲ τῆς Θρᾴκης ὀπέρας πεντακισχιλίους καὶ κεραμοποιοὺς διακοσίους· καὶ ἐπέστησεν αὐτοῖς ἄρχοντας ἐργοδιώκτας, καὶ ἕνα τῶν πατρικίων. καὶ οὕτω τελεσθένος τοῦ ἔργου εἰσῆλθεν τὸ ὕδωρ ἐν τῇ πόλει.

Nikephoros (c. 85, p. 160f) only refers to the different craftsmen as "artisans" (tekhnitai), but adds that they were paid from fiscal resources, and hence even if performing compulsory labor, were well compensated: "In the 5th indiction there was a drought ... For this reason the baths remained idle, since the reservoirs were empty. Consequently Constantine decided to renew the aqueduct which had been built by Emperor Valentinian and had been destroyed by the Avars in the days of Emperor Herakleios. He collected from the Roman dominions a great number of artisans skilled in construction (καὶ πλείστους ἄνδρας τεχνίτας εἰς οἰκοδομὴν ἐμπείρους ἐκ τῆς ὑπὸ Ῥωμαίων ἀρχῆς συναθροίσας), on whom he lavished many allowances from the public treasury and so completed this work."

[145] For example, Treadgold 1997: 341 calls them "peasant irregulars" sent to relieve the siege.

supplies, but they were able to defeat the relieving army and use their supplies to press on. This left the inhabitants of Tyana with no alternative but to surrender.

While the story appears to reveal Byzantine incompetence, it takes on new meaning in light of the above. For political reasons—Justinian II was never popular among later historians—the story has been compressed and manipulated to leave out significant details. As opposed to Theophanes' version where the semi-military crowd makes a disorganized attack, Nikephoros calls them a λαὸν ἄγροικόν τε καὶ γεωργικόν (a rustic and peasant crowd) sent to *assist* (ἐπαμυνόμενος) the besieged. What Justinian had sent was not a peasant militia, but a crowd of workers who were to arrive on the heels of the withdrawing Arabs and begin repairing the battered walls of Tyana as soon as possible. Moreover, the Arabs may have done more than considering withdrawal; a siege required a large army and parts of it may well have been crossing the Taurus back to Arab territory, hence the "intention" recorded by Nikephoros. However, distances were short, and the Arabs could have returned when they discovered the works going on in order to exploit the opportunity. Tyana is only a 24 hours' hard ride across the Taurus from the main Arab bases in Cilicia.[146] If they had not begun withdrawal, a large crowd assembling somewhere in Cappadocia with tools, supplies and equipment would soon become obvious to Arab scouts. What we have here, then, is a case of normal late Roman and Byzantine military organization, namely extensive use of civilian craftsmen, which is only revealed to us because one or two chroniclers could use the story for propaganda purposes.

It is mostly from the narrative sources that we can find this kind of information. The *Strategikon* in c. 600 and the anonymous *De re strategica* in c. 800 reflect common practice to a certain extent, as both recommend civilian participation in siege defense, but provide little information as to exactly how they can be of assistance. It is only in around 900 a Byzantine source, the *De obsidione toleranda* (§ 10), spells out for us which civilian skills were most useful in conducting a siege defense: "arms manufacturers, engineers, siege machine operators, doctors, bronzesmiths, saddlemakers, bridlemakers, shoemakers, tailors, ropemakers, ladder climbers [roofers], oarmakers, builders, sailors, caulkers, architects, mill stone cutters, astronomers..." Similar lists are indeed extant for the 7th century, but in unlikely

---

[146] I thank the muhtar and inhabitants of Kayasaray (Halkapınar), nestled at the northern end of a high mountain pass across the Taurus, who can organize trips across the mountains and kindly provided me with this information.

places: among the Slavs assaulting *Thessalonica in 662, as described by *Miracula sancti Demetrii* (see also 6.2.3 and 7.2.2), or the extensive papyrus records documenting Coptic craftsmen conscripted for the dockyards of the early Caliphate (see 7.3). In the former case, their organization and skills were assimilated from Byzantine models over time; in the latter, taken over wholesale by conquests and used for new purposes. The composite image clearly points back to a complex late Roman practice of marshaling not only armed forces but also very large segments of society in its conduct of war.

2.5 Conclusion

The East Roman or Byzantine army survived as a centralized institution despite immense challenges. The army maintained most of the infrastructure and expertise necessary for siege warfare, even during times of severe stress and reorganization, but was at varying times supplemented by other institutions and social groups. In addition to estates and private forces noted in the previous chapter, 6th-century bishops and landowners had both improvised and formalized roles in local defense. Similarly, civilian craftsmen were increasingly mobilized not only to build defensive works, but also to participate in urban defense and even in offensive operations. While socio-economic and political developments in the 7th century eliminated magnates and private forces and limited the role of bishops and landowners, civilian craftsmen continued to supplement the regular army. The wealth of the church was instead appropriated outright by Herakleios, probably inspiring the Franks to do the same. East Roman innovations in provincial, military and fiscal administration also had a significant effect on Western developments. Through wars, client relations, and acculturation, Byzantine practices and reforms had a significant impact on Western kingdoms, especially the Visigoths and Franks, in devising a method of exploiting all available wealth and administrative talent as well as counterbalancing the power of aristocratic families. East Roman manpower was frequently supplemented by allied or client troops, which ensured mutual acculturation to fighting styles and military technology. The Ostrogoths and Lombards as a result developed kingdoms from client armies that emerged in the 5th and 6th centuries, respectively, formed by Byzantine styles of fighting and closely resembling the Byzantines in organization and equipment. What made Byzantium different from the West, and particularly the Franks and Visigoths in the 7th century, was not the day-to-day running of affairs and redistribution of goods, which for historical, practical, social and economic reasons were quite similar, as they were all derived from a system of public burdens supplementing or in lieu of regular taxes. The difference lay in the fact that the Byzantine state chose to retain ultimate control of all administrative records, while in the West, the kings could use aristocratic competition in order to maximize benefits to the state—basically the better an aristocrat could exploit his resources in royal service, the more he stood to gain from it, while this spared the Western governments substantial costs and difficulties in administering their territories.

CHAPTER THREE

THE SUCCESSOR STATES IN THE WEST: OSTROGOTHS, VISIGOTHS, AND LOMBARDS

The Ostrogoths, Visigoths and Lombards all took shape as peoples in the Roman frontier region of the middle and lower Danube. In their early years, they might also be described as Roman client or even field armies, since they were often in Roman service, large segments of these people stayed loyal to the East Roman Empire, and there was at times little to distinguish them from other field armies in the Balkans that took to arms against the central government during the 5th and 6th centuries. They should therefore be treated together as products of the Balkans military culture, but due to their inability to find satisfactory settlement in the East, they migrated into the chaotic West where they finally established the independent kingdoms with which we are familiar. The survey of East Roman developments in the previous chapter will show that there was more to unite the Mediterranean than to divide it, and that patterns of military organization could change at a similar pace throughout the former Roman world.

3.1 *The Ostrogoths, 493-554*

Theoderic's Ostrogothic kingdom lasted only two generations, from 493 to 554, but during its heyday, it was the most successful and thoroughly Romanized of all the successor states. There is a general consensus that ancient social structures, such as a high degree of urbanization and a complex economic system, survived very well during this period. The Ostrogoths absorbed surviving Roman administrative structures and collaborated closely with the Roman senatorial class. Boethius and Cassiodorus are only the most famous of these, while the latter's official correspondence is the most important source for the inner workings of the remarkably Roman Ostrogothic administration.[1]

---

[1] The Roman character of the administration has been well known for a very long time, see e.g. Hodgkin's introduction to his translation of Cassiodorus' *Variae*. See further Goffart

3.1.1 Ostrogothic Ethnogenesis

The Ostrogothic state in Italy, Pannonia and Dalmatia originated from a combination of Roman and semi-Roman field armies in the Balkans.[2] One group had been integrated into the East Roman army at least since the 460s as *foideratoi*, but may have been settled in Thrace since the 420s with subsequent new additions in the following decades. Another group, previously Hunnic clients, sought settlement in the Roman Empire after Attila's empire disintegrated in the 450s. By the 470s, the two were competing for Roman patronage. The Thracian Goths had been enjoying this for decades, and their leader, Theoderic Strabo, at times held the office of *magister militum*, but conflict with the emperor Zeno (474-91) opened the way for the "Hunnic" Goths. From 474, these were led by Theoderic the Amal, later called the Great (c. 451-526), who had been given as a Gothic hostage to the Romans and therefore educated at Constantinople in the 460s.[3] In the following years he held a royal Gothic title as well as Roman commands, and was used by Zeno to keep Strabo at arm's length. All of these Goths depended on payment and supplies from the Roman government, and were willing to take up arms against the Roman state when this was not forthcoming.[4] They even joined forces against Zeno in 479. When his foremost Gothic rival died in an accident in 481, Theoderic the Amal gained control over both Gothic groups, and soon forced Zeno to give him high command in the Roman army in 484-86, when he was appointed to lead Gothic and Roman troops against the rebels Illus and Leontius in Anatolia. He was removed early in the campaign when Zeno feared disloyalty, and Theoderic openly revolted against the Roman government again in 487.

1980 and 2006 for his contribution to the debate on how the Ostrogoths were settled in Italy (cf. chapter 1.2.2).

[2] For the early history of the Ostrogoths, see Heather 1991: 227-308 and Wolfram 1988: 248-78.

[3] The degree of Roman influence from his decade at court is debated, but it appears from Theoderic's achievements that his educators were very successful in fostering his ambitions: cf. Theoderic's letter to Anastasios (*Variae* 1.1): "We above all, who by Divine help learned in Your Republic the art of governing Romans with equity ... Our royalty is an imitation of yours, modeled on your good purpose, a copy of the only Empire." The issue is treated extensively in Heather 1996 and mentioned briefly in Wolfram 1988: 262f and Moorhead 1992: 14 and n. 34 with references to more specialized discussions. The ethnographic perspective is treated exhaustively in Amory 1997. Heather 1991, 1996, and 2007 has a different conception of the Gothic "core" population, but agrees that the Roman environment played a fundamental role in Ostrogothic ethnogenesis.

[4] See e.g. in CO: *Singidunum (captured 472 from the Sarmatians), *Ulpiana, *Stobi, and other Illyrian cities (in 473, although *PLRE* 2 s.v. Theodericus 7, p. 1080 dates this to 479).

Had he been successful, he might have become the Gothic version of Zeno, who had come to power backed by Isaurian troops, or, like Vitalian, secured a permanent commission as *magister militum* (see 2.1.2). In the event, Zeno persuaded Theoderic to turn to Italy, where Odovacar was attacking kingdoms allied to the East Roman Empire. The Goths then crossed the Balkans, storming Gepid fortifications and fighting off Sarmatian raiders on the way. Entering Italy in 489, they defeated Odovacar's troops at the river Adige. After some complex maneuvering over the next year or so, Odovacar was blockaded at *Ravenna until a treaty was negotiated in 493, but Theoderic personally murdered his rival and assumed rule of Italy. He received formal Eastern recognition from Anastasios in 497.

The army that followed Theoderic into Italy became the Ostrogothic people, although Gothic ethnicity was still associated with the Illyrian and Thracian field armies for the next two generations before fading away.[5] Theoderic's Goths numbered perhaps 20,000 men (around 100,000 people if families were included), but this number must have increased significantly by Justinian's invasion.[6] The Ostrogoths were a composite people at all stages of their history. Led by a Roman-educated king, they were, as a product of Balkans military culture, a conglomerate of Germanic groups (various Goths, Gepids, Rugians) and probably Roman provincials. To this must be added former West Roman clients and servicemen who survived the purge of Odovacer and his close supporters; later on the Goths absorbed Roman and Moorish troops with apparent ease. In effect, they fought against, alongside or *as* Romans for well over a century. Indeed, Rance noted that Ostrogothic battlefield tactics were not of a particular "Germanic" type, but in many respects indistinguishable from East Roman tactics. While Wolfram's observations of surviving elements of Roman military organization in Theoderic's kingdom goes some way towards explaining this similarity,[7] many scholars have been reluctant to admit that

---

[5] See above, chapter 2.1.2.

[6] Wolfram 1988: 279 and Heather 1996: 164 for the numbers. An important difference between Amory's and Heather's models is whether they brought families with them to Italy. I agree with Heather (esp. 2007a) that many of them probably did have families due to the explicit evidence he cites from Procopius, as well as the fact Roman military men often had their families with them, especially in their permanent garrisons (cf. chapters 1 & 2 above). This is more in line with Amory's argument of the common Roman military background of the Goths. The conquest of Italy is treated in Wolfram 1988: 278-84 and Moorhead 1992: 17-31.

[7] Rance 2005; see Wolfram 1988: 300-306 for references to older literature and discussion of surviving Roman elements, including *limitanei* in Rhaetia and military colonists subsumed under Gothic ethnicity.

this affected military organization and practice, choosing instead to interpret Ostrogothic military history as an extension of ancient Germanic practices, albeit on a grand scale.[8] Ostensible failures at *Rome (537) and *Rimini (538) have been regarded as symptomatic of Germans reaching above their limited technological (and implicitly, intellectual) capabilities. Everything that does not fit into this framework is simply ignored, a position that clearly jars with their administrative and cultural sophistication.[9]

3.1.2 *Strategic Situation*

The Ostrogoths under Theoderic the Great took over Italy and Dalmatia from Odovacar in a war that revolved around the control of the old West Roman capital *Ravenna (490-93), which was besieged for over three years. By the time the new regime was fully established, it controlled the whole Italian peninsula. To the northwest, its territory extended to the Alpine passes against the Burgundian kingdom. To the north, Ostrogothic control reached into the old Roman provinces of Rhaetia and Noricum (almost) to the Danube, which was the eastern border down to the fortress city of *Sirmium, captured from the Gepids in 504.[10] From there a line ran south to the Adriatic, leaving western Illyricum and Dalmatia under Ostrogothic control. While Theoderic to begin with limited his ambitions to Odovacar's old kingdom, with a few additions, he found himself in an extensive war against the Franks, Burgundians and East Romans in support of the Visigoths in 507-08. The Visigoths were effectively driven out of Gaul, but held on to the Mediterranean coast with Ostrogothic assistance. Interestingly, only one set piece battle may have been fought, but the Ostrogoths relieved

---

[8] The classic statement is Thompson 1958; Burns 1984: 184-201 also subscribes to a Germanizing framework. The rationale behind the Gothic cavalry charge against well-positioned Roman infantry at Busta Gallorum in 552 has been taken as evidence of brute (and rather dim) Germanic aggression, but such charges were, under the right conditions, encouraged in later Byzantine military manuals in order to minimize the impact of foot archers, who would only have time to fire two volleys against a charging enemy. Indeed, the Roman forces operating around *Sisauranon 541 were nearly defeated by the Persians (who were famed for their archery skills), and only saved by a well-directed cavalry charge by Ostrogothic troops that had only recently arrived from Italy.

[9] This is the approach of Halsall 2003: 224, who in effect dismisses Procopius, of all authors, for being an eyewitness who wrote down his account ten years after the events based on notes taken at the time.

[10] Notethat this conflict has particular interest for ethnicity in the Balkans at the turn of the 6th century, as it involved Bulgars in Roman service, Huns, Gepids and Ostrogoths. Depending on the identification of the Huns and Bulgars, loyalties seem to have been very hazy in this particulary region during this period.

the cities of *Arles and *Narbonne from Frankish and Burgundian sieges, respectively.[11] When Theoderic annexed southern Provence to his kingdom, he had to defend his new acquisitions from intense Frankish and Burgundian attacks. Subsequently, an Ostrogothic governor, Theudis, administered the Visigothic kingdom on behalf of Theoderic, but beyond regular tribute and some administrative reforms imposed from Italy, Theudis was effectively independent. For a short period, Theoderic also exercised overlordship over the Vandal kingdom while arbitrating the affairs of other minor kingdoms. Around 520, the Ostrogoths dominated the western Mediterranean basin.

Ostrogothic power struggles after Theoderic's death in 526 precipitated the decline of their state. Within a few years, Justinian found reasons to meddle in the succession, and after the successful occupation of Vandal Africa, decided to intervene directly. The course of Justinian's war is well known and can be followed in the relevant entries in Corpus Obsidionum (535-54).

3.1.3 *Military Organization*

Goffart's argument on barbarian settlement has won most recognition in the Ostrogothic case. However, the issue is not crucial here since, firstly, the settlement probably took many varying forms based on available Roman precedents, and secondly, any original settlement would soon give way to individual and regional developments, as outlined in chapter 1.2.2. Gothic troops settled in three main clusters, where they lived with their families and owned property: Liguria in the northwest, Veneto in the northeast, and Abruzzo on the central Adriatic.[12] This is also where most of the heavy fighting occurred during Justinian's invasion. In addition, there were substantial garrisons in the Alps and, after 508, in Gothic Gaul, that protected Italy from Frankish and other incursions; these troops were permanently settled there with their families. Procopius describes the mobilization of the Goths and the distribution of equipment after the Roman invasion: "and only the Goths who were engaged in garrison duty in Gaul he was unable to summon, through fear of the Franks."[13] They were

---

[11] *Chronica Gallica* no. 689, MGH AA 9 (p. 665) ostensibly has: "Tolosa a Francis et et Burgundionibus incensa | et Barcinona a Gundefade rege Burgundionum capta." See however CO for Thompson's correction and further discussion of the campaign in chapter 1.3.4 above.

[12] See Heather 1996: 237ff for a brief discussion of their strategic disposition.

[13] Proc. 5.11.28.

however withdrawn within a year due to pressure from the newly established Frankish Burgundian kingdom (see chapter 4.1.2).

In addition to border forces, there were Gothic urban garrisons in major towns in central and southern Italy, Sicily and Dalmatia.[14] In Sicily there were small Gothic garrisons in *Panormus (Palermo), Syracuse and *Lilybaeum (535). In the major towns of the north, Gothic presence meant that they provided the bulk of the defenders during war, but in many cases, especially in the south and Tuscany, the urban population was expected to assist in defensive operations. At *Naples in 536, the citizens, including Jews, defended sections of the city walls alongside a garrison of about 800-1,000 Goths, but at *Salona (536), the Goths were uncertain of local support and had to withdraw. At *Milan (538) and other Ligurian cities the citizens actively turned against their Gothic overlords and invited in a small Roman force, but when besieged the Romans abandoned the citizen militia and general population to their fate.

There is no good reason to postulate a two-tiered system of expeditionary forces and lower-quality border troops among the Ostrogoths.[15] Heather argues that such a system existed based on the old *comitatenses-limitanei* distinction and his interpretation of Ostrogothic social structure: there was a significant distinction between a "core" of Goths, whom he regards both as a dominant social group and as the institutional equivalent of the *comitatenses* or field troops of the late Roman army, and remaining troops, in effect *limitanei*, whose quality (and political attachment or identity) was more questionable. However, the military interpretation seems unlikely, since the field-frontier troop dichotomy has been demonstrated to be false, and cannot be identified in the extensive narrative sources.[16] During the Justinianic war, there was no distinction whatsoever between where the Gothic troops came from: the border garrisons noted above were

---

[14] Christie 2006: 357-69.

[15] See *contra* Heather and Barnish in the discussion sections in *Ostrogoths*; they maintain the artificial distinction between different troop qualities based on the old, but flawed idea that *limitanei* were inferior troops.

[16] See chapter 2.1.2 for the postulated distinction and its refutation. Heather bases much of his argument on the presence of "notables" among the Goths, described by Procopius as *dokimoi, logimoi* or *aristoi*. These numbered around 5-6,000 (a quarter or so of the fighting Goths, cf. below) and Heather alleges that Gothic resistance only crumbled when this group was decimated at *Ancona 551 and the battle of Busta Gallorum. While his argument for a large, but distinct social group is plausible, it is not based on a broad evaluation of Procopius' use of the term in other contexts; e.g. at *Naples 536, *logimoi* describes local Roman notables responsible for policy decisions, and from the aftermath of the siege, it seems that this group only involved a handful of people.

apparently the "most noble" among the Goths and were called up for expeditionary service; additional border territories were ceded to the Franks (most of Gaul in 534, Provence in 536, Veneto in the 540s) in order to free up troops. When some of the Gothic garrison troops in the Cottian Alps did decide to surrender in 539, they were hindered by others from the same garrisons who were on field service, who were the last to give up, and only did so when the Romans captured their families.[17] Their *modus operandi* was thus similar to both that of the Roman frontier units as well as the field armies raised in the Balkans for expeditions against Africa, Italy and the East, and their concerns and behavior were very much like those of the Romans.

How large the total Gothic military establishment was is difficult to say. Procopius claims that 150,000 Goths besieged Rome, which is clearly far too large for a single force. Heather follows Hannestad in estimating 25-30,000 men for the whole Gothic army during the 530s and 540s.[18] While this fits well with Heather's estimate of Theoderic's followers in 493 (thus excluding any substantial non-Gothic additions), it is on the small side considering the very extensive operations documented during the war. To begin with the siege of *Rome (537f), it is possible to provide a reasonable estimate of the Gothic force based on numbers given at various points during and after the siege. Altogether, the Goths marched away with *at least* 20,000 men after they gave up the siege.[19] Heavy losses and disease over a year of fighting may bring the number to an original high of 25,000. Furthermore, Gothic garrisons had been captured at *Palermo, *Syracuse and presumably *Lilybaeum during the occupation of Sicily in 535; 2,000 men for the whole island when war was looming seems to be a reasonable minimum. The Gothic garrison at *Naples, along with those who surrendered in Samnium during the first year would make at least 3,000. Thus, we have at least 30,000 available for fighting in central and southern Italy in 537-8 alone. There were still substantial garrisons in the Alps, and large forces, including both Goths and Sueve clients, were sent into Dalmatia against *Salona in 536 and 537. Most of the Dalmatian Goths surrendered to the Romans, so the Gothic reserves fighting in the Po valley must have

---

[17] Proc. 6.28. 28-35; see also *Alpine forts 539.
[18] Heather 1996: 164; cf. Hannestad 1960.
[19] This is in part dictated by the internal logic of the Gothic siege camps at *Rome 537f; for troops afterwards, see especially *Rimini II, 538 and Proc. 6.11.1 for catalog of 9,800 soldiers left to garrison cities in central Italy; a similar number would have engaged in the siege of Osimo.

come from elsewhere.²⁰ These forces are difficult to estimate, but from Procopius' description of the extensive fighting in Dalmatia, several thousand must have been engaged there; similar numbers are likely for the garrisons and reserve forces. Thus we are speaking of an army approaching 50,000 soldiers (or well in excess of that when client troops are included). A fairly regular influx of Roman deserters and prisoners could help keep up numbers during the 540s, when Romans and even Moorish federates are reported in Gothic service.²¹

A complicating factor is the Ostrogothic fleet, established under Theoderic the Great. Cassiodorus preserved among his letters instructions for the construction of a fleet of 1,000 warships, a huge number by any standards, requiring at least 50,000 rowers.²² This brings the total of the Ostrogothic armed forces to well over 100,000 men. If the garrisons sent (presumably permanently) to Spain in 508-11, various classes of arms producers, those in the military administration, and servants to fighting men are included, Procopius' figure of 150,000 might actually be a reasonable East Roman estimate of all who were theoretically associated with the Ostrogothic military establishment, including soldiers, clients, naval personnel, arms makers and other support functions.

Although the fleet has received a poor reputation from the great loss at the siege of *Ancona (551, see also below), this is not quite representative of its wartime performance. At the beginning of the war, the Adriatic division conducted a coordinated land and naval expedition against *Salona (537), where the Romans had made extensive preparations to withstand a

---

[20] See Amory 1997: 168f for a list of Gothic garrisons and army divisions that went over to the Romans.

[21] See Heather 1996: 327f for a brief list; he miscalculates one instance (220 for 320 who went over to the Goths at *Rossano 548). The list suggests that somewhere around 2,000 Romans went over to the Goths during the late 540s, assuming a similar number defected from each of the smaller garrisons. However, Heather ignores some interesting implications. The 700 Isaurians who surrendered at *Rome (549) had observed how well other Isaurians in the Gothic army had done for themselves. This indicates that this group may have been somewhat larger and more important than that suggested by the three garrisons in the list by that point. Furthermore, many began to drift back to the Roman side when it was rumored that Germanus (subsequently Narses) was assembling a large Roman army in the Balkans and bringing cash for their salaries that were in arrears. Finally, after the battle of Busta Gallorum, a great number of formerly Roman soldiers *still* in the Gothic army were executed (Proc. 8.32.20).

[22] The estimate of the rowers is given by Pryor and Jeffreys 2006: 14, who express disbelief at the organizational ability of the Ostrogoths, although recognize that a fleet *was* completed. What kind of ships Theoderic had built determined how they were manned. Cassiodorus only mentions *dromones*, in effect rowed warships. Cosentino 2004 believes that the task *was* accomplished in full, but that many were transports rather than warships.

siege. The Goths were able to surround the city completely both by land, building a stockade around the city, and by sea with their fleet, despite a partly successful Roman naval counter-attack.[23] At the height of their naval power, the Ostrogoths operated two large divisions with extensive reach, and if these are added together (to a total of nearly 750 ships), Theoderic's original establishment may have been revived or was still been in operation. The Adriatic fleet was fully capable of expeditions not only to Dalmatia, but during the later stages of the war, 300 ships intercepted Byzantine supply ships and even plundered mainland Greece.[24] At the same time, another division of 47 ships attempted to assist the Gothic siege of *Ancona, but the Romans proved superior in seamanship and naval tactics.[25] The Gothic Tyrrhenian fleet, based in Naples and the Aeolian Islands, intercepted Roman supply ships from Sicily headed to Rome.[26] The Goths also used it to invade Sardinia and Corsica successfully,[27] while 400 ships were mustered for the invasion of Sicily which was only abandoned upon the arrival of Narses' large expeditionary army.[28]

3.1.4 *Logistics: Adminstration, Labor and Supplies*

As in the East, the Ostrogothic authorities organized labor, materials and supplies for military purposes through a combination of public duties, tax exemptions, and private initiative. Strictly military production is poorly attested in the surviving sources, but we do have some indications of how the Ostrogothic government supplied their troops with arms and armor. There are strong reasons to believe that at least some of the late Roman *fabricae* survived in Italy.[29] Arms could be issued by the government, or

---

[23] Proc. 5.16.8-18; further on the following operations, see Pryor and Jeffreys 2006: 13-19.

[24] Proc. 8.22.17-20, 30ff, Gothic fleet of 300 ships crossed to Cercyra, raiding the island and parts of mainland Greece, even intercepting supply ships intended for Narses' army.

[25] Proc. 8.23.1-3, Goths besieged *Ancona land and sea with 47 ships. 4-8, reinforcements arrive from Salona (38 ships) and Ravenna (12 ships). 9-28, forces draw up, various harangues. 29-34, Goths less than competent, crowding or spreading too much, trying to fight a land battle or ram individually. 34-38, Romans more competent, ramming single ships and overwhelming crowded ones with archery; only eleven Gothic ships escape while the rest sunk or taken, while many Goths perish. 39-42, Goths burn their ships and give up the siege, Ancona is relieved while reinforcements return to their bases.

[26] Proc. 7.13.5-7.

[27] Proc. 8.24.31-39; cf. *Caranalis 551.

[28] Proc. 7.37.5ff; see also *Messina (549) for the Gothic crossing, and *Sicilian forts (551) for Roman recapture.

[29] Wolfram 1988: 303 and n. 295 (at p. 511) and Christie 2006: 352 believe the *fabricae* were still operational. S. James 1988 discusses the evidence, arguing that the expertise no

government decree (e.g. at Salona), and Cassiodorus explicitly refers to arms-makers, *factores armorum*, presumably still organized in *fabricae*.[30]

Wartime food supplies were particularly well organized, and one of the main reasons that the Ostrogoths held out for so long. The Ostrogothic fleet was most effective as a logistical tool, carrying supplies to Ostrogothic garrisons and civilian populations from Provence to Dalmatia, and carrying troops to Sardinia, Sicily, Africa and Greece.[31] Depending on needs, supplies to the cities in the Po valley were supplemented from Istria by the Adriatic fleet or Liguria through a system of river transport boats. When the Romans began to station themselves around *Ravenna in 539 for a siege, they seized the Gothic grain boats when they stranded on the Po because of low water levels.[32] The Ostrogothic field armies operating in Gaul and the garrisons stationed on the River Durance were supplied via the Rhône from granaries at Marseilles while the fleet at Ravenna could supply Liguria.[33] During a famine in Gaul, private shipmasters along the whole western coast were encouraged to bring supplies in return for profit.[34] In the major cities there were state granaries to store foodstuffs, as Cassiodorus described at the beginning of the Gothic war.[35] Their importance is illustrated at *Ravenna (539f), which surrendered in 540 only after Belisarius had bribed a citizen to burn the granaries, hastening the shortage of supplies.[36] Finally, the fleet was fundamental to supplying the last Gothic field army in Campania in 552, and its defection to the Romans determined the fate of the Gothic state.[37] We also see a system of pur-

---

longer had to be organized in *fabricae*, but that it was not inferior to previous forms of production.

[30] Cassiodorus, *Variae* 7.18-19, but see the above n. and cf. *Salona 536.

[31] For a comprehensive survey of the Ostrogothic fleet in action, see Pryor and Jeffreys 2006: 13-19

[32] Cassiodorus, *Variae* 12.22, 24 for supplies from Istria; *Ravenna 539f for transport boats on the Po.

[33] Cassiodorus, *Variae* 3.41, from the granaries at Marseilles to the Durance garrisons; 3.44, used to supply *Arles in the context of the siege in 508; 2.20, ships at Ravenna to help supply the royal household while in the region of Liguria.

[34] Cassiodorus, *Variae* 4.5; while the shipmasters similarly commissioned to bring supplies from Spain to Liguria on another occasion (*Variae* 5.35) committed fraud by selling their cargo at other locations.

[35] Cassiodorus, *Variae* 3.29, old state granaries at Rome turned over to private individuals for repair; 10.27, state granaries at Dertona and Ticinum (Pavia) to supply Liguria, while those at Treviso and Trent were to supply Venetia; 12.27(f), the granaries at Dertona and Ticinum to be used to relieve famine under the supervision of Datius, bishop of *Milan; however, he led the defection to the Roman side which led to the siege in 538.

[36] Proc. 6.28.1-6, 25-27.

[37] See *Gothic fortified camp, 552, which may have relied on supplies from Sardinia.

chases for units marching across the kingdom to another front; thus a Gepid division marching from Dalmatia through Italy to Gaul was organized into units for commissariat purposes; these received cash stipends for purchasing supplies from local populations instead of requisitioning drafts animals, equipment and foodstuffs.[38] The ability to transport supplies over long distances and keep stores for garrisons and civilians under siege made it possible to provide for large forces in wartime and to send large field armies to Spain, Burgundy, Sicily, Pannonia and Dalmatia.

The need to maintain permanent military installations in the Alpine passes and the Po valley required direct royal administration in conjunction with local troops and landowners. Thus Goths and Romans were ordered to fortify Dertona near the Po and build houses inside to ensure that it could function as a refuge in time of war; this was also enjoined at the fort of Verruca.[39] Since there is good archaeological and written evidence that fortifications and city walls were regularly maintained at an advanced level, indistinguishable from late Roman structures,[40] it is reasonable to conclude that elements of the former Roman system of using specialist military engineers in garrisons and *fabricae* to direct repair and fortification works survived. Certainly the administrative practice of requiring landowners to construct or repair a specifically assigned length of walls (*pedatura*, cf. 2.4.2) was the normal Ostrogothic procedure, e.g. at Feltria,[41] as it was in the East. In Ostrogothic Italy, the public and royal architects took over an important aspect of the office of the *ballistarii* as chief organizers of fortifications. The *ballistarii* were chiefly responsible for organizing this procedure in the East Roman Empire, but there, too, public architects were acquiring an increasingly important role.[42]

During the war of 507-08, the repair of walls and towers at the newly acquired *Arles were paid for by the government. Such repairs had to be organized locally by the *possessores* in cooperation with the Ostrogothic expeditionary armies. The administrative practice would again be a combination of assigned *pedaturae* overseen by local architects and probably

---

[38] Cassiodorus, *Variae* 5.11. This is discussed extensively by Goffart 1972.
[39] Cassiodorus, *Variae* 1.17 (Dertona); 3.48 (Verruca).
[40] Halsall 2003: 222 passes over the Italian defenses in one short paragraph, effectively reducing the Alpine defensive system of the Ostrogoths to *one* fortified site. See Christie 2006 and articles in *Ostrogoths* (e.g. Brogiolo 2007) for a better assessment.
[41] Cassiodorus, *Variae* 5.9 for the *possessores* of Feltria, who were assigned *pedaturae* they were responsible for building, but at the same time their costs were offset by subsidies paid out by the government.
[42] Cassiodorus, *Variae* 7.5 (and 7.15) for the *architectus publicorum*; cf. 2.2.2.

Gothic military engineers. In return for their loyalty (and expenses) in supporting the Ostrogoths and carrying out defensive works, the *possessores* received tax exemption and were relieved of the burden of paying and supplying the troops, which indicates the extent of wartime destruction and cost of defense, even in areas not directly ravaged by invasion.[43] Outside the main Gothic settlements and military installations in the north, the maintenance of city walls was left to the initiative of the local curia. In Catania, for instance, the local council of *possessores, defensores* and *curiales* wrote to king Theoderic asking for permission to tear down the old amphitheatre so as to provide materials for a new city wall, probably in response to East Roman naval operations against southern Italy in 507-8.[44]

The day-to-day working of the system, as far as it can be reconstructed from peacetime activities, is familiar from the later empire and the contemporary East. The government could impose public duties or use taxes, or a combination of both, in order to organize large-scale monumental work and public infrastructure. Theoderic ordered materials for public works at Rome and Ravenna, such as monuments, aqueducts, cloacae and walls. While spolia from decaying buildings were available in most cities and could be used for a whole range of purposes, as we have seen, the walls of Rome required 25,000 custom-made tiles a year for maintenance, presumably from industrial-scale workshops organized on the model of late Roman *fabricae*. Stones for repairs of Rome's walls should also be collected from nearby fields, while timber was levied on *possessores* at Forum Julii.[45] The manual labor required was either paid directly from public funds or levied via the landowners, who mobilized their tenants and clients with appropriate skills. Theoderic himself sent experts and craftsmen to perform public works. These were presumably urban guild members paid from the royal fisc, or perhaps servants on the royal estates (*domus divina*); the latter were certainly assigned their *pedatura* just as the landowners at Feltria (see above). The planning and execution of works were often done

---

[43] Cassiodorus, *Variae* 3.44 (*Arles 508); 3.40 and 3.42 on tax relief, at first only to areas directly affected by the war, but later extended to the whole of the province.

[44] Cassiodorus, *Variae* 3.49. Procopius calls the city unwalled, so the work was never completed. The letter was probably sent in response to East Roman naval raids in Apulia in 507-8 (Moorhead 1992), but since the situation was soon resolved, the curiales probably decided to abandon the project.

[45] Cassiorodus, *Variae* 1.28 on collecting stones; 4.8 on timber at Forum Livii; this was also levied for the fleet, cf. below. Although pay was offered for collecting stones, Cassiodorus concedes that it was low. While such labor formed part of the burdens imposed on landowners and tenants, cf. the *pedaturae* above, the cash was probably offered as an extra incentive to a task with high priority.

under royal supervision by public architects, while land surveyors and specialists in mosaics and stone-working are also attested in various capacities.[46] There is thus an astonishing degree of administrative continuity (cf. 1.2.5), but the Ostrogoths seem to have spent government revenue and given tax exemptions more freely, perhaps on the model of Anastasios (cf. 2.4.1).

As opposed to the detailed narratives in Joshua the Stylite or other Eastern sources, we rarely see Ostrogothic military engineers and civilian craftsmen in the field, only the outcome of their labors. The organization of labor and materials for military installations differed little from normal public works in the East. The relation between mobilization of labor and military effect is most clearly seen in the establishment of an Ostrogothic naval organization. Here we have good evidence how Theoderic ordered landowners as well as the administrators of the royal estates to assemble timber, and *saiones* to recruit sailors and clear waterways of fish traps.[47] The fleet consisted of 1,000 *dromones*, and would have required immense material and manpower resources. During the war with Byzantium, the Ostrogoths were able to maintain this very large naval establishment, since the number given by Cassiodorus is consistent with the fleets enumerated in Procopius. This presupposed effective administration and organization of labor even in time of war.

These skills carried over into field engineering. In many instances, Procopius explicitly states that Gothic machines and field works were made with great ingenuity. Fortification technology in Italy was of great sophistication, and can hardly be distinguished from earlier West Roman and later Byzantine forts unless there are datable finds such as coins to identify the occupants. The narrative sources give little to distinguish Ostrogothic from Roman military engineering. In terms of labor, materials, finance and organization, then, East Roman and Ostrogothic warfare were fundamentally similar.

---

[46] Payment from the fisc is attested in several *Variae*, e.g. 1.21, where citizens of Rome contribute along with royal funding, which is however to be strictly audited, so that funds actually go to works, not private pockets. For experts, see *Variae* 3.52, on land surveyor (*agrimensorem*) to be sent to adjudicate a property dispute between two Roman *spectabiles* before it erupts into violence. Further *Variae* 1.6, marble-workers (*marmorarios*), i.e. experts in mosaic, ordered to be sent from Rome to Ravenna by Theoderic. Marble-workers proper, also sent by Theoderic, but this time *to* Rome, were called by circumlocution in *Variae* 2.7: "quibus hoc opus videtur iniunctum in fabricam murorum faciat deputari" in order to use various spolia from public buildings for the embellishment of the city.

[47] Cassiororus, *Variae* 5.16-20; the detailed instructions support the argument of Cosentino 2004.

3.1.5 *Ostrogothic Siege Warfare*

Just as in battlefield tactics, military organization and logistical capabilities, Ostrogothic siege warfare differed little from East Roman practices. They regularly constructed siege camps or full-scale siegeworks to hem in the besieged,[48] and were also adept at other sorts of field fortification, such as the fortified bridge mounted with *ballistae* that they constructed near Vesuvius at the end of the war.[49] Despite many comments on the late or rare appearance of Ostrogothic artillery during these wars, it is possible to argue that the Ostrogoths had defensive artillery from the outset, where the Ostrogothic defensive barrages during the siege of *Osimo in 539 seem to indicate as much. However, secure identification is obscured by Procopius' writing style and emphasis. Furthermore, the Ostrogoths had a formidable arsenal of offensive siege engines that have not yet been taken seriously. Artillery was in fact rarely used (or at least hardly ever described) in offensive siege operations during the early 6th century, so engineering capabilities must be evaluated on the basis of other machines.[50] During the first assault of *Rome (537), at the northern Salarian gate sector, they brought up four powerful rams, which Procopius described in great detail, noting that they were very destructive. In addition the Goths had built large moving towers that terrified the civilian defenders. Procopius relates how Belisarius dismissed the assault with a laugh, since they were using oxen to haul the towers forward. Many scholars have taken this anecdote as proof of barbarian ineptitude, or at least as a classicizing *topos* on the same, but a close reading of the text reveals that there is more to these events. The Ostrogoths were in fact well prepared and knew what they were doing.

Elsewhere Procopius describes how the Ostrogoths had large shields that made them impervious to regular archery fire. When they assaulted Hadrian's mausoleum, they could take cover under the colonnade of St. Peter's to get so close that the defenders were unable to operate the *bal-*

---

[48] At *Rome in 537 (Proc. 5.19.1-5, 11f), the Goths established seven camps around half the city: "And the Goths dug deep trenches about all their camps, and heaped up the earth, which they took out from them, on the inner side of the trenches, making this bank exceedingly high, and they planted great numbers of sharp stakes on the top, thus making all their camps in no way inferior to fortified strongholds."

[49] The *Gothic fortified camp (552) had amongst various other engines *ballistrai* mounted on towers, making any approach impossible.

[50] See discussion in chapters 5.2.2 and 8 *passim*; in effect, the general diffusion of the traction trebuchet in the late 6th century greatly increased the use of offensive artillery in siege warfare.

listae against them. The angle was too steep and the range too short to use machines, which only left the use of bows. When archery failed against their large shields and the Goths were about to scale the wall with ladders, the defenders survived by breaking up the statues on the mausoleum and dropping them onto the heads of the Goths.

The effectiveness of their shields makes the Gothic approach with oxen more comprehensible. The oxen were protected by serried ranks of armored infantry: "Belisarius, seeing the enemies' formation marching slowly *with the machines* (...)."[51] Instead of expending fire on well-prepared infantry marching slowly in formation and protected by large shields, Belisarius deliberately let them get close to the moat. He then built up the confidence of his troops with two well-aimed bowshots that took down two of the armored leaders of the formation, to roars of approval from the defenders. Only "then did Belisarius signal to the whole army to set in motion all the archery, but those around him he ordered to shoot at the oxen only."[52]

From this it is clear that the strategy was planned out beforehand, since simultaneous firing began from the whole wall on a given signal. While Procopius uses the generic *toxeumata* (archery), bows alone would have been insufficient to break up the Gothic formations. From Procopius' description of the *ballista*, however, where he compares it to a large bow, it is possible that his use of *toxeumata* included the *ballistae*, which could wreak havoc on the infantry protecting the oxen. This was what Belisarius was waiting for; he personally led those who were to fire on the oxen, taking advantage of the gaps and chaos caused momentarily by massive archery and ballista fire against the infantry. The oxen fell "immediately" (αὐτίκα) when the whole operation was set in motion. At that point the Goths gave up an outright storm at that sector, but kept up pressure with continuous archery fire against the parapets.

Meanwhile assaults took place at Hadrian's mausoleum to the west, mentioned above, and Vivarium to the east. Here the "machines," which possibly included towers and certainly rams, worked as planned, and the defenders were terrified by the Gothic assault before Belisarius arrived with substantial reinforcements. The Goths assaulted with machines along a large sector, and were able to reach the outer wall of the Vivarium, which

---

[51] Proc. 5.22.2; my emphasis: Βελισάριος δὲ βαδίζουσαν ξὺν ταῖς μηχαναῖς ὁρῶν τὴν τῶν πολεμίων παράταξιν ...

[52] Proc. 5.22.7: καὶ τότε μὲν Βελισάριος τῇ μὲν στρατιᾷ πάσῃ κινεῖν τὰ τοξεύματα πάντα ἐσήμαινε, τοὺς δὲ ἀμφ' αὑτὸν ἅπαντας ἐς μόνους τοὺς βόας ἐκέλευε βάλλειν.

was less heavily fortified than other parts of the wall. They put their rams to good use, breaking through (διορύσσοντας) the wall, but here Belisarius kept his cool, holding the reinforcements in reserve at the gates. He ordered one group to hold up those who had broken through the wall, while he himself led a sally out the gates that caught the Goths outside by surprise. At the same time, a sally at the Salernian gate drove off the Goths at that sector as well. Most of the Gothic siege engines were burnt that night. It had been a close call on two of three sectors, only warded off by quick thinking, good leadership and an extremely well-equipped and highly motivated expeditionary army. A similar deployment of a large siege tower at *Rimini (II, 538) has also been used to dismiss Gothic engineering capabilities, but again, a close reading of the text makes it clear that the Goths actually handled the engine well and adapted to circumstances. Furthermore, Gothic approaches were similar to that used by the Romans at *Amida (503 and 504), and while military means were ultimately fruitless, they won *Rome by treason (545f, 549f) after hard fighting, and took a host of Roman cities by surrender or storm (see CO s.aa. 541, 542, 544-50).

3.2 *The Visigoths in Spain, 508-711*

Collins recently argued that the greatest problem for the Visigoths in military terms was the fact that they experienced, besides fairly limited power struggles among the nobility, long periods of general peace, and may have had little need for a large, cumbersome and expensive military establishment.[53] However, the threat from the Merovingian Franks remained immense, and required strong defenses in northern Spain and southern Gaul. Campaigns against the Sueves in Galicia and the Romans in the south consumed great resources for over half a century (c. 570-625), and the Visigothic kings had to campaign regularly against the Basques who were prone to raid great swaths of northern Spain. The Visigoths can in fact be shown to have possessed most of the necessary infrastructure for carrying out siege warfare, offensive and defensive: there were permanent garrisons in many cities, in suburban forts, and border fortifications. Labor was organized within the *obsequium* of the king, secular magnates, and high-ranking clergy, who could mobilize dependents with necessary skills. While not particularly impressive compared with imperial standards, they

---

[53] Collins 2004.

were sufficient for repairing walls, building fortifications and siegeworks, and operating siege engines, the main elements of siege warfare.

3.2.1 Strategic Situation of the Visigothic Kingdom of Toledo

After they were driven out of most of Gaul except Septimania in 507, the Visigoths reestablished a kingdom in Spain centered on their capital Toledo, which gradually extended control to the whole Iberian Peninsula.[54] The Visigothic military model has received some attention, but the lack of narrative sources has made it difficult to determine how the system worked in practice, and especially how it survived the transition from Aquitaine to Spain and further how it evolved in the kingdom of Toledo.[55] Many elements developed simultaneously in Frankish Gaul, while others heralded many significant developments in the Frankish world that were only brought to fruition under the Carolingians. These were still recognizably late antique in their origins and fundamental organization, since the Franks took over the structures left behind by the Visigoths in southern Gaul. Involvement of clerics in warfare was regularized by Visigothic law a century before the Carolingians formally did so, although fighting bishops was a common practice already in 6th-century Gaul (see chapter 4.2.2). There were also significant parallels in manner of recruitment such as estate-based troops with high levels of technological expertise and fixed numbers of troops according to wealth, though measured differently). Visigothic Spain was, furthermore, an important conduit of Ostrogothic and East Roman influence to the West, and any adaptation by the Visigoths to Eastern practices could have repercussions in the Frankish area.

After the Visigothic loss of Aquitaine in 507-08, Theoderic the Great sent his generals Ibbas and Theudis to stabilize the situation in the Iberian Peninsula in 508 and 511, respectively, and through the latter acted as regent for the minor Amalaric (r. 511-531).[56] After 531, Theudis assumed the kingship until 548, and was in effect independent of the Ostrogoths, though he continued to send tribute. Under Ostrogothic tutelage, the Visigoths began to consolidate control over the Iberian Peninsula, where large territories were either under the Sueves (Galicia) or effectively independent (northwest and south). This consolidation is first shown in the establishment and

---

[54] Basic narratives are provided by Thompson 1969, Heather 1996, and Collins 2004.
[55] The most recent monograph is by Pérez Sanchez 1989, but this has not completely superseded the sections on the army in Thompson 1969 and King 1972. Halsall 2003: 59-63 provides a brief but more recent overview.
[56] See chapters 3.1.2 and 1.3.4 above for this.

maintenance of a frontier command in the Pyrenees and Septimania that was able to keep the Franks at bay and even reconquer some territory.[57] The East Roman invasion and occupation of the southeastern coast of Spain in 554 was possible due to a Visigothic civil war and the virtual independence of many southern cities, many of which remained outside both Visigothic and Roman control through most of the 6th century.[58] The Visigoths mounted significant campaigns involving siege warfare during the late 6th and early 7th centuries under the kings Leovigild, Reccared and Sisebut in order to gain control of the rest of the peninsula.[59] In addition, there were civil wars, suppression of Basque raids, and some frontier conflicts with the Franks. When the Arabs extinguished the Visigothic kingdom during the invasion of 711-18, they may have taken over and continued significant aspects of Visigothic military organization with the aid of Visigothic magnate families who became clients (Theodemir in the southeast) or converts (the Banu Qasi in the Ebro valley),[60] and also took over some of the logistical infrastructure of the Visigothic state, using it for their own campaigns in Gaul.[61]

3.2.2 *Visigothic Military Organization*

For about a generation after the battle of Vouillé in 507, we know little about Visigothic military organization, except that it worked to keep out the Franks and consolidate control over the central plateau of Hispania. By the late 7th century, legal sources describe a well-established system of raising and organizing armies that were based on the followings of great landowners.[62] The "proto-feudal" model championed by Spanish historiography has largely been rejected.[63] Rather, the 5th-century model from

---

[57] See *Saragossa 541; this is further discussed in chapter 4.1.2 below.

[58] See Ripoll Lopez 2001 for a good recent assessment. While she correctly rejects the concept of a fortified *limes*, her interpretation of the East Roman province as mere coastal enclaves may be a little too far in the other direction.

[59] See *Cordoba 549, *Málaga 570, *Medina Sidonia 571, * Cordoba and other cities 572, *Orespada 577. The siege of *Sevilla 583 is also analyzed below.

[60] See Collins 1989: 39ff and *passim* for Theodemir; ibid. 204f for the Banu Qasi or Cassii.

[61] See the discussion of Charles Martel's campaigns in chapter 4.3.3 below and the relevant entries in CO.

[62] King 1972; Wickham 2005: 98ff.

[63] Spanish historiography (see e.g. survey by Pérez Sánchez 1989: 83-103) has used this material to postulate a "proto-feudal" system that looks forward to feudalism proper in the Frankish world. This is rejected by most international scholars with the decline of feudalism as a valid model. See e.g. Collins 2004, Kulikowski 2004.

Aquitaine prevailed in the early 6th century, when Theudis recruited a private army of 2,000 professional soldiers (στρατιῶται) as well as a force of bodyguards from the estates of his Hispano-Roman wife.[64] Some of them may have belonged to the original Ostrogothic expeditionary army, others were survivors from Vouillé (or their descendants, depending on how late in Theudis' regency this actually happened). The practice of maintaining professional troops from estate wealth was probably not unusual among the Hispano-Roman nobility. They had largely been independent in the late 5th century until the establishment of the Visigothic capital at Toledo, and they had access not only to immense wealth, but also troops in the form of slaves (see below), barbarian invaders, federates, and decommissioned Roman soldiers, as argued in chapter 1.

There were two categories of Visigothic troops discernable to us in the 6th and 7th centuries: free Goths and slaves. Most of the former and all of the latter were organized in the *obsequium* of a magnate, royal officer, or served in the king's own guard. Free, professional soldiers were the institutional (and perhaps biological) descendants of free Gothic and Roman troops in the 5th century. Tax laws indicate that some were free landowners with inherited military obligations, as in contemporary Ostrogothic, East Roman, and later Carolingian society.[65] Extensive royal estates supported the royal following of *gardingi*, the Visigothic equivalents of contemporary Lombard *gasindii* and Merovingian *antrustiones*.[66] This royal *obsequium* must have been the largest military following in the kingdom with divisions stationed as border garrisons and in larger cities as reserves, serving alongside the followings of great magnates and local officials. From *Historia Wambae* we have evidence of one who apparently was commander of a division of the royal *obsequium*, i.e. analogous to the Carolingian *vassi dominici*.[67] According to Isidore, Gothic troops drilled regularly, were ca-

---

[64] It is often dubiously asserted that these were slaves; e.g. Halsall 2003: 45, Collins 2004: 43. The text of Procopius 5.12.5of is quite unambiguous: Theudis' wife "not only possessed great wealth but also owned a large estate [or district, region] in Spain. From this estate he gathered about two thousand soldiers and surrounded himself with a force of bodyguards ..." ...περιβεβλημένην μεγάλα χρήματα καὶ χώρας πολλῆς ἐν Ἰσπανίᾳ κυρίαν οὖσαν. ὅθεν στρατιώτας ἀμφὶ δισχιλίους ἀγείρας δορυφόρων τε περιβαλλόμενος δύναμιν ... It is clear that the *doryphoroi* came in addition to the 2,000 *stratiotai*, which is the technical term for professional soldiers.

[65] King 1972: 72.

[66] Thus Halsall 2003: 48f.

[67] The failure of the Visigoths to establish a successful royal dynasty in Spain that lasted more than two generations may indicate that most military resources were on magnate hands. Indeed, some kings tried to establish a system by which *gardingi* should continue

pable of fighting on horse and foot, and used lances and javelins in battle.[68] From this information it seems that their battlefield tactics around 600 were very similar to contemporary East Roman and Lombard practices.

The second category, slave troops, had been used in Spain since the time of Honorius,[69] but is best known from the late laws of Wamba and Ervig. These troops were regularly called up on royal authority, and could together with free soldiers form very large armies with great range and diverse capabilities. However, the decentralized organization proved highly fractious in local magnate conflicts and civil wars, and royal power had to be asserted forcefully with every new reign. Wamba complained that nobles only turned up with about one in twenty of their slaves. He required that a full one in ten should accompany their master on royally commanded or mandated expeditions, fully equipped for war.[70] This may be a realistic assessment of the actual power of magnates, and this provision is even more interesting in light of 7th-century legal provisions dealing with internoble squabbling and small-scale fighting in an agricultural and estate-centered environment. Neither slaves nor free men had any responsibility for their actions when commanded by their master, even for arson and pillage. The law also describes people being shut up in their houses and courtyards (i.e. of their estates) or having their estates attacked while on expedition. It also mentions slaves committing crimes while their masters are away, hence Wamba knew well the consequences of leaving too many armed men at home.[71] Magnates did this to protect their holdings from competitors, to exploit opportunities to pilfer their neighbors and rivals who were on campaign or other royal business, or even avoid abuses from royal officers who might act similarly. Thus Wamba's estimates of military strength are very interesting figures. Magnates were fully capable of arming

---

in office after a royal election. This means that they were normally disbanded when a royal dynasty died out, which normally happened within two generations. Establishing a permanent royal army would strengthen the royal position considerably, but as far as we can tell, this never happened. Newly elected kings remained dependent upon their personal estates; even if they could use fiscal resources to expand their own followings, they had to have considerable magnate support in order to succeed. See King 1972.

[68] These are the soldiers called "Goths" by Isidore, *HG* 69f.

[69] Cf. the brothers who raised an army of slaves from their estates in 409 to face the Vandal/Alan invasion; see chapter 1.2.4.

[70] *LJ* 9.2.9.

[71] Most of *LJ* 8.1. is concerned with this problem; see especially laws 1-8. Note especially law 1 (from Reccesuinth), who distinguishes between free and slave followers, but holds that neither are responsible when following their lord's command to attack someone else's person or property.

and sending out one twentieth of their slaves on expeditions, but the Crown wanted to use their full military potential, which was closer to one tenth.

Theudis' following in excess of 2,000 men must be an absolute maximum for professional soldiers, but slaves were another matter, since they were mostly infantry and some fulfilled support functions. However, they were not random peasant serfs being armed and shipped off to the front, as it were, but rather a specific group with military training, on par with specialized craftsmen of similar social status, also attached to the estate and available for military service. This system provided much of the necessary labor for repairs, supplies, fortifications and field works, since the legislation envisages a division of labor even within the slave force between heavy and light infantry as well as some engineering specialists:[72]

> ... and in order that said slaves may not come unarmed, but may be provided with the proper weapons, whoever brings them must furnish a part of them with suitable armor, and the greater portion must be provided with shields, two-edged swords, lances, bows and arrows, and some even the equipment for catapults [probably trebuchets] and other arms, and he who brings them must parade them, armed in this manner, before the king, general, or commander-in-chief.[73]

If some of the later manuscripts may be trusted, one in ten means a straight fraction of the total slave population, as it calls for half of the (able bodied) male population over the age of 20 up to a maximum of fifty.[74] A lord would

---

[72] This closely resembles the Carolingian *obsequia* with their logistics teams, cf. chapter 4.3.2 and 4.3.4 below.

[73] *LJ* 9.2.9: "... ita ut hec pars decima servorum non inermis existat, sed vario armorum genere instructa appareat; sic quoque, ut unusquisque de his, quos secum in exercitum duxerit, partem aliquam zabis vel loricis munitam, plerosque vero scutis, spatis, scramis, lanceis sagittisque instructos, quosdam etiam fundarum instrumentis vel ceteris armis, que noviter forsitan unusquisque a seniore vel domino suo iniuncto habuerit, principi, duci vel comiti suo presentare studeat."

The translation is slightly modified to reflect the phrase *quosdam etiam fundarum instrumentis*; traditionally the *quosdam etiam* is ignored by translators and scholars alike, although it marks out a special group within the *obsequium*, while *fundarum instrumentis* is simply rendered as "slings." Obviously, *instrumentis* ("equipment," or better here, "engine parts") goes poorly with hand-operated slingshots, while there is significant evidence that *funda* around this time had begun to acquire the meaning of "trebuchet," cf. chapter 8.2.3 below.

[74] This is also argued by Thompson 1969: 265: "According to some manuscripts of the text, they had to bring one half of their slaves up to a maximum of fifty." The apparatus in the edition reads (377.22ff): "*Pro* decimam partem servorum suorum *praebent* V 15. 16: medietatem servorum suorum, de his, qui inventi fuerint a vicesimo anno et supra, id [est] usque ad quinquaginta singulatim unusquisque." "*For* the tenth part of their slaves, [man-

need a base of 500 people to provide for such a following; many Visigothic magnates were obviously far richer and more powerful than this.[75] Surplus wealth could then be used for even larger followings, including personally free men who were better equipped with horses, arms and perhaps some property.

The thrust of Visigothic laws with hints in the narrative material strongly suggest not only the overarching importance of military followings, but also the gradual disenfranchisement of followers vis-à-vis their lords.[76] Troops organized and equipped as armored cavalry were expensive to maintain, so it is unsurprising that the personally free retainers (*buccellarii, saiones*) became more dependent upon their lords over time.[77] This process may have obscured the distinction between the Gothic regulars and slave troops by the late 7th century, although some of the laws quoted above still distinguish between free and slave who performed violent acts on behalf of their masters. Furthermore, we must consider that arming one's slaves also changed the manner in which one could treat them: Ian Wood points out an instance in the *Vitae Patrum Emeritensium* (*VPE*), where slaves on a royal estate attacked their new, royally appointed master who was devoted to an ascetic, religious life; they seem not to have appreciated having a scruffy, unkempt patron, no matter how sanctified he was.[78]

Visigothic kings continued to levy the *munera sordida*, such as the obligation to maintain bridges and roads or provide horses to the *cursus publicus*.[79] As elsewhere in the post-Roman world, these obligations became militarized because of political circumstances. The East Roman invasion was supported by the Roman population of the south, who allowed a very small Roman force of about 5,000 men to occupy a significant part of the peninsula. During the civil wars recorded in the *History of Wamba*, the support of local magnates, clerics and royal officers stationed in the cities decided the success or failure of the revolt. Cities thus had considerable

---

uscript] V 15. 16 *provides*: half of their slaves, of those who are found from (their) twentieth year and above, that [is] up to fifty (men), each and every single one of them."

[75] See Díaz 2000 on the organization and wealth of Visigothic estates and the significance of cities. The Visigothic system compares well with the Carolingian *mansus* reckoning. Even though the basic unit of calculation is different (the Carolingians used number of *mansi*, or fiscal household units, rather than number of slaves), it seems that most military followings in the Carolingian and Ottonian eras were of the same order of magnitude.

[76] See chapter 1.2.4 for the discussion of Euric's law on *buccellarii* and cf. the equivalent law in *LJ*.

[77] Cf. the discussion in Sarris 2006: 174f.

[78] Wood 1999: 198.

[79] King 1972: 71.

resources and political power. The role of craftsmen, urban militias and rural populations is poorly attested in the few siege narratives, although there are some indications. From the general image of social structure and urbanism in Spain in the 6th century, we can conclude that civilian craftsmen continued to be of considerable importance.[80] In particular the role of the church in military organization is worth emphasizing. Bishops controlled vast landed wealth but were located in fortified cities, and were capable of organizing large-scale building projects.[81] They were also obliged to provide troops,[82] thus providing a kernel of urban defense. This is not attested in the law until 673, when Wamba imposed the same obligation of military service on clerics as secular nobles and free men. Although the law was rescinded by Ervig, it was probably not an end to ecclesiastical military followings, but rather a way of temporarily purchasing support. Nor could Wamba's law have been the beginning of this practice, since Visigothic bishops were very much like their secular counterparts. Both groups were concentrated in an urban environment, but had great landholdings in the surrounding countryside.[83]

The surrounding rural population, at least those personally dependent on the major landowners, could also be mobilized in certain circumstances. The Visigothic kings used these powers to organize building projects in conquered territories;[84] sometimes harsh measures were taken, e.g. by Leovigild when he retook cities in revolt or held by the East Romans; rural populations (*rustici*) were massacred in large numbers. In light of the laws above, they must in some instances have been the dependents of magnates who led the opposition to the king, although some may have revolted against the institution of an oppressive magnate regime.[85] Peasant reactions suggest that the king and his officers were (re-)establishing *munera* under royal control. Indeed, royal organizational capabilities are obvious

---

[80] For urbanism in 6th-century Spain, see Kulikowski 2004 chapter 12; for economic activity, Retamero 1999.

[81] *LJ* 5.1.5, regulating how bishops could extract financial support from their dioceses to conduct necessary repairs on church buildings.

[82] For a discussion of the church's military obligations, see Pérez Sánchez 1989: 138-45.

[83] See Wood 1999 on the evidence from the *VPE*; Retamero 1999 on the economic context of urban life; but note Wickham 2005 who argues for smaller estates in late antique Spain.

[84] See J.Bicl. 61, s.a. 581 (Wolf 1999: 69): "King Leovigild seized part of Vasconia and founded the city which is called Victoriacum." See further Squatriti 2002.

[85] See CO *Cordoba 572, which had a strong magnate presence, and was hence a likely example of the former; and *Cities in Orespada 577, which may have represented a general peasant revolt.

from the excavated fortified city of Reccopolis, founded by Leovigild in 578 and named in honor of his son, Reccared:

> With tyrants destroyed on all sides and the invaders of Spain overcome, King Leovigild had peace to reside with his own people. He founded a city in Celtiberia, which he named Recopolis after his son. He endowed it with splendid buildings, both within the walls and in the suburbs, and he established privileges for the people of the new city.[86]

Royally appointed officers were responsible for provisioning the army, and if they were derelict in their duties, were subject to heavy fines to be exacted from the properties of the responsible officer.[87] There was also a system to raise troops to man the border garrisons, although we do not know, with the exception of royal troops, how or by whom these garrisons were provided for regular duty. Either they were formed from men permanently stationed at a fort who could be moved around according to need, as in the Ostrogothic and East Roman models, or by magnates on a rotational basis as a public obligation, as under the later Carolingians. There are indications that local magnates and royal officers took care of needs wherever they were, but the considerable requirements of the border fortifications in the north were bolstered by royal troops. During threats of foreign invasion, nearby magnates and troops were obliged to come to assistance.[88]

---

[86] J.Bicl. 50, s.a. 578 (Wolf 1999: 67): "Leovegildus rex extinctis undique tyrannis et pervasoribus Hispaniae superatis sortitus requiem propria cum plebe resedit et civitatem in Celtiberia ex nomine filii condidit, quae Recopolis nuncupatur: quam miro opera in moenibus et suburbanis adornans privilegia populo novae urbis instituit." See further Olmo Enciso 2007 and chapter 6.1.1 below.

[87] *Liber Judiciorum* 9.2.6: "*Concerning those who Appropriate Army Rations, or are Guilty of Fraud in the Distribution of the Same:*" "We deem it advisable that, in every province and castle, some one shall be appointed as a collector of provisions, for the use of the army; and said collector, whether he be the governor of a city, or not, shall at once deliver all provisions collected by him in his district, to those who are entitled to receive the same. If it should happen, however, that the governor of the city, or the collector, should delay to deliver them, either because through his negligence, he has not taken possession of them, or because of his unwillingness to do so the officers of the army may lodge a complaint against him on account of his refusal to deliver said provisions to those charged with their distribution. The general of the army shall then give notice to the king, and the days which have elapsed since said provisions should have been delivered shall be computed. Said governor of the city, or collector of provisions, shall then be compelled to pay from his own property, four times the value of said provisions, for each day lost by his neglect. We hereby decree that a similar rule shall apply to all officers of the army who are charged with such duties."

[88] Most of this information can only be gathered from the *History of Wamba* (cf. the relevant entries in CO under 673); assistance to nearby frontiers is attested in *LJ* 9.2.8, which

3.2.3 Visigothic Siege Warfare

Despite the paucity of sources, we can reconstruct a coherent image of how the Visigoths waged war and besieged cities, especially when set in the context of their known enemies, such as the East Romans and the Franks.[89] While some cities were taken by treason, many involved major efforts of engineering and combat. Frankish-Visigothic border warfare was often concerned with the forts (*castra, castella*) and fortified cities of Septimania, Northern Spain and Aquitania. On several occasions we learn of storms, e.g. the Visigothic capture of the Frankish fortresses *Beaucaire and *Cabaret. The earliest well-attested siege took place at *Sevilla, which was occupied by Hermenigild in a revolt against his father, king Leovigild, in 579. Leovigild was slow to react, and actually organized an expedition against the Basques in the following year. Perhaps he used the time to negotiate a settlement, but eventually he prepared a large army that began a serious siege (*gravi obsidione*) in 583. Since the city was on a major river, the Guadalquivir, and close to East Roman territory, it appears to have taken some time to blockade it completely. This was achieved in 584, when Leovigild rebuilt the walls of the ancient city Italica nearby and blockaded the river. Through a combination of combat (*ferro*) and starvation (*fame*) he gradually subdued the city, which was taken by storm (*pugnando*).

We would of course like to know exactly how the fighting was conducted, but we have precious few details on how the Visigoths expelled the East Roman garrisons from Spain in the first quarter of the 7th century. However, we know that Roman garrison forces, even if outnumbered, were well equipped to defend against sieges. The Roman troops attested in Spain consisted of regular troops (*milites*) from the East who were well versed in siege warfare and were also led by Eastern officers who organized repairs of fortifications while on temporary assignment in Spain. Similarly isolated and exposed garrisons in Italy put up tenacious resistance in hard-fought sieges.[90] From the brief descriptions of how Sisebut (612-621) and Suinthila (621-31) conquered several unnamed *Roman cities and fortifications (c. 614 and 624, respectively), it is clear that it happened by force

---

requires all bishops, magnates and royal officers to come to assistance if within 100 miles of the frontier in case of enemy attack.

[89] See chapter 2 on the Byzantine and 3.3. on the Lombard situation; in light of the paucity of Visigothic source materials, we know few details, but there were others sieges, e.g. at *Saragossa 653, during another civil war in the mid-7th century.

[90] *PLRE* 3 s.v. Comentiolus; see further chapter 2 above and chapter 3.3 below for examples.

(*pugnando; proelio concerto*). Isidore of Seville's *Etymologies* provide some clues as to how the Visigoths fought. While his encyclopedia contained much information cobbled together from ancient authorities, his descriptions of siege engines conform remarkably well with contemporary practices. Furthermore, as opposed to preceding[91] and following sections[92] in the *Etymologies*, this excursus is not laced with quotations of ancient authors and has very few of his customary, bizarre etymological derivations (many are in fact correct, including Greek terminology). This supports the idea that he was drawing on contemporary information. He briefly describes the sling (*fundum*) for throwing stones; the *ballista* for throwing stones or spears; it also had the alternative form *fundibulum*, which may be a traction trebuchet.[93] To counter these threats, he describes the *testudo*, "an armored wall made by interlocked shields."[94] His description of the ram (*aries*) is only concerned with the iron-capped ram itself, but does not mention the protective framework.[95] Instead he goes on to describe straw-filled sacks used to lessen the impact of the rams,[96] protective screens of rawhide "set up as protection from the enemy when constructing siege-works,"[97] and finally the *musculus*, a small siege shed used to protect those who undermine walls.[98] Visigothic elites were thus familiar with virtually the whole arsenal of contemporary siege warfare (cf. chapter 5.2 below).

By the time the Visigoths occupied the south in the 620s, Roman Syria, Egypt and much of Anatolia was lost to the Persians, the Balkans to the Slavs and Avars, Constantinople was threatened by a joint Avar-Persian siege, while the nearest Roman provinces, Italy and Africa, were threatened and partially occupied by Lombards and Moors respectively. There would thus have been little hope of reinforcements or even pay from the government in Constantinople. Some of the Romans may have fled, but others, with no really attractive options, went into Visigothic service either as prisoners or deserters. This would only reinforce already existing similarities

---

[91] See e.g. *Etymologiae* 18.7. on spears.
[92] See e.g. ibid. 18.12. on shields.
[93] Ibid. 18.10.1.The *ballista* and *fundibulum* are also attested in Visigothic law; the latter in the less technically correct form *fundum*, as we saw above. See further chapters 8.2.1 and 8.2.3 for a discussion of this terminology.
[94] *Etymologiae* 18.10.2.
[95] Ibid. 18.11.1.
[96] Ibid. 18.11.2.
[97] Ibid. 18.11.3.
[98] Ibid. 18.11.4.

in military practice, and added another skill unknown to the Visigoths since the kingdom of Toulouse, a navy. Isidore wrote in the early 620s that:

> Until recently they lacked experience in only one aspect of fighting: they had no desire to wage naval battles. But ever since King Sisebut took up the royal sceptre, they have made such great and successful progress that they now go forth with their forces on sea as well as on land. Subjected, the Roman soldier now serves the Goths, whom he sees being served by many peoples and by Spain itself.[99]

The *Historia Wambae Regis*, a late-seventh century description of king Wamba's suppression of a revolt in Septimania (the Visigothic province of southern Gaul), is one of the most detailed military accounts in the West and gives precious details of how the Visigoths combined their technological and organizational capabilities.[100] Wamba, leading a punitive expedition against the Basques following his coronation in 672, received news of the revolt of his general Paul, who had originally been dispatched to suppress a minor revolt in Narbonne, but had taken over Septimania and extended the revolt to Tarraconensis. In 673, Wamba diverted his army from the western Pyrenees to the environs of *Barcelona, suppressing the representatives of Paul, before crossing the eastern Pyrenees in three columns, storming the *castra* of *Collioure, *Ultrère, *Llivia and *Clausurae on the way. We know from archaeological excavations that these fortifications were large, complex stone structures, some dating back to the late Roman period, others possibly modeled on Byzantine forts, yet it only took the Visigothic armies a couple of days to overcome these formidable obstacles.[101] Upon reaching the great walled cities of *Narbonne and *Nîmes, held by rebel Goths, Franks, and Gallo-Romans from Aquitaine, the Visigoths stormed the walls within a few days with barrages of arrows and javelins. Furthermore, although the terms are not mentioned in the text, it is clear from the description that they used heavy artillery to good effect.[102] Finally, the army was followed by a fleet, which supported at least one of the sieges and led to the surrender of *Maguelone.

The forces involved appear to have been large. Wamba sent a main division of 10,000 men to support the vanguard besieging Nîmes, who had

---

[99] Isidore, *HG* 69.
[100] See in general the introduction to the English translation by Martínez Pizarro 2005.
[101] Martínez Pizarro 2005: 49f with references.
[102] In addition to the relevant discussions in CO, note especially the important argument by Thompson 1969, which is supported by King 1972 and Bullough 1970.

encircled the city but were unable to storm it.[103] A rough guess would put Wamba's total force at about 20,000 men (a vanguard of 2-3,000 men, a minimum for blockading Nîmes, a similar rearguard under the king, troops to reoccupy border forts, 10,000 in the main force plus naval forces). In addition, there were the forces of the general Paul, probably substantially fewer but still at least 5,000 Visigoths and perhaps a similar number of Franks and Gallo-Romans to hold the fortified cities as well as the border fortifications. Also, the author of the *Historia* explicitly states that Wamba deliberated against a well-equipped expedition with all available forces, only taking those that were already on campaign and those that could join on the way. If he had chosen a different strategy of a slow, deliberately planned campaign, which was also an option, he would have been able to raise an even larger army, but that might allow the rebels time to consolidate their position. The very rapid response by Wamba appears to have frightened off potential Frankish and Gallo-Roman allies from whom the rebel Paul was expecting support and caught his Visigothic supporters off guard. The scale of the expedition, marshaled from only a part of the available potential, indicates that the Visigoths had a very large military establishment. The resources the king could raise seem to have topped at least 50,000 men, capable of defending borders, engaging in long-range expeditions, and taking part in complex siege and naval warfare as other states did throughout the Mediterranean basin.

3.3 *The Lombards*

The Lombards, despite their late arrival as major actors in the 550s, very soon became acculturated to Roman styles of war, as one would expect from a group largely composed of former Roman client peoples. While they remained politically more independent than most client groups, they display an astonishing degree of similarity to East Roman warfare and military organization throughout their existence.

3.3.1 *Ethnogenesis on the Middle Danube*

The Lombards are a people with an ancient name but unclear history before they entered the old borders of the Roman world in 508, when they defeated the small Herul kingdom on the middle Danube. It seems in fact

---

[103] Accepted by King 1972; Thompson 1969 is slightly more skeptical.

that in large measure, these "Lombards" were composed of the Germanic peoples, predominantly Sueves, who were clients of the Ostrogoths. The Heruls, in contrast, already displaced by the Ostrogoths as a mainstay of the fading West Roman military under Odovacar, were driven from their homelands and sought settlement on East Roman territory in return for military service around 512.[104] By 526, the Lombards began expanding well south of the Danube (and apparently absorbing many Sueves), where they came to the attention of the Ostrogothic kingdom and East Roman Empire. The Ostrogoths recruited large numbers of Sueves for their expedition against *Salona in 537, but as the war went in Rome's favor, these clients sought new sources of patronage. The Romans needed troops, especially after the plague of 541-42 had devastated Roman recruiting grounds in the Balkans and the Romans were embroiled in a two-front war against the Goths and Persians, respectively. In 547, Justinian granted the Lombards the use of cities and fortifications of the formerly Roman province of Pannonia, where a substantial Roman population lived in a diminished but still recognizably Roman urban and socioeconomic environment.[105] Due to their settlement and integration into a still-Romanized region, the Lombards became one of the last products of the Balkans/Pannonian military culture, before this collapsed with the arrival of the Avars.[106] As partly independent clients, however, they competed fiercely with their Herul, Gepid, and lesser extent Hunnic and Slav neighbors for subsidies and alliances with Rome.

The settlement in Pannonia stipulated that Lombards serve in Roman armies, and the first 1,000 Lombards were recruited by Germanus in 549 for his abortive Italian campaign, which was interrupted by inter-client squabbling, especially between Gepids and Lombards. This took some time to suppress, but finally, after years of drilling and operations in the Balkans with Heruls and Gepids, who served under similar conditions, and the regular Roman field army, a total number of 5,500 Lombards (2,500 regular troops with 3,000 armed retainers, perhaps from subject populations)

---

[104] Goffart 2006: 205-10. While there was a semi-independent Herul state, many Heruls found employment as regular soldiers (*stratiotai*) in the East Roman armies. In *addition*, the Romans could recruit Herul units under their own leaders directly from their homelands; these were normally dismissed at the end of a campaign, but many clearly found a career in the Roman army as regular soldiers and officers.

[105] See Christie 1995: 48-55 and 2000 on the archaeological evidence for a sub-Roman society; Pohl 1997 on the treaty with the Romans.

[106] On Balkans military culture, see chapter 1.3.2, 2.1.2 and 3.1.1 above.

came with Narses to Italy in 552.¹⁰⁷ Procopius claims that they were undisciplined off the battlefield, and had to be shipped off from the field army in Italy for unauthorized rape and arson.¹⁰⁸ Apart from their abuses, however, there was no problem with Lombard martial qualities. They fought bravely as dismounted cavalry against Totila at the battle of Busta Gallorum alongside Heruls and other barbarian mercenaries.¹⁰⁹ At the Persian siege of *Phasis (556), Lombards (again together with Heruls) held a long section of the walls flanked by Roman regulars in an extremely hard-fought siege. The integration went so far that in appearance and tactics, Lombard soldiers became virtually indistinguishable from regular East Roman troops in the decades around 600. Physical representations of Lombard soldiers accord well with the *Strategikon*'s description of Roman heavy cavalry, and it is generally recognized that the Lombards were equipped and fought as East Romans.¹¹⁰

If they were not in Roman service, some Lombards went on occasional raids into Roman territory, presumably under independent military leaders who were too impatient to wait for a suitable mercenary commission, or to pressure Roman authorities to provide subsidies and employment.¹¹¹ Otherwise the Lombards continued to feud with their neighbors east of Pannonia, the Gepids, who held the Tisza valley down to the Danube and the great Roman fortress city of Sirmium. The Gepids were defeated by the Lombards in alliance with the newly arrived Avars in 567 and partly absorbed by either group, but the formidable Avars soon terrified the Lombards into leaving Pannonia.¹¹²

---

¹⁰⁷ See Pohl 1997 for a general survey of early Lombard-Roman relations; Jarnut 2003 argues that Roman influences was fundamental to early Lombard development in the period 488-550s.

¹⁰⁸ Pohl 1997. However, this may cover up a political inconvenience, since the Franks threatened war over the presence of Lombards, whom they intensely distrusted and perhaps hoped to bring within their sphere of influence.

¹⁰⁹ Proc. 8.31.5ff.

¹¹⁰ See e.g. the catalogue in Menis 1990: 96 for the decorated crest of the helmet of Valdinievole (with discussion), 114 for detail of king Agilulf (r. 590-616) flanked by two guards equipped in East Roman style. Compare this with the cavalryman on the decorated plate from the treasure recovered at Isola Rizza, p. 229f with discussion at 231f, which remarks: "Questo tipo di armatura bizantina fu adottato dai Longobardi." Since the latter image probably predates the Lombard invasion (or represents a Roman soldier nonetheless), it can profitably be compared to Agilulf's troops. The similarity is striking. For a contemporary description of the late-6th-century Roman cavalryman, see *Strategikon* 1.2, including plumed helmets and small circular shields.

¹¹¹ Christie 1995, Pohl 1997.

¹¹² On the Avars, see chapter 7.2.3.

In the decades following their settlement in 547, the Lombards had gradually fused with the locals. By the time they migrated into Italy in 568-9, they were a composite people. Paul the Deacon lists Gepid, Bulgarian, Sarmatian, Pannonian, Suabian (Sueve) and Norican villages in Italy originating with the migration and still distinct generations later.[113] Notable in this list are the Pannonians and Noricans, who were former Roman provincials that had once provided extra manpower to the "barbaric" invasion of 406 as well as to the Huns and Ostrogoths.[114] The Sarmatians were of old a rich source of Roman military colonists settled in Italy and elsewhere since the 4th century,[115] but Paul might in fact be referring to 6th-century Hunnic groups (Utigurs and Kotrigurs), who were active in the Roman army alongside Bulgar contingents.[116] In addition, there was a large Saxon contingent that took leave of the Lombards and went on to Gaul in the 570s.[117] To a large extent, then, the "Lombards" were composed of, fused with or depended on groups that had an important role in Late Roman client system and successor military organization, from well before the Alan-Vandal invasion in 406 to the Avar wars around 600.

3.3.2 *The Lombards in Italy*

Since the evidence of violent conquest when the Lombards first entered Italy in 568 is limited, it has been suggested that they were installed in northern Italy by the invitation of the Byzantine government in order to settle land ravaged by plague and famine, and provide garrisons against the Franks.[118] However, the Romans knew well the risk of settling tens of thousands of militarized people in concentrated blocks. It is more likely that, just after the very last Ostrogothic resistance had been defeated in 562-3, a manageable number, perhaps a few thousand Lombards, were invited to provide garrisons to a certain number of cities and fortifications and bring along erstwhile Roman provincials to help repopulate an Italy

---

[113] *HL* 2.26. We do not know if Paul means in *his* own day or is simply lifting the phrase from an older source.

[114] See chapter 1.3.2 and Goffart 2006.

[115] Jones 1964: 85, 619f. Some of the names of Lombard dukes indeed sound rather Turkic (=Bulgarians, Huns) or Persian (=Sarmatians), such as Zaban and Rodan.

[116] The Bulgar component was bolstered by the addition of Alzeco's Bulgarians who joined the Lombards in the 660s. See chapter 3.3.3 below and further chapters 2.1.3 and 7.2.4.

[117] *GT* 4.42.

[118] Paul the Deacon provides a fanciful anecdote on how Narses invited the Lombards in order to avenge a slight against him by the Roman government, but there might be a grain of truth to this. See e.g. Christie 1995: 73 for the less legendary interpretation of events.

devastated by war and plague. This was common enough in Thrace, Illyricum, and the East.[119] Whatever the plan, the Roman authorities quickly lost control; within a year or so, the Lombards had taken control over cities and forts in Friuli and the Po valley from Venetia to Liguria, in fighting that sent local populations and official representatives fleeing.[120] While they were besieging *Pavia closely for three years (probably from 569), the immigration began in earnest: well over 100,000 people flooded from Pannonia onto the northern plains of Italy, and Lombard forces quickly began to expand southwards.[121] Some Lombards also tried their luck against the Franks, but were heavily defeated by the well-defended cities of the Burgundian kingdom, which could also boast of a formidable commander in the Roman patrician Eunius Mummolus.[122] In the following decades, the Lombards overran much of Italy, and by 600, they controlled most of the north and Tuscany along with the large independent Lombard duchies of Spoleto and Benevento in the south.

The Romans held on to various coastal enclaves and large areas around Rome and Ravenna, linked by a fortified military road across the Appennines. Due to pressures in the Balkans and the East, the Romans were unable to send substantial armies to Italy. Lacking the ability to form field armies large enough to challenge the Lombards head on, the Romans constructed a complex defensive system based on fortified cities, ports, *castra*, watchtowers and military roads that held together each of the separate, ever-diminishing provinces. These systems have received extensive attention in archaeological and documentary studies, and are hence the best known throughout the Mediterranean region in this period. Some sections were inherited from Roman defenses, some were Ostrogothic additions, while others were Lombard developments.[123] Wherever the Lombards con-

---

[119] E.g. Huns (from Attila's empire), Utigur/Kotrigurs, Goths, Heruls in the Balkans; Tzani, Armenians, Persians in the East. See chapters 7.2.1-2 and 2.1.2.

[120] See CO 569: *Vincenza, *Verona for Venetia; *Milan and Ligurian cities.

[121] Christie 1995 claims the migration might have counted as many as 150-160,000 people. If the narrative of Paul the Deacon is taken literally, the real immigration only began well into the siege of *Pavia (569-71) and the Roman soldiers were expelled from the Po valley; only *then* (tunc) Alboin brought the above-mentioned peoples to Italy (*ad Italiam aduxisse*).

[122] The Lombards were heavily defeated by Mummolus' *Improvised fortifications at Embrun in 571, but made another, initially more successful invasion in 574, when they besieged *Arles, capturing many surrounding cities; they were bought off from *Aix, and defeated after commencing sieges at *Valence and *Grenoble. See also chapter 4.1.2.

[123] For the condition of Byzantine Italy, see Brown 1984; for fortifications in the "Byzantine corridor," see Zanini 1998, Menestò 1999, chapter 4 of Christie 2006, and the discussion below.

quered territory, they obviously took over Roman infrastructure. Substantial Roman armies were only launched in 590 (in cooperation with the Franks), 663 (by Constans II), and 788 (with local Lombard support, but betrayed to Charlemagne). While the first two had some success in recapturing cities and territory, the lack of sustained offensives and reinforcements meant the gains were soon lost again, and warfare between Lombards and Romans became a slow process of reducing fortifications one by one. This type of long-term coexistence led to strong mutual influences, so that Roman and Lombard military organization and style of warfare in effect mirrored each other.

As an alternative to direct confrontations, the Romans bribed or hired as many Lombards as they could to relieve pressure on the Italian communities while strengthening their own army on other fronts.[124] This arrangement worked fairly well at times. Several Lombard units served with distinction in the East Roman army; this was so common that the Syrian population threatened the Persians that 60,000 fearsome Lombards would soon be arriving.[125] This was far from an empty threat. The Lombards were renowned for their martial qualities, and for a time dominated the Balkans field armies. Guduin and Droctulf (in Greek: Drokton) served as *hypostrategoi* (subordinate generals) in the Balkans.[126] Droctulf fought well on behalf of the Romans against the Lombard king Authari (584-90) at the siege of *Brescello (584), where he was forced to withdraw, and *Classis (584), which he helped recapture for the Romans from the duke of Spoleto.[127] He achieved further fame for his exploits when raising the Avar siege of *Adrianople (587). Another Lombard commander, called Maurisio, defended *Perugia (593) for the Romans, and was killed when the city was stormed

---

[124] See e.g. Men. fr. 22: Envoys from Rome beg for assistance from Constantinople against the Lombards, but the Emperor found it impossible because of the Persian war; instead he sent money which they used to buy support from "some of the rulers of the Lombard people ... with their followers" (τινας τῶν ἡγεμόνων τοῦ Λογγιβάρδων ἔθνους ... ξὺν τῇ κατ' αὐτοὺς δυνάμει). Then those willing could be brought to fight in the east; otherwise, they could spend the money on the Franks (dated by Blockley to 577 or 578). The policy was followed up; Men. fr. 24 shows the Romans unable to send a relieving army to Italy, but sent a small force as well as gifts and promises of rewards to win over Lombards. "Very many of the chiefs (πλεῖστοι τῶν δυνατῶν) did accept the Emperor's generosity and came over to the Romans." This occurred early 579, before hostility with Avars erupted; cf. *Sirmium 579.
[125] See chapter 2.1.2 above.
[126] A *hypostrategos* had special responsibility for organizing the artillery under the overall command of Herakleios the Elder when he was stationed in the East in the 580s (see *Persian fort 587); it is however difficult to prove that this was one of the normal functions of such officers. See chapter 2.2.3 above.
[127] On Droctulf's career, see *PLRE* 3 s.v.

by Agilulf (590-616). These men probably commanded military followings which ensured them high position (and pay) when in Roman service, and independence if circumstances warranted it. For instance, Grasulf, the duke of Friuli, was (unsuccessfully) wooed by Roman and Frankish diplomacy to participate in the joint grand campaign of 590.[128] The duke Nordulf mobilized his own independent following (*homines ... suos*) when he entered into imperial service in time for the same campaign, during which he participated in the storm and capture of several cities under Lombard control.[129] While he went to Constantinople to advise the emperor Maurice on Italian affairs, his troops mutinied for lack of pay and went over to king Agilulf, who even demanded arrears on their behalf from the Emperor's representative, Pope Gregory the Great.[130]

Another alternative was to call in the (Austrasian) Franks, whom the Romans tried to entice through bribes and diplomatic pressure.[131] The Lombards were severely harrowed by Frankish armies in the last decades of the 6th century and forced to concede substantial territory in the southern Alps in the 580s. The closest call came during the joint Franco-Roman campaign in 590, when the Franks forced over a dozen fortifications to submit before demolishing them and forcing the inhabitants into slavery. The Lombards maintained their precarious independence by withdrawing into walled cities and fortifications,[132] whence they ambushed Frankish armies on the move or waited for dysentery and supply problems to force them out. For instance, the Frankish capture of the fort *Anagnia (575) by surrender prompted the nearby Lombard count to ravage the newly acquired Frankish territory, but the count was himself killed in a Frankish counterattack. When the Franks ventured beyond Anagnia to ravage the district of Trent, the duke of the city, Eoin, defeated them and averted further threat.

---

[128] See Pohl 1997 for the context; since he accepts Bachrach's late date, 588-91, for the Frankish diplomatic letter on this affair (often dated to c. 580), it must have been during the preparations for the 590 campaign.

[129] See *Roman invasion of Italy 590 for the context.

[130] See *PLRE* 3, s.v. Nordulf, and Pohl 1997 for a discussion of these events.

[131] For basic treatments of the diplomatic entanglements, see Goffart 1957 and Goubert 1965. Bachrach 1994 provides an interesting case study of one particularly involved affair that has some bearing on the Lombards.

[132] In addition to the examples in CO under 590, see the explicit statement in *HL* 3.17, where, on a previous occasion, the Frankish king Childebert had accepted money from Maurice but made peace with the Langobards. The Lombard strategy was simply formulated: "The Langobards indeed entrenched themselves in their towns" before negotiating for peace.

The Lombards benefited from the Merovingian division of realms, which caused endemic competition between the kings of the *Teilreiche*: soon enough, internal squabbles would force Frankish attention northwards, unless dysentery, malaria and supply problems did the job. Under Dagobert I, however, the united Frankish realm appears to have enjoyed formal hegemony over the Lombards, but we know little of how this was achieved or exercised. We can infer that lack of internal political diversions, such as civil wars and outright sabotage, allowed Dagobert to carefully plan campaigns over several seasons, as he was able to do in Visigothic Spain and against the Slavs.[133] This lesson was well learned by the Carolingians, and king Pippin refused to become permanently involved in Italy while he still had business with the Aquitanians. Thus after coming to the Pope's aid when *Rome was besieged by the Lombards in 754 (not in CO) and 756, he immediately withdrew most of his forces. It was his son Charlemagne who, within four years of the conquest of Aquitaine and a year and a half after the death of his brother and co-ruler Carloman, destroyed the Lombard kingdom in 773-74 through a massive military operation.

3.3.3 Lombard Military Organization

As a result of their origins in sub-Roman Pannonia and later symbiosis with the imperial provinces of Italy, Lombard military organization was very closely modeled on that of the East Roman Empire. In the migration days, the Lombards were organized into *farae*. Once believed to be clan groups, these were probably military followings patterned on the large *obsequia* attested elsewhere in the late Roman world.[134] They were led by individual *duces*, who were in effect independent military leaders. Their specific objective was to gain income through mercenary service or conquest and settlement. While booty was useful and attractive, it was no alternative to regular pay and sustenance—as we saw above, Lombard troops in Roman service (Nordulf's men) reacted much the same way as Isaurian, Thracian and Illyrian regulars in Italy did when their pay was in arrears. To provide for their men, Lombard leaders established themselves individually in around 35 cities, where they exacted taxes and/or tribute from the Roman

---

[133] See chapter 4.1.2.
[134] See chapters 1.2.3-4. Military followings were often called *familia, solacium, therapeia* and the like—the latter was used by Procopius for Lombard retainers in the 550s—hence the association with clan or family groupings. At *HL* 2.9, Paul calls *farae* "families or stocks of the Longobards"—"faras, hoc est generationes vel lineas".

population. The settlement as garrison forces in major cities under regional dukes and the emphasis on securing regular income reflected East Roman military organization.[135] At first, they did so under the auspices of the kings Alboin (560-72) and Cleph (572-74). Subsequently Lombard kingship lapsed for a decade, as the dukes pursued their own ambitions in Italy, Gaul, or in Roman service. It was probably in response to formidable Frankish and Roman threats in the 580s that the Lombards decided to elect Authari king (584-90) and endow him with the wherewithal to lead the Lombard army by donating half their estates to create a royal domain.[136]

In the following century and a half, to the final capture of Ravenna in 751, Lombard and East Roman military activity became in effect a perpetual continuation of the Gothic wars, where fighting consisted of raid, counter-raid, blockades and sieges, punctuated by long periods of relative peace. Lombard and Roman organization continued to mirror each other closely. The original arrangement with powerful dukes, having their own independent power bases and followings, lasted well into the 7th century, when royal control extended to the detriment of ducal power as royal officers (*gastalds*) took over judicial and military functions. Royal power seems to have been particularly boosted by increased prestige (and the settlement of a large Bulgar contingent) after the failure of Constans II's *Italian campaign in 663 or his death in 669.

In the course of their settlement and in the following generations, it seems that many Lombard soldiers acquired land and settled down, but retained their military obligations. With the decline of ducal power, these were now claimed by the king, and provided a pool of semi-trained, well equipped manpower into the eighth century.[137] By then, these obligations were extended to the whole free population with economic means to equip themselves, probably in response to formidable Frankish threats.[138] This is

---

[135] Brown 1984 and Christie 1995; see further *HL* 2.32 on the brutal treatment of Roman population, exaction of regular taxes/tribute, and the fragmentation of the Lombard kingdom into enterprising dukedoms—"unusquisque enim ducum suam civitatem obtinebat." For Goffart 1980, this phrase is a linchpin in his argument that the Lombards were the last "barbarian" group to use Roman techniques of administration in their new conquests.

[136] *HL* 3.16; see Christie 1995 and Pohl 1997 for the political background.

[137] For a basic discussion, see Halsall 2003: 81-84. However, he treats the Lombard military in isolation from the nearby East Roman developments. This very closely resembled the East Roman garrison forces in Italy in the 7th and 8th centuries, as the now garrisoned field armies evolved into an elite of modest-to-large landowners who still formed traditional military units of considerable military effectiveness, while semi-professional militias were organized in the larger cities, especially Rome. For this, see Brown 1984.

[138] The fundamental study is Bertolini 1968, who assembles all the legal and charter evidence to uncover the intricacies of social structures. However, this evidence in particu-

reflected in later Lombard legislation: a free man is called an *arimannus* (i.e. army-man) or in Latin, *exercitalis*, "one of the army (*exercitus*)." While this arrangement may seem "tribal" or "Germanic," the Latin form (*exercitalis*) is attested well before the Germanic *arimannus*.[139] In fact, this system seems to have arisen during the 7th century, modeled on the organization of the Byzantine army in Italy, which to a large extent consisted of local landowners who had also become the social and administrative elites in their provinces but retained their military expertise (cf. 2.1.3).

These free *arimanni* were useful for local defense and bolstering expeditions, as elsewhere in the early medieval West. This was especially useful for the defense of cities and fortifications, and Lombard troops probably ranked somewhere between Byzantine urban militias (e.g. at Rome) and professional units. The Lombards still had considerable magnate followings, whose importance in the 7th and 8th century has been overlooked.[140] Even if most of the dukes came under royal control, they still retained military forces in the form of the *gasindus*. Royally appointed judges, *gastaldi*, and great landowners without official office also had their own *gasindus*. Again, these forces are only obliquely known through legislation, but the stipulations are remarkably similar to the legislation of Euric from the Visigothic kingdom of Toulouse.[141] It is impossible to determine the ratio of *arimanni* (individual troops) to *gasindii* (those in a *gasindus* or *obsequium*), but individual landowning *arimanni* must have been relatively more important than their Visigothic and Frankish equivalents. The simple reason is that large-scale estates were fairly uncommon in Italy under the Lombard kings, as opposed to the estate-based military organization common in Gaul and Spain; hence the socio-economic basis for magnate followings was lacking in Italy.[142]

---

lar is by nature slanted towards free, propertied individuals attested in surviving 8th-century charters, whereas documentary evidence is lacking for the earlier periods. The basic grades of equipment, depending on estate size for landowners (specified according to acreage) and wealth for merchants (only relative indication—rich, middle or low) are found in the late laws of Arichis.

[139] Delogu 1995 believes in the ancient and intrinsically Germanic, democratic character of the institution of the *arimannus/exercitalis*, but recognizes that they would have limited military value due to their economic concerns and lack of training. Halsall 2003 points out the novelty of the institution in light of the contemporary political situation.

[140] Halsall 2003: 64f argues that Lombard *obsequia* were more important in the 7th century than the apparent "horizontal" recruitment (i.e. of free men by the king) the 8th-century material implies, but then reverts to the standard pejorative "bands" for describing them. See also Christie 1995: 114 on their prevalence as late as king Ratchis (744-49).

[141] Both laws stipulate that arms were to be kept by the free retainer, but property returned, if the retainer left the service of his patron.

[142] Wickham 2005, 2009: 145; see further chapter 3.2.2 above and chapter 4 below *passim*.

Most impressively, the Lombards were able to reduce Roman fortifications, sometimes head on through storms, sometimes through lengthy blockades, both of which required significant engineering and logistical capacities. As most East Roman troops, we can infer that Lombard soldiers possessed basic engineering skills from the very start, especially having trained and fought with the Romans in several siege campaigns in the econd half of the 6th century (see 3.1.1). As commanders of cities with substantial numbers of suburban fortifications, dukes probably had engineering experts in their households or stationed in their cities, as was the common East Roman practice. Droctulf, Nordulf and Maurisio, the commander of *Perugia (593) most likely did, since they defended walled cities against assault and themselves attacked and demolished Lombard fortifications.

We can also identify civilian craftsmanship skills available to Lombard rulers that we know had military application from elsewhere. The evidence we have indicate that the Lombards continued Ostrogothic and East Roman practices. Civilian craftsmen were obviously available in the surviving cities and fortified settlements, where architectural remains and luxury objects demonstrate small-scale, but quite refined skills.[143] Lombard legislation shows great concern for the *magistri comacini*, master craftsmen recruited from the Como region.[144] This area had been an important link in the East Roman defenses of Italy, and the Roman fort at Lake Como held out for nearly two decades after the Lombard invasion. There is therefore reason to believe that craftsmen in the Como area had skills with particular military application.[145] We can also infer the importance and capabilities of Lombard craftsmen from their frequent assistance to the Avars; as will be argued in chapter 7.2, Lombard (or subject Italian) skills were fundamental to the Avars' ability to assault and capture Roman cities in the Balkans.

The activities of craftsmen are difficult to pinpoint within Lombard Italy, but the framework set out in the preceding chapters provides some good indications where to look. Lombard ability to construct fortifications is well attested in the archaeological evidence. These were substantial con-

---

[143] See e.g. Menis 1990 for examples of Lombard architecture; Christie 2006: 129, 144 for craft traditions in individual Lombard cities; J. Mitchell 2000 for Lombard patronage of craftsmen.

[144] Rothari 144f, though we only hear of them in the context of compensation for damages during construction.

[145] See chapter 4.1.5 on how Italian craftsmen in the early 6th century were sent to Austrasia.

structions in dense concentrations across Italy, and come in addition to fortified urban centers throughout Italy.[146] Although urban civilization became poorer and more primitive, the fortifications were obviously constructed and maintained by someone. In light of the close contacts and similarities between Byzantine and Lombard military organization, we can assume that the Lombards used similar methods to raise specialist craftsmen and provide logistical support for their armies. Early evidence comes from a letter by Gregory the Great (590-604) preserved in Paul the Deacon's *Historia Langobardorum*. Gregory elsewhere provides ample evidence for how he went about organizing defenses for Rome against the Lombards c. 593 (not in CO); the pope's letter to the Lombards shows how they controlled and mobilized labor much the same way as he did. The Lombards cooperated with the Pope on logistical projects during a time of truce: Gregory the Great asked the duke Arichis of Spoleto to assist his representatives to cut timber in Calabria: "... we ask, saluting your Highness with paternal love, that you should charge your managers who are in that place to send the men who are under them with their oxen to his assistance ..."[147] The *Liber Pontificalis* documents how the 8th- and 9th-century popes mobilized local rural communities, ecclesiastical and secular estates, and urban or regional corporate bodies to maintain particular lengths of wall (in effect, *pedaturae*) at Rome, down to burning lime, in accordance with typical late imperial organization.[148]

---

[146] For Lombard defenses, see Zanini 1998, Christie 2006: 383-99 and Brogiolo 2000.
[147] *HL* 4.19.
[148] Delogu 1988 provides a detailed survey of the popes' building activities based on the *Liber Pontificalis*, which shows that restoration and new construction began in earnest with Sergius I (687-701) and continued steadily until it exploded under Hadrian I (772-95) and Leo III (795-816), due to fiscal stability and Carolingian patronage. It should be noted that the popes used recognizably late antique methods of organizing such work (paying and supplying laborers from the Pope's own resources, organizing laborers from the city and surrounding territories), and that substantial building activity began even earlier, during the pontificate of Honorius (625-38), who built many suburban churches and other structures (*LP* 72.3-6). Furthermore, Delogu ignores substantial military (walls) and civilian (aqueducts) infrastructure; for walls see *LP* 89.1 (burning of lime for the walls, under Sisinnius, 708), 91.2 (burning of lime to restore walls of Rome under Gregory II, 715-31), 92.15f (the same, with details on pay and restoration of walls of Centumcellae under Gregory III, 731-41), 97.52, 92 (substantial rebuilding of walls at Rome under Hadrian I); Davis 2007: 143 n. 90 also has details on a survey over walls, towers, battlements conducted at the time; the various groups involved were recruited for restoring particular measured stretches of wall *per pedicas*, or by the foot; this is the same system as the *pedaturae* noted in late Roman (1.2.5; Roman authorities also required burning of lime and other related services), East Roman (2.4.2) and Ostrogothic (3.1.4) warfare above.

A similar system would have been in operation in the Lombard territories, as the Lombards imposed strict public duties, similar to *munera*, which are only obliquely visible in the laws. King Liutprand ruled in 720 that merchants (*negotiatores*) and master craftsmen (*magistri*) were not supposed to be abroad for more than three years without valid excuse, and anyway had to inform their judges of their whereabouts. If they failed to do so, "Whatever sales *or other obligations* [*oblegationes*] the children undertake from their father's property after the expiration of the time period set, shall remain permanent and they shall pay their own or the debts of their father." The king ensured absolute control by confiscating property if the children returned it to the father or his wife remarried without royal permission.[149] Wall-building was a universal obligation in late antiquity from Mesopotamia to Mercia,[150] but especially well attested in contemporary Italy, where the complex conflict lines required a similarly complex defensive system. Hence it was probably one of the major burdens envisaged here. Finally there was always the option of calling out dependent labor from estates and workshops through ducal officers, as duke Arichis obviously could do, or simply paying skilled craftsmen.[151] The military potential of Lombard craftsmen was so great that it was desired abroad; the Avars, fabled for their conquest of cities, were provided with skilled shipwrights by the Lombards to construct riverboats, which were used during the Avar sieges of *Singidunum (588) and possibly *Sirmium (579ff).

3.3.4 *Lombard Siege Warfare*

The Lombards had neither a Procopius nor a Julian of Toledo to provide the same kind of details as we have for the Ostrogoths and the Visigoths. Hence we know little of exactly how siege warfare was conducted, so we must collate the available information from a variety of sources. Capturing cities by siege was a significant feature of Lombard warfare, from the Roman surrender of *Pavia (569ff) right up to the Lombard capture of *Ravenna (731) and siege of *Rome (756). Indeed, there is little doubt that siege warfare was the predominant form of war throughout Lombard history, as few battles are recorded and only slightly more raiding. Both raids and battles tended to occur in conjunction with siege campaigns. Furthermore, Paul's information is scarce, and in some cases, what he calls "ravaging"

---

[149] Liutprand 18.
[150] For Mercia, cf. Offa's Dyke, built in sections by designated teams (Squatriti 2002).
[151] The *magistri comacini* were certainly hired labor, and explicitly envisaged to work alongside or lead the (skilled) slaves of the commissioning lord. See Rothari 145.

may in fact have been one of the many possible stages of a siege, or the actual blockade itself. For example, the fighting that affected several Lombard cities in 670, including *Nimis, *Cividale and *Forlimpopoli, is impossible to define precisely.[152] The Lombards "carried off" (*abstulerunt*) *Narnia and other cities (756) while using siege engines at *Rome. In contrast, the Franks "conquered" (*conquisivit*) *Ravenna; the stylistic variation clearly carries some value judgment, and is not meant as a precise technical vocabulary.

Paul uses the terms *invadere* or *pervadere* quite often; at *Cumae (717) this involved the capture of the actual fortification, which was retaken by a Roman surprise assault at night. *Classis (723) was returned (*reddita*) to the Romans by the Lombard king Liutprand after one of his dukes had attacked (*invasit*) and thus taken the city. At *Narnia, the *civitas* was "conquered" (*pervasa est*). We know nothing else, as the *civitas* could refer to the surrounding territory as well as the fortified city, but during a *Lombard campaign in 731, these words are used in a string of sieges, apparently with little variation in meaning. King Liutprand "besieged" (*obsedit*) *Ravenna, "attacked and destroyed" (*invasit atque destruxit*) *Classis, "attacked" (*invasit*) *Fregnano and a string of other fortified cities in Emilia (*castra Emiliae*), including Bologna, before he "took" *Sutri "in the same manner" (*pari modo ... pervasit*), but this was again returned to the Romans. This string unambiguously refers to the fortifications, *castra*, which were the object of the fighting. Thus, Paul's terminology is fairly clear in most instances, and even more so in the following: a Lombard army "besieged" (*obsedit*) *Perugia (593); another "besieged ... took ... and razed to the ground" (*obsedit ... cepit ... et ad solum usque destruxit*) *Montselice (602); they likewise "took" (*cepit*) *Ligurian cities in 643. Such "taking" was clearly violent: at *Oderzo (643) the Lombard army "assaulted and destroyed" (*expugnavit et diruit*). Similarly *Lodi (701) was "attacked [and] captured" (*expugnata ... capta*); in the same year the Lombard king Aripert "assaulted and soon took" (*expugnans mox cepit*) *Bergamo.

Paul does however provide some episodic notices on how cities were stormed; we thus have evidence of the use of rams at *Cremona and *Mantua in 603 as well as *Bergamo (701) and several cities during a *Beneventan campaign in 702. While artillery may be inferred at *Padua in 601, where

---

[152] Cf. chapter 5.1; for example, during blockades, there were often small-scale skirmishes, sallies or attempted storms that immediately raised the siege into another category, but whenever these produced no result, a brief chronicle would take no notice and simply refer to the siege, ravaging, and/or surrender phases.

fire was thrown (*iniecto igni*) into the city, artillery terminology is only attested later, when the Lombards in *Benevento barely survived a Byzantine campaign in 663. While *Lucera and most of the southern Lombard cities fell, they were able to defend *Acerenza, and a royal army from Pavia relieved *Benevento, where a Roman *petraria* was used to throw the head of a captured Lombard messenger into the city. "Other war machines" (*diversis belli machinis*) distinct from battering rams (*arietibus*) were used at *Bergamo (701); these were almost certainly artillery. Fredegar Continuatus, the *Liber pontificalis* and letter collections give some more information. When the Franks attacked the Lombards in 755 in support of the pope at Rome, the Lombards prepared to defend *Susa "with weapons, machines of war and a mass of supplies/equipment" (*cum telis et machinis et multo apparatu*). The Lombards must have had a similar arsenal when defending *Pavia and other strongholds after the Franks broke through, in 755 and 756 as well as during Charlemagne's final conquest in 773-4. When the Lombards besieged *Rome in 756, they encircled the city, encamped at the gates to prevent exit and entry, while assaulting the walls "incessantly with a variety of engines and multiple contraptions" (*incessanter cum diversis machinis et adinventionibus plurimis*). It was ultimately their ability to defeat the Byzantines and threaten Rome by assault on fortified positions that brought on the fury of the Franks.

The nature of warfare can also be inferred by other means. The Franks demolished Lombard fortifications after they were forced into submission or captured in 590. Their Roman allies fought their way into several Lombard cities by rupturing the walls in intense fighting in the same year. Sometimes fighting is unattested in literary sources, but the complex defensive systems provide a strong hint as to the nature of the fighting, and at Invellino and S. Antonino, archaeologists have found evidence of artillery used in the fighting over the fortification walls.[153]

3.4 Conclusion: From Clients to Kingdoms

All three successor kingdoms closely reflect relations to surviving elements of the Roman state at the time. The Ostrogoths arose as a Balkan client army that eventually became freed from the East Roman orbit. However, due to their origin as invaders and the ideological separation from much

---

[153] See CO for *Frankish and Roman campaigns; see Christie 2006: 349 for references to the archeological reports.

of Italian society (Arianism as well as "Gothicness"), they preserved a form of military organization and practice that very closely reflected traditional Roman forms, making little use of the innovations noted in chapter 1; indeed, apart from their name, the creative distribution of taxes noted by Goffart, and perhaps one or two tactical elements (the lack of horsed archers is commonly quoted but this is probably a fallacy), there was very little to distinguish them from the Roman regulars they fought in the 530s to 550s, many of whom shared a similar ethnic background. The Visigoths, in turn, seem to have preserved the organization devised in the 5th century, basing their forces largely on the recruitment of professional soldiers and slaves into the retinues of great magnates. This type of organization was sufficient to support siege warfare and drive out the Romans in the south, and also provided a framework for absorbing Roman troops who were captured or deserted. The Lombards closely mirrored the long-term developments in the East Roman Empire. In the 6th century, their troops were largely organized as the retinues of independent dukes who drifted in and out of Roman service; in the 7th and 8th centuries, Lombard military organization closely resembled that of the remaining Byzantine armies in Ravenna and Rome. All three kingdoms fought against the Romans and were intermediaries for Roman influence to the Franks throughout their existence.

CHAPTER FOUR

THE LAST LEGIONS ON THE RHINE: SIEGE WARFARE IN THE FRANKISH KINGDOMS

The scale and complexity of Merovingian warfare is often severely underestimated. As a result, the development of an effective military machine under the Carolingians remains poorly explained. However, during the 6th century, the Frankish kingdoms achieved military parity with the societies examined in chapters 2 and 3. The problems experienced by Frankish expeditionary armies outside of the *regnum Francorum* had more to do with the partition of Clovis' kingdom in 511, leaving the Frankish *Teilreiche* at loggerheads with each other, rarely able to concentrate on foreign ventures for more than a few campaigning seasons at a time before being distracted by civil wars and other problems. The exception was under the united Frankish kingdom in the early 7th century, which established a far-flung hegemony. This lesson the early Carolingians took to heart. They never undertook a campaign to conquer territory outside the old *regnum Francorum* until they had established full control over all former Merovingian *regna* and thus eliminated debilitating distractions. Throughout, late Roman forms of labor obligation and estate management provided the logistical basis for Frankish armies, and the Carolingians systematically built upon the structures of their Merovingian predecessors.

4.1 *Frankish Warfare and Military Organization in the 6th Century*[1]

Identity shifts were clearly important in Frankish ethnogenesis. Movement of people also entails movement of knowledge, and as much of the Frankish military establishment descended so directly from Roman institutions and a Roman(ized) population, it is only reasonable to assume *a priori* that later society would also be highly affected by Roman practices as they existed in the 5th century. Indeed, the effect of this has been demonstrated in chapter 1.3, where we saw that the earliest Franks fought and were organized in a recognizably late Roman style. This is especially the case as the

---

[1] For context, see Ewig 1976: 114-71; E. James 1988; Wood 1994: 33-139; Kaiser 2004; van Dam 2005.

institutions that supported warfare had been were reinvented and adopted for more constrained economic and social conditions. An estate-based military economy and troops raised as private followings were immensely important 5th-century innovations. The financial and administrative structures of the Merovingians also rested on late Roman labor obligations and substitution of services for direct taxes, which were gradually abandoned around 600.

4.1.1 *The Problem of Gregory of Tours*

While older models of Germanism, feudalism or migratory-tribe grand narratives have largely been discarded, many of these advances have taken place since Bachrach's basic study on *Merovingian Military Organization* argued for strong continuities in military organization and practice from the Roman era. While largely accepted by historians of late Antiquity, his conclusions have not been followed up by scholars of the Merovingian period despite the great number of studies that point out the continuing *romanitas* of landholding, ecclesiastical organization, urban and local administration, taxation, mentality and social structures.[2] On the contrary, in the model recently proposed by Halsall, social, economic and cultural constraints prevented significant continuity from Roman military practices. He argues that sieges had a negligible, near ritual, place in Merovingian warfare and only became marginally more important under the early Carolingians. Logistics and the maintenance of troops boil down to ritual handovers of gifts that "oiled the cogs" of social relations among lords and retainers, and "large" armies (in the range of 5-6,000 men at the most) survived by essentially plundering their own territories.[3] In a similar vein, van Dam argues that the Merovingians, in contrast to the Romans,

> ... kept no large standing army and instead relied on local levies, garrisons and armed retainers for each campaign. By Roman standards their armies were comparatively small, often only a few thousand or a few hundred men; and because their campaigns, whether within Gaul or against neighbours, produced so much booty and so many captives who could be ransomed, the armies were virtually self-supporting. Unlike Roman armies, the armies of the Frankish kings and their magistrates were furthermore not intended to fortify outlying frontiers. Instead, military campaigns were too often sim-

---

[2] Bachrach 1972. Wood 1994: 64, E. James 1988 and others refer to Bachrach but have not applied the implications of his research to their own work. See Kaiser 2004 for extensive discussion of and bibliography on continuity.

[3] See Halsall 2003 (*passim*); for his model of siege warfare, ibid. 215-27.

ply manifestations of royal (and aristocratic) concerns about manliness and prestige, and the armies of the kings and their magistrates frequently terrorised their own subjects.[4]

Although van Dam's summary of troop types is to a certain extent correct, his model tells us equally little of maintenance, training and remuneration of troops between campaigns. Nor does his model adequately explain supply and campaign logistics when they occurred—post-campaign gains were very often *not* forthcoming, and anyway of little help before or on the march. Furthermore, there is no explanation whatsoever for Merovingian engineering skills, which are ignored or ridiculed by minimalists. Finally, if the mere distribution of plunder to ephemeral armies were the basis of warfare, it is difficult explain how Frankish pressure or influence provoked the rise of kingdoms capable of large-scale military engineering among surrounding peoples from the late 6th to the early 8th centuries, and the fact that Charles Martel and the early Carolingians could face the Arabs on equal terms in the early 8th century on the basis of essentially late Merovingian military organization. Even in the 6th century, most of the Franks' foreign adventures to Italy and Spain revolved around the control of cities and forts in competition with Ostrogoths, Visigoths, Lombards and East Romans, whose abilities in siege warfare have been chronicled in the preceding chapters. In the minimalist model, the Franks would be seriously disadvantaged facing such enemies, especially far from home territory.

The reason for the prevalence of minimalist interpretations lie, firstly, in the very limited source materials for the Frankish realm, and secondly, the even more limited approaches to these sources by modern scholars. We rely almost exclusively on Gregory of Tours and Fredegar and his continuators far into the Carolingian period. How these sources are read imposes heavy restrictions on how we understand the Merovingian military, which again affects interpretations of later Carolingian military organization. We have already seen that Gregory is a highly unreliable source for the period around 500, but even closer to his own day, the same lack of reliability persists. His greatest recent commentator, Walter Goffart, found that he was in fact an accomplished writer of Christian satire, but was completely disinterested in portraying other features of his life and times.[5] Instead, his anecdotes skillfully set virtuous saints up against the atrocities of those in power, who often happened to be his personal (or his king's

---

[4] Van Dam 2005: 210ff on the economic basis for warfare; 211 for quote.
[5] Goffart 1988.

political) enemies.[6] He sought to chastise or ridicule his cruel and brutal contemporaries for fighting incessant civil wars, but only occasionally provides an insight into mundane, everyday business. Unless a vivid description of a military event or social feature is incidental to his main argument, he completely lacks a "pictoral sense" of his own society. Then it is only to be expected that he would focus on the ridiculous, flawed, brutal and failed aspects of civil wars. On this basis, however, secular incompetence is taken as the default position in minimalist assessments.[7]

Gregory's dislike of war, except for successful expansion abroad, has been noted in the introduction. However, even when dealing with foreign adventures, he systematically telescoped events, exaggerated the achievements of Frankish armies in cases where he was ill informed,[8] and emphasized failures in his own day to prove a moral point. This especially applied to civil wars, where Gregory was occasionally involved, and the only period of Merovingian history (575-594) to receive significant coverage of his history (in books 5-10); the rest of the 6th century is given short shrift in just over two books.[9] There are some useful siege anecdotes from the mid-6th century, mostly of an episodic nature, and too brief to provide us with a "thick context," i.e. a coherent image in and of themselves. His frequent, but critical and brief references to sieges during later civil wars improve only slightly upon the situation. However, one should not exaggerate the problem: although Gregory's treatment is taken by many scholars to mean that nothing much actually happened, such an interpretation is possible only by ignoring interesting allusions that make much sense in light of the preceding chapters.

---

[6] Gregory owed his position to the Austrasian king Sigibert, who died in 575, and subsequently came under his brother, Guntram, king of Burgundy, who lived to 592. Although he did oppose some of Guntram's policies, he was far more interested in telling about all the bad things done by his nephew Childebert and other opponents, both local and national, thus giving a skewed picture of Merovingian administration, politics and warfare. In Gregory's mind, *any* example of forceful government with regards to taxation, exaction of supplies, or raising of troops is often lambasted as abuse of power or outright plundering.

[7] Loseby 1998a points out that it is difficult to correlate current knowledge of the early Frankish economy, based on archaeology and other sources, with Gregory's few incidental remarks, but at the same time evaluates the military function of Frankish cities based squarely on the evidence provided by Gregory. Bachrach 2002b has a more positive assessment of Gregory as a military historian once his biases are identified.

[8] Cf. his exclusion of *Arles 508 from the *Vouillé campaign. It would also be profitable to compare his treatment of Buccelinus' invasion of Italy with the image presented by Greek sources, for which see chapter 4.1.3 below.

[9] For the effect this has on our picture of Merovingian society, see Weidemann 1982 for prosopographical information on office holders, which shows a sharp and massive increase around 575.

Instead of rehearsing the same quotes from and debates over Gregory,[10] I propose to follow a different approach. Interaction with the Visigoths, Burgundians, Ostrogoths, East Romans and Lombards provides a basic framework of contemporary administrative and military practices that can be used in two ways: 1) By examining the Franks interacting directly with their neighbors, from sources rarely used, we can evaluate what skills they actually possessed, and on what scale. 2) The basic template presented in chapters 1-3 provides the "thick context" against which to evaluate the scantier, anecdotal information for internal affairs in Frankish sources and thus how Frankish armies were supported. If the Frankish kingdoms were not exceptionalist in the relevant aspects of religion, ideology, economy or administration, one could ask rhetorically why this should be the case in warfare.

4.1.2 *The Establishment of Frankish Burgundy and Wars with the Visigoths, 534-89*

A generation after their defeat of the Visigoths in Aquitaine, the Franks consolidated their control of Gaul by conquering Burgundy in 534. We only learn that the Franks accomplished this *obsedentes* (besieging) *Autun.[11] To gain an impression of what Franco-Burgundian siege warfare was like, we can take into consideration events at the beginning of the century that are somewhat better known. The Franks had first become embroiled in the *Burgundian civil war of c. 500, in support of Godigisel against his brother Gundobad. The Frankish-supported side was at first victorious, and Godigisel entered Vienne in triumph. The Franks followed up with a siege of Gundobad at *Avignon, but Clovis was persuaded for political and strategic reasons to abandon the siege and receive tribute, although it is clear that an extensive campaign had been planned, and was in fact under way, against all of Gundobad's territories. Gundobad was then free to attack his brother Godigisel at *Vienne, who was supported by a Frankish garrison. The fighting as attested by Gregory involved defensive archery from the walls which held the attackers at bay. However, due to lack of supplies, a large part of the population was expelled from the town, indicating a

---

[10] As well as Fredegar and other later sources; see 4.2.1 below.

[11] For the background see chapter 1.3.4. The sources are very thin (cf. *Autun 534). The conquest itself only receives a brief reference in the Chronicle of Marius of Avenches (s.a. 534), and a stylized account in dramatic form by Gregory (3.5-6) with the invasion and siege of Autun briefly mentioned in 3.11. For a brief discussion of the conquest, see Wood 1994: 51-54, who also notes how unsatisfactory Gregory's knowledge of the whole affair is.

lengthy siege. Among those expelled was the engineer (*artifex*) responsible for maintaining the aqueducts, who may have been inadvertently thrown out of the city. He helped the besiegers take Vienne, showing the way for sappers through the aqueducts, as it took some engineering with crowbars to make an entry. When attacked from both inside and outside, the Frankish garrison blockaded itself in one of the towers, but the Burgundians did not wish to provoke a war and sent them to the Visigoths instead of enslaving them.

From the survey of the wars in 496-508, it is clear that the Franco-Burgundian sieges occurred in a socio-economic, administrative and political environment that still had recognizably Roman features. They were also similar to contemporary East Roman siege practices: immediately recognizable features include the existence of urban engineering specialists, expulsion of population, importance of archery in wall-fighting, use of sappers and aqueducts to gain entry, and defenders banding together in individual towers when the wall fell.[12] Not only Burgundians fought over cities and forts; Frankish rebels also took to fortifications and had to be subdued. The *Auvergne campaign (524 or 532) had begun as a joint Frankish invasion of Burgundy, but was sidetracked by king Theuderic of Austrasia who used the opportunity to suppress a revolt against his rule in outlying territories. The campaign involved a siege of *Clermont, during which the walls were breached. This was then used as a base from which to storm the fort of *Vollore and besiege *Chastel-Marlhac, which was forced to pay a ransom. At the same time or shortly after, *Vitry-le-Brûlé was the base of the rebel Munderic, and the fort was subjected to a hard-fought siege involving massed projectile weapons, possibly including artillery. It ended when Munderic was treasonably killed during negotiations.

When the Franks finally absorbed the Burgundian realm in 534, another urbanized and economically advanced area came under Frankish control. They also gained important features of late Roman military organization which only reinforced existing tendencies. Autun had been a center of Roman military engineering where a unit of *ballistarii* had been stationed. It is uncertain how well the formal institution had survived, but it was common practice for federate armies, such as the Burgundians, to be supplied from the regular Roman military infrastructure.[13] The Burgundian kings descended from a line of *magistri militum*, or commanders-in-chief, established during the latest phase of the Western empire, and

---

[12] For a general discussion of siege tactics, see chapter 5 and cf. 2.3.
[13] See chapter 1.1.2 and 1.3.2 above.

competed with the Italian army under Odovacer. This office gave them control over any existing military installations (which were probably not automatically dismantled with the deposition of the last emperor); furthermore, Romans remained important in their military organization, a practice that continued under Frankish rule.[14] In the late 6th century, the chief military officer under the Frankish Burgundian kingdom, the patrician Eunius Mummolus, belonged to a family of Roman landowners. His record as commander during the Frankish civil wars and the defense against the Lombards is impressive.[15] There is therefore every reason to assume some level of continuity as demonstrated for Visigothic Aquitaine, and the Roman element may have been reinforced by the acquisition of Roman fortifications in the Alps, now isolated by the Lombards.[16] The territories of the Burgundians became the core of the Frankish kingdom of Burgundy, which existed in various manifestations throughout the period considered here. It was the Frankish Burgundian kingdom that for the most part expanded against the Visigoths, although other Frankish kings with territories in southern Gaul were sometimes involved in the fighting.

Frankish intervention into Visigothic territory after 511 was rare until late in the 6th century, but involved major efforts. Some ventures appear rather opportunistic in Gregory's narrative, and seem to have been poorly planned due to short time span available when a suitable incident occurred, or the fact that debilitating political distractions arose in Gaul.[17] Their first major offensive after the conquest of Aquitaine was in 532, when Childebert I of Neustria in alliance with Theuderic embarked on their *Septimanian campaign. Childebert was unsuccessful in attacking Rodez, but the army of his ally Theuderic stormed *Dio and subdued *Cabrières. In 541, the kingdoms of Austrasia and Burgundy again united to invade across the Pyrenees via Pamplona to *Saragossa:

---

[14] Indeed, Wood 2003 notes that the Burgundian settlement was the creation of a Roman general, Aëtius, in the first place. Afterwards, the Burgundian rulers acted as imperial legitimists in exercising the office of *magister militum*; their royal title is in fact only attested in 494, but as late as 516 king Sigismund received his ancestors' title *magister militum* from Anastasios, describing himself as *militem vestrum* (your soldier) in his correspondence with the emperor. The Burgundian system was in effect a surviving Roman regional command which was maintained without interruption from the 450s and formed a significant part of the Frankish Burgundian army after 534, cf. the involvement of Burgundian troops at *Milan 538, sent by the Franks to assist the Ostrogoths in order to avoid breaking treaty with the Romans.

[15] See chapter 3.3.4 above for the context of Lombard siege warfare.

[16] Christie 1995: 81f (Susa and Aosta), 84 (Comacina).

[17] Thompson 1969, Wood 1994: 169-175 and James 1988: 92f note difficulties the Franks had in invading Spain.

This year two [text: five] of the Frankish kings entering Spain via Pamplona came to Saragossa, which was besieged for forty-nine days and they devastated and depopulated almost the whole province of Tarraconensis.[18]

This time, Visigothic defenses were well developed, and the Frankish expedition ended in failure when the Visigoths were able to close off the passes and trap the withdrawing Franks in an ambush.[19] It should be noted that the siege itself lasted 49 days, indicating significant Frankish logistical capacities. Although it was recorded by Gregory of Tours, we only acquire "military" information from the *Chronicle of Saragossa*; Gregory entirely focused on how the appropriate religious rituals saved the population, but also indicates that the Frankish army was camped directly outside the walls of the city.[20]

The civil war between king Guntram of Burgundy and Sigibert of Austrasia in 566-67 resulted in extensive campaigns centered on *Arles (566, 567) and *Avignon (567), in or close to Visigothic territory. This alarmed the Goths sufficiently to establish a special frontier command in Septimania in response.[21] Guntram invaded Visigothic territory on several occasions from 585 to 589; the purpose may have been to reestablish his authority after Gundovald's revolt (see 4.1.4 below). The army sent in 585 against *Carcassonne abandoned the effort when a Frankish commander was betrayed and ambushed after failed negotiations. Another column raided the countryside around several unnamed cities, captured one of them, and began a siege of *Nîmes, but Gregory alleges (without providing any details) that the Frankish army was unable to take the strongly fortified and well supplied city. A Visigothic counterattack took the Frankish border forts *Beaucaire and *Cabaret, the first by surrender, the second by a hard-fought storm. There was renewed fighting around *Carcassonne, where the Franks were defeated near the city walls, and another Visigothic assault on *Beaucaire, both in 587. The Visigothic frontier command worked well, but the attacks were also meant as a means of applying political pressure rather than an attempt at outright conquest, as the Visigoths sent ambassadors

---

[18] *Chronicorum Caesaraugustanorum Reliquiae*, MGH *Chronica Minora* II, p. 223: "Hoc anno Francorum reges numero V [read: II] per Pampelonam Hispanias ingressi Caesaraugustam venerunt, qua obsessa per quadraginta novem dies omnem fere Tarraconensem provinciam depopulare attriverunt." See Thompson (following note) on the emendeation from "five" to "two" kings.
[19] Thompson 1969: 14f; see further the entry in CO.
[20] This is discussed in more detail by D. Bachrach 2003: 21f.
[21] Thompson 1969: 18f, dated to 569. See further GT *HF* 4.30, entries in CO, and chapter 3.2.2 for more details.

whenever they captured Frankish fortifications. Inspired by a Gothic revolt in 589 and family grievances, Guntram sent a new army to support the Gothic rebels, but this was crushingly defeated along with the revolt.[22] Again the goal was *Carcassonne.

The result of these expeditions was clearly not what Guntram expected, and in stark contrast to the performance of Burgundian armies against the Lombards, who had been driven back from their assault on Burgundian cities in the 570s, and the pretender Gundovald, who had been crushed in an intensive siege campaign.[23] Guntram responded to the first defeats in 585 by organizing a council to rectify the incompetence shown by the Frankish commanders, who had been expected to conquer Septimania. He also ordered at least 4,000 extra troops into his southern provinces in order to hold forts and cities against the Visigoths (see *Cabaret 585), perhaps to replace the 5,000 casualties in the failed campaigns. Such poor results continued as long as the kings of Burgundy operated on their own and were hampered by treason. Many in the south of his kingdom had supported the pretender Gundovald, and one way to get back at Guntram was obviously to betray or sabotage his subsequent expeditions. Even Guntram's son, Childebert, may have been supplying the Goths with intelligence on Frankish movements in 589, causing the loss of 5,000 men fallen and 2,000 captured against *Carcassonne.

The Burgundian Franks and the Visigoths had in effect reached military parity: as long as the Frankish kingdoms were disunited, the Franks had limited resources to put into the field for longer periods of time, but could in return effectively defend their borders, which meant that, large battlefield losses notwithstanding, the Franks did not lose much territory to the Visigoths, despite the latter's demonstrable fighting qualities against the East Roman Empire. The tables were turned when Dagobert, as sole ruler over the Franks, established Frankish hegemony over the Visigoths in the early 7th century.[24]

4.1.3 *Austrasian Interventions in Italy, 538-90*

While the kingdom of Burgundy was most often embroiled with the Visigoths or one of its Frankish neighbors, it was the Austrasian kingdom that pursued expansionist ambitions in Italy. Situated along the Rhine, but con-

---

[22] Thompson 1969: 92ff, 103; see further *GT HF* 9.31; *VPE* 5.12.3; J.Bicl. s.a. 589 (ii 218).
[23] For the Lombards, see chapter 3.3.2 above; for Gundovald, see chapter 4.1.4 below.
[24] See 4.2.1 below.

trolling the upper Danube and the central Alpine passes to Italy, Austrasian armies regularly intervened from 538 to 590.[25] The evidence is far better than in the Burgundian case, as East Roman authors chronicled Frankish involvement in some detail. It is necessary to be circumspect about this evidence, since it has been shown that Agathias manipulated his narrative due to political considerations, while both Procopius and Agathias included some untenable and self-contradictory ethnographic *topoi*, such as half-naked barbarians.[26] However, once these distortions are taken into account, a detailed examination of the evidence shows that there is nothing particularly barbaric about Frankish warfare. Accounts center heavily on logistics and the ability to win and hold fortifications and fortified cities. A close examination of several of the battles fought by the Franks also shows that they utilized tactics compatible with those of their neighbors.[27] At the battle of Derthon (Tortona) in 539, the Austrasian expeditionary force routed two separate armies, one Gothic and one Roman, in a single day.[28] Roman armies greatly feared the more numerous Frankish troops, and only willingly engaged in battle when they had a distinct advantage, especially when the Franks were suffering from dysentery and had been deprived of supplies. The Franks in turn were systematically disadvantaged by inherent logistical and political problems. Despite this, their successes were significant: for some years from 545 until around 562, the Franks controlled the southern slopes of the Alps and substantial territories in the Po valley, and at one point threatened to take over the whole of Italy.

The Austrasian Franks' main invasion route through the central and eastern Alps allowed them to establish a strong presence in the Trentino and Veneto, where they maintained control of a large number of cities and forts for many years, causing the Romans serious problems. Procopius acknowledged that these fortifications were too difficult and time-consuming for the Romans to reduce as early as 552, when they were preoccupied

---

[25] The army that was sent to assist the Ostrogoths at *Milan 538 consisted of Burgundians, but was sent at the behest of the Austrasian king Theudebert in order to cover up his involvement.

[26] For assessments of Agathias' general views on the Franks, see Av. Cameron 1968, who argues that when he wrote in the 570s, the Franks were allies of the Byzantines and had to be excused for some of their excesses. For the ethnographic excursuses by both authors on Frankish warfare, see Bachrach 1970.

[27] See Agathias 1.21.4-8 and Bachrach 2001: 178-87 for an analysis of the Battle of Rimini, where the Franks fought in good order; see also below.

[28] See the background to the siege of *Fiesole 539; the Roman army at Derthon was there to shield those besieging Fiesole from the Ostrogoths assembling at the military base located there.

with defeating the Ostrogoths in central and southern Italy.[29] At their height around 554, the Franks in addition controlled most of inland northwestern Italy and also held some cities as far south as Tuscany. Frankish armies were led by two Alamannic *duces*, the brothers Butilinus (Buccelinus) and Leutharis.[30] Agathias claims that Butilinus coveted the Gothic crown, and may have been operating independently of the youthful king Theuderic, but when the brothers were killed in battle, Ammingus, a Frank, led the Frankish troops in the north of Italy. He was able to block Roman progress for some time, but was finally defeated in battle and killed around 561/2. The downfall of Butilinus is well described by Agathias, but our main source for later events, Menander Protector, is only preserved in fragments, while Gregory of Tours and Paul the Deacon only present a highly telescoped version of events.[31]

During their dominance north of the Apennines, Frankish armies made Roman movements very hazardous. Since the Franks were pursuing long-term territorial ambitions, they evidently planned their logistics very well. Due to the distances across the Alps, they relied on local resources when garrisoned. In the field, they foraged efficiently and used garrisoned cities or fortified camps to store their supplies and protect their forces. Large foraging teams were active not only plundering and collecting livestock, but also collecting timber for fieldworks or other purposes. On one occasion, Roman troops skirted the more numerous Franks by marching in small detachments at night, but could hear the din caused by peasants wailing and trees crashing down.[32] Since many cities and parts of the Gothic army were collaborating with the Franks, we can also assume that they had access to whatever remained of the Gothic logistical system when they garrisoned cities.

Frankish garrison forces were willing to take on the Romans on several occasions. The Frankish garrison at *Parma (553) at first preempted a siege when they received accurate intelligence on an approaching Roman force,

---

[29] See chapter 3.1. above for context.

[30] Their relations with the reigning king of Austrasia, Theuderic, are somewhat ambiguous, as he was still very young, and some of his other magnates may not have wished to be directly involved in the affair. This may have been a deliberate distortion on the part of Agathias, however. Further on their careers, see *PLRE* 2 s.vv. Butilinus and Leutharis.

[31] See Wood 1998b: 239-42 and E. James 1988: 97f for analysis. Gregory has famously compressed the whole affair by conflating different events, exaggerating Frankish success and placing it in the wrong reign. The sources for Ammingus are Menander fr. 3; Paul the Deacon *HL* 2.2. For the dates, see *PLRE* 2 s.v. Amingus (*sic*).

[32] This incident occurred in the context of *Parma 553.

and ambushed it in the vicinity of the city, using an amphitheater as an ambush site. At this stage, many Italian cities went over to the Franks. At *Lucca (553), a Frankish garrison led the defense of the city against a besieging Roman army for three months. They resisted Roman attempts at storming the city with rams supported by archers, slingers and fire-throwing artillery. Just as the Romans did against the Ostrogoths and Persians, the Franks led sallies out to fight off the besiegers when they were threatening to break through. What brought the Frankish garrison down was not military incompetence, but a severe political disadvantage. The citizens of Lucca had been closely aligned with the Goths, and when the Goths were increasingly pushed back, a significant segment of the population at Lucca decided to call in the Franks rather than submit to the Romans. However, a pro-Roman element existed in the city, and as the siege dragged on, they managed to bring more and more of the population over to their side, especially after Narses had resorted to psychological warfare against them. In the end, the pro-Roman Luccans had convinced most of the population to abandon the Franks at the most devastating moment. During a Frankish-led sortie supported by the local inhabitants, the citizen militia deliberately withheld support from the exposed Franks. This led to their defeat and a negotiated surrender in which the Franks were allowed to withdraw. It is important to notice that even a fairly isolated Frankish garrison south of the Apennines was willing to fight on even after the Romans had managed to make a breach in the walls and only gave in when abandoned by the population alongside whom they were fighting.

Frankish expansionism came to a halt in 554 when the Frankish army overextended itself in a gamble for the whole of Italy instead of securing their hold on the north. Butilinus and Leutharis had good knowledge of the complex and diverse loyalties in Italy, and probably believed that a devastating show of force would bring over former Gothic sympathizers all over the peninsula to the Franks,[33] and perhaps even cause Roman soldiers to defect. They organized a massive raid that split up in Samnium and then followed the eastern and western coastlines respectively. Butilinus, leading the western column, reached all the way to Calabria across from Sicily, but was defeated on his way back at the famous battle of Volturno (Casilinum). He had set up a large fortified camp whence camp followers spread out into the surrounding countryside to gather supplies and plunder. Agathias describes the camp thus:

---

[33] For instance, there was still a large Gothic garrison at *Conza, which was besieged by the Romans in 554-55.

204   CHAPTER FOUR

Having stationed his army there he had a strong line of earthworks built around them (χαράκομά τε περιεβάλετο καρτερὸν), the effectiveness of which, however, depended on the nature of the terrain, since the river which flowed to his right seemed to constitute a natural barrier against attack. He had brough great numbers of wagons with him. Taking off their wheels and fitting them together rim to rim in a continuous line he stuck their felloes into the ground and covered them with earth right up to the hubs, so that only a half circle of wheel protruded above ground-level in each case. After barricading his entire camp with these and numerous other wooden objects (τούτοις δὴ οὖν καὶ ἑτέροις ξύλοις πολλοῖς ἅπαν τὸ στράτευμα ἐρυμνώσας) he left a narrow exit unfenced, to allow them to sally forth against the enemy and return again as they wished. The bridge over the river constituted a possible source of trouble if left unguarded. So he seized it in advance and built a wooden tower on it in which he placed as many as he could of best armed soldiers and his finest fighting men so that they might do battle from a safe point of vantage and repel the Romans should they decide to cross over.

However, there was an epidemic outbreak of dysentery, the Romans had stripped the countryside of supplies on his projected route of return, and then destroyed the fortified bridge from the Frankish camp to their foraging areas. This forced Butilinus to meet Narses in a decisive battle on highly unfavorable terms.[34] His brother Leutharis reached Otranto on the heel of Italy, but his vanguard was ambushed on the way back and the army returned to the north severely affected by disease. Leutharis died soon after. With the defeat of Ammingus a few years after, the first great Austrasian venture to Italy came to an end.

Austrasian ambitions remained, however, and when the Lombards had taken over northern Italy in 568, the Franks tried to reestablish their hegemony. They also attacked the Lombards at the behest of the Romans on several occasions.[35] Without local support, it is clear that the Franks faced considerable difficulties supplying armies across the Alps, while the Lombards shut themselves up inside their forts and fortified cities. The Franks were however successful in systematically reducing many Lombard strongholds and some were subsequently garrisoned, others demolished (itself a feat of engineering).[36] Cooperation with the Romans failed in the decade 580-90, but not for military reasons. The Austrasians were acting in their

---

[34] See Agathias 2.4.5ff for the camp; ibid. 2.6.3-6 for the foragers.

[35] The Burgundians stopped after turning back Lombard raiders and taking over Byzantine forts at Susa and Aosta by treaty; the Austrasians intervened in Italy in 576, 584, 587, 588 and 590. For the context, see Christie 1995: 86f and Goubert 1965: 22-26.

[36] On the difficulties and dangers of demolishing late antique fortifications and buildings, see p. 282 of the discussion section following Loseby 1998a.

own interests. On some occasions, they were compelled to fight by East Roman diplomacy, but chose to withdraw when the situation became precarious, such as when their Frankish neighbors threatened—the Burgundian kings would not have been too happy to see the Austrasians in control of Italy.[37] It also seems that the Franks preferred to have the Lombards as clients and buffers as long as the Romans were strong, since Frankish hegemony in the early 7th century was undisputed.[38] The Austrasians were most energetic when they could revive claims to north-eastern Italy from contiguous territory in Alamannia. In 590, Childebert II sent an army under 20 *duces* via the Brenner Pass. The largest contingent of 14 *duces* reached to a point north of Verona. On this occasion they were supposed to cooperate with a Roman army that was very successful on its part, but both sides claimed that the other failed to live up to the terms agreed upon. In fact, the Austrasians were probably forced to renege on the agreement due to pressure from the Burgundian realm (Lombard ambassadors passed through Guntram's kingdom on the way to Austrasia) as well as dysentery and lack of supplies.[39]

By the 590s, the Franks had been in close direct contact with the East Romans for well over half a century, either as adversaries or partners. They met with spectacular defeats when they pushed their luck too far, but whenever they put their military and political muscle into their efforts, they were successful, especially in north-eastern Italy where they eventually maintained a more or less permanent presence during the latter half of the 6th century. This means that they had to adapt to local forms of war, which they seem to have done very well—unless they were already well prepared from their home provinces. Indeed, the Roman commander in 590 was both surprised and furious at the Frankish withdrawal; the East Romans knew well how to assess enemies, had fought alongside or against the Franks on numerous occasions, and were well aware what capabilities they had. In fact, the contemporary *Strategikon* treat the Franks and Lombards as essentially the same when it comes to organization and tactics.[40]

---

[37] Cf. Christie 1995: 80ff, who points out that the Lombard and Saxon invasions affected Guntram's territory, while Sigibert of Austrasia hired the Saxons and had them march through his brother's kingdom, wreaking havoc on the way.

[38] See chapter 4.2.1 below.

[39] See CO, *Frankish invasion of Italy and *Roman invasion of Italy, both in 590, for details. The best source is in fact *Epistulae Austrasiacae* (40), which demonstrates the nature of the campaign (wall-breaking sieges), in contrast to Gregory's critical account. See further chapter 6.2.1 for the relevance of these campaigns to 7th-century Frankish siege warfare.

[40] *Strategikon* 11.3.

Lombard military organization and practice, in turn, as we have seen, derived from the Romano-Gothic-Byzantine infrastructure and the Lombards first rose to prominence as "subcontractors" for the East Roman military. Postulating a radically different kind of warfare once Frankish armies were back across the Alps does not make much sense. A survey of Frankish siege methods during civil wars, the technology used and the logistical structures that underlay them will amply demonstrate that the Franks fought in a manner similar to their neighbors.

4.1.4 *Frankish Civil Wars*

Many sieges were fought in Gaul due to incessant civil wars in the latter half of the 6th century. Cities and forts were used as refuges for people and movable goods, a course of action advocated by saints as well as kings.[41] Although their movable wealth may not compare with classical times, walled cities (*urbes*) were still valuable as administrative centers for their hinterland (*civitates*), and whoever held the city also controlled the collection of taxes from the surrounding *territorium*.[42] Thus, siege warfare was often about ensuring the political control over the flow of revenue. This made raiding an effective means of applying pressure, as it deprived both the opposing king and his local supporters of vital income, in addition to the obvious logistical benefits of depriving one's enemy of supplies and destroying their morale. Despite the frequent allusions to raiding leading up to surrender, the exact cause for submission is actually indeterminable based on the meager narratives in Gregory. Ravaging of the economic basis of the city was clearly effective, but it also seems that the prospect of a formal siege terrified populations into submitting.[43] If this interpretation is correct, this means that negotiations for many apparently "peaceful" or "political" submissions were conducted with the display of the early medieval equivalent of the proverbial big stick, a full siege train arrayed outside the city. Ravaging in late antique practice was only the common and necessary prelude to destroying the defenders' supplies. Indeed, sometimes even the threat of ravaging was enough to bring about submission, but this too

---

[41] GT 6.6, 6.41.
[42] See Loseby 1998a for Merovingian urban functions.
[43] *Rheims; *Soissons and other cities changed hands for unrecoverable reasons during a civil war in 562. Similarly the nature of *Tours 567 is impossible to determine, although it is clear that the ruler Clovis was scared away by the approach of the joint armies of Guntram and Sigibert under Mummolus. For an elaboration of this argument in light of contemporary examples, see chapter 5.1.

was a common occurrence elsewhere.[44] But Gregory himself gives us several clues that a simple "raid" might have much more to it. The siege of *Clermont (524) is actually completely unknowable from Gregory's *History*, which only states that Theuderic's army devastated (*devastat*) the region and came/gained access to (*accedens*) the city, which was then garrisoned. The nature of the fighting was actually a formal siege involving a large siege-camp and some means of wall-breaking that led to the capture of the city. This is only revealed as incidental details in several of his hagiographic writings, and should give us reason to be wary of other "uncomplicated" surrenders—hard fighting may indeed have preceded their submission without anyone bothering to record it.

We do however have brief reports that mention sieges with unambiguous vocabulary: some ended in storm,[45] others were blockades,[46] some involved sallies,[47] or ravaging that brought about submission, as well as the enslavement of the defeated populations and massacres.[48] Preparations were made by defenders for siege, although Gregory is maddeningly unspecific.[49] However, once when trouble was brewing at Tours, Gregory lets us know that the walls and gates were guarded, and on another occasion, a garrison of 300 men was mobilized by the count to man the walls.[50] Thus, the means of defense certainly existed. The inhabitants of Frankish cities (as well as others attacked by the Franks) engaged in religious rituals to invoke heavenly favor and avert disaster, including processions on the walls, fasting, carrying relics and so on.[51] Earthly assistance was also avail-

---

[44] Thus *Tours submitted to the demands of Roccolen after he had threatened its countryside in 575. See chapter 5.1.1 for the context, degrees of threat, and common parallels.

[45] *Chalon-sur-Saône 555; *Poitiers 567; *Tournai 575.

[46] *Clermont 555.

[47] *Arles 567.

[48] *Rheims 556/7; the civil war of 573 saw Chilperic conduct extensive ravaging, taking of captives and massacres throughout the *territoria* of his targets *Tours, *Poitiers, *Limoges and *Cahors, all of which submitted. *Paris was similarly ravaged by Sigibert the following year, when Chilperic handed back the cities he had taken, acknowledging they had been forcibly subdued. *Tours submitted to Guntram, who then helped the men of Bourges take *Poitiers in 584, both of which submitted after ravaging.

[49] Thus the citizens of *Toulouse (585) had prepared for war (*bellum parantibus*) against the usurper Gundovald, but were brought over by the sight of his huge army; similarly the rebels Ursio and Berthefried had entrenched themselves in an old *Fortified estate church at Woëvre in 587; there they "fortified the basilica with their weapons." It might only be poetic for preparations to fight in the open, but the description of the general context (an old fortification on a naturally defensible site) indicates that it was suitable for a proper siege defense.

[50] GT 7.29. This was only a skeleton garrison for guarding the gates.

[51] E.g. *Clermont 524 and *Dijon 555; see further chapter 6.1.3 below.

able, and sieges could be ended by relieving armies that took on the besiegers in open battle.[52] Alternatively, cities submitted when their king or his army was defeated in battle.[53] Sometimes sieges were settled by bribes and treachery (see below), but that was very common elsewhere. The complex intertwined politics of the period blurred the lines as there were often competing factions and aristocratic groupings within each city; in extreme cases, one side in a local conflict could end up besieging the other.[54] At other times, faction could be exploited by besiegers. Thus, even if technology and tactics are rarely given in detail, there is little to separate the social aspect of warfare in Gaul from other regions surveyed.[55]

The prevalence of large-scale logistical efforts would not only require military manpower, but also a large number of skilled craftsmen and engineers. Unfortunately, Gregory gives us little to go by, as he is averse to mentioning details such as technology or people of subaltern status, but we cannot conclude from this that all the above mentioned sieges were simply armies waiting in front of a city wall. On a few occasions, he mentions that a siege was conducted with siegeworks, using the verb *vallare* ("surround with a wall").[56] On some occasions, he goes further. When Mummolus fell out with his king, Guntram, in 581, he improved the defenses of *Avignon (583) by diverting the Rhône in a channel around the city and added traps in the riverbed. This was also practiced by the Romans in the East. Gregory's emphasis on the personal and political circumstances obscures the rest of the siege,[57] but the importance of technology, supplies and engineering skills in Merovingian siege warfare is abundantly clear from his only detailed description of a siege, that of *Convenae* (St.-Bertrand-de-Comminges) in 585. The main protagonist of this event, Gundovald, claimed he belonged to the Merovingian dynasty and tried to establish a kingdom in southern Gaul with the support of Mummolus and other prominent aristocrats. Guntram sent a large army to deal with the revolt, which ended when Gundovald and his key supporters were confined to the fortification of Convenae and defeated in a hard-fought siege.

---

[52] *Soissons 576.

[53] Thus *Périgueux, perhaps *Agen, and other cities in 582.

[54] As when a contested episcopal election ended in the siege of *Uzès in 581, which, to boot, was ended with a bribe to the weaker incumbent.

[55] Further on this, see chapters 5 and 6.

[56] Thus *Arles 567 and *Bourges 583; cf. Jordanes' vocabulary on *Rome 537f, discussed at ch. 6.4.

[57] The account in GT 6.26 is highly person-focused and episodic; of the siege we only learn about the negotiations, where Mummolus fooled Guntram Boso to wade into the river and into the trap. See the analysis in CO and the Roman parallel at *Phasis 556.

Since our understanding of Merovingian siege warfare to a large extent hinges on this siege, an accurate understanding of the key events is necessary.[58] Halsall's assessment of the first storm is worth quoting in full:

> Again the picture is fairly crude. The attackers attempted to make battering rams. These had shelters constructed from wagons, and roofs cobbled together from planks, old saddles and bits of wickerwork. These makeshift contraptions were, unsurprisingly, not very effective, being crushed by rocks dropped from the ramparts of the city. The defenders' tactics consisted largely of dropping things from the walls; we are told of boxes of stones, rocks and, interestingly, barrels filled with pitch and fat, which were set alight.[59]

This essentially rephrases part of Thorpe's translation ("protected with wattle-work, old leather saddles and planks of wood," given in full in CO), compresses somewhat the storm itself, and ignores the rest of Gregory's description. Thorpe includes not only the original *cletellae*, which he translated as "pack saddles," although the classical form for this is in fact *clitellae*; but also adds its interpretation in the notes of Krusch and Levison's MGH edition, as *crates* (pl.). There are very solid philological and lexical reasons for deriving *cletellae* from *crates*. The term means wicker-work, and it is often used in classical literature to describe defensive wicker screens and earth-filled wicker baskets used during sieges.[60] Thorpe presumably aimed at adding color, but hardly provided a scientific rendering. The edition provides the following text:

> Quintus et decimus in hac obsidione effulserat dies, et Leudeghiselus novas ad distruendam urbem machinas praeparabat, plaustra enim cum arietibus, cletellis et axebus tecta, sub qua exercitus properaret ad distruendos muros. Sed cum adpropinquassent, ita cum lapidibus obruebantur, ut omnes adpropinquantes muro conruerint. Cupas cum pice et adipe accensas super eos proicientes, alias vero lapidibus plenas super eos deiciebant.[61]

---

[58] Bachrach 1994: 93-153 deals with the military aspect of the Gundovald affair in great detail, and has an extensive analysis of the siege, including the complex maneuvers that led up to the siege, the establishment of the besiegers' camp, logistics, the geography of the site (which is important to understand the narrative provided by Gregory. Google Earth has provided some help in checking his analysis of the fighting, although here I focus on the technology, organization and morale).

[59] Halsall 2003: 225.

[60] Bonnet 1887: 747 n. 4 refers all the way back to DuCange, who demonstrated that this variety was common in late Latin. It derived from *crates* as a diminutive form *cratellae*, which alternated with Gregory's *cletellae*. See e.g. Caesar, *De bello gallico* 5.40.6, where *crates* are used to construct battlements and parapets in preparation for an enemy assault.

[61] GT 7.37, ed. p. 359.1-6. See also ibid. n. 2 (and the preceding note) for references to the philological studies relevant to *crates*.

Accepting Krusch's explanation for *cletellae*, a literal translation would be:

> The fifteenth day of this siege had come, and Leudegisel was preparing (or: began to prepare) new engines for destroying the town (*urbem*), wagons [equipped] with rams, covered with wicker screens and planks, under which the army could advance to destroy the walls. But when they approached, they were so pelted with stones that all those who were approaching the wall fell. [The defenders] threw [not only] barrels burning with pitch and fat upon them, but [also] threw down other [barrels] full of stones.

The wagon-borne rams appear somewhat less ramshackle with an accurate rendering: planks and wicker-work were fairly standard protection gear for late antique siege engines.[62] They were not impervious to everything, however, and Romans, Persians and Huns experienced having their siege sheds crushed at close range by skilled defenders.[63] It is also worth pointing out that the late antique site of Convenae, modern St-Bertrand-de-Comminges, was situated on a low hill with difficult access along a few narrow paths cut into the steep hillside. As Bachrach pointed out, the assaulting teams had to negotiate an uphill attack with very limited space. In the East, the Romans' Sabir Hun mercenaries developed lightweight portable sheds to protect the ramming teams in similar terrain; the use of wagons in this manner indicate similar resourcefulness in adapting to difficult situations.[64]

Furthermore, unless we are deceived by poor syntax (which is of course possible), Gregory seems to be describing *two* distinct phases of defense against the storm, first a barrage of stones against those who were approaching (*cum adpropinquassent* should not be taken as "when they had arrived," since no point of reference is provided), which indicates that this occurred at some distance from the wall. Only then were those who reached the walls bombarded from close range with burning or stone-filled barrels. Finally, a possible interpretation of the phrase *ut omnes adpropinquantes muro conruerint* (given here as "that all those who were approaching the wall fell") would be "so that all those who were approaching rushed to the wall" when the barrage of stones began, leaving the men exposed to the barrel traps prepared above. This interpretation lends credence to the possibility that Mummolus and his allies were deploying some sort of

---

[62] See chapter 5.2 *passim*. However, even the much ridiculed packsaddles, if made of leather or perhaps folded canvas, would be more than acceptable to resist arrows and fire.

[63] See chapter 1.1.1 for the siege of Aquileia (361, not in CO) for similar tactics in Roman civil wars; the defenders of *Naissus (442) crushed a large number of Hunnic rams, but were overwhelmed by their numbers. Cf. Vegetius 4.16 and 23, who describes the very same means.

[64] See the discussion of rams in chapter 5.2.5.

stone-throwing artillery,[65] and that this surprised the besiegers, who would expect to advance slowly, protected from smaller projectiles by the superstructure of the rams. In that case, the plan for the besiegers would have been to engage the defenders at a distance with archery and the like, and try to suppress them before approaching the wall, in order to avoid precisely the scenario that befell them. It is thus not a description of the besiegers' incompetence, but rather an indication that the defenders were particularly skilled and able to neutralize the normal mode of assault.

The siege involved several other interesting features. Fighting on the walls was clearly important, as we have seen; even bishop Sagittarius, one of the conspirators, was clad in armor and threw stones onto the heads of those assaulting (but the instance was specifically noted by Gregory due to his moralizing bias against armed clerics). Frustrated at the direct approach, the besiegers began to fill in a ravine with fascines to approach from the eastern side, but Gregory claims that this "did not harm anyone," without giving an explanation. However, it may be that it was this threat that caused some of the defenders to lose their nerve, as they opened secret negotiations to betray the city. How long actually passed between the first assault and the betrayal of the city (and how long it took to make the engines in the first place, before the assault began, or how much time was spent filling in the ravine) is never revealed, but Bachrach argues that the whole affair took several months.[66]

4.1.5 *Military Organization and Siege Logistics in the 6th Century*

Small magnate retinues or "warbands" that fought for glory and plunder,[67] then, can hardly have provided an adequate basis for the 6th-century Frankish armies that fought over fortifications on equal terms with their neighbors. The size of armies is the first issue. The individual siege not only required overwhelming force on the part of the besiegers,[68] but in many cases, several sieges were conducted at the same time, while other forces protected supply routes, garrisoned forts and cities, raided enemy territory,

---

[65] This is also inferred by Bachrach 1994: 139f. For a discussion of this possibility, see chapter 8.3.2.

[66] Bachrach 1994.

[67] As we have seen in the introduction to this chapter, Van Dam 2005 and Wood 1994 have extremely simplistic views of what motivated early Merovingian armies. Halsall 2003 is more sophisticated, but still unconcerned with how the Merovingian armies were actually supplied and maintained.

[68] I have found a minimum ratio of 2:1, but the norm was preferably 4:1, and often as much as 10:1 or even higher. See discussion in chapter 6.2.2.

and shielded against relieving armies.[69] Gregory provides many interesting figures for late 6th century Merovingian armies, most ranging from garrison forces of 300 professional troops guarding the gates of Tours,[70] around 4,000 for garrisoning a number of fortifications on the Visigothic border, to field armies numbering 10-15,000 on a single campaign.[71] Bachrach has used the latter numbers to extrapolate individual field armies on the scale of 20,000 men during serious inter-kingdom conflicts, but this is beyond what many scholars are willing to accept, and many opt for much lower numbers.[72]

East Roman estimates of Frankish strength provide a useful check on these numbers. Diplomatic correspondence shows that in 538, Justinian asked for a Frankish mercenary division of 3,000 men from Theudebert of Austrasia when the Romans were hard pressed in Liguria.[73] The Romans only had 1,000 men in the whole province at the time, and 300 were besieged at *Milan along with her citizens. The force requested was only a fraction of the troops available to the Austrasian king, as it was to be sent as an auxiliary force to serve under Roman command, *in solacium Bregantini patricii*, who was in charge of the local defenders at *Milan.[74] It was not an army that would operate independently during joint operations, as in the late 6th century: the Romans only needed to strengthen their garrisons in Liguria until reinforcements could arrive, and did not want to give the Franks the opportunity to exploit the situation. This is nevertheless what happened. Theudebert politely excused himself to Justinian for the current campaigning season, blaming the late arrival of the Roman ambassador. However, he surreptitiously had 10,000 Burgundians join the Goths at *Milan, and openly sent his own army the next year.

Procopius provides the highly improbable 100,000 men for Theudebert's army in 539, but it nevertheless destroyed an Ostrogothic as well as a Roman army in the course of a single day. It must have been quite large to take on such a challenge with confidence and win so spectacularly. Since the Roman army that had moved into the area at the time numbered close to 10,000 men, and the Goths were presumably as numerous, we can esti-

---

[69] This is discussed in detail in chapter 5.1.
[70] The garrison at Tours only involved some of the troops available to the count; see GT 7.29.
[71] For a survey of the numbers reported by Gregory, see Weidemann 1982.
[72] Most notably during Guntram's suppression of Gundovald, cf. Bachrach 1994; for the minimalist side, see chapter 4.1.1 above.
[73] *Ep. Austr.* 19.
[74] *PLRE* 2 s.v. Bergantinus (sic).

mate that the Franks matched them combined, i.e. forming a total of 20,000.[75]

Agathias claims that an army of 75,000 men invaded Italy in the 550s, and 30,000 of them were defeated by Narses at Volturno in 554. This force appears far too large at first, but an inspection of Frankish activities shows that it was actually on a similar order of magnitude. If Agathias' figure of the Roman army at Volturno, 18,000 men, is correct, the Frankish army was about the same size or slightly smaller, i.e. 15-20,000 men. Leutharis' army would have been about the same size or smaller. Thus perhaps about 30,000 men for the whole raiding force would be a reasonable estimate (which is given by Agathias as the number of Franks at Volturno), but this may have included some Goths who joined on the way. There were still enough Frankish troops in the north to hold fortifications; a smaller force of around 10,000 would suffice including some Gothic and other local assistance.[76] A reasonable, conservative estimate of the Frankish force, then, would be 30,000 soldiers from north of the Alps, including a large number of Alaman clients. These were in addition assisted by (a guesstimate of) up to 10,000 local Italian troops of indeterminate nature such as Goths and disaffected Italians.

Finally, considering the extensive regional responsibility and large personal military resources of the Frankish *duces* (see below), the Frankish army that was sent to aid the Romans in 590 under 20 *duces* could hardly

---

[75] On the various armies, see the entries in CO for the year 539, especially *Fiesole. The Frankish army experienced logistical problems quite fast, while the rapid outbreak of dysentery in their camp and its effects are further indicative of a very large body of men. This first foray into Italy seems to have failed, in the end, simply for being too large, and was probably an opportunistic attempt to gain booty and territories during what appeared to be the death throes of the Ostrogothic kingdom—the Franks had only recently taken over Gothic Gaul, in 536, and had probably not fully established themselves in the region yet. Hence logistical problems would be natural if they had not fully organized their most recent conquests.

[76] Several Frankish detachments garrisoned major cities in Tuscany, including *Lucca and *Parma. The Goths normally garrisoned important cities with 1,000-4,000 men, depending on size and strategic importance. In addition several raiding and foraging parties were active in the north. The one defeated by Narses at Rimini counted 2,000 men, while another large foraging party was active around *Parma 553. A much larger army raided the length of Italy, splitting in two: one group followed along the eastern coast, another along the western. The eastern army was so large that it had a scouting van of 3,000 men. The Franks marching on the west coast were said by Agathias to number 30,000 men; they built a very well fortified camp as well as a fortified bridge to ensure safe passage for foraging parties to the surrounding countryside. It was only after the bridge was destroyed and his troops were heavily afflicted with dysentery that Buccelinus decided on battle. The Romans however followed a strict doctrine of not engaging the enemy unless overwhelmingly superior.

have numbered less than 20,000 men. In light of these rather consistent numbers, we must conclude that the Austrasian Franks could raise expeditionary armies in the range of 20-30,000 men *across the Alps* without excessively taxing royal resources. It is impossible to say whether these numbers included the camp followers who helped with logistics and construction, or whether such individuals came in addition. That would of course add to the grand total. I would hazard to guess that the very large figures given in East Roman sources were in fact sober diplomatic estimates of the total potential manpower resources of one or more of the Merovingian kingdoms at different times.[77]

While such numbers explain the extent of Frankish activities in Italy, they are in serious conflict with much current historiography, and beg two important questions: on what basis were they raised, and how were they supplied? As we saw in chapter 1.3, the Frankish armies of Clovis and his sons were dominated by professional troops settled between the Rhine and the Loire, who were the direct descendants of Roman legions, in large part of Frankish stock, as well as other categories such as *laeti* and federates.[78] For reasons of supply and political control, they were widely distributed on estates belonging to the Merovingian ruling families and their close allies. While opulent villa centers were abandoned in the 5th century in northern Gaul, this may only indicate a shift in patterns of exploitation that were related to the needs of the army, similar to common 5th century-developments in (informal) East Roman and (formal) Visigothic military organization, where estates had a significant role. In fact, Aëtius had a strong position in northern Gaul due to his great estates there, and after his successor Aegidius broke with Rome in 461, all fiscal lands would have fallen under local military control.[79] He also had to maintain large forces on the Loire in order to face his Roman enemies and their Visigothic allies. Personal wealth combined with former fiscal lands provided much of the

---

[77] This is fairly common in Byzantine historiography. If correct, Procopius' figure may be an early-6th century estimate of the total army of Clovis' kingdom (i.e. before the conquest of Burgundy); Agathias of Theudebald's Austrasian kingdom *including* manpower from the client kingdoms—Alamans, for instance, were important in Butilinus' campaign, while he also seems to have had substantial influence over Burgundy.

[78] Bachrach 1972; Halsall 2003: 40-46; chapter 1.3.3 above.

[79] A number of scholars, such as James 1988 and Wickham 2009, suggest as much but there is little direct evidence. However, Devroey 2003: 245 identifies an example at Tournai, where the Roman *praetorium* and textile works survived as units of the Merovingian and later Carolingian fisc. Aëtius had extensive estates in Gaul (?—clearly outside Italy, Dalmatia and Pannonia); see Prosper s.a. 432 and *PLRE* 2 s.v.

power of the lesser rulers Syagrius, Paul, Arbogast and Childeric in the late 5th century.[80]

When the latter's son, Clovis, gained full control over the north, he also gained all of these resources, in addition to at least some elements of the traditional form of taxation for remaining land.[81] Direct taxation by the government is in fact well attested throughout most of the 6th century, especially in the Loire and Seine valleys—indicative of the distribution of troops requiring support—and was only gradually suppressed and became obsolete by the early 7th century.[82] Within this framework, Roman unit structure survived in recognizable form in the early 6th century. Procopius' famous description of recognizably Roman units in the Frankish army confirms that the Merovingians were also quite conservative in their military administration.[83] The soldiers who served the early Merovingians were nevertheless called Franks, and had tax exempt status in return for their military service. A "Roman" in Salian law was whoever still paid taxes, but in the course of the 6th century, the extension of military service among "Romans" and complications caused by property acquisition by "Franks" blurred the distinction, and Frankish identity (and military service associated with tax exempt status) became universal north of the Loire.[84] As we have seen, the Merovingians also absorbed Visigothic and Burgundian military organization, and in the course of the 6th century gained control over a wide belt of client kingdoms east of the Rhine and along the upper Danube (Thuringians, Alamans, Saxons) that added to their potential manpower.

At a certain point in the early 6th century, trusted officers and cadet lines of the Merovingian dynasty began to organize these Franks within

---

[80] See Kaiser 2004: 117-22 for a summary of the debate on the origins of the Merovingian fiscal resources and estate structure. He refers to work by Ewig, who demonstrates that the Merovingians drew heavily from Roman fiscal lands. Sarris 2004 effectively demonstrates how estate organization was of Roman origins and not a late Merovingian or early Carolingian innovation, but it is uncertain how early the bipartite structure arrived in the north of Gaul. See further the discussion in chapter 4.2.1 below.

[81] Daly 1994 discusses the contemporary evidence for Clovis' policies. For the continuity of Roman fiscal practices, see Goffart 1972, 1982 and 2008.

[82] Goffart 1982; Wickham 2005 and 2009.

[83] See the quote from Procopius in chapter 1.3.3 above.

[84] For the latest assessments of the spread of Frankish identity, see Halsall 2003: 46ff and 2007 *passim*. Taxes had been exempt for certain lands held at the time concessions were given (now unrecoverable), but new land acquired from tax-liable individuals would not be included. This is the reason for much Frankish protest in the late 6th century, since they felt that *any* lands they held, no matter how it was acquired, should be tax exempt. For this see Goffart 1982: 14.

the framework of their personal households, but the process is highly obscure. We have an early example in Sigisvult, a royal relative who was sent to garrison *Clermont (524) with his *familia*. Otherwise, the transition from a tax-based army to an estate-based conglomeration of military followings is hard to trace, and can only be established with the hindsight provided by Gregory of Tours, whose information is most detailed for the last decades of the 6th century. This process, and the constant divisions and reshuffling of territory of the divided Frankish kingdom, resulted in the structure familiar from the later 6th century.[85] Within the royal household(s), by far the largest and most widespread,[86] there was a distinction between at least two categories of royal troops, analogous to the *doryphoroi* and *hypaspistai* in East Roman military followings.[87] Some of these were called *antrustiones*, of higher status, while the bulk of soldiers in the king's *obsequium* were simply called *pueri regis*, "the king's boys."[88] Both were maintained by the households of the kings and their families (i.e. living off the proceeds of any one of a large number of estates, or taxes still collected). To ease the supply situation outside the campaigning season, they were probably settled or garrisoned in very small groups such as those attested in contemporary Egypt. The troops within the royal household were administered by his *maior domus*, who took direct control during regencies and became more prominent during the 7th century.[89]

Royal troops in outlying districts were led by regional military commanders, *duces*, who "bear a close resemblance to the *duces* found at this same time in Lombard and Byzantine Italy or Visigothic Spain."[90] In the north and east, the *duces* led fixed districts (e.g. Champagne, Burgundy) that probably reflect late or sub-Roman military organization;[91] otherwise, their commands could fluctuate depending on changes in the political geography or served as extensions of the royal household. The early *duces* may in fact have had humble backgrounds as officers in the early Merovingian military establishment or the royal household (cf. the high prevalence of Germanic names among them), but soon became synonymous with the

---

[85] E. James 1988 provides a series of good maps on the divisions after 561; see 4.2 below for further discussion.
[86] For the distribution of Merovingian fiscal lands (palaces, estates, villas etc.), see Ewig 1965.
[87] See chapter 1.2.3.
[88] On the Merovingian royal household, see the discussion in Halsall 2003: 49f.
[89] On this, see Kaiser 2004: 115f and chapter 1.2.3 above.
[90] Lewis 1976: 389.
[91] Bachrach 1972 *passim*.

high aristocracy.[92] When not in charge of a division of the royal household troops, late-6th-century *duces* with estates of their own had substantial military followings in their own right, which may have numbered several hundred men.[93] This came in addition to their official commands, which included subordinate counts, who were in charge of the *civitas* and its military resources. Counts are normally believed to be of "Roman" origins and also had their own followings, which may also have numbered in the hundreds.[94] Aquitaine and the immediately surrounding *civitates* preserved a military organization that was taken over from the kingdom of Toulouse, strongly based on private military followings. During the 6th century but probably a survival from the gradual transition to Visigothic rule a century earlier, troops were organized *civitas* by *civitas* due to the political divisions of the day. Merovingian kings often only held scattered city territories in the south and southwest, and regional commands were only created when a large number of cities could be grouped together.[95]

The exact composition of individual Merovingian armies is often difficult to determine, as in most cases they are only referred to as an *exercitus*, army, of a region or kingdom. At a lower level, Gregory refers to the *homines*, men, of a particular *civitas*.[96] A close analysis of the narrative sources reveals that the lower-level *civitas*-organization had two tiers. The largest group consisted of able-bodied poor civilian men (*pauperes*), organized by the landowners or royal officers upon whom they depended. This group

---

[92] In general on their function and distribution, see Lewis 1976. The most updated prosopographical survey is found in Weidemann 1982. Halsall argues in several publications that the high aristocracy in northern Gaul only became settled towards the end of the 6th century, and that internal competition is reflected in grave customs, which he believes reflect competitive display among the rising aristocracy in the early 6th century, but the practice disappeared when the situation stabilized.

[93] See GT 5.14, who shows that over 500 men accompanied *dux* Guntram Boso when he escorted Merovech, a Merovingian prince, who had gotten entangled in a difficult political situation. Some of the men may have been Merovech's *pueri*, who are attested intermittently, but it seems reasonably clear that most of the men belonged to the duke. Mummolus the *patricius* (equivalent of or higher than *dux* in the Burgundian realm) defied king Guntram with his own personal following, while Guntram appointed another *patricius* to take command in Burgundy.

[94] E.g. the count at Tours had 300 men at his disposal which came in addition to ecclesiastic military resources, since they were used to pen in Gregory. While some of them may have been royal troops, landowners with military obligations or the like, a large proportion were at the count's immediate beck and call. Similarly, when Sigisvult was sent as governor (presumably count) of *Clermont in 524, he brought his whole *familia*, or following, to hold the city.

[95] See e.g. GT 8.18; Lewis 1976: 390.

[96] On the latter, see e.g. Bachrach 1972, Heather 2000.

was essential for logistical purposes (see below) and could also provide extra manpower for defending cities and fortifications, but did not normally fight.[97] The revolt of Munderic at *Vitry in 524 was accompanied by throngs of the common people, presumably his personal dependants mobilized in this fashion. The (far) narrower group, and the basis for expeditionary forces, was formed by professional troops, *homines* proper, who served in the retinues of local magnates,[98] sometimes supported on campaign by sections of the general "militia" for logistical purposes.[99] Gregory gives us a hint of this composite structure: when Guntram ordered the *homines* of various cities to attack the Bretons in 584, most of the men of Tours seem to have taken part (such as the troops under the count's authority). However, the 'poor citizens' (*pauperes*) and the 'young men' (*iuvenes*) of the cathedral failed to show up for the campaign, citing the traditional exemption from expeditionary duty.[100] The "young men" were clearly the military members of Gregory's *familia*,[101] while the *pauperes* provided support functions. Merovingian armies, then, consisted of conglomerations of military followings and divisions of the royal household troops.

The retinues of bishops and lay magnates are mostly extras and props in Gregory's drama (they were the ones who actually exercised "aristocratic" violence), but they accompanied their lords in all their affairs, and are thus ubiquitous in all his writings. They were hence a large and important social group. They must be regarded as professional, full-time soldiers, because they never seem to be involved in any other sort of business; indeed, they seem to have been more engaged in fighting (due to internal conflicts and aristocratic feuds) than most Roman soldiers normally were. In the narrative and legal literature, they go under a vast array of names,

---

[97] This is somewhat analogous to Bachrach's general levy in many publications (e.g. 2001), although this example is particularly pressed into serving his argument in his 1993b article. However, Durliat 1993 argues that these men were not called out to fight, but to provide transportation services and workforce during sieges and other conflicts requiring access to labor. Similarly, I do not accept their usefulness for battle and argue with Durliat, on the basis of the structures identified in chapters 1-3, that *pauperes* called out in support of armies were indeed the standard late antique *modus operandi* inherited directly from the Roman Empire. See further below.

[98] In structural terms, i.e. the social and military significance of military retinues, if not scale and capability, I am in agreement with Halsall 2003.

[99] Perhaps they also included well-equipped, individual landholders, as envisaged by Bachrach's "select levy," but they are for structural reasons less important for the argument presented here.

[100] GT 5. 26.

[101] Elsewhere, Gregory refers to his *pueri* who were involved in scuffles, e.g. GT 8.40.

including *pueri, vassi, satellites, antrustiones* for individuals, but as groups were known as *trustis, contubernium, obsequium, familia*.[102] The size of such followings is in most cases difficult to gauge, but as we have seen, several hundred seems to have been normal for the most powerful dukes and counts;[103] in effect, they were the same size as the military followings of East Roman generals, but far more ubiquitous because all magnates, officeholders and most bishops (see 4.2) had such followings.

According to Halsall, large armies were impossible to sustain because few cities in Gaul had more than 5,000 inhabitants, and many villages only around 50.[104] What is often forgotten in such arguments, however, is that a very large number of these villages belonged to much larger estate complexes, whose cultivators paid dues and/or performed services for their lord (cf. the *pauperes*), depending on the nature of the estate organization.[105] The diversity of the estate economy, even in northern Gaul, is clear from two documents from the early 6th century: the testament of St. Remigius and the *Pactus Legis Salicae*. Remigius willed his personal property, which at his death consisted of the portions of four estates and other scattered holdings inherited from his father, a typical medium-range northern Gallic landowner of the mid-5th century.[106] It is sometimes pointed out that Remigius' holdings were rather small, but as a cleric, he may already have disposed of much of his property long before the will was drawn up, and at any rate it was only a portion of a substantially larger complex that still functioned, but had been shared with his relatives. It will be recalled that Genovefa kept *Paris (490) supplied from her estates for over ten years; similar logistical abilities were common around 500.[107] The *Pactus Legis Salicae* confirms the image of a medium-sized, but quite diversified estate economy in northern Gaul, which only became more complex and large-scale the further south one looks, and far better attested in the 7th century.

---

[102] The terms are too ubiquitous to be surveyed here (see searches of these terms in the relevant volumes of MGH at dmgh.de); I plan to return to the problem of retinues in a separate article.

[103] GT 5.24, Merovech was escorted by Guntram Boso who had over 500 men.

[104] Halsall 2003: 125f.

[105] The exact nature of agricultural organization is still debated. Wickham 2005 and 2009 argues that between the great cases of the late Roman Empire and the bound Carolingian peasantry, many villages were largely independent, but the evidence adduced here shows that great landowners exercised considerable social control over their dependents. See chapter 4.2.1 and *passim* for this.

[106] Wightman 1985: 261.

[107] Chapter 1.2.5 above.

Since soldiers were dependents of a lord, they were supplied through the estate structure of their patrons in peacetime.[108] However, on campaigns, it was the personnel and agricultural surplus from villages and estates near the marching route that provided for an army's logistical needs. Foodstuffs could be assembled in advance, and were levied from the general population as a tax. This was immensely unpopular, at least in Gregory's presentation, but seems to have been fairly routine in the 6th century. The vast throng accompanying princess Rigunth (4,000 of the "common people" plus her personal escort and the retinues of prominent officers who accompanied her) was supplied at depots.[109] An alternative was to shift produce from the royal, aristocratic and ecclesiastic estates whose forces were directly involved in a specific campaign (and presented as the proper alternative by Gregory) instead of burdening them on the people,[110] who had immense labor obligations anyway. There is good evidence that foodstuffs were prepared in advance for ambassadors and their retinues according to detailed lists, ordering what should be stored in specific quantities at specific locations.[111] Estate managers had assembling and shifting supplies as their regular daily business, and are known to have supplied cities in preparation for sieges (*Convenae 585). Since troops were scattered in small numbers and only occasionally brought together for specific purposes, such as hunts, valuable for training, or *publicae actiones* to provide security and enforce the law (or, of course, squabble with political rivals), the logistical operations were quite simple considering the scale of estate organization, and rarely noticed by any texts. On a larger scale, armies were preceded by officials who went about collecting necessary foodstuffs, which could be deposited in granaries; from the tone in Gregory, it seems clear that they were zealous going about their business. A final alternative, however, was to buy supplies.[112]

Early Frankish engineering was much more sophisticated than commonly thought, and was possible thanks to the ability to organize labor on

---

[108] If their patron was the king himself, or royal officers serving away from their personal resources, they could live off royal estates and the taxes still levied by the government.

[109] GT 6.45, 7.10.

[110] See the previous note; however, as the following on purchases shows, Gregory always found grounds for criticism.

[111] E. James: 1988: 188f; see also Marculf I, 11 for the original text.

[112] While Gregory uses this as an occasion for criticism, it seems clear that the troops returning from the *Frankish invasion of Italy in 590 (see GT 10.3) used their arms (probably gained as booty from the Lombards, although this is not mentioned by Gregory) to purchase supplies from the local population on the return trip. See further Goffart 1972: 176 for similar purchases made by troops on the march in Ostrogothic Italy.

THE FRANKISH KINGDOMS   221

a massive scale. The Franks were quite adept at building field fortifications, such as the one built at Volturno, or in Burgundy for stopping Saxon and Lombard invasions.[113] They could also bridge rivers, a particularly difficult task that required highly trained specialists in the East Roman Empire.[114] Civil engineering was quite substantial; the course of rivers were diverted on several occasions, one known example to protect the city from being undermined by the current, the other to provide extra protection during a siege.[115] There was clearly an ability to build stone fortifications; thus bishop Nicetius of Trier had a heavily fortified residence built in the mid-6th century, while Gregory of Tours marveled at the fortifications of Dijon.[116] Chilperic, when threatened with an invasion by his brother in 584, ordered his magnates to repair city walls and bring their relatives and movable goods inside. He recognized that their lands and immovable goods risked being destroyed during an enemy invasion, and therefore guaranteed that they would be reimbursed for any losses.[117] There was thus an *obligation* to repair city walls on the part of the landowners, who could again draw upon their dependants to perform these tasks. It was also in their self-interest, since power struggles among magnate factions often involved military action.

Indeed, Merovingian kings had the same mechanisms available as Valentinian III, Theoderic and Anastasios to impose burdens of military logistics.[118] The well-known labor requirements that descended from ancient *munera* had become the traditional seigneurial obligations of the dependant agricultural population, mobilized by their patrons on royal orders.[119] While the "Franks" vociferously protested against taxation, providing military and logistical service was not an issue. As demonstrated by 7th-century immunities granted to monasteries, common obligations required by

---

[113] For the former, see 4.1.3 above and Agathias 2.4.5-7; for the latter, see chapters 4.1.2 and 3.3.1 above.

[114] GT 10.9; the Avars forced Roman engineers to bridge the Danube for their siege of *Sirmium (579). I argue that these were also the ones responsible for transmitting advanced siege technology to the Avars, cf. 7.2.3.

[115] For the river diversion, see Coates 2000: 1122; for the added siege protection, see *Avignon 583.

[116] GT 3.19.

[117] See Bachrach 2002b for the general context and analysis of the strategy that lay behind.

[118] See chapters 1.2.5, 2.2, 2.4 and 3.1. for this.

[119] See Kaiser 2004. Goffart 1972 is the fundamental study on the extension and exploitation of labor burdens in lieu of taxes from the 4th century to the Merovingians and Carolingians; the problem is addressed in chapter 4.2.1 below.

the king, administered through his officers and landowning subjects, included transportation and bridge-building. *Civitates* and *castella* are specifically mentioned as places where such labor was normally called out.[120] No immunities were given for repairs of fortifications, however. The exact method of organizing repairs must have been the assignment of *pedaturae* to the landowners in question, as was the case with Ostrogothic *possessores* or East Roman social units and corporate bodies. Although labor obligations were also universal, in e.g. Roman Mesopotamia and Ostrogothic Gaul, as we have seen, extraordinary burdens or expenses were sometimes defrayed through tax relief or cash payments. The decline of direct taxation in Gaul meant that magnates had to shoulder far larger military burdens in the form of retinues, expeditionary service, and garrison troops whenever called upon, as well as routinely supplying labor for logistics and engineering. Thus, while military service and the burden of repair was mandatory on landowners (and apparently not an issue), Chilperic had to make sure that they would support him even if their estates were being ravaged. If they risked losing their economic basis, a negotiated settlement with his rival would soon become more attractive, as we saw above.

During the Merovingian era, most cities (discussed further in chapter 6) still had economic activities useful for military purposes, and were also the homes of at least part of the *familiae* of kings, bishops, counts and sometimes other magnates. Where their craftsmen and specialists actually resided is more problematic and probably varied from case to case.[121] As early as the *Pactus Legis Salicae*, the Franks highly valued their dependent labor: not only were there detailed punishments for stealing or damaging a wide range of crops and livestock, it also lays down heavy fines for the theft of skilled slaves. Indeed, the range of craftsmen available and degree of specialization under the Merovingians is rarely addressed by military historians, whatever their views, but they are in fact quite ubiquitous in the original sources, while recent archaeological surveys show that their skills in many key crafts were neither inferior to, nor more narrowly distributed than, those of Roman craftsmen.[122]

---

[120] Wood 1994: 215.
[121] See Claude 1981, Nehlsen 1981 and chapter 6.2.3 below for further discussion.
[122] Craftsmen are listed in X.6: "He who loses [i.e., steals and sells] ... a ... metalworker, miller, carpenter, or groom (*stratore*), or any other craftsman worth twenty-five solidi ..." Those responsible for the running of the estate and keeping of horses are also attested: 6. He who steals or kills or sells and overseer (*maiorem*), steward (*infertorem*), butler (*scancionem*), horsekeeper (*mariscalcum*), groom (*stratorem*), a metalworker, goldworker or carpenter ..." Further: "7. He who steals a slave who keeps young horses (*puledrum*) ..." For

All of these groups have actual or potential military applications, and could be summoned at will by their lords whenever their services were needed. A certain number of craftsmen joined any major expedition as camp followers to perform various tasks as need arose, forming a specialized segment of the *pauperes* (noted earlier in this section). Thus Mummolus had his servant *faber* (probably one of several—he was only mentioned by Gregory for being so huge) brought from *Avignon (583) to *Convenae (585). In addition to destructive traps, the defense may also have involved artillery. Bishop Nicetius of Trier's large fortified estate center was defended by a *ballista*.[123] These were complex machines requiring specialist operation (*tekhnitai* or *ballist(r)arioi* in Greek sources), and unless imported from East Rome, they were trained in a local tradition. It just so happens that Mummolus had been commander of a region that had extremely strong Roman traditions, and that craftsmen there could maintain military skills over several generations. We can recall the *artifex* at *Vienne in 500 who played a vital role during the siege. Nicetius, in turn, was bishop in the region that had one of the highest concentrations of Roman arsenals and *fabricae* during the early 5th century, and where self-consciously Roman officers were still active until at least 480. It is possible that Franks had picked up ballista-operating skills on an Italian expedition. If this is the case, it reveals that once in Gaul, the experts would have to be maintained by a magnate's household, which basically proves its suitability as a valuable form of military infrastructure. Indeed, in the *Epistulae Austrasiacae* there is preserved a letter from bishop Rufus of Turin to Nicetius, explaining how he finally has the opportunity to send the *portitores artifices* that Nicetius asked for. The combination of terms seems to be highly unusual, but they were presumably boat (barge) builders, as Nicetius' estates were on navigable Rhine tributaries. Another explanation is that military skills survived along with military organization, and was gradually reorganized according to political developments, with more and more of the logistics and resource allocation devolving on great magnates in return for tax exemptions and immunities.[124]

---

a survey of craftsmen and their social relations in Merovingian Gaul, see Claude 1981. Henning 2007 provides a very illuminating survey of the level of specialization and skill among various key crafts, finding that the Merovingian (and Carolingian) craftsmen were highly competent.

[123] See chapters 1.2.5 and 8.2.1 for further discussion.

[124] For Rufus and Nicetius, see *Ep. Austr.* 21. See chapters 1.2.5, 8.2.1 and 8.3.2 for further discussion of this point.

4.2 The 7th Century: Ascendancy of Military Followings and Proprietal Warfare[125]

Urban civilization seems to have reached its nadir sometime around the late 7th century, but it never disappeared completely. At the same time the estate economy became relatively more important. This led to the rather peculiar situation that while cities remained significant political and administrative centers, refuges, and worthwhile military assets (and thus targets), most of the forces by the early 8th century used to capture cities, and even their siege trains, appear to have been raised and supplied increasingly from royal, ecclesiastical and aristocratic estates.

4.2.1 Fredegar, the Liber Historiae Francorum, and the United Frankish Kingdom

The source problem is most acute with the only connected narrative, the chronicle of Fredegar,[126] which is even more brief than Gregory: 60 years of Merovingian history from 594 to 654 are covered in the space of 69 pages in the standard edition; the next 73 years to 727 (mostly taken from the *Liber Historiae Francorum*) in 11. From our discussion of the 6th century, it is clear that Gregory did not give a fair description of warfare and military organization; only incidental details interpreted in the light of a few Merovingian documents and external sources allow us to piece together a more revealing image. Fredegar covers more time in fewer pages, provides far fewer revealing anecdotes, and what must have been large, complex campaigns involving battles, raids and the fall of great cities barely merit descriptions of a line or two. Fredegar's handling of the Gundovald affair in 585, where he used Gregory as a source, amply illustrates this problem. Whereas the siege of *Convenae receives a reasonably full description in Gregory's account, including composition of forces, their commanders, logistical arrangements, technology used, tactical difficulties met by defenders and attackers, as well as the aftermath, Fredegar only informs us that the city was taken and the pretender Gundovald was thrown off a cliff.[127]

Thus, the next stage of civil war ended with the fall of two kings and their formidable grandmother, Brunchild, and the sole rule of Chlothar II

---

[125] For the context, see Ewig 1976: 172-230, Wood 1994: 140-272, and Fouracre 2005.

[126] See Goffart 1963 and Wallace-Hadrill's introduction to his edition and translation for older debates; for a new assessment, see Collins 2007.

[127] Compare Gregory's original account with abbreviation in Fredegar 4.2; see CO for references to Gregory.

by 613, effectively ending civil wars for two generations through the reigns of his son Dagobert I (r. 623/29-639) and grandsons to 656. During this war, there were at least two great campaigns revolving around cities, in 600 and 612. Since the wars of this period still largely involved the people who were active during the later years of Gregory's narratives, there is no reason to assume a fundamental change in style of war. Indeed, the text of the first campaign, conducted jointly by Theudebert and Theuderic against Chlothar in 600, is quite explicit that the *Cities (*ciuitates*) along the Seine were *ruptas*. The word *ciuitates* should here be taken to mean the walled cities (normally *urbes*), as it is set in opposition to the surrounding *pagus* (sic for acc. pl.) that were also ravaged. An absolutely minimal explanation would be that the walls were breached, and we know from elsewhere how that could happen.[128] The conflict was deadly serious as the populations of these towns and their surrounding districts were enslaved. During Theuderic's campaign against Theudebert in 612, *Naix and *Toul were captured (*ceptum, cepit*) by force, and after a disastrous battle, *Cologne submitted. The first two cities probably involved major efforts, but we know little of the modality, especially at *Cologne. Since it involved the fall of a king, we must assume that it took some effort.

Chlothar II's *Edict of Paris*, issued in 614, shortly after the war, has been seen as a major turning point in royal-magnate relations. Since it stipulated that a count had to own land within the area where he held office, it has been regarded as an important concession to the aristocracy, who thus gained more local control than before. In addition, a large number of immunities against remaining royal rights and exactions survive from the 7th century. The traditional view was that this provided local magnates the wherewithal to usurp royal prerogatives within their territories, setting the stage for magnate infighting in the late 7th and early 8th centuries (and by extension the Carolingian breakup in the late 9th century). However, Murray showed how this edict was actually based on Byzantine policy enacted in Italy in 554 in order to ensure uncorrupt government: an officeholder with lands could effectively be punished with confiscation if he was derelict in his duties. While this is largely accepted, it is still common to emphasize the relative decline of royal power through the decline of the

---

[128] See the phrase *rumpendo muros* ("by breaking walls") used by East Roman commanders to describe their activities during the *Italian campaign in 590 in alliance with the Franks; see chapter 5.2 for various methods of breaching walls.

residual Roman tax machine which operated until around 600.[129] Wickham spells out the process in detail but argues that the decline of a centrally administered tax system did not matter too much, since the Merovingians had extensive lands of their own and anyway their armies were too small to need a tax base for their upkeep.[130] As we have seen, this view can hardly be correct—the Merovingian armies were both large and professional. Thus, another explanation must be found. Murray also pointed out that immunities from royal intervention on estates were predicated upon enforcing peace and discipline, i.e. a requirement to participate in the defense and internal policing of the kingdom. Goffart found that while the direct taxes were gradually eroded, as noted above, the extensive obligations (*munera*) required by the Roman state survived. Although they were to be administered by the landowners themselves, the obligations were painstakingly exacted by the Merovingian kings.[131]

If we relate the obligations that were regularly applied to landowners in return for tax concessions in the late West Roman and the contemporary East Roman Empires to the change in Merovingian fiscal policy between c. 550 and 600, we reach some interesting conclusions. With the gradual militarization of the entire landowning population throughout the 6th century, the Merovingian kings were more than happy to relinquish tax income as long as they received *the equivalent in services*, the most expensive of which was of course military service. This had been a regular feature of late Roman fiscal administration that continued to be applied in the East Roman Empire and the Caliphate whenever necessary, but interestingly, Justinian again provides a practical model: his novel of 545 abolished something of the arbitrariness and harshness in the late Roman fiscal regime, as extraordinary levies were subtracted from the same or following year's tax bill (see chapter 1.2.5). This formally established the principle that services could substitute for payments, if exacted on an equivalent scale. In the West, service obligations did become the guiding principle of military administration in the last half of the 6th century, and remained an inescapable condition for immunities in other matters. Following from Goffart's arguments on a two-tiered tax structure, where only taxes in gold

---

[129] Goffart provides a series of studies on the fate Roman taxation; see e.g. 1972, 1982, 2008.
[130] Wickham 2005 and briefer 2009: 120f; see also section 4.1.1 for similar views. However, Wickham describes the capabilities and complexity of Merovingian administration in other respects. The ability to keep records would obviously be useful for enforcing military and other obligations.
[131] Murray 1994 and next note.

or in kind were resisted, early medieval rulers were in effect able to exploit a larger proportion of the wealth in their kingdoms than the Roman emperors ever had by delegating lower-tier exactions of labor and services to landowners.[132] Justinian's model explains not only how taxes became defunct by 600, but also how large armies were maintained without a system of direct taxes even in substantially poorer, less urbanized areas. Wickham demonstrates how this process happened at different rates in different Frankish regions, indicating a process of negotiation and gradual consolidation. Royal authority, then, to a large extent was focused on enforcing these obligations, while Gregory of Tours demonstrates how this transition was perceived. They were never allowed to atrophy the same way as taxes, however, because the services were still maintained by magnates for their own benefit and in competition against other aristocrats under weak kings. Inter-magnate competition was largely based on the resources mobilized for royal service as well.

The pattern of military organization remained the same as in the 6th century, with some modifications. The mobile range of some offices gave way to more fixed geographic structures, as landowning alone replaced direct taxes as the basis for power.[133] Large-scale landowning is very well attested during this period: the testament of the bishop Bertram of le Mans from 616 shows that at the time of his death, he owned estates and lands throughout most of Gaul, estimated at 300,000 hectares (3,000 km^2), producing corn, wine, salt, pitch and providing pasture and forest areas, which could be used for livestock, horses and construction.[134] His estates were thus able to produce a very wide variety of foodstuffs and materials, much of it by tied labor (slaves or *coloni*).[135] Although I have found no argument about the manner of organization, it is clear that the transition from ancient forms to bipartite estate organization was under way and may have been spurred by the complete militarization of landowner's obligations. Even before Sarris argued for the introduction of the bipartite estate to the West around 400, it was recognized that this particular form of organization began to spread in Gaul perhaps as early as the late 6th century. An increasing number of ecclesiastic and royal agricultural complexes were

---

[132] See especially Goffart 1972 and 1982, although some of the implications are only stated in passing.
[133] See Lewis 1976 for the evolution of a more fixed structure; for a discussion of the role of landowning versus taxation, see chapter 4.2.2.
[134] Wood 1994: 207-10, 214f.
[135] For Bertram, see Weidemann 1986: 102-12; for other examples of episcopal estates, see Lebecq 2000.

systematically reorganized as bipartite estates throughout the 7th century, the argument being that increased aristocratic competition and royal pressure required ever more rational and productive forms of management.[136] The role of the church hierarchy in military organization became entirely regularized, and was also extended to the rapidly multiplying monastic houses.[137]

Under united rule, Frankish hegemony was established all along the fringes of the *regnum Francorum*: Fredegar claims that the Lombards paid tribute to the Franks.[138] After the debacle in 589, the Burgundian Franks left the Visigoths alone, but Dagobert I, as ruler of a united Frankish kingdom, intervened on behalf of the pretender Sisebut against king Sisenand in 631. This invasion was very successful, reaching all the way to Saragossa and installing Sisebut as king. The Franks were rewarded by a huge Visigothic payment of tribute.[139] Dagobert's reign marked the high-water mark of Frankish influence in the Visigothic realm. Frankish hegemony also extended over southern England, while Frankish dukes were imposed on Gascons in the southwest and several eastern peoples. This led to further integration of Saxons and Slavs into Frankish political culture, economic structures and military organization and practice. For example, the Slavic Wends, who gained independence from the Avars under a Frankish merchant named Samo, began organizing a state in the 620s that built massive fortifications.[140] When Dagobert wanted to subdue him, he set in motion a large-scale invasion in four columns, two from Frankia proper, one from Alamannia, and one from Lombard Italy. The most notable of the fortifications prepared by the Slavs was *Wogastisburg (630).[141] These were difficult to storm and took a major effort by the Austrasian army, which failed nonetheless. Some of the Frankish invading columns still took many

---

[136] Sarris 2004; for the reorganization of estates into biparte structures, see Goffart 1972: 373-76.
[137] See chapter 4.2.2 on this.
[138] Fredegar 4.45.
[139] Thompson 1969: 171f.
[140] Fredegar 4.68.
[141] The site is yet to be identified, but large-scale earth and wood fortifications were the norm under Avar hegemony (cf. Squatriti 2002) and have been identified on a massive scale in 9th-century Slav polities with known antecedents in the 8th century. The practice of fortress-building among western Slavs probably originated the 7th-century context discussed here; see Brather 2008: 62-66 and 119-40 for an introduction to the 3000 currently known early and high medieval Slav fortresses constructed (mostly) in earth-and-wood. At the same time the Balkan Slavs and Bulgars were beginning their processes of slow assimilation to Roman structures; cf. the relevant sections in chapter 7.2.

slaves, but the Slavs responded by extensive raiding and were joined by the Sorbs. Slav invasions occurred in 631 and 632. The rise of a powerful Slavic polity disturbed the Frankish client system, and the Franks struggled to maintain control over their tributary frontier peoples. The Slav incursion threatened Frankish control over Saxony, which the Franks invaded. Having submitted, the Saxons asked to guard the border instead of paying annual tribute. The Thuringian Duke Radulf also sought independence upon the death of Dagobert; his *castrum at Unstrut (639) was subject to an uncoordinated Frankish attack. Since Radulf was himself a Frank, he had several supporters in the Frankish camp who held back their assistance. A combination of skilled defense and betrayal led to great Austrasian losses and Thuringian independence.

These earth-and-wood fortifications could withstand large Frankish armies that invaded in multiple columns. Their scale and significance should therefore not be underestimated; these were the works of nascent states with significant administrative and organizational abilities. Given the location of these states in Central Europe and along the middle and upper Danube, inspiration may be due to Avar, Bulgar or even Byzantine influence, but in light of common Merovingian practices demonstrated above, and the fact that the Wends and Thuringians were periodically ruled by the Franks (as were, probably, the Saxons), there seems to be little doubt that their military organization and practices were most strongly influenced by their huge Frankish neighbor to the west.[142]

4.2.2 Bishops, Magnates and Monasteries

After the death of the great Dagobert in 639, and especially after the reign of his sons, the devolution of power and control onto great magnates come to the fore in Fredegar's first continuator, the *Liber Historiae Francorum*. This source emphasizes the great civil strife that took place after 656. Indeed, Merovingian politics during 656-80 was characterized by Wood as "The Failure of Consensus."[143] Little is known of late 7th-century politics and foreign interventions, but this does not signal an end to Frankish power.[144] The famous *History of Wamba* in 673 illustrates that rebels still counted on Frankish support for holding cities and fighting sieges (see chapter 3.2.3). The extremely swift response by Wamba precluded Frankish inter-

---

[142] See chapter 7.2. *passim* for a discussion of the various possibilities.
[143] Wood 1994: 221.
[144] For an assessment of Frankish power in the 7th century, see Fouracre 2005: 377-80.

vention on a large scale, but he was careful to let the Franks in the rebels' service return home unharmed so as to avoid political consequences. Although the Lombards heavily defeated the Franks in a war in 663, it is clear from other evidence that the Lombards still feared their northern neighbors, despite the apparent decline of a strong, centralizing dynastic power. During this period, the evidence of warfare and military structures in *LHF* and its derivation Fredegar *Continuatus* 1-10 is very meager, as this source covers 70 years in a couple of pages. The factional struggles actually seem like proprietal warfare centuries later, i.e., there was a close correlation between possession of landed resources and military power, rather than formal office, which foreshadows Charles Martel's rise to power two generations later and warfare until well into the high middle ages.[145]

Another direction in this direction was the complete militarization of the church. Many Western bishops had militarized their offices since the 5th century, as most of them came from a secular background as senatorial aristocrats, and we saw a particularly egregious example of this with Hilary of Arles in the 440s.[146] Maintaining household troops was not yet universal among the clergy, however, and at any rate, bishops and clerics were expressly exempt from military service in the mid-6th century. It is not clear to what extent all their dependents (e.g., large-scale tenants) were affected by this.[147] Yet a large number of bishops clearly did have military followings, and the practice seems to have become universal by the late 6th century. Since they were exempt from military service, king Chlothar I (511-61) wanted to impose a tax of 1/3 upon the church, but the whole affair was averted by the wrath of bishop Injuriosus.[148] Afterwards, kings seem to have applied different approaches, instead demanding military service. This is one of the reasons why Gregory of Tours kept getting into trouble with the king and his officers. Not only did he refuse to let Tours pay taxes, which the Merovingian kings let slide towards 600 because they were substituted by military service; but he also weaseled his way out of supplying troops and joining royal expeditions, the most important royal prerogative. In his defense, a difference in interpretation may lie behind Gregory's behavior: the king believed they were close enough to be called

---

[145] For the term and its applicability in the high medieval period, see France 1999.
[146] See chapter 1.2 for the general context. The standard work for the militarization of the church from late antiquity to the high middle ages is Prinz 1971, who places the regularization of this practice to Charlemagne, although here it is dated to the 7th century. See further Gauthier 2000, who ignores the military role of bishops.
[147] *Conc. Aur.* 541, c. 13.
[148] GT 4.2.

out in full force for the (defensive) expedition to nearby Brittany, but were kept at home by Gregory himself in order to protect the city in case of Breton counter-raids or other problems.[149]

The incident indicates that this was a period of transition for the church but it was completed shortly after. While the attempt to tax church lands was fiercely denounced and apparently resisted successfully, it seems that by the early 7th century, the compromise was that the church, just as the secular magnates, should provide military forces on the same conditions as all landowners. This means that the manpower and logistical capacity under the bishops' control was also directly transferable to warfare. The ability to organize large-scale labor also existed among secular magnates, but few secular records survive. Again this is only well attested for a bishop, Desiderius of Cahors (d. 655), who not only came from a family of southern Gallic magnates and secular officeholders, but also had a distinguished career at the courts of Chlothar II and Dagobert before his election. At Cahors, Desiderius dedicated himself to urban improvement. He organized the building of several churches inside and outside the walls.[150] In order to improve the water supply to the city, he asked his colleague, Caesarius of Auvergne, to supply craftsmen (*artifices*) who were to construct underground wooden piping.[151] Most importantly, he also organized extensive repairs of the city walls,[152] which is very interesting for a time when urban civilization hit an absolute low point in the West. These repairs were not only for show. Bishop Leudegar had a similar program at *Autun, where he embellished the baptistery, built a new atrium and repaired the *domus* (pl.), presumably the buildings used by his household. Significantly, he also restored the city walls (*murorum urbis restauratio*); his career ended when he was besieged by royal troops in c. 679 (see CO and below). The technology used for building was quite advanced. In *passio Praejecti* (c. 11), relating events in the 660s, bishop Praejectus of Auvergne (Clermont) organized the renovation of a building in a nearby village or suburb (*in vico*). The

---

[149] Incidentally, during the same campaign another church was miraculously saved from being fined for not participating; see GT 7.42.
[150] *Vita Desiderii*, c. 16 (at p. 574f): erection of basilicas inside and outside the city walls of Cahors. Also mentioned in c. 20.
[151] Desiderius I, 13 to Caesarius bishop of Auvergne: "[...] Praeterea credo, quod nec vobis lateat, qualem egestatem de aqua, quam fons prebeat in hac Cadurcina civitate, habemus. Sed voluntas nobis inest, si possis permittit, ut per tubos ligneos subterraneo officio ad ipsa civitate aquam ducere debeamus. Proinde, quia novimus, quod peritos ex hoc artifices haberes, precamur, ut conpendium nobis de ipsos faciatis. Et unde iubetis, non plenissime reservimus."
[152] *Vita Desiderii* c. 17 (p. 575).

builders (*structores*) operated a treadmill crane which they used to hoist stones 60 feet up; the miracle of course involved a catastrophic machine failure in which a *structor* against all odds emerged unharmed from the rubble.

While the episcopacy had become thoroughly militarized, the increasing number of monasteries founded in the 7th century meant that large tracts of land risked escaping traditional burdens. This was rectified by Dagobert, who imposed heavy military obligations on monasteries as well.[153] This was clearly a Merovingian innovation, but the precedent was set when the militarization of the episcopacy had become regularized.[154] Thus, by the early 7th century, most of the institutional features familiar from the Carolingian era were well in place, and the rapid extension of the bipartite estate, which was suited for maintaining and supplying private retinues, as we have seen, is attested on episcopal and monastic lands in the 7th century.[155] Again we might see an Eastern precedent for these policies: as noted in chapter 2.4, the East Romans in the first half of the 6th century formalized a series of obligations on bishops to organize local defenses, among other things, perhaps inspiring Clovis II. Herakleios imposed massive levies on the church in the 7th century in order to raise money for his defensive wars. If the Merovingians followed East Roman cues for how to control their counts, they surely did the same for their bishops and monasteries.

4.2.3 *Late Merovingian Siege Warfare*

Due to the poor source situation, we have very few detailed accounts in chronicles of how warfare was actually waged in the late 7th century. However, the institutional framework of large estate complexes with extensive military obligations allows us to understand how armies could be raised and supplied. Furthermore, a few illuminating descriptions in saints' lives survive that we know were composed during this period. These *vitae* and *passiones* do not, unfortunately, give the same level of detail as the *Miracula sancti Demetrii*, but that is because they seek to exculpate their main protagonists, the bishops of various cities, who were engaged in brutal struggles for power with their magnate peers; thus their military efforts are very much toned down. Still, illuminating details survive. There we see the

---

[153] *Miracula Martinis*, c. 7 (p. 571f).
[154] Thus Lebecq 2000; for the argument that Charles Martel reimposed these obligations, see introduction to chapter 4.3. below.
[155] Sarris 2004.

violent deposition of two bishops, with one bishop's city surrounded by ravaging enemy troops and another siege fought out in quite a lot of detail.[156]

In the *Acta Aunemundi*, the bishop Aunemund of *Lyons (662) incurred the king's displeasure, and the king wanted him taken captive or dead. Aunemund had been accused of possible treason, specifically wanting to call in foreign peoples, most likely the Lombards.[157] The king dispatched three dukes, apparently with overwhelming force, to quickly resolve the issue. The information in *Epistulae Austrasiacae* and Gregory provides some indication of strength. If one duke hade in his personal following 500 men, and could draw on counts, clerics and others with military obligations when on expedition, so that 20 *duces* lead a force of about 20,000 men back in 590, we can deduce that three *duces* led an army of 1,500-3,000 men (or even more, since this was within the kingdom and royal troops as well as all other categories were presumably available). When the force arrived and began to besiege Lyon, Aunemund apparently found, despite fervent prayers and alms, that he did not have enough support to withstand the army that began to set fire to the countryside and lay waste the suburban villages (*oppidis*). He subsequently gave himself up as a sacrifice alongside the abbot Waldebert, from whom he had sought advice and support. The army returned with their captives, camping near the estate of the church of Chalon. While Waldebert was released, Aunemund was killed later that night. The site where the army camped is very interesting. Such a large body of men would have some difficulty in carrying all their supplies, but having set up a camp near a production center of the church, they could presumably be supplied from their stocks.

A decade later, bishop Leudegar of *Autun (679), who had made preparations defend his city (see 4.2.2), found himself in a similar scenario. Again the issue was treason against the king, who raised and dispatched an army against Leudegar. However, he had far better local support than Aunemund, and it took hard fighting over the walls (*supermurale*) to beat him into submission. This was a world of wall-building bishops who fought their kings from the ramparts of their sees; it is also the environment in which Arnulf of Metz, Pippin the Elder, and Pippin (II) of Herstal established the Carolingian dynasty and its power base.

---

[156] The political context to be found in Wood 1994: 224-38.
[157] Geberding and Fouracre 1996: 178f.

4.3 The Carolingian Ascendancy in the 8th Century[158]

Those familiar with the debate surrounding early Carolingian warfare will immediately have recognized many "essentially" Carolingian features in the preceding discussion. While the Merovingian system to a certain extent fragmented around 700, most of the resources previously mobilized by the Merovingian kings remained in the hands of and were continually utilized by the great magnates, of whom the Carolingians became the most prominent through incessant warfare under Charles Martel (714-41), Pippin III (741-68), and Charlemagne (768-814). The lack of a formal, centralized bureaucracy actually worked to the advantage of the Carolingians. Major proprietors had always been required to provide public services since the Roman Empire. The regular substitution of military service for direct taxation had already begun under the Visigoths in Toulouse, and was completely regularized for secular magnates by the early Merovingians. Bishops were involved in military affairs as great landowners in their own right since the 6th century, a process that was completely regularized by the early 7th century. By then, the foundation of monasteries had transferred vast tracts of land to institutions outside the system. During the 7th century monasteries were also obliged to provide troops. With the regularization of military service from church lands, a process begun by the Merovingians but completed by Charles Martel, *all* lands were subject to this obligation.[159] The Visigothic system in Toledo demonstrates how service should be performed according to wealth, assessed in numbers of slaves, while the Carolingians (and their Ottonian successors) defined military obligations in units of cultivated land. Performing such service was rarely contested, though the terms were continuously renegotiated accord-

---

[158] For the Carolingian period, see *KdGr*, *NCMH* 2, Bullough 1970, Collins 1998, Fouracre 2000, Verhulst 2002, Barbero 2004 and Innes et al. 2011.

[159] Thus Fouracre 2000: 122-26; similarly, Devroey 2003: 283 argues that the church escaped direct military obligations in the 6th and 7th centuries. However, as we have seen, the military forces of ecclesiastics clearly existed and were quite formidable in the late 6th and 7th century, and that some 7th-century Merovingian kings had imposed their will on the church. Charles Martel was reestablishing central control over these resources by forcing ecclesiastical institutions to relinquish some lands to secular lords in order to counteract their potential independence; indeed, by the mid-8th century, there is more evidence for clerics in war again (see e.g. *Ihburg 753 and *Rome 756), which would indicate that the Carolingians had imposed their will effectively. Ecclesiastic military service is far better known in the 9th century, when reform-minded clerics began to complain of the obligations imposed by them and the uncanonical living that followed among many abbots and bishops.

ing to changing needs and circumstances. Greater freedom from taxation, judicial immunities and so on appear to have given greater power to magnates and weakened the state. However, the magnates were engaged in an arms race with their peers: they had to use their surplus resources on military expenditure in order to defend their position in society and their private property from rivals. When royal power was strong, it could extract the magnates' military resources in the form of service against external objectives. Their "arms race" thus benefited royal power, as it kept bureaucracy at a minimum while maximizing military resources. Only when kingship was weak or divided could the aristocracy choose to follow one of multiple (or excluded) heirs to engage in internal conflicts.

4.3.1 *The Debate Revisited*

The feudal paradigm has been universally abandoned as an explanation for Carolingian warfare. The central feature of feudalism was the institution of vassalage, which involved temporary land grants, fiefs or benefices (*beneficia*), from a lord, to a vassal (*vassus* or *vassalus*), who swore fealty to his lord and performed military service in return. This model was heavily promoted by the giants of Carolingian history, especially Bloch and Ganshof, and used to permeate most aspects of Carolingian and high medieval history.[160] During the last generation, however, its basic assumptions have been reexamined by a number of scholars who have concluded that that benefices and military service were rarely linked, and anyway played a small role in military organization.[161] The technical and tactical corollary to the feudal paradigm, the Brunner-White thesis of the ascendancy of heavy cavalry (thanks to the stirrup), requiring vast landed resources for upkeep,[162] has been completely refuted.[163] Still, there are major discussions of several aspects of Carolingian warfare that have not been resolved; some questions have not even been raised, far less properly discussed. We can broadly distinguish two camps in the debate, simplified as minimalist-exceptionalists and maximalist-continuists. Minimalists and exceptionalists believe in little continuity from late Antiquity or correlation with

---

[160] See the classical studies *Feudal Society* (English tr. 1989) and *What is Feudalism* (3rd ed. of English tr. 1976). Much classical military history has been written based upon these assumptions, for which see the following notes.

[161] Bachrach 1974, Reynolds 1994, Halsall 2003, Fouracre 2000.

[162] For a summary of the Brunner-White thesis and the ensuing debate, see DeVries 1992: 95-110.

[163] See e.g. Sawyer and Hilton 1963 and Bachrach 1980.

Byzantine and Islamic military organization. Maximalists and continuity-proponents believe in a high degree of continuity and large-scale operations. It must be emphasized, however, that many scholars not particularly interested in military history take an agnostic view of these matters, or reflect shades in between.

1) SIZE AND ORGANIZATION: Minimalists take it as axiomatic that Carolingian armies were rather small (some even quote a "canonical number" of 5,000), arguing from a diverse range of historical parallels or socio-economic constraints, including contemporary city and village size, size of armies in more bureaucratized states, contemporary (Byzantium, Caliphate) and later (e.g. England and France during the Hundred Years' War), and finally, simple disbelief that armies could be comparable to earlier Roman or later medieval forces. In place of feudalism, proponents of this view have recently tried to establish their view of Frankish "otherness" through emphasizing peculiar cultural traits, which again are mostly interpreted with anthropological models of gift-giving, especially the redistribution of tribute and booty as the main cementing factor in military organization.[164] Maximalists argue from known figures in Carolingian history, such as number of royal estates, counties, bishoprics and monasteries, which all carried military obligations, and aggregate numbers from this basis. This generates a vast reservoir of manpower that could be raised, at least in theory. From narrative and legal sources, they argue that the Carolingian armies seem quite diversified in composition, and were capable of handling many tasks, such as battles, raiding, siege warfare, fortification building, and other engineering.[165]

---

[164] These arguments can in varying degrees be found in Lot 1946, Contamine 1984, Verbruggen 1997, Ganshof 1968, Reuter 1985 and 1990, and Halsall 2003. Squatriti 2002 refers to the "canonical number" of 5,000 men in Charlemagne's army based on several of the preceding scholars. Note that the numbers estimated are the same as those proposed by older generations of scholars who accepted varieties of the Brunner-White thesis. Since low numbers were based on extremely high costs of maintaining heavy cavalry and the underestimation of the Carolingian estate economy, a different army composition and supply system will radically affect estimates.

[165] The "high count" for Carolingian warfare is provided by Werner 1968. Bachrach 2001 (and multiple other works) relies very heavily on Werners estimate, but argues that the number provided by Werner is the total available to put into the field at any one time. Bowlus (2002) has similar views. France 1985 argues for diverse capabilities in Carolingian armies, and 2002 criticizes Reuter (1985 etc) for failing to account for the immense costs and challenges of maintaining military followings. He also, rightly, shows that Charlemagne's power to a large extent depended upon magnate consensus, and when this was not forthcoming, the military potential of the Carolingians was severely circumscribed. However, he settles arbitrarily for the number 20,000 for Carolingian field armies. Goldberg 2006 and

2) OBJECTIVES AND METHODS OF WAR: Minimalists argue that the smallish Carolingian armies moved rapidly, mostly by horse (thus maintaining important segments of the older Brunner-White thesis), and fought by raiding and skirmishing. Objectives included plunder, tribute, and new offices as rewards for the Frankish aristocracy, which then turned on itself when it ran out of territories it could plunder. Evidence of sieges and other large-scale efforts are ignored, explained away or reinterpreted in the otherness-framework.[166] Maximalists believe that the Frankish armies fought over objectives that required large armies, including sieges and building fortifications on a significant scale. This in turn required immense efforts that could hardly be performed from horseback.[167]

3) LOGISTICS: This is the weakest point in the minimalist argument. Since they use lack of bureaucracy and generally restricted views of society to demonstrate the armies were small, most do not even bother to consider how even small cavalry armies were actually supplied on campaigns of hundreds or thousands of kilometers. Indeed, even an army of 5,000 horsemen would require a huge number of remounts and depots along campaign routes in order to achieve the marching speeds postulated for Charles Martel and Charlemagne.[168] Furthermore, it is now well established from quantifiable engineering projects in the Carolingian empire and contemporary societies, such as the *fossa carolina*, Offa's Dyke, Danevirke, and the Avar and Bulgarian ditches that early medieval societies could organize thousands upon thousands of laborers to perform large-scale tasks in a single place over weeks, months, and maybe even years.[169] Here too, maximalists have not followed up arguments strongly presented under the preceding points. While demonstrating that the Carolingians in fact did have vast potential manpower resources, work focusing on how they supplied their supposedly huge forces in the field from an estate-

---

D. Bachrach 2012 show in detail how Carolingian structures functioned much the same way in the East Frankish realm into the 9th and 10th centuries, respectively.

[166] Most poignantly Reuter 1985 and 1990, followed by Halsall 2003.

[167] For strategic maximalism, see Bachrach 2001; Bowlus has similar arguments. While France can hardly be termed a maximalist, he does support a certain scale of operations and level of sophistication, especially under Charlemagne and later. Similarly Verbruggen believes that Carolingian armies were small, but argues for sophisticated strategies and advanced tactical skills.

[168] Various estimates are provided in the works cited in the previous notes for this section (4.3.1).

[169] See Squatriti 2002 and 2004 for the most recent treatment; Hofman 1965: 446 calculated that it would take 6,000 men (7,200 with support functions) working 55 ten-hour days to excavate the *fossa Carolina* based on the remaining physical evidence.

based economy has only begun to appear in the last few years.[170] As argued above, estate management derived from Roman forms of indirect taxation provides an explanation,[171] and the functioning of its daily mechanics begins to be visible in the reign of Charlemagne, if not earlier.

4.3.2 *Size, Composition, and Distribution of Carolingian Military Forces*

Goffart recently argued that Carolingian military obligations as attested under Charlemagne stemmed from the conversion of Roman taxes to military service. In his model, land units with their tax assessments were distributed to individual soldiers who were in return obliged to serve upon royal summons. These land units were still used by the Carolingians to assess the scale of obligation, now under the term *mansus*. The *mansus* was long taken by scholars to be a fiscal unit that was later assigned a military obligation, but Goffart turns this premise on its head.[172] The mechanics of his model, i.e. the reassignment of taxes as the rendering of military service, is the basis of the argument of this study, but it is clear that Goffart describes a situation that could not far outlast the 5th century when it was presumably first applied. As we have seen, by the mid to late 6th century, a large proportion of Frankish troops were organized in magnate retinues. In practice this left the magnates to administer the military obligations on behalf of the Merovingian and later Carolingian kings. This also means that there was no great change in the recruitment, status, or organization of troops between the Merovingian and early Carolingian periods. Thus, outside of the campaigning season (which lasted 3-6 months a year), Carolingian troops were widely dispersed throughout the Frankish empire, either on estates belonging to their lords,[173] or on garrison duty in one of the

---

[170] Brühl 1968 blazed a trail with his study on how the Frankish kings and their successors supplied their private retinues in peacetime, but this has not been followed up until recently; see Verhulst 2002: 127, Gillmorr 2002, Bachrach 2007b, Campbell 2010.

[171] See chapters 1-3 above and sections 4.1.5 and 4.2 in this chapter.

[172] Goffart 2008.

[173] Although feudal doctrine demanded that they be provided with fiefs, realities were far more complex. See e.g. Innes 2000: 145-53 for evidence of a specter from landholding troops of some standing to others concentrated in the households of their lords, and more specifically in military villages. France 1985 has found evidence for a *vicus militare* where 110 soldiers lived together in a village belonging to the estate of a 9th-century monastery. These were clearly neither part-time farmers nor *rentier* warriors concerned with administering fiefs. For the rare case of an individual professional soldier, named Gundhard, see Innes 2000: 146f. Despite identifying evidence of professional soldiers, Innes argues that it was indeed possible to create "a viable army from a mass of free and half-free peasants." For a different perspective on the role of peasants, see chapter 4.3.3 and 4.3.4 below.

castra or *civitates*. Most of these groups were very small, comprising the following of an individual magnate, abbot or bishop, numbering from a few score to a couple of hundred, and forming natural tactical groups.[174] The widespread distribution of military manpower meant that much of the time (6-9 months a year), troops were able to consume locally produced goods. When on expedition, they brought up to 3 months of supplies for use beyond the frontiers. While marching through the empire, they lived off goods gathered up in advance according to plans laid months or even years ahead.

Cumulative evidence indicates that the later Carolingian (and Ottonian) rulers had large military followings even when going about on peacetime business. Brühl and Werner argue for royal followings in the region of about or over 1,000 men in West and East Francia respectively in the 9th century; presumably a slightly higher number would accrue to the senior middle kingdom in 843.[175] I would therefore argue that Charlemagne and Louis the Pious disposed of very large personal military resources. Before the division of the Empire in 843, they had around 4,000 men in their own *immediate* followings, settled on their core lands in Austrasia and Neustria, to enforce their will, form the core of expeditionary armies, act as a strategic reserve for frontier commands or as a rapid reaction force in case of emergency. These were called *vassi dominici* (*non casati*), i.e. "royal boys

---

[174] The base numbers used in the following calculations (counties, bishoprics, royal palaces and other fiscal units) are drawn from Werner 1968. Other numbers, such as the total royal following, or size of individual followings, are indicated as they occur.

[175] For the most recent assessment of the Carolingian court (without estimates of size), see chapter 3 of McKitterick 2008. Werner 1968 assesses the retinue of the German kings at 1,000 men at least; Brühl 1968: 70f, 170, 175f uses records of the German royal retinue's daily consumption to show that it was supplied with enough foodstuffs to feed 4,000 men a day. Brühl adduces further evidence for the size of retinues, which he also believes numbered in excess of 1,000 men. The sources make it impossible to reconstruct the exact size of the royal retinue, but some are quite suggestive: Brühl points out that at the colloquy of Mersen in 870, each king was allowed to have 6 bishops, 10 *consiliarii* (the "Gefolge des Gefolges", or most intimate advisors in the kings' personal households [= senior *vassi dominici*?]) and 300 *ministeriales et vassallos*; at the Carolingian conference of three kings at Savonnières in 862, there were 200 *consiliarii* altogether.

If we assume that the *ministeriales* were only servants and counted a third of the retinue at Mersen, this gives a ratio of 20 (junior) *vassalli* per *consiliarius*: altogether 4,000 men for the immediate personal following of the three kings. If all had a military role, this would make it 2,000 per king and 6,000 altogether. Furthermore, the *Normannenfürst* Harald had more than 400 men in his retinue when he was baptized at Ingelheim in 826. "Harald konnte sich an Macht und Ansehen bei weitem nicht mit dem Frankenkönig messen, dessen Gefolge wir daher unbedenklich auf mehrere Hunderte, sehr wahrscheinlich sogar auf über tausend Mann veranschlagen können." The following that accompanied Rigund in 584 numbered 4,000, "so handelt es sich da natürlich um Heeresgefolge."

(without land grants)," and were supplied by Carolingian estate managers who reported directly to court several times a year.[176] Royal resources dwarfed those of most magnates until divisions and sub-divisions of the kingdom in the late 9th century began to level the playing field. Until then, the only chance for successful revolt was to form vast coalitions of magnates who used a royal candidate to win over royal troops. Subkingdoms ensured closer royal control of how such troops were used when the distance to core Frankish lands were great, e.g. in Aquitaine, Italy and Bavaria.

In addition to those who were maintained directly by the royal household, there were also *vassi dominici casati* ("royal boys" provided with estates). They were established on distant royal or ecclesiastical estates granted as *beneficium* (precarial tenure in return for service) in various provinces, and came in addition to the troops that were supplied directly by the royal estates. Senior *vassi dominici* had large enough *beneficia* to provide for their own retinues, which must be added to the royal following. We know relative sizes of the followings of *vassi dominici casati* quite well. According to the *Capitulare Episcoporum*, issued c. 780, each of these *vassi dominici* who were stationed outside the palace had resources equivalent to or only slightly smaller than those of middle-level local counts (on which see below), whose personal followings seem to have numbered around 50 men.[177] They were distributed along with local magnates' troops to occupy newly conquered territory, guard established borders, or engage in minor warfare and raids along the frontiers. Their social status was very low, as

---

[176] See Bachrach 2007b for the Carolingian supply system; see further the discussion below on the same and the status of the *vassi*. The word *vassus* is derived from the Celtic *gwas*, the semantic equivalent of Latin *puer* (boy); hence originally designating a very low social status.

[177] The size of the Carolingian military following is rarely stated explicitly in narrative and administrative sources. Fortunately, the *Capitulare Episcoporum* (dealing with alms to be given and fasts to be performed by royal officers during a famine) provides not only an indication of the relative wealth of counts and *vassi dominici*, but also numbers from which we can deduce the potential size of their followings: "... Comites vero fortiores libram unam de argento aut valentem, mediocres mediam libram; vassus dominicus de casatis ducentis mediam libram, de casatis centum solidos quinque, de casatis quinquaginta aut triginta unciam unam. Et faciant biduanas atque eorum homines in eorum casatis, vel qui hoc facere possunt; et qui redimere voluerit, fortiores comites uncias tres, mediocres unciam et dimidiam, minores solidum unum." A literal translation, leaving technical terms, would be: "The more powerful counts [should give] one pound of silver or the same value, the middling [counts] half a pound; a *vassus dominicus* of two hundred *casatae* half a pound, of one hundred *casatae* five solidi, of fifty or thirty *casatae* one ounce. And they and their *homines* in their *casatae* shall make a two-day fast, or [at least] those who can do this; and

many seem to have been semi- or unfree.[178] Hence it was a quite spectacular humiliation for a great magnate to be forced to undergo a "vassalic" submission to the king in the 8th century.[179]

Prosopographical research gives a good indication of their numbers, which were very large. 1,000 *vassi dominici* are known; if only a tenth of that number held substantial *beneficia*, they would each dispose of a following of anywhere from a dozen to a few score men (perhaps 50 on average, as middling counts), making a total of 5,000 men. In addition were some *casati* who were provided with modest estates for their own personal upkeep or as rewards for their services. Whenever on royal missions, whether *casati* or *non casati*, the *vassi* could draw on supplies from royal *fisci*.[180] Altogether, then, Charlemagne personally had some 9,000-10,000 professional soldiers at his disposition. Even though the number seems large, it would still only make the burden at about 12 men per fiscal unit or palace known

---

to redeem himself [from the fast], the more powerful counts should pay three ounces, the middling ones an ounce and a half, and the lesser count one solidus."

The ratio between military obligations and property size is very explicitly regulated in Carolingian capitularies. Four *mansi* (another term for *casatae*, i.e. the holdings of *casati vassi*) provided for one well equipped infantryman (if the *mansus* is taken as a household of 5.5 people, this means one in 22 served), but in this case only applied to the dependent households, cf. Verhulst 2002: 24 for the household ratios. Although the numbers are somewhat later, it was based on very long experience in what could be expected from the resources available to landowners and their tenants. In addition, there were lands belonging to the demesne that provided additional resources but were excluded from the *mansus* reckoning, cf. Goffart 2008. What follows are therefore minimum numbers: a *vassus dominicus* of two hundred *casatae* could provide a following of at least 50 men; of 100 *casatae*, 25 men; of fifty, 12 or 13, and thirty, seven or eight. The latter (the men of 30-50 *casatae*) may only have been servants and armed escorts for individual service by a *vassus dominicus*, but the former (the men of 100-200 *casatae*), supported by more substantial demesnes, gifts from the court (e.g. horses) and industrial estates belonging to monasteries (e.g. arms and armor), could both be more numerous and better or more heavily equipped. In order to avoid further speculation on composition and equipment, however, the base number of 50 men (mostly well equipped) for the followings of the most wealthy *vassi dominici* is sufficient for our purposes. This means that middling (*mediocres*) counts of equivalent wealth would also have similarly sized followings, while the wealthier (*fortiores*) could have 100 men. Hence the average of 75 presented below.

[178] See the relevant studies in *Spätantike*.

[179] E.g. Tassilo, duke of Bavaria. Indeed, later "vassals" were in effect independent lords who used the ritual to reinforce political bonds or recognize hegemony without conceding political control; these bonds were very different from those between semi-free *pueri* and great magnates.

[180] See Bachrach 2007b on accounting practices that demonstrate this during the reign of Charlemagne; the exact relationships between and distribution of burdens among the various other fiscal units under royal control, ecclesiastical lands, and local magnates' resources have received little attention, but see Brühl 1968 for peace-time exactions by the royal household, including his retinue, when on the move.

to be under Carolingian control, and even fewer (say ten men) if some of these men were granted benefices from ecclesiastic lands. Even the smallest of known royal estates could probably support such numbers without too much difficulty,[181] and Gillmor has recently calculated that the royal estates indeed produced enough horses every year to support a force of 10,000 cavalry.[182]

Counts, drawn from the upper echelons of great Carolingian landowners,[183] would on average have more men in their personal followings than the *vassi dominici*. If middling counts could raise 50 men, great counts (like bishops, see below) should have retinues of about 100 men. A very careful average, then, would be 75 men for each count in at least 500 active coun-

---

[181] Cf. the little estate at Annappes (*Asnapius*) surveyed in *Brevium Exempla* 25, which had "a royal house, well built of stone, with three chambers; the whole house surrounded by galleries, with 11 rooms for women; underneath, one cellar; two porches; 17 other houses inside the courtyard, built of wood, with as many rooms and with the other amenities all in good order; one stable, one kitchen, one bakehouse, two barns, three haylofts. A courtyard with a strong palisade and a stone gateway with a gallery above from which to make distributions. A smaller courtyard similarly enclosed with a palisade, well ordered and planted with various kinds of trees. Household linen: one set of bedding, one tablecloth, one towel. Equipment: two bronze bowls, two cups, two bronze cauldrons and one of iron, one cooking pan, one pot-hook, one fire-dog, one lamp, two axes, one adze, two augers, one hatchet, one chisel, one scraper, one plane, two scythes, two sickles, two iron-tipped spades. Wooden equipment: a sufficient quantity. Produce: nine baskets of old spelt from the previous year, which will yield 450 measures of flour; 100 *modii* of barley. In the present year there were 110 baskets of spelt: of these 60 baskets have been sown, and we found the rest [in store]; 100 *modii* of wheat: 60 have been sown, and we found the rest 98 *modii* of rye, all of which has been sown; 1,800 *modii* of barley: 1,100 have been sown, and we found the rest; 430 *modii* of oats, one *modius* of beans, 12 *modii* of peas. From the five mills, 800 smaller *modii:* 200*modii* were given to the workers on the home farm, and we found the rest. From the four brewhouses, 650 smaller *modii.*From the two bridges, 60 *modii* of salt and two shillings. From the four gardens, 11 shillings, two *modii* of honey, one *modius* of butter in payment of dues; 10 sides of bacon from the previous year, 200 sides of new bacon, along with the offal and lard; 43 measures of cheeses from the present year. Livestock: 51 head of older horses, five three-year-olds, seven two-year-olds, seven yearlings, 10 two-year-old colts, eight yearlings, three stallions, 16 oxen, two donkeys, 50 cows with calves, 20 bullocks, 38 yearling calves, three bulls, 260 older pigs, 100 piglets, five boars, 150 ewes with lambs, 200 yearling lambs, 120 rams, 30 she-goats with kids, 30 yearling goats, three he-goats, 30 geese, 80 chickens, 22 peacocks."

[182] Gillmor 2008 calculates from the *Capitulare de Villis* and the *Brevium Exempla* that 4,500 horses were produced per year by royal estates, and that each horse could serve an average of seven years. I would suggest that these horses mostly went to the *non casati*, and that the *casati* were expected to provide for themselves from their own properties or other sources such as royal monasteries or other ecclesiastical estates granted as *beneficia* for such purposes. They would probably be assisted by royal resources if they incurred extraordinary losses while on campaign.

[183] For a survey of the most significant noble families and their functions, see e.g. Werner 1965.

ties (of 700 known but including multiple countships, thus some counts would be vastly wealthier and thus have far larger retinues than "middling counts" noted above), the total being 37,500 men.[184] How many men a count would command of the free, propertied arms-bearing population or magnates without public office is difficult to say. Given the importance of this group in legislation, we can double the count's retinue; that would give a total of 75,000 men. In case of emergency, e.g. hostile raids, the general population could be pressed into service but I believe this was a very rare occurrence during the expansionary phase, only becoming common again after the division of the empire in 843.[185] A similar number of monasteries (500) plus 190 bishoprics, averaging 100 men each (but subtracting some land going to *vassi dominici casati* and other exemptions), would put the number of troops recruited from church lands at about 60,000.[186] The chronicles, letters and capitularies clearly demonstrate that these magnates actually showed up for service as required, often at very great expense, both for offensive and defensive warfare.[187]

Altogether the Carolingians at their height in the early 9th century had potential military resources numbering about 145,000 men, most of whom

---

[184] County numbers as in Werner 1968, who also uses the *Indiculus loricatorum* to calculate an average comital following of around 50 heavy cavalry, but it is clear that this document involves the peculiar situation that troops were detached for long-distance service (according to some unknown factors, such as ratio of infantry-cavalry, and political reliability or importance/competence of magnate concerned) from the total following, which could be substantially larger and also included some specialized infantry. I put the size of the average comital following higher on the basis of 9th-century information (e.g. *Annales Bertiniani* and *Annales Fuldenses*) and the information from the *Capitulare* analyzed above, which provides a base number of 50-100 men depending on wealth.

[185] There are many references in Carolingian polyptychs (estate surveys) to individual peasants who also had guard duties and the like, see e.g. the *Polyptych of Wissembourg* 11.2 (from 819) for six men who owed guard duties (*qui vigilare debent*)—although for what is not specified. Bachrach in his works (e.g. 2001) argues that direct military obligations involved *all* able-bodied males, but it is clear from the polyptychs that a few particular individuals in a fairly large population had particular competence and possibly even some equipment to perform guard service; the general population was *theoretically* available for military purposes, but mostly levied for labor and logistics, and very rarely for defensive service during the period under consideration here since the Carolingians were almost consistently on the offensive.

[186] Some episcopal followings were huge, approaching royal sizes. The archbishop of Ravenna, for instance, had a following of 500 men in the 9th century; see Reuter 1985: 83 for references. Otherwise, *Indiculus loricatorum* indicates *divisions* of the followings of ecclesiastical magnates numbering from one to several score. In excess of 100 men was probably the norm (cf. the *vicus militare* in France 1985), hence an average of 100 seems to be a safe low estimate.

[187] See e.g. Lupus of Ferrières, *Epistolae* 25, who meekly asked Charles the Bald (in 840) that his *homines* be allowed a few weeks to rest between campaigns.

were professional, full-time soldiers. This is roughly equivalent of the number of troops found in the Roman provinces around 400 AD that came under Carolingian control, and larger than the contemporary Byzantine army.[188] The numbers were of course smaller in the early 8th century, but at that time, many of these same resources were under the control of local magnates or former client states, so that the aggregate military resources of the Frankish realm(s) were not much smaller, and probably never dropped much below 100,000 men throughout our period. According to the lowest current population estimates of around 10 million inhabitants in Charlemagne's empire at its largest extent, soldiers would only account for about 1,5 per cent of the population, again analogous with the Byzantine military, which did not exceed 2 per cent.[189]

The military followings were full time troops with little else to care for. Their leaders were soldiers by profession and reared for the job from childhood.[190] They only had room for independent political maneuver during civil wars or *interregna*. The troops they commanded appear to have lived together and could thus train as units; in addition, there was heavy emphasis on individual training within the household.[191] Units and combinations received regular training during hunts, which normally took place each fall after the campaigning season but before going into winter camp. This was a favorite pastime of all kings and magnates, but was specifically recommended as useful for training bodies of troops by Byzantine military man-

---

[188] Estimates of the Byzantine army around 800 vary somewhat. Treadgold 1995, a Byzantine maximalist, argues for an army of over 80,000 men in 773 which increased due to fiscal and military reforms to 120,000 men in the early 9th century, but commits the fallacy of regarding the bulk of the army, the *themata*, as part-time soldiers. Haldon 1997: 101ff is more skeptical (especially of the size of the *tagmata* around Constantinople, which Treadgold believes numbered around 24,000 men), but recognizes that Treadgold's number may reflect the total potential strength of the army at time. Since Haldon (see e.g 1984) is on better empirical ground on the *tagmata*, it seems that 100,000 men would be a reasonable estimate for the total number of professional troops in the Byzantine army in the early 9th century.

[189] McCormick 2007 gives a range of ten to 20 million for current estimates. The Byzantine estimates are from Treadgold 1995: 162 and hence within a maximalist range. Using available lower estimates for the Byzantine army (such as Cheynet or Haldon) and higher numbers for the population around 1000 (Laiou and Morrisson 2007) produces a ratio of merely 0,5 per cent during the height of the Byzantine Empire.

[190] The letters of Lupus of Ferrierès attest to tutor for orphaned child of military commander, while even clerics who had had a secular upbringing continued military exercises after taking office. I plan to return to the 9th-century evidence at some later point.

[191] Bachrach 2001: 84-107. While somewhat conjectural, his conclusions based on Vegetius and Rhabanus must generally be regarded as correct.

uals.[192] Training on a larger scale occurred during army assemblies that were arranged in the spring, though we have little concrete evidence of this until later. We do however know that this was the late Roman practice, was continued by the Byzantines,[193] and at least on one occasion do we know of the large scale games at Strasbourg in 841.

4.3.3 Objectives and Means: Charles Martel and Pippin the Short

When and why the Carolingians formulated their expansive policy is contested amongst scholars. There is little doubt that they fought in order to impose their authority on the old Merovingian *regnum* (most of which was still organized as outlined in 4.1-4.2 above) and eliminate dynastic threats; in the process, they also conquered much additional territory.[194] In most theaters of war, this resulted in sieges of fortifications that sometimes had to be starved into submission over months, or stormed with the full range of tactics and technology used by contemporary societies. Continuing the practices of Merovingian civil wars, the Carolingians frequently attacked cities fortified in Roman times; furthermore, urban populations were again on the rise from the 8th century onwards. This was characteristic of fighting in southern Gaul, northern Spain, and most of Italy. In Aquitaine, the local nobility also used a range of lesser fortifications that either survived from suburban Roman forts, or were more recent constructions. While some were improvised from natural defenses, others were substantial structures. Outside former Roman territory, across the Rhine (and to a lesser extent

---

[192] E.g. *Strategikon* has a separate section (12 D) among other drills on the value of coordinated mass hunting as a means of acquiring cohesion, and recommends that it be done even while on the march. Further see chapter 2.1.3.

[193] Again see the various drills prescribed by *Strategikon*, both individual and unit (books 3, 6 and 12 *passim*), as well as the training recommended in *De re strategica*, now dated to c. 800. These drills were clearly used in the field; thus Roman armies headed for Italy often stayed in the Balkans for one or two winters to drill; Narses had his soldiers train at Rome (Agathias 2.1.); while the annual *adnoumion* was used for equipment checks and drills before going on campaign, as attested by the 9th- and 10th-century Byzantine military manuals.

[194] Older scholarship, such as Verbruggen 1965, tends to downplay the strategic efforts of the early Carolingians before Charlemagne. Bachrach 2001 argues that the objective of early Carolingian warfare was to reconstitute the *regnum Francorum* of the Merovingians, while Charlemagne did nothing less than reconstitute the West Roman Empire after a lapse of some three centuries. Most scholars have little problem with the first suggestion, but choose to focus on contingent problems that led to the next stage of conquest; for example, Costambeys, Innes and Maclean 2011 have pointed out how the 8th-century conquests tended to be directed against territories that supported potentially legitimate Carolingians or were ruled by the closest cadet line of the Carolingian house.

along and north of the Danube), the Carolingians faced a different type of fortification: large earth-and-wood structures that could be just as challenging as any stone fortification. The wars against the Saxons, Slavs and Avars were dominated by campaigns against their strong points, while the Carolingian conquests were held and defended by constructing massive fortifications that became the kernels of the first cities beyond the Rhine. Regarding this conquest as driven by "plunder and tribute" is far too simplistic. Large territories had to be organized, administered and defended; political settlements had to be made with local elites who were willing to cooperate. Maintaining a military following was not cheap, but sending it out on constant campaigns was extremely expensive in sheer material cost, and the nature of warfare required even more manpower be mobilized for logistical and engineering tasks that produced no plunder whatsoever.

Charles Martel's career began inauspiciously with a defeat at the hands of the Frisians in 716-17. He learned well, however, and never lost a battle again until his death, although a few sieges probably failed.[195] One of these early victories, in about 721, over a rival at *Angers, included a siege of the city, though our main source, Fredegar Continuatus, says nothing of the outcome or means employed in the siege itself, only that he raided the territory and brought back a great amount of booty.[196] In light of his later exploits and solidly Merovingian upbringing, it would probably have resembled the conflict presented in the *passio Leudegarii* (see chapter 4.2.3). His most famous victory, of course, is the battle of Tours/Poitiers in 732/3 when he defeated a large Arab raiding army. What has usually been passed over since the days of Brunner's theory on the origins of feudalism, however, is the fact that he besieged many cities in the south. This happened within a decade after concerted Arab campaigns that revolved around sieges of great cities in southern Gaul in the 720s. Just to give an example from a contemporary observer: at *Toulouse (720), the Arabs used heavy artillery (probably trebuchets, see chapter 8.2.3) and other machines against the city, but were driven back by an Aquitanian relieving army under Duke Eudo.

---

[195] On his career in general, see Fouracre 2000.
[196] See also CO: "Afterwards the *Princeps* Charles pursued Ragamfred, laid siege to the city of Angers, laid waste the neighborhood and then returned home with rich booty." || His ita euulsis Carlus Princeps insecutus idem Ragamfredo Andegavis ciuitatem obsedit; vastata eadem regione, cum plurima spolia remeauit. || Here however the word *obsedit* may mean the same as *occupavit*, as is the case in the *Vita Genovefae*. For a similar instance, see *Paris, 490. Nonetheless, the context and the use of *vastata* indicates that the process that made the situation so was quite violent.

This means that when Charles followed up his great victory in the battle of Tours, he was prepared to fight the Arabs on their own terms. Charles himself attacked Aquitaine in c. 735 after the death of Eudo, occupied (*occupavit*) *Bordeaux and *Blaye, and subjugated (*subiugavit*) them and other "cities and suburban fortifications" in the region. Soon after he effected the integration of Burgundy, subjecting the great men in the district of *Lyons (738) with a large army. In the meantime he faced the Arabs at *Avignon (736), where he defeated an Arab garrison, installed since 734, with a massive artillery bombardment followed up by a storm.

> Then as once before Jericho, the armies gave a great shout, the trumpets brayed and the men rushed in to the assault with machines and *restium funibus* [probably trebuchets] and attacked above the walls and {into} the intramural buildings; and they took that strong city and burned it with fire and they took captive their enemies, smiting without mercy and destroying them, and they recovered complete mastery of the city.[197]

Immediately afterwards he proceeded to their provincial capital, *Narbonne, which he surrounded with continuous earthworks and assaulted with rams.[198] The Franks thoroughly defeated an Arab relieving army before moving on to the cities of *Nîmes, *Agde and *Béziers. By what means Charles reduced these cities is not mentioned, but apparently he was quite thorough: "... funditus muros et moenia destruens igne subposito concremauit, suburbana et castra illius regionis uastauit." In light of the bombardment at *Avignon and the rams used at *Narbonne, it is not unreasonable to conclude that he achieved this level of destruction by using similar means.

At first sight, it might appear that Charles Martel's reign saw a massive increase in siege warfare, but we only know so much of Charles Martel's sieges in this area because the last continuation of Fredegar was commissioned by his brother, duke Childebrand. He played an important role in establishing Carolingian control of Burgundy, led many of the campaigns

---

[197] See CO for source references: "In modum Hiericho cum strepitu hostium et sonitum tubarum, cum machinis et restium funibus super muros et edium moenia inruunt, urbem munitissimam ingredientes succendunt, hostes inimicos suorum capiunt, interficientes trucidant atque prosternent et in sua dicione efficaciter restituunt." Since the text itself is difficult, the translation by Wallace-Hadrill is problematic and rendered differently here. For example, *funibus* should probably be emended to *fundis*; the phrase *funibus restium* (with ropes of ropes) does not translate as "rope ladder," as Wallace-Hadrill suggested. Also, attacking "over the walls and intramural buildings" probably makes better sense with trebuchets hurling projectiles over the walls and hitting the buildings inside. See the entry in CO and chapter 8.2.3 for further discussion.

[198] The outcome is not mentioned, but it was in Frankish hands by the time the Aquitanians attacked the city in 763.

in the south from his base at Lyons, and his extensive participation in sieges allows us to designate him as a siege specialist. His skills came in handy in the years following the death of Charles in 741, whose control over Aquitaine was limited. While Provence was pacified in the last year of Charles' reign, his sons fought a long series of brutal wars to establish Frankish control in Aquitaine. The first conflict was a revolt against Frankish control, and took advantage of an abortive Frankish civil war which ended when Pippin and Carloman besieged their brother at *Laon in 741. In the following year, the brothers invaded Aquitaine, supported by their uncle Childebrand. They defeated the defenders and ravaged the territory of the cities of Orléans and *Bourges before they stormed the fort of *Loches, whose garrison (*custodes*) were taken captive. The Carolingians stopped at this point due to fierce resistance and other concerns. Pippin's main enemies during the next years were the Saxons, who, during the reign of Charles Martel at the latest, had begun building substantial fortifications and provided serious resistance to the Franks.[199] He also famously intervened in Italy against the Lombards, who were besieging the pope in *Rome on several occasions, best attested in 756.[200]

It was only in 760 that the Aquitanian war began again, apparently deliberately provoked by Pippin, who organized a series of raids that left most of the region burning. The Aquitanians under their duke Waiofar responded with raids against Carolingian marching routes and royal villas.[201] However, this was clearly not enough to deter Pippin, who began systematically to attack Aquitanian forts and fortified cities. In 761 he took by storm (*per pugnam coepit*) the *castra* of *Bourbon, whose garrison was taken captive, *Chantelles, and *Clermont, where the population perished when the city was incinerated. As a result of these atrocities, many other fortifications (*castella*) surrendered by treaty (*per placitum*). In 762 the campaigns continued with a Frankish siege of *Bourges, which was surrounded completely by a double set of siegeworks: first a *munitionem fortis-*

---

[199] The adaptation of more advanced military practices in Saxony, such as fortification building, is dealt with by Hardt 2001 and discussed briefly in chapter 7.2.5. The campaigns of Pippin were against the fort of *Hohenseeburg (743), taken *per placitum* (through negotiations); despite an ostensibly successful campaign where Pippin had a *magno apparatu* (a large siege or supply train), the Saxons stormed *Ihburg (753) and killed the bishop Hildegar from Cologne—his men presumably formed the core of the garrison at this fort. The *Saxon strongholds at Sythen (758) were taken by storm.

[200] In 755, Pippin broke through at *Susa, besieged *Pavia, and took a number of other Lombard strongholds before the Lombards submitted; in 756, Pippin again besieged *Pavia.

[201] Fredegar Continuatus 42.

simam against the city, and secondly a *uallo* that surrounded all the siege engines, *machinis et omni genere armorum*. Subsequently there was a brutal fight over the walls, leaving many wounded, before the Carolingians breached the walls and took the city. From Charles Martel's campaigns we know that the most important *machinae* available to the Carolingians were rams and trebuchets. The garrison left by Waiofar was dismissed, while the Gascons and count Chunibert and his men were forced to swear an oath of loyalty to Pippin, who sent their wives and children to Francia to ensure their good behavior. The fighting was so destructive that Pippin had to order the repair of the city wall before proceeding to besiege and capture the *castrum* of *Thouars.

The year 763 began with Aquitanian counter-assaults. An ambush against the Frankish garrison at *Narbonne that was returning from patrol failed miserably. Aquitanian raids (*ad praedandum ambulare*) from Auvergne against Frankish staging-posts at Lyons were defeated by the Frankish counts guarding the Loire border. Likewise the Aquitanian count of Poitiers, raiding the area of Tours (*Toronico infestando praedaret*), was defeated by the men of the abbot Wulfard of St. Martin (*ab homines Vulfardo abate monasterio beati Martini*). With the failure of these counter-raids, one of Waiofar's closest supporters, his uncle Remistanius, went over to Pippin. At this point Waiofar decided he could no longer hold all the great cities in Aquitaine, and razed the walls of "Poitiers, Limoges, Saintes, Périgueux, Angoulême and many other cities and fortified places." Pippin had these cities occupied, organized repairs of the walls, and followed up with a raid deep into Aquitaine, where he was faced by Waiofar with a great army, including Gascons from across the Garonne. Fredegar reports a great Carolingian victory, but the battle is not mentioned in the *Annales Regni Francorum*, only the raid. It appears that the outcome was not all that positive for the Carolingians. Fredegar indirectly admits as much, having Waiofar ask for peace, offering formal submission in return for his cities. This is not a request to be expected after a catastrophic defeat. Carolingian lack of success can also be inferred from the lack of military activity reported for 764 and 765.

Pippin used this time to plan new campaigns. Rebuilding city walls and *castra* destroyed by siege or by Waiofar also must have taken a long time. The last important one was Argenton, built in 763 according to Fredegar, but dated to 766 by the *ARF*. It is possible that the construction took several seasons. Based in his newly garrisoned fortifications, Pippin raided much of the region so thoroughly that "the Gascons and the magnates of

Aquitaine now saw that they had no option but to come to him: many there swore an oath to him and submitted to his authority." Pippin may have thought that the war was practically over. During a winter campaign in early 767, he marched from Narbonne to capture *Toulouse and accept the surrender of *Albi and *Gevaudan. Subsequently he brought his wife and court to Bourges, before setting out on a summer campaign to the Garonne, where he captured "many rocks and caves" (*multas roccas et speluncas*, called *castella* by the revised annals) and the *castra* of *Ally, *Turenne and *Peyrusse. The final conquest of Aquitaine was seriously threatened in 768 by the defection of Remistanius, who raided Pippin's new acquisitions south of the Loire so thoroughly "that not a peasant dared work in the fields and vineyards." Pippin solved the problem by using the Loire as a supply route and attacking Waiofar along the Atlantic coast. Remistanius was captured by another division and executed, and shortly after Waiofar was assassinated by his own men. When Pippin died later that year, he bequeathed a formidable war machine to his son Charles.

4.3.4 *Organization and Supplies: Charlemagne and Louis*

Several authors have suggested that the Carolingian estate economy was geared towards the logistical needs of the Frankish army.[202] This demonstrably applied to siege warfare as well. The marching routes for the campaigns in 4.3.3 above as described in *Fredegar Continuatus* and the *Annales Regni Francorum* demonstrate that troops were assembled at major known estate centers, especially those belonging to the kings, and were supplied by estates on the route of march. This applies particularly to the logistical and engineering teams that accompanied the troops. The labor necessary for sieges and all other logistical tasks were raised as labor corvées from estates,[203] whose resources and personnel now covered all the tasks usually performed by the late Roman army of the 4th century, or a combination of civilian and military participants, as in the East Roman Empire. According to the famous *Capitulare Aquisgranense*, issued in 811:

> The equipments of the king shall be carried in carts, also the equipments of the bishops, counts, abbots, and nobles of the king; flour, wine, pork, and victuals in abundance, mills, adzes, axes, augers, slings [i.e. trebuchets], and

---

[202] For surveys of the Carolingian estate economy during this period, see Ewig 1965, Verhulst 2001 and Devroey 2006: 443-583, with discussions of military use *passim*. Further on the formidable logistical capabilities of monasteries, see Lebecq 2000. For military logistics, see Bachrach 2007b and Gillmor 2008.
[203] See below for how this worked in the 9th century.

men who know how to use these well. And the marshals of the king shall add stones for these on twenty beasts of burden, if there is need. And each one shall be prepared for the army and shall have plenty of all utensils. And each count shall save two parts of the fodder in his county for the army's use, and he shall maintain good bridges and good boats.[204]

Although fortified cities were defended or used as springboards for further campaigns, it is thus abundantly clear that all Carolingians and their immediate successors maintained siege expertise attached to their estates, and that these could be redistributed according to military needs. Charlemagne completed the conquest of Aquitaine in 769 by building a fort at Fronsac, using the men and materials (*Francos cum omni utensilia et praeparamenta eorum*) stationed at Angoulême by his father Pippin. This expertise was a direct continuation of Charles Martel's and Pippin's practices, which are evident from the ability to build fortifications, repair city walls, and construct siegeworks. Charlemagne also used this basic infrastructure to perform several impressive engineering tasks, such as building fortified bridges across the Elbe, and attempting to excavate a canal, known as the *fossa Carolina*, to link the Danube and the Rhine.

Saxony was an area without any trace of Roman urban civilization, but had long been acculturated to Frankish forms of war. This meant that great forts of wood and earth became increasingly important in the Saxon wars, and many large, well-excavated fortifications were repeatedly fought over.[205] While some fringe peoples, such as the Bretons and Avars, already possessed these skills,[206] others were also influenced by Carolingian forms of warfare, acquiring both the ability to storm fortifications and build similar structures of their own. Thus the Slavs, whose acculturation began in the West in the 7th century, were fighting with the Carolingians on equal terms by around 800,[207] as were the Danish kings (rather than Viking raiders).[208]

Frankish warfare was highly flexible, as the same forms of organization were also applied against more economically and technologically

---

[204] *Capitulare Aquisgranense* 10. See further chapter 8.2.3 for the terminology.
[205] *Eresburg (772, 776); *Büraburg (773); *Syburg (775); *Bockholt (779), abandoned at the Frankish approach; *Saxon fortifications (776, 785); see also chapter 7.2.5 and Bachrach's 2013 work on *Charlemagne's Early Campaigns*, vol. 82 of this series.
[206] Thus see the *Breton fortifications (786) and the *Avar fortifications (791), which were apparently massive, but abandoned due to the vast size of the Carolingian army. For the Avars, see also chapter 7.2.3.
[207] Thus see chapter 4.2.1 above for the 7th century; for the age of Charlemagne, *City of Dragawit (789), *Slav fortifications (808), and *Hohbuoki (810).
[208] For a brief discussion of western Slav and Scandinavian acculturation, see chapter 7.2.5.

advanced enemies in the Mediterranean. Louis the Pious was trained by Charlemagne's advisors, and was at an early age put in charge of a very difficult border against the Umayyad emirate of Spain, against which Charlemagne had failed in 778.[209] His biographer, *Astronomus*, tells how Louis attacked *Tortosa (809) with *arietibus, mangonibus, vineis et ceteris argumentis* when he was sub-king of Aquitaine.[210] It also mentions that he received four estate complexes that should cover the needs of his household, and received support from Frankish magnates organized in the usual style to besiege a number of other cities.[211] In Italy, Charlemagne led armies in person, or delegated them to his sons, who even reached the southern Lombard duchy of Benevento.[212]

Some have maintained that Charlemagne's late capitularies on military affairs represented innovations and even unfulfilled ambitions. From the analysis of Charles Martel's and Pippin's campaigns, however, we can see the systematic manner in which troops raised from estates provided the manpower and expertise necessary for conducting sieges and building fortresses. Every magnate with military obligations was by definition also required to have the equipment and expertise to do anything that may be required on campaign, whether that be fighting in battles, conducting raids, defending borders, garrisoning cities, repairing walls, constructing fortifications, or besieging an enemy city. Organization and practices become better attested in the 9th century, but all the features are recognizable from Eastern parallels and Western precedents. While all this makes the Carolingians sound rather "Byzantine," we cannot argue that "Byzantine bureaucracy" surpassed the Carolingian administration: the population of Istria, conquered from the Byzantines by Charlemagne in 804, was appalled by the heavy exactions and labor dues imposed by the new duke:

> At the time of the Greeks, we were never forced to provide fodder, we never worked free for public estates, we never fed dogs, and we never had to raise money as we do now. We never paid for flocks, as we do now, having to hand over sheep and lambs every year. We have to provide transport services as far as Venice, Ravenna, and Dalmatia, and along the rivers, which we did not have to do before. When the duke has to leave for the emperor's war, he takes our horses and forcibly leads our sons away with him. He makes

---

[209] See Bachrach (forthcoming).
[210] Astronomus 16 (p. 330, a. 808); the siege is dated by *ARF* to 809.
[211] Bachrach 1974; see *Lerida (797), *Huesca (797ff), *Barcelona (801), *Umayyad *castella* (802).
[212] For context, see West 1999; sieges were fought at *Pavia (773), *Cividale and *Treviso (776f), *Beneventan *castrum* (792), *Chieti (801), *Lucera (802), and *Ortona (802).

them bring carts and then takes everything and sends them home on foot. He leaves our horses behind down there in France or he shares them out among his men. At the time of the Greeks, they took one sheep for every hundred from those who had that many, for the needs of the imperial envoys. Now anyone who has more than three must hand over one every year.[213]

The tone is clearly polemical, and some of these impositions probably levied illegally, but they are all well documented in Carolingian history and clearly derived from late Roman *munera*. Many of these obligations were subsequently revoked in order to retain the loyalty of Charlemagne's new subjects. While similar exactions are well known in Byzantium as well, this little ray of light on the daily business shows us that we cannot be dogmatic about the efficiency of bureaucratic systems and assume that the Carolingian administration was somehow lighter. One could say that the more personal the administration, the closer the supervision. It can rather be argued, then, that the military establishment consumed the lion's share of the surplus production of the agricultural economy of the Carolingian age. This in turn gave little room for a heavily urbanized society throughout much of Merovingian and Carolingian history.

Thus, in 818, the monastery at Wissembourg required annually from its estates "eight cavalry horses, thirty-six oxen, eight carts and twenty men," similar to practices at royal estates.[214] Furthermore, these exactions were only imposed upon the dependent population, who were basically the support team for Wissembourg's military *familia* on campaign. Finally, the walls of Wissembourg were maintained (and thus presumably defended) by the surrounding population. The walls were divided into sections by towers and gates, and each section was maintained by a different social unit, enumerated in a document c. 900: "groups of villages from the rural hinterland," townspeople or *"urbani,"* outside settlers such as "Frisian merchants," and finally the *"familia* of the abbey of Murbach." Innes argues that this is evidence of the localization of Carolingian administrative structures, which had earlier operated on a more regional scale.[215] However, it should now be clear that parallels to this method of organizing construction can be found from Byzantine Mesopotama and Rome to Ostrogothic Gaul and Anglo-Saxon Mercia; the *modus operandi* of building walls by *pedaturae*

---

[213] Translation quoted in Barbero 2004: 194; see ibid. 193ff for discussion and further references to the original.
[214] Innes 2000: 146; see further ibid. 159-62 for the extensive corvée labor called out to royal palaces.
[215] See Innes 2000: 162ff for the list of "social units" involved.

254   CHAPTER FOUR

and asupporting troops would be familiar to bureaucrats and landowners in 4th- and in particular 5th-century Roman Gaul. The antecedents lay in late Roman *munera* and the survival and militarization of Roman public burdens.

4.4 Conclusion

In stark contrast to the views of most historians, we must conclude that the Frankish armies of the mid-6th century may have differed little from Justinian's army. Both the Franks and the Romans had professional soldiers organized in late Roman fashion—from equipment and tactics to unit structure. Others were organized within private military retinues, but this was also a feature of East Roman military organization. Logistical resources, especially production of weapons, were controlled by magnates, but this not only developed from officially sanctioned late Roman practices (1.2.5), but seems also to have remained the case in the East at least until the mid-6th century. In the 7th century, royal control of Frankish military officers was deliberately modeled on practices Roman Italy in the mid to late 6th century. Furthermore, the gradual atrophy of direct taxes may stem from East Roman models, as services substituted for taxation, albeit on a new, systemic scale, whereas Constantinople chose to retain control over taxes by using service substitution only intermittently. The administration of the military resources of bishops and monasteries similarly followed East Roman cues: bishops were thoroughly militarized by 600, but Anastasios and Justinian had long before established a system where bishops were responsible for organizing local defenses. Even Dagobert's requirement that all church land, including monasteries, contribute militarily, may be related to Herakleios' emergency measures demanding forced loans from the church during the Persian wars (chapter 2 *passim*).

A practical model also existed among the nearby Visigoths. The 7th-century evidence proves how estate-based armies could provide both the manpower and technical expertise to conduct sieges. The decline of the large estate in the East at the same time makes it appear to later observers that we are dealing with completely different and unrelated systems by the time we reach the Carolingians. However, the reconstruction provided here shows how adaptations to late Roman military organization in fact naturally evolved in different directions following socio-economic and political developments, especially as they began to diverge in East and West around 600. Furthermore, there should no longer be any doubt about the effectiveness of the Frankish descendant: the resources available to the Merovingian kings, and reestablished by the Carolingians in the 8th century by relentless campaigning, were formidable, despite the ostensible lack of a centralized bureaucracy and use of direct taxes. Indeed, it allowed them to fight on equal terms against armies backed by centralized, highly bureaucratic states in Aquitaine, Spain and Italy.

CHAPTER FIVE

THE ANATOMY OF A SIEGE: TACTICS AND TECHNOLOGY

Siege warfare was one of the most difficult and resource-intensive forms of warfare in antiquity and the middle ages. A siege campaign rarely revolved around one city or fortified place alone, but often had to deal with a string of fortifications of varying sizes, often over months or even several campaigning seasons. Conducting just one siege took skill, training, planning, manpower, technological knowhow and logistical backup. Conducting several sieges over months and years, while constantly being prepared to face sallies as well as relieving armies, was one of the most difficult feats for an antique or medieval army and the infrastructure that supported it. The same can be said for siege defense; it took years of preparation, planning, competent workmanship and every conceivable technological and military skill.

5.1 *Siege Strategy and Tactics: Basic Definitions*

The fundamental siege methods as defined by *De re militari*, which distinguishes between storm and blockade, is applicable throughout the period under consideration here:

> ...note that there are two types of siege (*obsidendi duas esse species*)—one in which the adversary deploys forces in suitable positions and attacks the besieged by means of unremitting assaults (*continuis insultibus impugnat obsessos*), the other in which he prevents those under blockade (*inclusis*) from getting water, or hopes for a surrender through famine (*deditionem sperat a fame*) since he has stopped all supply-lines. By this strategy he himself remains at leisure and safe, while he wears down the enemy.[1]

While the basic distinction holds good for some analytical purposes, we have already seen examples of interchangeable approaches during one siege, with armies shifting between blockade and storm, sometimes back and forth several times, depending on circumstances. Often several cities and fortifications were attacked during the same campaign. This not only multiplied manpower requirements, but such campaigns also demonstrate

---

[1] Vegetius, *DRM* 4.7.

great variety in siege methods even when all other factors (region, period, actors, climate) are similar.[2]

The blockade could require very large forces to establish, since manpower was needed to construct fortified camps and other siegeworks necessary to cut off the besieged. Furthermore, it could rapidly develop into a storm if some opportunity arose.[3] The storm in turn often required the same basic infrastructure as the blockade, as well as vastly superior forces in order to overwhelm the defenders. This meant that the besiegers had to carefully decide what approach to take, and sometimes this was only done when the fortifications were inspected firsthand.[4] If both parties persisted, continuous storming could drag out over weeks or months, so that again, the distinction between blockades and storms is further blurred.[5] On a strategic level, both approaches often involved shielding forces that protected the besiegers, distracted relieving armies, and ensured lines of communication; several armies could thus be involved in a single siege.[6] Even supporting fleets took part in transporting supplies, enforcing blockades, or carrying out assaults and relief operations.[7] Often, the besiegers needed to defeat the enemy's field armies *before* they could begin attacking a fortification; at other times fierce battles were fought simultaneous with sieges.[8] Due to these manpower requirements, attackers were just as likely to run out of supplies as the besieged.

---

[2] See chapter 2.3 above, and especially the examples of *Constantina, *Harran and *Edessa 502f for raiding as prelude to sieges; the Roman sieges of *Amida 503 and 504f demonstrate different approaches according to strategic priorities, opportunities and failures. At *Reggio 542 the Ostrogoths attempted to storm the city several times before settling on a blockade. The opposite was the case when the Romans besieged *Lucca 553, where they progressed from blockade to storm, but at *Cuma 552f a Roman storm was successfully repulsed even when the wall was undermined, leading to a lengthy blockade.
[3] Thus at *Tzacher/Sideroun 557, the Romans discovered a secret approach after tightening their blockade, which led to a successful storm.
[4] Thus the Romans at *Orvieto 538f decided to blockade the Ostrogoths due to the strong position of the city; the Avars at *Cividale 610 inspected the site to determine where to attack, but the ultimate result was determined by the betrayal of the city by a prominant Lombard noblewoman.
[5] *Naples 536 saw 20 days of continuous storming; *Constantinople 715 experienced 6 months of daily fighting during a civil war, but finally fell to treason.
[6] See *Fiesole 539, *Kamakhon 766 for good examples of shielding forces; *Ravenna 539f and *Constantinople 717f illustrate the complexity of operations, which involved not only several armies but also naval forces.
[7] See e.g *Naples 542f, *Rome 545f and the previous note; further discussion in chapter 5.2 below.
[8] Thus the Ostrogoths took *Singidunum 472 from the Sarmatians by siege *after* defeating them in battle; the garrisons at *Florence and other cities were confined to their forti-

Conversely, the defenders had to conduct repairs, assemble supplies, bring in reinforcements, prepare weapons, and evacuate people and livestock. Since an enemy's intention might be difficult to gauge (distinguishing between a raiding party and a scouting vanguard must have been difficult), preparations had to be made even when a full-blown siege never materialized, or the speed and success of an invasion meant that such preparations were impossible.[9] Thus, in many instances a political settlement was preferred by both parties, or the stresses of the siege caused ruptures in the ranks of one of the parties.[10] In fact, negotiations were conducted at any point of a siege, from the approach of a vanguard, through actual fighting, even to the brink of collapse.[11] Otherwise, the mere threat of a siege (even a bluff) could be enough to win the desired concessions.[12] The sheer complexity of possibilities is poorly reflected in most of the laconic chronicles, which only mention the fact of surrender, raid or otherwise.[13] A far more apt definition, then, is the one provided by the extant work of Aeneas Tacticus on siege warfare from the 4th century BC, which eschews the straight blockade-storm dichotomy of Vegetius to set the strategic and tactical maneuvering of a city under siege within a wider socio-economic, political and cultural context, all of which could affect the fate of a *polis*. His approach is fundamentally similar to the realities outlined above, and explains the range of events included in the *Corpus Obsidion-*

---

fications after a field battle in 542; similarly the Persians did not invest *Archaeopolis 550 until they had chased off the Roman field army encamped at Phasis.

[9] The complexity of a rapidly developing situation is well evidenced by *Thessalonica 615 and 618; especially at the latter, raiding parties preceded the full army which both blockaded and stormed the city for 33 days. A similar Avar use of a large vanguard is found at *Constantinople 626. The citizens of *Edessa 484 successfully took to their walls under threat.

[10] Thus the rebels penned in at *Papyrius for four years (484-8) were betrayed by their own men; see further chapter 6.2.1 below.

[11] The Lombards were bought off when they began preparations to invest *Aix in 574. *Ravenna 489-93 ended in a political compromise after several rounds of blockade; *Edessa 502f is a good example of negotiations during several stages of a siege; the Romans tried different approaches, including stratagems and assaults, at *Sisauranon 541, but were only able to negotiate a settlement when a raiding party discovered by accident that the Persian garrison was out of supplies; at *Semalouos 780 the Romans surrendered to the Arabs only after long and very hard fighting.

[12] Thus the Vandals at *Carthage 533 relied on political means in their attempt to regain their capital, which they seemed loath to destroy through raiding or siege. Even the Persians tried to impress, e.g. at *Theodosiopolis 576, where they had troops and arms drawn up before the city. See further chapter 6.1.3 on the effect of morale.

[13] Thus e.g. *Hohenseeburg 743 and *Chantelle (and other forts) 761, all taken *per placitum*, but see discussion at 4.3.3 and cf. *Semalouos 780.

um. While this chapter focuses on the strategic, tactical and technological aspects of a siege (both Vegetian and Aenean), chapter 6 deals with the socio-economic, cultural and political (i.e., largely Aenean) aspects in detail.

5.1.1 *The Blockade*

Raiding must be regarded as fundamental to early medieval warfare. Not only a means of acquiring "plunder and tribute" in Reuter's memorable phrase, it also played an important role in siege warfare, since raids were often deliberate preludes to formal sieges. Since the purpose of a blockade was to deprive the defenders of supplies, the besiegers often began by ravaging the surrounding countryside. This could occur well in advance of the total encirclement envisaged by Vegetius, thus more in line with Aeneas Tacticus' understanding of siege warfare. Systematic raiding of the countryside destroyed the economic basis of a city (for which see chapter 6.1.2). There therefore was a deep anxiety among soldiers as well as civilians over the consequences of raids. Most obviously they were concerned with the loss of life, captives, livestock or movables and destruction of buildings. However, at inopportune moments, raiders could steal or destroy a whole harvest or disrupt essential agricultural activity.

We must be careful to distinguish between, firstly, opportunistic raiding, conducted by nomadic tribes, bandits or marginal groups for immediate material gain or honor (i.e., "plunder and tribute"), and, secondly, strategic raiding, which had a long-term political or military goal. Strategic raids were aimed at two objectives. The first was political, to force specific concessions, such as ransom or recognition of political hegemony (short of outright conquest). In the former case, the appearance of an army was sometimes enough: the Persians threatened numerous cities in Syria in 540 with sieges or with raids of their territories, often achieving considerable ransoms in the process.[14] Sometimes outright submission was achieved; although such political submissions were frequently reversed, they established useful precedents for later campaigns. One cannot assume that such threats were carried out by merely showing up; at *Apamea in 573 the Persians established a tight blockade, hemming in the city with elaborate siegeworks, which were an essential component to their negotiations.

---

[14] Cf. *Hierapolis, *Chalkis (both 540).

The other strategic objective was to destroy the enemy's ability to perform military operations by hitting at his supplies and infrastructure,[15] paving the way for conquering armies by depleting an enemy's reserves and will to resist.[16] Since supplies were often stored up in fortified points, it took very little for such raids to change character from a *chevauchée*, or attack on the open countryside, to an assault on a fortified position. We see all forms of raid in the campaigns that have been analyzed in chapters 2.3, 4.1.4, and 4.3.3. Arabs, Huns and Kurds were let loose to plunder indiscriminately. Arab tribes in particular were renowned for their capacity for plunder during the sixth century.[17] While they were acting in their own interest, the Persians who had organized them could exploit the chaos and distraction to focus on long-term objectives. The Persians more systematically raided areas they intended to return to next season during the Anastasian war. The Aquitanians attempted to break Pippin's stranglehold on their territory by raids against Frankish staging posts. Most of them were defeated by Frankish border guards, but when Remistanius defected and attacked from the inside of Frankish defenses, he was able to make all agricultural activity impossible that year.[18]

If a district was disrupted this way for more than one agricultural season, it meant certain famine and ensuing epidemics, so everyone from great landowners to poor shepherds had much to fear from a prolonged enemy presence.[19] Indeed, the small fortified places and accompanying inscriptions in northern Syria show the concerns of the local population in the face of Arab and Persian incursions, and how little the central government could do outside of providing larger settlements and forts with defenses and garrisons.[20] This ability the Arabs put to good use during the Islamic conquests of the 630s. Raids were so persistent and systematic that they in effect were blockades of the cities of Palestine, but freed the Arabs

---

[15] Some also argue that the objective was to force the enemy to fight in open battles (or lose prestige), cf. Halsall 2003; see also the debate in the first issues of *JMMH*.

[16] This remained standard Byzantine operating practice on the Arab frontier until the end of the 10th century. See McGeer 1995.

[17] In addition to Procopius' statement at *Sisauranon 541 (see also chapter 7.3.1 for this), the Syriac authors, especially Joshua the Stylite and John of Ephesus, convey this in great detail during the Anastasian war and in the 580s, respectively.

[18] See chapter 4.3.3 and especially the various *rocks, caves and fortifications taken in 767.

[19] Thus the Goths starved at *Osimo 539 when cut off from their own lines of supply and foraging parties.

[20] For an analysis of inscriptions from Syrian fortifications, see Trombley 1997; for further discussion, see chapter 6.1.1 (infrastructure) and 6.1.2 (relations with the countryside).

from having to tie down forces to camp outside the city walls.[21] The territory of *Thessalonica 662 was subject to fierce and continuous raiding for two years by the Slavs before they made a brief attempt to storm the city. As a result, the population was utterly emaciated. Thus, following the more flexible definition in Aeneas Tacticus, we have to postulate an intermediate stage of persistent raiding, which made movement outside of fortifications hazardous to the extent that it constituted an effective blockade, but complicated the task of contemporary chroniclers and modern historians.

Seasonal raiding was often the deliberate prelude to a siege in the form of a close blockade, which physically sought to deny entry and exit from all gates. Most commonly this was achieved by establishing fortified camps or a fortified perimeter around the city. The Arabs are depicted in various traditions as effectively blockading *Damascus, *Emesa, *Tiberias, *Caesarea and *Jerusalem (see 634-35, 640), in the first case by establishing camps by the gates, in the last by ravaging the surrounding area to the extent it was impossible to leave or enter the city safely for months. This would obviously disrupt trade, communications, and agricultural activities, depriving the city of its essential necessities. The Lombards established blockades against *Pavia (569ff) and *Rome (756, where a papal letter enumerated the camps of the various Lombard contingents according to gate), but were in turn blockaded in *Pavia on several occasions (755, 756, 773f) by the Franks. The Carolingians honed this skill against the Arabs in the early 8th century,[22] but it was common in the Merovingian period as well.[23] The Romans regularly relied on blockades, especially if a storm failed or seemed too risky, e.g. at *Amida (504), *Cumae (552f), *Parma (553). The Ostrogoths blockaded the Romans at *Salona (537) and *Rome (537f; 545f). The Visigoths put this skill to use during civil wars, when Leovigild blockaded his rebel son Hermenigild at *Sevilla (583f), even rebuilding an old Roman fortified city as a siege camp and establishing a blockade on the river Guadalquivir.

In addition to controlling the countryside and establishing camps or siegeworks to block access to foodstuffs, it was also necessary to cut off the supply of water. Aqueducts were regularly (and easily) cut, even when the siege in practice only amounted to a storm. Whereas food stores could last for months or even for years, lack of water immediately terminated resistance. The Ostrogoths at *Urbino (538) extended their hands in surrender

---

[21] See appendix I.
[22] see chapter 4.3.3, especially *Avignon, *Narbonne 736/37.
[23] This is extensively discussed in chapter 4.1.4 above.

when they realized their well was dry after only three days of siege, just as the Romans were moving up the hill to assault the fortifications. *Beroea (540) had to surrender to the Persians when the horses and livestock brought into the acropolis consumed scarce water resources. Due to this weakness, besiegers followed several approaches at the same time: the Vandals at *Carthage (533) cut the aqueduct while they tried to subvert some of the defenders; the Avars at *Constantinople (626) ran out of supplies and support long before the city was affected, but the cisterns had to cover the needs of the city for the next century and a half. In the cases of *Vienne (500) and *Naples (536), the ruptured aqueducts allowed the besiegers to enter within the walls. The Persians cleverly installed three layers of underground piping to fool the Romans at *Petra (550); the Romans never found the bottom pipe, and only discovered the ruse after capturing the city by other means. Even the Merovingians installed subterranean pipes when repairing their city walls, as we saw with Desiderius of Cahors in the mid-7th century, and Gregory of Tours noted the importance of good water supply to the defenders of *Chastel-Marlhac (524).[24]

Cut aqueducts left defenders dependent on pre-existing wells and cisterns, which also proved useful when ancient infrastructure fell into disrepair or a site was relocated. At *Naples (536) and Constantinople, large intramural cisterns could cover the needs of the city. However, sometimes these were located just beyond the walls, so that the besiegers had to approach the walls in order to destroy it; thus the forces of Aegidius broke a well outside the walls of *Castrum Cainonense in 463, while the Romans attempted to do the same at *Osimo in 539, although they had to settle on poisoning the well since resistance from the walls was fierce and the cistern well built. However, even these resources sometimes failed. Thus *Thessalonica was left without water when the rains failed during the Perboundos affair (662). The lack of water could just as well affect the besiegers. The Romans nearly gave up besieging the Moors at *Toumar (544) when their water supply ran out, while the Roman garrison at *Sergiopolis (542) was about to surrender to the Persians when a Christian Arab in the Persian army revealed that they would have to withdraw soon due to lack of water.

A blockade could last anything from a few days to several years, but normally took weeks or months, and rarely lasted over a year. *Narbonne was besieged over the winter of 436-37. It seems that Roman garrisons and populations in particular could be extremely tenuous if well led and pre-

---

[24] See chapter 4.2.2 above.

pared. It is unclear how long the siege lasted, but the Romans at *Perugia seem to have withstood an Ostrogothic blockade that lasted up to several years from 545, but the city was finally taken by storm. The Ostrogoths in turn could also hold out for a long time, such as at *Conza (554f) and many other occasions. At *Comacina (587), the garrison was besieged for six months by the Lombards; at *Dara, the Romans held out for six months in 573, while in 603-04, the city held out for nine months. *Sirmium resisted the Avars for two years from 579, and was only abandoned on imperial orders when it was clear that no relief could be organized. At *Mardin (606ff), it appears that the Roman and garrison held out for two and a half years, and may only have fallen as the result of Herakleios' revolt, which drew Roman forces from Syria. Long sieges were common throughout our period; Charlemagne invested *Pavia (773f) for nearly a year.

Starvation and thirst were common causes for surrender. The Roman garrison at *Crotone held out extreme starvation while waiting for reinforcements over the winter of 551-52. Cannibalism occurred at *Amida (504f), when the Persian garrison abandoned their Roman concubines to their fate (the women reportedly ambushed the few emaciated men left in the city and ate them), while the Roman garrison at *Piacenza (545) resorted to cannibalism before surrendering to the Ostrogoths. Otherwise, bad weather put an end to operations. Roman troops besieging *Amida (503) simply drifted to nearby cities or even went home when winter set on, and on a few occasions, torrential downpours may have ended sieges.[25]

Inclement weather and lack of food led to epidemics, and sickness was a common problem on both sides of a conflict. The inhabitants of *Clermont suffered from an epidemic when they were shut up inside their walls in 555. Exhaustion probably caused the many Ostrogothic sick abandoned in their camp at *Rimini (II, 538). The sweltering heat caused illness among Roman troops before the siege of *Sisauranon (541). The same problem seems to have affected the Frankish armies operating in Italy in the 6th century, a problem attributed to the unfamiliar climate. However, Arab armies operating in Anatolia encountered the same problem. The Umayyad army besieging an unnamed *Fort in "Asia" (741/2) suffered an epidemic in their camp before they were defeated, while the Abbasid army besieging *Kamakhon (766) was struck with dysentery and other intestinal diseases when eating fruit and vines in excess. Malnutrition and climate in combination with cramped quarters seem to have caused such out-

---

[25] Thus *Castrum Cainonense 463; also see the discussion of the Visigothic siege of Adrianople, chapter 1.1.2.

breaks. Epidemics are reported among the population of Palestine in the aftermath of the Arab invasion, presumably due to starvation and the weakened state of the population after long periods of raiding against the countryside.[26]

The physical means by which close blockades were instituted are further treated below (5.2), but it must be noted that sometimes, a blockade could only be partially established, allowing some supplies to be brought in; this only lengthened the siege. Most famously the Ostrogoths could only enclose the northern half of *Rome (537f), but attempted to close the blockade, albeit unsuccessfully.[27] Otherwise, they began complete blockades, such as at *Salona (537), where the Ostrogoths instituted both land and naval blockades. While their fleet was defeated by the Romans, they still maintained the landwards blockade for some time. The naval Arab attempted siege of *Constantinople in the early 670s probably only cut off access from the Aegean and parts of the Sea of Marmara, while the Romans could bring in supplies from Thrace and over the Black Sea; conversely, in 717-18, the Arabs established a complete land blockade, but were defeated at sea and subsequently suffered terrible logistical problems.

5.1.2 *The Storm*

A storm could take place at any time during a siege, from a surprise attack by a rapidly advancing vanguard to desperate attempts to break the deadlock of an unsuccessful blockade. Sometimes storming tactics were simply necessary in order to complete a blockade, such as the destruction of the extramural cistern at *Osimo (539). It was of course possible to take a fort by stratagem, without any obvious threat of an army nearby. A Roman who had deserted to the Persians reappeared with 400 Persians at *Martyropolis in 589. Claiming that they were deserters from the Persian army, he gained access to the fort and managed to take it over. A battle or raid could also develop into a storm of a fortification if the situation warranted it.[28]

---

[26] Appendix I.

[27] Although they managed to blockade *Rome very successfully in 545f.

[28] Thus a Slav raiding party attempted to take *Thessalonica in 604, but mistook a suburban fort for the city itself, giving the garrison time to mobilize. The Roman assault on the village of *Anglon (543) followed from a battle where a Roman invading army encountered and defeated the Persians in the field. Victorious, the Romans pursued them into the village under the fort, which had been prepared for ambushes and sorties, thus leading to an ultimate Roman defeat in siege-like conditions. However, successful battles were generally conducive to successful sieges, such as the *Cities along the Seine that were stormed after battle during the Frankish civil war in 600.

While such occasions may have arisen without a specific plan, capturing fortifications by surprise was certainly possible. *Viminacium and other forts (441) along the Danube were taken by surprise storm by the Huns. The Romans' surprise capture of *Imola (538/9) led to the occupation of the whole of Emilia. Sometimes surprise attacks could have bizarre consequences. When the Romans had just taken over *Verona (541) from the Ostrogoths through treason, the latter lingered in the hills overlooking the city after escaping. When they noticed that the Romans were disorganized (most of the army was well outside the city squabbling over the distribution of booty), the Ostrogoths charged back in through the still open gate, forcing the Roman troops to defend themselves on the wall from an assault from the inside; in the end, the Romans had to jump *out* of the city to escape. As the Huns, the Avars seem to have excelled in surprise assaults. Thus they took *Singidunum by surprise in 583, but suffered heavy losses; later that same year *Augustae and *Viminacium fell, presumably by similar means. When the Romans were under pressure, they sometimes excelled at surprise assaults. When *Amorion had fallen to an Arab invasion, the Romans sent a "special ops" force that took the city by surprise in the dead of winter (666), while *Cuma (717) was retaken from the Lombards by a night surprise attack. Several Visigothic sieges were resolved by treason or surprise, but this was far from the only method employed.[29]

A storm might be conducted for political reasons. The Persians stormed *Sergiopolis in order to scare other cities on their Syrian campaign in 540. Similarly they assaulted *Theodosiopolis (610/11) before engaging in negotiations, which they seem to have conducted from a position of strength, since they had demonstrated that they were willing and able to fight (but also to preserve their forces if fighting could be avoided). The Saxons tried to talk the Frankish garrison at *Syburg (776) into surrendering, and attempted to storm the fortification when they refused. However, the very threat seems to have been efficient at *Eresburg, which had surrendered shortly before. Even more compelling than political issues were problems of supply; at *Sura (542), the Persians were forced to storm since they were running out of water and could not engage in a blockade.

If surprise, stratagem or other approaches were impractical, a formal siege had to be instituted, during which a storm took place at any stage. The most desired objective was to overwhelm the defender at first onslaught. At *Naissus (442), massed towers, rams, archery and ladders were

---

[29] See e.g. *Medina Sidonia 571, clearly treason; *Cordoba 572, a night attack but treason is not specifically mentioned; for other methods, see chapter 3.2.3 above.

made possible by a bridge constructed across the nearby river; here overwhelming resources took some time to amass (preventing surprise), but were clearly enough to defeat any opposition. The Arab naval assault on *Constantinople in 654 had a similar aim, but was defeated by the forces of nature within sight of the walls. Opportunity sometimes arose during a blockade. Several Roman assaults on *Amida (504f) began as minor incidents, but such tactics were banned by the commanders after heavy losses: at one occasion, a Roman assault "in a dense mass" led to 40 Romans fallen and 150 wounded, while the Persians on the walls were well covered by protective sheds, and only lost nine men. *Toumar (540) was stormed when a particularly brave Roman patrol was able to defeat Moorish guards in a mountain pass, thus opening a new direction up to assault. A similar scenario occurred at *Tzacher/Sideroun (557). Otherwise, storming took place at intervals. The Ostrogoths made many attempts on the wall during their lengthy blockade of *Rome (549f), but were only able to storm the city when they subverted some of the Isaurians in the Roman garrison, who opened a gate while the Ostrogoths diverted Roman troops by assaulting from another direction.

Finally, a concerted effort was made to end blockades if they dragged on too long. Thus *Perugia (545ff) was taken by storm after a lengthy blockade. The Persians had been besieging *Dara for six months in 573 when they ended the siege with a full-scale assault, including the use of a tower. While they exploited the Roman defenders' inattentiveness (the Romans had become overconfident, were tired of the cold, and had become lax in their guard duties), it was a well-planned assault. This seems also to have been the case when Leovigild defeated the rebels at *Sevilla (583f); despite a long, complex blockade, the city was taken by storm. *Caesarea Maritima was possibly blockaded for years, certainly for months, before it was taken by a concerted effort combining artillery, ladders and treason around 640.

As with the length of a blockade, the length of a storm (as far as it can be defined as such) varied immensely, and sometimes alternated with lengthy periods of blockade. The Visigoths stormed *Narbonne (673) in only four hours or less, apparently upon the day of arrival; the Slavs spent three days trying to storm *Thessalonica in 662; the Persians used 19 days to effect a breach in the walls of *Jerusalem in 614. In contrast, the Arabs engaged in naval battles outside *Constantinople daily for months every sailing season during much of the 670s. Due to the inherent dangers and immense losses, continuous storming or "wallfighting" could not be sustained indefinitely in the same manner as a blockade.

5.2 Siege Tactics

Vegetius contains a brief description of the standard late Roman arsenal around 400: he describes the tortoise, which can carry a hook for pulling out stones or a ram for smashing; vines and screens used to let attackers approach the walls to undermine the walls or fire against their opponents; mounds used to overtop walls (e.g. using archery to shoot down on defenders); mantelets (*musculi*), small shelters for troops to remove obstacles, such as stockades (*sudatum*) and fill in ditches (*fossatum*) to allow siege towers to advance.[30]

The great variety of tactics used for subduing a city remained essentially the same for the next centuries, and show only marginal regional differences, although differences become slightly more pronounced over time. Due to the interchangeability of blockading and storming tactics, I have rather chosen to group tactics according to how labor intensive they were. A siege often began with archery and ladder assaults straight from the march. Although ladders could cause logistical problems, this approach required the least preparations (5.2.1). Some types of artillery could also be made ready within days (5.2.2). These tactics were often used in combination with or in order to cover the remaining approaches. Fortified siege camps and siegeworks (i.e. any form of encircling fortification) were often set up first, unless the assault was opportunistic or a deliberate surprise; these were useful both for blockades and storms (5.2.3). The next stage would be to neutralize defensive advantages by overtopping the walls with mounds or towers, which required immense labor and engineering skills (5.2.4). This was sometimes complemented with the use of battering rams and other types of siege sheds that aimed at breaching gates and walls, mostly under the cover of one or all of the above (5.2.5). Towers and wall-breaking engines often required considerable efforts just to approach the wall, as many fortifications were surrounded by wide moats that had to be filled in under enemy fire. A slightly less risky but more labor intensive approach was to undermine the foundations of a wall (5.2.6). Most of these tactics were met with similar countermeasures from the defenders (archery, artillery, extension of walls and towers, countermines) and are treated as they occur, but some special defensive devices and approaches are dealt with in the next section (5.3).

---

[30] Vegetius, *DRM* 4.14-16.

5.2.1 *The Basic Approach: Archery and Ladders*

The basic but most efficient method in storming walls was the use of volleys of arrows, javelins, and stones hurled from hand operated slings against defenders on the walls and towers; the defenders responded in kind.[31] Such missile duels could last for hours, days or even weeks before one side gained the upper hand. Fierce missile exchanges are well attested at e.g. *Antioch (540), *Cumae (552f), and *Thessalonica (615, 662). The Romans made failed attempts at *Naples (536) for twenty days with heavy losses before gaining entrance through an aqueduct; although the time is not specified, a long archery exchange from the walls occurred between the Franks and Burgundians at *Vienne (500). If the attackers had overwhelming firepower, the barrage would be so extensive that the defenders kept their heads down or were scared off the parapet. Sometimes the barrage was so intense that most defenders fell, were wounded or panicked and ran away. The attackers could then approach with siege ladders, climb the walls, and defeat any remaining enemies on the walls in hand-to-hand combat.[32]

There were also intermediate stages from first volley to an assault on the walls. The Persians gradually increased the intensity of their archery volleys in order to test the defenses and lure out the Romans at *Phasis (556), nearly overwhelming the Romans who were caught by surprise at the sheer density of arrows. At *Narbonne (673), there was a long exchange of arrows (*sagittae*), and the besieging Visigoths additionally used javelins (*telorum iactu*) against those on the ramparts until they gave in to overwhelming fire. Having scored a first victory, the besiegers then moved up to the gates, where fighting continued until they "leapt" (*insiliunt*) over the walls, presumably by the means of ladders. Indeed, defenders had little recourse if their artillery and archery was overwhelmed by the besiegers, except to surrender or sally out to break up the barrage (see 5.3.2). In one excep-

---

[31] At *Noviodunum 437, the Romans assaulted with arrows (*eballon*) and javelins (*ēkontizon*). The fight over the extramural cistern at *Osimo 539 involved *belos, toxeumata, lithous*; at *Edessa 544, Roman soldiers and civilians defended with *sphendonas, toxa*; *Vitry-le-Brûlé 524 saw hard fighting with *iacula*. For further examples, see the following.

[32] This practice was universal, but most frequently attested for Persians. See for the Persian capture of *Amida 502f with ladders; the Persian capture unmanned walls of *Beroea 540 with ladders while defenders withdraw to acropolis; the Persians scaled the walls of *Antioch 540 with ladders after the flight of the defenders; there was a failed Persian attack with ladders and archery against *Petra 541 and *Edessa 544. The Goths as well were used to storming with ladders along with towers and rams, cf. the multiple attempted storms at *Rome 537f, for which see also Proc. 5.21.2-5; 5.22.10f; 5.22.12-25.

tional case, however, Roman clients, the Rubi (possibly Rugi), revolted and took control of *Noviodunum (437); when overwhelmed by archery and javelins, they took children from the population to set up as human shields on the wall. The Romans immediately stopped their assault and negotiated a settlement.

Occasionally besiegers used *testudo* (or functionally equivalent) formations to get within range of the walls. Sometimes this is mentioned explicitly, but it can also be inferred from the context.[33] At *Rome (537f), the Ostrogoths had large shields that allowed them to approach the walls despite heavy defensive fire. They then used archers protected by shields to tie down large numbers of defenders to the north while assaulting Vivarium at the eastern sector.[34] It was only sallies and the large number of *ballistrai* at Rome (see chapter 5.2.2), operated by skilled engineers, that scared the Ostrogoths out of firing range after long and extensive exchanges of archery fire. The Romans used portable screens on a framework of poles, called a *stoa* ("colonnade") or simply *mekhanē* (engine), to cover assaults. The Slavs made simple covers of planks and hides to protect against missiles.[35] Archers were regularly stationed on siege towers.[36] Otherwise, it seems that troops advanced behind or under cover of other engines, and very often, ladders are mentioned in conjunction with rams.[37] Defenders in turn set up "Cilician" mats made of goat's hair or other padded materials above the parapet to cover against archery fire.

Archery regularly accompanied or covered the approach of siege engines.[38] Even if it did not drive the defenders completely off the wall, covering fire made countermeasures difficult, such as operating ram-crushing devices, traps, throwing stones, pitch or other flammable materials (cf. below). Thus a continuous barrage allowed attackers to bring engines up close, or could provide protection for the building of mound and siegeworks close to the fortifications. However, these approaches required planning, coordination and training. If the defenders were well prepared, fire had to be coordinated against defenders on both the walls and on the

---

[33] Approaching wall with shieldbearers at *Amida 504f; tilting uphill *testudo* against Tzani at *Rhizaion 558; against *Beïudaes 587.
[34] Procopius, Wars 5.22.10f
[35] Thus *Urbino 538; *Thessalonica 615.
[36] Cf. *Naissus 442 and 5.2.4 below.
[37] At *Convenae 585, Frankish troops seem to have trailed the battering rams; similarly the Huns assaulting the *Chersonese 559 brought up rams and ladders at the same time.
[38] E.g. Roman assault on *Tzacher (557) in the Caucasus, at Agathias 4.20.3.

towers,[39] and attackers had to choose where to best assault. They also had to be sure that they could keep up the pressure while men and machines were exposed. In order to achieve this, it was necessary take to risks by assaulting inaccessible sections, so that the defenders would be unable to concentrate their forces, such as during the Persian assault on *Archaeoplis (550), where Daylami infantry were assigned to rough terrain, while Persians and Huns assaulted on more level ground with engines and elephants. Good organization and planning was required by a surprise assault, since defenders could be quickly roused.[40] Thus at *Rome (537f), the Ostrogoths used archery and ladder combinations on many occasions, either to cover for and distract from approaching engines, in full-on frontal assaults, or as the vital element of surprises or subterfuge.

If a section of wall could be cleared (even only for a short while), the besiegers could then set up ladders to scale the wall, but these had to be very well prepared in advance. At *Naples (536), the Roman engineers had miscalculated the height of the walls, so when they were about to scale the walls while the defenders were under heavy suppressing fire, they found that the ladders reached only well below the parapet. The delay could have ended in disaster, but the situation was saved by lashing ladders together two and two.[41] Often troops got as far as the walls, set up ladders, and reached the crest of the wall only to be repulsed by defenders that had crouched down behind the parapet.[42] Due to the advantage in elevation, defenders could simply hurl stones by hand onto approaching troops and scaling parties.[43] Fighting must have been extremely difficult in these circumstances, and Khusro himself had to approach the base of the ladders at *Amida (502f) and threaten execution to anyone who descended without wounds. When the Romans assaulted *Petra in 550, the general Bessas himself joined his troops in mounting a ladder, but he fell down, and as he was rather old (70 years) and fat, he had to be dragged by his leg to safety by his *doryphoroi* who formed a *testudo* over him, before he charged again.

---

[39] E.g. at the siege of *Lucca 553, where the Romans directed archers and slingers to shoot at infantry on the ramparts, while engines and artillery were directed at the towers (at Agathias 1.18.4).

[40] Thus the Roman defenders of *Philippopolis and *Adrianople, taken by surprise by Avar assaults in 587, gave hard resistance. See further chapter 7.2.3 for Avar attempts to surprise Roman cities.

[41] Proc. 5.10.5, 21-24.

[42] Thus *Ancona 538, where the Ostrogoths were repulsed by one of Belisarius' personal retainers, or *doryphoroi*.

[43] E.g. bishop Sagittarius at *Convenae 585.

In the end, the wall was taken by some nimble Armenians who climbed up a precipice at a spot that was regarded as impregnable, and from there onto the walls.

Archery and missiles frequently caused great losses on both sides. Commanders seem to have been particularly prone to injury or death, since they had to direct operations and thus expose themselves, or lead charges in order to boost morale, as we just saw. Thus Roman commanders were killed defending *Sura (540), *Petra (541), and *Dara (573). A Frankish *dux* was killed while assaulting a fort during the *Frankish invasion of Italy in 590. Although they eventually were unable to take *Petra (549) from the Persians due to failed mining operations, Roman archery was extremely efficient in suppressing the defenders and allowing sappers to approach the wall. The Persian relieving army later only found 150 men unhurt of the original garrison of 1500, while 350 were wounded. The rest had been killed by archery.[44] The casualty rate of 90% is instructive, and although probably exceptionally high, it means that, in normal circumstances (e.g. fewer failed mining operations by the besiegers), defenders under continuous assault would either have to sally out to drive off their opponents or negotiate terms. Roman assaults killed many Moorish defenders at *Zerboule (540). However, losses at *Sisauranon (541) convinced the Romans to institute a blockade instead of continuous storming.

Roman, Ostrogothic and Persian skills in combined archery and ladder assaults are very well attested. The Persians, famed for their besieging sophistication and ability to undertake complex engineering operation, often used this seemingly simple approach with great effect, especially when they were advancing deep into Roman territory. It was a costly approach in terms of manpower and casualties, but less demanding in logistics than the vast siegeworks constructed at *Dara (573) and elsewhere. Even at *Dara in 540, however, when the Persians repeatedly assaulted the walls with archery fire, they were able to drive the defenders off the first circuit wall, but were afraid to follow up their success by scaling due to the double walls. They chose to tunnel under the outer wall instead of exposing themselves to fire from the higher inner walls. After the Persian wars, the Romans rarely had the chance to go on the offensive, but Theophanes reports

---

[44] See Proc. 2.29.34-36; 2.30.15 for the details. A similar event in Roman North Africa is reported by Procopius at *Zerboule in 540. No figures are given, but the Romans prepared to withdraw, unaware of the heavy casualties they had inflicted on the Moors over the course of three days of archery fire, before they found that the Moors themselves had fled the night before.

that the Romans attacked the Arab garrison of 5,000 stationed at *Amorion (666) with "planks" during the winter, taking the garrison by surprise and massacring them to a man. The advantage of this tactic was that it could be used by peoples and groups that had limited logistical ability to settle a siege rapidly. Slavs stormed dozens of forts and cities in the Balkans this way, but continued to use this approach even after acquiring significant engineering skills. The Avars also resorted to these tactics, both on their own and with client peoples who did the hazardous work for them.[45]

Frankish sources in the 6th and 7th centuries rarely mention engines, but often the use of bows, javelins or slings; interestingly, the situation is reversed in the 8th century, when engines were comparatively frequent but individual weapons are hardly mentioned at all. They do however refer to heavy casualties when breaching the walls at *Bourges (762); it would not be unreasonable to infer that this was caused by missiles.[46] We can conclude the same about the Lombards; they mastered the full arsenal of siege engines and had a proven ability to storm Roman-held cities, but we never hear specifically of weapons wielded by individuals. The same applied to the Visigoths in most of the 6th and early 7th centuries, although they clearly had advanced logistical abilities and, as the Lombards, stormed Roman-held fortifications. Indeed, they proved extremely skilled at approaching and scaling walls with archery fire and ladders during the late 7th century. Wamba's campaign in 673 resulted in the storming of several major fortifications where archery and javelins played a major role, and the fighting was settled by assaults with ladders.[47]

5.2.2 *Artillery*

There were two major types of artillery in use to the end of the 6th century: the *ballista* and the *onager*. A third artillery weapon, the traction trebuchet, became prominent during the latter decades of the sixth century and is discussed extensively in chapter 8.

---

[45] E.g. at *Topeiros in 549; the city was defended by its citizens after the garrison had been lured out and ambushed. Despite heroic resistance, the inhabitants were driven back from the walls by massive archery. The Slavs subsequently set up ladders, storming the fortifications, slaughtering male inhabitants and enslaving women and children. Se further the discussion in chapter 7.2.2-3.

[46] For Frankish examples, see Charles Martel at *Avignon 736; *Narbonne in 736 or 737. Wallace Hadrill translated the phrase "restium funibus" as rope ladders, a translation supported by Bachrach 2001: 106, but the wording makes little sense and should clearly be emended. See comments in chapters 4.3.3 and 8.2.3 and in CO.

[47] The best descriptions are found at *Narbonne, *Nîmes, both in 673.

The *ballista* (Gr. *ballistra*, βαλλίστρα) was a very large tension crossbow with a winding and trigger mechanism. It shot long heavy bolts with great accuracy over a long range: two bowshots according to Procopius. The ballista was most consistently used by the Romans and Western successors for urban defense. As we saw, the Romans had specially trained military and civilian engineers (*tekhnitai/ballistarii*) for the task. Its use was also associated with the navy, e.g. at *Phasis (556), while river patrol boats were also probably armed with *ballistrai*.[48] It is fairly well attested in Ostrogothic use during the Justinianic wars, while there is incidental information on its use among the Franks and Visigoths in the West.[49] Interestingly, it is hardly mentioned in Persian use at all: once defensively at *Amida 504f, but there the Persians may have appropriated Roman machinery. The Romans kept using it throughout the period. It is attested on several occasions at *Constantinople (626, 663, 715), and large, bow-powered, bolt-shooting artillery is well known from 9th- and 10th-century texts.[50] It seems that it was complimented with, and perhaps partly superseded by, a smaller, handheld derivation that became the progenitor of the crossbow. This may explain the rather dramatic drop in incidences in the sources; indeed the Visigothic *ballista* seems to have been a rather small device apt for hunting game.[51]

In most cases *ballistae* were mounted on towers, whence they could be aimed in different directions. They were highly efficient in keeping enemies at a distance, as well as destroying their morale. *Ballistae* were especially well suited to stop general assaults or protecting garrison troops returning from a sortie. At *Rome (537f), *ballistrai* were effectively used to save several sorties, and utterly demoralized the Ostrogoths on at least one occasion as a Goth was pierced by a *ballistra* bolt and stuck to a tree, still standing. They were nevertheless able to approach the walls many times. The same was the case at *Constantinople (663). Unfortunately there is little information on the number of *ballistae* to be used at any one siege, but at major urban centers there seems to have been a great number. For example, Procopius tells us that there was (at least) one *ballistra* on each tower at *Rome (537f), and this can be inferred from the sheer effect reported in the sources and from the physical infrastructure of surviving late antique walls.

---

[48] See chapter 2.2.2 for references to the Danube flotilla.
[49] Venantius Fortunatus 3.12 for the *ballista* at bishop Nicetius' private fort on the Moselle and 8.2.1 for a discussion of the meaning.
[50] See e.g. Haldon 1999: 134-38, 189 and *idem* 2000 *passim* for discussion and references, although in chapter 8.2.2 below I take issue with his interpretation of *manganon*.
[51] See Rihll 2007 for this argument.

The other capital artillery piece in late antiquity was the *onager*, a one-armed tension catapult for throwing stones, incendiary bombs, or anything that could damage or lower the morale of the enemy.[52] It was also used mainly defensively from city walls, though it is implicit in many descriptions of offensive siege warfare and perhaps caused the type of damage described in Menander (see below). The Huns at *Aquileia (and similarly *Milan and *Pavia) in 452 used heavy artillery to cover the approach of their wallbreaking engines. Defensively, onagers were placed on the ramparts, probably because they were heavier, had a powerful recoil that could cause damage to the base upon which they stood, and were more difficult to aim in different directions than the *ballistae*.

Unambiguous descriptions of the traction trebuchet appear at *Thessalonica (586), but it was probably distributed throughout most of the former Roman world by that date. We shall leave problems of dating and terminology until chapter 8. However, it quickly became an important improvement to the ancient arsenal, and by the early decades of the 7th century, if not earlier, it seems to have completely displaced the *onager*.

With a strict literal reading of the sources, the *ballista* and *onager* were mainly (and regularly) used for defensive purposes, as the absence of defensive artillery was worth special mention.[53] I have found no instances of the offensive use of ballistae during this period, but this is probably misleading. Sources very often refer to "machines", sometimes specifying one type but leaving the others undefined; thus when rams and the like are mentioned explicitly, other "machines" are likely to be artillery, and vice versa. Machines that were used defensively from walls and towers were probably also artillery.[54] Furthermore, Latin and Greek terms for "throwing" or "missiles" can be ambiguous and cover a wide semantic range.[55] This can be demonstrated with certainty with defensive artillery, where the engines are unnamed, but their effects and projectiles were clearly

---

[52] Incendiary weapons: *Edessa 544, *Petra 550, *Lucca 553, *Phasis 556, and *Padua 601, although some of these may have been traction trebuchets. See the discussion in chapter 8.2. *passim*.

[53] E.g. *Sergiopolis 542.

[54] Thus the Romans used machines *apart* from sheds etc. to assault *Tzacher/Sideroun 557; machines and *ballistrai* on large wooden towers are mentioned at the *Gothic fortified camp that faced off a Roman army in 552.

[55] Cf. the discussion of *toxeumata* ("archery") in the Roman and Ostrogothic arsenals at *Osimo 539; cf. also the Frankish siege of *Vitry-le-Brûlé, where the Frankish army *iacula transmitteret* ("hurled" or "shot" javelins?), maybe with the help of *ballistae*.

those of artillery.[56] The same is sometimes the case with offensive artillery; at *Lucca 553, "flaming ... missiles" were thrown at the towers, while the engines used for hurling missiles against fortifications begin to be mentioned explicitly from *Onoguris (555) onwards.[57] The trebuchet was arguably far more efficient than its predecessor, having greater range, destructive power and accuracy, and was therefore mentioned far more often in the sources; the problem is a long transitional period in the 6th century before terminology became standardized, so its diffusion has required special attention.

Used together, these weapons formed a formidable field of fire through which a besieger must pass.[58] At *Theodosiopolis (421/2), the huge stone-thrower (likely an *onager*, but perhaps an early trebuchet) set up under the direction of the bishop crushed the head of a blaspheming Persian commander, but presumably also worked against more worldly threats from siege towers and other engines. The Roman "Crusher" certainly worked against the Persian ram approaching the city walls of *Amida (502f). Artillery could stop assaults in their tracks, e.g. against the Avars at *Diocletianopolis (587), the Avaro-Slav assault on *Thessalonica in 618, and again at *Constantinople in 626 and 663. Artillery barrages were useful against machines, ships and personnel; such barrages are reported on numerous occasions at *Alexandria (608, 609, 642f).

Probably before the introduction of the traction trebuchet, and certainly afterwards, attackers systematically directed artillery against the defensive devices, artillery and personnel of the besieged. If both sides were well armed, a siege would open with formidable exchanges of artillery before one or the other side was overwhelmed. Artillery duels are reported at e.g. *Sagalassos (664), *Thessalonica (618, 662). In the former case, a single engine was set up against a particular gate. Against well defended cities, however, batteries necessarily had to be large. 50 trebuchets are reported facing *one* wall at *Thessalonica (586),[59] while the Arabs are reported to have set up 72 trebuchets against *Caesarea Maritima around 640.

Attackers used artillery for several purposes. It does not seem to have been capable of destroying the circuit wall itself during this period, but it

---

[56] Cf. *Naissus 442, *Petra 541, *Edessa 544, *Cumae 552f (can be inferred on both sides), and *Convenae 585.
[57] This was probably a traction trebuchet; the problem is discussed in chapter 8.3.1.
[58] For the complexity of such fields of fire, see Marsden 1969-71.
[59] See *Miracula St. Demetrii* 154.

was used against those manning the walls and their engines. Artillery was also useful for destroying any screens the defenders might have set up against small missiles, and most importantly, the breastwork itself, which would be far weaker than the rest of the walls. A remaining fragment of Menander describes how "the battlements were shattered by the blows and the whole wall of the tower was weakened by their force."[60] Since the fragment speaks specifically of projectiles (*bolais*) weakening the wall, it is safe to assume it concerns the effects of stone-throwing artillery.

This effect is also demonstrated by an anecdote in Theophylact concerning the siege of *Beïudaes in 587. A Roman soldier named Sapeir was able to scale the wall of a Persian fort, ramming spikes into mortar and cracks. The first time he climbed up, a Persian defender was able to push him off. Sapeir was caught on the shields of his comrades and immediately made another attempt, when the Persian desperately kicked the broken parapet onto Sapeir's head, as it had been weakened by the preceding artillery fire. The same happened again, and the third time he could easily mount the rampart and kill his opponent. Thus, if deprived of a breastwork by artillery fire, defenders would be exposed to a barrage of arrows and sling stones, while besiegers could climb up ladders unhindered and walk onto the parapet. Since artillery clearly had such an effect on stone parapets, artillery could also be used in order to destroy extra protective measures set up by defenders, such as Cilician mats or wooden superstructures.[61]

Setting up engines took some time, since it was necessary to reconnoiter the site and bring up the sufficient parts and materials. The Persians could explain away the large amounts of timber being transported to build a fleet at *Petra (549) as materials for defensive artillery for their garrison there. At *Sagalassos (664), the Arabs assembled materials at the spot to build a large trebuchet under the direction of a captive Roman engineer, while at *Kamakhon (766), they used "Armenian wagons" to haul cedar timbers for their traction trebuchets. The particular process of "setting up" machines is attested briefly at *Chlomaron (578). Depending on complexity of engines and availability of materials and labor, it took from a couple of days to over a week to set up a battery of trebuchets.[62] Since they could be constructed according to variable specifications, it took some time to acquire

---

[60] Menander, fr. 39: τὰς μὲν ἐπάλξεις καταρραχθῆναι ταῖς βολαῖς καὶ τῶν τείχων ἅπαντα τοῦ πύργου σαθρωθῆναι τῇ βίᾳ

[61] See 5.2.1 and chapter 8 *passim* for examples of artillery use, although the destruction of superstructures are rarely mentioned.

[62] *Thessalonica 615: the Slavs use traction trebuchets during a three-day storm; *Constantinople 626: the Avar Khagan arrived with the main force on July 29, an assault with

ANATOMY OF A SIEGE: TACTICS & TECHNOLOGY

aim depending on the dimensions of the engine and skill of the crew. This problem is not reported among defenders; Roman military engineers and civilian populations seem to have been well practiced in using their trebuchets, and could hit targets with their first shot. Once aim was established, traction trebuchets could be used for continuous barrages day and night. This could be aimed against the defenses, in order to clear the wall, or against the city itself, in a form of terror bombing which proved highly efficient.[63]

On several occasions, artillery pieces were captured and turned against the enemy; thus the Romans captured some sort of artillery from the Persians defending the outworks at *Beïudaes (587) and turned them against the main fortifications. The Persians drove a Roman besieging army away from *Nisibis (573), and then transported the Roman trebuchets and towers to *Dara, which fell six months after. The Arabs also did this on at least one occasion (*Fort near Antinoe, 642), while many groups, especially the Huns, Avars and Arabs, used the services of deserters or prisoners of war to build machines for them, or lead construction teams.[64]

Arab siegecraft peaked dramatically around 650, when massive fleets armed with artillery were sent against Cyprus (*Lapethus 650), *Arwad (649) and *Constantinople (654). After the mid-7th century, detailed narratives are in most cases lacking with the notable exception of *Thessalonica, where siege artillery is well attested in 662, and *Narbonne (673), where heavy stonethrowing artillery can be inferred. However, the terminology for the traction trebuchet can be traced at regular intervals throughout the period. Thus we can document continued use of artillery among the Romans, Lombards, Franks, and Arabs to the end of the 8th century and beyond, and wherever we have sufficient details, the problems and capabilities were the same as those reported around the turn of the 7th century.[65]

---

some engines was made on July 31, while the main battery was ready on August 1; *Drizipera 588: *helepoleis* ready on the seventh day.

[63] The problem of establishing accurate aim is reported at an unnamed *Persian fort in Arzanene 587, where teams of trebuchet operators kept the barrage going in relays, but the crews clearly took some time to establish accurate aim. Similarly, the Avaro-Slavs at *Thessalonica (586) could barely hit the walls of the city, but the more practiced Roman crews had no problem with their aim. *Edessa's Persian garrison and Jewish defenders were beat into submission by a Roman trebuchet attack against the buildings of the city in 630.

[64] See in general chapter 7.

[65] In general see chapters 2.2., 3.2.2, 3.3.3 and 4 *passim* for the institutional context and non-narrative evidence and chapter 8 for a detailed discussion of diffusion; for examples in context, see *Damascus 690, *Tyana 708, *Constantinople 717/18, *Toulouse 720, *Nikaia

5.2.3 Siegeworks: Camps and Encircling Fortifications

Whether a besieging army wished to storm or blockade a fortification, they would in most cases have to set up a physical barrier to prevent egress from the city and to protect the besiegers' encampment against sorties and relieving armies. Fortified camps and siegeworks were essential to this end. Storms with missiles and artillery could on occasion be organized from the march, so to speak, so that at *Thessalonica (586), the author of *Miracula St. Demetrii* commented on the lack of barbarian palisade (*kharax*, χάραξ) and outworks (*proskhōma*, πρόσχωμα), which was offset by their huge numbers and perhaps the use of shields positioned as a temporary defense in the style of Roman and Byzantine marching camps. More complex besieging methods, however, such as towers, mounds, tunnels and undermining (5.2.4-6), needed safe bases for men and materials. Even if it did not immediately result in capture by storm, successful engineering very often convinced garrisons and civilian populations that it was better to negotiate terms than to face the imminent risk of a sack; thus the Goths at *Todi and *Chiusi (538) surrendered when the Romans began constructing fortified camps, *kharakōmata* (χαρακώματα), and thus made it apparent that they were settling in for a long siege.

A close blockade presupposed siegeworks, either in the form of fortified camps at selected sites or a continuous encirclement. Since this was such a standard practice, the nature of the encampment hardly receives any notice: a Frankish siege camp is reported at *Clermont (524), but little is said of its structure, which has to be deduced from other types of Frankish field fortifications (see below). The use of fortified camps instead of encirclements had much to do with the terrain, availability of men and manpower (on both sides), and the relative position of the camps vis-à-vis the gates. Thus the Romans established two camps, one on each side of *Urbino (538), while against *Osimo (539), there was a series of camps arranged according to unit. The Goths partially encircled *Rome during their first siege of the city (537f) with a series of seven heavily fortified camps around the northern half of the city, facing the major gates. They nearly closed off the city by establishing another siege camp to the south of the city a few months later. Those to the north were built in the same manner as Roman marching camps, surrounded by very wide and deep trenches, with the earth heaped up onto a large rampart topped by palisades. Pro-

---

727, *Avignon 736, *Narbonne 736 (probable), *Rome 756, *Kamakhon 766, *Barbad 776, *Syburg 776, *Semalouos 780, and *Nakoleia 782.

copius notes how well built they were, but does not provide any measures or details of any special features. Although the Romans were capable of performing sallies and often managed to beat Gothic forces in the field, they were unable to storm the camps, which allowed the Goths a safe place to prepare for storm and to send out soldiers to harass Roman supply lines.[66] The Goths used siege camps or *kharakōma* (χαράκωμα) in a similar fashion in 545f, this time more successfully, even defeating relieving armies under Belisarius while holing up the defending Roman garrison. The Lombards as well used a fortified camp (*castra*) when besieging *Trent (680), while at *Rome (756), a Lombard army was encamped in divisions outside the major gates, preventing egress and providing a secure base for continuous assaults on the walls.

The only certain method of holing up a besieged enemy was to surround their fortification with a continuous ring of siegeworks. This was a particularly labor-intensive method, but as long as the siegeworks were well guarded, it ensured that a city had no hope of relief or way of smuggling in supplies. At *Orleans (453), the besiegers circled the city with *magnis aggeribus* (large ramparts), while the Franks at *Vitry-le-Brûlé (524) set up some sort of encircling fortification (*vallat*). This also occurred at *Bourges (583; *vallant*). The Goths completely encircled *Salona (537) with a continuous stockade (the other meaning of *kharakōma*, which is clear from the context) on land along with a strong naval blockade. The Romans blockaded *Osimo (539)—at first in a rather lax manner, but then tightening it to cut off all supplies, presumably a transition from scattered camps to more continuous earthworks. This is well documented at *Parma in 553, where the Roman detachment assigned to taking the city failed to encircle the walls with continuous earthworks. Narses had to send one of his own officers to organize a full blockade with the proper methods, as he had also done at *Cumae (552f) shortly before, creating a complete ring of earthworks with guards set up against the city gates.[67] Roman siegeworks are further attested at *Nisibis (573), which was surrounded by *qalqūmē*, the Syriac form of *kharakōma*, and at *Martyropolis (589). The Persians could also construct large siegeworks and captured several cities in this manner, although they are only explicitly mentioned in a few instances, e.g. *Edessa (544), *Dara (573, including a continuous brick wall), and *Apamea (573). Siegeworks effectively stopped sorties. At *Tzacher/Sideroun (557),

---

[66] For details, see Proc. 5. 19. 1-5, 11f; 6. 3. 1-11
[67] For important details, see *Salona: Proc. 5.16.8ff; *Osimo: Proc. 6.23.9-39; *Parma: Agathias 1.17.2f, 1.18.1-3; *Cumae: Agathias 1.11.5

the defenders made a sortie to destroy the "works" (ἕρκη), but were repulsed, while the Lombards besieged at *Benevento (663) apparently overran the Roman fortifications (*castra*) outside, although they were driven back and the siege was only raised by a relieving army.

Field fortifications and fortified bridges were constructed by the Romans, Franks and Ostrogoths on several occasions to secure supply routes to their own armies and prevent relief to the besiegers. Since these are described in some detail, they may give some indication as to the appearance of fully developed siegeworks. The Ostrogoths constructed a heavily fortified bridge to block Roman relief up the Tiber, supported by nearby siege camps but containing its own garrison, apparently 200 men in each of the two towers. Although Belisarius devised ingenious floating towers, armed with incendiary projectiles that set ablaze one tower (killing 200 Goths inside) and nearly broke the Gothic blockade, poor coordination on the Roman side and efficient Gothic reinforcements from the fortified camps ended in defeat of the relieving army and the Romans evacuating the city.[68] The Huns built a bridge to reach *Naissus (442) with their siege engines, while the Avars used Roman engineers to build large bridges across the Sava in order to blockade *Sirmium (579ff).[69]

The Ostrogoths also built a large *fortified camp (552) that had towers mounted with artillery; the Romans had to respond in kind, and only managed to dislodge the Goths after their supply fleet had gone over to the Roman side. A similar Frankish fortification was set up under Butilinus' expedition in 554. This too seems to have been a formidable construction. Indeed, the Persian *kharakōmata* built at *Edessa (544) were burnt when they withdrew, indicating a substantial structure perhaps including towers, breastworks, covers and other superstructures. These hints are only reinforced by the massive earth ramparts that have been identified as belonging to this period; it seems that Avars and Bulgars in particular were masters at moving earth on a massive scale.[70] Thus, the Avars set up a fortified camp (χάραξ) at *Tomi (597f), while at *Constantinople (626), their siegeworks consisted of a palisade (σουδᾶτον), ditch (σοῦδα) and fortified camp (φωσᾶτον).

---

[68] See *Rome 545f for details on the Gothic fortified bridge near Rome. For other Ostrogothic field fortifications that also included bridges, see *Gothic fortified camp 552.

[69] The description in Menander (fragment 25.1) is ambiguous but implies some sort of fortification.

[70] See Squatriti 2002 for a summary of the archaeological evidence.

The Arabs frequently encamped before city gates during the great conquests (e.g. *Damascus 634/5), but soon acquired even more substantial engineering skills: during their great siege of *Constantinople (717f), they built a double set of siegeworks from the sea of Marmara to the Golden Horn, cutting off the city overland, while protecting their camps against the Bulgars. These consisted of deep trenches and embankments crowned with stone walls, stretching for a length of at least 12 km (6 km each way from the Sea of Marmara to the Golden Horn), but probably more to accommodate for topography and the large army. The Franks under Charles Martel are attested building continuous earthworks around besieged cities during his campaigns in Aquitaine and Septimania during the 720s and 730s, with special stations for artillery, e.g. at *Avignon (736). *Pavia was completely encircled in 755 (*castra metatus est undique*), while continuous siegeworks are well attested during the Aquitanian campaigns of Pippin the short in the 760s; at *Bourbon (761), "he placed encircling fortifications/camps" (*in giro castra posuisset*) before storming the city. Although Frankish siegeworks are difficult to discern in the sources for the sixth and seventh centuries, it is clear from the above that the skills involved in building siegeworks were essentially the same as building field fortifications and marching camps.

5.2.4 *Siegeworks: Firing Platforms—Mounds and Towers*

A successful storm presupposed overwhelming firepower from the besieger. Due to the difference in elevation, this approach was extremely hazardous and often ended in heavy casualties, and in order to even the odds, attackers regularly made use of some sort of mobile protection (screens, mantlets, etc) to compensate for the defender's angle and elevation. Sometimes attackers constructed a mobile or stationary platform that allowed the attackers to shoot level with or down on the defenders.

One favorite method of the Persians was heaping up huge siege mounds to overtop city walls. The early 6th-century wars between Rome and Persia provide some of the best attested examples. The Persians built a great mound at *Amida in 502f. The Roman response was a combination of undermining, well-directed artillery, and building counter-towers upon the walls. When the Persians besieged *Edessa (544), they began building a mound from the eighth day of the siege, having gathered materials and scouted the surroundings. The mound was so great that it threatened to overtop the wall. The Persians preserved this skill and used it with great success during later wars. Their assault on *Dara in 573 was a particularly spectacular affair. As well as encircling the city and diverting the river, they finally captured the city by using mounds and towers to overtop the city walls. The resources used by the Persians on that occasion must have been enormous, taking months of preparation as well as months of hard work, depending on close support from nearby Persian bases. Further advances were halted when the Romans were able to mobilize more fully.

Another method, used occasionally by the Romans but often by Persians, Avars and Goths, was to set up mobile siege towers. Most of the time, these towers were mounted on wheels and propelled by human muscle power (although the Goths experimented with using oxen), had a framework of sturdy beams that could be reinforced with iron bracing, and covered in hides to protect against arrows and incendiary projectiles. This device was a risky undertaking, since it presented the defenders with a huge target for artillery and incendiary missiles, and often took hours or even days to get into place; it could also be sabotaged by quick engineering on the part of the defenders. Thus the Huns brought forth large engines with beam frameworks on wheels and carrying archers (i.e., presumably towers) against *Naissus 442. The Romans built three large towers bound with iron against *Amida (503), but had to abandon the attempt and burn the towers due to a change in strategic circumstances. The Goths nearly took *Rome (537f) by using towers, as explained in chapter 3.1.5 above; they were also close to success at *Rimini (538), although the Romans managed to sabotage it only a few hours before it would have been used for the final assault.

The sheer sight of such huge machines terrified defenders, especially civilians who had not been previously exposed to them (*Rome 537f), and if successfully brought up to the wall, they could wipe out the defenders from above (*Naissus 442). However, the large resources required to build them (the Avars built the largest reported number, 12, at *Constantinople 626) and their cumbersome nature meant that defensive artillery had to be silenced before the towers approached, otherwise the results could be disastrous. The Romans had to abandon their towers set up before *Nisibis (573); these too may have been transported by the Persians to *Dara, but there they seem to have settled for fixed towers built close enough to overtop the walls; two of them were set alight by the Romans, but the last one covered the final Persian assault that led to the fall of the city. The Avaro-Slavs had huge towers (*púrgous hupermegétheis*) that overtopped the walls of *Thessalonica (618), only to see at least one of them collapse by "divine intervention" upon approaching the walls, killing those inside. Following the logic of the hagiographer, divine assistance did not preclude prosaic instruments such as trebuchets and other engines.

Towers set on ships seem to have been a frequent occurrence and of more use than land-based towers, since they could be maneuvered and shifted to other sections more easily. Belisarius did this against the fortified Gothic bridge on the Tiber with great effect.[71] The Romans could also hoist skiffs from ships' masts. At *Palermo (535), these were instrumental in taking the city, while at *Phasis (556) the Roman fleet hoisted skiffs to the top of the masts; these even carried artillery, and were hence equally instrumental in defending the city from Persian assaults. Finally, a sailor at *Constantinople (626) set up a rig carrying a skiff that was used to burn the approaching twelve Avar towers, which seem to have been very well covered by the massive battery of Avar trebuchets and hence threatened to overtop the city walls. The Arabs are reported to have built vast siege-towers on their invasion fleet against *Constantinople in 654. They can hardly have crossed the eastern Mediterranean, Aegean *and* Marmara seas bearing this equipment, so it was probably installed on one of their forward bases. The presence of the towers may in fact have contributed to the demise of the fleet when the weather suddenly turned, so that the Arabs suffered far greater losses than would a conventionally equipped fleet.

The last explicit report of towers in the sources examined here took place at *Thessalonica 662, when a large Slav tower with several stories (one for a ram and sappers, one for artillery, and one for archers) was left incomplete when its engineer was divinely struck with madness.

5.2.5 *Wallbreaking: Machines*

The battering ram (here only "ram" for short) was basically a large beam tipped with an iron head, suspended from a framework in the form of a shed or house, which is most often *not* mentioned separately; while the "ram" is properly the ramming device used against a wall, the term should normally be taken to mean the whole machine including its protective shed. Rams in the hands of a skilled besieger could effect a breach by repeatedly battering the walls (requiring further undermining) or breaking through the gates, and thus settle the conflict. They should be regarded as a capital weapon, but in modern analyses are in fact quite often ignored in favor of artillery weapons. The Huns used a large number of rams to overwhelm the defenders and breach the walls of *Naissus (442). At *Orleans

---

[71] *Rome 545f, esp. Proc. 7.19.1-22.

(451), the Hunnic rams were about to break through when the relieving army arrived, but at *Aquileia, *Milan and *Pavia (452), unspecified *"machinae"* were responsible for breaking through the walls under the cover of artillery.

Persians at *Amida (502f) used a large ram that was brought up the siege mound. Large, efficient rams were used by Ostrogoths at *Rome (537f). Procopius was thoroughly impressed by their performance, and described them as a sturdy framework of beams, with a pointed roof like a house, mounted on four wheels, operated by 50 men. The Ostrogothic tower- and ram tactics almost broke through at the Vivarium sector of the siege, and the attack was only stopped by a Roman sortie led by Belisarius himself. The Romans had to use similar tactics against a Persian ram assault at *Petra (541). Due to their heavy construction, rams depended on level ground. At *Petra (550), the Romans' client troops, the Sabir Huns, constructed three lightweight protective frameworks of long poles and hides that were equipped with rams' head from the Roman arsenal. The combination was extremely efficient. It took 40 men to carry and operate, could be maneuvered up steep hills, and otherwise functioned as a regular ram. In addition, the men inside were equipped with grappling poles to pull incendiary materials from the roof of the shed and perhaps also pry out cracked stones that came loose during the attack. The Persians asked the same Sabirs to build similar rams for them at *Archaeopolis (550), but in far larger number. The wagon-mounted Frankish rams at *Convenae (585) were clearly the result of a similar attempt to adapt to difficult terrain.

In many instances, the use of rams may be inferred, as machines operating close to the wall effect breaches: thus at *Edessa (544), several types of machine were probably used at the same time to cross the moat and break through the forewall (προτείχισμα); at *Cumae (552f), rams were probably used, while at *Lucca (553) the walls were breached (διετέμνετο). Several cities besieged during the *Roman invasion of Italy (590) were taken *rumpendo muros*, i.e. by breaking the walls. The Romans again used *diversis machinis* against *Benevento (663) alongside artillery, probably rams, and in 711, a ram was among the weapons sent to besiege the rebels at *Cherson. In 595, the Avars broke down the walls of *Singidunum, but it is unclear at which stage this occurred (possibly after occupation, but most likely during the fighting). The Lombards certainly used rams regularly: they broke the walls of *Mantua (603) *cum arietibus* (with rams), and in the same campaign took *Cremona and *Valdoria, while a royal Lombard army took rebel-held *Bergamo (701) *cum arietibus* and other machines, apparently "without difficulty." Rams or the equivalent can be inferred in Frankish use during the 7th century, while it is explicitly mentioned in use by Charles Martel in Southern Gaul and Louis the Pious on the Spanish March.[72] Even more interestingly, it is mentioned in Arab use only once (at least in the sources examined here, at *Kamakhon 766), but on many occasions brief sources speak of breaches caused by the Arabs; thus *Caesarea Maritima (640) was breached, but probably not by the mentioned trebuchets; similarly, the walls of Kīlūnās (642) in Egypt were "cast down" without any details given; at *Nikaia (727), the partial destruction of the walls is observed.

Although rams made a great impression on contemporaries and were sometimes used in large numbers, at *Thessalonica (586) the "iron rams" needed some support, as ramming against the walls would cause much debris and require many hands to clear a path and undermine or burrow through. Therefore, rams were often accompanied by siege sheds that protected sappers who did this work; alternatively, the sappers could operate their sheds on their own (provided that the walls were at least partially cleared). Some may have been light enough to be confused with the screens used by infantry to approach the walls (e.g. *Urbino 538), or were more permanent (but movable) devices to protect sappers constructing mounds and other siegeworks. Otherwise, specific kinds are attested.

Khelōnai (χελῶναι) seem to have been shaped as rams' sheds and of a similar size, but were full of sappers armed with pickaxes, crowbars and grappling poles; they were present in very large numbers at *Thessalonica in 586 and again in 618, where they were made from wickerwork and hides (ἐκ πλοκῶν καὶ βυρσῶν); at *Constantinople (626), the *khelōnai* supported the towers and trebuchets.

Spaliōnes (σπαλίωνες; also attested in the form παλλίωνες; cf. *Strategikon* 10.3) may have been a lighter version of the same; at *Onoguris (555), they supported rams against the walls. At *Phasis (556), teams of Persian sappers could hack away at the wooden fortifications under their cover. Some portable sheds were referred to as "little houses" and may have been integrated into or detachable from the offensive works, which seems to be the case at *Tzacher/Sideroun (557); in return, the defenders used a *spalion* in order to attack the Romans at work filling the moat and attacking the walls.

---

[72] See chapter 4.3.3 and the examples of *Toulouse 720, *Avignon 736, *Narbonne 736, and *Nîmes 737; rams are only explicitly mentioned at *Narbonne, but the level of destruction and the term "other machines" or the like alongside use of artillery indicates that at the other sieges as well, the Franks used rams or the equivalent, for which see the following.

A fragment of Menander has a very detailed description that adds to Agathias' descriptions:

> In Menander *spaliones* (σπαλίωνες) are machines (μηχανήματα), screens made of stretched ox-skins raised on beams (ξύλοις) to the height of a man. Soldiers (ὁπλῖται) go inside this and, sheltering beneath the skin, approach the wall. Wielding tools for cutting stone or breaking through walls, they make channels beneath the ground, pressing forward and digging through, and they either tear down a part of the wall or by some other means get themselves within tunnels. Then they do one of two things: they either break up through the ground and penetrate within the circuit wall or they reach the reservoirs of the spring inside the place and immediately empty them by drawing the water off into a hollow tunnel.[73]

At *Kamakhon (766), siege sheds of some sort were used to cover sappers who were working to approach the walls; this may reflect the return of Persian practices as the Arabs had relied more on artillery under the Umayyads (see chapter 7.3).

5.2.6 *Wallbreaking: Engineering*

Two major engineering projects remain: the filling of ditches to bring up wallbreaking engines, and the undermining of walls. Both required moving immense amounts of materials under difficult conditions, and are hence to be characterized as some of the most difficult and labor-intensive siege methods in common use during the period. Mining could also be used by the besieged against the besiegers, either to undermine their mounds (e.g. at *Amida 502f and *Edessa 544), or to intercept the enemy's tunnels (e.g. *Dara 540). On occasion, it took betrayal to intercept a mining attempt (e.g. *Constantina 502f), while a failed attempt at undermining at *Amida (504f) revealed an existing mine into the city, although this was betrayed by the excitement of a Roman woman within the Persian-occupied city, overjoyed at the sudden appearance of Roman soldiers crawling out from the ground. Collapse of the walls was one of the worst things that could happen to a fortification, since morale often hinged on the security they provided. The defenders of *Antioch (540) panicked when part of the interior scaffolding they had built to hold more defenders broke down. The noise was such that the defenders thought that the enemy had successfully undermined the walls.

---

[73] Menander fr. 40.

The filling of ditches is vastly underestimated both by modern and medieval authors, and often not reported. We have no details on this at *Constantinople (626), for instance, but the waves of Slav and Avar infantry sent across the killing field were probably not expected to climb down into the deep moat and then back up again in the face of triple walls. At *Phasis (556), the Persians had large teams of porters who transported materials and filled in the ditch surrounding the Roman fortifications. This allowed the Persian infantry and *spaliōnes* to approach the walls when the general assault began. The Romans used similar tactics against *Tzacher*/Sideroun (557), while the Franks did the same at *Convenae (585). Although Gregory is critical of the efforts, the prospect of engines approaching over a filled-in ditch convinced some of the rebel defenders to open negotiations and betray the city. Sometimes there was a contest between ditch-fillers and ditch-excavators: at *Rimini (II, 538), the Goths were filling in the ditch with fascines to allow their tower to approach, but the Romans snuck out at night to expand the ditch again.

Undermining could be done in two manners: firstly, from up close, by bringing up siege sheds to the base of the walls. These covered sappers who dug down to the foundations and hacked away stones and filling with crowbars and pickaxes. A great range of siege engines were used, improved or invented for these purposes. Rams were frequently used but nearly as cumbersome as siege towers to handle; more lightweight versions became common during the course of the sixth century, and could easily be constructed by tribal or nomadic groups with rudimentary engineering skills, as we have seen. *Spaliones* effectively covered the besiegers at *Onoguris (555), allowing the men to excavate the foundations, which they then loosened with hammers and crowbars.

Alternatively, and safer from enemy fire, was to dig tunnels from the outer siegeworks all the way into town or to the base of the walls or a tower. The purpose would be to get into the city, capture a section of the walls from within (to allow your own troops to ascend with ladders) or even better, gain control of one of the gates. The Persians as well as the Romans occasionally dug tunnels to get within the walls. At *Dara (540) the Persians dug a tunnel to get within the outer circuit wall, but were stopped by a counter-trench; the Romans recruited nearby miners to do the same against *Amida (504) and *Nisibis (573), but were on either occasion discovered or driven back by the Persians. In sum, under-wall tunnels were a rather risky undertaking, but entering through aqueducts could be more efficient: at *Vienne (500) and *Naples (536), crowbars were used to expand aqueduct tunnels and thus gain entry; at *Constantinople (705), when Justinian II reclaimed power entering through an aqueduct, no engineering is reported, but it is hard to believe that there were no obstacles to such an obvious threat.

Most of the mining efforts attested were directed at undermining walls and towers, whether done from tunnels originating in the siegeworks or done under the protection of a siege engine at the base of the wall. Often mining was combined with a number of other approaches, such as archery and artillery to keep defenders from destroying rams and penthouses or discovering where tunnels were directed.[74] By removing the stones of the base, replacing them with wood, and firing it, whole towers or sections of wall could be brought down, with devastating effects to morale. The Romans defending *Petra in 541 surrendered to the Persians when the latter managed to undermine and bring down a tower.[75] The Romans turned tables on the Persians shortly after when they in turn besieged the Persian garrison installed at *Petra (549), undermining and bringing down a large section of wall. The Persians were only saved by a very large building standing behind the fallen section of wall, preventing the Romans from following up with a storm.[76] When the Romans finally did take the city (see *Petra 550), it was the result of a failed mine they had dug earlier that finally caved in. The Romans were assaulting from another direction when the wall collapsed. This led to a scramble where the Romans managed to exploit the breach and capture the city.[77] This practice can be observed further West as well, though in a somewhat circumspect manner. Gregory of Tours describes how the walls of *Angoulême (508) collapsed like those of Jericho in the face of Clovis' army. In light of the above, it seems clear that one does not need to invoke divine intervention to explain the events. However, during the *Auvergne campaign in 524, there is good direct evidence that the Franks undermined the walls of *Clermont. The Persians kept using the tactic well into the 7th century, undermining walls of *Jerusalem (614).

Even a successful undermining might not end in capture. When the Romans undermined *Cumae (552f), the debris left by the fallen section of wall was so rough as to prevent a Roman storm. The Goths were able to organize resistance and hold the breach while the Romans struggled to get across the fallen wall.

---

[74] E.g. *Rome 537f, *Chlomaron 578.
[75] Proc. 2.17.23-28.
[76] Proc. 2.29.34-36.
[77] Proc. 8.11.11-21, 54-58.

5.3 Defensive Responses

In addition to reciprocal responses to individual threats as noted above (archery against archery, artillery against artillery, mining against mining etc.), defenders had to display particular inventiveness in order to survive. Sometimes this involved using creative engines not otherwise attested, e.g. the rigs improvised at *Constantinople (626) or the "Wolf" at *Rome (537; see below) and other inventions, but for the most part, if threatened by starvation or storm, the defenders either had to make a sortie against the besiegers, hoping to break off a particularly dangerous assault or engineering project, or hold out until relief could be organized.

5.3.1 Technological Responses

Sometimes it is difficult to distinguish between purpose-built engines and creative *ad hoc* responses. Thus, it was common to stretch mats and screens of goats' hair or other materials over the parapets (cf. 5.2.1). These were often highly improvised; even mattresses and bedding covers are attested. Nevertheless, they could give some cover against enemy arrows and sling stones. Similar materials were used by the defenders of a *Persian fort in Arzanene (587) to soften the blows of catapult projectiles against the walls, while at *Amida (502f), the Romans used oblique beams and/or bundles of reeds to lessen the impact of the Persian ram. Alternatively, hooks were used to snag rams' heads or overturn siege sheds; the Avaro-Slav assault on *Thessalonica (618) saw a large number of their *khelonai* overturned this way, exposing their crews to archery fire from the walls.

Both military manuals and historians record the use of various contraptions, stones, logs, columns, boiling oil and other suitable objects to throw onto the heads of assailants, and especially their engines. At *Naissus (442), huge boulders were tipped from the parapet onto the oncoming rams, crushing a great number of them. From the description of their size and effect, they must have been hoisted onto the walls by some sort of crane. The same could be argued for the barrels of stone or burning pitch and fat that were hurled against the besiegers and their rams from the parapet at *Convenae (585). Sometimes the effect was less than desired, as when the Goths at *Osimo (539) rolled wagon wheels down the steep hill at the Romans who were charging against them, only to miss every single man. At *Rome (537f), however, the Romans at Hadrian's mausoleum were almost overrun by the Ostrogoths, who evaded both ballista and archery fire and began scaling the walls with ladders. The desperate defenders resorted to an act of cultural vandalism by breaking up the classical statues that adorned the monument and hurled the broken pieces onto the heads of the ascending Goths. The Roman defenders of *Kamakhon (766) had large beams with stones in one end that were swung against those attempting to storm the walls.

Gates were particularly vulnerable, even when flanked by bastions on both sides. At *Rome (537f), Belisarius' troops set up a device called a "wolf," essentially a reversed trap door (like a winding bridge in an upright position), with spikes protruding outwards. This was released from the walls and fell down onto anyone trying to rush the gates. Later on, when Belisarius had to rebuild the demolished walls of *Rome (546f) and was defending against an Ostrogothic army that suddenly appeared, the unfinished gates were protected by a large number of *triboloi* or spikes that made passage impossible.

Finally the physical structures themselves had to be modified. Thus, Belisarius had his men extend the left flank of the merlons inward (at 90 degrees) on the parapet at *Rome (537f). This protected them from flanking fire on their left side, so that they could hold their position longer than otherwise (although returning fire to the left must obviously have become more difficult). The parapets at *Antioch (540) were extended with wooden scaffolds, in order to allow more defenders to participate. Walls were sometimes extended upwards to face towers or mounds (e.g. *Amida 502f, *Edessa 544), although this could result in typical problems of hastily performed works that proved impossible to maintain under fire.[78]

5.3.2 *Sorties*

Sorties were very common during sieges. Procopius famously records a total of 69 engagements outside the walls between the Romans and the Goths at *Rome (537f); at least two of these would qualify as major set-piece battles, while a large number were serious skirmishes involving hundreds if not a couple of thousand men. Since the Goths were never able to quite blockade the city, the Roman army could use open spaces to engage in running battles and return to the fortifications and the protection of artillery and archery; otherwise they could go on longer forays in order to disturb Gothic supply lines or even capture lesser fortifications. This begs the question of the physical boundaries of a siege, since some parts of it could apparently be carried out beyond the sight of the walls. Indeed, at

---

[78] For a further discussion of the physical infrastructure, see chapter 6.1.1.

*Ancona (538), the Roman garrison went out to chase off the Ostrogoths *before* they could begin a proper siege, but were themselves surprised by the huge number of Goths. The Romans successfully made a sortie against the Slavs at *Thessalonica in 604 before the siege could begin, and used the same tactics against the invading Arab armies in the 630s.[79] The stakes could be very high: at *Messina (549), the Roman garrison tried to prevent an Ostrogothic invasion of Sicily, while at *Rimini (552) the Ostrogoths tried to stop Narses' invasion of Italy, and the garrison at *Singidunum (588) went some way in order to burn the fleet that was being prepared by the Slavs. Similar considerations may lie behind the battles reported near *Carcassonne (587), where the Visigoths emerged victorious over a Frankish invading army. At *Mopsuestia (772), the garrison seems to have surprised an Arab army returning from Anatolia by laying an ambush or otherwise attacking unexpectedly out in the open.

Not all cases are as extreme, however, and most of the time, a sortie would be conducted somewhere between the shadows of the walls and the encampment of the enemy. The objective was often more immediate, to destroy or fend off dangerous engines; thus *Rimini (II, 538), where the Romans went out to prevent the Goths from recovering their siege tower, which was about to be hauled back from the filled-in moat for the night. The Romans at *Edessa (544) sent out Hun soldiers to stop the Persians from completing their mound; later on, the Romans won a fight on the mound itself. At both *Rome (537f) and *Archaeopolis (550) the Romans were able to stop dangerous ram assaults with sorties; in the latter case, they even managed to catch the unarmed ramming crews and Persian archers out in the open.

Sometimes the fight was over supplies; thus at *Osimo (539), Goths and Romans fought on the hillside below the walls, apparently over lush pastures for their horses. At *Fiesole (539), however, the Gothic position was more desperate, and there they ventured out before they starved to death, although the sortie was unsuccessful and they were confined within the walls until they surrendered. Similarly, the Romans holding Hadrian's mausoleum after *Rome fell in 549f decided on a desperate sortie, as they lacked any supplies, but were convinced to accept a peaceful surrender.

Sorties could also be parts of elaborate ruses. The Persians used the fortress of *Anglon (543) as a safe base for a sortie that was the last element in an elaborate ambush of the Roman army, while the Romans at *Phasis

---

[79] For this, see Appendix I.

(556) managed to smuggle out a large force that returned shortly after, posing as a "relieving army," and ultimately causing a full Persian collapse. Objectives could also be much simpler; the Roman garrison at *Harran (502f) managed to catch a Hunnic chief during a sortie. Although this infuriated the Huns and their Persian masters, it was a strong bargaining position that ensured that the city and its territory were left unharmed.

As a complex operation, the sortie required good intelligence and loyalty on behalf of the defender in order to not fall into a trap, which would be devastating. Belisarius often saw his plans at *Rome 537f disrupted by deserters who passed on intelligence to the Ostrogoths, but on some occasions was able to maintain secrecy or use well-timed sorties to disrupt Ostrogothic assaults. Successful maneuvers were carried out by the Romans from their encampment at *Rhizaion (558), which was assaulted by Tzani. The Romans exited from several gates simultaneously, catching the Tzani by surprise and finally subjugating them to Roman rule in the process. At the *Chersonese (559), a sortie combined with a naval assault chased off the Kotrigur Huns, while the garrison and militia at *Alexandria (609) combined a sortie with an artillery barrage to chase off the army of Bonosus, Phokas' general.

Failed sorties were common, but they were often meant to probe the besiegers' fieldworks and perhaps cause diversions for other purposes, as the Lombard attacks on the Roman *castra* at *Benevento (663). Otherwise, they were provoked by the besiegers who hoped to weaken the garrison's morale and ability to fight. Thus, the Romans managed to lure out some Persians at *Amida (503) by sending sheep close to the fortifications, although their attempt to force a captured Persian officer to convince his men to surrender ended in failure (and impaling). At *Lucca (553), the Frankish garrison correctly went out to break off a successful Roman assault, but were betrayed by the city's militia in the middle of the battle. At *Arles (567/9), the newly installed garrison was betrayed by the population during a sortie. However, at *Chastel-Marlhac (524), the defenders had to pay ransom for captured troops, who had gone out in hope of capturing booty. Otherwise, there are reports of multiple failed sorties at *Sicilian forts held by the Goths (551) and at *Conza (554f), which virtually ended Gothic resistance in Italy.

Despite their complexity and inherent risks, sorties were often attempted, since successful ones could end sieges. The Goths at *Rome (546) were thus driven off, but the tables were turned when the Goths at *Caranalis (551) broke up a Roman blockade. The Lombard king Perctarit was defeated during a civil war due to a successful sortie from *Trent (680). The best approach was to coordinate a sortie with the arrival of a relieving army. The Romans failed miserably in this respect due to disagreement among commanders at *Rome (545f), but at *Orleans (604), the besieged garrison went out when a relieving army arrived. Although Bertoald, the commander, died, the relief was a success, and this was by far the most common means of ending a siege.

5.3.3 Relieving Armies

In many instances, sieges were avoided due to the danger of relieving armies. The Ostrogoths under Totila only began besieging *Fermo and *Ascoli (544) when it became clear that no relief was forthcoming. Defenders were equally anxious for relieving armies to arrive, and sometimes gave warning to their rulers or generals that they would only be able to hold out for another week or month before supplies ran out. Thus the Romans holding *Rimini (II 538f) notified Belisarius that they would surrender in seven days due to lack of supplies (despite successful defense), unless reinforcements arrived; similarly one month's warning was given at *Spoleto (545). This could be a ruse, however; the Persian commander at *Petra (549) tried to buy time for a relief army by promising surrender within a set date. However, if a relief army was defeated or failed to appear, the only option would be surrender. The citizens of *Emesa (635) told the Arabs that they would not surrender until the Roman army was defeated, but when none appeared and the Arabs began to storm the walls, they finally gave in.

While the besieged may not have had large enough forces to defeat more numerous besiegers, a besieging army rarely had the ability to maintain a siege when a relieving army arrived. A notable exception is the Arab siege of *Constantinople (717f), where double walls were built to protect the Arab camp against Roman sorties from the city and Bulgar raids from Thrace. Thus, if not very well prepared for the prospect, a besieging army would simply pack up and leave if relief was on the way. Even Attila's Huns left *Orleans (451) just as they were about to break through due to the arrival of Aëtius. Later on, Mummolus relieved *Valence and *Grenoble (574) from Lombard sieges, and a relief army broke the siege of *Soissons (576). Simply the presence of a potential relief army could be enough to raise the siege. For example, the Persians gave up on *Theodosiopolis (576) due to a nearby Roman army, and the Lombard army approaching from Pavia convinced Constans II to withdraw from *Benevento (663).

Unless caught unawares or during a withdrawal, the alternative was a set-piece battle. Droctulf, the Lombard duke in Roman service, lifted the Avar siege of *Adrianople (587) and defeated the Avars in battle by using feigned flight. Shortly after, a Visigothic feigned flight combined with a successful ambush ended in total defeat for the Frankish army headed for *Carcassonne (589). Theophylact attributes an Avar withdrawal from *Drizipera (588) to a miraculous relieving army, although humans were sometimes acceptable instruments for divine intervention in hagiography, which is surely the source of Theophylact's narrative. In 589, however, a Persian relieving army defeated the Romans before *Martyropolis in battle. On rare occasions, however, the relieving force was surprised; the Arabs had nearly given up *Tyana (708) when they learned of a large Roman army approaching with supplies and materials to rebuild the walls. The relieving army was caught by surprise by the Arabs, its defeat leading to the prompt surrender of the city. Due to their very offensive style of war, the Arabs frequently conducted sieges deep in Roman territory. At *Syke (771), the Arabs were trapped by Roman armies, but in desperation they managed to defeat some of the Romans and escape. However, this sometimes went spectacularly wrong: at *Synnada (740), the Arab besiegers were surprised and utterly defeated.

A relieving army often had the added complication of having to bring in extra supplies to the besieged city. *Narbonne was relieved after a year (436f) by Litorius, who also brought supplies into the city. The Roman fleet was useful in bringing relief and supplies to *Otranto (544) and broke the sieges of *Ancona (551) and *Crotone (551f). Belisarius also brought supplies and extra troops to *Rome (545f) by ship and riverboat, but the failure to coordinate with the garrison allowed the Goths to prevent any relief from coming through. However, the Roman Danube flotilla worked well at *Singidunum (595). Indeed, throughout the 7th and 8th century, the Roman fleet was probably responsible for much of the political geography of Italy and the Balkans. Even *Alexandria (642f, 646) could be relieved or retaken by Roman fleets, and *Thessalonica (662) survived as a Roman enclave in large measure due to the Roman fleet.

The size of a relief army had to be substantial. Reportedly at *Clermont (474), 19 men were enough to break the siege. Some scholars actually take this number literally;[80] it is however clearly an episode focusing on the exploits of Ecdicius and his immediate bodyguards among a substantially

---

[80] See entry in CO for this.

larger force, as the heaps of Visigothic fallen and the other events described during the relief of the siege would make little sense otherwise. A similar number is reported at *Cividale (670), when Wechtari with 25 followers is said to have defeated 5,000 Slavs. At *Rome (537f), the slow and incremental arrival of relief forces meant that Belisarius could not raise the siege at once, but he could use his troops to increasingly harass the Goths. Even at *Osimo (544), 1,000 men with servants were not enough to break the Gothic siege; although they did manage to get into the fortifications somehow, they had to leave before they consumed scarce resources. The Persian army that was sent in relief of *Petra (549, 550) arrived too late to help the garrison there, but was large enough to capture Phasis without a fight and commence a siege of *Archaeopolis (550). At *Onoguris (555), the Romans besiegers learnt of a Persian relieving army, but apparently did not have enough troops to keep up the pressure *and* turn back the relief. A halfhearted attempt with only 600 men was defeated by the Persian army of 3,000, which went on to raise the siege and chase off the rest of the Roman army.

Relief could also be indirect. The emperor Maurice threatened the Avars besieging *Tzurullon (588) with a Turkish invasion of Avar lands, and the Persians besieging *Chalcedon (615) withdrew when they learnt that a Roman army threatened Persian territory.

5.4 *Conclusion: Towards a Thick Description of Sieges*

The nature of the siege during this period has recently been harshly characterized as near-ritualistic, and siege methods caricatured as "basic", "a few simple techniques", "with a minimum of finesse," "of a basic nature," "tragicomic," "uncomplicated," and "fairly crude."[81] This view cannot be sustained in light of the numerous examples of complex siege operations attested and the sophisticated institutional framework possessed by the successor states. While the classification in this chapter is necessarily somewhat artificial, several well attested sieges show in detail possible combinations and the order in which they occurred. Most complete, classicizing narratives include details on most of the available engines and tactics, as well as a host of details on social and political issues dealt with in chapter 6. The most famous of all sieges during this period, *Rome (537f), is too often dismissed or treated as the quintessential late antique

---

[81] Halsall 2003: 224, 226f.

siege from which most recent scholarship has worked. While it provides enough evidence to elucidate tactics in great detail, the problem is that relying too heavily on a single narrative prevents us from understanding how prevalent these tactics were, and allows some to explain it away as a piece of classicizing historiography. Southern and Dixon have a good survey of late antique siege warfare, including engines and tactics, but their work is based largely on Ammianus, Vegetius and Procopius, and thus open to criticism for using too few and possibly classicizing sources.

However, there exists a wealth of other narratives that essentially confirm the complexity of sieges, while episodic anecdotes and schematic notices in fact reinforce this impression once the evidence is assembled systematically. Thus *Theodosiopolis (421/2) shows the range of machines used by the Persians: *helepoleis* (ἑλεπόλεις: normally the primary siege engine of the day, in this context probably battering rams; see chapters 5.2.5 and 8.2.1 below), *mēkhanais* (μηχαναῖς: any variety of artillery and siege sheds), *pyrgous* (πύργους: towers); *Naissus (442) similarly the range of engines available to the Huns. We have seen this variety obliquely in the Franco-Visigothic war (chapter 1.3.4, with further evidence from the *Burgundian civil war in 500 and later at *Arles 508) and more directly in the East during the Anastasian war. Complexity is the key word throughout the 6th century; in the East, at *Edessa (544), and *Onoguris (555); in the Balkans, *Thessalonica (586), and in the West, at *Sevilla (583f) and *Convenae (585). While sources are generally less detailed in the following centuries, similar complexity is demonstrable through the Arab conquests, Roman *invasion of Italy (663), further conflicts in the Balkans (*Thessalonica 604, 615, 618, 662; *Constantinople 626, 654), in Visigothic Spain (esp. 673), and even in Merovingian Gaul. By the 8th century, there is little sign of simplification; the Franks used most of these tactics against Arabs and Aquitanians (chapter 4.3.3); the Arabs still amassed vast fleet-borne armies against Constantinople (see below), but also used heavy machinery when assaulting Anatolian fortresses, such as *Tyana (708); *Kamakhon 766 illustrates the logistical and tactical capabilities and difficulties of Abbasid armies.

The use of naval assets have been noted in the context of organization, logistics, and blockades; however, fleets were often used to storm fortifications from the sea, or alternately protect them; they could also break sieges (see further 5.3.3). Thus the Roman fleet facilitated the storm of *Palermo (535) and *Classis (584/5), and at *Phasis (556) helped defend the city; at *Ancona (551) a naval battle ended the siege, while the Danube flotilla relieved *Singidunum (595). Naval assets were used during the Visigothic civil war in 673, and the Roman civil war in 608 (e.g. *Alexandria), while the Arabs relied heavily on fleets against *Constantinople (654, 670s, 717f) and elsewhere. Incendiary devices were often used in sieges, but are dealt with in the context of artillery or undermining; the famous Greek Fire was in this period mostly confined to naval use and has been left out of the discussion unless directly related to a siege narrative. Finally, "chemical" warfare has not been examined in detail here, but should be noted: the Arabs used smoke against *Dvin (640) and the underground city of *Kasin (776); James has recently proposed that the mines from the Persian siege of Dura-Europos in 256 provide evidence that poisonous gas was used to stop a Roman countermine.[82]

It is clear that siege warfare was no simple matter, and that there was no one single favored method of besieging. It rather depended on contingent conditions, some of which we know or can reconstruct, but most of the time these are unknown. However, once a siege was undertaken, it resulted in a complex of practices, where different approaches were used simultaneously or alternated over time. While the detailed empirical evidence seems to indicate that sieges were more significant in the East Roman Empire in the 6th century, this corresponds quite exactly with the survival of the late antique genre of classicizing historiography. Similar events, but with more "simplified" evidence, can be well documented throughout the period and area in question here. A solution could be a very detailed philological examination of the semantic content of each verb in a "thick" context, but this soon boils down to a problem that may be dubbed the *Fredegar complex* (cf. Fredegar's handling of *Convenae 585): most of the sources used here are second- or third-hand epitomes that telescope complex events down to a neat phrase, a process that is observable in detail where several versions of the same basic story can be identified and traced over time through different recensions.[83] Indeed, *every* siege was a dreary process of assembling materials, food supplies, weapons, building siege engines, setting up camp, endless patrols, fruitless attempted storms, abortive sallies and any of the above mentioned features. Even the rich narrative provided by the contemporary Joshua the Stylite on *Amida (502f)

---

[82] See chapter 1.1.1 and S. James 2011. I thank Jan Frode Hatlen for the reference.

[83] See Kretchmer 2006: 269-92 for a detailed philological study of the simplifications made in the different versions of *Historia Romana*, with examples of "simplification of warfare" and political events at 269-74. Kretchmer demonstrates the various authors' tendency to omit minor wars as an important principle of abbreviation.

compares unfavorably with Ammianus' "classicizing," but incisive and autoptic account of the Persian capture of the same city in 359.[84]

What *is* clear, is that neither the Ostrogoths, Visigoths, Franks nor Arabs shied away from long blockades, but willingly combined them with other methods. For the most part, their sieges were characterized by formidable logistical abilities that matched those of the Romans and the Persians. This was achieved by applying their considerable (but occasionally ignored) administrative apparatus to their conduct of war. Yet societies with primitive economies and fairly crude administrative apparatus (even when compared to the contemporary West), such as the Slavs and Avars, took part in siege warfare on equal terms with the Roman Empire. While this will be explained in chapter 7, we will first see how all these societies shared in the same political, socio-economic and spiritual world.

---

[84] Lenski 2007.

CHAPTER SIX

THE ANATOMY OF A SIEGE: ECONOMY, SOCIETY AND CULTURE

Late antique states were dominated by common socio-economic realities, cultural and mental outlook, and similar practical concerns. While different urban models evolved in the 5th-8th centuries, the new developments in military organization (chs. 2-4) and similarity of technology (ch. 5) made sieges possible in areas where only a rudimentary Roman infrastructure survived. Furthermore, civilians, urban artisans, city-dwelling farmers, and people from the surrounding countryside were significant to the defense of cities and supplemented regular troops, whether they belonged to a local garrison, a magnate's retinue, or were installed by a distant ruler. Thus, even thinly settled cities remained important military and political centers.

During a siege morale was variable, depending on levels of loyalty and antagonism within the population and towards allies. Logistical and psychological factors were also highly significant; perceived threats were often enough to cause a surrender. In order to build up morale, civilian and military leaders sought to ensure supplies, rewards, and repairs, while bishops and clerics developed stories of miraculous protection from God and the saints during previous sieges. These stories were used in processions and services performed by the clerics and populations when sieges were imminent or even while they were in progress. This practice is reflected in the sources: apart from the classicizing historians, hagiographies and miracles stories provide the most detailed accounts, in contrast to extremely brief chronicles and annals.

6.1 *The Topographies of a Siege*

With increasing insecurity and the proliferation of political boundaries crisscrossing the former Roman world, the structures of the empire provided a basic infrastructure for defense. As borders solidified, new fortifications were built. This infrastructure required constant upkeep. Furthermore, most urban communities lived in a close symbiotic relationship with the countryside. Preparing for a siege meant storing up months or even years of food in advance. Unless this had been planned well in time, simply stripping the countryside of food could break the morale of an urban population. The people who lived in and around them were not only concerned with their physical protection, however, but also took every opportunity to ensure divine assistance.

6.1.1 *Defensive Topography: Infrastructure and Fortifications*

Any kind of fortified or defensible site might be subject to a siege, ranging from the great late antique *metropoleis* of Constantinople, Antioch, Thessalonica and Rome, via lesser provincial cities, down to the smallest hilltop fortification. Although some cities were partly or completely abandoned in the period under consideration here, in most places urbanism was significant enough for states and elites to fight over them. Sieges fought on the great city walls constructed by the Romans from the 3rd to the 5th centuries thus dominate the narrative sources. Similarly, lesser fortifications (*castra, castella*, φρούρια) built either as border forts or as suburban refuges during the late empire continued in use. For example, *Ratiaria (447) was taken along with nearby *phrouria*. Suburban fortifications have not received enough attention by either historians or archaeologists, but appear quite frequently in the sources: at *Naples (536), a suburban *phrourion* was captured as the Roman army was settling in for a siege, while at *Thessalonica (586), the invaders mistook the nearby fortification of Matrone for the city itself, giving the inhabitants another day to prepare. The Franks regularly contended with suburban fortifications, both at home (see below) and on foreign adventures. For instance, during the *Frankish invasion of Italy in 590, they besieged and took a number of suburban fortifications around Milan, border forts in the Alpine passes, and demolished over a dozen forts in the region of Trent.

In the East, the tradition of new construction continued on a large scale in the 6th century. New foundations such as Dara and Iustiniana Prima were small cities with a civilian population but with very strong fortifications.[1] On some exposed frontier areas, however, fortifications were dedicated solely to military purposes, and had little or no civilian population or infrastructure, although fighting over them could be very intense.[2] In some regions, forts were designed with large circuit walls, covering open spaces that could accommodate a large number of refugees in time of cri-

---

[1] Henning 2007.
[2] See e.g. *Aphum 578 and *Chlomaron 578 for fighting over lesser fortresses on the Roman-Persian frontier.

sis.³ With 7th-century invasions, cities were moved to more defensible, hill-top sites, which were provided with substantial walls; in effect they acquired the characteristics of 5th-6th century border fortifications.⁴ Old *kleisourai*, fortified mountain passes, were revitalized, especially on the new Taurus frontier that arose between the Romans and the Arabs around 700.⁵ New fortified towns were founded in Thrace the late 8th century, when the Romans began expanding into territories formerly held by Slavs and Bulgarians. The practice was so common that Roman officers set down detailed principles of fort-building and city-founding in military manuals.⁶ It is likely that the 8th- and 9th-century Roman foundations were prototypes for some of the well fortified Bulgarian cities built in the 9th and 10th centuries.⁷

In the West we can see the foundation of cities on a smaller scale among the Ostrogoths and Visigoths, but they also founded a number of substantial fortresses (*castra*) in their northern provinces. Theoderic set up fortifications with permanent lodgings at Verruca and near Tortona. Since they were constructed in densely populated areas, they were designed to function as refuges but could also double as more or less permanent settlements.⁸ Alongside urban fortifications, they backed up the interior of the *clausurae* (the Latin equivalent of *kleisourai*, fortified border passes) that crossed the Alps. This system was well developed in the 4th century but appears to have been partly abandoned in the chaotic 5th century. The Ostrogothic revival was probably inspired by the East Romans. The *clausurae* were protected by further fortifications. Substantial ones have been identified at Susa and Aosta at the western end of the Alps. These fortifications were used in turn by the Late Romans, the Ostrogoths, the East Romans, the Franks and the Lombards.⁹

The Visigoths founded or rebuilt several cities. The reconstruction of Italica was important in the siege of *Sevilla (583f), but the campaign also involved the siege and capture of a number of other unnamed cities and

---

³ See Dunn 1994 and studies in *TLAD* for examples from the Balkans; see Liebeschuetz 2001 *passim* for other examples.
⁴ This is the argument of Niewöhner 2007, who synthesizes most of the existing archaeology for Anatolia.
⁵ Haldon and Kennedy 1980.
⁶ *Strategikon* 10.4; *De re strategica* 9-12.
⁷ See *PRT* 2 for the archaeology and chapter 7.2 *passim* for the historical context.
⁸ This process was followed up during the Lombard-Byzantine wars and may signal the prelude to the famous process of *incastellimento*, although the evolution of antecedents is very complex; see in general Christie 2006 for the situation to the early 9th century.
⁹ Christie 2006: 357-69.

forts. Leovigild protected his conquests with new fortifications, but his best known achievement was the construction of Reccopolis in Tarraconensis, which was on the route from Toledo to the northeastern border fortresses in the passes of the Pyrenees.[10] The Visigoths were also inspired by East Roman practices, as they established fortifications in the most important mountain passes between Septimania and Tarraconensis. This meant that any invader from the north had to pass not only the fortified cities of Arles, Nîmes, Narbonne, Carcassonne, and their suburban fortifications (which now also doubled as border forts), but then had to traverse fortified passes before descending onto the Ebro valley, again dominated by fortified Roman cities, while the plateau beyond was protected by the new foundations of Leovigild.

There is some dispute as to the fate of urbanism in the north of Gaul, especially in Rhine (and tributary) valleys the 5th and early 6th centuries.[11] This did not keep the Franks from fighting over cities, however. Most of the Frankish expansionary wars and virtually all of the civil wars were fought inside former imperial boundaries. Roman infrastructure dominated and was kept in usable order. During a Frankish civil war in 584, the Neustrian king Chilperic ordered his magnates to repair city walls within his realm and garrison forts by while he remained in the field with a large army.[12] Substantial urban repair was undertaken by Merovingian bishops in the 7th and the Carolingians in the 8th century. There were also some additions. Some magnates built substantial fortifications outside of the traditional urban infrastructure, including Nicetius' fortified residence near Trier, and the fortress of Dijon, which was built as the residence of a bishop but which Gregory thought of as equivalent to a city.[13] The *fortified estate church at Woëvre (587) is particularly intriguing. Some of these fortifications may have originated in suburban forts or other installations that were of Roman origins, and thus still usable. However, others were original constructions, demonstrating that Frankish practice was similar to that of the Ostrogoths and Visigoths. Cities were frequently used as refuges, but could only accommodate a limited number of people. This was especially the case in northern Gaul, where the relative low number of cities in a densely populated countryside meant that distances to the nearest forti-

---

[10] For the general context, see Collins 2004: 52-56, for Reccopolis see Olmo Enciso 2007.
[11] See especially Halsall, 1995, Ward-Perkins 2005, Liebeschuetz 2001. For a critique of Halsall, see Bachrach 2003 and Wickham 2005; for further literature, see the notes for chapter 6.1.2.
[12] GT 6.41; see discussion in Bachrach 2002b.
[13] Ibid.

fied city were great and the capacity to accept refugees limited. This may explain why suburban fortifications have received little attention and give the impression of abandonment: they were only used intermittently during civil wars, and they had little or no economic function.[14]

We know very little of Frankish border fortifications under the Merovingians. In the southeast, the Burgundian and Austrasian realms took over parts of Ostrogothic, East Roman and Lombard infrastructure at various times, but in the region facing against the Visigoths, Frankish forts probably originated in suburban fortifications.[15] The narratives of warfare in the early and mid-8th century give the impression that the great fortified cities of Septimania and in northern Aquitaine functioned as border forts, although at *Bordeaux and *Blaye (735), both the *urbes* and the *suburbana castra* were the object of sieges. In the north, the Merovingians based themselves in the fortified cities and forts on the Rhine, such as Cologne and Duisburg. With renewed Frankish expansion outside of imperial borders, however, the early Carolingians began an ambitious program of fortress-building. These were very substantial constructions that were used both as bases for further expansion and defensively against Saxon and Slav raids. Eresburg had to be rebuilt after it was destroyed by the Saxons, and Charlemagne strengthened the defenses with another fortress on the river Lippe.[16]

While conquering empires such as the Avars and Huns largely deported populations in order to exploit them and create a wasteland between themselves and the Romans, the Arabs maintained the urban fabric in many of the regions they conquered.[17] Due to the expansionary nature of their empire, the Arab core territories were not threatened by external enemies during this period. This allowed them to establish large, un-walled garrison cities in Egypt and Iraq. In border areas, however, they took over pre-existing infrastructure. Thus, in Septimania, they took over the Visigothic logistical and fortification system and used this as a base for raids

---

[14] Halsall 1995 is quite dismissive of these structures, although he identifies several around Metz.
[15] See examples cited in chapter 4.1.2. The standard works on Aquitaine and Toulouse by Labrousse 1968 and Rouche 1979 are unconcerned with suburban fortifications in their surveys of the regional topography.
[16] *ARF* 776 and *Eresburg, *Syburg in CO s.a.
[17] A notable exception: Carthage was laid waste, but had already been in decline; it was replaced by new foundations; Qayrawan and Tunis, garrisoned with troops and provided with (Coptic) craftsmen from Egypt. See chapter 7 for more on the differences between conquering peoples and their treatment of conquered populations.

into Gaul. In northern Syria and Mesopotamia, Roman and Persian forts and fortified cities were garrisoned and used for bases for further expansion. The same also occurred in Khorasan. By the end of the 7th century, the Arabs also expanded their protective shield by re-founding abandoned Roman cities, so that the Byzantine-Arab border became the most heavily fortified zone in the early middle ages.[18] By the late 7th century, also, integration into local urban society meant that city walls began to take on a much more significant role in internal Arab strife.

The infrastructure of individual fortifications varied immensely. The sources sometimes describe a combination of fortification surrounded by a settlement, which occurs on several occasions in the Caucasus, e.g. at *Anglon (543) and *Tzacher/Sideroun (557). In most cases, however, defensive works surrounded the central built up area (but recall the qualification about urban structure noted above). The first line of defense was normally a moat. At both *Salona (537) and *Rome (537f), the Romans improved the fortifications by adding a moat (τάφρος), and the latter was said to be particularly deep. At *Phasis (556), a deep moat with stakes planted inside was filled with sea water. At *Avignon (583), the defenders excavated a moat that had deeper pits, and submerged it all by drawing water from the Rhône. Walls and outworks normally followed the patterns laid down by the late Roman Empire. This included a circuit wall, ideally tall enough to prevent easy storming, thick enough to withstand artillery and siege engines, and well enough founded to protect against undermining. The specific measurements however varied immensely from place to place. In quite a few cases, late antique cities had double walls, the inner larger than the outer; otherwise, the outer wall took the form of a *proteikhisma*, apparently a raised breastwork high enough to prevent assault with engines and infantry but not as imposing as full-scale walls (e.g. the double walls of Constantinople had an additional *proteikhisma* that rose up from the moat).[19]

Siege campaigns very often resulted in the razing and repairing of walls. Due to the long-term strategies used to capture a city (5.1), repairs were often undertaken at first sight of trouble. Thus *Edessa (502f) began preparations for a siege when the first Persians raiding parties appeared in 502,

---

[18] Baladhuri devoted a separate chapter to reconstructed fortifications and new foundations. For a survey of the frontier wars, see in general Haldon and Kennedy 1980.

[19] Double walls: *Dara 540; *proteikhisma*: *Edessa 544. The standard work on Byzantine fortifications, including Constantinople, is still Foss and Winfield 1986, but see now especially Niewöhner 2007 and other studies cited at the beginning of next section that cover the region surveyed here (6.1.2).

although the formal investment only began the following year. When it became known that the Arabs were organizing an expedition against *Constantinople in 715, reparations began immediately, although the Arab army only arrived in 717. The Romans abandoned *Batnan (503) without a fight to a Persian raiding party since their walls were "broken down" (to conduct repairs?); similarly at *Sura and *Beroea (540), the uncertain state of the walls induced the population to negotiate ransoms with the Persians. The Ostrogoths had to abandon *Salona (536) partly due to the dilapidated state of the walls, which were immediately repaired by the Romans when they took over. This was a systematic Roman policy in order to hold newly conquered territories; they did the same at *Carthage (533) and *Rome (537f), where they also modified the merlons (5.3.1). This took time and effort, hence the Ostrogothic policy of breaking down walls in order to prevent Roman garrisons from using them as bases, especially as the Ostrogoths began to find popular support unreliable.[20]

The Aquitanians began to destroy walls when they realized that they could not motivate the population and resist Pippin; even Charlemagne himself resorted to this in 778 when he took *Pamplona. The explicit rationale was to prevent rebellion. At *Antioch (540), there was some debate whether one should fortify an outlying rock that the Persians could exploit. Since the time to do the work was too short, and ongoing works would only alert the Persians of an opportunity, it was decided to do nothing. At *Kallinikos (542), the Persians could exploit ongoing works to take the city and raze the remaining walls.

Towers held out longer, and often had to be cleared one by one.[21] Here too late Roman advances dominated military architecture, as many curtain walls had regularly spaced, protruding towers that could carry artillery and ensured that no one could approach the curtain walls without being subject to enfilading fire. Many fortifications also had an acropolis, either inherited from classical times or added as an extra bastion. At *Beroea (540) the citizens fled to the acropolis when they were unable to pay ransom to the Persians. Hadrian's mausoleum fulfilled a similar function at *Rome (549f), when 400 Roman soldiers withdrew there after the breach of the main walls; the Ostrogoths did the very same thing in 552, storing their valuables there but trying to hold the main circuit as long as possible. The

---

[20] *Benevento 542, *Pesaro 544, *Rome 545f (1/3 razed by Ostrogoths) and 546 (rebuilt by Romans), but failing in an assault on the city, the Ostrogoths withdrew to Tibur which they rebuilt.

[21] The Franks garrisoned at *Vienne 500 banded together in one of the towers.

Persians at *Petra (550) made a last stand in the acropolis but were trapped and killed by fire when they refused to surrender.

Finally, we must also note natural fortifications, where the sheer physical features of the terrain added to man-made structures or were strong enough to protect the site itself; otherwise, some late antique structures built for other purposes were massive enough to be converted to *Ersatz* fortifications. The steep sides of *Petra (550) made an assault with wheel-borne engines impossible. The Roman assault on *Acerenza (663) was hampered by "the highly fortified position of the place," including not only man-made fortifications, but also its situation on a high-rising hill. *Ersatz* fortifications include amphitheaters, modified bridges, aqueducts and so on. Belisarius had a tower (πύργος) built on the *Milvian bridge (537), although there the size of the Ostrogothic army scared away the garrison left behind to defend it. At *Spoleto (546), the Ostrogoths fortified the amphitheater instead of the city. However, Narses apparently found the population of the city more reliable, and chose to rebuild Spoleto and garrison it (after capturing *Narnia 552). Substantial fortifications could be made of earth and wood; the old fortifications at *Phasis (556) were still usable according to Agathias, but were reinforced with massive ramparts. The Frankish general Mummolus built extensive *improvised fortifications at Embrun (571) with wooden ramparts. Such wooden fortifications remained very common in central Europe into the 9th and 10th centuries, and were the basis of East Frankish expansion into Slavic territory (see ch. 7.2.5).

On a few rare occasions, fighting is even reported inside the city, either on rooftops or from street to street. During the storm of an *Abasgian fortress (549), the population fought from the rooftops while the Romans controlled the gates. At *Antioch (540), as well, there was fighting in the streets. The physical layout of the city and its defenses were so massive that even if the Persians took control of the acropolis on Mt. Sulpicius, it took some time to descend into the city proper, which allowed the civilians who did not flee to organize defenses in the streets. The population of *Nisibis (573) fought from house to house for seven days, even forcing the Persian besiegers back onto the walls, but finally lost their nerve and surrendered. Similar events are reported at *Amida (502f) and *Damascus (634f).

Water supply was fundamental and often targeted; thus at *Rome (537f) aqueducts were physically cut by the Ostrogoths. However, due to the topography, defenses, or ingenuity of the defenders, this was sometimes impossible. *Osimo (539) had a water supply outside the walls which proved impossible to break, so it was poisoned instead. The Romans failed to cut

ANATOMY OF A SIEGE: ECONOMY, SOCIETY & CULTURE   307

the water supply at *Petra (550) because the Persians had built three layers of piping underground; the Romans only discovered the top two. Bridges were fundamental to logistics and often necessary to capture, build or break down. Thus at *Naissus (442), a bridge was built and used to bring up engines; at *Rimini (552), the Ostrogoths broke a nearby bridge to prevent Roman progress; while the Persians at *Kallinikos (542) could surprise the citizens partly because they rapidly built a bridge and re-crossed the Euphrates. The Avars built bridges (using Roman expertise) across the Sava to capture *Sirmium (579ff). Charlemagne built two fortified bridges that helped him capture the *City of Dragawit (789). Finally, the offensive use of fortifications deserves comment. The Romans unsuccessfully attempted to take *Thebothon (573) in order to block of *Nisibis, but the Arabs successfully pursued this strategy when they constructed a fort in order to blockade *Tripoli (644) from land.

6.1.2 *The Topography of Settlement: City and Country in the "Dark Ages"*

In the period considered here, there were only a handful of very large cities with over a hundred thousand inhabitants, such as Constantinople, Rome (although both declined to around 50,000 by the 8th century) or Baghdad. Most cities numbered a few thousand; a few dozen may have reached tens of thousands, especially in the early period or in the Caliphate. In the vast majority of cases, the urban population had a close symbiotic relationship with the surrounding countryside, and frequently, only part of the urban area had in fact been walled. This was the deliberate decision of late Roman authorities and urban communities, since the total built-up area was often too large to protect.[22] Furthermore, the gradual development of villages in favor of villas and scattered farmsteads in the post-Roman West led to a settlement pattern that had previously been more typical of the East. This produced a rather peculiar form of urban settlement that existed from Mesopotamia to the Rhine valley in the 6th and 7th centuries, where walled urban centers (whether thinly populated or not) were surrounded by a number of villages clustering just outside the wall, a few hundred meters distant, or further away but otherwise associated with the central settlement.[23]

---

[22] Liebeschuetz 2001: 82ff. For an evaluation of Roman fortification policies and practices in Gaul in the 3rd and 4th centuries, see Bachrach 2010.

[23] The exponential increase in archaeological materials on late antique and early medieval urbanism and its countryside means that I have had to limit myself to the available works of synthesis. For this and the following paragraphs, these studies have been most

The agglomeration of a partly inhabited center and a cluster of nearby villages in effect made an urban community, though of a somewhat different character from classical urbanism. Although the city centers themselves may have had small populations, the evidence from Paris shows that around 10,000 people were within a day or so of the walled enceinte. This demographic and economic pattern meant that suburban areas were legitimate objectives of attack before, during and after sieges; the campaigns of Charles Martel and Pippin the Short illustrate this well.[24] Such strategies only reinforced the incentive to take refuge in the nearest city, where the rural population defended themselves as any other urban dweller. In Gaul in 584 this happened on royal orders; at *Autun (679) it was the local bishop who organized the flight of the surrounding population to the city. Both *Thessalonica and *Constantinople were periodically filled with refugees, and in Anatolia the many newly built or revamped fortifications provided security from Arab raiders, although the population still lived in villages in areas more distant from the most exposed frontier zone. It became standard practice to use fire signals and messengers to warn local authorities to evacuate the countryside and bring people and flocks into fortresses.[25]

Since city centers were used as refuges during war, keeping some of the walled space free of buildings may have been useful to provide space for grazing, cultivation, and temporary refugee camps. This was done on a large scale at Constantinople, where the Theodosian Walls included a large space (simply called *khōra*, or countryside) that was never built up. At *Rome (549f), such space was actively used to cultivate crops even during sieges. Due to the political circumstances, walled areas should be defined as royal, aristocratic, or ecclesiastic space—areas were kings, counts, bishops lived with their entourages, military retinue, and immediate providers of services, although much of their political and economic power came from their estates in the surrounding countryside. The population from the surrounding *pagus* took refuge inside *Castrum Cainonense (463), *Thessalonica (615, 618) was full of refugees from Naissus and other Balkans cit-

---

useful: *Städtewesen, Landscapes, TRW* 4, *TRW* 9, *LAA* 2 and *PRT* 1-2 for general surveys; *PRT* 1, Verhulst 2000, Loseby 1998 and 2000, Hill 1988, Christie 2006, Kulikowski 2004, Périn 2002, Gauthier 2002, Halsall 1995 and Ripoll Lopez 2003 for western Europe; *PRT* 2, Niewöhner 2007, *TLAD*, Curta 2001: 120-89 and *Constantinople* for the Balkans and Anatolia; for the early Islamic East see Walmsley 2007.

[24] See 6.1.1 and 4.3.3 above; especially *Nîmes 737, *Loches 742 and the Aquitanian campaign of Pippin. In all these cases, suburban areas were particularly targeted by the participants.

[25] Haldon 1999: 56-60, 78-83, 150.

ies that had fallen to the Avars, and as we saw, the dependants of bishop Leudegar of *Autun (679) were assembled inside the city walls.

This pattern was more common in the West, especially away from the Mediterranean coastline. In part it reflected the lower degree of urbanism from the late Roman era, but also the greater degree of relative security. For example, the core areas of the Frankish empire in northern Gaul were hardly ever threatened by external foes. A similar argument could be made for the Visigothic kingdom of Toledo, where the north was densely fortified but otherwise there was a move towards a more rural lifestyle.[26] When threats were constant, such as in the Balkans and Mesopotamia in the 6th century and Anatolia in the 7th and 8th centuries, the whole agricultural population lived more or less permanently inside the walled area, producing small but densely settled cities where the inhabitants were mainly agriculturalists. When times were more peaceful, however, urban settlements evolved organically to fit the pattern of economic activity. This resulted in fairly large villages from only a few hundred meters to a few kilometers from the fortified area of the city. Such villages and fields could be used as bases or for materials and pasturage by a besieger (e.g. *Clermont 524), so often the population prepared for a siege by clearing suburbs of the surrounding countryside of crops, materials and buildings.[27]

Many urban inhabitants had small farms or gardens close to the city wall, while more large-scale activity (such as magnate estates with vineyards and the like) could be undertaken up to a few days distance from the walled area. Urban inhabitants and those living in suburban villages spent much time tending their fields, and often camped out at harvest or vintage. Very good sources for such activity are narratives of sieges and raids that describe how not only villagers, but also urban populations, were surprised when out tending to their fields. As diverse cities as *Narbonne (413), *Singidunum (583), *Thessalonica (586), *Theodosiopolis (576) and Euchaïta (644) experienced surprise attacks while a large proportion of the population was out in the fields. When *Constantina, *Harran and *Edessa had their territories raided by the Persians in the autumn of 502, a total 18,500 people working and sleeping in the fields were surprised and captured. The extensive estates of the Papacy are well attested during the Lombard siege of *Rome (756). The Lombards devastated suburban areas, taking captives

---

[26] Kulikowski 2004; Ripoll Lopez 1999.

[27] E.g. *Phasis 556, forcing the Persians to haul supplies and materials over great distances; the Persians returned the favor against the Romans at *Nisibis 573. See also *Constantinople 626.

from the *familia beati Petri* (the household of St. Peter) who lived on his estates throughout Latium.

Arab raiding skills were recognized and used systematically by the Persians, who sent Arabs against Syria while the Persian army besieged *Jerusalem in 614. Again in the environs of *Jerusalem (634f), the Arabs killed some 4,000 "poor villagers" during the raids that brought about the submission of the city. At *Antioch (638/9), the Arabs exploited the festival of St. Symeon to take captives from the surrounding villages. The Visigothic contender Froia had Basques raid the countryside and prevent egress from *Saragossa (653). At *Thessalonica (586), the Slavs systematically raided around the suburban forts (*phrouria*, φρούρια), suburbs (*proasteia*, προάστεια) and fields (*agrous*, ἀγρούς) around the city; they collected prisoners and consumed everything edible found outside of the city walls within two days. Many of the inhabitants were away for the harvest season and thus unable to return (or taken prisoner). After the siege, the garrison sent out cavalry scouts to search the countryside and verify that the besiegers in fact had left the region. This was essential: in 618, the city was hit by a cavalry raid coordinated to arrive from several directions at the same time. Due to the total surprise, again citizens were caught out in the fields and taken captive or killed.

The Slavs had to clear out due to the size of their armies, lack of logistical system, and underdeveloped state of agriculture in the inner Balkans due to persistent raiding. Most of the successors had no such problems. The estate and village economy was very resilient and could, in times of peace, support quite large forces. Thus after the siege of *Tzurullon (588), the Roman troops assembled to fight the Avars were billeted in the villages of the Thracian countryside. The enormous force of 52,000 men gathered to face the Persians during the Anastasian war was partly supplied from Egypt, but the troops were billeted in and fed by the towns and cities of Mesopotamia and Northern Syria. The Persians paid local inhabitants from the countryside to provide supplies to *Amida (504f); even the Abbasids at *Kamakhon (766) could conscript some labor but they also attracted large numbers of farmers and merchants with cash to bring in supplies. Frankish armies in Italy operated for years (553-62) or months (e.g. three months during their *invasion in 590), and at *Saragossa (541) the Franks spent 49 days raiding the surrounding countryside while besieging the city. The Frankish army besieging *Convenae (585) was too large for the local resources to support, so supplies were carted in. The Lombards at *Arles (574) exploited local Frankish resources by setting up camp at a suburban villa, which would have had significant stocks. Besieging armies were also supported by fleets. The enigmatic Visigothic fleet that supported Wamba's campaign in 673 has not been satisfactorily explained, but for the Arab fleet supporting *Constantinople (717f) there are extensive papyrus documents that provide valuable details on the preparations made in 714 and before (cf. chapter 7.3).

The close integration between city and countryside explains why urban populations were so susceptible to pressure on their livelihoods. As we saw in chapter 5.1.1, raiding commonly preceded a formal siege, and indeed, in many cases *was* an efficient method of siege. *Thessalonica (esp. 662) suffered greatly at the loss of her countryside to raiders. In 615 and 618, the city was full of refugees that put an extra strain on resources. Authorities resorted to draconian measures of house searches and seizure of foodstuffs in order to prevent hoarding, but at the same time officials were often corrupt and used such opportunities for their own gain. The arrival of supply fleets ensured that the city remained a Roman outpost. Naval supplies certainly ensured the survival of many Roman enclaves in the Balkans, Italy and southern Spain, while Roman resistance against the Arab conquest dragged on for 10-15 years on or near the coast of the Levant despite the ostensibly rapid Arab progress.[28] The supply situation, then, was far more important than any other single factor, and when a city was well supplied, any threat had to be backed up by realistic force, as at *Convenae (585). Otherwise, raiding could prepare the ground for a return the next year, which seems to have been the intention when an attempt to besiege *Nîmes (585) failed or was given up.

Since dependence on the countryside was so great, raiding of the *territorium* forced garrison troops out in the open to chase off raiders. Normally raiding parties were quite small detachments from larger armies, so that an individual garrison or a mobilization of nearby forces would suffice. As these partial numbers are frequently reported in the sources, they are sometimes taken as the benchmark of the total size of armies.[29] However,

---

[28] The Arabs controlled the countryside around Gaza from February 634 and the region of Bethlehem and Jerusalem was impossible to traverse safely by Christmas that year; most of the inland cities in *Palaestina* were in Arab control by 635 and Greater Syria by 637-8, although resistance at *Caesarea continued at least until 640 (possibly 641), at *Tripoli until 644, and at *Arwad until 650. See further in Appendix I-II.

[29] Halsall 2003: 120f makes much of a Frankish cavalry raiding party of 600 men that caused great problems for the Romans in 355. In comparison, 2-300 Cheyenne Dog Soldiers (most of whom were equipped with bows and arrows) practically contained US expansion into Kansas during the period 1863-69. At the same time, both the North and South routinely

small raiding parties moved extremely fast on horse, were hard to detect, and could cause immense destruction, especially due to their ability to surprise. Thus small detachments of a few hundred to a couple of thousand men were constantly moving around the countryside during the Anastasian wars (ch. 2.3 *passim*). A large raiding party of 2,000 Roman cavalry devastated the Ostrogothic heartland in Picenum and took *Rimini (538), while 1,000 Romans on the east bank of the Euphrates was more than enough to convince the whole Persian army to adhere to a temporary truce until they had crossed safely and felt free to assault *Kallinikos (542).

Unless intelligence was very good, it would be difficult for a local commander to decide whether the raiders were just interested in collecting booty, were a prelude to a determined siege, or deliberate ruses to lure out and ambush garrisons. At *Constantina (502f), units from the field army systematically cleared nearby villages of Persian raiders, but one such unit was drawn into a well-prepared ambush and heavily defeated. The newly installed garrison at *Arles (567/9) went on a sortie to protect the territory of the city, but had arrived as part of an occupying regime and was betrayed by the population. Balkan garrisons made aggressive sorties against Slav raiders. At *Topeiros (549) the Romans were drawn into an ambush, with tragic consequences for both garrison and inhabitants, but the strategy of sending sorties into the countryside worked successfully at *Thessalonica (604), where the Slav raiders were massacred. This was also the policy of the garrisons in Roman Palestine and Syria, who routinely confronted Arab raiders in the field until the Persian conquest. The first phase of the Arab conquest in 633-35 began with large-scale raiding of the countryside, and at first the garrisons responded in the regular manner, going out into the field to chase them off. The result was a series of hard-fought minor battles, but due to local Arab superiority in numbers, many garrisons were defeated or forced to flee, leaving cities without access to the countryside and bereft of many of their defenders.[30]

In times of extreme need, the government organized supplies, or facilitated the ability of cities to do so. Recognizing the lack of preparations before the war, Anastasios had public magazines set up in cities on the eastern frontier at the end of hostilities in 506. The Ostrogoths did likewise under Theoderic and in the 530s, and were hence well prepared when the

---

fielded individual armies ranging from 20,000 to 100,000 men. For the Cheyenne Dog Soldiers, see Grimes 2000; for the size of American Civil War armies, see McPherson 1988 *passim*.

[30] See Appendix I-II.

Romans invaded.[31] Their logistical system played an immense part in their ability to resist the Romans. They both operated a taxation system that could collect taxes either in money or in kind, and remit taxes in return for requisitions, thus keeping the fiscal control over all taxable resources. When besieging *Naples in 542f and *Rome in 545f, Totila even ordered his army *not* to raid, but ensure that the farmers kept up production and hence taxes; since no Roman field army could challenge Ostrogothic control of the countryside, it would be counterproductive to harry one's own tax and supply base while the Roman garrisons were withering away inside their fortifications. The fighting around *Parma (553) between Romans and Franks was over the right to extract labor and resources; for some time, the Franks held the upper hand and thoroughly exploited their prerogatives. The Romans had similarly concentrated on consolidating their control of the countryside in Lazica after the capture of *Archaeopolis (550); Procopius criticizes the local commanders for this decision, arguing they should have concentrated on the border forts that were difficult to supply, but establishing a proper machinery for taxation may in fact have been a necessary prerequisite for garrisoning such sites.

The Visigoths had legal provisions for collecting supplies that were stored inside cities.[32] Estates were fundamental to Frankish logistics, which operated either through the old *annona* taxation, being redistributed by royal officers, or as a burden upon landowners, who were then responsible for supplying their own retinues and dependents. Gallic landowners carried a great responsibility that had evolved from *ad-hoc* solutions in the late 5th century. Ecdicius supplied a starving city with produce from his estates; St. Genovefa did likewise. This was regularized under the Visigoths and the Franks, alongside the obligation to provide troops and organize public works. Production centers had great capacity for storage, and goods could then be transported according to need. The public granaries at *Convenae (585) had been filled by a local landowner; there is great debate about the administrative mechanisms that lay behind, but the imposing logistical abilities of Merovingian and later Carolingian magnates have been demonstrated above.[33]

Despite functioning supply systems, raids made it impossible to stock up on supplies, or the siege could extend beyond what had been stored in advance. When authorities were aware that supplies were at risk, they ex-

---

[31] See *Ravenna 539f and chapter 3.1.4.
[32] See chapter 3.2.2 above for an extensive discussion and references.
[33] Brühl 1968; Verhulst 2003; Wickham 2005; chapters 1.2.5 and 4.3.4 above.

pelled large parts of the population. During the Roman preparations for the Arab siege of *Constantinople (717f), Anastasios II in 715 expelled all the citizens who were unable to supply themselves for at least three years. This was done more brutally during the middle of the siege at *Vienne (500) when the Burgundians threw out poor people during the siege (to their own detriment), but at *Rome (537f), the Romans took advantage of a lull in the siege to bring people to safety in an orderly way. At *Convenae (585), Gundovald and his supporters tricked most of the inhabitants to go out of the city in order to stop an enemy raid, and refused them entry afterwards. The concern was not only supplies, but also the loyalty of the locals (cf. chapter 6.2.1).

Lack of preparations was regarded as a valid reason to submit to an enemy without a fight; Anastasios forgave the citizens of *Martyropolis (502) for their rapid submission to the Persians, while later surrenders of Roman cities hardly ever had repercussions for the local leadership. During the Frankish civil wars, cities that had gone over to another legitimate Merovingian ruler were forgiven after the war was over. In many cases, surrender was preceded by serious raiding, so there was no doubt about the attacker's intentions.[34] This is not a Frankish peculiarity, since the Persians used the same approach many times, e.g. against *Amida in 578: the siege of the city lasted only three days but raiding continued for 15. In light of the history of Persian operations in the region, it would have been unwise for the defenders to assume that the Persians were bluffing (which was probably the case), and at any rate the raiding would have had serious economic consequences.

Enemy destruction and rumors of their atrocities could cow both civilians and their garrison (see 6.1.3), and in such circumstances, an alternative to fighting was simply to buy off the enemy. The inhabitants of *Edessa and *Hierapolis (540) bought off the Persians. They were confident that they were able to withstand a siege, but were loath to let the Persians destroy the countryside. For political reasons, paying ransom was favorable to both parties. The besieged avoided having their lands destroyed or a foreign garrison installed in their city. The besieger could construe it so that the city had paid tribute and even formally submitted. Thus he could present the transaction as a victory that resulted in booty or tribute.[35] When the Avars failed before *Sirmium in 568, the Avar ruler explicitly asked for a

---

[34] This is discussed extensively in chapter 4.1.4; see e.g. *Rheims 556/7, *Tours and other cities in 573; *Tours again in 575f; *Tours and *Poitiers in 584.

[35] For similar events, see e.g. the Persian campaign in Syria under the entries for 540.

token ransom of the city so that he did not lose face before his subjects. When the Roman commander refused, the Avars withdrew but let one of their subject peoples, the Bulgars, loose on Illyricum. Although a token concession might seem reasonable, the payment of tribute, even if only nominal, produced some ambiguous results. The same invader could return and occupy the territories of cities that had previously submitted *pro forma* or paid a ransom. Whether faced with tribute or actual occupation, the population had to resist; if not, the arrangement could become permanent. This perhaps lies behind the many "double sieges" of the early Islamic tradition, where a city seems to have been captured on at least two distinct occasions within a year or so. Rather than a result of historiographical confusion or recycling of *topoi*, there are in fact quite well attested scenarios where formal or token submission may have been a temporary expedient, resulting only in a seasonal payment of tribute and a brief recognition of sovereignty, which was later exploited by an invader to press new and harsher claims. During the Persian wars, token submission at some point in the past could result in later claims to sovereignty. Khusro had personally received the temporary submission of *Apamea in 540 in return for not ravaging its territory. The population was sure that it would receive the same treatment when his army returned again in 573, but not only did the Persians raid and destroy a number of fortifications in *Apamea's territory, but the population itself (his theoretical "subjects") was deported despite performing the same submission ritual as in 540. Altogether, 273,000 people are said to have been taken from the city and countryside during that campaign. Indeed, many urban populations in Syria and Mesopotamia accepted the Arab conquerors' demand for submission and tribute, but often did not admit garrisons and administrators during the first conquest phase.[36]

The sight of enemies destroying fields, extramural churches, barns, suburbs and so on, while carrying off livestock, movable goods, and perhaps even friends and relations, occasionally had the opposite effect of what the invaders might have intended. Instead of being cowed, civilian populations were sometimes enraged to the extent it was difficult for military commanders to hold them back.[37] In these cases, defense could be extremely spirited, and on a number of occasions civilian defenders fought just as well as professional troops (see 6.2.2 for this).

---

[36] See Appendix I-II.
[37] The inhabitants of *Rome (537f) were furious at the Ostrogoths; farmers at Constantinople were angry for Hun depredations in 559 (see *Chersonese for context).

Recent scientific evidence shows exactly how devastating warfare could be during the 7th century, although the evidence is not in the form of datable ruins. Pollen analysis shows an abrupt disappearance of all anthropogenic flora in southern Cappadocia during the years 664-678, followed by a quarter millennium of shrubs and forest growth, with virtually no evidence at all of agricultural activity. This evidence tallies well with known military events. The campaign of Khālid ibn al-Walīd, the first large-scale expedition to winter in Anatolia, began in 664 and extended for two or possibly three years.[38] The (certainly intended) effects of these invasions were to destroy the economic basis for urban civilization and hence defense of fortified centers; hence the fact that some cities in the region appear to have been abandoned "over night."[39]

6.1.3 *Cultural Topographies: Morale and Ritual under Siege Conditions*

In 559, the Kotrigur Huns sent a raid in force into Thrace in order to extort tribute payments from Justinian. Agathias, who was himself present at the time, recorded the irrational fear of the population of Constantinople. The anecdote probably reflects the horror experienced by any city during the period facing a similar situation:

> With the enemy encamped at such close quarters the citizens of Constantinople were terror-stricken and were already conjuring up the horrors of a siege, the burnings, the scarcity of foodstuffs and finally the walls being breached. And so it frequently happened that even in the central thoroughfares of the city crowds of people would suddenly break out into a run, pushing and jostling in an unaccountable fit of terror, as though the barbarians had already forced their way in, and a tremendous din was raised in the shops as doors were violently slammed. And not just the common people but even the authorities had succumbed to the prevailing mood of anguish and fear. Even the Emperor himself was, I imagine, impressed with the gravity of the situation.[40]

The reaction is hardly surprising. A siege caused immense human suffering. The terror of the population of Thessalonica in *Miracula sancti Demetrii* grew in the face of Avar invasions, as refugees from cities further north told of the invincible Avar besieging armies.[41] In Greater Syria, the Arabs

---

[38] See Appendix III.
[39] See particularly modern Viransehir, possibly ancient Mokisos, which seems to have been abandoned very rapidly. Private communication with Albrecht Berger, whom I thank for sending me his 1998 article on this problem.
[40] Agathias 5.14.6ff.
[41] *Thessalonica 586, 604, 615, 618, 662.

even developed an apocalyptic tradition that presaged the long-feared Roman return, since Roman fleets regularly conducted raids along the coast, and were capable of carrying substantial forces that shook the Caliphate's control in Lebanon in the 680s.[42]

Maintaining morale was therefore one of the most important factors in siege warfare. The threats were immense and the will to fight was often very fragile. Military developments had an immediate impact on morale. Most cities could sustain quite heavy losses during a siege, but morale was effectively broken if a commanding officer was killed. The personal leadership of Roman generals as well as the Persian shah sometimes decided the outcome of sieges, since they could spur their troops to brave action.[43] While Roman sources tend to emphasize Persian use of threats and terror against their own troops, this may be a reflection of ancient ethnographical stereotypes that painted the Persians as servile and motivated by fear. The Romans themselves certainly dreaded the shah's presence, and at *Kallinikos (503) his personal threats to kill the whole population convinced them to hand over a Persian general whom they kept captive. Perhaps more devastating than the death of a commander, a supporting field army defeated in battle often convinced garrisons to give up. Thus the Roman garrisons in *Florence and other Italian cities were severely rattled by a series of defeats in the field (542), while *Centumcellae (552f), *Florence, *Volterra, *Luni and *Pisa (553) surrendered after the Ostrogothic defeats at Taginae and Mons Lactarius in 552.

The pressure upon the population could be even greater than on professional soldiers. The Romans garrisoning *Milan (538) struck their own deal with the Ostrogoths, leaving the Milanese to massacre and slavery. Otherwise, they suffered from low rations and the never ending threat of a sack. At *Rome (545f), the population was under such severe stress that many committed suicide before civilians were allowed to escape. The city ultimately fell due to a breakdown in discipline among the garrison troops. Otherwise, long watches, boredom, cold weather and lack of sleep all led to indiscipline. Overconfidence could also have the same result; thus *Amida (502f) and *Vollore (524) were captured when the besieged thought they were safe and were beginning to be lax in their guard duties. The Roman negotiator at *Dara (573) was so confident in the city's defenses that he failed to relay a reasonable offer for ransom to the rest of the city council;

---

[42] See the relevant studies in *FCIW* 8.
[43] See e.g. *Petra 541; also chapter 5.2.1 above for further examples of the role of commanders. See also the charge of Bessas at *Petra 550.

he was personally scolded by the shah after the capture of the city, when the immense riches captured were piled up, far beyond the meager ransom demanded. Finally, the knowledge that relief was on the way could make garrisons and populations endure terrible conditions and very long sieges. *Centumcellae was besieged for many months (549f), but held out in the knowledge that a large Roman army was being organized in the Balkans; the Ostrogoths at *Taranto (552) decided to hold out when learning of the election of Teïas.

Both besiegers and besieged tried to enhance morale through a variety of means, while at the same time demoralizing the other party. Taunts were frequent and often directed at an enemy's manhood and military skills. The Goths reviled the Roman troops as Greek "actors of tragedy and mimes and thieving sailors"—which probably accurately reflected the attitudes of many Westerners, since that may have been the only Greeks they had met before the arrival of Belisarius' army.[44] However, when the Goths were first defeated in 540 and the Roman army marched into *Ravenna (539f), the Gothic men were reviled by their wives for being defeated by such "Greeklings."[45] Being reviled for lack of manliness by their women probably does fully not explain later Gothic resistance, but it may have shamed many Goths into action when the opportunity arose later on. Another form of abuse was to mock the opponent's military skills, which established superiority and boosted morale, especially if followed up by successful military action. The Romans abused the Arabs' lack of skill in handling military equipment at *SYLWS (664) before they used their own trebuchet to destroy the one mishandled by the Arabs. Despite their fear of the Persian shah, the Romans would on occasion verbally abuse him if he was present, inviting him to try to storm the walls himself after the failure of his soldiers; the enraged Khusro ordered another attack, in part to revenge insult, in part to take advantage of the elated and disorderly mood of the besieged.[46] At *Melitene (576), the Roman envoys scolded Khusro for burning the city, calling it a "piece of mischief" unworthy of a great king. The Visigothic rebels holding *Narbonne (673) similarly cursed king Wamba from the walls, but this only provoked a fierce assault from the angry besiegers.

The practice of taunts and counter-taunts was often enhanced by psychological games. Prostitutes rather crudely showed their private parts to

---

[44] Proc. 5.18.40.
[45] Proc. 6.29.34.
[46] *Amida 502f; see further Whitby 1994.

the Persians at *Amida (502f). It may have had a great impact on the besiegers, since one of the prerogatives of the conqueror was to rape women in a city taken by storm (see 6.3.2 below), and we know that it caused great frustration among the Ostrogoths when Totila denied them this right at *Naples (542f and 6.3.2 below). Totila brought about the surrender of the city by giving three months rather than the one requested by the defenders to wait for reinforcements; when they realized that there was no hope of relief, the inhabitants immediately surrendered.[47] Narses in return played a trick on the inhabitants of *Lucca (553). Having captured some of their leading citizens, Narses used them to force a deal with the citizens to surrender if reinforcements failed to arrive. However, they failed to live up to the deal, and Narses had his hostages led out to be decapitated in sight of the city walls. Unbeknownst to the people of Lucca, however, he had replaced the hostages with common criminals, and the citizens were astonished to receive their living relatives back into the city. The hostages later played an important role in convincing the citizens to abandon the Franks and surrender to the Romans. At *Phasis (556), the false rumor of a Roman relieving army was used both to boost the morale of the defenders and demoralize the Persians, who were forced to send part of their army against the new threat.

Terror tactics were systematically used by both sides of a siege. When a captured Persian officer failed to convince his comrades at *Amida (503f) to surrender, the Romans impaled him within sight of the city wall. Terror was also used to keep one's own forces in line: a prominent citizen at *Carthage (533) was impaled in public view outside the city for colluding with the Vandals, while the Romans besieging *Osimo (539) burnt one of their own alive when they discovered that he had been relaying messages between the besieged garrison and Ostrogothic authorities. At *Thessalonica (615), Slavs attempted to attack the sea walls in *monoxyla* but failed, leaving many dead on the shore. Roman soldiers went out and beheaded them, placing their heads on the walls in full view of their comrades on the landward side. Impalement remained in use in the Caliphate. During the great civil war of 680-92, administrators who had supported the wrong side were impaled in the aftermath of *Damascus (690) and *Mecca (692).

If siege engines were produced in large numbers or unfamiliar forms, they terrified the defenders. Accurate firing from defensive artillery broke the morale of the attackers, but besiegers' artillery could in return "terror

---

[47] Proc. 7.7.

bomb" the dwellings inside the walls (5.2.2). The Ostrogothic garrison at *Osimo (539) finally surrendered when Belisarius had the Goths captured at *Fiesole paraded past the fortifications. However, sometimes only the realization that a siege was in preparation was enough to break morale; thus the Ostrogothic garrisons at *Todi and *Chiusi (538) surrendered when they learned that the Romans were to institute formal siege. The Persians attempted to convince the Romans at *Dara (573) to surrender by using the terror of siege engines, but when this failed, they instituted a tight blockade that led to the final capture of the city.

Concern for family and civilians was pressing for most defenders, and determined their actions. When the Romans were besieging an *Abasgian fortress (549), Procopius comments on the defenders' brave resistance, since they were fighting for their homes. Despite Roman brutality, they had serious concern for their own civilians. At *Noviodunum (437), the children from the population were used as human shields, prompting Roman troops to cease hostilities. The Ostrogothic field army recruited from garrisons in the *Alpine forts (539) surrendered when they learned that the Romans had taken their families. Illyrian troops who had taken several *forts around Bologna (544) returned home when they heard that a Hunnic invasion had struck their home region, prompting concern for families and property. On the other hand, families and children in the East prevented the Roman soldiers holding *Centumcellae (549f) from defecting (as many other Romans did in the 540s). In a rare instance, the Romans besieging *Petra (550), when on the brink of victory, practically begged the surviving Persian garrison in the acropolis to surrender with reference to Christian piety. The Romans only resorted to force when the Persians refused, using incendiary projectiles to incinerate their last stronghold.

Civic, military and religious ritual was another important factor in building morale.[48] Procopius records the use of music to cheer up the defenders of *Rome (537f), and his troops sang military victory songs when they successful defeated Ostrogothic assaults. So did the civilian defenders of *Antioch (540) when they drove back the Persians during heavy street fighting. These songs included hymns to the reigning emperor and prayers for his victory over the enemies of Rome. A small group of Romans managed to enter *Petra (549) during an assault on the walls, proclaiming the emperor triumphant. When Roman communities refused to accept orders from regional commanders, they often resisted in the name of the

---

[48] See McCormick 1986 and D. Bachrach 2003 for good surveys.

emperor. The military cries *Deus adiuta* (*Romanis*), "God bring help (to the Romans)" and *Deus nobiscum,* "God with us" are well known in battlefield manuals and probably not unheard of during sieges. *Kyrie eleison* was also common, its use explicitly mentioned at *Thessalonica (618) and *Kamakhon (766). When the inhabitants at *Theodosiopolis (421/2) were terrified by the approach of the Persian army and siege engines, the bishop had a large catapult set up and aimed at a blasphemous Persian commander, having it shoot as he said "in the name of the Lord who ha[s] been blasphemed." Similarly, one of the civilian defenders of *Thessalonica (618) devised a rhythmical chant that was used by trebuchet pulling teams: "In the name of God and Saint Demetrius!" It was probably pronounced something like (with stress marked by acute accent): *is toúnoma theoú ke ayíou dhimitríou*.[49] Indeed, some of the liturgical elements in the *Miracula sancti Demetrii* were, in effect, prayers for the catapults to work effectively and smite the pagan foes (especially in 618). The Muslim element in Arab armies used the *takbir,* i.e. the war-cry *Allahu akbar.* It is reported by a Syriac source as early as a skirmish outside *Constantinople (663), and was consistently used according to Islamic chroniclers (see e.g. *Barbad 776). At *Kamakhon (766), however, the premature use of the *takbir* during a surprise attack ruined the surprise, allowing the defenders to organize themselves. The Franks used a combination of war-cries and trumpets when assaulting fortifications, e.g. at *Avignon (736).

The ritual aspect was also backed up by specific religious motivation that begins to be articulated in the late 6th century. The citizens of Apamea were willing to fight a "holy war" against the emperor Justin II in order to keep their relic of the cross. Justin wisely averted disaster by praising their zeal and taking only half of their piece of holy wood.[50] At *Drizipera (588), this motivation even extended to dying and receiving recompense in the next world. This kind of idea might rather reflect the *milieu* in which Theophylact wrote around 630, as such ideas became ever more common in Herakleios' rhetoric against the Persians, and may have inspired at least some elements of early Islam. This willingness to die in battle in return for heavenly rewards is a well known feature of early *jihad* theology.[51] Interest-

---

[49] Tarver 1995 points out during modern trials of a traction trebuchet that a rhythmical chant would be very helpful for the pulling teams, but found no evidence of such in medieval sources.

[50] Menander fr. 17.

[51] For martyrdom in Islam, see Cook 2007; for *jihad, EI*² s.v. (*djihad*) and relevant studies in *FCIW* 8; for historical context, see the relevant studies in Howard-Johnston 2006 and 2010.

ingly, the pope expressed a promise of heavenly rewards to Carolingian troops fighting to protect *Rome in 756.

Considering the importance of otherworldly assurance and assistance, religious processions were frequent during sieges. When Artabasdos held *Constantinople (742f), he may have begun the cycle of iconodule reaction to iconoclastic emperors by restoring icons in the city. Icons had previously an important role in the defense of *Constantinople; the defenders paraded the icon of the virgin painted by Luke (Hodegetria) on the Theodosian walls against the Avars in 626. Some, including the enemy, even reported the sight of the Virgin herself on the walls, while St. Demetrius had apparently manned the walls of *Thessalonica himself in 662, killing Slavs with his mace. Many cities had relics or even living saints who could help boost morale. Caesarius of *Arles (508) was met with a formal reception (*adventus*; see below) when he returned from Burgundian captivity, and immediately afterwards he oversaw defenses of the city. When the Franks unsuccessfully besieged *Saragossa in 541, Gregory of Tours believed the inhabitants survived because they arranged a procession on the walls donned in sackcloth. The clergy of *Dijon (555) refused to meet the rebellious prince Chramn inside the city, but instead arranged a formal reception outside the city. During the meeting, they performed a service to find "random" scripture texts that prophesized against Chramn, essentially validating their refusal to admit him. Bishop Leudegar arranged an elaborate procession that stopped at each gate at *Autun (679). Bishop Anianus of *Orleans (451) led prayers for a relieving army to arrive, while St. Maximus at *Castrum Cainonense (463) led public prayers for rain when the besiegers had cut off their only water supply; in both cases, the sanctity of the bishops ensured relief—the rain produced at the behest of St. Maximus even scared off the besieging army. Whereas the inhabitants at *Amida (502f) had employed improper taunts, the citizens of *Constantina (502f) were fortified by their bishop Bar-Hadad (Baradotus), who held services on the walls and gave the Eucharist to those on watch. This practice continued right through this period; at *Kamakhon (766), the Romans celebrated the liturgy on the walls.

All sieges contained some element of military pageantry and display. It was common to parade troops and even siege engines. A good show of discipline and equipment had an obvious psychological impact, as did the sight of engines and supplies. The *Strategikon* even recommended parading the best-looking troops in front of the enemy, while taking care to keep shoddily clad soldiers in the rear. Indeed, threats had to be credible: there would be little incentive to surrender to a hundred scraggy men waiting tamely below with ramshackle engines that collapsed of their own accord. Sometimes the passage of vast armies, even if only in the same general region, prompted immediate surrender. Sebeos relates that the whole of Anatolia surrendered to the Arab armies converging on *Constantinople (654). This was nothing out of the ordinary, as *Edessa and other Mesopotamian cities surrendered to the Persians in 609 for similar reasons. The Romans holding the *Milvian Bridge (537) were utterly terrified by the size of the Ostrogothic army, and ran off without even notifying the rest of the garrison at *Rome. To increase the terror, besiegers lined up in serried ranks at the beginning of a siege in order to impress the besieged. The size of Gundovald's army convinced *Toulouse (585) to surrender immediately, despite having prepared for a siege. Similar considerations probably lay behind the frequent surrenders in Frankish civil wars, in addition to the political aspect (6.2.1). The Avars deliberately lined up before the walls of *Constantinople (626), and the Avaro-Slavs did likewise at *Thessalonica in 586. There it took the defenders two days to regain their confidence, but by then they began to taunt the Khagan, inviting him in to take a bath—depending on the dating, perhaps this was a spoof on his wives' bathing at *Anchialos (588), or perhaps implied that he really *needed* to take a bath.

These rituals had an important function in ensuring divine assistance, and were prescribed in detail by military manuals.[52] They also reinforced hierarchies of authority or established new ones. The formal *adventus* or *introit* arranged for magistrates, military officers and foreign representatives (whether ambassadors or conquerors) was important, especially in politically sensitive contexts. Joshua the Stylite describes the unwilling but politically necessary reception of a Roman commander befor the beginning of the Anastasian war. Caesarius of *Arles was clearly a controversial figure, perhaps even a traitor, but was nonetheless received (or arranged to be in order to cement his authority) with an appropriate *adventus*-ceremony before the Franks and Burgundians attacked in 508. Such receptions and rituals were also held for enemies. The Avar Khagan simply helped himself to imperial robes deposited at *Anchialos (588), claiming legitimacy and demanding taxes from cities within his reach. Khusro was far more sophisticated. He had the local citizens arrange circus games in

---

[52] See *Strategikon* 2.18 on battlecries; 7.A.1 on the blessing of the standards one or two days before battle; Charlemagne's campaign that led to the capture of *Avar fortifications in 791 was preceded by a three-day fast and elaborate liturgical rituals.

*Apamea in 540, where he presided as the legitimate ruler, even supporting the Greens during the races against the Emperor's traditional support for the Blues. Again in 573, a similar *adventus* for Persian envoys was exploited by the Persians to take the city by surprise. We can only imagine the pomp (or ritual humility) that surrounded bishops and dignitaries as they offered to ransom or surrender a city. Joshua alludes to this when he describes the reception the Persians received at *Martyropolis (502) and the "dignified bearing" of bishop Bar-Hadad who went out to appease the Shah during the siege of *Constantina (502). Procopius tells of several occasions when formal surrenders were arranged by the local bishop. Christians on the Persian side of the border also appealed to common Christianity. When Maurice led the siege of *Chlomaron (578), Christians in the region of Arzanene went out in liturgical procession to meet the Romans and even offered liturgical vessels as ransom. This was of course an offer Maurice had to refuse. Such formal receptions are reported by Arab historians. Baladhuri describes the *adventus* arranged by the locals for 'Umar at *Adhri'at in 636. These events are routinely misinterpreted by modern historians. While the "welcomed as liberators"-argument has long been proven unfounded (cf. 6.2.1 and 7.3.2), such stories are occasionally employed in polemics, while some scholars use them to provide a "balanced" account. For example, Kennedy argues that the Arab conquest was accompanied by a "carneval atmosphere," clearly implying that the population was staging a positive reception for the conquerors. Indeed, local dignitaries arranged formal receptions for the Arab conquerors on a few occasions, which included processions and music. These receptions were a form of appeasement and submission rather than a joyful expression of welcome, and at *Adhri'at the citizens' efforts were brusquely rejected by the Caliph himself. He only put up with the ceremony when his advisors explained that a rejection would be interpreted as a refusal to accept the city's submission.

The besiegers of course had their religious rituals, although this is less well attested. The Visigoths attacking *Septem (547) were taken by surprise when they were celebrating Sunday mass. Attila and the Huns used pagan rituals; at *Toulouse (439), the Roman commander Litorius used haruspices to divine the future before assaulting the peaceful Goths, who had even sent an embassy led by clerics. In Salvian's view, there was no wonder he was killed in the ensuing fight, but paganism and sub-pagan rituals seem to have been common in the Mediterranean well into the 6th century. The Avar Khagan swore pagan and Christian oaths during negotiations over the siege of *Sirmium (579ff).[53] Zoroastrian religion played an important role in the Persian wars. On campaign, the shah brought his own personal fire temple, which was left behind in the confused fighting that (ultimately) led to the capture of *Melitene (576). An omen from the shah's horse convinced the Persians to attack *Sura (540), while the Magian priests convinced him to continue the rather lackluster campaign in 576, which this time led to the capture of *Sebasteia. While Islamic rituals presumably dominated Umayyad and Abbasid armies, this cannot be taken for granted in light of the large numbers of clients and subject troops regularly serving well into the 8th century. Hindus and Zoroastrians in the Abbasid armies performed their sacrifices as late as the siege of *Kamakhon (766), to the great astonishment (and revulsion) of the Zuqnin chronicler. For Christian polemicists, Islam was more legitimate than religions that offered sacrifices to idols and multiple gods. In a few instances, human sacrifices are also recorded. Early 6th-century Franks still sacrificed humans, at least during their 539 invasion of Italy; Sidonius relates a similar practice among Saxon pirates in the late 5th century.[54] There is also a non-Christian undercurrent in some Roman practices into the 8th century. At *Rome (537f) itself, the population attempted to open the doors of the temple of Janus, an ancient ritual for a state of war. An astrologer predicted a Roman victory at *Markellai (792—Theophanes' subsequent narrative clearly demonstrates his error), while a very bizarre event is recorded at *Pergamon (716). Apparently, the Roman defenders cut up a pregnant girl, boiled the fetus, and dipped their sleeves into the pot. It is difficult to know what to make of the story, unless the originator of it wished to blame the Arab capture on some sort of sin on the part of the inhabitants.

Some of the best sources for siege warfare, hagiographies and miracle stories, were composed in order to reassure people that divine assistance was forthcoming in times of trouble. They would have had liturgical functions, being read out during mass on feast days, and perhaps even used specifically during times of threat. The *Miracula St. Demetrii* indeed has long sections of a liturgical nature, especially from the sieges of *Thessalonica in 615 and 618. While they provide a whole mass of incidental details,

---

[53] He may have just wanted to hedge his bets, as it were, but he probably had significant numbers of Christians in his ranks, even if many were captives; see chapter 7.2.3 for further discussion.

[54] Frankish sacrifices in the context of *Fiesole 539; Saxon sacrifices mentioned in Sid. Ap. *Ep.* 8.6. The evidence adduced from Illerup for Scandinavian military developments shows that this was common practice in northern Europe in pre-Christian times.

hagiographies are also heavily manipulated, mostly by omission and exaggeration. Manipulation increased with time and space. The miraculous intervention of the bishop Anianus at *Orleans (451) was well established a generation later, at the time of Sidonius Apollinaris, and became the basis for the *Life* of Anianus and Gregory of Tours' version. In contrast, the contemporary Priscus, as summarized by Jordanes, focused on relations between the defenders, logistical works, and the rituals that Attila performed in order to divine the outcome of the siege. The *Miracula sancti Demetrii* was written some time after the events. It could take some liberty in its descriptions (for instance, the threats posed by the Slavs and Avars may have been exaggerated in order to emphasize the miraculous intervention of the saint), but since the intended audience was the city's population, and some of the participants of previous sieges apparently were still alive at the time of composition, there were clearly limits to how great this manipulation could be, and there are indications in the text that a too miraculous interpretation of events was sometimes met with blunt criticism from the congregation. This reveals a skeptical spirit among populations that had been tested too hard. While religious motivation could be fierce, and religious explanations for events dominated in late antiquity (or at least in our preserved sources), we cannot assume that everyone thought in the same manner. The critical questions that were posed by the congregation and recorded in the *Miracula sancti Demetrii* are quite reasonable, while Dionysios preserves, apparently drawn from a hagiographical source, the blunt question posed by the population in and around *Antioch (638/9) after a particularly brutal Arab raid: "Why does God allow this to happen?" The stock answer could have provided little comfort; it was in effect "because of our sins." As Sarris has shown, skepticism against such stock clerical answers was probably far more widespread than most scholars normally think.[55]

6.2 The Urban Community at War

Maintaining internal cohesion was a difficult task, especially if a city had a diverse population, but religion and ethnicity had little to do with this. There are for instance many anecdotes of Jewish betrayal of Roman or Ostrogothic cities, but most of them can be shown to be literary *topoi* or exaggerations. *Any* corporate body or individual could at any time choose

---

[55] Sarris 2011.

ANATOMY OF A SIEGE: ECONOMY, SOCIETY & CULTURE 327

to betray their city, for any number of motives. Religious disaffection between Christian groups was insignificant. Arianism was not an issue in Western sieges, while the cities that provided the fiercest resistance to the Arabs (and suffered the most severe consequences) were predominantly monophysite cities in Roman Mesopotamia, northern Syria, Upper Egypt and Armenia. Rivalries between local magnates were far more important and devastating. While internal divisions sometimes came to the surface during foreign invasions, they caused completely unpredictable outcomes during civil wars, when betrayal, double dealing and surrender were more common.

6.2.1 *The Politics of a Siege: Loyalty and Dissension*

Maintaining cohesion in an urban community was always complicated by diverging loyalties.[56] For the most part, the damaging divisions occurred between the representatives of the central government (high-ranking military commanders; governors) and local dignitaries (landowners and officials) who had different priorities. For instance, Justinian had his officers forbid local city councils from ransoming their cities from the invading Persians in 540, and the Roman general Bouzes even refused to allow the population of *Dara to ransom captives that the Persians had taken from *Antioch and other cities. However, quarrels among officers were all too common and led to failure at *Rome (545f), a confused withdrawal and rout at *Nisibis (573). When the conqueror was completely alien to the local population, it was mostly fear or personal agendas (revenge, personal gain) that induced people to betrayal, but even in such cases, local parties would try to gain political advantage in their internal power struggles.

Political loyalty amongst a complex set of actors was a determining factor. Urban defense depended upon cohesion between garrison forces, magnates, clerics and officials whose primary interest lay with the city and its territory. Although they might squabble amongst themselves, the picture was further complicated by dignitaries, officers, and any forces sent by the central government in reinforcement. Sometimes garrison forces and the local population could not find common cause. The situation might change within a few years; while *Kamakhon was fiercely defended by its garrison

---

[56] For particularly good examples of the complexities of urban communities under siege, see *Rome 537f; *Antioch 540; *Archaeopolis 550; *Tours 575f; *Convenae 585; *Thessalonica 682.

in 766, it was simply handed over by an Armenian garrison in 793. The Ostrogothic garrison at *Salona (536) decided it could not rely on the local population and abandoned the city to the Romans. However, power struggles between locally rooted parties were just as damaging. Cities had up to several prominent individuals or families who had their own agendas and could sway public opinion through patronage networks, and some were willing to bring in foreign forces to achieve their ends. The Vandals tried to split public opinion among the citizens at *Carthage (533; see also 6.1.3 above), and even attempted to subvert Roman soldiers by appealing to common Arianism the single *potential* incident involving Arianism ever recorded during a siege. They also approached Massagetae (i.e. Hun) mercenaries, playing on their fear of being settled in Africa. A fair share of booty and a guaranteed return to their homelands reassured the loyalty of all the soldiers at Carthage. For the Merovingians, it was often sufficient to apply pressure to the power base of wavering magnates, e.g. at *Rheims 556/7, but even the Franks received some Lombard support during their *Invasion of Italy in 590. The Ostrogoths were initially forced to abandon *Rome (536) when the population had decided to invite the Roman army in, but during the siege of 537f it is clear that they still had some residual support amongst some in the city, while others could simply be bribed. Belisarius had civilians who took part in the defense "mingle" (see 6.2.2) with regular troops to ensure loyalty and solidarity; he even had to expel the pope himself due to dubious loyalties. The same happened to the bishop of *Antioch (540). Indeed, bishops were as likely to engage in power struggles as any magnate, and in the West consistently behaved like their secular counterparts.[57]

A single individual or small group of people had the power to decide local policy. Thus a Roman official betrayed *Theodosiopolis (502). Matasuntha, the Ostrogothic queen, actually invited the Romans into *Rimini (538), while her husband Vittigis was desperately attempting to hold the Appennines. According to Procopius, the queen was driven by personal motives. *Milan (538) and other Ligurian cities likewise invited the Romans under the auspices of their bishop and leading citizens. A Lombard noblewoman invited the Avars in at *Cividale (610). Besiegers tried to exploit this during negotiations, and could apply extreme pressure on envoys. The Persians captured and tortured the bishop of *Sergiopolis (542) for failing to follow up an agreement he made with the Persians at *Sura (540) to

---

[57] See e.g. the bishops' involvement at *Narbonne and *Gerona 673; *Autun 662 and chapter 4.2.2 *passim*.

ransom captives. The Arabs sent a bishop to negotiate at *Arwad (649), but the population refused to let him back to them, inducing the Arabs to return next year with an even bigger siege train. This was shortly after *Emesa (635) and some other Syrian cities had managed to negotiate very extensive privileges in return for a surrender. Later on, the Arabs used a bishop to betray the island of Cos, perhaps in a similar manner.[58] However, public opinion could be hard to sway: the population of *Alexandria (642f) nearly stoned the patriarch for negotiating a surrender to the Arabs. Complicating the issue there were still ongoing power struggles after the death of Herakleios (see below).

In many cases, collusion between invaders and local population was for simple material gain. Instead of raiding, invaders paid local populations to bring in supplies. This problem is widely reported on the Roman eastern frontier throughout this period (e.g. *Amida 504f, *Kamakhon 766). Turncoats gained power and brought dangerous knowledge and leadership. A former *doryphoros* of Belisarius went over to the Ostrogoths, leading their fleet to capture *Mourikion and *Laureate in Dalmatia by surprise in 542. A simple way to ensure loyalty was to pay money and distribute booty; king Wamba was certain to reward his troops when they captured a number of Pyrenean forts (*Collioure, *Ultrère, *Llivia) in 673. Otherwise, garrisons simply decided to change sides or accept bribes from the enemy, especially if their pay was in arrears or other political conflicts had shaken their loyalty. This was a perennial problem in Italy, where some Ostrogoths went straight over to the Romans at the outset of the war, whereas unpaid Romans frequently ended up in Ostrogothic armies in the 540s. However, they might return to their old allegiance, especially if back pay and supporting armies were forthcoming. Many urban communities were torn over which side to support.[59] This was particularly acute in some frontier regions where daily relations crossed political boundaries. Integration with nearby Slavic groups at *Thessalonica (662) was considerable and blurred boundaries between Romans and Slavs that had been very sharp a few generations earlier; by 682, boundaries were even further blurred and a group of descendants of Roman captives in Pannonia nearly took over the city under Bulgar leadership.

The most extreme expression of disloyalty was to betray a city under siege. The motives were again the same as for "pre-siege" betrayal (above),

---

[58] See the *Aegean campaign that followed the fighting at *Tripoli 653.
[59] The examples are too numerous to elaborate in the text; see e.g. *Spoleto 546, *Rimini 549f, *Perugia 552, *Lucca 553, etc.

but normally it was only a single individual or a small group that acted in this manner. Of all the groups involved at *Vienne (500), it was the local hydraulic engineer who showed the way in through the aqueduct after he had been thrown out of the city. Some Italians were ambivalent about both Roman and Ostrogothic garrisons. *Verona (541) was betrayed to the Romans by a local guard, but at *Tibur (544) the locals betrayed the Isaurian garrison after a quarrel. While the Isaurians managed to escape, the civilian population was massacred by the Ostrogoths. Other Isaurians went over to the Ostrogoths when they saw erstwhile comrades doing well in the army that was besieging them at *Rome (549f). Several Roman strongholds in Spain were betrayed to the Visigoths (e.g. *Medina Sidonia 571), but we never learn by whom or what the context was. During the Roman siege of Lombard *Benevento (663), a royal army from Pavia came in relief, but it was abandoned by a large number of troops on the way, perhaps bribed through Roman diplomacy or unwilling to accept a successful royal expedition. Besiegers too could be abandoned or betrayed by their own. Those besieging *Thessalonica (586) were plagued by bouts of infighting, and many actually deserted *to* the city. The multi-ethnic Avar-led army before *Constantinople (626) virtually collapsed when the largest component, the Slavs, had had enough and left after many of their own had been ambushed and massacred in the Golden Horn. Persian armies had numerous Christian troops whose loyalty was sometimes ambiguous. Arab client troops in Persian armies even divulged information to besieged Roman cities.[60] Armenian *naxarars* (militarized nobles) fought for Roman as well as Persian armies, but from the 7th century they had an alternative in the Arabs.[61] This meant that no matter the period, operations in Armenia could be complicated by local feuds and ambitions manipulated by foreign intervention—and *vice versa*.

All societies systematically resorted to siege warfare during civil wars, which could be fought just as bitterly as any other conflict. Maurice began preparations for a siege, but had to abandon *Constantinople (602) due to lack of support. However, there were two hard-fought sieges at *Alexandria (608, 609) during Herakleios' revolt, and *Constantinople was besieged on many occasions by Roman forces (698, 705, 715, almost in 718, but severely in 742f). Lombard dukes at *Bergamo, *Comacina and *Treviso (591) resisted royal power and had to be subjugated by siege, while the revolt

---

[60] E.g. *Sergiopolis 542.
[61] Garsoïan 2004: 117-25 and the relevant chapters of Sebeos with Howard-Johnston's commentary.

against the Visigothic king Wamba was defeated by a campaign of sieges (see chapter 3.2.3 and CO entries for 673). Similar recalcitrance was a common cause of Merovingian civil wars (see below). Even the Carolingians Pippin and Carloman had to besiege their rebellious brother at *Laon in 741.

Civil war produced many conflicting loyalties that made cohesion impossible at times. During the incessant West Roman civil wars of the 5th century, barbarian auxiliaries and clients had a hard time choosing whom to serve, resulting on different barbarian groups on each (Roman) side at *Bazas (413). The Visigoths were essentially party to another Roman civil war when they were handed over *Narbonne (461) in return for their support, yet again to one "Roman" side against another. Due to the fragmentary nature of the sources and unclear political boundaries, few scholars can determine who belonged to which party fighting for *Angers in 463. As late as 473, however, the Ostrogoths may have invaded *Naissus, *Ulpiana and *Stobi at the behest of the West Roman government.

Civil wars shook the East Roman Empire in 602-3, 608-13, and 641-44, all critical junctures during the Avar, Persian and Arab invasions, respectively. Hard fighting is recorded in Egypt.[62] Examining the conflict in 608-13, Olster found that the party lines on the ground had very little to do with religious adherence, but depended on local networks of patronage, political power, and simple ambition or antagonism.[63] If larger issues were involved, they had more to do with dynastic legitimacy. This was why *Edessa (603) accepted a boy claimed to be Maurice's son, Theodosius, as rightful emperor with Persian support, and at *Theodosiopolis (610/11) the Persian army used him to convince the city to surrender. Just like other groups, Roman rebels colluded with foreign invaders. For example, mutinous Roman troops took *Hadrametum (544) with Moorish help, but a Roman army retook the city with help from the local citizens. *Carthage (544) mobilized in case the city would be besieged by Moorish troops in league with Roman rebels, although no siege materialized. Some of the 8th-century sieges over *Constantinople (715, 742f) and *Cherson (711) involved long blockades, heavy fighting and siege engines, indicating very sharp divisions within Roman society.

Frankish civil wars amply illustrate the same points, but can be more difficult to interpret since there normally were three or four legitimate

---

[62] In addition to the two sieges of *Alexandria 608 and 609, see also *5 Egyptian cities 608 and *Manuf 609.

[63] Olster 1993.

kings at the same time and endless magnate faction. Such divisions existed within every region, but a royal show of support for one faction was often enough to settle a dispute. If no such support was forthcoming, magnate feuding could descend into outright war, such as when one party besieged the other after a contested episcopal election at *Uzès (581). However, one should not exaggerate magnate independence. Internal competition among them benefited the Merovingian dynasty in general, and Merovingian kings had no compunction about depriving magnates of office, whether by surreptitious, legal or military means. At *Poitiers (585), the bishop actually had to bribe the royal garrison when they turned on him. Magnates revolted for unstated reasons, but may have been expecting support from the king of another *Teilreich*; this was indeed the only "legitimate" way to revolt that could attract support from other magnates.[64] Thus Theuderic had to intervene in the *Auvergne in 524 in order to establish his authority over a local faction that was in revolt against him. Gregory never explicitly states that this faction was supported by another Merovingian king, but it cannot be ruled out. Furthermore, Gregory often mentions how an army from one kingdom arrived and somehow threatened a city that belonged to another king. As argued in chapter 4.1.4 above, these threats were in many instances backed up by the prospect of an actual siege. We see the process at *Poitiers (567), where an army arrived to extract an oath. The means of extraction was clearly violent: it "overpowered" (*oppressit*), "conquered" (*obruit*), "killed" (*interimit*); the citizens were only willing to give oaths to someone "thus ... coming to" (*sic ... accedens*) the city. Incidentally, the vocabulary (apart from *accedens*, used by Gregory as the key word) is consistent with the stages of siege warfare.[65] Given the level of violence, "support" for a legitimate Merovingian could be forgiven (in most instances), but support for pretenders such as Gundovald or magnate rebels without legitimate royal backing led to wars of annihilation. Furthermore, forcibly extracted oaths also led to treason if the opportunity arose.[66] There is every reason to believe that magnates took just as much "convincing" under the Merovingians as they did under the Carolingians, as evidenced by the brutal sieges during Pippin the Short's conquest of Aquitaine in the 760s (see chapter 4.3.3).

---

[64] See Fouracre 2005 on the strength of the Merovingian dynasty in Frankish political culture.

[65] Similar "oath-extracting" expeditions are found at e.g. *Tours 567; *Perigueux 582 may have involved oath from several distinct parties.

[66] Forgiveness: *Paris 574 after the oath was extracted with fire and sword; annihilation: *Convenae 585; betrayal: *Arles 566, 567/9, *Carcassonne 589 (probably).

In contrast to modern preoccupations, class divisions and "community cohesion" (i.e. minority relations) seem to have caused little concern during sieges, at least to the successor states. Networks of magnate patronage lay behind most "popular" unrest, although sometimes a particular group (such as circus factions or the Jews) was singled out in later sources. Jewish communities were accused of betraying cities on a few occasions. This seems only to have been a concern on the Roman eastern frontier, however, and possibly evolved into a *topos*. Olster has shown that the degree of Jewish initiative and participation in the civil war of 608-10 is highly exaggerated in some sources.[67] In light of the number of sieges considered here, and how widespread Jewish communities were, it is remarkable how *little* attention they receive in siege narratives. At *Thessalonica (615) they are mentioned as witnesses of the miracles performed by St. Demetrius, and must have participated in the defense as any other civilian group (cf. 6.2.2). The Jewish community guaranteed supplies at *Naples (536), defended their own sector, and were indeed the last to surrender. At *Arles (508), the Jews were singled out by Caesarius for attempting to contact the Franks and Burgundians, offering them a safe section to set up their ladders. The offer was made by throwing a stone with a message attached to it against the Frankish lines, but was found by a patrol from the city that investigated the no-man's land between the walls and the besiegers. This discovery was conveniently made just after one of Caesarius' relatives (and fellow citizen of Chalons) had crossed the lines and offered to surrender the city. It must therefore be dismissed as a rather obvious attempt to deflect attention from an embarrassing incident that could not be hidden. As with other corporate groups in a city, in most cases the Jews seem to have done their job, fighting loyally for their local community and otherwise drawing little attention to themselves. There was only one egregious example of betrayal at *Tella-Constantina (502f), but it is difficult to say why this happened at this particular occasion.

For the next century there were no incidents and apparently no connection between the Jewish and Samaritan revolts and Romano-Persian warfare. This changed dramatically with the last Persian war of 603-28, when Jewish communities collaborated actively with the invading Persians on at least some occasions. At first, the Persians supported the Jews, who arranged a formal *adventus* for the Persians at *Caesarea in Cappadocia (611) while the Christian community fled. However, the brutality at *Jerusalem

---

[67] Olster 1993: 102ff.

(614), when the Jews massacred the population in retaliation for Christian atrocities, was one step too far, and instead the Persians began to cultivate Monophysite bishops. Again at *Edessa (630), the Jewish community supported the Persian garrison that refused to leave in accordance with the recently signed peace between Rome and Persia. Jewish collaboration during this conflict is reported too widely to be dismissed as a *topos*, but it is hard to discern a rationale behind their actions.[68] However, the *perception* of Jewish treason (even if largely a polemical invention) may also lie behind some of the restrictions imposed on the Jewish population by Herakleios, Dagobert in Francia, and Sigibert in the Visigothic kingdom at the time, although this is controversial.[69] There was little reality to these concerns; in the outskirts of *Jerusalem (634f), Jewish and Samaritan villagers were massacred alongside Christians, while contemporary Jewish scholars apparently viewed the rise of Islam with great suspicion.[70]

In some interpretations, Jewish irredentism during the Persian invasions lay behind the rise of early Islam and led to the first Arab assaults on Palestine.[71] However, only one Jewish community, at *Emesa (635), is said by Baladhuri to have actively assisted the Arabs during the conquest phase, but this may simply be a reflection of incidents that happened during the Persian invasion; in every other respect it is an isolated event *not* included in his description of the actual siege.[72] Most communities responded with resistance if at all possible, only surrendering when hope for reinforcements was cut off. There are in fact striking parallels between the behavior of Syrian cities against the Persians in 502-03, 540, 542 and 603-14, and the response in the face of the Arabs in the 630s. The population at *Martyropolis (502) meekly submitted with an *adventus* in the same way as the inhabitants of *Adhri'at (636) and many other cities.

The Persians were far more insecure of their (non-orthodox) Christian communities than the Romans were of either heretical groups or Jews. On

---

[68] It appears to have had an irredentist goal of reestablishing Israel, cf. Sebeos 42; this passage underpins much of the argument presented by Cook and Crone 1977. For a recent discussion, see the commentary in Thomson and Howard-Johnston 1999: II, 238ff and what follows here.

[69] See Bachrach 1977 for a discussion of the context.

[70] See especially the *Doctrina Jacobi nuper baptizati* and the discussion in Kaegi 1968.

[71] Most notably Cook and Crone 1977 and the preceding notes.

[72] See Baladhuri 211, but this is in the context of the battle of Yarmuk, a battle account riddled with historiographic and source critical problems;contrast his treatment of the siege at 200f. A much later instance, the Jewish betrayal of *Neocaesarea, is reported by the Zuqnin chronicler in 729, but I suspect that this may be a reproduction of a far earlier (6th-century) event.

several occasions in the 6th century, in fact, when the Romans were on the offensive in Mesopotamia, the population of Persian cities appealed to common Christianity in order to stave off Roman armies.[73] Furthermore, internal divisions within Christianity had little effect on the course of warfare. East Roman armies were composed of Monophysites, Chalcedonians, and Arians, but antagonism between these groups virtually never arises in siege narratives. In fact, apart from the Arian non-incident at *Carthage (533), I have only found evidence of "infighting" (which was limited to a bitter exchange of words) between Monophysite Armenians and Monophysite Syrians in the Roman army during the Arab invasions;[74] otherwise, adherence to a particular Christian rite or theological position seems not to have been an issue in warfare; indeed, the most fierce opposition to Constantinople's religious policies was to be found in staunchly Chalcedonian Italy and North Africa.[75] More serious, but often ignored, is that the Romans, even during the midst of Arab invasions, engaged in civil strife (especially 641-44) which severely weakened resistance in Egypt, as support for the two royal factions divided Egyptian society just as any other province.[76]

Before the first civil war (656-61), most of the Arab conquering elites and soldiers were settled in the garrison cities and ruled through Roman or Persian administrators. Civil wars were mostly settled in the field, since there was little attachment to classical urbanism. This gradually changed

---

[73] E.g. the population in Arzanene during the siege of *Chlomaron (578).

[74] For this incident, see Dionysius 80 (Palmer 1993: 164); Brock 1982 shows that many Nestorians were just as bitter on the Monophysites as the latter were on both Chalcedonians and Nestorians, but from the actual narrative evidence this never seems to have been an issue for soldiers in the field or populations under siege. Indeed, the two protagonists in Dionysius recognize each other as Christians serving under a Christian king without further ado.

[75] The interpretation of "oppressed churches" liberated from Chalcedonian persecution is a construction of later polemical tracts which found their way into the standard chronicles of the time (cf. chapter 7.3.2). To a large extent these were produced to make sense of the bitterness and incrimination resulting from Persian appointments and Roman reappointments of bishops (and secular officers, although this is conveniently "forgotten" by ecclesiastical historians) during their last war. Another motive was to explain the humiliation of Arab rule over Christians. See Dionysius 21 for the Persian expulsion of Chalcedonian bishops after their conquest of Mesopotamia; ibid. 40 for Herakleios' restoration of the same after the Persians had been defeated. For how authors made sense of their apparent humiliation under Arab rule, see Whittow 2010: 93. This thread is taken up again in chapter 7.3.2 with some examples of secular protagonists who probably shaped events more than the theologically minded clergy.

[76] The dichotomy between "Greek" and "Copt" had nothing to do with this and has indeed been repeatedly refuted; see most recently Sijpesteijn 2007: 442.

under Umayyad rule, when Arabs settled especially in the Syrian cities and civil wars began to turn on fortified cities in the 680s and 690s (e.g. *Damascus 690 and *Mecca 692). A number of Syrian cities were besieged and many had their walls razed around 745 (e.g. *Emesa). This not only brought siege warfare to the heart of early Islamic politics, but also resulted in Christian subjects being directly involved in urban defense and having to choose sides in Arab civil wars, a choice that could have serious repercussions and led to streams of refugees to Byzantium, especially in the early 9th century (see also ch. 7.3).

6.2.2 Societies at War: Garrisons and Civilians

Ideally, any fortification should have a garrison of professional soldiers who did the heavy fighting. This was unquestionably the case in most of the Roman Empire. As argued in chapter 1.2.1, most of the regular West Roman army survived well into the mid-5th century (e.g. *Aquileia 452), while in some locations or regions recognizably Roman formations still existed into the early 6th century. In the East, regular army units continued to fulfill garrison duties throughout this period. Professional garrisons were often very small, numbering in the low hundreds; the Roman garrison at *Tella (639) was only 300 strong, while 300 men of a somewhat larger force escaped from *Tendunias (641). During the 6th century, even major cities such as Constantinople, Alexandria, Antioch and Jerusalem appear to have had only small garrisons, if any at all. These normal, peace-time postings of the army were bolstered in times of war, but supporting troops had to be brought in from stations in other cities or even other provinces.[77] Still, reinforced garrisons in major cities or strategically important border fortifications were rarely above a few thousand men.

The second largest defending force on record during the 6th century was 6,000 men at *Antioch in 540, but this number was only achieved by stripping the Lebanese provinces of their regular troops, whose homes were undefended and the troops therefore became concerned about the safety of their families and property. This may have led to the breakdown of their morale at a critical juncture, as the professional reinforcements escaped, leaving the fighting to the militia. *Rome (537f) was initially defended by the field army under Belisarius, numbering only 5,000 men. Ad-

---

[77] When the Romans prepared to defend *Salona 537 against an Ostrogothic counterattack, they had to denude the garrisons of surrounding fortifications in order to hold the city.

ditional reinforcements would have brought the total up to 6,600 (but we have no knowledge of casualties, so it could have been several hundred less), while further reinforcements (around 3,000 men) allowed Belisarius to send out 2,000 cavalry to raid the communications of the Goths and capture fortifications. At the most Rome could thus have had around 7,000 (at shorter intervals even 10,000) professional troops, in addition to sailors and camp followers.[78] In 549-50, however, only 3,000 men were available to defend *Rome. The Roman garrison at *Rimini (II, 538) was 1,200 cavalry and 400 infantry; while useful for strategic purposes, the large cavalry component was difficult to supply once the Ostrogoths began their siege.[79] In 626, *Constantinople had a very strong garrison of 12,000 men cavalry, but this was fairly little for manning the 6 km of double walls on the land side, not to mention the sea walls that needed at least some defenders. 7,000 men reportedly defended *Caesarea (640), probably a special case since it was the last major port in Palestine and would be full of troops who had fled from elsewhere and/or reinforcements sent to keep the city in Roman hands.

The Persians appear mostly to have relied on professional soldiers, since most of the population in the Caucasus and Mesopotamia were Christians who were deemed unreliable. The heavy presence of professional garrison troops to compensate for this appears to have made Roman counter-attacks more difficult; often they faced strong Persian garrisons of 1,000-3,000 of highly motivated and disciplined troops when besieging the most important strongholds in the Caucasus and Mesopotamia.[80] We know next to nothing about Arab garrisons, although 5,000 Arabs were reportedly massacred at *Amorion when the city was recaptured by the Romans in 666, and they certainly had enough troops to match or even outnumber Persian and Roman garrisons if necessary.

---

[78] The garrison at Rome was only 5,000 men—large for the period, but not large for the size of the city and scope of the threat; see Proc. 5.22.17. It was reinforced by 1,600 Hun, Slav and Antae cavalry in the spring of 538. This allowed Belisarius to engage in spectacular running battles; see Proc. 5.27.1-5. After the arrival of further reinforcements and supplies, Belisarius sent out troops to take fortresses in Latium; see Proc. 6.7.3-34. The 2,000 freely moving cavalry able to take several fortresses, even *Rimini—led to Goths giving up Rome, and they were defeated while withdrawing; Proc. 6.10.

[79] A similar problem arose at *Beroea 540.

[80] Some of the examples are discussed below; e.g. *Amida 503, 504f, held by 3,000 men; *Petra 549, held by 1,500 men, replaced by 3,000, but only 2,300 available for defense in 550; some were probably assigned to hold lesser fortifications and other necessary tasks, such as scouting, foraging, or escort duty.

Ostrogothic garrison forces in time of war numbered from 400 to 4,000, depending on the size, strategic importance, and loyalty of the city.[81] 800 men were taken captive at *Naples; before losses, the garrison may have been around 1,000. Indeed, the Romans had 1,000 troops to hold the city against the Ostrogoths again in 542-43. The Ostrogoths garrisoned *Portus (537) with 1,000 men when they had captured it from the Romans, who had been unable to station a garrison there. Similarly, 1,000 men held *Orvieto (538f). Lombard garrisons were on a similar order of magnitude, even into the 8th century, when *Cumae (717) was held by 800 men.[82]

At some point between c. 450 and 550 privately recruited troops, i.e. magnates' military followings, began to comprise the bulk of professional soldiers in the Frankish and Visigothic kingdoms. Smaller fortifications and border forts were guarded exclusively by professional soldiers, often settled permanently in the region they were defending. Garrisons were normally composed of the military followings of local magnates and dignitaries. At *Chastel-Marlhac (524), 50 men were captured during a sortie but they were ransomed by the population. The loss of these men was clearly not enough to affect the outcome even of a moderate fortification. When Gregory of Tours was falling into royal disfavor, the local count mobilized 300 men only to guard the gates and ensure that Gregory could not flee (see 4.1.5); the number of professional troops could probably be doubled or perhaps even tripled in time of war by Gregory's *familia* and the retinues of other magnates. At *Orleans (604), the mayor of the palace was surprised by his enemies with a following of 300 men while on routine peacetime business. This means that larger forces were available in times of war; individual magnate followings could number up to 500 men, and a city was normally held by up to several such magnates.[83] As in the East, troops could be transferred from other regions in time of crisis; thus after the fall of *Cabaret and *Beaucaire (585), 4,000 men were sent to hold border forts against the Visigoths.

Recent calculations of the garrison forces on frontiers where the size, date and use of fortifications are well attested provide very large aggregate numbers; thus Curta estimates that the 6th-century Romans garrisons in the Iron Gates sector alone of the Danube frontier numbered 5,000 men, while a list of Balkan forts with known dimensions generate a total of at least 32,390 men. Christie similarly argued that the Roman garrisons in

---

[81] See the discussion at *Rimini II 538.
[82] *LP* 91.7 (Gregory II, 715-31).
[83] See chapter 4.1.5 above for the retinue size of Merovingian magnates.

Ligura in the 7th century numbered up to 13,000 men. Both may perhaps operate with garrison complements on the high side: Christie assumes 100 men per *castellum* and 500-1000 men in major centers, while Curta assumes that tiny fortlets had about twenty men, each small to moderate fort would have a *tagma* of 100-500 men, while 2,000-4,000 would serve in large fortified sites. However, apart from the largest garrisons, which were probably rare, these estimates are quite consistent with the numbers uncovered in narrative sources, so this means that early medieval landscape was densely populated with professional garrison troops.[84]

Such professional troops did heavy fighting, especially at the gates and during sorties. Regular military units were stationed unit by unit along the walls. We have a good enumeration at *Phasis (556), where the walls were held in sections by the different client units (Moors, Tzani, Lombards and Heruls), regular Anatolian divisions (elsewhere called Isaurians), other eastern regiments, and the private retinues of several commanders. Civilian militias performed support roles and were assigned specific sectors of their own or were assigned among the professional troops, such as at *Rome (537f). The practice is also specifically recommended by the *Strategikon*.[85] Like military units, civilians were normally divided into corporate groups with responsibility for their stretch of *pedatura*. Recognized groups include circus factions, monks and Jews.[86] For instance, when Maurice mobilized to defend *Constantinople in 602, he had 1,500 Greens and 900 Blues enrolled. They had also been mobilized in 598 when the Avars threatened the *Long Walls. Otherwise neighborhoods and guilds seem to have performed the same functions alongside rural refugees. Magnate (and generals') followings seem also to have assigned particular sectors. At *Thessalonica (586), the defense normally involved the military servants of wealthy men and the military following of officials. Other sources also enumerate the various groups that participated in the defense of the city. Antioch was defended by the "demes" and youths—i.e., neighborhoods and circus factions, or circus factions and private retainers, depending on

---

[84] For the numbers see Curta 2001: 181ff; Christie 2006: 372. For example, Curta generates his estimates based on surface area available to the garrison held against standard legionary camps and modern military usage of space, but one of his prime examples, Nicopolis ad Istrum, was probably built as a refuge with much open space for the surrounding populations, while many soldiers would have had families living with them.

[85] *Strategikon* 10.3.32ff; the quote is discussed in chapter 2.4.2.

[86] E.g. Jews hold particular stretches of wall at *Constantina 502f, *Arles 508, and *Naples 536; Jewish defenders witness miracle at *Thessalonica 615; monks held a stretch at *Amida in 502f and Mardin, for which see 7.3.2.

the difficult interpretation of the meaning of "deme" and "youth."[87] At *Alexandria (608), according to John of Nikiu, there were "army regulars, barbarians, citizens of Alexandria, the Green Faction, sailors, archers." Many of the same groups participated in the fighting in 609, while "youths" provoked the Persian siege of *Jerusalem (614).

Civilians would perform basic guard-duties when the threat is less imminent, such as patrolling designated stretches of wall, guarding gates and manning towers within their area of responsibility. If they were deemed less than reliable, they shared such duties with professional troops, but in many instances seem to have carried much responsibility on their own. In time of peace or during lulls they would spend as much time as home as possible, but were mobilized for wallfighting whenever necessary. This was obviously the case when threats first arose (e.g. *Thessalonica 604) or during storms, but they also had to be put on alert when professional troops made sorties. While such details are reported only in the best attested sieges, we have evidence for active civilian participation in many other instances.[88] Civilian men were expected to throw stones with sling-staffs, slings or their hands, push away siege ladders, and hurl objects onto enemy rams and penthouses. For example, shepherds and farmers at *Edessa (544) fought bravely on the walls and even drove back Hun sheep rustlers during a sortie. An attentive farmer on guard duty in a tower discovered a Persian sneak attack at night and raised the alarm. During the same siege, some Edessenes helped Roman troops drive back a Persian force that had breached the wall. At *Alexandria (608, 609), civilian participation in the fighting was extensive.

Women, children and the elderly could undertake various logistical tasks, such as carrying missiles and stones, boiling oil, and so on (*Amida 502, *Edessa 544). At *Topeiros (549), the civilian population had to defend the walls after the garrison was lured out and ambushed by a Slav raiding party. Belisarius improvised but probably not without precedent, enrolling the civilians on the army paylist and assigning them to battalions alongside professional soldiers during the lengthy first siege of Rome. At *Tibur (544), the population shared guard duties with Isaurian troops (until they fell out with each other). The distinction between professional and civilian defenders was also clear in the West. The royal garrisons were distinct and had to be expelled when a city changed sides, as at *Poitiers (584). These

---

[87] For the demes, see Al. Cameron 1976 and Astachova 1995.
[88] *Ancona 538, the civilians hold the wall while professional troops go on sortie; cooperation between *Loukanoi* and Franks at *Lucca 553.

were clearly distinct from the "mob" that was assembled to defend the city in 567, while at *Arles (567/9), Gregory distinguishes between the bishop and *urbani* on the one side, and the garrison (*exercitus*) on the other. A large number of *rustici* were killed by Leovigild when he had taken *Cordoba and other cities (572). This can surely be attributed to their active resistance.

As we saw in chapter 6.2.1, civilian-garrison relations were complex, and either group (or factions within a group) could betray another. Sometimes this was not done out of malice or desire for gain, but for simple survival. Hence the Roman garrison installed at *Milan (538) made a deal that sealed the fate of the population, but conversely, the population at *Ashparin (503) was terrified by the Persians into handing over Roman soldiers and their commander who had just taken refuge there. In most cases, however, there was strong solidarity amongst the besieged. The population of *Chalkis (540) refused to hand over their garrison and commander to the Persians. During occupation, a newly installed garrison would have to find some way of collaborating with the locals if they had not been massacred or deported. The worst-case scenario occurred at *Amida (504f), where the Persian occupying garrison had tied up many of the males in the city and left them to starve in the amphitheater. At first they had taken the women as their concubines and provided them with food, but when rations became low, they simply abandoned the women to starvation as well and stayed on the walls day and night.

Estimating besieging forces is much more difficult and controversial during this period, since it necessarily has to confront the "how big were medieval armies?" question. Some scholars have suggested a ratio of 1:4 between besieged and besieger if the latter wanted to successfully take a city by storm, and furthermore postulated a ratio of one man per 1-1.3 meters of wall.[89] Using the first ratio on the information given above, it should be a simple task to estimate average besieging armies, while remaining walls can be used to calculate the necessary garrison size. However, there seems to have been much variation in practice from such ideal figures. One methodological problem is identifying the number and function of civilian defenders, a notoriously difficult task. Another difficult issue is the density of defenders on the walls. While the basic ratio is a useful guide, it would vary with the strength of the wall, the natural accessibility of particular stretches, equipment and skill, and the numbers on opposing sides. For

---

[89] See Bachrach 1990 for a series of figures applied to Anglo-Saxon Burghs.

instance, when a besieger had superior forces, their archery and artillery volleys could simply overwhelm the defenders, who would then have to be concentrated in greater density in response. There are also cases where, even during a storm, long stretches of wall were only lightly defended, while others were fought over with great intensity. Finally, it also neglects the fundamental issue of labor. Many besieging methods were labor intensive. While soldiers most certainly participated in much digging, hauling and building, the requirements were often so great that additional labor was recruited. This obviously complicates the issue.

On other occasions, we have reliable numbers that simply do not conform to the basic ratio. The Romans managed to storm *Petra (550) with 6,000 men when the Persians had a garrison of 2,300, while the Persians failed to take *Sura (540) with 6,000 men, although the garrison only consisted of 200 professionals alongside an unknown (but not particularly large) civilian population.[90] Besieging armies numbered from one or two thousand to 10,000 or more, but many such armies were active at the time. Small armies could take a number of lesser fortifications. A Slav raiding party of 3,000 men, divided in groups of 1,800 and 1,200 men were able to take a number of Balkan fortifications in 549, including *Topeiros.

However, for sieges of large cities, and campaigns that involved several cities and other fortifications, much more substantial forces were needed. The largest army ever assembled in the 6th century counted 52,000 Roman soldiers gathered in Mesopotamia in 503, but not all were located in the same place, and the army required special consignments of grain from Egypt and the imposition of bread-baking duties (i.e. *munera*) on the urban population in the region. When engaged in the war, it operated in more manageable units. One such unit of 12,000 men is mentioned as besieging *Nisibis, while a substantially larger force invaded deeper into Persian territory. Another large force is the army that invaded Persian territory in 543 (*Anglon), numbering 30,000 men. Single field armies in Italy rarely exceeded 20,000 men in one location, perhaps apart from the Frankish invasions in 539 and 553-54. Even such large forces normally broke up to pursue various objectives. The Romans took *Verona (541) with a detachment from a larger army numbering 12,000. This was a case of treason, so an overwhelming force was not necessary. The rebels marching on *Carthage (536) numbered 8,000 men and were supported by 1,000 Vandals and some slaves. At about the same time, somewhere in excess of 9,000 Romans (plus

---

[90] At *Onoguris (555), I estimate around 1-2,000 Persian defenders and 5,000 Roman besiegers; a Persian relief force numbered 3,000 men.

the crew of the fleet) took *Palermo (535), and a similar number was involved in the siege of *Naples (536). The Ostrogoths probably had 25-30,000 men at *Rome (537f). At *Ancona (538), a force of 4,000 men supported by another army besieged the city. The evidence from Visigothic Spain suggests the same order of magnitude, although there are few points of reference in administrative or narrative sources outside the *History of Wamba*. For the Frankish realm, I argued in chapter 4.1.5, based on various narrative and administrative records, that the global military potential was very large, and that individual field armies may have reached the 20-30,000 mark, even on Italian expeditions, and were certainly large enough to conduct full-scale sieges. In most cases, the actual besieging army would be smaller, as in the Roman and Italian case, since a siege required many supporting functions. Finally, truly gargantuan forces were amassed on occasion. The Avars assembled something like 80,000 men at *Constantinople (626), according to Howard-Johnston, while Syriac sources claim that the Arab forces involved in the siege of *Constantinople in 717-18 numbered 200,000 men, which might be correct if all naval crews from Egypt and Syria through the Aegean are included alongside the besieging army, laborers and engineers, and additional Arab field forces in Anatolia.

While small cities with locally recruited garrisons were common, the presence of allied forces, reinforcements or integration into a new political structure could complicate the composition of defenders. *Vienne (500) was defended by Burgundian and allied Frankish troops. *Conza (554) was the last stronghold of the Ostrogoths in southern Italy, but may also have included Franks and Huns. Similarly, most large armies were multi-ethnic conglomerates, drawing on many sources of manpower. At *Sisauranon (541), before and during the siege, the Roman force included Ostrogoths (who drove the Persians away with an effective cavalry charge), *hypaspistai* (personal retainers of Belisarius), and Arabs. Persian armies could be just as diverse; at *Archaeopolis (550), the Persians had 4,000 Sabir Huns and Daylami infantry in addition to their own substantial cavalry (20,000 horses starved during the campaign). Avar armies are discussed in chapter 7.2.3 and *Thessalonica 618, *Constantinople 626.

6.2.3 Specialists at War

There appear to be woefully few descriptions of civilian specialists in West involved in siege warfare. In narrative sources, an *artifex* is attested at *Vienne (500), and a *faber* at *Convenae (585). However, this has more to do with constraints of genre and the exclusion of low-status individuals from narrative history in general, and lack of interest in military events in particular (0.3.6). As the preceding chapters show, the institutional, socio-economic and cultural context was similar or at least very comparable to that of the East Roman empire and can be explained in some detail based on a common Late Roman background. Civilians were expected to participate in defense (1.2.5), and are fairly well attested also in the West in similar capacities to equivalent groups in the East (6.2.2). From the narrative evidence examined in chapter 5, it is clear that technical expertise was very common during sieges, although the few explicit mentions of engineers, *tekhnitai* (τεχνῖται), hardly do justice to the complex infrastructure that lay behind East Roman siege warfare, whether the expertise was military or civilian (cf. chapter 2.2.2). Evidence from the eastern frontier as well as the Balkans does however demonstrate that many of the experts recruited were also specialists in civilian construction. *Strategikon* explicitly mentions several types of craftsmen who were or could be drawn from civilians, an image confirmed by narrative evidence and further strengthened by the provisions in Justinian's novels. From the late 6th century, civilian expertise was also mobilized to use on offensive campaigns. Civilian engineers, craftsmen and specialists, then, were systematically mobilized for warfare alongside military expertise (2.4). It is therefore imperative to understand the role and distribution of craftsmen and engineers in East Roman society in order to explain how the logistics of siege warfare were organized in the East.[91]

The Persians, like the Romans, had specialist military engineers of their own, but they also relied on the same sources of civilian labor. For instance, the trebuchets and towers built by the Roman engineers (*myqnyqw* (pl.), i.e. μηχανικοί) at *Nisibis (573), were taken over by the Persians, who had the required skill (*ūmānwān*) to handle the Roman engines (*ūmānwātā*) and use them against *Dara. Like the Romans ever since the Anastasian war, they also recruited masons (*pāsūlē*) who diverted an aqueduct and must have had considerable labor for a host of other tasks, such as towers and siegeworks.[92] Even the Slavs at *Thessalonica (662) employed woodcutters, carpenters, ironworkers, missile and armor makers, and specialist engineers (μαγγανάριοι) who could build complex siege engines, in a man-

---

[91] In addition to chapters 1 and 2 above, see Ruggini 1971, von Petrikovitz 1981 and Cuomo 2007 for late Roman and early Byzantine craftsmanship; Ousterhout 1999 traces further developments into the middle Byzantine period.

[92] For *tekhnitai*, see *Naples 536, *Rome 537 and 546f (where their absence is noted), *Edessa 544 (Persians), *Petra 550 (both Persians and Romans); other expertise in the Roman army, *Osimo 539; among the Persians, *Phasis 556 (build bridges and barriers).

ner that resembles East Roman military recruitment of specialist craftsmen (2.2 and 2.4 *passim*). Most sources simply do not divulge this kind of information, but it has been reconstructed in chapters 1-4 above, where it was shown that an important explanation lies in the involvement of civilian craftsmen in urban defense alongside engineers from the regular army, a model which persisted in the East Roman Empire and was adopted *in toto* by the early Islamic conquerors. This class of specialists provided for the needs of the conquering armies: conscripted civilian craftsmen in Egypt and Syria contributed in every conceivable manner: they built fleets, repaired fortifications or built them up from the ground, and provided engineers for siege campaigns (7.3).

Therefore, it is important to look for the same categories of craftsmen in the West, because we have established that civilians participated in most aspects of siege warfare there too (6.2.2). Furthermore, wherever attested, military skill in the West was drawn exclusively from civilian craftsmen in magnate retinues or in surviving cities. On the basis of common Roman origins, subsequent East Roman analogies, and the evidence assembled in chapters 3 and 4, it is clear that most craftsmen and hence their military capacities fell under magnate control by the 6th century, and were distributed on both estates and in surviving urban centers. While sources are sparse, such craftsmen have been identified in written sources and archaeology for most of the early medieval West.[93] In quite a few instances it is possible to demonstrate how bishops built, supplied and defended fortifications. While secular magnates are more elusive, their economic power base and control over resources was similar to that of bishops, and can be effectively traced from Charlemagne onwards.[94]

In the West we do however have some very good examples from before 600 that demonstrate continuity from the Roman empire and parallels with the contemporary East: Ostrogothic and to a lesser extent early Lombard Italy organized public works in Late Roman fashion (3.1.4; 3.3.3). In Gaul, Genovefa as well as Caesarius of Arles organized large-scale building

---

[93] While Marazzi 1998: 148 is critical of post-Roman craftsmen skills, emphasizing decline and their simple organization, Henning 2007 shows that there is very good archaeological evidence for the survival and widespread distribution of advanced craftsmanship in Gaul and beyond from Roman into Merovingian and Carolingian times; for the written evidence see the articles collected in *Handwerk*, especially Claude 1981 and Nehlsen 1981, who show that a very large proportion of specialist labor and craftsmanship was unfree and thus at the beck and call of their magnate patrons.

[94] See the discussion in chapter 4.2.2 on the bishops and 4.3 *passim* for evidence relating to secular magnates.

activities using the same people, infrastructure and authority that formed the basis of late Roman *munera* (cf. 1.2.5); Gregory of Tours refers to new churches and construction in his miracle stories (4.1 *passim*). Landowners as a class were also associated with maintenance of city walls, in Italy and Gothic as well as Frankish Gaul. Nicetius of Trier had his ballista and recruited specialist craftsmen from Italy. We know from laws and other sources that landowners controlled specialist labor, or could afford to hire it, as they had their own military following, it is not unreasonable to assume *a priori* that civilian dependents were mobilized in support functions.

In the 7th century, evidence seems very thin in West indeed, but with the knowledge of previous period and contemporary East, we can establish a framework of skills to look for. Byzantines and Lombards mirrored each other closely in Italy. The papacy maintained organizational skills and social control typical of the Late Roman Empire; indications are that Lombards did the same (3.3.3). It is abundantly attested in Gaul under Desiderius of Cahors, who organized several building projects: churches, aqueducts, and city walls. He also had specialist expertise (crane operators) and advanced technology: altogether, it is very clear that Frankish bishops (and other magnates) possessed the technology, manpower, and organizational skills to defend and attack walled cities (4.2.2-3). The building capabilities of Visigothic bishops present a similar reality; archaeological evidence (Reccopolis) and narrative sources (especially Wamba's campaign in 673) confirm that this image is applicable to secular magnates as well.[95]

The situation in the West becomes much clearer thereafter. Late Visigothic laws required magnates (intermittently bishops) to mobilize their retinues including engineering specialists, who were evidently available to Wamba; by the reign of Charlemagne, we have detailed lists of tools and engineering equipment (and people able to use them; 4.3.4). The obvious link between the two periods can be found in the campaigns of Charles Martel and Pippin (4.3.3), which demonstrate the capacity to use advanced technology (rams, trebuchets, field fortifications) and fight Arabs on equal terms. A final hint as to early Carolingian military success can be found in Pippin's last military venture, when he established a base full of "equipment" for further expansion into Aquitaine. Charlemagne possessed a formidable heritage that he implemented to the full. His reign demonstrates

---

[95] *LJ* 5.1.5; see further chapter 3.2 for discussion of examples and context.

a phenomenal ability to organize labor for both civil engineering (e.g., *fossa Carolina*), fortifications, and siege warfare, all of which were supported by an estate economy and craftsmen called out by their lords according to need. Carolingian realities reflect solutions envisioned by the Theodosian Code, or attested in 6th-8th century Byzantium, where urban and rural craftsmen were regularly called out to supplement the army for engineering needs.

Beyond this, the navy provided a valuable supplement to urban defense as well as offensive siege warfare in the East Roman Empire. Sailors are seen partaking in offensive and defensive siege warfare.[96] In the West, the Ostrogoths (and Vandals) also developed significant naval capacities that were often used in conjunction with sieges. Even the Visigoths took over some East Roman naval infrastructure around 600, and in 673 had a fleet that could assist in Wamba's campaign. Since sailors played a key role in many sieges and were trained to operate rigging, sails, cranes and so on, we know that wherever large fleets were stationed and sailors recruited, many of the skills relevant to siege warfare would obtain. In addition we can add the skills obviously relating to repair, maintenance and outfitting of ships associated with the major ports and naval facilities.

6.3 Ending the Siege

If the Romans had emerged victorious at the siege of Dura-Europos with which we began chapter 1, the evidence of destruction and siegeworks would probably not have survived until today. However, the remains vividly demonstrate how much labor it would have taken for the survivors to clear the debris and what engineering skills were necessary. Extensive repairs would have to be made, first and foremost reconstructing towers and large sections of wall damaged by undermining, rams and artillery. The partly undermined sections would have to be carefully demolished and rebuilt if the ground was (or could be made) stable; if not, a new course for the wall would have to be found. The citizens and garrison would also have to be sure that all tunnels had been identified so no weak spots could dam-

---

[96] E.g. at *Panormus 535, *Rome 537, *Phasis 556, *Alexandria 608, *Thessalonica 618, and *Constantinople 626. See Pryor and Jeffries 2006 for a philological-technical history of the Byzantine navy. Despite its bulk, it has extremely little to say on the infrastructure (ports, dockyards, arsenals, recruitment etc.) behind the Byzantine fleet and virtually nothing to say about the people who built the ships. McCormick 2001 has a bit more on the infrastructure of waterborne traffic during late antiquity.

age the walls or be exploited by besiegers on later occasions. Such repairs took the collective labor of civil and military engineers, artillery experts, skilled military and civilian craftsmen, and an enormous amount of unskilled labor. In addition, enemy siegeworks had to be demolished. Removing the tons of earth and wood in a Persian mound, Gothic siege camps, or Frankish earthworks was another labor-intensive task that had to be performed. Furthermore, the defenders often made modifications to defenses, extending moats, digging ditches, countermines, and countermounds, or demolishing structures near the walls. Often these modifications only had ad-hoc functions and disrupted normal civilian life, so they had to be removed when the fighting subsided.

6.3.1 *Consequences of Survival*

There were naturally victory celebrations, award ceremonies, thanksgiving services, and funerals after surviving a siege. At *Marseilles (413), the population rejoiced at the return of Boniface when he ended the siege. The 200 Roman fallen at *Phasis (556) were given an honorable burial. When Theuderic's army successfully relieved *Orleans (604), he followed up with a triumph at Paris. Triumphs remained common; a triumph took place at *Beritzia (772) after a Roman victory. A thanksgiving service was held in the church of St. Demetrius after the Slavs abandoned *Thessalonica (615), and at the church of the Virgin at Blachernae after the Avars abandoned *Constantinople (626).[97]

The end of a siege was the time to settle the scores. At *Constantina (502f), a pogrom began against the Jews, but was stopped by the bishop and the commander. The offending army might have the tables turned on them. The Persians routed at *Phasis (556) lost 10,000 men. The Tzani who assaulted the Roman camp at *Rhizaion lost 2,000 to 40 Roman dead. The Franks who had besieged *Saragossa (541) were themselves caught and massacred in a Pyrenean pass. On a few occasions, captives were executed; thus the women of *Thessalonica (615) led a Slav chieftain into the city and stoned him to death.

Even when victorious, there was often some cost to the besieged. Starvation and ensuing epidemics were frequent when a blockade or siege stretched out over weeks and months. The Roman garrison at *Otranto

---

[97] This is quite well studied; see McCormick 1996 and D. Bachrach 2003 for victory celebrations; further examples may be found at *Edessa 502f and the siege of Amida in 359, with discussion and references in chapter 1.1.1.

(544) was emaciated after a long siege and had to be cared for back at Salona. This also happened to *Thessalonica on several occasions, but most severely in 662. The *city of the Slav chieftain Dragawit (784) had to give hostages to the Carolingian army. The citizens of *Harran (502f) escaped a siege, but had lost many inhabitants to the Persian raiders and even had to pay a ransom of 1,500 goats in order to placate them. The citizens of *Beroe (587) had to pay a small ransom to get the Avars to leave, but at *Singidunum (588), the Romans had to pay a more substantial sum (2,000 gold darics) and other valuables after a seven-day siege. When the citizens of *Thessalonica refused to pay the besiegers to leave in 618, the Avars renewed hostilities, destroying the countryside and surrounding shrines, and threatened more raids in the future. In the end the population decided to come to terms, and as part of the settlement, the besiegers came up to the walls to sell captives and booty back to the Thessalonians. It seems that *Germanikeia did not fall when besieged in 778, but suffered large losses through bribes, booty, and captives. The economic consequences of a siege were indeed wide-ranging; at *Kamakhon (766), the large injection of coin into the local economy as the Arabs paid the local population for logistical support spurred a sudden increase in forgery.

When possible, a failed besieger tried to destroy most of his infrastructure so that it would not fall into enemy hands. Thus the Avars had a rearguard set fire to their camp, engines and the suburbs. However, at *Nisibis (573), the equipment left behind by the Romans was in good order. The Slavs who fled *Thessalonica (615) abandoned machines and booty they had taken from the surrounding countryside. The Ostrogothic camp at *Rimini (II, 538) was full of equipment that had to be disposed of, as well as sick Gothic soldiers that were unable to flee and had to be cared for. The Persian reinforcements at *Petra (549) had to rebuild a large section of wall with sandbags as well as care for 350 wounded men and 1,000 dead. The burial of the dead was more often a concern if a besieging army had been heavily defeated. The numerous Visigothic losses at *Clermont (474) had to be buried, while a very large number of Slavs had to be picked out of the water at *Constantinople (626) before their bodies could be disposed of. Little is said in other instances, but the large number of fallen reported in various conflicts must have taken much time and effort to bury or burn.

Due to the complexity of siege campaigns, even a raised siege would lead to new fighting. There might be losses despite victory. The new Roman garrison at *Otranto (544) began raiding out from the city, but was ambushed and lost 170 men. The extremely high Persian losses at *Petra (549) have been noted (above and chapter 5.2.1). In other instances, a spectacular victory led to instant turns of fortune. The Avar failure at *Constantinople (626) led to such a loss of prestige that they hardly figure on the political map again until their core territories were absorbed by Charlemagne. The Arabs had overrun all of Anatolia in 654, but when their whole fleet was destroyed before *Constantinople, the Romans regained all of Anatolia and began counter-offensives in Armenia. The consequences may have been dire enough to provoke the first Arab civil war from 656-661, which started with discontent among Arab troops in Egypt. At *Kamakhon (766), the consequences were not as dramatic, but the Abbasid army experienced an immense humiliation, as they lost large values, a very large number of troops, and much prestige, which is not reflected in the standard Islamic histories.

6.3.2 *Consequences of Fall*

In most cases, a storm ended in a combination of atrocities. Thus the population of *Amida (502f) was subject to massacre, executions, ritual humiliation, rape, captivity, and occupation, during which the new "administration" treated the population abysmally. Cities were rarely deliberately destroyed, although their walls might be razed (6.1.1). However, burning may have had some specific function: *Toulouse (507) was *incensa* and plundered. *Sebasteia and *Melitene burnt in 576, but they may have been abandoned. Carolingian warfare was often a mix of terror and politics; hence *Avignon (736) was stormed and burnt while the population was massacred; in Aquitaine, the same policies were pursued against recalcitrant magnates who soon lost the support of urban populations (see further 4.3.3).

The most common fate was captivity, since it served as an incentive to the conqueror and terrified civilian populations. The population of *Vollore (524) was taken captive after the city was taken by storm. The families of Ostrogothic troops were enslaved at the *Alpine forts (539); presumably they were released when their men surrendered shortly after. The population of *Beaucaire (587) was taken into captivity by the Visigoths. The early Carolingians also took captives, e.g. at *Loches (742). Some were taken captive by ruse; thus the Persians at *Sura (540) feigned negotiations over ransom, but used the celebration and embassies to gain access to the city and take most of the population captive. They were ransomed by the bishop of *Sergiopolis, who promised to pay for them in the future, but he was himself taken for failing to live up to his agreement and tortured by the Persians when they returned in 542. The Persians negotiated ransom for prisoners to gain time to bring them over to their own territory before the Romans could do anything. This was the fate of the population of the campaign in 540, when the inhabitants of *Edessa offered a large ransom. The objective of Persian wars often seems to have been to take captives; thus even poor farmers were taken at *Kallinikos (542).

That rape was accepted as the conqueror's prerogative may be inferred from Totila's ban of this customary right (see below). However, in most cases rape is euphemistically framed as the "distribution" of captives, while conquering garrisons formed liaisons with local women with varying degrees of willingness. The Persian garrison at *Amida (see 502f, 503, 504f) kept the women in the city as concubines in return for food, but the alternative was cannibalism or starving to death. After *Milan (538), the men were massacred, while the women were distributed to the Burgundians who had provided manpower to the Ostrogoths. The Avars massacred the men at *Cividale (610) but divided the women and children amongst themselves. While the distribution of women to the conqueror was fairly common, there are actually few explicit mentions of rape before the 640s (see below). Whether this change is only an accident in sources or a change in practice is uncertain.

Captives were often transported over great distances. The Persians brought large numbers of iron and wooden fetters to bind their captives, and presumably they were transported this way back to Iraq. This was the fate of the population of *Antioch (540), who had a new city built for them by Khusro. *Apamea and its surroundings yielded 273,000 captives in 573. The Persians again deported the population of *Emesa (610/11). In both cases, the population had already surrendered, so the Persians broke trust. They seem to have been motivated by economic and military needs that required large population transfers back to Persian territory.[98] The *Franks who invaded Italy in 590 brought the captives back to Gaul, but many of them were ransomed afterwards, and the Franks seem to have used this leverage to begin reestablishing their hegemony over the north of Italy. The Arabs routinely took large numbers of captives, both during the early conquests and on subsequent raids. Only to provide a handful of examples, the populations of *Turanda (712), *Mistheia (712), *Galatian forts (714), *Sardis (716), *Palozonium (740), and *Laodicea Combusta (769) were led into

---

[98] On this, see chapter 7.1.2. For the economic and military functions of such population transfers, see e.g. Liebeschuetz 2003, who argues that control over populations equaled political control.

captivity. The population of *Tyana (708f) was deported to Syria and driven into the desert—whether this means that they were killed by exposure is uncertain. However, the sight must have been disturbing, and deeply affected the conquered populations in Syria and Egypt. After the capture of *Amasia (712), a Syrian chronicle records that "an endless train of booty and slaves" headed for Syria, while the Coptic pope in Alexandria, who had to help administering taxes and labor that were directed at further Arab raids and conquest, spent great sums on ransoming Christians who had been enslaved by the Arabs and brought to Egypt.[99] On other occasions, populations were deported and resettled without being completely deprived of their freedom. In 769, the Arabs removed *Germanikeia (i.e., its population) to Palestine. The Romans in the 8th century did the same; the populations of *Germanikeia (745), *Melitene and *Theodosiopolis (750) were deported to Thrace in order to provide defenders and settlers for newly founded cities.

For professional troops, a surrender was sometimes an opportunity to find new employment. The Ostrogoths at *Petra Pertusa, *Todi, *Chiusi and *Urbino (538) were recruited into the Roman army. Some were immediately sent to the East or Sicily and Naples, while others seem to have stayed with the new Roman garrison at Petra. At *Osimo (539), they got to keep half their property and were enrolled into the army; the remaining property went to the Roman soldiers. Right to the end of the war Ostrogothic soldiers went over to the Roman side *en masse* when they had had enough; thus Aligern at *Cumae (552f) decided to change sides when the Franks invaded, while the last Ostrogothic fort in southern Italy, *Conza (554f), surrendered after a long siege and its 7,000-strong garrison was integrated into the Roman army. Roman soldiers occasionally changed sides; thus most of the garrison at *Beroea (540) went over to the Persians after a negotiated surrender, while after a series of failures around *Florence (542), Roman prisoners joined the Goths. Of 400 men at *Rossano (548), 320 went over to the Goths. The Visigothic king Sisebut ransomed Roman captives after taking several *Roman cities in Hispania (614f), apparently in order to employ them in his own forces. Similarly, the Persian garrison at *Sisauranon (541) went over to the Roman side and was immediately dispatched to Italy. Evacuation was also an option to get rid of enemies who refused surrender. The Lombards allowed Roman garrisons to escape back to safe territories, e.g. from *Padua (601) and *Mantua (603).

---

[99] Trombley 2004.

ANATOMY OF A SIEGE: ECONOMY, SOCIETY & CULTURE 353

A successful assault often ended in massacres. At *Amida (502f), this was not only in the frenzy of the storm, but also survivors were taken out and executed by various means *after* the fighting was over. This also happened after the Persians took *Nisibis (573), since the Persians wanted to punish the Romans for not surrendering when they had the chance. The Persian garrison occupying *Theodosiopolis (502f) was massacred when the Romans took the city. At *Ashparin (503), the Roman garrison was massacred but the population was spared. The Romans who stormed *Naples (536) began massacring the population and take slaves in revenge for deaths among their comrades and relatives, but this was stopped by Belisarius, since the city was to be reintegrated into the empire. However, the population themselves took vengeance on two notables who had advised resistance. One was impaled, the other died (by suicide?) before they got to him. Women and children were usually spared for captivity. At least Agathias was indignant at Roman atrocities at *Tzacher/Sideroun (557), where Roman soldiers killed many women and children early on in the conflict.

The actions of a new occupation power were often determined by political objectives. A successful besieger might distinguish between garrisons and civilian populations, either dependent on contingent factors (e.g. betrayal), or on long-term political goals. The Slavs who massacred the civilian men after *Topeiros (549) did so deliberately with horrifying means. The intention may have been to terrify other garrisons and populations, and ensure compliance among the women and children they had taken captive and intended to bring all the way back across the Dabube. New garrisons committed atrocities at *Amida (503, 504f). So did the Frankish garrison installed in *Clermont after Theuderic's *Auvergne campaign in 524. This might be deliberate policy to break the will of the population (cf. chapter 6.1.3). Otherwise, terror was used against political enemies. Thus Justinian II hanged representatives of the former regime when he regained power at *Constantinople in 705, and when *Emesa fell in 745 during the third Arab civil war, the walls were razed, values were confiscated, and political enemies were massacred and the corpses of dead enemies ritually humiliated.

Kindness and a good political settlement were more common when the occupation was expected to last long and loyalties could be swayed by less brutal means. The Avars also sometimes pursued a policy of kindness. When the Romans were finally forced to evacuate *Sirmium (579ff), the population was emaciated. The Avars tried to ensure that they were fed, but many gorged themselves and died.[100] Belisarius famously failed to reach a lasting settlement at *Ravenna (539f), which led to new bouts of war. Most of the cities in Italy ultimately surrendered peacefully, but were reintegrated into the empire only after years of hard fighting (*Florence and following entries for 553). The Ostrogoths pursued the same policy, since their aim was to rule Italy, not gain booty. Thus the population of *Naples (542f) was given extremely lenient terms. Not only were they gradually and carefully fed to avoid gorging themselves to death, but Totila banned rape and even executed one of his own men for raping a Roman girl. Furthermore, the Roman garrison was given free passage and fraternized with the Ostrogoths (presumably including a large number of former Roman soldiers) before they went to Rome. Totila followed the same policy at *Rome (545f), and despite some vacillation, finally decided to rebuild and resettle *Rome after his final capture of the city (549f).

Persian policies after 610 were similar. They tried to ensure local support and negotiated surrenders in return for tribute and submission with local dignitaries, and worked seriously to establish a lasting political structure in the Roman provinces they conquered. Thus the leading men at *Theodosiopolis (610/11) came out to present themselves to the Persian general Koream outside the city before opening the gates; *Damascus surrendered in peace in return for tribute. Originally, the Persians organized the peaceful submission of *Jerusalem (614) where the population was left alone and local officials continued to work. Unfortunately, this also meant that local factions were allowed to continue scheming. Some youths began a revolt and persecuted the Jews. It was this that led to the Persian siege and subsequent massacre, in which 17,000 were left dead and 35,000 were taken captive according to Sebeos, but other sources claim as many as 90,000 dead. Persian massacres and taking of captives are also reported at *Caesarea in Cappadocia (611), *Alexandria and the East (619), and *Ancyra (622).

The Arab conquests were quite similar to the Persian wars. As we have seen above, the reaction of the Roman population to the conquerors varied, but seems in fact to have been *less* accommodating to the Arabs than to the Persians. The Arabs therefore had to use the full range of possibilities. When besieging and capturing cities, they vacillated between gener-

---

[100] A similar fate befell the Persian garrison at *Akbas 583, who gorged themselves on water after they surrendered, causing the death of many.

ANATOMY OF A SIEGE: ECONOMY, SOCIETY & CULTURE 355

ous terms (what later became the *dhimmah*)[101] and atrocities, which included rape,[102] captivity,[103] and massacres.[104] They could also make other deals depending on political and military progress.[105] Sometimes pacts were given *after* hard fighting, as in Palestine (Appendix I), *Lapethus (650) and *Arwad (650), and often these include provisions for the population to go to Roman territory. The later Arab sources distinguish between *Rūm* (i.e. ethnically Greek representatives of the Roman state as it existed in the 8th and 9th centuries and hence some sort of alien presence) and the local population. However, this is a later construct in both Islamic and Christian sources to explain and justify the conquests. Indeed, contemporary sources do not operate in similar terms (cf. chapters 6.2.1 and 7.3.2). Instead, the Christian Syriac sources show populations making agreements with the Arabs in the same terms (the most commonly used term is *melltā*, lit. "word" but here meaning "promise" or "agreement") as they had previously made with the Persians; the same terms are even used by Persian garrisons when they surrendered to the Romans.[106]

---

[101] E.g. *Palmyra 634, *Tiberias 635.
[102] *Constantia on Cyprus 649, *Euchaïta 644.
[103] *Antioch 638/39, *Dvin 640, *Aegean Islands 653.
[104] *Bahnasā 640/41, *Kīlūnās 642, *Khram 643.
[105] The garrison at *Babylon 641f was allowed to escape massacre if it handed over military stores to the Arabs.
[106] Thus the Romans surrendered *Emesa 610/11 to the Persians who gave their *melltā*; the inhabitants of *Damascus surrendered, giving their *melltā* to hand over tribute; the Persian garrison at *Edessa (630) gave in and accepted the Roman offer of *melltā* after being bombarded with trebuchets; the patriarch Sophronius of *Jerusalem (634f) received a *melltā* and oaths (*mawmātā*) on behalf of all Palestine when 'Umar came to the city, probably a year or two years after the siege; only tribute had been conceded in the meantime. At e.g. *Palmyra 634 Baladhuri uses the Arabic *amān*; for further examples of and comments on the Arabic vocabulary, see Hill 1971. I would argue that Hoyland 2011: 65 n. 84 oversimplifies in postulating that Theophanes' terms *polemō/logō* reflects the Islamic terms *sulḥ^{an}/'anwat^{an}*, transmitted through the Syriac intermediary *b-ḥarbā/b-melltā*, since "the late Roman terms in Greek were *kata kratos/homologiā*." However, Conrad 1990, who first suggested the connection, only identified four instances of the use of the terms after the Islamic conquests: *polemō* (twice) and *logō* (twice), although Agapius uses only *amān* in both instances where Theophanes uses *logō*. Hoyland additionally makes the point concerning events in the pre-Islamic period, where however Agapius has no term (*Ancyra 622) or again uses *amān* (*Edessa 630). The opposition *b-ḥarbā/b-melltā* more likely goes back to (near-)contemporary Syriac notices, similar to those preserved in e.g. *Chr. 640*. The apparent shift in terminology has more to do with shift in genre from classicizing or ecclesiastic history to chronicle (see ch. 0.3) rather than influence from the emerging Islamic historiographical tradition. Furthermore, siege/surrender terminology in Greek was far more variable among classicizing historians than this simple dichotomy allows, as a quick perusal of 6th-century entries in CO will demonstrate; for instance, Procopius uses the term *es logous* in addition to a flurry of other terms.

Brogiolo observed that cities captured in war could experience three levels of destruction: material (i.e. the physical destruction, e.g. burning and razing of walls and buildings), demographic (Brogiolo emphasizes massacres, but captivity would have a similar effect), and institutional. The latter could be achieved in a number of means: often cities were deprived of their administrative privileges, such as control of their *territorium*; to this can be added the installation of garrisons, deprivation of trading rights, demotion as an administrative center, removal of élites, and a number of other sanctions that could be instituted by a conqueror.[107] Only the first of these options is possible to see in the archaeology with any precision, but that also requires the complete abandonment of the site without attempts to clear the rubble and rebuild. However, burning was an extremely rare event, and hardly any of the cities examined here were completely abandoned after a siege. Italian urbanism experienced a crisis in the mid-to-late 6th century, but here it was the result of a combination of factors: warfare drawn out over generations, plague, and not least, socio-economic change. Foss provided a strong argument that several Anatolian cities went in to rapid decline as a direct result of Persian expeditions in the 610s and 620s, but in light of new evidence from a number of sites as well as weighty methodological considerations, Niewöhner has recently shown that the process was far more drawn out and may perhaps be more profitably compared with the Italian situation. Otherwise, only cities in the Balkans under Hunnic attack in the 5th century show a radical decline or shift in settlement pattern that can be interpreted as the result of the Hun policy of population transfer.[108] While we may suspect that the decline of cities was spurred by specific events, then, it would be methodologically unsound to assume that a conquest would have to be comparatively peaceful even if the archaeological record shows little clear evidence of destruction at a specific time period, especially as the most effective forms of siege warfare affected the countryside or infrastructure that normally was repaired.

---

[107] Brogiolo 1999: 111f.
[108] Italy: Christie 2006; Anatolia: Foss 1975, *contra* Niewöhner 2007; Balkans: *TLAD*. Visible destruction: only Dura Europos with which we began chapter 1.1 well before our timeframe, in 256, and Amorion, captured by the 'Abbasids in 838, are known to me outside the Balkans (see articles in *TLAD* for examples and discussion of urban change or abandonment); see the literature cited in 6.1.2 for individual cities; for the assumption that conquest must be "seen" in the archaeological record, see Schick 1994 on Palestine and the contrast presented in Appendix I and II.

6.4 Conclusion: Deconstructing, or Reconstructing, Thin Sources

Despite the complaints of poor sources, it is possible to demonstrate a fairly coherent image of social, political and military problems faced during a siege. Again, the weight is of course on the 6th century East, but enough remains from other places and periods to recognize that we are still operating in the same world. Perhaps some parts are a little less wealthy and literate, but they are still conversant with each other. We may understand the source problem better if we were to put ourselves in a Western medieval annalist's slippers. Provided he understood Greek, and having in front of him Procopius' ample description of the siege of *Rome (537f), he had to figure out a way to sum up the action in the one line for the entry of that year. The siege included several rounds of wallfighting, the use of machines and archery on both sides, fighting over nearby fortifications such as *Portus, *Ostia, *Centumcellae, 69 battles outside the walls, and a host of other military events. In addition, there was the expulsion of the pope, starvation among the population, extensive negotiations between the parties, and long periods of truce. On top of this, the siege ended with a *battle* where the Ostrogoths were heavily defeated. Indeed, Jordanes provides us with the key, as it were, in his two summaries of the siege. In the *Getica* 312 (138.4-14), the context is clearer:

> "... where Magnus the count with a small army was evicted and completely destroyed. When Vittiges heard this, like a furious lion he assembled the whole Gothic army, and having marched out of Ravenna, he tired the defenses of Rome with a long siege. However, his boldness was frustrated, and after fourteen months he fled from the siege of Rome and prepared himself to subdue Rimini." || ubi dum Magnum comitem cum parvo exercitu ipsi evulsi et omnino extincti sunt. quod audiens Vitiges ut leo furibundus omnem Gothorum exercitum congregat Ravennaque egressus Romanas arces obsidione longa fatigat. sed frustrata eius audacia post quattuordecim menses ab obsidione Romanae urbis aufugit et se ad Ariminensem oppressionem praeparat. ||

In the *Romana* 374 (49.14-17), there is slightly more technical detail:

> "However, in the following the same Vittigis encircled the fortifications at Rome and assembled his engines and towers, with which he tried to enter the city, [but as they were] consumed by fire he failed, laboring for nothing for a whole year." || secundo vero ipso Vitigis Romanas arces vallante congreditur machinasque illius et turres, quibus urbem adire temptabat, igne consumptis per anni spatium quamvis inaedia laborans deludit. || [109]

---

[109] My own guess, made independently of Jordanes, is something like this: "In this year, the Emperor's men came to Rome and expelled the Pope. The Goths encamped against it

Jordanes' style is representative of most medieval chronicles. There are only hints at the technology used in *Romana* (*machinasque ... turres*), very little indication of the logistical problems (*vallante* may represent the vast siege camps housing something like 25,000 Ostrogothic soldiers), and nothing of the extensive fighting on and outside the walls nor anything to indicate the political problems faced by both sides as well as the extensive fighting ranging from Portus to Picenum. The only elements that receive some attention are the resources mobilized (*omnem Gothorum exercitum congregat; congreditur machinasque ... turres*) and the siege as a year-long failure (*frustrata eius audacia post quattuordecim menses ab obsidione aufugit; igne consumptis per annis spatium quamvis inaedia laborans deludit*). Such phrases are normally treated as clichés, but in light of the information from Procopius we can be reasonably certain that most of the resources of the Ostrogoths were mobilized and that the siege was very arduous and did last over a year. In addition, the two versions represent the siege quite differently in terms of vocabulary. While *Getica* uses unambiguous siege terminology twice (*obsidione*), both *Getica* and *Romana* also employ diffuse circumlocutions that tend to be ignored by scholars who use other chronicles to reconstruct military history (*oppressionem; vallante; urbem adire temptabat*). Jordanes' summary is fair when we know what happened from Procopius, but if he had been the sole surviving source, his testimony would surely have been used to prove the futility of siege warfare to achieve military objectives during this period.

This is far from the only case where one type of source seriously misrepresents events: we have seen in the introduction how Gregory failed to mention *Arles (508), and only gives away the actual breach of the wall at *Clermont (524) in one of his hagiographical writings; otherwise *Clermont would be only known to us as a "stylised" city takeover typical of Merovingian politics.[110] Even the prime source of siege details in this period, Procopius, treats the Hunnic siege of *Aquileia (452) as an entertaining marvel story, but here Jordanes preserves invaluable details from Priscus' original account. The tendency to abbreviate beyond recognition

---

for a year, but achieved nothing, and were defeated in a battle at the Milvian Bridge."|| Hoc anno, homines Imperatoris venerunt usque ad Romam et Papam expulerunt. Gothi vero per unum annum contra eam castra metati sunt, sed nihil potuerunt, et expugnati sunt in proelio iuxta Pontem Milvium. ||

[110] For *Arles 508, see CO and chapter 0.2.1; for *Clermont, see CO and chapter 4.1.2; Halsall characterizes Merovingian siege warfare as "stylized" even in relation to the besieging skills of other societies, whose capabilities he describes in rather pejorative terms (see conclusion at chapter 5.4 for references).

is rife throughout the period under examination. The *Annales Regni Francorum*, the official annals of the Carolingian court, do not match up to the Continuation of Fredegar during the Carolingian conquest of Aquitaine (4.3.3), while Theophanes and Tabari, the bedrock of Byzantine-Islamic history, scarcely provide a hint at the fierce contest and complex events at *Kamakhon in 766. We must therefore be wary of taking one-line annal entries as the sole basis for reconstructing late antique and early medieval siege warfare.

Incidentally, it should be noted that Jordanes wrote at Constantinople in an environment conducive to full-fledged classicizing historiography with detailed siege descriptions and where there were enough generals, bureaucrats and soldiers who could inform him of current realities, but his choice of genre left virtually no room for this. Even if most other early medieval authors were more poorly informed, they were just as deliberate in their choices.

CHAPTER SEVEN

APPROPRIATION OF MILITARY INFRASTRUCTURE AND
KNOWLEDGE

Transmission of technological knowledge and military practices was relatively easily between the successor states since they ultimately derived from the late Roman state. However, it is also clear from many of the examples cited that several political, tribal or ethnic groups external to this late Roman tradition achieved many if not most of the same capabilities, especially from the later 6th through 7th centuries. This applies particularly to the Slavs in the Balkans, the Avars, and the Arabs. While the Slavs and Avars only seem to have possessed these capabilities for little more than two or three generations, the Arabs quickly developed a fearsome war machine unparalleled in military history. Considering that all these groups can be termed conquest societies, took control of large swaths of Roman or formerly Roman territory, and can also, with some justification, be termed "tribal" groups originally having a fairly simple socio-economic organization, without many of the institutional bases described above, we must explain the discrepancy.

A major problem in diffusion studies has been the fact that little has been done to study the context of a particular transfer: a particular invention or skill is divorced from its environment and traced without examining its institutional framework. From chapter one and two it is abundantly clear that there was a common foundation for use and maintenance of knowledge and technology throughout Mediterranean area in late antiquity, and that there was a surprising degree of homogeneity in practice as a result. Competition, contact and conflict led to a rapid dissemination of new knowledge. As we have seen in chapter two, the trebuchet was always part of a larger ensemble of poliorcetic knowledge and practice. Thus, in order to understand the diffusion of e.g. the trebuchet to groups outside the late antique system (ch. 8), we must first trace the creation of a context in which it could be appropriated and then used. Only then can we plausibly reconstruct the most likely routes of dissemination.

7.1 The Hunnic, Persian and Visigothic Templates

The concept of co-evolution means that polities or societies in (military) contact over longer periods of time would by necessity adapt to each other. However, the pace and efficiency of this process was tempered by the institutional framework that was to adopt innovations in practice or technology. Thus clients of the Roman Empire effectively adopted individual equipment and combat tactics. The former could be adopted through simple means by small-scale craftsmen, while the latter required training, cooperation or conflict typical for client troops. More complex capabilities, such as mass production of diverse weapons for extended campaigns, siege and logistics trains, and technological support for engines and engineers required an institutional framework that had to be evolved over centuries or acquired through conquest.

7.1.1 Client Integration and State Formation: The Visigoths

Client kingdoms outside the Roman Empire, although dwarfed by the resources and power of Rome, to a certain degree underwent *co-evolution* (see 0.2.2), as continuous cross-border diplomacy, military service, and conflict shaped these polities from rather crude and primitive tribal structures to small, but more economically and socially diversified polities (1.3.1). Gothic political structures by the 4th century is under debate; while some argue for fairly large political structures dominated by an ethnically and socially homogenous class of free men, others believe that Gothic political structures were somewhat less developed and coherent. While I am skeptical towards the rather maximalist position of Heather, and believe that the Goths were not politically or ethnically coherent, there certainly existed a stratified society with a military/political upper class some way towards state formation and thus capable of adapting to Roman impulses.[1]

In such a context, there are several ways in which technological and cultural transfer occurred: Roman captives (e.g., Ulfila) brought (Arian) Christianity, language, ideology, and the low-level technology that was necessary for production of luxury goods and military items on a Roman model. Service in Roman armies is poorly attested but for the Goths began in the 3rd century, when they sent troops to serve against the emerging

---

[1] The debate over Gothic ethnic coherence and early state formation is dominated by Heather, on the one hand, and Halsall, Kulikowski and Lenski on the other. See bibliography under these authors for major contributions and cross-references below to relevant sections for more detailed references.

Sassanids. Military service is better attested from the 4th century, and was mostly required for the Persian frontier. In addition, there were individual recruits about which we know little but who may have participated in continuous personal transfers for generations. Overall, familiarity with Roman fighting styles, technology and organization evolved over a century and a half. Thus by the late 4th century, Gothic client troops could be expected to contribute substantially to the Roman field armies operating in the East. Usurpers (Procopius) and legitimate emperors (Valens) clearly thought they would be useful as federate or permanent garrison troops and military recruits settled on Roman soil. The re-evaluation of their poliorcetic skills shows why the empire thought so. Rather than dismiss Gothic skills out of hand, we must keep in mind that if the Goths were to fail against a Roman city while fighting without any logistical support, it would certainly have to be against Adrianople, a center of arms and artillery production and a station for elite field troops (1.1.1-2).

During their sanctioned migration to the Roman Empire, "Goths" included Germanic-speakers as well descendants of Roman provincials, but they were also joined by Alans and Huns. After the revolt, this motley group was again joined by Roman subalterns: slaves, provincials and some deserters. Dispersed in the Balkans 382 partly on their own accord and partly as a result of imperial pressure, many were recruited into regular army. Some of the "Gothic" elements revolted intermittently after 395 but it has proven very difficult to pinpoint either continuity from the participants at Adrianople or the ethnic composition of those in revolt in 395.[2] The group certainly contained a strong Gothic element attested in the survival of an East Germanic language, and were joined by other Germanic-speaking client or federate groups in 406. However, throughout this period, those who became Visigoths were already heavily mixed with Roman provincials and other barbarian groups, served in army, were supplied by factories, and received their ethnic name from a Roman unit designation. Other, similar groups not in revolt had very diverging fates at the center of Roman politics or simply fade away, presumably due to complete assimilation. In appearance, behavior, fighting style and ethnic composition, these "Vesi" appear indistinguishable from Roman units, and should in effect be considered Romans by the time quite different political developments caused them to look more barbarian or "Germanic" (see 1.2.1-2).

From their settlement in Gaul in 418 to the end of the Western empire, the Visigoths tried to support whomever they thought were (or should be)

---

[2] Lenski 2002 and Halsall 2007 for this emphasis.

legitimate emperors. A separate kingdom is only attested securely in 439. It was a recognition of failure for both imperial and Gothic policies, according to Halsall, mostly a result of the Vandal takeover of Africa. Furthermore, Halsall emphasizes that their political objectives, though at times indicative of independent policy vis-à-vis central authorities, almost always aligned with the interests of Gallic aristocrats, who themselves were often at odds with Roman authorities in Italy (see 1.3.2). A third stage of technological integration, probably accelerating around 439-40, was the large-scale, imperially legitimized usurpation of military and civic administration by this precise class of landowning aristocrats, who piecemeal, willingly or by coercion, brought their capabilities over to the Visigoths. The subsequent incarnations of a truly independent Visigothic kingdom, first in Toulouse, then in Toledo, had a military organization largely created by Roman authorities in an effort to compensate for shortfalls in revenue around 440 (1.2.5). However, since these measures were administered by the same aristocratic group of landowners who just as often were in opposition to imperial policies, the reforms had two unintended results: they provided the means for independent kingdoms to emerge without necessarily being supported by the central imperial bureaucracy, while simultaneously ensuring that late Roman military organization survived, in a modified form, well into the middle ages.

Similar trajectories may be reconstructed for most barbarian groups who formed their own kingdoms: a period of client assimilation through co-evolution outside of the empire, resulting in troops and political leaders who could easily be integrated inside imperial structures as military units, military officers, or even aristocrats. At the same time, centrifugal tendencies due to internal Roman dissention caused many Romans or Romanized barbarians to be defined outside Roman identity. Once securely established between c. 440 and 480, the emerging successor kingdoms reversed the process so that Romans were assimilated into barbarian identities, but the erstwhile Romans' knowledge and military organization remained. The changes that brought about "barbarian" or "medieval" military organization was a Roman creation around 440, *after* most barbarians had already been Romanized, were integrated into, and had become dependent upon Roman military institutions.

7.1.2 *Inter-state Transfers: The Sassanids*

The Sassanids, like former Persian dynasties, were the greatest house among a highly militarized aristocracy, and their empire was in effect a coalition of semi-independent Iranian regional rulers under overall Sassanid suzerenity. The aristocracy had their own private forces, fought as heavy cavalry with retinues composed of landowning gentry (*dehqans*), and were supported by peasant infantry (*paygān*) that proved very versatile in both battle and siege warfare. For instance, the *paygān* were those doing all the heavy labor at *Phasis (556). Most of the explicit evidence on them comes from Ammianus, but their functions remained essentially the same through the 6th century: to fight as archers and light infantry, storm walls, build siegeworks, mounds and engines, undermine enemy fortifications, and do all the supporting logistical labor such as collecting materials, building bridges, and clearing roads.[3] Subject or client peoples provided additional manpower as light cavalry (Kurds, Huns, Arabs) or heavy infantry (Sabir Huns, Daylami), but especially Caucasian peoples, like the Armenians and Caucasian Albanians, had a military aristocracy that seems to have mirrored that of the Iranian nobility. Procopius mentions Persian engineers (τεχνῖται) on two occasions, once at *Edessa (544) constructing a siege mound, and then at *Archaeopolis (550), referring obliquely to Persian engineers that would normally construct siege engines. Otherwise we know very little of Persian military engineers. Evidence from the Arab conquests indicates that some aristocrats had particular expertise in siege warfare and probably had staffs of craftsmen and engineers within their households. The competence of Persian engineers is beyond doubt in light of the enormous amount of evidence on engines and siege methods assembled in chapter 5.2, where Roman-Persian warfare provides much of the evidence. In addition, they could draw on the technical expertise from the same sources as the Romans: firstly the civilian population in Mesopotamia (cf. 6.2.3), and secondly, the militarized Armenian nobility.

In order to strengthen their position, the early Sassanid rulers took measures to establish military officers and standing forces under their control. Pursuing a policy first instituted in the 3rd century, the Sassanids continued to systematically deport Roman populations throughout the 6th and early 7th centuries. This was probably not a prerequisite to advanced technological knowledge, as it is sometimes claimed. The evidence from Dura-Europos (see 1.1) shows that the earliest Sassanids had complex besieging capabilities *before* they systematically began to deport Roman captives. It was rather a policy of diversifying and strengthening the Persian economy in general while weakening the Romans, thus also fulfilling a symbolic

---

[3] *EncIr*, online edition: "Army, i, Pre-Islamic Iran, section 5. *The Sasanian period.*"

function of establishing superiority. Furthermore, Pourshariati argues that the Sassanian shahs also needed to strengthen their hands vis-à-vis the great regional sovereigns in Iran, some of whom had imperial ancestors and harbored commensurate ambitions. This was accomplished by concentrating deportees in non-Iranian Mesopotamia, which already provided the shah with the bulk of his revenue.[4] Nevertheless, transfers of technology and organizational skills did certainly result from such deportations, which had two major advantages. Firstly, deported Roman populations were settled in colonies that were modeled on Roman cities and preserved, to some extent, familiar social structures and institutions, so that the deportees could perpetuate their way of life and their knowledge over time.[5] Secondly, the Persian Empire was already a sophisticated, urbanized society with advanced administration and an immense capacity for organizing labor and supplies in support of their military ambitions, as well as large-scale civil engineering.[6] New knowledge traveled quickly within the empire, and once established, it could be preserved by new centers outside the original colonies. Thus, while it is difficult to determine whether their poliorcetic skills were survivals from the ancient empires of Mesopotamia, influenced by Central Asian neighbors, or appropriated from the Romans by population transfers, there is little doubt that once acquired, the Persians had no problem maintaining and redeveloping them over the following centuries.

7.1.3 *Conquest Appropriation: The Huns under Attila*

As we have seen, the Huns had a formidable poliorcetic ability after 442. If we base our argument on the fragmentary sources analyzed in chapter 1.1.3, the conclusion must be that the Huns learnt their besieging skills from the Romans, as most other Roman clients did. However, little used Syriac sources show a Hunnic incursion against Mesopotamia in 395 that took the fort of Ziatha by siege.[7] Furthermore, Howard-Johnston has revived the old argument that the Huns were related with the the Xiung-nu, steppe enemies of the Chinese in the 1st century BC, and by the second half of the 4th century, they are attested north of the Caucasus.[8] They may therefore

---

[4] Pourshariati 2008.
[5] See the articles on the Parthian and Sassanian periods in *CHI* 3(2), Rubin 1986 and Lieu 1986.
[6] Howard-Johnston 1995; Rubin 1995.
[7] Greatrex and Greatrex 1999.
[8] Howard-Johnston 2007 argues convincingly that the Huns recently arrived from Central Asia. Scholars have been skeptical of this due to the vast time span and distances

have been familiar with poliorcetics and advanced technology from China and central Eurasia over several centuries. By all accounts, in the age of Attila they had certainly been in contact with the Persians and Central Asian polities skilled at siege warfare for over a century.[9]

While I have taken issue with some of Thompson's interpretations on Hunnic and Germanic warfare (e.g. *Naissus 442), he is fundamentally right in that both groups were disadvantaged in their lack of infrastructure. My main contention is that client service fundamentally conditioned "Germans" for Roman-style warfare (see 1.3 and 7.1.1). The Visigoths had not managed to gain control of the Roman logistical system nor enough experts in the short period at their disposal when they failed to take Adrianople, but were certainly not as incompetent as most modern scholars claim. Furthermore, when the Visigoths were integrated into the Roman military infrastructure, they very rapidly became fully proficient in siege warfare. The Huns clearly grasped the value of poliorcetic knowledge and may have had far older traditions for it in light of Howard-Johnston's argument. However, Ziatha was taken by a large-scale raiding party who probably did not use siege engines, so Hunnic invaders had the same problem as the Visigoths, namely implementing siege warfare on such a large scale as they did in 441-42 without support from a complex infrastructure. This is particularly pressing because the Huns employed the standard Roman siege engines of the day, such as battering rams, siege towers, and artillery to take a very large number of cities.

After their arrival in Europe, Huns regularly served in Roman armies for nearly two generations in all types of warfare, including sieges.[10] One of these occasions had occurred as recently in 439, when Huns participated in the (possible) Roman siege of the Goths at *Toulouse under command of Litorius. They could thus have received extensive training through service with the Roman army and seen the value of advanced technology in practice; or at least adapted their practices to current Roman methods. The Huns acquired the necessary expertise to implement siege warfare them-

---

involved and also what we do have preserved in the sources. Militating against this, known Chinese innovations, such as the traction trebuchet and the stirrup, make their way westwards a century later through different routes. For the context see chapter 7.2., for the trebuchet, chapter 8.

[9] See Needham et al. 1994 on Chinese poliorcetics. While Thompson has been extremely skeptical towards Hunnic poliorcetic skills due to their "stone-age" technology, as he put it, Heather in the afterword to the second edition of Thompson 1948/1996 is more accepting. This is discussed further by Heather 2005: 146-49.

[10] For a survey of Hun auxiliaries and client troops, see Maenchen-Helfen 1973: 255-58.

selves directly from the Romans. In 441, *Viminacium and other forts were taken by surprise during market season, while a large proportion of the garrison forces in the area were away on the expedition against the Vandals. The combination of extensive client service, familiarity with Roman strategic dispositions, and surprise attacks on forts that were open to trade allowed them to take a very large number of Roman captives or deserters from the militarized borderlands of the Balkans and Pannonia. Many of these fortifications had arsenals and trained engineers,[11] while their civilian populations practiced the normal range of trades and crafts. While on an embassy to Attila's court, Priscus met a Greek-speaking Roman merchant who appeared to be completely integrated into Hun society by the looks of his haircut, dress, and general appearance; indeed, archaeologists would have identified him as a Hun. Priscus recounts their debate over the pros and cons of civilized life versus the freedom of the steppes. The merchant had been taken captive at *Viminacium (441), but won freedom through bravery on the battlefield, and explicitly stated that the Huns *did* force Romans to perform dangerous military service for them.[12] Siege warfare must certainly qualify. One issue that is often overlooked in this context is that Attila was appointed as *magister militum*, i.e. general of Roman troops, by the Western Empire. This possibly gave him legitimacy in the eyes of Roman subjects and captives, and there is in fact evidence that he was obeyed by Romans.[13]

The most important Central Asian heritage of the Huns was the sophisticated political culture that allowed them to exploit the talents of their multi-ethnic empire. Skilled manpower for storming cities was not a problem, since the Huns had themselves served as clients and later subdued a large proportion of the Roman client system.[14] Thus, most of their troops were Goths and other barbarian groups who had also seen service in the Roman Empire. In addition to prisoners and clients, Roman or Romanized officers were often in Hun service. Aëtius spent significant amounts of time

---

[11] Maenchen-Helfen 1973: 136, 201 simply takes the presence of Roman military engineers for granted, as opposed to Thompson 1948, who spent much energy in disproving the evidence for Hunnic siege warfare. In light of the normal Roman infrastructure described in chapter 1.1.1, the Huns must have captured quite a few specialists during their campaign in 441.

[12] Priscus fr. 11.2.495-98.

[13] Priscus fr. 6.

[14] The takeover was hardly voluntary, and many clients preferred to remain in Roman service; cf. Priscus fr. 2, where client groups flee to Roman territory. On the Rhine, which was on the fringes of Hunnic control, Attila and Aëtius competed for control, supporting separate contenders to succeed as king of a Frankish client polity (Priscus fr. 20.3).

at the Hun court, relied heavily on Hun auxiliaries, and had close personal relations with Hun rulers. Due to his use of Huns in the fierce struggles during the 420s and 430s for power inside the empire, he would have had strong incentives to train them in necessary skills, including siege warfare, but if so, his policy evidently backfired with the rise of Attila. There were several other high-ranking Roman bureaucrats and advisors at Attila's court who could provide intelligence, advice and logistical organization for complex warfare, e.g. Orestes, later *magister militum* in Italy and father of the last Roman emperor in the West.[15]

Furthermore, other peculiarities show Hunnic dependence on Roman engineering. The bridge built at *Naissus (442) must have been constructed by Roman engineers; constructing bridges to carry heavy siege towers is indeed a rare nomadic skill. Attila's second-in-command Onegesius had a captured Roman bath master build a fully functioning Roman bath for him. Unfortunately, the poor bath master did his job too well; instead of being released, as he had hoped, he was forced to remain in attendance to keep the bath operating.[16] These problems demonstrate that the Huns faced a fundamental dilemma. Since they conquered little territory and reportedly destroyed cities and deported their populations, the knowledge they appropriated depended directly upon the people that possessed it. While the Huns could import and exploit the necessary labor and materials to build complex structures that were the hallmark of classical civilization, they had not figured out a way to maintain them over time. This can be extended to military knowledge as well. While they achieved some success by rooting up populations, transferring them far from home, and treating many of them poorly as slaves and subject peoples,[17] such populations had few incentives (beyond threats) to participate in the Hunnic imperial venture. The fate of such captives was a recurring and extremely contentious problem in Hun-Roman negotiations. The Huns adamantly demanded the return of all those who fled to Roman territory, threatening war if they were not forthcoming.[18] While some were "political" refugees, others were important to the Hun economy and war effort. Indeed, the Hunnic

---

[15] *PLRE* 2 s.v. Orestes; see also the discussion in MacGeorge 2002: 276.

[16] Priscus, fr. 11.2.364-72.

[17] Priscus, fr. 11.2. 415ff describes most Roman captives as having tattered clothes and filthy hair, living in effect as beggars on the fringes of the Hunnic camp.

[18] Thompson 1948: 196; see further e.g. Priscus fr. 2 on a Hunnic demand of back-pay of ransom for captives who had already fled; ibid. 9.3 for negotiations over refugees—the Huns demand that the Romans at Asemus hand over any refugees in their city, but the Asemuntians simply lie and hide them away.

empire itself was institutionally frail, dissolving upon the death of Attila. Subsequently, most of his achievements disappeared with it—even some of his heirs sought refuge inside the East Roman Empire. Anyone else with significant skills returned to the Roman Empire (e.g. Orestes) or found employment in any number of petty successor steppe states, whose resources and range were far inferior to the great Hun Empire.

7.2 The Balkans, 530-825: From Client Assimilation to Conquest Appropriation and Back

While some of the client kingdoms on the northern border of the East Roman Empire sometimes appear to be more aggressive and independent than in the 4th century, the Romans were able to manage them until the Avars began causing trouble around 580; even after the collapse of the Danube frontier and the establishment of a Bulgarian khaganate, however, both Slavs and Bulgars still sought trade, recognition and service from the Byzantines. Indeed, according to recent research, the Slavs took form as an ethnic group not in the Pripet marshes, but "in the shadows of Justinian's forts," or even of the walls of Thessalonica. This environment and some particularly good contemporary sources allow for a detailed examination of processes of assimilation, technological transfer, and the problem of institutional frameworks among states that did not preserve as much Roman infrastructure as the Western successors.

7.2.1 Huns as Clients: Utigurs, Kotrigurs, Sabirs and Bulgars in the 6th Century

From the lower Danube to the Caucasus, a series of Hunnic tribes established themselves as successors to the great Hunnic Empire during the last half of the 5th century. Sometimes they can be distinguished as Utigur, Kotrigur and Sabir Huns, with the former two based north of the lower Danube and Black Sea and the latter north of the Caucasus. All groups interacted closely with Romans, while the Sabirs also had close relations with the Persians. The difference between these groups, other Huns, and the earliest Bulgars is unclear (perhaps also to themselves), but after c. 550 the Utigur, Kotrigur and Sabir Huns seem quite distinct in Greek sources. The Bulgars crop up before 500 and appear intermittently in the 6th century and more commonly in the 7th century after the collapse of the Avar khaganate. "Bulgar" was Turkic for "disturber," an apt name for an artificial conglomeration of soldiers that tried to establish itself among other tribes or groups as military contractors.[19]

The Utigurs and some Bulgar groups were (at least intermittently) Roman clients. Offshoots were settled on Roman territory; for example, surviving Kotrigurs humbly asked for permission to settle after Justinian had stirred up the Utigurs against them in 559. Those that lived across the Danube varied between raiding and providing federate or mercenary troops, as Roman clients had done all along the frontier for centuries. Due to their connections across the river, raids could exploit good knowledge of Roman military practices and dispositions, since they were timed with Roman expeditions that drew many soldiers from the Balkans to Italy or the eastern frontier. The objective was not to conquer territory, but force concessions out of Roman authorities, much in the same way as 4th-century clients tried to do.[20] An exceptionally successful Hunnic raid against *Illyria in 539 resulted in the capture of 32 lesser forts and some larger fortifications. Their success depended on a solid knowledge of Roman military dispositions (much of the Thracian field army was away in Italy) and practices. Procopius states that they had not engaged in wallfighting before, but Hunnic raiding divisions were able to bypass the fortifications at the pass of *Thermopylae and the walls of the *Chersonese, engaging in diversionary assaults on the walls, and take *Kassandria by storm.[21] Unless it was a holdover from the glory days of the Hunnic Empire, this newfound experience was probably acquired in Roman service and perhaps cooperation with some of the other client groups, such as the Gepids, who held some Roman infrastructure, and Heruls, who were used extensively by the Romans.

Certainly in the 550s, Huns used the common siege methods of the day. In the Caucasus and Mesopotamia, Huns were regularly used through most of the 6th century, especially by the Persians. This gave them extensive expertise in siege warfare. At *Petra (550), the Sabir Huns in Roman employment even invented a light-weight ram's shed that could be carried

---

[19] The mixed or "artificial" character of these groups is clear from the careers of Mundo, the mercenary of indeterminate ethnicity who ended up in Roman service, and Vitalian (see chapter 2.1.2 for the Balkans military culture in which he operated).

[20] For most of these raids, we lack explicit information on any goals, but that of the Kotrigurs in 559 certainly had the objective of pressuring the Romans for tribute. See Curta 2001: 116f for a summary of Hunnic raids.

[21] Proc. 2.4. Whitby 1988 notes that this might actually point to the effectiveness of the Balkan fortification system, since this would have prevented them from capturing booty or prisoners among the well-prepared and well fortified settlements of the northern and central Balkans, a region that had been used to raiding for well over a century. They thus had to risk penetrating deeply to the southern Balkans to find less prepared areas.

uphill by a team of 40 men, and still support a regular battering ram. They constructed similar devices for the Persians in large numbers at *Archaeopolis (550). In the Balkans, the Romans tried to balance Utigur and Kotrigur Huns against each other. Kotrigur raiding caught Roman attention, which is borne out by their employment in the Roman army after a raid in 551,[22] but Justinian then supported the Utigurs, which brought about Kotrigur jealously and the raid in 559. The raid ended in disaster for the Kotrigurs, as they were stopped before Constantinople, the walls of the *Chersonese, and at Thermopylai without achieving anything, while Justinian arranged for the Utigurs to attack their homes and deprive them of all their horses. What is most interesting for our purposes is the means the Kotrigurs used to attack the walls of the *Chersonese, including the elusive *helepoleis* (possibly tortoises with battering rams or trebuchets), ladders and archery, and even small boats to skirt the defenses by sea. All these efforts were met with efficient Roman countermeasures, but the besieging abilities of the Kotrigurs were fair, considering the fact that they were in an extremely difficult situation, far from any logistical base. Those of the Sabirs were excellent for a nomadic people, but heavily conditioned by long-term service for Romans and Persians.

Within a short time all these Hunnic groups were absorbed by the Avar Empire, providing the Avars with valuable military services, including intelligence and manpower. It is also likely that the Avars employed them for siege operations along with other subject groups. Hunnic siege capacity had the same origins as that of the Slavs, but Huns served more frequently, for more extended periods, and in a greater number of capacities than the Slavs in the Roman army. They thus acquired a somewhat more sophisticated poliorcetic capacity than the Slavs at an earlier date (mid-6th century).

7.2.2 Slavs and Appropriation

The Slavs come into our attention in the 6th century along the lower Danube, "in the shadow of Justinian's forts," as Curta puts it, but only settled in the Balkans in the 7th century. From the 530s until the establishment of the Bulgarian empire (680s), the political map, as far as can be ascertained, was dominated by very small chiefdoms and a fairly primitive material

---

[22] Proc. 8.21. In fact, Narses seems to have purchased the services of a great number of these, as they are found in great numbers in his army in Italy the next year, cf. Proc.8.26.7-17.

culture that was beginning to evolve more complex forms when the Bulgars disrupted this development. The Slavs have generally been assigned a very small role in the diffusion of technology, since they appear to have had a very primitive social organization as well as a low level of technological development, especially in the 6th century.[23] In contrast, it can be argued that the Slavs were very quick learners and absorbed military knowledge from deserters, prisoners of war, or their own service in the Roman army to the extent they were systematically exploited for their skills by the Avars.

In the mid-6th century, Slav troops had a distinguished record of service in the Roman army, well before they began to infiltrate the Roman provinces in the Balkans.[24] Procopius relates how a contingent of Hun, Slav and Antae cavalry, 1,600 strong, reinforced *Rome during the spring of 538. These troops increased the garrison by a third and allowed the Romans to sally out aggressively against the Goths on many occasions.[25] Slavs and Antae are attested in Italy, including the Roman siege of *Osimo (539) and the Roman defense of Lucania.[26] Agathias mentions them on the eastern front, at *Phasis (556) and *Tzacher/Sideroun (557), where they were on both occasions involved in siege warfare which included massive use of artillery.[27] The Slav threat to the Balkans, however, only began materializing in about 545, when a Slav raiding army was fortuitously defeated by Narses in Thrace, who had recently brought newly recruited Herul troops to winter in the area.[28] The timing of this Slav attack and their avoidance of a large field army when invading some years later[29] indicates that the

---

[23] See in general, Vryonis 1981; Whitby, 1988 *passim*; Lemerle's commentary to *Miracula St. Demetrii*; Curta 2001 and 2006; and Kobyliński 2005 for historical background and development; for an assessment of early Slav military organization, see Malingoudis (1994). I generally follow Curta, whose treatment is most comprehensive.

[24] Cf. the story of Childebudius, commander of Danube from 531 killed in battle against Slavs; in the context, Procopius (7.14.) mentions many Roman captives among Slavs and Antae.

[25] Proc. 5.27.1-5.

[26] Proc. 6.26 and 7.22.1-6.

[27] Agathias 3.21.6 for Dabragezas, an Ante taxiarch who led "his own troops" on the Roman patrol boats up the river Phasis during the Persian siege of the city; ibid. 4.20.4 for the Slav Suaranas who stopped a Misimian sally from their fortification at *Tzacher 557 with a well-directed javelin. On both occasions the Romans engaged the enemy with large compliments of artillery. It is difficult to say how involved Slav and Antae warriors were in the operation of siege engines, although the position of Dabragezas as a taxiarch with his own following, operating patrol boats normally equipped with *ballistrai* and perhaps other engines is highly indicative. On patrol boats see e.g. *Strategikon* 12.B.21, where boats used for crossing rivers are mounted with small *ballistrai*.

[28] Proc. 7.13.21-26. For an exhaustive list of raids, see Curta 2001: 116f.

[29] Proc. 7.29.1-3.

Slavs were by then familiar with Roman military dispositions and practices, possibly after the return of some who had recently served in Roman armies. This is further confirmed by their raid down to the Aegean coast in 549, during which they divided their army, defeated several field troops on the way before attacking a number of Illyrian and Thracian cities. On one occasion, they lured out the garrison of *Topeiros, killing all of them in an ambush, before storming the walls with barrages of archery and scaling ladders.[30] The speed and skill with which this happened apparently astonished Procopius, who relates that:

> ... both [Slav] armies captured many fortresses by siege (πολιορκία), though they neither had any previous experience in attacking city walls (τειχομαχήσαντες), nor had they dared to come down to the open plain, since these barbarians had never, in fact, even attempted to overrun the land of the Romans. Indeed it appears that they have never in all time crossed the Ister River with an army before the occasion which I have mentioned above [in 545].

These tactics were difficult to master, as we have seen, but the confidence, strategic skill and distance over which the raid was carried out indicates that the Slavs had learned well from service in the Roman army, especially in Italy. Their success appears to have prompted another, larger invasion the year after (550), but this time they were blocked at Naissus by the huge army amassing under Germanus at Serdica for the projected invasion of Italy, so they turned on Dalmatia instead. Only when the Roman field army reached Salona did the Slavs make another attempt on Thrace, wintering in 550-51 and receiving reinforcements from across the Danube. This army was initially successful, defeating a Roman army at Adrianople, taking prisoners and capturing their standards, but eventually a part of their army was defeated by the Long Walls, leading to the liberation of captives and restoration of booty and standards.[31] With Gepid assistance, Slavs also raided Illyricum in 551, and again took a great number of captives, while the Roman forces were only able to deal with stragglers from the main army.[32] These invasions were disruptive but did not threaten Roman control of territory, and the threat seems to have ended shortly after as a result of Justinian's diplomacy. Indeed, the Slavs and Antae are not mentioned as threatening at all in Agathias, where the Utigur and Kotrigur Huns appear as the great bogeymen. Only well after the beginning of the Avar men-

---

[30] Proc. 7.38.
[31] Proc. 7.40.
[32] Proc. 8.25.1-6.

ace, in 578, did another large raiding army appear in Thrace.[33] The new round of invasions culminated with the great raids against Thessaly and Greece in the 580s and the great Slav-led assaulted on *Thessalonica in 586.

Something had gone wrong with Roman client management. Curta proposes that trans-Danubian trade and subsidies decreased c. 545-60 due to Justinian's great project of fortification. This forced the elite within loosely organized Slavic groups to find new means of acquiring prestige; firstly, through intertribal warfare which helped solidify nascent political structures around chieftains, and secondly, through new bouts of raids after the arrival of the Avars in 568.[34] However, Slavs clearly served in the Roman army during this period, so that the recruitment aspect of client management was still in operation. Thus an important source of income was available, as attested by coin finds, some as far away as the Baltic shore.[35] It is possible that the strong defenses, a "most favored"-policy of recruiting troops from certain tribes, and other forms of recognition and subsidies in kind maintained stability and helped the formation of more consolidated political units. Perhaps the plague also played a part.

What prompted the new raids is uncertain. The Avars have traditionally been blamed for pushing the Slavs across the Danube, but the Romans resorted to traditional client management techniques and bought Avar support immediately after the Slav raid of 578, ferrying a huge Avar army across the Danube to attack Slav homelands.[36] This did not have the intended effect, as the Slavs continued raiding the next year, the Avars took advantage of perceived Roman weakness to besiege *Sirmium (579ff), and the frontier was completely breached for most of the 580s and much of the 590s. I would rather suggest that client management had been too successful. There were now a number of Slav chiefdoms that could be manipulated in normal circumstances, but this no longer worked with the Avars consolidating control over some of them and pressuring others. The first named Slav leader, Dauritas, rejected Avar claims of suzerainty with boasts of his own prowess, and he or his colleagues may have raided Roman territory in order to demonstrate this in the face of Avar claims; indeed, such raiders killed Avar envoys whilst traversing Roman territory, perhaps as explicit revenge for the Avar raids against their homelands. The events

---

[33] Menander, fr. 20.2.
[34] Curta 2001 assembles the archaeological evidence; the issue is treated more succinctly in Curta 2006: 57-61.
[35] See Curta 2001: 176ff for his interpretation as well as coin finds.
[36] Menander, fr. 21; see further Curta 2001: 91f.

after 578-79 then acquired a dynamic of their own: unable to play the Avars and Slavs out against each other, the Romans had to face attacks from Avars, Slavs under Avar suzerainty, and independent Slavs with their own objectives.

The next we hear from the Slavs is in the early 580s during the battles for *Sirmium (579ff) and *Singidunum (588), when Slavs built boats for the Avars to cross the river on two separate occasions (probably similar to *monoxyla*, log boats, reported at *Thessalonica in 615). In 586, while the Avars were threatening the Danube plain and eastern Thrace, the Slavs took the opportunity to make an attempt on *Thessalonica, although the dating is controversial.[37] Interestingly enough, it appears that mostly Slavs were involved in this effort, with little or no assistance from the Avars, although they probably encouraged it. The siege was a short but intense affair, including massive batteries of trebuchets and hordes of tortoises and rams assembled against the walls. The tactics involved were those traditionally used in late antiquity, although the hagiographer John provides the very first clear description of a trebuchet (see below and chapter 8 *passim*). What is important to note here is the fact that the Slavs were able to build and handle each instrument individually, but actually seem quite incompetent in arranging them to coordinated batteries, aim accurately, or use them to provide cover fire for their tortoises and rams—attempting this would have been counterproductive anyway, considering their poor aim. Thus, what seems to have been lacking is not craftsmen who were able to build the individual machines but a framework for the highly specialized experts who could train engine crews and coordinate fire in the field, The great number of prisoners of war taken over the decades could provide some specialists, as many Roman captives were soon integrated into Slav

---

[37] See Lemerle's commentary and Whitby 1988, favoring 586, and Vryonis 1981 favoring 597 (see CO entry for further discussion). The first date is based on context, as the Avars invaded along the Danube in 586 and might have prompted the Slavs to attack Thessalonica. Vryonis has a good case for 597 in that the Thessalonica's bishop at the time, Eusebius, is attested in Gregory the Great's letters from 597, but mostly his argument relies on the Bousas story, which in the current context in Theophylact Simocatta occurs in 587, as a *terminus post quem* for the transmission of advanced artillery technology. This story is demonstrably misplaced from a chronicle and perhaps occurred in 586; cf. the discussion in CO. Postdating the siege still proves nothing, though, as military technology often took quite some time to adopt, especially in the scale showed by the siege of Thessalonica, and implies a longer period of trial and error. Earlier Avar engineering on the Danube and Sava, as well as great numbers of Roman prisoners of war taken over the previous decades probably provide a better explanation, and Bousas, who I do believe existed, was one of very many such experts pressed into Avar service.

society, and the Slavs could probably provide the mass of carpenters (cf. their boatbuilding skills).[38] However, this means of appropriation could not replace an institutionalized framework for providing training and organization over time.

Slav raids continued during the early 7th century, but now their mode shifted.[39] At *Thessalonica in c. 615, various Slav tribes are distinguished, and they brought their families for settlement. While the assault was brief and ended in defeat, the Slavs used all the customary methods, such as siege sheds, trebuchets, archery and ladders, using archery volleys to cover their attempt to fire the gates, and a fleet of *monoxyla* to attack the harbor. According to the *Miracula sancti Demetrii*, most of the equipment was built in the vicinity. Since this took a few days, the defenders had time to prepare for an assault and repulsed them effectively. In 618, however, the same tribes had called in Avar assistance, which involved a large, multi-ethnic army, including subject Slavs and Bulgars. The equipment and tactics used by this army was essentially the same, but with the addition of a large wooden tower and heavier siege sheds. Furthermore, the bombardment seems to have been more intense. While the assault again failed, barbarian siegecraft now differed little from that used by Romans, Persians and Ostrogoths. What seems to have been lacking was a logistical system that could maintain armies in the field over longer periods of time.

For a long time afterwards, we know little of Roman-Slav relations. Slavs were a major component of the Avar army that besieged *Constantinople in 626, but after the Avar failure became completely independent. Some Slavs were defeated and forcibly recruited into the Roman army around 660 but defected to the Arabs. However, the Perboundos affair, which probably took place in 662, illustrates a new reality. The Slav chiefdoms (the *Sklaviniai*), now settled in Macedonia, Thessaly and Greece, had consolidated under rulers who were heavily influenced by Byzantine customs. Perboundos was one such ruler who spoke Greek and dressed in Byzantine manner. Accused of plotting against *Thessalonica, he was arrested and brought to Constantinople. Other Slav rulers were welcome at the court in

---

[38] For the great numbers of prisoners of war and their integration into Slav society, see *Strategikon* 11.4; indeed, a great many former Romans were regarded as unfaithful and unreliable by the author of the *Strategikon* (11.4.31). This is further discussed by Curta 2001: 348f and Kobyliński 2005: 530.

[39] The problematic raid on *Thessalonica 604 has been left out of the discussion. While I have retained Lemerle's date for the purpose of the Corpus Obsidionum, Curta presents a good argument for c. 584. Similar raids probably occurred at the time of either possible date, however, so it does not affect the argument too much.

APPROPRIATION OF INFRASTRUCTURE & KNOWLEDGE 377

Constantinople and successfully spoke his case, so the Emperor decided he was to be released after he had attended to the current war with the Arabs. Perboundos thus seems to have had much support among some Byzantines and was only executed after he escaped and was recaptured. This provoked the other kings to organize raids around *Thessalonica, before an assault on the city was organized by the "kings of the Drougoubites" with support from others. Their arsenal was again impressive, but this time the organizational support was far stronger. Not only was there a specialized engineer who could built a large siege tower with different engines for each storey; the kings also put at his disposal a large crowd of lumbermen, carpenters, ironworkers, fletchers and arms makers who were clearly under their authority. They also built "carpentered" boats, i.e. an evolutionary step from their traditional *monoxyla*. Specialist craftsmen, then, existed in abundance in the small Slav kingdoms of northern Greece and Macedonia, ruled by kings who spoke Greek and had close ties with Roman society. Such assimilation was very important for adoption of social features, military organization and technology. This also made them highly useful to greater polities in the late 7th century. The Romans settled conquered Slavs in Anatolia and employed them as soldiers, although with mixed success; on at least two occasions, large Slav contingents defected to the Arabs and were subsequently settled in the Caliphate. Similarly, the Bulgars deliberately settled various Slav tribes on their Danube and Thracian frontiers in order to provide defense.[40]

In sum, the early Slavs were good at storming fortifications on their own initiative, especially if the communities were isolated and surprised. Their skills were a combination of the evident martial prowess of all adult males in early Slav society, paired with specific techniques for siege warfare as universally practiced in late antiquity. These skills, along with knowledge of Roman military infrastructure, dispositions and defensive practices, would in the model proposed here derive from two specific sources: firstly, service in the Roman army; secondly, from a great number of prisoners of war who were assimilated into Slav society. Although the sources do not allow us to accurately pinpoint named individuals or places for transfer, the events cited above provide a sufficient context for the transmission of such knowledge during the 530s and 540s, culminating in the great raids of 549-51. Interestingly enough, although prisoners of war conceivably could have taught Slavs to build and use siege engines, there seems to be

---

[40] See Curta 2001: 110f for these examples.

no evidence whatever of independent Slav tribes using engines or engineering during this period. After a lull in the Slav threat during the 550s, with a few known examples of Slav/Antae service in the Roman army during that period, they again became a major threat after the arrival of the Avars. We know little of the Slavs' military capabilities at the time, since our major detailed source, Menander, is fragmentary. However, they were used frequently as shock troops as well as craftsmen and logistics specialists by the Avars, perhaps even used as "trebuchet-fodder" during the storming of heavily fortified and well defended centers such as Thessalonica and Constantinople. While besieging abilities in general were a result of service in the Roman army, I would argue that their technological skill attested 586-626 must be ascribed to a large influx of Roman captives, some of whom were assimilated into Slav society. However, after the final settlement in Greece, the Slavs developed a society not too different from the surviving European provinces of the Byzantine Empire which was capable of supporting specialist craftsmen. Continued military contact and service in Byzantine armies would ensure that they were still familiar with current forms of warfare, while their own social and political evolution had ensured the stability and complexity required to conduct siege warfare without resorting to predatory means. A possible evolution into larger and more prosperous Slavic states was however destroyed by the arrival of the Bulgars on the lower Danube in 680.

7.2.3 *Avars and Appropriation*

The story of the Avars is well known. They arrived in the Carpathian Basin in 568 as refugees from the Turkic Khaganate,[41] entering an alliance with the Lombards to defeat the Gepids. The Lombards fled to Italy, however, allowing the Avars to claim all former Gepid and Lombard territory, including *Sirmium, which the Romans had recaptured as the Gepid kingdom collapsed. Having inherited Gepid claims to *Sirmium, the Avars strove hard to gain the city. When they failed, they allegedly sent 10,000 Kotrigur Huns to invade Dalmatia while the Avars themselves withdrew to the recently conquered Gepid territories.[42] They spent some time consolidating their rule, and apart from a raid in 574 were not a direct threat to the Ro-

---

[41] Menander fr. 10.1 relates negotiations between Turks and Romans, in which the Turkish representatives explain the flight of 20,000 of their Avar "slaves." The number appears to be credible, coming from a diplomatic source. See Blockley's 1985 commentary for further discussion. For the general history, see Pohl 1988.

[42] Menander, fr. 12.5.

mans until 579, when they began a long siege of *Sirmium. From the capture of Sirmium around 581 (possibly 582), Avars raided and captured fortifications throughout the northern and eastern Balkans until they were systematically driven back by Roman forces from 597 to 602. It is now recognized that the Danube frontier held through the reign of Phocas and some years into the reign of Herakleios. The Avars therefore shifted their attention towards Italy, with diplomatic contacts attested in 603 and an invasion in 610, but on the Danube they were quiescent until Roman armies had to withdraw to face the Persian invasions, sometime around 615 or later. They had direct leadership over multiethnic armies at *Thessalonica in c. 618 and led the siege of *Constantinople in 626,[43] but their loss of prestige meant that their hegemony collapsed and was further reduced by the Bulgars c. 680, leaving a rump state in present-day Hungary, which persisted until conquered by Charlemagne.

The Avars have a formidable reputation as besiegers in modern scholarship, in large part because they have been credited for introducing the traction trebuchet, a major new innovation, to western Eurasia.[44] Its importance in the history of technology and siege warfare cannot be overestimated, so its origins are of great interest. However, there are three very weighty issues against this theory. Firstly, although it is difficult to trace because of unclear terminology in the sources, the trebuchet had probably already been in use in the Mediterranean world for generations by the time the Avars arrived, a point to which we shall return below. Secondly, the Avars display an astounding lack of engineering expertise in virtually every field related to siege warfare and logistics. They can instead be observed browbeating, paying or threatening various subject peoples and captives into performing tasks for them. Previous inhabitants, e.g. Slavs and Huns, demonstrably had long traditions in craftsmanship and engineering, such as boatbuilding, bridging, or use of rams, tortoises and artillery. In no case can the Avars be seen as innovators. It is important to note that the trebuchet required great expertise in building as well as handling, and was often misused by inexperienced crews, while the logistical demands were huge—another point of frequent Avar failure. Finally, all the preserved Greek authors who wrote on the subject have absolutely no knowledge of such a spectacular innovation at the time (it is of course possible that Menander said something in his lost sections), although they had no

---

[43] On the complex nature of the Avars and their relations with Slavs, Bulgars and Huns, see Pohl 2003 and for the Slavs in particular, see 7.2.2 above and Curta 2001.
[44] See the discussion in chapter 8 for this.

qualms about crediting barbarians with a host of innovations, literary stereotypes notwithstanding. Procopius was impressed by the clever Sabir rams, while the *Strategikon* freely acknowledged that Avar and Slav-inspired equipment was standard issue in the Roman army. Avar cavalry tactics were regarded as superior to Roman and Persian practices.[45] The *Strategikon* also speaks of the trebuchet, but has no mention of any foreign origins. As we shall see in the last chapter, several 6th-century authors provide clues that the trebuchet was probably used on the eastern frontier of the Roman empire well before any Avar arrival, though Procopius and Agathias had no clear idea what it was (though it was not new to them), while to Theophylact, writing about 630, it was so common it merited no particular comment.[46]

A more likely scenario is that the Avars learnt using the full range of late antique siege technology from their subjects and from the Romans, only after their arrival in Pannonia. As the Huns before them, they were certainly aware of the importance of siege warfare and had both the courage and individual skill to assault fortifications without other documented support, just as the Slavs at *Topeiros (549). However, for this they relied on speed, surprise, and subterfuge, as indicated by a peculiar story in *De administrando imperio*: the Avars had surprised a Roman patrol, taken their equipment and uniforms, and used the knowledge of patrol routines to arrive at their home base at the appointed time, completely fooling the inhabitants, who thought their soldiers had returned.[47] They also brought with them a sophisticated political culture and ambitions far beyond those of normal client groups, again as the Huns; they thus had mechanisms to organize and exploit the labor of subject peoples. Hence, the use of siege engines, whenever the sources document this, was always handled by subject peoples whose knowledge of them predated the Avars' arrival. Sabir and Kotrigur Huns, both of whom were absorbed by the Avars, were proficient in storming city walls and using rams and trebuchets. Slavs had the same capacity. Gepids are also attested in Avar armies. Their independent client kingdom controlled Sirmium and possibly other fortifications. Both Gepids and Lombards had until recently controlled large sections of the former Roman infrastructure in Pannonia with imperial "blessing."[48] It is

---

[45] *Strategikon* 1.2.19f (Avar lances and neckguards); 1.2.38 (Avar-style horse armor); 1.2.46ff (Avar tunics); 1.2.60f (Avar tents, regarded as both practical and appealing); 2.1.19ff (superior Avar and Turkic cavalry tactics); 12.B.5 (Slav javelins recommended for infantry).
[46] See chapter 8.
[47] *DAI* 30.
[48] Christie 1995.

possible that the Avars took over some of this in a partially functioning state, so that Gepid elements were absorbed by the Avars and provided logistical and technical services to the Avars. This is not the end of it, however.

Roman fortifications at *Singidunum, *Augustae and *Viminacium were taken by storm in 583, while eight or nine fortified cities were possibly taken in 586, giving an impression of highly competent besieging skills. However, both campaigns were determined by surprise and speed (e.g. *Singidunum 583). When they had to move further inland in 587, the Avars failed at *Beroe, *Diocletianopolis, and *Philippopolis, and abandoned the siege of *Singidunum in 588 after only seven days and a small ransom. During the rest of that year, they succeeded at *Anchialos with the support of a Slav-built fleet. At *Drizipera, they set up a fortified camp and used *helepoleis* (trebuchets) from the seventh day of the siege, but again abandoned the city, possibly because the arrival of a relieving force. They similarly besieged *Tzurullon, but abandoned it as well. The Slavs had been involved in a major project of timber-cutting and boat-building before the garrison at *Singidunum destroyed their ventures with a raid, prompting the siege in 588. However, the Avars were most keen to finish preparations so did not press the siege when the operations had been moved to a safe spot. From there on the successful attack of *Anchialos and the use of siege engines and fortified camp at *Drizipera must be attributed to Slav laborers. The Avars were in fact consistently dependent upon external sources of labor. The Lombard king Agilulf (590-616) sent craftsmen to build boats for one of the Avar sieges: "At this time also king Agilulf sent to Khagan, king of the Avars, craftsmen (*artifices*) to make ships, with which the same Khagan conquered a certain island in Thrace."[49] The chronology is problematic, because at this time, the Romans were on the offensive and had effectively beaten back Avar attacks. It should rather be associated with the siege of *Singidunum in 595, an otherwise unknown conflict after Agilulf became king in 590, or perhaps Agilulf did this as duke on his own or on behalf of king Authari (584-90), thus at the time of *Singidunum in 588.

The shipwrights—who were probably Italians and maybe even former Roman experts—provided the expertise to direct the efforts of Slav laborers. Other specialist expertise certainly came from captured Roman engineers. The story of Bousas, captured at *Appiaria in 586 (traditionally 587),

---

[49] *HL* 4.20: "Hoc quoque tempore misit Agilulf rex cacano regi Avarorum artifices ad faciendas naves, cum quibus isdem cacanus insulam quandam in Thracia expugnavit." *PLRE* 2 s.v. Agilulfus places this in 602.

is often taken as a literary device based on assumptions of barbarian inferiority, although others have taken it quite literally and built elaborate chronological arguments upon it. Both approaches are unnecessary. The role of Roman captives is well documented elsewhere in the conflict by John of Ephesus; he relates that specialist engineers, Syriac *myknyqw* (mīkanīqū) and builders, Syriac *bānāyē*, were forced to build the bridge across the Sava, which sealed the fate of *Sirmium (579ff). The class called *myknyqw* in Syriac sources are precisely those who were involved in building siege engines and earthworks and palisades in Mesopotamia at the time (cf. chapter 2.4.2). Furthermore, garrison artillerymen were common in the Balkans. In light of the large number of border fortifications taken by the Avars, it would be surprising if they did *not* find some specialists among their captives. Other elements of the Bousas story may have been invented, but the fundamental elements of a captive engineer forced to build effective engines were not.

Clearly, subject peoples were fundamental to Avar besieging abilities, and the Slavs most of all. When the Khagan returned the favor to Agilulf in 603, he sent Slavs who helped take *Cremona, *Mantua and *Valdoria by storm. During the campaign, the besiegers used battering rams to breach the walls of *Mantua, "in like manner" as *Cremona and possibly also as *Valdoria. The siege of *Thessalonica (586) may have been instigated by the Khagan, but was wholly carried out by Slavs; in 615, local Slavs tried to take the city on their own initiative, while in 618, the Khagan sent an army consisting of subject Slavs and Bulgars. The great army assembled against *Constantinople in 626, according to the information assembled by Howard-Johnston, numbered some 80,000 men and included Slavs, Huns, Scythians, Bulgars and Gepids. The Slavs predominated as heavy and light infantry and launched their *monoxyla* in the Golden Horn, but all the other groups ("Scythian" could mean any nomadic group at the time) had well demonstrated besieging capabilities predating the Avar arrival. The Avars however enforced heavy labor obligations upon their populations. The vast earth ramparts which crisscross much of the Balkans bear eloquent testimony to this: while their function is debatable, the sheer manpower and organizational ability necessary to build them is undeniable.[50]

In conclusion, most of the labor, logistical work and engineering within the Avar Khaganate derived from a combination of previously existing knowledge and forms of organization that were harnessed and directed

---

[50] For this, see Squatriti 2002.

under Avar leadership. Some of this was maintained into the late 7th and 8th centuries in the reduced Avar state in Pannonia, which Charlemagne only conquered after several major campaigns (see e.g. *Avar fortifications in 791). We see some evidence for the internal workings of Avar administration in one of the last stories in the *Miracula St. Demetrii*, which shows how the Avars ruled their subject populations. The descendants of Roman captives were organized into a separate group under a Bulgar leader, and seem to have had a significant degree of autonomy. However, with the collapse of Avar power, and perhaps encouraged by the establishment of the Bulgars on the Danube in 680, the Romans broke away around 682 and escaped to *Thessalonica, which their Bulgar leader attempted to capture by subversion. The plot was discovered, but despite the threat, some of the leaders and the whole population were reintegrated into the Byzantine state and army. If the plot had been successful, we might have seen a "Bulgaria" on the Aegean, even though most of the population would have been a mix of Slavs, Romans and other groups; conversely, the Bulgars that settled on the Danube were probably far closer to Byzantine society than normally recognized.

7.2.4 *The Bulgars, 680-825*

The origins of the Bulgars that settled on the Danube in c. 680 are obscure, but the topic has been treated above and needs not detain us here.[51] It is clear that vacuum after the Avar and Roman collapse gave the Bulgars room to maneuver when they came under Khazar pressure, taking over areas that had been under Avar hegemony. Once entrenched, the Bulgars applied many of the same methods of rule and warfare as the Avars had.[52] Bulgar military organization is evident from the first Byzantine expedition against the Bulgar division under Asparuch in 680/81, when *Bulgar fortifications near the Danube estuary were besieged by Constantine IV, indicating a significant ability to organize labor and build earthworks. At that time or shortly after, Slav subjects provided much of the manpower, and

---

[51] See chapter 2.1.4 on origins; further Beševliev 1981 on the earliest Bulgar state; Browning 1975 for a comparison of Byzantine and Bulgarian economy; Curta 2006 for a recent assessment of the region; the studies on Pliska and the Balkans in *PRT* 2 for a wealth of new archaeological evidence on the earliest Bulgar state, including perspectives on both their settlements and their earthworks; the latter is also treated by Squatriti 2002; finally, Sophoulis 2012 has the most recent synthesis although his focus is on the decades surrounding 800.

[52] Other Bulgar or Bulgar-led offshoots ended up in *Thessalonica (682), Austrasia (around 631), and Italy (660s), but were in these cases assimilated or subjected by the host country.

were settled by the Bulgars as border guards against both Byzantine territory and the rump of the Avar khaganate in Pannonia. Within their area of rule, there were also substantial Greek-speaking populations, evident from middle Greek inscriptions. The Bulgar leadership thus succeeded in an Avar-style form of rule.

Rather than confrontation, the Byzantines attempted to reestablish traditional client relations. Archaeological evidence indicates that Constantine IV had indeed attempted to use economic means before resorting to his failed military expedition. This policy was resumed after 681, with trade attested from 690 at the latest. Justinian II was helped back to the throne in 705 by the new khan, Tervel, who raised his Bulgar and Slav subjects in his support. Tervel was richly rewarded and acknowledged as "Caesar" (i.e. junior emperor) by Justinian, but publicly only called himself *archon*, the contemporary term for provincial governor, in his inscriptions at Madara. He thus recognized his place within the imperial hierarchy, and despite some double dealing on the part of Justinian (*Anchialos 708), supported him against Philippikos (*Cherson 711), and only attacked Roman territory again when Justinian was overthrown and new tribute (rather, subsidies) were not forthcoming. However, after a new peace treaty in 716, relations were good, trade well regulated, and the Bulgars supported the Byzantines during the great siege of *Constantinople (717f), attacking the Arabs who camped in Thrace. The Bulgars had clearly been brought within the Byzantine political orbit and caused no trouble in the following years. A change in policy only came in 755, when the Bulgars felt provoked by the arrival of Syrians and Armenians in newly fortified settlements in Thrace. Instead of negotiating a new treaty, Constantine V organized a series of successful campaigns by both land and sea which brought devastation and chaos to the Bulgars, although expeditions in 766 and 774 failed due to storms that destroyed the fleets with their men and crews. A Bulgar counterattack against *Beritzia in 772, intended to carry off the population, was stopped by the Byzantine army. Instead, the Byzantines began resettling several long-abandoned Thracian cities, encroaching on territory claimed by the Bulgars. In 792, the Byzantines built *Markellai and used it as a base for an unsuccessful attack on Bulgar border fortifications. The high point was reached under Nikephoros I (802-11). Byzantine territory now reached all the way north to the Haemus mountains as far west as Serdica (Sofia), taken, resettled and garrisoned in 807.

So far in the conflict, the Byzantines had not subdued the Bulgars, but the Bulgars were unable to stem the Byzantine advance. Although the Bulgars had much manpower, were skilled at building earth and wood fortifications on a large scale, and were by know very familiar with Byzantine political culture and military dispositions, Byzantine organization far surpassed that of the Bulgars in the building of fortifications and siege warfare. Nevertheless, the Bulgars went on the counteroffensive and managed to take *Serdica by treason in 809, massacring most of the garrison and population. The circumstances are very unclear. The behavior of several Roman officers after the fall was clearly suspicious to Nikephoros, who refused to grant them an amnesty. Fissures within the army and elite probably led some to treasonous behavior; there had already been a mutiny during his campaign that spring. Nikephoros was able to continue his campaigns, however, and by 811 had almost destroyed the Bulgars, who humbly begged for peace. Here, however, the Bulgar earthworks served their purpose: the Byzantine army was forced to camp in a mountain pass due to such an obstacle, and during the night they were ambushed and Nikephoros was killed, his skull serving as the Bulgar khan Krum's drinking cup. This led to the dramatic reversal of fortune allowed Krum to capture a number of heavily fortified Thracian cities and took the fight as far as the gates of *Constantinople in 813.

In addition to the massive loss of morale after Nikephoros' defeat, the Bulgar advance was possible thanks to newfound expertise. At least two military engineers defected to the Bulgars; the first, one of those denied an amnesty after *Serdica in 809, and the second, an Arab who had been baptized, joined the Byzantine army and was stationed at Adrianople (see *Mesembria 812), but was now dissatisfied after being underpaid and beaten for complaining about it. Between them, they had a significant impact on Bulgar besieging abilities: *Debeltos and *Mesembria were taken along with their populations and arsenals, *Adrianople besieged with a wide array of engines, and there, the Bulgars began preparing an even larger battery for assaulting Constantinople. This was never achieved due to the sudden death of Krum and subsequent infighting over the succession. The preparations were abandoned, and shortly after, Leo V was able to reestablish the balance, which was maintained through most of the 9th century.[53]

The failure to attack Constantinople was an accident of fate. One could think that once the succession was settled, the Bulgars could again begin to organize more expeditions and use their new-found expertise. After all, their population had grown, a number of Christians from the Empire had

---

[53] For the most updated treatment of these wars and their context, see Sophoulis 2012.

been resettled in the north, and they could conceivably have conquered the rest of the Balkans if not Constantinople itself. However, there is reason to believe that they still had an inadequate infrastructure to maintain siege technology over time, and that their expertise disappeared with the experts that had introduced it. During the revolt of Thomas the Slav, a Byzantine pretender, in 821, the two Byzantine sides engaged in many sieges against each other. However, when the reigning emperor, Michael, was besieging Arcadiopolis in 823, he carefully avoided advanced siege technology which could be observed by the nearby Bulgars, instead only constructing a ditch and palisade (which the Bulgars already knew how to do quite well) to contain the defenders.[54]

Only a decade after capturing a number of heavily fortified cities by storm, then, the Bulgars had apparently lost this ability. This may be explained by a number of factors. During their succession disputes in 813 and afterwards, the Bulgars turned on their own Christian population, killing many who refused to abjure their faith. This was certainly not a climate conducive to encouraging any defectors or captives to provide services for them. Sophoulis points out a number of other factors that may have played a part. Large numbers of Christian subjects, who probably possessed most of the necessary competence, were sent to serve on the northern frontiers beyond the Danube. The Bulgars were already dependent on Slav manpower for their numerous fortification projects, and pressure from the Carolingians seems to have forced them to focus defensive efforts in that direction. Furthermore, peace negotiations immediately after the war stipulated the return of captives and deserters; our sources are quite explicit that individuals from precisely these groups were involved in technology transfer during the wars of the early 9th century. Finally, the level of the Bulgarian economy was still not advanced enough to provide the necessary environment or infrastructure for absorbing such knowledge and maintaining expert craftsmen over time independently of Roman or sub-Roman populations. The growth observable in the late 8th and through most of the 9th centuries, the conversion to Christianity in the 860s, and other stimuli laid the foundation for large-scale construction projects, including fortifications on Byzantine models, and made tsar (i.e. Caesar) Symeon's ambitions possible in the 10th century.

---

[54] Treadgold 1988: 241; the reason is explicitly given by Genesius 2.8.

7.2.5 On Northwestern Peripheries: Western Slavs, Saxons and Danes

The client mechanisms observed on the late Roman frontiers in the 4th and 5th centuries, and the East Roman frontiers in the following centuries, also operated on the northern borders of the Frankish world. In the 6th century, Clovis reestablished a client belt that included all transrhenan Franks, Alamans and Thuringians; to this was added Saxon and Slav territories in the late 6th and early 7th centuries, which resulted in an early bout of state-building. Perhaps an agent of the Merovingian kings' client management apparatus, the famous merchant Samo established the first Wendish (west Slav) state in 623/4, in opposition to both Frankish and Avar hegemony. Many have argued that this could only have happened after the Avar Khaganate had collapsed, so sometime after c. 630. However, the Avar invasions of the Balkans after c. 615 must have drawn much of their manpower and attention away from their western borders. Avar armies were active in Thrace in 623, while the huge expedition against *Constantinople in 626 must have been in the works for up to several years, thus giving time for an early separation. Indeed, just as the Avars had used a Bulgar to rule over an ex-Roman population in the mid-7th century, there is every reason to suspect that Samo might have been a similar creation installed to organize defenses on their western frontier. The Wends were afterwards difficult to subdue in their great earth and wood fortifications, despite the resources available to the 7th-century Merovingian kings. Dagobert's expedition against *Wogastisburg in 630 was not successful, but apparently impressive enough to ensure that the Slavs were not a significant threat for some time. However, the Thuringian duke Radulf, who was of Frankish origins, exploited the death of Dagobert to establish his independence and withstood a retaliatory siege of his *castrum at the river Unstrut in 639, in no small part because of his close connections with some of the aristocrats leading the Frankish army. The Slavs again began to make an impact on Frankish history during the reign of Charlemagne, who had to subdue the *city of Dragawit, a Slav ruler, in 789.[55]

The Saxons seem to have been cast adrift by the collapse of the Roman client management system, as historic Saxony had close relations with northern Gaul in the late 4th and early 5th centuries; this may have

---

[55] Great earth-and-wood fortification became a hallmark of Slav polities, e.g. the 9th-century Moravians, whose rulers established sophisticated fortified capitals with extensive economic activities and crafts production on the Carolingian model. For the excavated fortified settlement at Pohansko, see Macháček 2007; for bibliography on the famous Moravian capital at Mikulčice, see Poláček 2007; further discussion, chapter 4.2.1.

prompted the many "freelance" Saxon groups that were active as mercenaries and raiders in Britain, Gaul and Italy well into the 6th century. At home, the loose conglomeration of tribes was periodically under Frankish hegemony, and sometimes also under Frankish dukes. They paid tribute to the Merovingians, but this was changed to military service by Dagobert, probably in response to the rise of the Wends and other Slavs at the time. The ethnic difference between Saxons and Franks were small; they spoke closely related dialects, some border tribes passed as either Frankish or Saxon, depending on the period and source, and in the 8th century, Saxon social structure and military practice seems to have become hard to distinguish from that of the Franks. The Saxon aristocracy had at that point become a militarized, landowning elite, who led resistance against Frankish encroachments by building large-scale earth and wood fortifications that mirrored those on the Frankish side of the frontier. This led to the endemic raiding and wars over fortifications in the late 8th century.[56] During the reign of Pippin (741-68), the border was reasonably stable, although there were several tit-for-tat attacks. Saxon *Hohenseeburg was taken by the Franks through negotiations (*per placitum*) in 743. However, in 753, Saxons killed the bishop charged with defending *Ihburg, so presumably they took it by storm. The Franks in return stormed *Saxon strongholds in 758. Charlemagne had to subdue a large number of Saxon fortifications before he gained control. The Saxon ability to build fortifications, strike back, and use siege engines of their own indicate a quite sophisticated military structure; for instance, at *Syburg (776), they used a combined assault of *petrariae* (trebuchets) and fascines to cross the moat and assault the fortification, but it seems that they were outshot by the Franks. Once subdued, however, the Saxons quickly became fully integrated into the Carolingian empire and a pillar of further expansion eastwards, as well as providing manpower on many other fronts.

To the north, the Danes were influenced by this process as early as 737, at the latest, when the vast Danevirke was built; by 808, they could assault *Slav fortifications and provoke Carolingian intervention.[57] Afterwards, Godfred, the Danish king, organized his army into teams that were to take

---

[56] For the Ostrogothic and Frankish influences north of the Rhine and Danube (i.e. on the Saxons and other client peoples), see Wood 1998; for the integration of the Saxons and their similarity to the Franks in the 8th century, see Wood 2000.

[57] The background is of course much "deeper;" as emphasized in chapter 1.3.1, Roman influences had a profound effect on Scandinavian society already in the first centuries AD; for the transitory period, see Näsman 1998, who traces the shift from "late antique" to Frankish influences in the 6th century and the concomitant consolidation of larger polities;

APPROPRIATION OF INFRASTRUCTURE & KNOWLEDGE   389

responsibility for repairing or rebuilding designated sections of Danevirke. This is remarkably similar to the method that must have been used to build Offa's Dyke. Indeed, the Pope was at the time organizing the repairs of the walls of Rome, assigning sections of wall to various social units and corporate groups.[58] This points to an interesting conclusion: the "dark age" kingdoms in England and Scandinavia had learnt, through Frankish intermediaries, the ancient practice of organizing corvée labor and assigning the teams to building fortifications by *pedaturae*. This must also lay behind the Slav and Saxon fortifications. Only the materials they used may have been less durable.

7.3 The Arabs and Islam: Appropriating and Domesticating the Late Antique System

Throughout the period from 502, when Arab tribes participated in Roman-Persian warfare in the east, until around 614, when Arabs are said to have raided monasteries around the Dead Sea (*Jerusalem), the inhabitants of the northern Arabian Peninsula appear to have had little capacity for systematic conquest, but an enormous appetite for booty.[59] Procopius explicitly stated that Arabs were inept at siege warfare when discussing *Sisauranon in 541, though skilled like no other at raiding. Indeed, the extensive raiding that dominated the very first Arab attacks under Islam in 633 and 634, and caused many Roman cities to surrender in 634-35,[60] conforms well with the image later Muslims liked to project onto their forbears, and is the one favored by modern historiography. However, a close analysis of Christian sources in particular reveal that the Arabs excelled in siege warfare as early as 640 at the latest,[61] were mobilizing greater resources than the Romans or Persians ever had done by the 650s, and continued to do so for several generations afterwards.[62] This was achieved by skillfully appropriating Roman (and Persian) methods of administration and organization as the conquest progressed. The ground had been prepared by centuries of Romano-Persian client politics, which had much the

---

Näsman 2000 provides good backround on the Danish economy during the period treated here (8th-early 9th centuries).
[58] See chapters 1.2.5, 2.4.2, 3.3.3 and 4.3.4 for this.
[59] For Arab involvement in Roman and Persian armies, see e.g. chapter 2.3 *passim*; for the raids against the Dead Sea region in 614, see Dionysius 24.
[60] For this, see Appendix I.
[61] Two examples are set into context and analyzed in Appendix II.
[62] For an unknown attempt to conquer Anatolia outright in the 660s, see Appendix III.

same effect in the Syrian and Arabian deserts as Roman policies had against the northern barbarians. Conversely, the Arabs soon established client systems of their own, which in many respects continued Roman and Persian military organization.

7.3.1 *Background and Early Events*

The traditional assumptions about early Islamic warfare, whether late antique, Islamic, or modern, are based on ethnographic stereotypes, although contemporary sources reveal a complex reality. Arabs in the Syrian and north Arabian deserts were long-time clients of the Roman and Persian empires. The Persian Arab clients, the Lakhmids, were often enrolled as auxiliary forces during Persian expeditions. Although they were chiefly charged with raiding, they seem also to have had stations in Persian siege camps, so were at least familiar with Persian besieging technology if not the foremost participants in Persians siege warfare. Their equivalents on the Roman side, the Ghassanids, lived in present-day Jordan and inner Syria, in or near several cities that had all the facilities of late antique urbanism, ranging from public squares and buildings to complex fortifications.[63] Unfortunately, I have found no narrative evidence for Arabs defending any of these cities from threats until the Islamic invasions, but on the basis of later anecdotes and archaeological evidence indicating that urban society there differed little from the general picture of late antiquity, we must assume these cities possessed the same infrastructure as elsewhere.[64]

The rise of Islam in the shadow of the great Roman-Persian war of 603-28 can only be studied from the much later and controversial Islamic sources. These also provide the only evidence for how the earliest Islamic armies fought before the expansion into the Fertile Crescent began. While much of the literature emphasizes raiding and Bedouin-style warfare, there were also significant conflicts over permanent settlements that involved

---

[63] On the relations between the Arabs and their imperial neighbors, see in general Shahîd 1989, 1995 and Retsö 2003, who more strongly emphasizes the empires' transformative effect upon the Arabs, and the problematic nature of their identity, even well into Islamic times.

[64] The nature of urbanism in Syria before the Arab conquests has been much debated; Kennedy 1985, 1990 proposed that urban economy and society in the Roman east were in decline, and were only revived by the Islamic conquests. However, many scholars have criticized this as far too simplistic, pointing out much regional variation and that the evidence points more towards continuity from the Roman to Islamic periods; see e.g. Walmsley 2000.

some besieging skills. The use of a catapult is attested at Ṭaʾif in 630. There is thus prophetic guidance on the morality of using such devices against settlements where women and children live (in the *hadith*, non-combatants are explicitly mentioned in the context of night raids), while the prophet himself is observed receiving advice from Salman the Persian on how to dig a large trench and embankment as a defensive device. These anecdotes may be historically true; they certainly provided useful moral and practical advice on certain operational and logistical issues for later generations of Muslims. However, there was a world of difference between operating a catapult and digging a ditch on the one hand, and on the other, conquering the great fortified cities of the Roman and Persian empires by the dozen.[65]

Indeed, in 640, a century after Procopius' quip on Arab poliorcetic incompetence, Sebeos describes an Arab assault on the walls of *Dvin (see Appendix II; *Dara was also assaulted on the same campaign) that clearly demonstrate exceptional abilities at the archery-and-ladder approach, which not only required great skills in coordination and training, but was also supplemented with a sort of chemical warfare not seen elsewhere. Probably during the same year, the Arabs also stormed the heavily defended port of *Caesarea using a combination of massed artillery along with ladders. Reinterpreting some of the early conquest accounts in Baladuri in light of common late antique practices (see Appendix I-II), it is clear that the Arab ability to raid had the same effect as blockades on the cities of Palestine and Mesopotamia, but they could also follow up with the storm of cities. The combination brought about most of the surrenders recorded, since defenders rarely gave up until they were convinced no rescue would come. The besieging methods became even more complex; at *Tripoli (644) the Arabs built fortifications of their own to blockade the city.

By 649, large fleet-based besieging armies descended upon Cyprus and *Arwad; in 653, the Arabs raided and established bases on the *Aegean Islands, and in 654 they reached the sea walls of *Constantinople. These fleets came from occupied Syria and Egypt, and carried vast arsenals of catapults and sundry siege engines. At the same time, Arab armies raiding into Anatolia and Armenia from Syra and northern Mesopotamia relied on the tactics displayed during the first conquest phase: speed, surprise, and storming walls from the march without much of a logistical train. Thus several *Armenian forts were attacked and one taken by storm in 643, while

---

[65] For the nature of pre-conquest Islamic warfare, see Landau-Tasseron 1995; for artillery in early Islam, Hill 1973.

the inhabitants at *Euchaïta (644) were taken by surprise and caught in the fields, believing that the approaching Muslim Arabs were Christian Arab allies of the empire. Further expansion was halted by the first civil war of 656-61. Apart from naval-borne siege trains, the use of siege engines on land expeditions was comparatively rare. After a revolt at *Tripoli in 653, massive losses in 654, and the chaos of civil war, use of siege engines within the former boundaries of the Roman Empire, both by sea and land, are not recorded for some time. At *SYLWS in 664, the Arabs instead forced a Roman master carpenter to build a large trebuchet, but otherwise the only reports are from internal conflicts; during the second Arab civil war (680-92), a trebuchet was used against the Ka'ba in *Mecca in 692, and several trebuchets manned by Christian inhabitants were used to defend Mardin around 690. In the eastern Caucasus and Iran, however, Arab armies did deploy catapults right through the 650s and beyond.[66]

7.3.2 *The Sources of Expertise*

The pattern of besieging methods employed in different regions, as revealed by contemporary and later sources, stand in contrast to the widespread distribution of related engineering skills in the early Caliphate. Both fortification and besieging technology are well attested from the early 8th century, and from then on frequently mentioned in both Islamic and Christian sources.[67] The temporal and geographic gaps could be ascribed to an accident of the source materials, which are very late or fragmentary for this period. However, these gaps correspond remarkably well with some peculiarities in how the Arabs administered their new conquests, and may be further explained as the long after-effects of the Roman-Persian wars.

Firstly, according to the reconstruction made in Appendix I and II, the Muslims attacked many cities in Palestine and Mesopotamia without siege trains. They instead relied on the basic blockade or storming tactics in order to subdue the inland cities in the south and at first avoided those on the coastal plains. It was only by incidental defections that they began to acquire more advanced technology. Persian garrisons had handled local defensive needs during the occupation of Syria (613/14-628/30). Even if

---

[66] Mardin is not in CO; see chapter 7.3.2 for discussion and references; for continuous use of siege engines in the former Persian territories of the new Islamic Empire, see Hill 1973 and the works of Chevedden (et al.).

[67] In general, see Haldon and Kennedy 1980, to which the *Chronicle of Zuqnin* adds much information on the organization and conduct of warfare. This is analyzed in the following.

they discontinued some aspects of Roman military organization, Foss has recently shown that the Persians upheld local administration and economic life in a manner that may also have been conducive to support the infrastructures of local self-defense in the interpretation offered here. A Samaritan unit of 5,000 men involved in defending Palestine might indeed have been a Persian creation.[68] There is however no notice of defensive use of siege engines in present-day Palestine, Jordan or southern Syria against the Muslims in the 630s despite fairly copious siege anecdotes examined here, although they are attested in Mesopotamia and Egypt, which had more time to prepare and received large reinforcements. Furthermore, Persian garrisons that wished to enter Roman service had been granted the right to settle in Syria by Herakleios (see below). These troops provided a loyal defense of the province until the Roman field armies had been defeated and most cities had surrendered. Thereafter, most of them made a deal with the invaders. In this they followed their brethren in Iraq, where we know of several large Persian army formations that went over to the Muslims wholesale, bringing with them Persian tactics, technology, institutions and strategic knowledge. One such Persian defector taught the Arabs to build a battery of 40 trebuchets against Buhasir in Iraq in 637. The frequent defections of Persian troops to the Arabs in the early years of the conquest, and later support of Persian engineers and soldiers, may explain the use of siege engines in the eastern Caucasus and further east in the following decades.[69] Furthermore, when the Arabs established client kingdoms detached from Persian or Roman control, such as Armenia, Caucasian Albania and for a brief spell, the Armeniac command (later *strategis*) in the 650s and 660s, military service would include their technological and logistical capabilities.

In Syria the situation was more complicated. No large Roman army formations went over to the Muslims, although some individual defectors are known; the Christian Arabs resisted fiercely, many fled if they could. The

---

[68] Foss 2003 for the Persian administration and Dionysios 49 for the Samaritans; cf. chapter 6.2.3 and 1-4 *passim*.

[69] See Morony 1984: 194-99 in general and 210ff for this particular incident (in Tabari II 2424, 2427); ibid. and 2004: 278 more generally on the defection of Persian military units to the Arabs, and the adoption of Persian traditions, such as the employment of the class of servants and youths (*ghulām*, pl. *ghilmān*) who also accompanied Arab armies. Although Arameans (mostly Christians) are recorded as assisting the Arabs as spies and on the battlefield in Iraq (Morony 1984: 175f), they are much less prevalent than in Syria and Egypt since most of the military responsibilities in the Sassanid Empire rested upon the Persian nobility.

local population in southern Syria, Jordan and Palestine may not have fully reinstituted the traditional framework for maintaining siege engines and other defensive measures between the final evacuation of most Persian garrisons in 630 and the first Islamic raids in 633-34. Even if the Persians had preserved the knowledge and institutions of urban defense to some extent, effective use required cooperation with professional troops.[70] The small, newly installed Roman garrisons were all too often forced to engage the Muslims out in the field, since that was the only way to break their blockade on the countryside. The decimation or escape of garrison troops could only further have weakened any defensive infrastructure, but also made any expertise unavailable to the Arabs. The Persian defectors, however, had certainly preserved defensive expertise as part of their garrison duties; such forces or Persians brought from Yemen (below) must have enabled the Arabs to besiege *Caesarea (640) with the large battery of 72 trebuchets.

The conquest of Egypt presents an even more complex image. Garrison forces had no choice but to face the Arabs in the open simply to prevent famine, but the Arabs used local superiority to overpower them.[71] It must also be remembered that Egypt was rent by civil war at the time (641-44), much as Syria was during the Persian invasion. One side evidently believed that they could use Arab support. In this confused context, a great number of cities were stormed, but siege engines only had a secondary role during the first year or so, as far as we can tell. The garrison at *Antinoe (642) handed over their arsenal of engines to the Arabs as part of their surrender agreement, but at *Alexandria (642f), the Arabs were at first driven back from the walls by defensive artillery. Alexandria only surrendered after a lengthy siege.

As we have seen, the first Islamic forces had arrived without much of a siege train or logistical support. However, in contrast to the armies that attacked Syria a few years earlier, by 640 the invading forces already comprised a multi-ethnic force, including Romans and Persians from Yemen, whose progress in Egypt was further strengthened a few high-ranking individuals in the administration went over to the Muslims outright, convert-

---

[70] See chapters 5.2. *passim*, 5.3.1 and 6.2.2; Foss 2003.
[71] For the most recent reconstructions of the Islamic conquests of Egypt, see Hoyland 1997: 574-90, Sijpesteijn 2007 and Howard-Johnston 2010 *passim*; for the problems facing the Roman garrisons who ventured out into the open, see e.g. *Bahnasā 640f, *Tendunias 641 and the final sections of John of Nikiu, *passim*.

ing to the new religion within the first year or two of the war.[72] With the aid of local administrators, the early Islamic armies received massive logistical support from the Egyptian population, who were required to perform labor services and provide resources in kind for their conquerors even as the Arabs were storming cities in Fayyum and lower Egypt.[73] From 643-44, evidence from surviving papyri and the narrative of Pseudo-Sawīrus shows exactly how well the Arabs had taken control over the local administrative apparatus: they already had river boats, crewed by Christians, to transport troops, supplies and booty, administered by former Roman officials.[74] They employed traditional Roman methods for procuring supplies and labor, only on a larger scale and directed towards new purposes; while levies of horses were common enough, the conquerors imposed labor obligations to open the canal between Fustat and the Red Sea in order to send grain shipments to the Hijāz, which was struck by famine.[75]

The wholesale control over the labor and resources of Egypt was directed towards further conquest and most notably the establishment of a navy. The expeditions in the second half of the 7th-century are largely known from narrative evidence. The massive fleets sent against Cyprus, the Aegean, and Constantinople from 649 to 654 required a vast infrastructure in the form of dockyards and arsenals staffed with conscripted labor.[76] The workings of this infrastructure in the early 8th century is known from preliminary preparations for *Constantinople (717f) and other expeditions, where the archival records of the governor Qurra b. Sharīk (709-14) have survived. Enormous exactions were placed upon the population, requiring forced labor, taxes and supplies in kind. Sijpesteijn points out how the Arabs dramatically tightened and enhanced existing administrative practices and exacted far greater resources than the Romans ever had. As a result, refugees (φυγάδες) were a persistent problem in early Islamic Egypt. The tax and labor obligations were therefore enforced by tight ad-

---

[72] See Sijpesteijn 2007: 439 and al-Qadi 2013 for the composition of this army; J. Nik. 114.1 for unspecified Egyptian converts to Islam; 120.29ff for high-ranking Roman supporters and how they administered taxes and labor; also 121.3f and 121.10f.

[73] J. Nik. 113.1-4.

[74] For this and the following, see Trombley 2004: 209 and *passim*; Sijpesteijn 2007: 440, 444ff, 448; according to the latter, the inspiration for this may have come from Persian administrators, but the basis was largely local personnel and practices.

[75] Thus Sijpesteijn 2007: 447f and Finster 1982. See also al-Qadi 2013: 45-61, who rightly argues that this constituted compulsory service and not "recruitment" or other voluntary means.

[76] See *Constantia and *Arwad 649, *Lapethus and *Arwad 650, *Tripolis and *Aegean Islands 653, and *Constantinople 654 for the operations.

ministrative control and draconian punishments. All travelers were required to have passports, which were chiefly issued in order to allow people to find work for paying mandatory taxes, or specialist laborers to reach the site of their labor obligations. Monks were branded in order to prevent refugees from settling in monasteries and thus escape the pressures of taxation and labor duties. The Christian administrators at province and village level were threatened with huge fines and capital punishment if they failed to send the required laborers, while specialized prison barges equipped with wooden blocks were used to transport captured refugees to their site of labor. At the dockyards, specialist craftsmen (τεχνῖται) such as caulkers, carpenters and ironworkers are attested. The extensive use of coercion notwithstanding, at least some laborers received wages, and all those involved with the fleets received rations (which, however, necessitated further burdens of taxation and exaction).[77] Christian craftsmen were sent to newly established naval bases in order to provide the necessary infrastructure for further expansion and repair of ships; thus 1,000 Coptic Christians were sent to Carthage in 701.[78] Carthage provided a safe base for assaults on the islands of the central Mediterranean and support for the armies that invaded the Maghreb and Spain in the following years.

The fleets were manned with Arab Muslim "emigrants" or *muhājirūn* (μοαγαρῖται), client troops or *mawālī* (μαυλεῖς), and Christian sailors. Many of the latter were conscripted, and also received poorer quality rations than the first two categories. Coptic Christian crews were apparently just as reluctant in their service as conscripted craftsmen. Many crewmen fled in 703 when an expedition against Sardinia failed, but Qurra was still seeking information on their fate and the whereabouts of survivors in 709. Expansion into Spain was not a priority at this point; the first governor of Spain was recalled for not obeying orders, while the North African fleet was directed towards *Constantinople in 717f. The Coptic crews of the fleet that arrived at *Constantinople in the spring of 718 defected en masse to the Emperor. This reluctance to serve stands in contrast to the official attitude of Christian bishops in Egypt, who justified the Arab conquest as a relief from Byzantine persecution of the "orthodox," i.e. Monophysite Coptic Christians. There is good evidence that this view was a deliberately constructed justification for Christian participation in Arab warfare against other Christians. There was a longstanding tradition in Egyptian ecclesias-

---

[77] Trombley 2004; Sijpesteijn 2007: 449f.
[78] Pryor and Jeffries 2006: 28 for the Coptic craftsmen; ibid. 24-34 for the context; further al-Qadi 2013: 46.

tic writings of emphasizing, almost to a bizarre degree, the role of the Coptic Church as a persecuted community no matter the actual historical circumstances.[79] In fact, Herakleios had removed collaborators, even members of the illustrious Apion family, after the Persian withdrawal in 629. It seems that many of the animosities that affected the course of the Islamic conquest ten years later had to do with such political measures,[80] while most of the Monophysite clergy in both Egypt and Syria were willing to accept communion with Constantinople. Indeed, the actual troublemakers during the 630s and 640s were largely Italian Chalcedonians intolerant of attempts at compromise.[81] Thus, the Monophysite portrayal of itself as persecuted was ahistorical, but a necessary fiction as long as the Coptic Church was used in the administration of the newly conquered territories.

Syria also had her naval facilities, although the lack of documentary evidence means that we know much less than we do about Egypt. Nonetheless, Syrian fleets participated in the naval expeditions from 649 to 654, when the institutional framework for naval-based siege warfare from Syria may have temporarily collapsed. This was a result of losses incurred against *Constantinople in 654, the flight of engineers and prisoners in the run-up to the expedition in 653, and the chaos of the first civil war from 656 to 661.[82] We know that a group of Romans, who had joined the Arabs early on, revolted at *Tripoli around 653, where they destroyed the fleet and arsenals that were preparing to go to Constantinople, and after killing the Arab garrison, fled by boat. The first Arab civil war probably allowed many other Syrians to flee; perhaps Kallinikos (see below) was one of them. Baladhuri explicitly states that the "industry", i.e. all the infrastructure needed for naval and siege warfare, was confined to Egypt at the beginning of the 660s. Caliph Muʿāwiya's (661-680) subsequent transfer of soldiers and engineers to the coast of Syria was apparently necessary because the Syrian "industry" had been damaged somehow.

---

[79] See Sijpesteijn 2007: 443 on the Coptic pope Benjamin's exhortation to the Copts that they should join the army of ʿAmr b. ʿĀs in the 640s; for the construction of a history of persecution, see Papaconstantinou 2006.

[80] See Palme 2007: 271 for the removal of Apion in 629; the sieges attested by John of Nikiu in the years 640, (*Bahnasā), 641 (*Tendunias, *Babylon) and 642 (*Kīlūnās, *Roman fortress near Antinoe, *Alexandria) show local parties acting in much the same way as they had during the Herakleios' revolt in 608-10.

[81] Herrin 1989: 206-19; see also chapter 6.2.1 above.

[82] For instance, an Arab raid reported in 663 against Thrace even reached the walls of *Constantinople, which necessarily presumes a naval venture. The Arabs had camp followers, either clients or Christian conscripts, but they did not attack the city itself and they turned back from the city when they came within ballista range. See also Appendix III.

Indications are that the Muʿāwiya systematically worked to strengthen or reestablish naval facilities along the Syrian coast in order to prepare for new naval raids along the Anatolian coast and the attempted siege or raid on *Constantinople (670f). This was best achieved by moving those who possessed the necessary expertise in military engineering and craftsmanship from inland Syria to the coast. The number of craftsmen and engineers with military competence inland was still substantial. When Herakleios defeated the Persians in 628, the reorganization of the defenses of the cities of northern Syria and Mesopotamia, close to the Persian border, must have had a high priority. For instance, large Persian contingents under the son of Shahrbaraz, the Persian general who had occupied Syria and Egypt, had remained in Syria in return for providing defenders; indeed, they fought the Arabs until the main Roman field armies were defeated. Afterwards, they reached an agreement with the Arabs and were settled in Emesa;[83] this might explain the significant presence of Persians in Syria for a generation afterwards. Furthermore, the hard resistance against the Arabs offered by several cities, such as the Roman garrison at *Dara (640), means that garrisons and arsenals were well established and functioning. Similar preparations must be presumed for several north Syrian and Mesopotamian cities. Thus, Muʿawiya transferred Persian soldiers, descendants of those who had settled in Emesa, from the inland garrisons to the coast in AH 42 (661-62). A little later, around 667 (perhaps connected to a census held in 668, although Umar reportedly made a census as early as 639), Syrian "craftsmen and carpenters" (ṣunnāʿ wa-najjārīn) were sent from the inland to the coastal cities in order to provide labor for the navies that were being organized for expeditions into the Aegean and the central Mediterranean.[84]

One of the craftsmen who was transferred in this manner was probably the famous Kallinikos, the inventor of Greek fire around 670 (see *Constantinople 670f). He was a Roman engineer from Heliopolis in inner Syria, but escaped to the Byzantine Empire. He might have been among those transferred by Muʿāwiya around 667, but his story may have been misplaced from events in the 650s or early 660s. At any rate, his escape clearly demonstrates that the Arabs used Roman engineers with military competence even a generation after the conquests: in Greek he is referred to as an

---

[83] Dionysius 54.
[84] For the census, see Michael the Syrian 11.XII.435/450 and Hoyland 1997: 418 n. 104; for the transfer, see Baladhuri 117; for early Islamic censuses in general, see al-Qadi 2008.

arkhitektōn, and in Syriac *naggārā*, cognate with the Arabic *najjār*.[85] There are indications that he was a civilian master builder who continued his or his father's trade without much disruption since the mid-630s, but maintained a certain level of military knowledge, due to self-defense arrangements continued under Arab patronage (see below). Since reliable troops were busy on the great fleets, and due to the propensity of Christian engineers and craftsmen to flee if given the chance, the Arabs on their expeditions against Anatolia in the 660s and 670s relied on a combination of basic siege tactics in order to storm cities or political means to convince them to surrender. Otherwise, specialist manpower was drawn from Persian, Armenian or Roman clients.[86] Subsequently, the second civil war took much of the energy of the Caliphate, so apart from the anecdote at Mardin (see also below), there is little evidence. Formerly Roman urban craftsmen with military competence were still maintained in Arab service throughout most of Syria and Mesopotamia in the late 7th century, but this was still mostly confined to the frontier and coastal arsenals and dockyards.

At port or in the cities, engineers and craftsmen could be controlled and exploited through a system similar to that attested in Egypt. We find an inland example at Mardin, probably during the second Arab civil war (680-92), where the Arabs even pressed the local clergy into manning trebuchets and defending the walls (presumably the cathedral and local monastery had their assigned *pedaturae*). This caused such concern for the clergy's spiritual well-being that they approached Jacob of Edessa with the following question:

> When our bishop of Mardin was attacked by those from the outside (*hānōn d-men l-bar*), the Arabs, who are ruling on the inside (*ṭayyāyē hānōn d-šallīṭīn men l-gaw*), ordered that everyone go out to the wall to fight and did not exempt anyone from going out, not even the priests. Then a priest or a deacon, when the battle was in full swing (*'šen*), threw a stone from the wall and struck and killed one of the fighters attempting to scale the wall. How is it right to deal with him as regards the canons? And I want to learn whether it is a sin for him [alone] or [also] for other priests and monks inasmuch as it was not of their own wish to be pulling the rope of machines (*māngānīqōn*) of war and to be throwing stones and killing fighters outside. And is it right that they serve in the priesthood, and is it right for a short time that they be subject to the law?[87]

---

[85] For Kallinikos, see Theophanes 353f and Michael the Syrian, 11.XIII, ed. p. 436, Hoyland 2011: 166ff.

[86] For the situation in c. 640, see Appendix II; for the 650s and 660s, see Appendix III.

[87] See Hoyland 1997: 606 for his translation and transcription quoted here; Jacob's answer (who leaves much of the judgement to the conscience of the individual priest and

It is thus clear that the Arabs maintained the obligation of civilians, organized in corporate (here the monks) and other social units, to defend and presumably maintain their *pedatura* of the city walls, including siege engines. It is certainly this type of environment that explains Kallinikos decades after the conquest. On land-based campaigns, however, it may have been too risky to bring the Syrian Christian craftsmen along. We know from the papyri that Egyptian laborers had a tendency to escape well into the 8th century, but in Syria, the situation is more poorly attested and has to be reconstructed from incidental remarks in different chronicle traditions.

In the north, soon enough the Arabs had to address defensive concerns, as the Byzantines began counter-offensives during the second civil war.[88] The whole frontier region against the Byzantines was heavily fortified, and must have had formidable defensive capacities, if we are to believe surviving siege descriptions. While this process is well known from Baladhuri, who gives details on places, dates, sizes of garrisons, and occasional military events, he hardly ever mentions those who actually built these fortifications or operated their engines.[89] For this we must turn to the Syriac sources, where we find that border fortifications were regularly built by Christian craftsmen recruited in northern Syria and Mesopotamia. Thus, the same people who had provided so much of the defense against the Persians, were now regularly involved in the defense of the Caliphate, although the examples only begin to multiply in the 8th century (for which see below). Sometime after 700, the competence of Syrian craftsmen was also used on offensive operations overland. Why this had become possible is uncertain. It seems that very strict administrative measures had been put in place in Syria during the censuses of 692 and 712, just as in Egypt. During the latter census, the whole population received neck seals in lead detailing their place of origin.[90] The closer control of Syrian craftsmen

---

ultimately to God) is quoted in n. 16. Hoyland suggests that "those outside" were the Byzantines, but there is nothing to indicate the killing of Christians here, so I would suggest that the second Arab civil war is a more suitable context, especially as this collection was compiled in the 680s. Since the text is only available in manuscript form, the historical context is uncertain, and also difficult to date precisely, I have not added this siege to the CO.

[88] The tit-for-tat warfare typical of the 8th century has been very well served by Lilie 1976 and Kennedy and Haldon 1980.

[89] Baladhuri organized his work geographically, so see his chapter 2.16, 3.3 and 4.1 for examples.

[90] See *Amasia 712 for references; although censuses had taken place in Syria since 668 at the latest and the census of 692 was perceived by contemporary Christians as the begin-

seems to have made it possible to bring them on campaign. To begin with, this was limited to the frontier, as Mopsuestia was built up by the same expedition that besieged *Taranton in 702. At *Tyana in 708-9, the Arab armies disposed siege engines that were powerful enough to break down (parts of) the walls, although the source of labor is not mentioned in either instance. However, the fleet that was sent against *Sardis and *Pergamon in 716 in order to establish bases on the way to *Constantinople (717f) carried soldiers and workmen (qalīgrē).

The great Arab siege of *Constantinople in 717-18 may have involved 200,000 men (Egyptian rowers, regular troops on ship and land, volunteers for jihad). The overland army was accompanied by 12,000 Syrian craftsmen (qūlāgrē), who had a train of 6,000 mules for supplies and 6,000 camels to carry weapons and trebuchet parts (zaynā saggī'ā, manganīqē). Islamic sources describe these engines in operation, while the Syriac sources confirm that the Arabs built a double set of stone walls to enclose their camp before Constantinople over the winter. The Arabs had no problem providing labor for menial tasks, especially in war, but certainly some of the engineers would have directed the efforts, since they were regularly employed in similar civil engineering projects in Syria and Mesopotamia, such as digging irrigation canals, building towns and fortifications, and repairing or building bridges.[91] In addition they were regularly recruited for constructing border forts. For example, bāb al-atrāk (Gate of the Turks, defending against Khazar invasions) in the Caucasus was built in 732 with ūmānē (craftsmen), naggārē (carpenters) and qalāgrē (laborers),[92] while Melitene was reconstructed in 761 by Syrian qalāgrē.[93]

---

ning of "Egyptian slavery" (i.e. in a biblical sense; see Hoyland 1997: 414, 418f n. 104), it seems that the census in 712 is the first where such strict measures of control had been imposed upon the Syrian population. See further Robinson 2005 on neck-sealing and al-Qadi 2008 on censuses.

[91] See e.g. Chr. Zuq. AG 1029 (Harrak 1999: 159f, AD 717-18) for construction of canals, forts, towns, villages and planting of farmland; AG 1052 (Harrak 1999: 165, AD 740-41), bridge across the Euphrates; AG 1053 (Harrak 1999: 165, AD 742-43), organization of craftsmen (ūmānē, qalāgrē) to rebuild bridge across Tigris ruined by flood, but prevented due to death of leader.

[92] Chr. Zuq. AG 1043 (Harrak 1999: 159, AD 731-732): "The same Maslama gathered a large crowd of craftsmen, carpenters and labourers and prepared everything necessary for building. They went in and rebuilt the Gate of the land of the Turks that he had torn down." Cf. ed. p. 169.8-11:

ܗܘ ܕܝܢ ܡܫܠܡܢܐ ܟܢܫ ܐܘܡܢܐ ܣܓܝܐܐ ܘܢܓܪܐ ܘܩܠܓܪܐ ܘܛܝܒ ܟܠ ܕܡܬܒܥܐ ܠܒܢܝܢܐ. ܘܥܠܘ ܒܢܐܘܗܝ ܠܬܪܥܐ ܕܐܪܥܐ ܕܛܘܪܩܝܐ ܗܘ ܕܣܚܦ ܗܘܐ.

[93] Chr. Zuq. AG 1072 (Harrak 1999: 201, AD 760-61): "'Abd-Allāh, son of Muhammad, the Persian Caliph, sent the son of Wahhab, along with numerous troops and craftsmen from

What is perhaps even more surprising, in view of later Islamic law and common scholarly perceptions, is that the Syrian Christians became directly involved in Arab (i.e. Muslim) internecine conflicts, and different factions seem to have supported different sides during the third civil war and subsequent 'Abbasid takeover (744-51). As late as 750-51, when the Arabs (representatives of the last Umayyads?) at Mayafarqin (Martyropolis) began mistreating the Christian population somewhere in or just north of Tur Abdin, the Christians revolted and organized armies that chased off all the Arabs, who resorted to recruiting other Christians to fight against the rebels. However, the rebels found employment with the newly established 'Abbasid regime so were clearly on the winning side. The militant response of the Christian populations to mistreatment, not to speak of their martial abilities, must derive from their continuous involvement in Umayyad border warfare.[94] During the third Arab civil war, Marwān, the Umayyad governor of the frontier province protecting northern Mesopotamia, mobilized resources (soldiers and craftsmen) that were normally used against the Byzantines and Khazars to march on and storm *Emesa and several other cities (745), where the walls were razed. It is reported that both sides pressed villagers armed with slings into service, and that 5,000 of them were killed in one of the battles between the parties. However, after the Christian rebels had joined the 'Abbasids, new crews of craftsmen and laborers were recruited to demolish the walls of Syrian cities where the Umayyads had strongest support.[95]

Conversely, parties to Byzantine conflicts sought support or refuge in the Caliphate. For example, a disaffected Armenian garrison surrendered

---

the whole Jazira, to rebuild Melitene of Cappadocia, which had remained in ruin for eight years. People and army were brought in and it became more inhabitable and prosperous than before." Cf. ed. p. 222.20-24:

[94] An early instance is provided by the *Canons and Resolutions of Jacob of Edessa*, where one question concerns a Christian deacon who during a severe famine found outcome by joining some soldiers, but then deserted after the famine was over. It is unclear which soldiers are meant, but considering the geographic location and timeframe, the deacon seems in fact to have joined the Arabs, perhaps during the second civil war (680-92). See Hoyland 1997: 605f for a translation and further references.

[95] *Chr. Zuq.* AG 1064 (Harrak 1999: 190, AD 752-53): "The Persians ('Abbasids) destroyed the cities to their foundations. After the Persians returned to the country for the second time, subdued it, strengthened it and ruled in it, they destroyed all the walls of the Syrian cities to their foundations at the order of the Caliph. Many laborers and workmen gathered and tore down all the walls, burned their gates and took away their plated copper and iron.

*Kamakhon in 793 although it put up fierce resistance in 766 (see below). Such military colonists could be put to work for the Caliphate, as were the independent Armenian nobles throughout the 7th and 8th centuries.[96] In return, the Byzantines exploited the Arab civil wars to deport Syrian and Armenian populations from *Germanikeia (745), *Melitene and *Theodosiopolis (750) in the Arab frontier provinces and settle them in newly founded cities in Thrace to defend against the Bulgars.

As late as 766, Syrian craftsmen were recruited for the siege of *Kamakhon (Qamh). From the Chronicle of Zuqnin, written by a Mesopotamian in c. 775, we have the following testimony:

> After the whole army marched in, they besieged a fortress called Qamh that was located on the borders. But among the many craftsmen (*qalāgrē*, ܩܠܐܓܪܐ) from all of the Jazira who marched with the army, some were left by 'Abbas to rebuild a fortress called "Ziad", while others marched with the army. 'Abbas brought Armenian wagons, in which he pulled numerous cedar beams. He ordered carpenters (*naggārē*, ܢܓܪܐ) to make mangonels (*mwndnyqn*, ܡܘܢܕܢܝܩܝܢ) with them, which he placed on a peak facing the fortress so that they might hurl stones with them inside the fortress. The Romans, who were inside the fortress, also mounted mangonels against them.

From elsewhere in the Zuqnin chronicle, we know that the majority of the population in the region was still Christian (cf. above), although by the late 760s, Arab (i.e. Muslim) landowners were increasingly common, and the Islamization of the local population began to gain pace as the 8th century progressed.[97] Probably by around 800, the Muslim population had become large and diversified enough to take over more of the responsibilities that had rested upon the Christian population, although the transition must have been gradual and is partly hidden from our view. However, by 812, an Arab engineer, who had converted to Christianity and found service in the

---

What the strong and wise kings had built at high cost for the protection from enemies, the Persians tyrannically destroyed and threw on the ground." Cf. ed. p. 207.17-26:

[96] See Garsoïan 2004: 118-36 for the very variable policies of Armenians and Arabs vis-à-vis each other.

[97] Sijpesteijn 2007: 451ff notes that the large-scale islamization and arabization of Egypt only began in the mid-8th century; the situation seems very similar to Syria. For evidence of both Christian and Muslim craftsmen in the first Islamic centuries, see Finster 1982; there must have been a long period of overlap as locals converted and descendants of the first Muslim garrisons began to engage in civilian trades.

Byzantine Empire, had defected to the Bulgars and taught them how to build the advanced siege machinery necessary to take *Mesembria.

The continued participation of Christian craftsmen in direct warfare against fellow Christians in the East Roman Empire can be explained by a number of factors. The Arabs had from the beginning sought to exploit divisions within the societies they conquered; conversely, among Romans, Visigoths, Armenians, Persians and Franks there were often parties to internal struggles who were willing to either call in the Arabs as external allies, or immediately supported them when the conquest began in order to preserve their own positions. Often these were very powerful people, such as landowners, military officers, or high-ranking clerics with much to lose. When supporting the Arabs, however, they could call upon traditional networks of patronage or means of administration and gain even further authority. In some instances, the Arabs sought to create internal divisions between the conquered populations. While it was argued above and in chapter 6.2.1 that religious divisions cannot explain the course of the Islamic conquests, we know that e.g. Mu'awiya set the Monophysite and Chalcedonian churches up against each other by having them compete for the privilege of collecting taxes and administering the Christian population. Similarly, the Arabs moved Christians from different provinces around to areas where they had no previous networks, ensuring a degree of impartial oversight where local officials tended to be lax in their enforcement of taxes and labor obligations.[98] Furthermore, local populations had strong incentives to defend against Khazar (and possibly Byzantine) raids, and since they suffered the consequences of internecine fights amongst the Arabs, they had little choice but to participate in their wars as other groups in the Arab empire. Finally, there was the simple monetary incentive. While the Arabs resorted to fear, threats, and exploiting or extending traditional obligations from unwilling populations, they also paid laborers; at least some of the specialists employed in the Egyptian dockyards received quite substantial wages, while many of the villagers who carried supplies for the Abbasid forces at *Kamakhon (766) were lured by the promise of money payments.

---

[98] See e.g. the debate arranged by Mu'awiya between Chalcedonians and Jacobites (Monophysites) in the surviving fragments of the *Maronite Chronicle*; for the crossposting of Christian administrators, see Dionysius 132ff.

7.4 Conclusion: From Appropriation to Domestication

Modern theorists distinguish between appropriating and domesticating technology. The application of its modern emphasis on consumer meanings and cultural perceptions is difficult to trace with the available sources for this period. However, it is possible to distinguish analogous developments on a more basic level: whether a technology is merely *used*, and thus vulnerable to loss, or integrated into society in such a fashion that it is self-perpetuating. The Visigoths and other barbarians adapted to a changing society so effectively that they are normally blamed for developments they were trying to keep up with. Other modes of transmission (co-evolution, client integration, conquest appropriation) are more obvious, but our view depends on the role of elites or social carriers, their relationship to practitioners, and which social group has left an imprint in the sources. For Syriac chroniclers, the exactions imposed on them by the Arabs extended obligations that were part of daily experience since Roman times. For early Islamic authors, military technology was of limited interest when it was only appropriated, or put to use for the benefit of the Islamic state; it only truly became domesticated when it was handled by Muslims. This largely explains the problem posed by Hugh Kennedy in the introduction (0.1.2): Islamic technology does not become visible in the Islamic sources until it came under the full political, social and epistemological control of Muslims. When exactly this happened is beyond the scope of this study, but it may be traced in outline. Large-scale conversion first occurred in and near the great garrison cities around 700, and made a certain economic impact, although its effect on warfare was still slight. By the end of the 8th century, however, there were telling signs that Muslims were becoming increasingly prominent in rural society as well as cities outside the old garrison settlements in Egypt, Syria and northern Mesopotamia, whether they were converts from the local population or descendants from Arab settlers. These are the regions where most of the early "Islamic" sappers had been recruited. Around 800 or shortly after, the change seems complete. By then, the Byzantines could recruit former Muslim military engineers who had mastered the full range of late antique siege warfare (e.g. the Arab at *Mesembria 812). The process may have been hastened by the Abbasid revolution in 750, which brought large numbers of Persians into mainstream Islamic society and ended the period of Arab dominance. Indeed, the Banu Musa brothers, who entered the historical stage as the first great Islamic engineers after 800, were of Persian origins.

CHAPTER EIGHT

DIFFUSION OF THE TRACTION TREBUCHET

The importance of trebuchet can hardly be overestimated. It represented a new type of military technology, as it was based on lever power rather than the classical means of torsion and tension. While it appears more ramshackle and primitive than the finely tuned torsion weapons of the Roman army under the Principate, it seems in fact to have facilitated a far more aggressive form of siege warfare when it began to spread sometime in the 6th century. Its diffusion demonstrates how the successor states were geared towards sustaining technology and associated logistical problems, and why some conquerors were more successful than others in appropriating this technology and successfully domesticating it.

8.1 *State of the Question: Historiography and Technical Aspects*

Despite recent advances in the historiography of the traction trebuchet, most models fail to take into account the socio-economic infrastructure of the societies that used it, or have applied social models that are not conducive to support advanced technology, especially not on the scale required for siege warfare. This has resulted in a historiographical impasse where the trebuchet can to a certain extent be traced through philological means, but most scholarship ignores the mechanisms by which it diffused into and throughout the successor states.

8.1.1 *Historiography of the Traction Trebuchet*

The origins, development and diffusion of the trebuchet have provoked much interest among scholars. In the last generation alone, well over a dozen of substantial articles and book chapters have appeared on the problem, while many military historians have discussed some aspect of this innovation in monographs or general surveys of the period.[1] The problem

---

[1] For the following discussion, see Huuri 1941; Hill 1973; Gillmor 1981; King 1982; Chevedden et al. 1995, 2000; Tarver 1995; Dennis 1998; Chevedden 1998, 2000; DeVries 1992: 133f (and further on other types of trebuchets). Between them, these articles summarize most of the last century and a half of scholarship.

is that most scholars have followed a strictly technical-philological approach without examining the society in which a particular type of technology is used and developed. Social diffusion mechanisms are also not discussed. Thus, although specific names can be traced over great distances, some scholars have great difficulties accepting an innovation at a certain point because it does not square with preconceived notions of society, or they simply cannot conceive how a certain innovation could reach so far or have such a great impact.

Most of the terminology associated with artillery in Greek, Latin, Arabic and Syriac sources has been assembled by Huuri. He identified the arrival of the traction trebuchet to Byzantine territory quite late, to the siege of Manzikert in 1071 based on the description of the pulling crews operating the Turkish trebuchets. The best description in later sources was the crusaders' siege of Lisbon in 1147; there was long a consensus that the traction trebuchet only came to the West sometime between these two dates but had been common in the Islamic world for some time. Hill argued for the introduction of the traction trebuchet to the Middle East from China in the 7th century based on Islamic sources. It was transferred via Central Asia to the early Islamic conquerors, who used the first trebuchet of their own at *Mecca in 692, thus dating the transfer to around 660 and placing it somewhere in northeastern Iran. While noting the early use of a catapult at Ta'if in 630, Hill argued that this must be a torsion type weapon. The Chinese hypothesis has received strong support due to Needham's monumental study on technology in Chinese civilization. As the traction trebuchet has an ancient pedigree in Chinese sources, the hypothesis is eminently reasonable.[2]

However, the dating of its introduction has been pushed back to the late 6th century by a number of scholars. White noted that the *Miracula sancti Demetrii* contains the first recognizable description of a traction trebuchet in the Mediterranean area at *Thessalonica (586). Vryonis, who first published an English translation of the *Miracula*'s description the traction trebuchet, did not actually recognize it as such, but both he and White argued that heavy siege machinery had to be transferred from the Romans to the Avars some point before this use. Vryonis preferred the late alternative date for the siege, 597, in order to link the transfer to a specific incident, the Bousas-affair at *Appiaria (586), which is normally dated to 587.[3] In

---

[2] See Needham and Yates 1994 for the standard survey and the literature quoted in the previous note for further examples of and references to Chinese artillery.

[3] For the debate see briefly DeVries 1992: 133f; Vryonis 1981.

contrast, Howard-Johnston has on numerous occasions argued that it was the Avars who introduced this innovation to the East Roman Empire, thus placing the transfer sometime after 568 but retaining a route of diffusion from Central Asia.[4]

Gillmor dated the introduction of the traction trebuchet to the West to the 9th century at the latest, based on bishop Abbo's poem on the Danish siege of Paris in 885-86, thus well before the siege of Lisbon in 1147. Although a diffusion route from Islamic Spain in the late 8th century is plausible, there are good philological grounds for proposing a route of diffusion from the Byzantine Empire through Italy in the mid-late-8th Italy.[5] Chevedden has in a number of articles proposed an elaborate but precise route of geographical and chronological diffusion based on the assumption that a specific set of terms all denote the traction trebuchet or some variety of it. In short, he follows a 6th century date for the introduction of the traction trebuchet to the eastern Mediterranean, arguing it was well known before the siege of *Thessalonica (586), although here he opted for the late date, 597, and a further westward expansion to Spain perhaps as early as around 600, but certainly by 673, and into the Frankish kingdoms in the early 8th century.[6]

Apart from Dennis and Howard-Johnston, most of these scholars have made no, or only a very limited, use of East Roman sources. Thus the introduction of the trebuchet to the East Roman Empire is problematic since its first arrival and subsequent internal diffusion have never been securely identified. Dennis has shown exactly how common the traction trebuchet had become in the last three decades of the 6th century by linking it to the term *helepolis* or "city-taker," which occurs frequently in Theophylact Simocatta and other Greek sources. His argument shows quite clearly that it was too widely distributed at an early date for this to be an Avar introduction; this is also borne out by the Syriac evidence (8.2.2 below). Furthermore, the great problem in most of the research so far is that technology has not been linked to its general socio-economic and specific institutional context. While Rihll has little new on the transfer of the trebuchet, she provides valuable insights into the connection between technology and (ancient) society, a topic often overlooked in the literature reviewed above. This also has implications for how technology was diffused, and will be discussed in chapter 8.3 below.

---

[4] Howard-Johnston 1984.
[5] Gillmor 1981 hints at the possibility on philological grounds, but see chapter 8.2.2 below.
[6] Chevedden 1998, 2000.

8.1.2 Technical Aspects of the Traction Trebuchet

The traction trebuchet displaced classical, torsion-powered artillery because it was simpler and required less competence to build, while maintaining comparable range and power, and it had far higher rates of firing and accuracy (when operated by a trained crew). Furthermore, it was probably *safer* to operate than tension weapons, whose bundles of taut sinews stored up huge amounts of energy even in resting state and were prone to catastrophic failure when in use.[7] Chinese and high medieval European and Islamic sources provide statistics on the phenomenal range, power and firing rate of traction trebuchets. These were made by experts with long experience and following traditions now lost to us; still, modern reconstructions with fairly small trebuchets provide good confirmation. Tarver made a trebuchet that was operated by crews ranging from a skeleton crew of four pullers up to a full complement of twenty-five people. It could shoot stones in the range of 1-5 kg from around 60 to 145 meters, depending on the crew/weight ratio.[8]

While somewhat simpler to construct than its ancient predecessor, the logistical requirements of the traction trebuchet were just as great or even greater, and it needed even more trained crews and engineers to operate effectively, as well as a large number of expert craftsmen. These not only had to choose the correct wood and cut it correctly for the rotating beam and other important parts, but also overcome a host of problems in creating the struts, framework, axle, sling pouch and release hook. The chief carpenter/architect (*rabb naggārā*) who fooled the Arabs at *SYLWS (664) into operating a defective trebuchet could have manipulated any of these elements subtly to have the Arab trebuchet shoot progressively shorter. The machine also needs a large amount of rope, and if built with a protective framework (see below), required further wood, wickerwork and fresh hides. For instance, the dry hides brought to *Thessalonica 586 proved useless against burning projectiles and had to be replaced with the skins of freshly slaughtered animals. Aiming required a coordinated crew led by someone who held and aimed the sling. His responsibility would be to give orders to the crew, release the sling at the right moment, and check where the projectile hit before beginning the next cycle. Furthermore, experimentation shows that while teams with even small amounts of training

---

[7] Tarver 1995: 141.
[8] For the Chinese and high medieval figures, which are difficult to apply to this period since any reference figures are lacking, see the literature cited in chapter 8.1.1; further on modern experiments, see Tarver 1995.

soon shot four projectiles a minute, accuracy depended on using projectiles of even weight. This meant laboriously collecting suitable stones and hewing them to order. For the battery at *Thessalonica (586), which counted 50 machines, this could mean as much as 200 stones per minute and 12,000 per hour. Effective use thus presupposed an efficient infrastructure for logistics, training and maintenance. The social units who were assigned their *pedaturae* could be expected to have at least some experts that would train citizens, perhaps even in peacetime, although such an infrastructure has to be inferred from incidental information analyzed in chapters 1.2.5, 6.2.2 and 6.2.3.

The most basic type appears to have been a simple pole-framed trebuchet, which probably had a small pulling crew of a handful of men and was used for throwing fairly small stones over great distances. It was thus quick to set up, easy to use, and efficient as an anti-personnel weapon, either defensively or during attack. It might even be used as a field weapon in battle, according to the *Strategikon*, where it is described as mounted on carts. A larger weapon with a sturdier frame was also known, and the standard term after 900 appears to have been a *labdarea*, based on the lambda-shaped frame. Such a weapon could have larger pulling crews and carry a far heavier payload, certainly destroying parapets and enemy siege engines. The final type was a trestle-framed trebuchet, named *tetrarea* for its square base (see discussion in 8.2.3 below). It may not have been much larger than the *labdarea*, but would have had a sturdy, protective trestle-shaped frame to protect pulling crews, thus allowing it to be placed closer to the walls and delivering heavier payloads over a shorter distance. The very largest traction trebuchets are reported to have had pulling crews of scores or even hundreds of men.[9]

8.2 The Philological Evidence

In her study of classical artillery, Rihll pointed out that during periods of innovation and change in technology, multiple versions of the same or similar machines often coexisted. Furthermore, innovations rarely had a clear terminology, as old terms were used with new meanings or new terms were invented without necessarily following precise technical refinements. Similarly, the terminology employed for traction trebuchet is very difficult to establish, since it is attested in half a dozen late antique languages and

---

[9] See in particular Dennis 1998 among the studies quoted above.

may have existed in (now indeterminable) different sizes and marks (cf. chapter 8.1.2). The confusion firstly arises out of the fact that none of the classicizing 6th-century historians had a good term for it. Furthermore, what became the standard term in the East during the 7th century, *manganon* and its derivatives, is used in our most authoritative source, the *Strategikon*, with two distinct (although related) meanings, besides other terms. The early confusion in terminology resulted from its recent innovation; it had no obvious classical predecessors, and also developed into at least three distinct types early on, with a great deal of overlapping of generic or misapplied terms for "machines," "artillery," and specific "mark" or "caliber." The range of terminology must have been especially difficult for those who were not directly involved in the everyday operation and maintenance of siege engines.[10]

8.2.1 *Generic, Classicizing and Uncertain Terms*

The term *helepolis* (plural *helepoleis*) or "city-taker" was originally used for siege towers and powerful rams in classical literature. It is still used in generic fashion for "siege engines" by the *Miracula sancti Demetrii* at *Thessalonica (586), where the term *helepoleis* introduces a string of machines: "*helepoleis*, both iron rams and huge stone throwers and the [engines] called *khelōnai*" (ἐλεπόλεις καὶ κριοὺς σιδηροῦς καὶ πετροβόλους ὑπερμεγέθεις, καὶ τὰς καλουμένας χελώνας).[11] However, Evagrius began to to associate *helepoleis* with artillery, particularly *lithoboloi*, which we know were used of traction trebuchets (see 8.2.3 below). Dennis argued that Theophylact Simocatta employed *helepolis* as a classicizing *Ersatz* term for trebuchet. This then stuck in literary contexts through Byzantine history long after the period treated here.[12] The generic meaning still attested in the late 6th century may cause problems for this interpretation. There is one more clue, however. Theophylact alternated *helepolis* with the very generic *katapeltēs*, but Evagrius explained that the *katapeltai*, a word by then belonging to the high register of literary Greek, were commonly called *lithoboloi*.[13]

Furthermore, *helepolis* occurs in Agathias on one occasion, at the Hun assault on the *Chersonese in 559. In light of Dennis' argument, and based on the increasingly specific use of this word in the late 6th century (Aga-

---

[10] For the classical situation, see Rihll 2007.
[11] *Miracula St. Demetrii* 139.
[12] Dennis 1998; cf. *Appiaria 586; *Persian fort in Arzanene 587.
[13] For Theophylact, see *Beïudaes 587 and *Diocletianopolis 587; for Evagrius, see *Dara 573, *Martyropolis 589, and *Akbas 590.

thias was writing in the 570s), it seems likely that Agathias uses this term to describe the traction trebuchet. He also uses another, extremely interesting term, *akontistērion* (ἀκοντιστήριον). While the further implications of this term are discussed below, it seems to denote an engine that had no good classical name. However, since he depended upon reports from military officers or other observers for his history, he probably would have received different kinds of reports from different regions which he filtered in his own peculiar way. Hence, he might not be aware of the fact that he was describing the same engines as the *helepoleis* when referring to *akontistēria*, which clearly were unknown to the classical arsenal; nor does he seem aware of a change in mode of warfare beginning with Narses' Italian campaign (see 8.3.1 below).

Finally, the term *ballistra* (βαλλίστρα) began to go through a shift in meaning in Greek around 600. It occurs in the *Strategikon* c. 600, denoting a light artillery piece that was carried on carts. While Dennis and Gamillscheg originally believed that was a bolt-shooting artillery piece driven by a large bow (as in Procopius, cf. chapter 5.2.2), Dennis later proposed a convincing argument that the *Strategikon* is actually describing a trebuchet.[14] Based on this, the term *ballistra* may be used with a new meaning at *Constantinople (626, 663), although the explicit mention of *toxobolistrai* at *Constantinople in 715 in addition to lever-operated machines (see below) means that tension-powered bow artillery (i.e. the *ballistra* of Procopius) was still a regular part of the East Roman arsenal. It is also used quite consistently from *Constantinople (626), and frequently set in contrast to stonethrowers whenever precision was required. In the West, however, the situation is more ambiguous. Isidore uses *ballista* of both stone- and bolt-shooter. Since the alternative *fundibulum* is also used for stonethrowers, perhaps a "stonethrowing" ballista was the old-fashioned, one-armed, torsion-powered *onager* (cf. chapter 5.2.2). Otherwise the term is rare. Venantius Fortunatus mentions the *ballista* placed on the fortified residence of Nicetius of Trier. While on the one hand Ventantius came from Italy and would presumably have been familiar with military terminology as it was used during the Gothic wars, the peculiar phrase *gemino ballista volatu*, literally "a ballista with twin flight," may hint at something else. The phrase poorly describes the arc of the simple one-armed *onager* as well as a bow-powered bolt-shooting *ballistra* as described by Procopius. It may instead reflect the movement of the whole lever beam in a traction trebu-

---

[14] See Dennis 1998 for discussion and references.

chet: as one side is pulled down, the other moves up, producing a distinct see-saw movement.[15]

8.2.2 Manganon *and Its Derivatives*

The word *manganon* (μάγγανον) is fundamental to understanding the diffusion of the traction trebuchet. In classical Greek denoting "magic" or a "trick," this semantic meaning passed to the derivations *manganeia* (μαγγανεία), which is frequently used in ecclesiastical writers to denote the trickery or wiles of the devil, and *manganeuō* (μαγγανεύω, verb) or *manganeuma* (μαγγάνευμα, noun), which is often used for more simple trickery or quackery.[16] The term *manganon* however became applied to lever-machines that were able to shift very large weights (thus appearing "magical" or at least clever) and used for several purposes. In the 1st century, Heron of Alexandria used the term as an element in a pulley system to operate other types of engines,[17] but it was also associated with construction (since it was used of technology employed to operate cranes) and the mechanical opening of doors (see below). A 5th-century hagiography records the use of such lever-machines (*mangana*) to tear the flesh of St. Euphemia with the aid of four stones as counterweights.[18] As early as the 4th century, *mangana* were associated with warfare: Epiphanius, a late 4th-century ecclesiastical historian, elaborates on how Joshua captured Jericho when "there were neither engines nor *mangana*, no rams, no *helepoleis*, but [nonetheless] the sound of the horn trumpet alone" brought down the walls.[19] He also used the word in adjectival form, *manganika*: thus μαγγανικαῖς μηχανίαις, but here with its meaning of magic tricks.[20]

In the late 6th century *Strategikon*, there were still two uses of *manganon*: one for the war machine (alternating with *petrobolos*; cf. 8.2.3), and one for a counterweight-operated construction engine that had been adopted for wall defense by dropping heavy objects (stones, sharpened logs

---

[15] E.g. as late as 552 a *Gothic fortified camp employed bolt-shooting *ballistrai*; see further 8.3.2 below.

[16] See the basic entries in LSJ and searches on the root *maggan-* in the Thesaurus Linguae Graecae, which yields (23 April 2011) 848 instances in the canon of classical and Byzantine Greek literature, the majority of which are forms of these two variations.

[17] Heron *Belopoeica*.

[18] *Martyrium antiquior sanctae Euphemiae* 13.a, in F. Halkin, *Euphémie de Chalédoine. Légendes byzantines* [*Subsidia hagiographica* 41. Brussels: Société des Bollandistes, 1965]: 13-33.

[19] Epiphanius, *Panarion* (*Adversus haereses*) 3.82 (p. 124.13f): οὔτε γὰρ μηχανικοὶ οὔτε μάγγανα ἦν, οὐ κριοί, οὐχ ἑλεπόλεις, ἀλλὰ μόνη φωνῇ σάλπιγγος...

[20] Epiphanius, *kata Basileidou d'* 2 (vol. 1 p. 258.9).

and the like) which were then pulled up again with the counterweight.[21] After c. 600, the adjectival form (neuter plural) *manganika* became common for the traction trebuchet. Originally, this would have been used with a word such as *mekhanemata* or *organa*, but in most instances it occurs alone. It appears thus in a number of sources, beginning with *Miracula sancti Demetrii* on the siege of *Thessalonica (618) and *Chronicon Paschale* on *Constantinople (626), where the description of the engines in use leaves no room for doubt.[22] Based on these descriptions and alternation with other terms whose meaning is secured (see 8.2.3 below), it is clearly a stone-throwing machine operated by lever power and not used in any other sense in this period.

Early on, there is a companion term *manganarios* denoting an operator of a *manganon*. The first mentions are decidedly civilian. Pappus, a 4th-century mathematician from Alexandria, enumerated those who might need a knowledge of leverage as the "guiding principle of construction" (μηχανικὴ προηγουμένη τῆς ἀρχιτεκτονικῆς), including *manganarioi* and *mekhanikoi*.[23] There are also some sporadic inscriptions: one at Sardis roughly around 400 mentions a hydraulic *manganareios* (μαγγαναρείου [sic] ὑδραλέτα),[24] while another from Bosra in Jordan, dating to 274, mentions a *manganarios* named Malik responsible for a building project, and hence probably an engineer specializing as (amongst other skills) a crane operator.[25] According to the 6th-century astrologer Rhetorius, the Sun in aspect with Aphrodite (Venus) and Ares (Mars), or Ares and Hermes (Mercury) at nadir would produce "ropedancers or *manganarioi*".[26] According to LSJ, here *manganarios* means a "conjuror," presumably deriving a definition from the classical meaning of its root *manganon*. However, variations with Kronos (Saturn) produced builders or potters (οἰκοδόμους ποιοῦσιν ἢ κεραμουργούς). Hence the art of the *manganarios* was closely associated

---

[21] For engines in the sense of catapults, see *Strategikon* 10.3.17-27, 40 and 10.4.8 where it appears alongside *petrobolos*; for the counterweight machine, ibid. 10.3.14ff. I thank Ernst Gamillscheg for generously discussing his German translation with me in Vienna.

[22] In addition, see *Thessalonica 662, *Cherson 711, *Constantinople 715, *Nikaia 727, and *Mesembria 812.

[23] Pappus, *Synagoge* 8.2, 8.4 (pp. 1024.14, 1028.16).

[24] Sardis 7,1 169 dated to 4th-5th century, at http://epigraphy.packhum.org/inscriptions/.

[25] *Inscriptiones graecae ad res romanas pertinentes*, vol. III no 1165: .... | ἔτου(ς) πτ´ | μη(νὸς) Ὑπερ|βερετέου [ε´]. || [Α]ὔξι Ζιζί|ους. | [Οἰ]κοδ|ομὴ Μ(α)|λίχου | μανγα||[ν]αρίου | ἐπὶ ἀρχῆ|ς Ἀντονί|νου Ἀνουνέ|ου κὲ Ἀνί(ν)α.

[26] Rhetorius, *Capitula Selecta* 213.1, 217.9, in: F. Cumont, *Codices Parisini* [*Catalogus Codicum Astrologorum Graecorum* 8.4. Brussels: Lamertin, 1921]: 118-225.

with craftsmanship, construction and by extension the use of machines. Rhetorius' grouping may also indicate that ropes were a defining feature, so his use of the term is probably the same as that of Pappus. At the Hippodrome, there were Blue and Green *manganarioi* who had a central role in the races, operating the machinery that opened the doors; there were also teams of *manganarioi* who participated in parades held when pay was due. Although a decidedly civilian role, I suspect that the competence of these *manganarioi* was transferable to defensive purposes, and that when the Blues and Greens were mobilized to man the walls, these *manganarioi* would be assigned to operating engines.[27] Indeed, by the early 7th century it had become the standard term for a military engineer: Slav military engineers are called *manganarioi* (*Thessalonica 618, 662); so is Bousas at *Appiaria (586) in Theophanes' rendering. 10th-century Byzantine sources refer frequently to *manganarioi*, who were regularly employed in siege defense and on expeditions.[28]

The terms survive in a number of derivatives in other languages, which also hint about the time and direction of the diffusion. *Manganum* or *mangonum* is the Latin variety of *manganon* and indicates the stone-throwing machine. It was in common use from around 800 (here exemplified by the *mangonibus* used by Louis the Pious against *Tortosa in 809), but since Greek had long used *manganika* as a standard term by then, we may posit an earlier borrowing of the short form that has not been recorded in surviving texts. *Manglion*, the Armenian variety of *manganon*, was used by Sebeos to describe one type of engine on board the Muslim fleet moving towards *Constantinople in 654. While it could have been borrowed from Roman field army in Armenia, the Bible translation from Greek to Armenian from the early 5th century uses this to render the Septuagint Greek *ballistrai*.[29] The early introduction of the term in this context is surprising, but possible due to the early usage of the term in a military context by Epiphanius.

---

[27] See chapter 2.4.2 for a discussion and references.

[28] See e.g. *De obsidione toleranda* 10.3 and *De Cerimoniis* on the Cretan expeditions in Haldon 2000.

[29] See *Constantinople 654 for the text and the commentaries in Thomson and Howard-Johnston's translation for further references to biblical passages. I thank his Excellency the Armenian Apostolic Archbishop of Vienna, Mesrob Krikorian, for taking me through the Armenian text and suggesting possible solutions to the problem. Most importantly, it is possible that the terminology was inspired by a previous translation made from the Syriac Bible in the early 5th century. I have unfortunately not yet been able to follow up this point, but hope to resolve the issue with the aid of a specialist in late antique Armenian history and philology.

While the early Syriac form *mngnwn* or *mngwn* (*manganōn* or *mangōn*; the latter was probably only a spelling error) occurred at *Nisibis and *Dara in 573, it was the Greek (neuter plural) adjectival derivation *manganika* that became the basis for the standard term for traction trebuchets in Semitic languages by the early 7th century. The shift seems to have occurred about the same time as in Greek. Hence *mangānīqā* (singular with regular masculine singular emphatic *-ā*; the regular plural is *-ē*) was the Syriac term from the early 7th century and was used consistently with the meaning traction trebuchet throughout this period. It is used thus by Dionysius at *Edessa (630), *SYLWS (664; *Chr. Mar.*), where the pulling at the ropes by crews of men is the main point of the story, *Mecca (692) and *Constantinople (717f). There were also a few varieties in Dionysius, perhaps betraying other sources: *mangānīqōs* (plural) occurs at *Caesarea (640) and *Lapethus (650). The form *māngānīqōn* occurs in the *Canons and Resolutions of Jacob of Edessa*, concerning an otherwise unknown siege at Mardin in c. 690, with a classical Arabic plural or the Greek neuter singular.[30] Finally, the form *m[w]ndnyqn* (plural) occurs in the Zuqnin chronicler's description of *Kamakhon (766), and derives from the Persian or Arabic version of this word, with the grapheme *d* representing the Arabic affricate *jīm* (or *dj*) in *manjanīq* (see below). The *waw* seems to be otiose, as it is only used occasionally, and may possibly represent a schwa. However, the ending *-n* does not fit with Syriac grammar, and I suspect it may be the Persian plural *-ān*, since military and technological terminology in Syriac tends to retain the plural form of the language from which it is borrowed.[31]

The word *manjanīq* is the ubiquitous Arabic term for traction artillery derived from the Syriac *mangānīqā* through entirely predictable phonetic and grammatical rules. Islamic historiographers record the use of this machine by the embryonic Islamic state as early as 630 (see 8.1.1 above), which must have been a trebuchet in light of the contemporary Syriac use of the same term with this meaning and the well-documented prevalence of this engine by the late 6th century.[32]

---

[30] See the discussion and reproduction of this anecdote in chapter 7.3.2 above.

[31] Cf. the plurals in *-ōs* and *-ōn* noted just above; furthermore, note e.g. the engineers (*mēkhanikoi* discussed in chapter 2.4.2, with Syriac plural marking rendered as *myknykw*, i.e. *mīkanīkū* or the like, which approximated contemporary Greek pronunciation). Otherwise, the phonetic evidence is inconclusive; while the affricate pronunciation of *jīm* may betray either Arabic or Persian influence, the rounding of /a/ towards /o/ in Persian, which I suspected due to the orthography with *waw* in the Syriac, only took place around 1300 AD. I thank Judith Josephson for sharing her knowledge of medieval Persian with me.

[32] See Hill 1973, Chevedden (et al.) 1998, 2000 and Kennedy 2001 *passim* for further references to this term in the Arabic literature; these were probably used at e.g. the sieges

8.2.3 Descriptive and Functional Terms

Other terms for the trebuchet were descriptive. The straightforward "stone-thrower" or *petrobolos* is frequently employed by *Miracula sancti Demetrii, Strategikon*, and other authors. Although a generic term, the earliest unambiguously described trebuchet, employed by the Slavs at *Thessalonica (586), was simply called a *petrobolos*. This was clearly a traction trebuchet, because, in addition to a number of other distinct features, it had a protective framework covered by skins (i.e. the trestle frame noted in 8.1.2). Furthermore, the *petroboloi* at *Thessalonica (618) had crews inside, were operated by *manganarioi*, and had protective frameworks that apparently were so large as to exceed the height of the breastwork of the city.[33] While this might be slightly exaggerated, the characteristics are in no way compatible with torsion-powered catapults. However, at *Thessalonica (662), the Slavs planned to mount *petroboloi* on top of a three-storey siege tower which would certainly overtop the walls.[34]

The term *petrarea* was used in Greek only by the *Miracula sancti Demetrii* and the *Chronicon Paschale*. At *Thessalonica (618), the citizens crewed the *petrareai* (where they were described as *inside* the framework) and were assisted by sailors who were described as *empeiromanganoi*, i.e. experienced in operating *mangana*, or lever and rope powered machinery. Like *petroboloi*, the Avars used *petrareai* at *Constantinople in 626, where the frameworks were covered in hides and bound together. Finally, at *Thessalonica, the Slavs assaulted with very high *petrareai* and wooden *manganika* "never before seen."[35] In Latin we find the term (as *petraria*) in Paul the Deacon (used by the Romans at *Benevento 663) and the Royal Frankish Annals (used by the Saxons at *Eresburg 776). The etymology is uncertain, but it seems to have been quickly confused with the word *tetrarea*. It might have sounded more sensible to Latin-speaking troops in Illyricum and Italy, or result from an amalgamation with the term *petrobolos*, firmly attested in use for large trestle-framed traction trebuchets in the *Miracula sancti Demetrii*.

Tetrarea describes the sturdy trestle-shaped framework that was typical of the heavy trebuchet; *labdarea* then denotes a lambda-shaped frame,

---

of *Barbād 776 and *Semalouos 780, although I have not worked with Tabari's original in these instances.

[33] *Miracula St. Demetrii* 206 and 203, respectively. There were also "huge *petroboloi*" at *Thessalonica 586.

[34] *Miracula St. Demetrii* 272.

[35] *Miracula St. Demetrii* 255.

thus somewhat more narrow than the *tetrarea*. These terms were little used in this period; I have only found evidence for *tetrareai* at *Constantinople (715), when Anastasios II was preparing for the Arab invasion force that arrived two years later, "placing *toxobolistrai* and *tetrareai* and *manganika* on the towers" (στήσας τοξοβολίστρας καὶ τετραρέας εἰς τοὺς πύργους καὶ μαγγανικά). However, both engines were standard elements in the Byzantine arsenal around 900.[36]

The term *funda*, "sling," occurs mostly in texts from Visigothic and Arab Spain. Its meaning extended from the original handheld sling to a term for the traction trebuchet by the early 7th century at the very latest, when used by Isidore of Seville in a list of heavy artillery (including *ballistae* and *fundibulum*;[37] see chapter 3.2.3). Wamba legislated that some specialists in the retinues of magnates be equipped with "equipment for *fundae*" (*fundarum instrumentis*; see chapter 3.2.2). Later Byzantine military manuals describe the standard hand-operated sling as tucked in the belt,[38] so the *instrumenta* must denote the parts of a large engine. This is clear from both the Visigothic sieges of *Narbonne and *Nîmes in 673 (conducted by the same Wamba), where the defenders were overwhelmed by massive barrages of artillery (respectively, *tantos imbres lapidum intra urbem concutiunt, ut clamore vocum et stridore petrarum civitas ipsa submerge aestimaretur; muros urbis petrarum ictibus petunt*). While the engines were not named, at Nîmes this barrage resulted in breaches in the wall that had to be repaired. At the siege of *Toulouse (720), the Arabs "were besieging the city and tried to storm it with *fundae* and other types of engines" (*eamque obsidione cingens fundis et diversis generum machinis expugnare conavit*) according to the Byzantine-Arab chronicle. This leaves little doubt as to the nature of the *fundae*, as they were clearly regarded as a type of *machina*. Furthermore, the Arabs at this time had perfected siege warfare, with the use of large batteries of trebuchets as one of their defining traits (see

---

[36] For this see Haldon 2000; he mistakenly identifies *manganika* with bow-powered engines (the *ballistra* of Procopius, cf. chapter 5.2.2), but it is clear from the above and the consistent use of *toxobolistra* to distinguish bow-powered bolt-shooting artillery that this cannot be correct.

[37] *Fundibulum* is used in Isidore's etymologies to specify a type of stone-throwing *ballista* (as opposed to a bolt-shooting one). Otherwise, it is rarely used in narratives. Chevedden 1998: 193f and nn. 37f adds a few references dating to this period, but mistakenly bases one instance on the apparatus in *HW*, where a 13th-century deacon added *fundibali* as a gloss. See the MGH ed. p. 516.

[38] *Praecepta militaria* 1.3.

chapter 7.3). Charles Martel stormed *Avignon in 736 with batteries of *fundis* that shot over the walls and into the city.[39]

A final term in Arabic merits our attention. The *'arrāda*, originally a loan translation of *onager*, was used in Arabic sources with little or no distinction from *manjanīq* (e.g. *Edessa 630, *Tella 639, Alexandria 646). If anything, it perhaps indicated a lighter version of the same engine. It had probably entered Syriac and Arabic vocabulary sometime in late antiquity, well before the Islamic conquests, when it still meant a one-armed tension catapult (*onager*), but underwent the same semantic shift as other catapult technology did in the late 6th century (cf. 8.2.1).

8.3 The Diffusion of the Traction Trebuchet: The Historical Context

Based on the philological evidence paired with explicit narratives, it is possible to demonstrate that most terms for artillery by the late sixth century referred to types of trebuchet. However, most of this evidence has so far been used to date the arrival of the trebuchet to somewhere around 580, thus likely after the Avars' appearance in 568. Hence, crediting them with introducing this innovation is not an unreasonable conjecture. However, as we have seen above (7.2.3), there are good *a priori* reasons to reject this interpretation. A further examination of the philological evidence in its historical context will clearly demonstrate that this view can no longer be sustained.

8.3.1 The Early Introduction of the Traction Trebuchet: Diffusion or Independent Invention?

Beyond what may be dubbed the current consensus of an introduction around 580, there also seems to be hard evidence of the existence of the trebuchet even before this. Joshua the Stylite described a huge engine used by the Amidenes (*Amida 502f) to lob vast stones, crushing the protective padding the Persians had placed over their siege mound and in the process killing the engineers working on it. It was, fittingly enough, named "the Crusher." The practice of naming large individual trebuchets became common later; many great trebuchets received poetic names due to their immense power. The custom can be dated to *Theodosiopolis in 421/2, where

---

[39] See the discussion in CO and chapter 4.3.3, especially on the emendation of *funibus restium* to *fundis restium*; perhaps also *fundibulis* would be an acceptable emendation in light of the above.

Theodoret describes a huge stone-thrower, named after the apostle Thomas, used against the Persians. Too little is recorded of its effects and use to be certain, but taken with the evidence from Epiphanius, this may be the first recorded instance of such a weapon.

There appear to be no clear descriptions of trebuchets in either Procopius or Agathias, while Menander, of course, is very fragmentary. Procopius, who participated at *Rome (537f), provides the standard description of artillery which is still familiar from the ancient arsenal.[40] However, a rocky outcrop that caused concern for the defenders at *Antioch in 540 had not been noticed by the original builders, nor dealt with during later repairs. This was a strange oversight in light of the great Persian threat, who could apparently threaten the walls from this point. It has been suggested that some sort of new machinery was causing this concern.[41] Indeed, Procopius does refer to an interesting incident where the Persian defenders of *Petra (550) built an additional wooden tower onto one of the bastions, whence they threw large stones at approaching rams. This "wooden tower" might actually have been the frame of a trestle trebuchet, protecting pulling crews from enemy fire. Procopius was by then finished with his military career, and seems not to have been aware of the trebuchet or its mechanics, so he would simply have misunderstood what was actually going on. In fact, the *Strategikon* advises that when siege towers approached the walls, counter-towers should be built on the facing walls without roofs, so that the defenders had enough room to operate their *mangana*. Here the lever mechanism is not attached to the trestle frame or tower, presumably so that both can be manipulated and moved separately according to need.[42] This seems to fit the description of Procopius.

Agathias as well had difficulties with certain technical innovations, but he knew his Procopius well. It is therefore very interesting that he described the Roman besiegers as setting up τὰ τῶν μεγάλων λίθων ἀκοντιστήρια καὶ ἄλλα ἄττα τοιάδε ὄργανα, "the hurlers of the great stones and certain other such machines" at *Onoguris (555).[43] Now Procopius knew first hand the traditional arsenal of ballistrai and onagers, which he had seen in use at Rome, and for which he would have been responsible for procuring supplies, at least on some occasions. Although he did not describe them too

---

[40] This is therefore the normal ending point for most modern studies on ancient artillery, e.g. Marsden 1969-71 and Rihll 2007.

[41] I have the suggestion from Whitby but the reference eludes me; perhaps the *LAA* conference at Oxford 2007.

[42] *Strategikon* 10.3.17-27.

[43] See Agathias 3.5.9 for the quote.

well, they were familiar and regular components in most sixth-century sieges, as we have seen (chapter 5.2.2). Agathias would also have known as much. However, it is clear that Agathias was here facing something unfamiliar for which there was no word, since ἀκοντιστήριον is a hapax that only occurs here (and in a dictionary entry in the Suda, which only quotes Agathias). Thus, rather than finding a classicizing *Ersatz* term for it, like Theophylact did, he tried to be creative (although recall that he also referred to a *helepolis* at the *Chersonese 559, cf. 8.2.1 above). The name unfortunately does not give us much to go by; it simply means a "thrower" or "hurler," derived from the verb to throw (ἀκοντίζω), which was mostly used about javelins (sg. ἀκόντιον). However, since it was used for "great stones" it is clear that we are not dealing with a giant ballista, and the expression "other such machines" hints at artillery of varying types and sizes. Furthermore, we have an instance at *Thessalonica (615) with *petroboloi* "hurling stones" (ἀκοντίζοντες λίθους), i.e. a quite similar context and meaning.[44] Finally, Agathias provides some evidence from Narses' invasion of Italy in 552-53 that indicates a more aggressive mode of siege warfare with more widespread use of artillery to capture cities, such as *Cumae and *Lucca.

Thus, on the basis of fairly hard evidence of unknown machinery in Joshua the Stylite and Agathias, as well as good indications of its construction in Procopius (especially when read against *Strategikon*), it is likely that the traction trebuchet had become known in the eastern Mediterranean area at the latest by around 500. The philological and (admittedly circumstantial) historical evidence may even support a date around 400. If the theory of diffusion from China is to be upheld, there are two possible routes via Central Asia. One is that the Huns or another nomadic group brought it with them well before the Avars, although this is problematic for reasons of logistics, organization, and lack of evidence. The other route is through the Persian Empire, which certainly did have the necessary infrastructure under the Sassanids, and engaged in warfare with sedentary Central Asian polities that were also in contact with China. However, there is no trace of this engine in any of the research conducted thus far on the wars of the 4th-early 6th centuries. Therefore we cannot rule out independent innovation. The basic mechanical principle of lever force was long known to the Greeks and Romans and apparently applied to military engines in the 1st century and used directly for military purposes in the late 4th. The early instances may have been experimental innovations spurred by particular

---

[44] *Miracula St. Demetrii* 185.

threats, so had not become widespread by the time Justinian's war of reconquest began in the 530s, but intense border warfare in the East from 540 and the regular involvement of civilian craftsmen and engineers in local defense provide a very plausible context for rapid diffusion.

8.3.2 *The Wider Diffusion of the Traction Trebuchet within the Former Roman World*

The traction trebuchet began to achieve widespread distribution throughout the Roman Empire (the original eastern provinces) by the mid-6th century, so that it became a familiar and regular part of the arsenal of Roman field armies as well as of local garrisons and even urban militias and urban engineers by the latter half of the sixth century. It would also presumably have spread to the reconquered Western provinces sometime after 540. In Italy a probable date is with Narses' army in 552. Herakleios the Elder was a commander in the East during the 580s, and as argued in chapter 2.2.3, he had subordinate officers in his army who were directly responsible for setting up and coordinating trebuchet batteries for assaults, and arguably a team of engineers who could direct the construction and use of the actual machines. It is a reasonable conclusion that he brought such knowledge with him when he was assigned to service in North Africa in c. 600, since most generals had extensive military followings and staff. We thus have one possible *terminus ad quem* for the arrival of the trebuchet in North Africa; tracing other commanders or units who served in both the East and in Africa will undoubtedly push this date further back, so a generation earlier is very likely.[45]

In view of the scarce evidence, Spain is surprisingly well attested when it comes to the diffusion of the trebuchet. It was certainly brought there by the Romans, thus after 554, and known to Isidore of Seville in his *Etymologies* around 600, so the Visigoths must have mastered its use before that. When the transfer actually took place to the Roman province is uncertain, but since it would have taken place to Italy with the arrival of Narses' army in 552 and Africa sometime before 600, a good guess would be the same range. Again, good textual evidence is lacking, but a Roman general heavily involved in siege warfare, like Herakleios the Elder, is well attested in Cartagena around 590, where Comentiolus, who took part in the clearing of Slavs and Avars from the Balkans, commissioned the rebuilding of the city walls of Cartagena and would also have been responsible for oversee-

---

[45] *PLRE* 3 s.v. Herakleios 3 and the *fasti* for the Exarchs and *Magistri Militum* in Africa.

ing other aspects of its defense. Although the transfer could have taken place earlier, Comentiolus would certainly have brought this technology with him if it were lacking. Thus the window is somewhat narrower: certainly after the Roman conquest in 554, most likely in the last quarter of the century, but certainly by the time of Comentiolus at the very latest. Shortly after, between about 615 and 625, the Visigoths overran the Roman province, storming and sacking a great number of fortified cities (see chapter 3.2.3). Unfortunately the sources do not tell us the means by which this was done, but in light of Isidore's information, it would have been a violent contest involving trebuchets on the defenders' side. Roman soldiers were recruited into Visigothic service; the Visigoths, in turn, certainly had the infrastructure to maintain this knowledge over generations. They certainly knew how to use it well when Wamba defeated the rebel Paul in 673.

Lombard military organization was compatible with Roman (or even created by the Romans), and as most fighting between the two was over the control of cities, we know that siege warfare played a major role during the long Lombard-Roman conflicts. As Roman clients, the Lombards may conceivably have introduced trebuchet technology to Italy, either circuitously via service for the Avars, or directly through service in the Roman army (see chapter 3.3.1). However, diffusion to Italy seems to have taken place by the early 550s, meaning that the Lombards would have gained this technology when they acquired control of the northern Italian cities. This may have even been encouraged by the Romans if they were actually installed there in order to defend Italy from the Franks and the Avars. As a very last alternative, the Lombards received support from Slavs when besieging Roman cities in the early 7th century. Thus, a date of around 580 (at the latest c. 600) may reasonably be postulated for Lombard acquaintance with this engine. Certainly when the Romans attempted a major offensive against the Lombards under Constans II in the 663 the Lombards were able to defend major cities against Roman trebuchets, so presumably they would have had a similar arsenal available by then at the very latest. Indeed, at *Lodi and *Bergamo 701, the Lombards assaulted "with battering rams *and other machines.*"

This brings us to the last Mediterranean power, the Franks. Again the narrative evidence fails us, so that it is only in the early 8th century we can positively identify trebuchet artillery during Charles Martel's campaigns in the south of Gaul, while it was most likely not known before it became common in Spain and Italy, i.e. after c. 550. This timeframe coincides with the Arab expansion, and since their poliorcetic capacity is well attested, they could presumably have been responsible for transmitting such knowledge during the first quarter of the 8th century. Yet Charles Martel's forces were extremely competent in the use of a variety of siege engines and in conducting siege operations in general (cf. chapter 4.3.3). This was a difficult and complex body of knowledge, and presumed an infrastructure to underpin it as well as time to practice and refine it. This the Franks had since the early Merovingian era; most of their infrastructural framework was in fact inherited from the Roman Empire and thus closely related to East Roman and other successors' military organization. It is therefore likely that the Franks became familiar with the new technology through their neighbors, the Visigoths and the Lombards, sometime in the seventh century, although an exact location or route of transmission is no longer recoverable. It may have occurred as early as Gundovald in 585, since he had Roman sponsorship, but this rests on a debatable reading of the siege description at *Convenae. However, around 550 (or 568), the bishop Rufus of Turin sent some craftsmen to Nicetius, who had *ballistae* on his fortified residence near Trier. Since urban craftsmen under magnate authority were the very type of people who could build siege engines by this date (cf. chapters 2.4.2 and 6.2.3), they may have been helpful in setting up the *ballista* for Nicetius (chapters 1.2.5 and 8.2.1). This leaves us with a very wide timeframe from the 550s to the 720s; Nicetius' *gemino ballista volatu*, "ballista with twin flight," leaves me inclined towards the earlier date. This leads to two possible conclusions: the Merovingians either preserved some aspects of Roman military craftsmanship that could readily accommodate new technology just as any other successor state, or they successfully created an institutional framework for maintaining siege engines within a generation or two after the disintegration of the Roman state in the West.

Due to its predominance in siege warfare from the 7th century onwards, technological diffusion largely has to be traced via evidence of the trebuchet and competence in fortress building. Hence the Muslim Arabs had some knowledge before 630; here contact with their Christian neighbors in Jordan, Syria and Iraq served as an early conduit. However, as shown in chapter 7.3, it was their whole-sale control of the late Roman and Persian populations and their successful appropriation of existing institutions that made it possible for them to conduct large- scale siege campaigns. Their genius lay in how they devised a social organization that maintained this knowledge while laying the foundations for a world civilization.

8.4 Epilogue

The traction trebuchet superseded ancient torsion artillery because each individual machine outperformed older equivalents, and could be used very effectively on a large scale (ch. 5.2.2, 8). However, the trebuchet required similar resources and infrastructural support to maintain; hence diffusion was only possible where ancient military infrastructure survived or was adapted in the 5th and 6th centuries. While the result towards 800 looks decidedly medieval and regionally diverse, it was in fact derived from a set of Roman institutional innovations that prevailed throughout the Empire in its last century before the fall of the West. The relative decline of urban civilization as compared with the high point of the Roman Empire in the 2nd century AD meant fairly little. Most Roman fortifications were still useful and hence fought over for centuries after their construction (ch. 6).

The decline of a centralized, tax-collecting bureaucracy used to support a professional army and its technical experts was also comparatively unimportant. All post-Roman polities developed mechanisms for maintaining troops, fortifications and technical expertise based directly on the vast complex of labor obligations known variously as *munera*, *angareiai* or liturgies, which gradually became militarized in the late 4th and most of the 5th centuries. Indeed, the decline of civic munificence in the West around 400 is directly proportionate with the extension of military obligations to the great landowning classes who were involved in the Roman civil wars of the 5th century, and we can actually trace the movement of soldiers and experts from Roman units to magnate service (ch. 1).

For two centuries, the East Roman Empire operated a dual structure that allowed both formal institutions (the army and state arsenals) and informal structures (estate-based private retinues, urban craftsmen and other corporate or social units) to share the burden of maintaining infrastructure, defending fortifications and engaging in offensive siege warfare. While the great landowners were eliminated by the early 7th century and a centralized bureaucracy continued to supply the army, urban populations remained militarized due to the constant threats to virtually every city inside the surviving Byzantine Empire for the next two centuries (ch. 2). Originally similar developments in the West took a different turn around 600. The ethnification of Roman armies (i.e. their adoption of various "barbarian" identities) has obscured this process until recently, but it is increasingly recognized that most of the early 6th-century Ostrogothic,

Burgundian, Visigothic and Frankish armies were still essentially organized as the Roman military units from which they descended or into which their barbarian ancestors had been integrated. Nevertheless, change in the manner of raising supplies and troops was particularly pronounced in the West, where it evolved from a centrally-managed system for the redistribution of foodstuffs, supplies and money around 400 AD to the imposition of localized labor services and decentralized management of supplies and soldiers. This had become more or less universal by c. 600, although East Roman influences remained strong, especially in Lombard Italy and Visigothic Spain. While the administrators were mostly great landowners (who also doubled as royal officials), the resources they administered were levied on the force of royal legitimacy. In this context, the end of direct taxation was unimportant as long as the services they had paid for were now rendered directly as a set of military and logistical obligations. This is most clearly visible in Visigothic Spain, while most elements of the "Carolingian" military system were in fact well established during the high point of Merovingian power in the early 7th century (chs. 1, 3, 4).

While this system was most prevalent in the successor that arose on former Roman territory by 500 AD, conquest peoples and neighboring empires were to varying degrees capable of absorbing military knowledge. Roman client politics created small, highly militarized polities along the frontiers whose purpose was to provide an extra layer of defense, keep other client polities in check, and send recruits or federate contingents to serve in Roman armies. Client management was actively maintained by the Ostrogoths, Franks and East Romans on the northern frontiers throughout the 6th century, resulting in the continuous integration of northern barbarians into (post-)Roman methods of organization and fighting. Eventually, this resulted in the rise of larger polities in the 7th and 8th centuries—Slav and Saxon tribes; Avar and Bulgar khaganates; Wendish and Danish kingdoms—capable of organizing large-scale labor projects and engaging in warfare with their southern neighbors on a more equal footing. However, the institutional aspects of warfare were more complex than the organization of labor and troops; it also required complex technical expertise, which required self-perpetuating mechanisms of support. In the successor states, this was possible because of urban survival, the estate economy, and the complex of military institutions that evolved by 600, whether in the form of magnate retinues or centrally organized army units (ch. 6). Similar arguments can be made about the Persian Empire. In contrast, most of the northern conquerors, such as the Slavs, Avars and Bulgars, acquired some of these skills through service in the Roman army. While this provided extensive familiarity with Roman practices, implementing expertise in siege warfare on a larger scale over time was achieved through the exploitation of captives and deserters. While sometimes these groups were large enough to form distinct client populations under a larger political umbrella, in most cases they were deprived of their urban or institutional context where their technical expertise had developed; hence the failure of the Avars and Bulgars to respectively maintain and acquire the complex of poliorcetic knowledge in this period. The Avars lost control over most of their client peoples in the 7th century, while the Bulgars only developed a solid institutional basis in the 9th century. In our period, the most successful to adapt were minor Slavic groups in the Balkans, who had begun to assimilate into East Roman political culture and economy in the late 7th century. Although their achievement may have hinged on the assimilation of a very large number of erstwhile Romans and was interrupted by the Bulgars shortly after, it demonstrates how peoples with very simple material culture and political organization could evolve over the course of a few generations. Similarly the Saxons, who were under heavy Frankish influence, became increasingly sophisticated in the 8th century, and around 800 Saxony was fully integrated into the Carolingian Empire, extending its influence deep into Scandinavia and Central Europe (ch. 1, 4, 7.1.-2).

The most successful conquerors were of course the Arabs, whose genius lay in preserving Roman (and Persian) provincial society, urban life, and even the political elite virtually intact, and exploiting this to their own purpose from the very first years of the conquest. This means that most of the administrative apparatus and technical expertise they employed during the conquest phase directly continued that of the East Roman and Persian Empires, down to the urban Christian craftsmen who built and operated their fleets and catapults. This also explains why their adoption is largely hidden from view in the Islamic sources, which were mostly concerned with the top echelons of the Muslim elite, and rarely mentioned the subalterns, whether Egyptian and Syrian Christians pressed into labor service or Persian Muslim client troops who were serving the new Arab political elite. By the time we have good descriptions of siege warfare in the 9th century Islamic sources, a century and a half of gradual client emancipation and the conversion of a substantial part of the local population to Islam had produced a new foundation for siege warfare supported by mainstream Islamic society (7.3).

A detailed examination of actual techniques and strategies shows that early medieval sieges were immensely varied, as methods shifted according to the objectives of the parties even in the course of a single siege. Furthermore, besiegers and defenders employed a wide range of field engineering, siege engines and other methods, again depending on circumstances and adapting to how their enemies acted (ch. 5). This common technological and military culture prevailed in an area that to a large extent also shared similar mentalities and related forms of socio-economic organization, resulting in a similar set of behaviors and responses among civilian populations and armies in times of war, whether ritual, political, or military, from Francia to Mesopotamia (ch. 6).

For most students of the high middle ages (even educated laypeople introduced to the horrors of the first Crusade), the model presented here will be very familiar. It could certainly be extended with ease throughout the 9th and 10th centuries. During the caliph al-Mu'tasim's great invasion of the Byzantine Empire in 838, the fortress city of Amorion came under siege. The Arabs built massive siegeworks, bombarded the walls with trebuchets, and could now also undermine the foundations of the wall under the cover of penthouses covered in hides. The Byzantines responded with artillery shot of their own, and dropped boulders onto those who approached the wall. Despite such intense exchanges going on for three days, the fortification was betrayed by a single defender. However, this was no simple matter. The "betrayal" was that he directed the Arabs' attention to a spot where the wall had previously been weakened by a flood and only poorly repaired; a barrage of artillery against the spot caused the wall to collapse, but only after the Byzantines for some time held off the collapse by building a wooden framework to support the wall. The Arabs were still unable to break through, and the fighting continued, as the Arabs tried to use towers and fill in the ditch around the walls. Even then, the Arabs only managed to capture the city on the twelfth day of the siege during an unofficial parley with the commander who held the section of wall that had been broken down. While the Byzantine representative was in the Caliph's camp, Arab troops slowly approached the breach in the walls, as the Byzantine troops had been ordered not to fight and only signaled them to stop. When negotiations were over and the Arab troops were close enough, their commanders gave the order to assault. This took the Byzantines entirely by surprise and the city fell without a fight.[46] Similarly, we have

---

[46] Vasiliev 1935: 160-173 contains extensive quotes and references to Greek, Arabic and Syriac sources; Treadgold 1988: 302f briefly summarizes the main points.

the Franks under Louis II besieging the Arabs at Bari 871 with siege engines, possibly with Byzantine help.[47] At Angers in 873, Charles the Bald diverted the course of the river and is reported to have brought new and marvelous engines that have caused much scholarly speculation. While many have argued for the traction trebuchet or more complex varieties of this,[48] I would conjecture that this saw the reintroduction of the siege tower to the West; as it was not mentioned after the fall of the Ostrogothic kingdom but common during the early Crusades. The integration of fringe peoples into mainstream, post-Roman military culture continued; at Paris 885-86, the Vikings used traction trebuchets with great effect.[49]

These brief anecdotes demonstrate that even throughout the 9th century, we can observe a common military culture from the Middle East to Scandinavia which was very much closely interrelated, where methods of siege were recognizable throughout and the mechanisms of diffusion continued to operate; Byzantines fought Arabs, Arabs fought Franks, and Franks fought Vikings. Even if innovations continued to be made in the East, both Franks and Vikings could apparently adopt these with comparative ease. The foundations for this common culture lay in the survival and transformation of Roman institutions and infrastructure in the 5th and 6th centuries, and their successful integration by conquerors and clients in the 7th and 8th centuries. Similarly detailed narratives that mention the use of engines, describe large-scale blockading operations, and breaching of walls, can easily be reproduced throughout the 9th century. Indeed, most of the details given by contemporary 9th-century authors, whether taken from the Arabic Tabari or the Latin poem of Abbo, bishop of Paris, could have gone straight into any of the 6th-century Greek and Syriac chronicles and histories without producing troublesome anachronisms. Such was the heritage bestowed by the Romans that only the ghosts of hairy barbarians roam among the ruins of their Empire.

---

[47] Kreutz 1991: 45 for the context and references to the sources.
[48] E.g., the hybrid trebuchet, as argued by Chevedden 1998.
[49] See e.g. Maclean 2003: 55ff for context and references.

APPENDIX ONE

RECONSTRUCTING THE ARAB INVASION OF PALESTINE AND SYRIA FROM CONTEMPORARY SOURCES AND THE IMPORTANCE OF ARAB SIEGE WARFARE

The Arab conquest of Palestine (the East Roman provinces *Palaestina* I-III) has long baffled historians, who have reconstructed a plausible sequence of events based on a partly implausible set of data drawn from the very late compilations of the great Islamic historiographers. While they provide enormous amounts of material, it has long been recognized that they provide incompatible dating schemes and lists of commanders, as well as much more serious problems such as the use of a small set of literary *topoi* to construct a large number of battles and other dramatic scenes whose complexity and level of detail only increase with the passing of time.[1] The same problem applies to many of the Christian sources, which are not only late but in some respects also influenced by the Islamic historical tradition (see Appendix III further on this). Therefore, the solution has been to select certain narrative traditions and apply them with common sense. Donner thus produced a reconstruction where the Muslim Arabs first overpowered the Christian Arab tribes in present-day Jordan, and only *then* proceeded to conquer Palestine; for similar reasons, Schick argued that the nature of the Islamic conquest must have been quite peaceful in contrast with contemporary descriptions since the Christian Arabs were building churches with inscriptions dating them to the very period (635-36) when Arab armies were active in Palestine and Syria. Kaegi rather selectively chooses narratives to support his views of the most useful military strategies.[2]

However, for the conquest of southern Palestine there are a handful of contemporary observations that provide exact information or good impressions of the situation. This can then be held against some of the siege narratives examined in CO, and may temper some of the criticism leveled against Islamic historiography. Firstly, it should be clear that the Arabs did not in fact attack Palestine from the east (dictated by the logic of conquering the Christian Arabs first), but rather from the south and north-east, completely avoiding the Christian Arab client tribes east of the River Jordan. The direction of assault is clear from three accounts, all of which were probably contemporary, and possibly even written down by eyewitnesses. Firstly, Theophanes seems to preserve a very old notice under AM 6122 (AD 630-31, or 631-32 with corrected chronology) which probably dates back to a near-contemporary account. Here the Christian Arabs, supported or led by Roman of-

---

[1] Thus Noth 1994.
[2] Mayerson 1964; Donner 1981; Schick 1995; Kaegi 1993; Howard-Johnston 2010 provides a model approach but ignores evidence from the *Chronlicle of 637* and Theophanes, who place the battle of Yarmuk /Gabitha in 636.

APPENDIX ONE: ARAB INVASION OF SYRIA & PALESTINE 431

ficers, fought against the Muslims and defeated them severely at *Mouchea*, probably Mu'ta in Arabic traditions, which lies in present-day southern Jordan. Meeting such stiff resistance, the Muslims found another way. Theophanes continues:

> Now some of the neighbouring Arabs were receiving small payments from the emperors for guarding the approaches to the desert. At that time a certain eunuch arrived to distribute the wages of the soldiers, and when the Arabs came to receive their wages according to custom, the eunuch drove them away, saying, 'The emperor can barely pay his soldiers their wages, far less these dogs!' Distressed by this, the Arabs went over to their fellow-tribesmen, and it was they that led them to the rich country of Gaza, which is the gateway to the desert in the direction of Mount Sinai.

The direction is noteworthy. The Arabs skirted the southern flank of the defenses, entered the Sinai Peninsula, and attacked Palestine from the south-west towards the hinterland of Gaza. This highly surprising direction meant that the Arabs avoided the dense layers of defenses to the east, beginning with client tribes and then a number of fortified cities on both sides of the River Jordan. The Arab incursion took place in the winter of 633-34, because the first Roman reaction is attested by a contemporary Syriac source, the *Chronicle of 640* (or as it is called by Hoyland, the *Chronicle of Thomas the Presbyter*), who records for the year 634:

> In the year [AG] 945, indiction 7, on Friday 4 February at the ninth hour, there was a battle between the Romans and the Arabs of Muḥammad (*ṭayyāyē d-Mḥmṭ*) in Palestine twelve miles east of Gaza. The Roman fled, leaving behind the patrician *bryrdn*, whom the Arabs killed. Some 4000 poor villagers of Palestine were killed there, Christians, Jews and Samaritans. The Arabs ravaged the whole region.

The ravaging and killing in the countryside was characteristic of the invasions. Thomas records that his brother Simon was killed along with many other monks in monasteries in the mountains of Mardin (in Mesopotamia) in the year AG 947 (635-36).

A fragmentary, but contemporary notice in Syriac, dating to c. 637, records that Arab raiders came from the north-east (and may have traversed the length of Palestine west of the River Jordan, depending on the reconstruction of some of the lacunae). There are also indications that the Arabs had moved in from the desert east of Emesa, which had surrendered, and then fanned out southwards to Damascus, Galilee and beyond, "and many villages were ravaged by the killing of {the Arabs of} Muḥammad (*Mūḥmd*) and many people were slain and {taken} prisoner as far as Beth ..."[3]

The general image is reinforced by Sophronius, patriarch of Jerusalem, who in his Christmas sermon for 634 records that the congregation was unable to reach nearby Bethlehem for the customary celebrations:

> We ... are unable to see these things, and are prevented from entering Bethlehem by way of the road. Unwillingly, indeed, contrary to our wishes, we are required to

---

[3] Preserved in Syriac is *byt zk* ... [*wt'*], with the last three letters tentatively provided by Brock; however it may possibly be Beth Zachariya, a monastery located outside Jerusalem, which would then indicate an expedition from Galilee to the south of Palestine. The general thrust of movements indeed seems to be *southwards* from Emesa.

stay at home, not bound by closely by bodily bonds, but bound by fear of the Saracens.[4]

While Sophronius prayed for the destruction of the Arabs and hoped that the emperors would receive divine assistance to carry it out, it is clear from this evidence that the Romans had not controlled the countryside of southern Palestine for nearly a year. This affected most of Palestine and Jordan, and must have made the supply situation for the cities intolerable. It seems that many of the "battles" recorded in Islamic historiography were the product of frequent sorties made by local garrisons (such as at Gaza) in order to protect local villages, rural populations, and agricultural production. Since the Arabs had no such concerns, moved rapidly, and had no defensive needs, they could concentrate far greater forces to ambush local garrisons or even armies composed of multiple garrison forces. Following the reconstruction in chapter 6.2.2 that most garrisons rarely numbered more than a *tagma* (100-500 men), Kaegi's reconstruction of the defenses established in the region, suggesting a presence of a mere 5,000 men to hold the cities of *Palaestina* I-III, is probably correct. Against this we must posit Donner's estimate of 24,000 men in the Arab invasion forces; even if attacking in two or three columns, which many reconstructions suggest, they would always have local superiority. Swarming the countryside and only using some forces to keep the Christian Arab federates busy, most towns were out of both supplies and garrison forces within a year of such indirect blockades, but in many instances, resistance dragged on into 635, while towns that had accepted a negotiated settlement (normally without admitting an Arab garrison) revolted as soon as reinforcing armies appeared.[5]

There are two main conclusions to be drawn from this argument: firstly, the contemporary sources clearly indicate that there were attacks on Palestine from *two* directions, the south-west and north-east. This goes against the grain of the traditional arguments presented above. Secondly, the first raids were specifically aimed at the weak points of urban communities, their agricultural hinterland. This was a deliberate strategy to force cities to surrender. If cities had strong garrisons, could be reinforced, or chose to revolt, the Arabs were always willing to take great risks to storm them (cf. *Caesarea Maritima, *Dara, *Dvin in 640 and the discussion in Appendix II), but preferred to set up deals that interfered little with local administration, elites, agricultural activity, and the flow of revenue as long as the threat of military counter-thrusts were eliminated. This was mostly the case after the battle of Gabitha/Jabiya (and/or Yarmuk) in August 636, when a large Roman relieving force was defeated on the border between modern Syria and Jordan.

This draws us to the trustworthiness of the Islamic sources, here represented by Baladhuri. At least some of the movements and tactics in contemporary sources actually confirms his version of events. Firstly, the Roman army that assembled at Gabitha probably did so to reestablish links with the Christian Arabs east of the

---

[4] For translation, references and discussion, see Hoyland 1997: 70.
[5] For combat in the field, surrender, and revolt, see CO entries *Jerusalem 634f, *Damascus 634/5 (although it probably occurred *after* *Emesa; cf. the fragmentary Syriac notice above), *Emesa 635, *Tiberias 635.

APPENDIX ONE: ARAB INVASION OF SYRIA & PALESTINE 433

River Jordan. The Muslims had skirted this region because of its strong defenses, and instead found ways to penetrate into the agricultural hinterland of the Levantine cities. The Muslim assaults on the Christian Arab cities only began when the Christian Arabs were cut off from direct links with Roman territory. Thus Baladhuri, amongst several other routes of march, also presents the Islamic armies coming from the north (the same recorded in Syriac sources as attacking *Emesa and *Damascus in 634-5), with one division splitting off to assault *Bosra and other cities in Jordan, probably late in 635 or early 636, prompting the arrival of a large Roman army to support their clients. The other element is the use of raiding; Baladhuri often resorts to catalogues of cities taken, but is specific about raiding (cattle-rustling and taking of captives is specifically mentioned) while the Muslims were marching towards Bosra. It should thus be clear that this tradition at least is in line with contemporary sources, and hence other information that can be gathered from the Islamic sources without conflicting directly with the reconstructions can be used to amplify our understanding of events.

APPENDIX TWO

'IYAD IBN-GHANM'S INVASION OF ARMENIA IN 640 AND THE ARAB CAPACITY FOR STORMING CITIES WITHOUT HEAVY SIEGE ENGINES.

In 640, 'Iyad ibn Ghanm led a campaign from the northernmost Arab salient in Syria across the Euphrates, into Mesopotamia (Jazirah), and deep into Armenia. This campaign is one of the earliest Arab campaigns that can be reconstructed in great detail by using Arab, Syriac and Armenian sources together. It illustrates the problem of relying too much on one single tradition in attempting a reconstruction of events, since no commentator has so far connected these disparate perspectives.[1] Separately, these traditions only give brief snapshots of apparently unrelated segments of the campaign. The Arab sources provide a detailed route of march through northern Mesopotamia to Dara, but only have sparse information on events in Armenia that does not extend much beyond the Bitlis pass. A Syriac source misplaces the campaign by one to two years but clearly link the Dara campaign with the campaign to Dvin. This is also the case with the Armenian source, which says that the Arab army came from "Asorestan" (taken to mean Syria west of the Euphrates by the most recent commentator), and then provides many details of events in Armenia proper, as well as the exact date of capture as 640, with the climactic siege and capture of Dara in October that year. It was lead by a commander who has received little attention, but who clearly was of great ability. 'Iyād was the commander of the *jund* (army) of Qinnasrin, and was detailed on a punitive expedition against the Byzantine commanders of Mesopotamia for breaking off the tribute payments they had agreed to make with the Arabs in 638. It is therefore possible that the early part of the campaign began in the autumn of 639, but most of the events took place in 640.

It is often said that the Arabs were met as liberators, but this is certainly not the case for the cities of Northern Mesopotamia, even in the triumphalist presentation given by Baladhuri.[2] There the Arabs systematically raided the countryside,

---

[1] Kennedy 2007 and Donner 1981 follow the Arabic traditions, which focus exclusively on a campaign against Mesopotamia (al-Jazīrah), while scholars of Armenian history (Manandean 1948, Howard-Johnston's commentary to Sebeos 42) have been concerned with the chronology as it can be reconstructed from Armenian sources. This is reflected in Canard's survey in EI^2, "Armīniya. 2. Armenia under Arab domination" vol. I p. 635f, where he notes both 'Iyād b. Ghanm's invasion in 640 as recorded by several Islamic historians (ending shortly after Bitlis) and treats the assault on Dvin as a separate incident undertaken by another army in 642. Kaegi 1993: 191ff, despite trying to exploit all the various traditions, gives the much later Arabic traditions pride of place and regards the contemporary Armenian information as "suspicious."

[2] For the following, see Baladhuri 269-77 (DeGoeje 172-77) along with the CO entries

APPENDIX TWO: ARAB INVASION OF ARMENIA, 640 435

perhaps for as much as two agricultural seasons, preventing the townspeople from going out to harvest. At Kallinikos (al-Raqqa), the peasants were out in the fields under guard of Christian Arab troops (simply called "Bedouin" by Baladhuri). The vanguard of 'Iyād's army carried out a successful raid that sent those who escaped fleeing to Constantina/Tella (*Tella 639). 'Iyād arrived at the city with his troops drawn up in formation before one of the main gates, clearly to intimidate the population. However, the inhabitants responded by shooting at the Muslim troops "for an hour," wounding some of them.

> In order to escape the enemy's stones and arrows, 'Iyād withdrew, and, after going round the city on horseback, he stationed horse-guards at its gates. He then returned to the main army and sent bands of soldiers who went around, bringing back with them prisoners from the villages and large quantities of food. It was the proper time for reaping the harvest.

The tactics seem to reflect those employed in Palestine, but here were are informed of the exact *modus operandi*: screening forces, in this instance cavalry divisions stationed near each gate, simply prevented garrison forces or messengers from exiting and refugees from entering, while the remainder of the army systematically cleared the surrounding countryside of supplies and captives. It was a very efficient approach, especially when the city was low on supplies (the harvest was not yet in), and a large number of the rural population were trapped outside. The peasants were preoccupied with the harvest but the Roman authorities were clearly aware of the dangers since they had been supplied with guards, but this was not enough to protect against the vanguard of a large invading army. The city authorities were faced with several means of pressure: with no hope of relief, cut off from supplies, and seeing people from the surrounding country (who had relatives, acquaintances and patrons inside the town) being taken into captivity every day, the local patrician agreed to a surrender within five or six days.

The objective was to capture the city without too much destruction in order to keep revenue flowing, but now for the benefit of the conquerors. The agreement detailed by Baladhuri stipulated cash payments as well as providing wheat, oil, vinegar and honey. A number of other cities were subdued on the same campaign. Amongst them, Edessa (al-Ruha), Carrhae (Harrān) and Samosata were obliged not only to make similar payments, but also provide labor services (in effect, *angareiai* or *munera*) to maintain roads and bridges, as well as guide Muslims and provide them with intelligence. However, conspicuously absent from these provisions are any mention of garrisons installed to hold these cities. A governor was established at Harran, presumably to supervise the payment of taxes from the region, but a local administrator (*'āmil*) and garrison was only installed in Samosata *after* the inhabitants revolted; what the revolt entailed is never explicitly revealed, but must have been the refusal to hand over taxes and provide intelligence and labor.

The course of events is highly revealing: the first Islamic conquerors established a new political leadership that forced taxes, supplies and labors from cities, but

---

*Tella; Mesopotamian cities 639: *Dara; Mesopotamian cities 640 and *Dvin 640, where references, quotes and transcriptions of the original may be found.

did not conquer them outright (i.e. establish direct control through local governors and garrisons) unless absolutely necessary. Furthermore, it seems that cities that were in a position to do so, *always* revolted once the Arab armies had passed. Samosata was on the outer fringe of the region traversed by 'Iyād. Across the Euphrates, it was less at risk of having its whole territory ravaged from the Arab bases at Edessa and Harran.

After the installation of 'Iyād's *'āmil* and garrison at Samosata, Baladhuri listed a long string of villages, regions and cities in northern Mesopotamia that were taken "on the same terms" as Edessa or other cities, although Ra's al-'Ain (Resaina) held out. The nature of the resistance is hard to gauge, but apparently played down by Baladhuri, who emphasizes the triumphal success. However, the defenders at both Kallinikos and Edessa had bombarded the approaching Muslims from the walls, convincing them to attack the countryside instead and settle for a blockade. Indeed, in other sources, the Arabs are reported as resorting to more forceful approaches: Agapius claims that the Arabs set up catapults (*'arrādāt*) against *Tella, which is a highly revealing contrast to the fast-moving cavalry armies in Baladhuri. Furthermore, the capture of *Dara (640), which Baladhuri only mentions in passing, is presented in a very different light by the *Chronicle of Zuqnin*:

> AG 952: The Arabs laid siege to Dara and attacked it. A great many people were slain on both sides, but especially among the Arabs. In the end they made an agreement and they conquered the city. From that moment onwards no human being was killed.

This affects our interpretation of the catalogues of surrendered cities in Islamic histories: it took hard fighting and many fallen merely to establish a political settlement that left most of the local economic life, society and administration intact. The local communities had demonstrated their ability to defend themselves, but came to terms before risking a storm that would ruin them utterly. Indeed, utter ruin befell *Dvin when the Arabs advanced into Armenia. There the intention was only to establish a foothold beyond the Bitlis pass in the mountain range that separated Armenia from the Mesopotamian plains, and this is reflected in Baladhuri, whose catalogue of cities ends in Akhlat (Khilat or Akhlat) on the northwestern shore of lake Van. Beyond this, the Islamic sources say nothing, but the *Chronicle of Zuqnin* continues:

> The same year they laid siege to Adavīn (Dvin), and in this city a large number of people were killed, as many as 1,200 of the Armenians.

This is confirmed by Sebeos (42), who also says that the Arab army came from "Asorestan" (i.e. Syria west of the Euphrates), then over the mountain passes (he does not specify Bitlis, but this is clear from Baladhuri) to "the land of Taron" which lies north of the Bitlis pass on the western shore of Lake Van. From there, we have a precise route of march where Sebeos details the regions that were occupied on the way along the northern shore of Lake Van. Rounding the north of the lake, the Arabs marched along the valley of Berkri, through the land of Gogovit to Mount Ararat, and from there descended into the Araxes valley to attack Dvin,[3] which

---

[3] I have not yet been able to establish the exact route of march around Ararat. The Arabs may have moved east (marching along the southern slopes of Ararat) at Dariwnk'

APPENDIX TWO: ARAB INVASION OF ARMENIA, 640

was taken on the fifth day after the Arabs appeared. They used barrages of arrows and smoke "all around" to drive the defenders off the walls and then ladders to storm the walls and open the gates. The capture of Dvin was a huge defeat for the Armenians, who lost 35,000 people captive and 1,200 dead. It is also a testament to the skill of the Arabs, who prepared for the siege in five days in the royal hunting preserve created by Khusro. Since they scaled the walls, they must have made ladders and presumably screens to cover their archers (approaching the walls without would be difficult, but large shields could prove effective as well). A hunting preserve would serve well to gather necessary materials. The smoke mentioned by Sebeos indicates an Arab innovation, since smoke does not seem to have been used in this manner previously.[4]

The intention here was neither to hold territory nor establish agreements, but to destroy the morale and defensive capacity of the large Armenian forces, which would become a bone of contention between the Caliphate and the Roman Empire for the next generations. The Arabs quickly exploited new political connections and their newfound administrative apparatus. On the way, the Arabs were assisted by an Armenian prince, Vardik of Mokk', which lies south of the Bitlis pass and thus within reach of Arab armies on the Mesopotamian plains. It may be that he only wished to avoid having his own lands destroyed, but there were always scores to settle between Armenian noble families. At any rate, his guidance allowed the Arabs to outmaneuver the Armenian forces, and furthermore, the invaders were able to cross a bridge (across the Araxes?) recently destroyed by the other Armenian nobles as they withdrew to Dvin and other cities. This could only have been undertaken with extensive repairs. In light of this, the provisions stipulated for the cities in Mesopotamia shows that they had their immediate needs in mind: supplies to maintain their armies, taxes to ensure distribution of wealth, and labor to make sure that the Arabs were not hindered by infrastructure destroyed by the retreating Armenian and Roman forces.

If this labor was taken from Mesopotamia and put to work in the Araxes valley, the Arabs may have employed it in the same manner against the cities they captured, although this is uncertain. Dara (in Mesopotamia) and Dvin (in Armenia) were both great fortified cities on the Perso-Roman border with complex defensive works and were well garrisoned. Indeed, they were more significant in Roman strategic thinking than many other regions as they were situated on the important Persian frontier. Furthermore, they had better time to prepare for the Arab invasions than the cities of Palestine. Nevertheless, the Arabs managed to storm both, but only after massive casualties, as pointed out by the contemporary observers. This evidence indicates that the Arabs were consummate but ruthless besiegers at an extremely early date, even when they did not use siege engines. The anecdotes recorded here also show that resistance was much more intense than re-

---

and then moved north up the Araxes valley to Dvin; otherwise north through the pass near Kolp to the Araxes upstream of Dvin, which they approached from the west.

[4] It might otherwise be a cryptic reference to undermining and firing the foundations of the wall. Even if it did not cause the wall to collapse, large amounts of smoke drifting up along the wall would have a rather unpleasant effect on the defenders.

vealed by the Islamic sources alone, and that the political settlements to a large extend left most of local society untouched. The creation of classical Islamic civilization was still some generations away. However costly, 'Iyāḍ's invasion was a step in this direction in two respects: firstly, it deflected attention away from the main thrust against the Roman bases on the coast of Syria (see e.g. *Caesarea Maritima 640) and the invasion of Egypt, which would soon be underway; secondly, it demonstrated to the Armenian nobility that the Caliphate had become a viable alternative to the Persian Empire, and as such they came to terms with it in the same manner as they had always done with the Persians.

APPENDIX THREE

ARAB GRAND STRATEGY, 663-669: 'ABD AL-RAHMAN IBN KHALID'S INVASION, SAPORIOS' REVOLT, AND THE BATTLE FOR ANATOLIA

Arab and Byzantine activities between the end of the first Arab civil war in 661 and the "first" Arab naval expedition against *Constantinople in the 670s have received little attention and lie in relative obscurity. According to older works on events in Anatolia, the raids reported those years were little out of the ordinary— rapid Arab advances across the Taurus range to take and destroy one or two urban settlements, a slightly unusual attempt to hold Amorion over winter, and a rather obscure Byzantine counterattack. The emphasis is on slow attrition, typical of Arab-Byzantine warfare during the Umayyad period, which inexorably resulted in the decline of urban civilization in central Anatolia. Lilie in contrast recognizes the ultimate Arab goal of capturing Constantinople, but following Greek and Arabic sources, his reconstruction of Arab strategy appears somewhat diffuse. Only Treadgold links Arab raids to political developments in Byzantium, most notably the revolt, traditionally dated to 668, of the *stratēgos* Saporios, general of the newly created Armeniac military command, later *stratēgis*. Following this dating, and placing an earlier Arab raid against Pontic Koloneia rather than Cappadocian Koloneia, Treadgold constructs an image of an exasperated *stratēgos* who turned to his tormenters for support.[1]

These reconstructions are based on the sources traditionally used for Arab-Byzantine relations, the large number of chronicles and compilations written in Greek, Arabic and to a lesser extent Syriac from the 9th to the 13th centuries.[2] Due to their bulk, number and relative coherence, the events, perspectives and chronologies taken from minor eastern chronicles written in the 7th-8th centuries are often dismissed out of hand as doubtful or even "fabulös".[3] This has created a long-standing historiographic distortion of events in the mid-7th century, where late and numerous sources take precedence over few contemporary ones. Recent advances in source criticism have turned this basic premise on its head,[4] since typically "corroborative" information, such as dating and the naming of commanders

---

[1] Lilie 1976: 57-96, esp. 68-74; also Brandes 1989: 48ff; cf. Treadgold 1997: 320.
[2] The most frequently quoted sources are Theophanes and Tabari, as well as the Islamic histories of Mas'ūdī and Ya'qūbī, and the eastern Christian sources deriving from the Syriac Common Source (see next but one note).
[3] Thus both Lilie 1976: 70 and Brandes 1989.
[4] The late Christian sources provide no independent confirmation of events reported in later Arabic sources. Theophanes and his cousins in Syriac (Dionysius, preserved in *Chr. 1234* and Michael the Syrian) and Arabic (Agapius of Manbidj) have been shown to derive in part from the emerging Arabic tradition, especially during these years. For this and the following, see Conrad 1992, Hoyland 2011.

in the Christian sources were in many instances influenced by the emerging Islamic historical traditions. Furthermore, dating schemes in Christian as well as Arabic sources are often demonstrably inaccurate. While the Islamic chronicles frequently provide multiple alternatives for the same event, in the Christian sources (especially those derived from the Syriac *Common Source*), events are rather arbitrarily assigned to a year in later derivations. In the original Common Source, most events were simply left undated but placed in a specific chronological context before or after dated entries. In later redactions, such as Theophanes, these events became assigned to a specific year, with the result that the same event appears to happen twice with slightly differing contexts or is simply off by one, two, three, four or even five years. Indeed, narratives were often broken up, reassembled, assigned to different dates and personalities, and related to new political constellations and religious developments, often mixed with materials from other sources. Segments relating embarrassing events, such as catastrophic defeats or religious views rejected by later generations, were simply scuttled and covered up by importing new events from other entries.[5]

Constans II's reputation—indeed the whole period of his reign (641-669)—has languished doubly. He suffered a near universal *damnatio memoriae* in Byzantine sources due to his promotion of the Monothelite creed, as he continued Herakleios' attempt to create a compromise between Monophysites and Chalcedonians. The complete silence of Nikephoros is an extreme case, but Theophanes and other Common Source derivates only provide disjointed facts, mostly concerned with failures, since his religious policies were condemned in 680 and his successes were ostensibly purged from historical records. Arabic sources, in turn, have reinforced the minimalist Byzantine perspective by ingenious distortions of their own. Instead of presenting complex, deliberately planned campaigns, events were broken up into discreet, yearly entries, with a single objective and assigned to single commanders.[6] This was patterned on the summer raids, *sa'ifa*, which became highly ritualized during the 8th century and with which the later Islamic historians were familiar. Much of this process could have been an honest attempt at making sense of apparent contradictions in these historians' eyes. For example, armies moving in several columns would have had several commanders, but in-

---

[5] See in general Howard-Johnston 2010 for a recent, more systematic attempt at re-dating events based on strict, testable criteria; for the structure and dating scheme (or lack thereof) in the original Common Source, now known as *The Chronicle of Theophilus of Edessa*, see Palmer's 1993 translation of "Dionysius reconstituted" (the lost Syriac intermediary source for Michael the Syrian's *Chronicle* and the *Chronicle of 1234*), Hoyland's 1997 concordance of the entries that were derived from the *Common Source* in Theophanes, Agapius, Michael the Syrian, and the *Chronicle of 1234*, and most recently, his 2011 reconstructed translation. Further observations and cross references (with multiple examples of misdating found *passim* throughout the 7th century) are to be found in Mango and Scott's introduction to their translation of Theophanes as well as Vassiliev's edition and translation of Agapius.

[6] See e.g. the volumes of Tabari covering these years and the analysis in Lilie 1976, who to a certain extent follows the same logic but is able to discern several long-term Arab strategic objectives.

APPENDIX THREE: ARAB GRAND STRATEGY, 663-669   441

termediate sources may have emphasized the activities of different commanders, providing a confusing set of alternatives. Thus, an army operating in e.g., three divisions under three different commanders over two years conveniently provided credible material for three years' worth of *sa'ifa*, even if the second year of the campaign was originally marked by a horrific defeat and the following year was quite uneventful as a result.

Distortions appear deliberate when there were power struggles or embarrassing defeats to cover up. The Arab naval expeditions against *Constantinople led to military disasters (654, 670f) that deeply affected the body politic and religious of the nascent Islamic empire. These failures set the stage for internecine conflict by creating crises of confidence in the leadership.[7] In the Islamic sources, these disasters were successfully recast or suppressed to the extent that they were effectively forgotten. The causes for conflict were completely different by the time Muslim historians compiled their works about two centuries after the events. In fact, the very nature of the campaigns that preceded the civil wars was misunderstood, perhaps deliberately. The result is demonstrably nonsense in the case of naval expeditions with hardly any trace of a historical core, perhaps because such expeditions so often ended in failure.[8] Due to such distortions, it is very difficult to recreate the course of Arab raids in from 650 to 670 based on Tabari and other Arabic sources, and virtually impossible to distinguish them from major invasions aimed at permanent conquest.

As a result of these distortions, all that remains in the Arab-Byzantine tradition on the reign of Constans II are, firstly, a number of disjointed and occasionally duplicated raids against Anatolia;[9] secondly, a literary exposé of a humiliating Byzantine defeat at the naval battle of Phoinix in 655, with only obscure hints and references to a vast campaign against *Constantinople (654) that was under preparation but never appears to have taken place;[10] and thirdly, an extremely colorful story of the confrontation between the eunuch Andrew, a representative of Constans, and Sergius, representative of the rebel Byzantine general Saporios, at the court of Mu'awiya.[11] The double *damnatio memoriae* erased several crucial developments from Byzantine, Islamic and much Eastern Christian historiography. Constans II's extensive Caucasian interventions from 652 to 660, as well as the near-miraculous salvation of *Constantinople from a vast Arab armada in 654, are known from the Armenian Sebeos and partly confirmed by snippets in other sources. Constans' partly successful Balkan campaign receives some notice, but

---

[7] This is the argument of Howard-Johnston when commenting on Sebeos' version of events from the failure of *Constantinople 654 to the first Islamic civil war.

[8] For instance, Conrad 1992 demonstrates that Arabic narratives of naval expeditions around 650 have been jumbled beyond any reasonable recognition already by the early eighth century, but contemporary papyri along with early narrative sources confirm the scale and importance of Arab naval warfare from the 640s on.

[9] For these, see the reconstructions offered by Lilie and Brandes (as cited in the first note of this appendix).

[10] For Common Source derivates concerning the battle at Phoinix, see Hoyland 2011: 141-44.

[11] Ibid., 153-61.

his *Italian campaign from 663 is only known from Latin *Liber Pontificalis*, which ignores military events, and the unsympathetic Paul the Deacon, who maximizes Byzantine losses. In the Latin and Armenian sources, then, Constans II's military strategy emerges as carefully planned campaigns to retain and extend Roman control from central Italy to the eastern Caucasus. Similarly, thanks to a series of recent studies on the 650s based on Sebeos, a contemporary observer, scholars have begun to reinterpret the policies of the Islamic empire, arguing that not only was their overall strategic goal to conquer Constantinople by a combined naval and land campaign far earlier than commonly accepted, but this was nearly accomplished. I will argue that a similar campaign directed at the conquest of Anatolia took place in c. 663-68, based on little used Syriac traditions, at least one of which is demonstrably contemporary. The political context, scale of Arab ambitions and their ability to organize such a campaign in the 660s requires a brief summary of the new interpretation of the assault on *Constantinople in 654.

The emerging consensus is that the Arabs began preparing for an immense invasion aimed at conquering the remnants of the Byzantine Empire immediately after they had defeated the Persians in 651. While Arab armies were moving westwards from Iran, the dockyards and arsenals in Egypt and Syria were producing ships and equipment at full capacity. Most of the personnel working on building ships and siege engines were former Byzantine subjects; prisoners of war, conscripted craftsmen and laborers, and clients who helped organize the fleet, administer supplies and provide intelligence. Some were former Byzantine officials, others were Persian military colonists settled in Syria by Herakleios around 630. Arab fleets, having recently eliminated threatening Byzantine naval bases at *Arwad and Cyprus (*Lapethus, *Constantia) in 649-50, began establishing logistical posts on the *Aegean islands along the southwestern Anatolian coastline in 653. Some formerly Byzantine subjects sabotaged preparations at *Tripolis around 653 and escaped to Byzantine territory, but that seems to have made little impact on the ensuing expedition.[12] Constans was in Armenia in 653-54 trying to secure his eastern flank, but was outmaneuvered by Arab military movements and internal Armenian political developments, as one of the chief nobles, Theodore Rshtuni, submitted to the Arabs and received 7,000 Arab troops to help him gain control over most of Armenia. Theodore proceeded to invade Roman territory, raiding up to the Black Sea, and seems also to have captured Trebizond. As a consequence, Constans was drawn further east to secure Armenia. Then the noose tightened, as word soon spread that the massive fleet heading for Constantinople was set in motion, and Constans had to hurry back to defend his capital. When Mu'awiya shortly after led the main Arab land forces into eastern Anatolia, occupying Theodosiopolis and Caesarea in Cappadocia in 654, the Armenian regions were effectively detached from Byzantium and organized as a client kingdom of the Arab empire.[13]

---

[12] See chapter 7.3.1 for this.
[13] Garsoïan 2004:122; Sebeos 49. Howard-Johnston reorders some of the events recorded by Sebeos; see II: 270-23.

So far there is little controversy, since most of the information from Sebeos is corroborated by allusions or explicit information in the sources traditionally favored by modern historians. However, most scholars have chosen to believe the traditional sources in that nothing much happened in 654. Despite the Arabs' overwhelming strategic advantage, their invasion apparently never materialized, and the whole affair seems to have gone out with a whimper. The famous Battle of the Masts (as it is known in Arabic) at Phoinix follows in 655, but the prelude to and context of the battle is far from satisfactory in the traditional narrative, as it simply seems to skip a year from the massive buildup in 653 to a naval battle in 655 where the Arabs strangely appear to be underdogs.[14] Recent, more convincing explanations follow Sebeos, who describes the massive Arab fleet ready to sail from Egypt and Syria in 654.[15] Clearly the *Aegean Islands occupied in the preceding years were logistical bases to facilitate the passage of such a vast fleet. Sebeos subsequently relates how Mu'awiya advanced overland from Caesarea to Chalcedon, while the Egyptian and Syrian fleets moved along the naval bases established on the way to take Constantinople from the sea. The Byzantine army withdrew in the face of this massive, double threat, leaving most of Anatolia to submit to the advancing Arabs. The complete submission of Anatolia appears to mirror the rapid submission of Syria and Egypt. This should make us pause and reconsider Byzantine reactions in the 630s, when there seems to be a "peaceful" transition to Arab rule.

The apparent collapse of Byzantine morale was reversed when the Arabs were themselves struck a devastating blow, but not by any human force. According to Sebeos, the Arab fleet was in sight of the sea walls of Constantinople, drawn up for assault directly from ship-borne towers, when a sudden storm utterly destroyed the fleet. The total Arab defeat had profound consequences for both Arabs and Byzantines. It would not be unreasonable to suggest that the scale of the defeat triggered the revolt against 'Uthman by Egyptian troops, at the very least by providing a fertile breeding ground for discontent.[16] In this interpretation, the subsequent Arab civil war was a direct result of massive loss of men (especially from Egypt), resources and prestige, but later distorted due to the extremely embarrassing, or to later generations, incomprehensible, context. The situation was barely shored up by the Arab victory at Phoinix and Mu'awiya's controlled withdrawal from most of Anatolia. The successful disengagement provided Mu'awiya with the military means to struggle for power after his patron and kinsman 'Uthman had been killed in 656.

While the Arabs soon engaged in an extremely brutal civil war that lasted five years (656-61), Constans had secured Anatolia and brought Armenia back to the fold already in 654. In line with the reinterpretation above, Constans' naval expedition in 655 was intended to take advantage of the Arab prostration with an im-

---

[14] Sometimes, alternative dating schemes (e.g. placing either the occupation of an Aegean island or the Battle of the Masts in 654) help close the apparent gap, but the fundamental problem remains. See also next note.

[15] For the following, see Howard-Johnston's commentary to Sebeos 50 (II: 273-76), O'Sullivan 2004, Cosentino 2008 and Howard-Johnston 2010.

[16] See Sebeos 52 and Howard-Johnston's commentary (II: 284-88) for the argument.

mediate counterthrust, but there were still substantial Arab naval forces that would have guarded the many ports and islands that had been conquered during preparations for the expedition against *Constantinople. These were assembled to a substantial fleet which stopped the reportedly larger Byzantine navy at the naval engagement at Phoinix.[17] There was a similar offensive overland against the remaining Arab garrisons in Armenia. Initially successful in recapturing *Dvin in 655, this expeditionary army was ambushed and had to flee to Caucasian Iberia, allowing the Arabs to take *Theodosiopolis the same year. Following these defeats, Constans must have concluded that he could not attack the Arab empire head on, and instead turned his attention to organizing a grander platform for counterattack. First there was an obscure expedition against Slav territories on the Balkans (c. 658), which produced a number of Slav recruits, but the Balkans were too poor and infiltrated by Slavs to provide the necessary strength and resources for his grand project.[18] He did however exploit the Arab civil war to return to Armenia, where he brought all of the Armenian nobles back into the Roman fold and even extended Roman control into Media in 659-60; this prompted Mu'awiya to ask for peace and pay tribute in 659.[19]

Constans drew inspiration from his grandfather Herakleios' successful strategy of assembling strength in the West in order to defeat an eastern foe. Thus, after returning from the eastern Caucasus in 660 and presiding over the trial of Maximus the Confessor for treason until early 662, he set out overland to Athens to reestablish Byzantine rule in the southern Balkans in mid-662. By 663, he landed in Italy and attempted to recapture and reorganize the Western provinces for a counterattack against the Arabs.[20] His absence in the West is the background for the events in the 660s that are under consideration here. As we noted above, the traditional narrative as well as the prevailing modern interpretation of the Arab incursions into Anatolia emphasize disjointed raids that makes it difficult to reconstruct their overarching strategic idea, except weakening Byzantium and keeping it on the defensive until the major naval effort that aimed at *Constantinople and the Anatolian coast in the 670s. By turning to the more contemporary Sebeos scholars have been able to put the events of the 650s into context. While his narrative ended with Mu'awiya's victory in 661, two little-used Syriac sources provide invaluable information on Arab raids during the 660s. The *Maronite Chronicle* was written in 664 (the last dated entry) or shortly after, and contains the first coherent, inside perspective of the emerging Caliphate seen from the viewpoint of Syrian Christians who still had some allegiance to Byzantium. Unfortunately the folios up to 657 (and one covering 661-3) are missing, but what survives is highly illuminating. This source provides a handful of well-dated anecdotes that allow us to

---

[17] Cosentino 2008 provides a slightly different reconstruction of events.
[18] Although the Slavs were now becoming much more sophisticated in military and political organization; cf. *Thessalonica 662 and ch 7.2.2. This explains the extensive use of Slavs as military colonists by the Romans in the late 7th century; see Sarris 2011: 295f.
[19] Sebeos 52 and Howard-Johnston's commentary, which also draws on later Armenian historians.
[20] For this, see CO entries for 663, under *Constans II's Italian Expedition and Zuckerman 2005.

grasp Arab strategy far better than the later sources, both Christian and Islamic. In addition, the Syriac *Chronicle of 819* provides an early notice that the campaign lasted for two years on Roman territory and was led throughout by 'Abd al-Rahman ibn al-Khalid, the son of the famous conqueror Khalid ibn al-Walid (dubbed "the sword of Islam"). He obviously had a much higher standing than the more obscure commanders credited with leading Arab armies in Anatolia during this period.

The information from the *Maronite Chronicle* and the *Chronicle of 819* allow us to reinterpret and re-date the traditional narrative of what took place in the 660s. While much of the Byzantine field army was in the West with Constans, his son Constantine was responsible for maintaining the defense of the East. Constans seems to have been confident in the peace agreement made at the height of the civil war in 659, and also that the defenses established in Armenia were sound, and the Arabs would not be able to move soon due to the civil war and its aftermath. Mu'awiya proved him very wrong. He clearly wanted to exploit Constans' absence in order to finish the work he had begun a decade earlier, and in 661 broke the peace treaty. This time he proceeded more deliberately and prepared the ground well by politics and intrigue. He managed to detach Armenia from Constantinople upon his victory in the Arab civil war, appointing Grigor Mamikonean as his client ruler.[21] This exposed central Anatolia from the east. The next move was to renew naval raids into the Aegean. In order to do this, he began a build-up of naval forces in Syria and Egypt. Again the Arabs turned to the conquered population, but it appears that sabotage, losses during the *Constantinople (654) campaign, refugees, and civil war had taken their toll on Arab industrial capacity in the coastal cities. Baladhuri reports that Mu'awiya transferred craftsmen from the cities of the Syrian interior to work in the dockyards along the coast. He also moved Persian forces, presumably for providing military manpower and expertise for projected naval expeditions.[22]

In 663, we have some indications of goals of this naval buildup. The *Maronite Chronicle* reports that an Arab force was in Thrace, ravaging *Constantinople's immediate hinterland. Such a force obviously needed a supporting fleet, but unfortunately the story of how they got there has been lost. It is however quite possible that Theophanes misunderstood one of his sources and misplaced information from other Arab naval expeditions to the context of *Constantinople 670ff; Howard-Johnston prefers 654, Jankowiak suggests 668, while I would hazard to guess that 663 is just as likely in light of the information from the *Maronite Chronicle*.[23] The raid may have been planned as much for political as for military reasons. The Arabs had set up a camp close to Constantinople, using it as a base

---

[21] Garsoïan 2004: 122; for a full evaluation of context, see Howard-Johnston 2010 and his commentaries to the relevant section of Sebeos.

[22] For this see chapter 7.3.

[23] Concerning the context, the Arabic tradition actually provides a clue: Elias claims that Busr invaded all the way to Constantinople (i.e. Chalcedon) and wintered, while an entry for the next year tells of a naval expedition to Constantinople; see Lilie 1976: 69 for references. On the "misplacing" of information, see references in *Constantinople 670f. For the following, see the translation quoted for *Constantinople 663.

for their plundering activities. At first, this enraged the urban population, and the youths (probably circus factions), refugees and soldiers in the city were eager to sally out against them. At a first attempt, they managed to catch the Arab camp unprepared with most of the troops out raiding; only clients and servants held the camp which the Romans proceeded to pillage. However, desiring to repeat their success the next day and planning another sortie, the junior emperor Constantine (IV) forbade them from doing so. When they insisted nonetheless, they were drawn away from the city by the now well-prepared Arabs and fell into an ambush, and were only saved by scurrying back to the Theodosian Walls, where the *ballistrai* protected them from their pursuers.

While the Arab force did not directly threaten to take Constantinople, its presence so close to the capital less than a decade after 654 was ominous, and ostentatiously drew attention to the impotence of the sitting regime and perhaps also forced it to transfer troops from Italy and Anatolia to Constantinople. Indeed, the next year, 664, was marked by a large-scale invasion by 'Abd al-Rahman ibn al-Khalid, that was far more dangerous than hitherto recognized, as it was assisted by an internal revolt that threatened to detach most of Anatolia from Roman control and bring it under Arab rule as a client kingdom.

With an exposed eastern frontier, a raiding Arab army that moved to within ballista range of the Theodosian walls, and much of the field army on an uncertain venture far away in Italy, some high-ranking Byzantines must have either have been quite unnerved, or wanted to exploit the situation for personal gain, or both, much as the Armenian nobility had done in the 650s. Saporios (Shapur) was the general of the newly established Armeniac command (see ch. 2.1.3) and likely the same person as the general Saburrus who accompanied Constans on his *Italian campaign in 663. He was of Persian origin and thus probably one of the military colonists recruited by Herakleios (or a descendant) who had been settled in Anatolia.[24] It is conceivable that he had family relations with Persians settled in Syria who were now in Arab service. His revolt has not attracted much attention, but appears to have been an integral part of Arab strategy. The traditional dating of the revolt is based on Theophanes, who places the key event in 667/8, when competing Roman loyalist and rebel embassies were in Damascus. This is, rather typical of the Common Source, both telescoped and possibly several years off: while Michael the Syrian provides the earliest date, 665/6, the *Chronicle of 1234* places it in 666/7 and Agapius offers the late date of 668/9. The revolt itself must have taken place over at least a few years, considering the diplomatic activity that went on between the three parties, and I suggest here that it began sometime between the raid up to Constantinople in 663 and 'Abd al-Rahman ibn Khalid's invasion in 664.

There are three reasons for this. Firstly, there is a very detailed report in the *Maronite Chronicle* of the first year of 'Abd ar-Rahman ibn Khalid's invasion of Anatolia. This source was written shortly after the events portrayed, and was exceptionally well informed about the invasion and its initial objectives, which include a city called *SYLWS (possibly Sagalassos), *Amorion, *Kios, *Pessinos,

---

[24] For his background, see Peeters 1933 and *PMBZ* s.v.

*Pergamon and *Smyrna. It explicitly states that the Arabs tried and failed to take *SYLWS and an unknown city in a lake (*Skoutarion?) by storm. The other cities were simply "subdued," a verb implying use of force, but no details are given. In contrast, Amorion surrendered by treaty (*mellta*), which appears strange for the most populous and best defended city in Anatolia. The explanation must be found in the collusion of high-ranking Byzantines—it would certainly not be the last time Amorion was involved in such schemes (cf. *Constantinople 717f).

This leads us to the second reason for re-dating the whole affair. The capture of *Amorion does not figure in the traditional Arabic sources as it ended in a terrible disaster, the whole garrison of 5,000 men being massacred in a surprise night attack in the dead of winter in 666/7. Lilie, following Theophanes, places this in c. 669 in order to synchronize it with a large incursion, recorded in CS and Islamic narratives, in support of Saporios (instead it raided to Chalcedon when he died). However, CS derivates also note several raids in the period 663-65, so we only have the very late conjecture of Theophanes, who chose to place the Roman recapture of Amorion *after* a late dating of Saporios' revolt.[25] There is good reason to believe that Theophanes is wrong on this point, and the (re-)capture of Amorion resulted from the raids reported around 664. It would certainly not be the first time Theophanes mistakenly inserted material into the framework provided by the Common Source. In fact, the Arab sources and CS derivates agree with the *Maronite Chronicle*'s general framework, as they show Arab commanders, including ibn Khalid, operating in the same general regions in central-western Anatolia. However, the slight historiographic distortions as described above have thrown most historians off the trail, as Amorion was unknown to or has been edited out of the different traditions. The *Maronite Chronicle* describes in contrast how ibn Khalid's campaign against *Amorion in 664 involved two fortifications which must be located in Pisidia.[26] While Amorion is ignored in the Arabic tradition or placed several years later, ibn Khalid is indeed described by Tabari as raiding in the vicinity of Antioch in Pisidia in 664/5. However, the invasion by all accounts (except Michael the Syrian, who however changed the detailed list of cities raided to general regions, in the process mistaking Cappadocian for Pontic Koloneia) studiously avoided the Armeniacs. The Arabs began raiding Hexapolis, on the southern frontier of the Armeniac *strategis*, only *after* the death of Saporios (see below).Thus there was a surrender of Amorion, not by force but by treaty, in 664; it was occupied for at least two years as part of a larger command of ibn Khalid (per *Chr. Mar.* and *Chr. 819*); and the occupation forces should be associated with the revolt of Saporios, whose territory was untouched by the invasion.

Finally, an early date for the revolt follows from Howard-Johnston's observation from Armenian sources how the Caliphate was engaged in diplomatic dealings not only with Caucasian rulers, but also Roman factions. Firmly dated to 667/8,

---

[25] See Lilie 1976: 72f; cf. the reconstruction in Hoyland 1997 for the original grouping of these events, and following notes.

[26] The vicinity to Amorion, presence of a large lake with inhabited islands (*'SQDRYN*), and the consonantal skeleton of *SYLWS all indicate a location in Pisidia, as all suggested reconstructions are very close to each other.

the Albanian prince Juansher came to Damascus where he met Roman envoys who were in opposition to Constans (i.e. Sergius). So far, it seems to confirm the traditional dating of the revolt, which is retained by Howard-Johnston.[27] However, Juansher, who fought alongside the Sassanids against the Arabs and later became a staunch supporter of Constans II, inexplicably exchanged his allegiance and submitted to Muʿawiya early in 665. Howard-Johnston only implies that a Hunnic raid across the Caucasus the winter of 664/5 prompted Juansher to seek protection, but another explanation springs to mind: in 661, as we saw, Armenia was detached from Roman hegemony. This had a domino effect, as Saporios must have returned from the *Italian campaign late in 663 or early 664 (certainly with Anatolic, possibly with Armeniac units) in order to deal with the threat presented by the defection of Armenia and the raid on the suburbs of *Constantinople in 663. Observing that his position was untenable, or using the opportunity to further his own ambitions, he instigated a revolt in 664 at the latest with support from ibn Khalid. With Roman support now cut off by Armenia proper, the Armeniac *strategis* in revolt, and Arab armies in west-central Anatolia, Juansher did what he could to shore up his own precarious position.

This reconstruction is further supported by some extremely interesting military details during this invasion. The Arabs did not bring a siege train. They relied on the ability to storm fortifications by the use of massed archery, ladders and various ruses and stratagems (cf. *Dara, *Dvin 640; *Euchaïta 644), building whatever they needed on the spot (*Skoutarion 664). This could be a logistically efficient approach, but costly in human lives, since a storm without long-term preparations like siegeworks and engines would expose the attackers to a punishing killing field before they were able to scale the walls or breach the gates. It might seem peculiar, since the Arabs demonstrably used siege engines successfully and on a large scale as early as 639-40, and consistently with their major naval expeditions from Cyprus (*Lapethus 649) to *Constantinople (654). There are several reasons for this choice. Most of the engineering and advanced competence resided with subject populations whom the Arabs still did not trust. They only began using Syrian engineers and experts on overland expeditions after 700, when they had in place good administrative mechanisms to control them, and the restructuring of their religious identity began distancing them from the Byzantine Empire. Otherwise the Arabs only used such expertise in naval dockyards or on fleets. At port such personnel were under strict control, but their loyalty on naval expeditions was a recurring and very serious concern for Arab authorities into the eighth century, and was only resolved by conversion and the creation of a large class of Muslim engineers and craftsmen by around 800. In the meantime, the Arabs apportioned their subjects' skills wisely (by sending many of them to the coast rather than on campaign), and used whatever resources they came across while on land-based campaigns (see ch. 7.3).

When attacking *SYLWS (Sagalassos or another Pisidian city), they employed the services of a Paphlagonian master carpenter (the *rabb naggārā* in chapter 2.4) who offered to build a large trebuchet (*manganīqā*) in return for the security of

---

[27] Howard-Johnston 2010: 119, 223-26.

himself and his family. His origins seem strange, as Paphlagonia, on the central Black Sea coast, was far from the area raided by the Arabs. However, part of it did lie within the Armeniac *strategis*; he might indeed have been assigned by the Roman rebels to assist their Arab allies in subduing cities further to the west. A similar argument may be made for the Slavs whom Constans had deported to Anatolia and who subsequently defected to the Arabs; if this conjecture is correct, they were only following the orders of their local officers as any other Roman soldiers. As we have seen, 'Abd al-Rahman ibn Khalid's army studiously avoided attacking Armeniac territory in 664, so there is every reason to believe that Saporios was not only involved from the start, but that he provided material support to the Arabs in the form of troops (including Slavs) and engineers. He also had the connections and credibility to help convince the troops at *Amorion to surrender, but perhaps not further away.[28] It thus came about that the Arabs established a base at the most strategically located and best fortified of all the Anatolian cities without a blow. Unfortunately, the *Maronite Chronicle* ends at this point, and we only have a brief notice in the *Chronicle of 819* that explicitly states (though in an extremely compressed fashion) that the invasion lasted two years and was led by 'Abd al-Rahman ibn Khalid the whole period. Thus, they had a very secure base for extensive attacks throughout Phrygia and Pisidia during the period 664-666.

With this framework in place, we can reinterpret evidence from CS and Tabari on apparently disjointed Arab raids during this period as well as Theophanes' description of the Byzantine counterattack. The main army from the *Jund* of Qinnasrin under the command of 'Abd al-Rahman ibn Khalid and supported by Saporios advanced into Cappadocia very early in 664 (either via Armeniac territory or via the Cilician Gates), with Koloneia as its first hostile objective. This army sought to exploit the distraction poised by the fleet-borne raiding army near Constantinople. It then proceeded to Pisidia, where it met with moderate success, and *Amorion, which submitted by treaty and became the main base for the subsequent Arab raids. Here the *Maronite Chronicle* is highly compressed, so that it is difficult to know the order in which events occurred. Perhaps there were two or more columns involved after Koloneia, as indicated by multiple commanders reported in later sources; however, after *Amorion, the major thrust was to the northwest, raiding and capturing cities in a wide arc from the south of the Gulf of Nicomedia along the coast down to Smyrna. Its movements thereafter are uncertain, but ibn Khalid's force remained in occupation of *Amorion and parts of Phrygia for two years.

The grand invasion thus led to Arab control of at least two thirds of Anatolia, where indirect control was exercised through the *stratēgos* Saporios over most of the Armeniac, and possibly parts of the Anatolic and Opsikian commands, while the main Arab forces were stationed in and around Amorion, which became the base of raids and attempted conquests throughout Pisidia, Phrygia and the north-

---

[28] Indeed, loyalties across Anatolia were divided, as the loyalist ambassador to Mu'awiya at the time, Andrew, had the support of at least some border forces who set up an ambush for the rebel representative Sergius upon returning from Damascus. See Hoyland 2011: 153-61 for the various CS derivatives.

western coast along the Sea of Marmara and the Aegean. What emerges, then, is an overarching Arab strategy that aimed at nothing less than the complete conquest of Anatolia, only leaving a medium-sized client kingdom comprising the Armeniac and perhaps parts of the Anatolic commands under Saporios. This would become a Byzantine equivalent to the Armenian client kingdom which the Arabs had established with the Armenian noble Grigor Mamokinean as its ruler in 661. Had Mu'awiya succeeded, Constans would have been left with little beyond Constantinople, a few exposed costal territories in Europe, and the province of Africa that was threatened from Arab land and naval raids from Cyrenaica.

'Abd al-Rahman ibn Khalid's invasion was not merely a "wintering" in Anatolia, then, but an existential threat to the last substantial region under Roman control. In light of its military achievements during the first campaigning season in 664, his army of must have been very large; certainly, the consequences of its advance through Cappadocia were severe in the extreme. Former Arab raids had only lasted weeks or months with fairly limited objectives, while major invasions took a more easterly route via Armenia to Theodosiopolis and Caesarea. After the raids of 644-45 (*Euchaïta, *Caesarea, *Amorion) and major invasion of 654-55 (*Constantinople, *Dvin, *Theodosiopolis) there was a decade of relative respite, so even though the wars bore hard on the Anatolian cities, the invasion of 664 was very different. We do not have narrative sources to tell us exactly what happened, but faced by such an invasion, any civilian population not supporting Saporios along the main invasion route must have fled, been massacred or taken captive and deported. We know for sure from the archaeological record that they did not stay. Pollen analysis taken from Southern Cappadocia show how, between 664 and 674, anthropogenic plants (i.e. cultivated by humans) began to disappear, being gradually replaced by shrubs and brush typical of abandoned agricultural lands. This was not a temporary development. Within a few years, trees began to appear, and within a few decades, all of southern Cappadocia was covered by dense forests, which dominated until the late tenth century, when plants associated with husbandry began to appear. This cannot be related to climatic or cultural changes as the area had been cultivated continuously since the Neolithic age and no similar change is attested in the intervening millennia. It was a human-made disaster which caused the complete abandonment of the area for a quarter of a millennium.[29]

Many Anatolian cities contracted during the seventh century, others shifted to more defensible sites, but few were abandoned outright. Mokisos, which was possibly located at present-day Virahşehir, is a notable example of abandonment. As Berger has noted, it appears as if the city was built on one day and abandoned on the next.[30] No accurate dating of this event is possible, but in light of its location in southern Cappadocia, we may have an accurate *terminus ante quem*. With the passage of ibn Khalid's army the region became effectively uninhabitable. During the two years or so ibn Khalid's invasion lasted, this was the main route for raiding armies on their way to Koloneia and Amorion, whence they fanned out to attack

---

[29] Haldon 2007.
[30] Berger 1998 and personal communication.

most of central and western Anatolia. While the Armeniac command was left alone due to the defection of Saporios, most of the Anatolic and Opsikian commands were not. Southern Cappadocia got the worst of it as hostile frontier territory. The population did not, and actually could not, come back afterwards. While some cities in the region survived due to their military function or were later revived as bases and garrison towns, southern Cappadocia as a whole never had the opportunity to recover. There was still some fluidity to the situation along the border following the Arab defeat, but the infamous *sa'ifa* became institutionalized during the late seventh and early eighth centuries, and most of the Arab raiding armies followed this route. As soon as the Arabs began to consolidate their hold on Cilicia in the early eighth century, any possibility of recovery was gone. Raiders could take several routes, and according to local informants in the Taurus foothills, it would take a skilled rider only a day to cross over from the Cilician plain, where the main bases for the *jihād* were established in the 8th century, to the Anatolian plateau. No one could survive such a permanent threat.

Despite these enormous upheavals, three developments hindered the Arabs from absorbing such large swaths of Roman territory, although the order in which they happened is uncertain. Firstly, 'Abd al-Rahman ibn Khalid had been recalled to court and died in 667, perhaps poisoned on Mu'awiya's orders.[31] He appeared to be succeeding where 'Uthman and Mu'awiya had failed so spectacularly in 654, and may thus have been an important threat to Mu'awiya's son Yazid, who was sent to lead campaigns in 663 and 668/9. Furthermore, as the son of the most successful Islamic general in history, Khalid ibn al-Walid (who was forcibly retired by 'Umar due to his popularity and prowess), he seems to have enjoyed immense prestige. In fact, several *hadith* record the regret of 'Umar for retiring Khalid and his premature death, as even 'Umar later said he would have been the most eligible to succeed as caliph if he were still alive.[32]

Secondly, Theophanes records the capture of *Amorion from the Arabs in the winter of 666-67 (a date indirectly confirmed by *Chr. 819*) by a commando-style raid in the deep snow. The Romans scaled the walls with "planks" or (perhaps better) siege ladders and massacred the 5,000-man strong Arab garrison. Brandes recently doubted this story, believing that the major features related in narrative sources, such as population and garrison sizes, the extent of fortifications and so on were incompatible with the archaeological data published until then.[33] This image has been dramatically overturned since Brandes wrote. It has been conclu-

---

[31] For this, see *EI*² s.v. 'Abd al-Rahman ibn Khalid.

[32] I have not been able to ascertain the exact reference for this tradition, but see *EI*² s.v. Khalid ibn al-Walid.

[33] Brandes 1999 expresses skepticism of siege accounts and large estimates of the population based on information from the 8th and 9th centuries, pointing to the preliminary archaeological reports available at the time. This has now been turned on its head; as shown by e.g. Ivison's 2007 synthesis (the actual excavation reports are by now very extensive), the lower city was densely populated and economically active, even in the middle of the 7th century. While visiting the site in the summer of 2007, Ivison also indicated to me that high estimates of up to 30,000 inhabitants may actually be correct, especially during market season.

sively proven that the lower town of Amorion was densely populated and heavily fortified during the seventh century. There is no longer any valid reason to doubt high figures or descriptions of elaborate fortifications. Theophanes' description of the Byzantine commando operation therefore deserves to be taken seriously. As a major hub of Arab military activity in the middle of hostile territory, the garrison obviously needed to be large. In similar conflicts in the 6th century, large garrisons were a matter of course in particularly important bases in contested or exposed regions. In light of the enormous manpower resources that the Arabs possessed, there is absolutely nothing incredible with a garrison of 5,000 men. In addition we would have to add at least some of the remaining population who would be enrolled for required service, such as logistics, construction and support, again based on late Roman *munera*, in line with common Arab practice of using existing administrative structures. Attacking a heavily fortified city defended by a garrison could only normally be done by an army two to four times the size of the defending force. Bringing up such an army unnoticed in the dead of night was a challenge. But this was, realistically, the best Byzantine option. Otherwise they would have to prepare a full logistical train for a blockade or a storm during the summer months, which was very dangerous since the Arabs then had access to relieving forces from Syria as well as from the Armeniac rebels. Trying to wait it out for months in a camp during winter was not an option, since heavy snow meant that armies could not normally move around and certainly not assemble supplies. This would include the Arabs, who in such circumstances would not be able to patrol the region properly for threats. Thankfully Theophanes took the trouble to include this anecdote from his source, whereas Nikephoros seems to have excised it as he regularly did with brief notices. It may originally have been preserved because the operation was under the command of Constantine, who reversed his father's religious policies at the Sixth Ecumenical Council in 680.[34]

According to this reconstruction, the position of Saporios must have become much more difficult after the death of his main patron and the loss of *Amorion to Constans, at the latest by 667. Mu'awiya did not let 'Abd al-Rahman's achievements slip completely, but used diplomacy to renew political alliances in 667/8 with both Roman and Caucasian actors, and subsequently sent new forces in support of the rebels. However, before they could arrive, Saporios was killed in an accident, and his soldiers appear to have defected back to the Byzantine side. This was the third factor that helped save Anatolia from Arab control. The dating of the recapture of Amorion and the death of Saporios is difficult; they may have occurred a year or so apart. Saporios' revolt is compressed to one entry in CS derivates, but clearly happened over several years. It thus began in 663/4, ending at the latest in c. 668/9. This led to renewed Arab attacks on the Armeniac command, in the new border region called Hexapolis (previously known as *Armenia IV*), but by now the Arabs were deprived of local support. Reinforcements were sent under Mu'awiya's reluctant son Yazid, perhaps to salvage something of the situation, but the ostensibly impressive raids across Anatolia in 668/9 had no permanent result.

---

[34] For the council, see Herrin 1987: 277ff.

APPENDIX THREE: ARAB GRAND STRATEGY, 663-669

The assassination of Constans in 669, the ensuing revolt in Sicily and its suppression, and Arab naval attacks on the Anatolian coast towards *Constantinople (670ff) until 674 formed parts of a grand scheme to dismember the empire. They were of course grand threats to the empire, but without local clients and a solid anchor in Anatolian land bases, they effectively became massive blows in the air.

PART TWO

CORPUS OBSIDIONUM

CONVENTIONS ADOPTED

The entries are organized chronologically, with one heading for each siege. In some instances, lists of cities without any specific information are given as a regional entry or under the first/best attested siege. Otherwise, unnamed forts and cities are included with one that is named, and in a few instances, whole campaigns have been given a survey heading in addition to individual entries. Depending on the nature of the source (e.g. if it is too compressed to untangle in a comprehensible manner) all the evidence for several sieges in a campaign is presented in the first entry; the other headings then only refer back to the first entry with an explanation or additional information if necessary. Within a year, sieges are organized as they occur in the sources, or if more precise information exists (which is rare), according to date within a year. When sieges in one year come from different sources and/or regions, there is no particular order, except that an attempt has been made not to break up campaigns within the same general region that stretched over years by inserting entries from other sources/regions. The basic heading/entry format is:

NAME OF CITY/FORT   length of siege   *terminology*   year
1. Sources and secondary literature; cross references to main discussion(s) of the siege in the first part of the book. The abbreviations are those used commonly throughout the book. Note the use of 'Dionysius' for the reconstruction of the *Chronicle of Dionysius of Tell-Mahré* in Palmer et al. 1993; when unstated, the source is the *Chronicle of 1234*, otherwise Michael the Syrian is specified. The same applies to quotations in the original Syriac. For other Syriac Common Source parallels with discussion, references and other information, I refer to Hoyland 2011.
2. Translations, original quotations, and/or summary with technical terms and important phrases in the original. Unless otherwise stated, these are taken directly from those sources listed in (1). Depending on the level of detail required to understand the geography, protagonists and objectives, brief explanations of the military/political context have sometimes been included. Otherwise, the secondary literature normally provides the necessary information along with further arguments or information about the siege that may be used in (3).

3. Discussion of sources, tactics, dating, or terminology; other sources that throw light on the siege (with further references); other studies; further cross references to instances where this particular siege is mentioned in the first part of the book.

Throughout (1-3) there are cross references to other sieges in the Corpus that impinge on the discussion; these are marked with an asterisk and date: Thus, the entry 'Naissus 442' refers to '*Ratiaria 447' in the discussion. If the date is the same as the main entry, the date is left out, but marked with (above) or (below) for ease of reference. There are also further cross references to chapter for discussion of specific points. However, in order to avoid excessive lists of chapter numbers, only significant discussions are cross referenced to, not every instance for which a siege has been used as an example in the notes or in passing.

In the headings, alternative names or additional sieges that do not receive their own entry are placed in (parentheses). Uncertain sieges (i.e. those that may only have consisted of a battle or raid near a city or fort with no follow-up campaign against the city) are marked by a question mark; thus '?Toulouse 439'. Spurious sieges are noted with the city name in [square brackets].

The length of siege is only indicated where the text explicitly provides this information, or it is easily recoverable from the context.

Similarly, dating in the period under consideration is notoriously difficult. The dates provided in the entries are based on the dates provided in the sources, standard editions and translations, or secondary literature quoted. Significant discrepancies have been noted and dates are discussed in detail when this affects the argument, but in general the dates should merely be regarded as a reference tool rather than a precise, absolute chronology. Approximate dates are marked with 'c.'; thus 'Sirmium c. 472'. Sieges that lasted from one calendar year to the next are marked with 'f' (following), or if over two years, 'ff'; e.g. 'Rome 537f'. The very few sieges that lasted longer are marked with a dash: Papyrius 484-88. Alternative dates are marked by a forward slash /, thus 'Thessalonica 586/97'. In all cases, the entry is placed according to the first possible year. Those beginning in one year and continuing into the next or may be assigned to alternative years are normally placed last within the *first* year indicated.

Siege terminology in the headings are meant to show the vast semantic range of commonly used vocabulary, the stability of certain terms (e.g. the Syriac *šrā 'al*), and equivalents in different languages describing the same events. It can also illustrate the sometimes surprising use of words or phrases to indicate what is clearly a siege, or should be regarded as such following the definitions in chapter 5.1. Note that the Greek expression (ἐν)στρατοπεδευσάμενοι ('having encamped, taken up position (in)') and related forms have on some occasions been abbreviated to '(ἐν)στρ.' in the headings, but the full form is mostly found in the relevant entry.

All translations have been marked by "quotation marks." Latin terms and short phrases are *italicized*; full sentences and block quotations are placed separately between straight lines thus: || ... ||. All transcriptions from Greek, Syriac and Arabic are also italicized. Translations or summaries of terminology from Ge'ez and Armenian are placed in [square brackets] on the heading. This is also used to indicate the nature of the siege or other relevant information when there is no simple phrase that directly indicates a siege, even if this is clear from the narrative evidence.

THE 5TH CENTURY

ROME 408ff

Wolfram 1988: 154-59; Kulikowski 2007: 1-10, 174-77.

Rome went through a series of three blockades during these years. All were meant by the Visigothic leader, Alaric, to pressure the Roman government in Ravenna to grant him and his followers settlement and integration into the Roman army. There was little fighting; most of the time, the Visigoths prevented foodstuffs from reaching the city, which still had a population of several hundred thousand and was thus very vulnerable to pressure on supplies. According to Kulikowski, the sack of Rome was done out of exasperation and was also an admission of failure on the part of Alaric, since peaceful integration was no longer possible.

ARLES 411

Olymp. fr. 17.2; GT 2.9; *PLRE* 2 s.v. Edobichus.

The usurper Constantine had been penned in at Arles by Honorius' forces: "Although the army of Honorius was still pressing the siege of Arles, he held out because Edobich had sent word that he was arriving with a large army of allies." Κωνσταντῖνος δὲ περικαθημένης τῆς Ὀνωρίου στρατιᾶς ἔτι πρὸς τὴν πολιορκίαν ἀντεῖχεν, ἀγγελέντος Ἐδοβίχου μετὰ πλείστης συμμαχίας ἥξειν.

Edobich's army was ambushed by Honorius' general Constantius. This led to the surrender of Constantine. Edobich fled to a friend, Ecdicius, a landowner (ἔφυγεν εἰς ἀγρόν τινα πρὸς Ἐκδίκιον κεκτημένον; the latter term renders the Latin *possessor*, or great landowner). Ecdicius clearly wanted to support the winning side and had Edobich beheaded, but although officially commended for his actions he was dismissed by Constantius without favor.

An important point downplayed by Olympiodoros that we only learn from Gregory is that Edobich's relieving army consisted of Frankish and Alaman client troops from beyond the Rhine. Although himself a Frank, Edobich's political allegiance was exclusively with the Roman world and he used his connections to support his patron Constantine. This illustrates well the problem noted in chapter 1.3 (*passim*), that "barbarian invaders" could just as well be participants in Roman civil wars; indeed they were here perceived as such by a Roman observer.

| VALENCE | *effringitur* | 411 |

Chr. Gall. a. 411 (71); *PLRE* 2 s.v. Iovinus 2.

"Valence, the noblest city of Gaul, was stormed by the Goths; Jovinus had gone there in flight." || Valentia nobilissima Galliarum civitas a Gothis effringitur, ad quam se fugiens Iovinus contulerat. ||

| NARBONNE | *ingressi* | 413 |

Hyd. a. 413 (55).

"The Goths entered Narbonne at the time of the vintage." || Gothi Narbonam ingressi vindemiae tempore. ||

| TOULOUSE | *capta* | c. 413 |

Claud. Rut. Nam. *De reditu suo* 1.496; Wolfram 1988:162 and n. 280 (p. 440).

Claudius tells how it was impossible for him to leave Tuscany because of *capta Tolosa*. Wolfram believes Toulouse was captured by the Visigoths at the same time as *Narbonne (413) and *Marseilles (413), but it may have been as late as 416/17.

| ?MARSEILLES | λαβεῖν ἤλπιζεν | 413 |

Olymp. fr. 22.2; Wolfram 1988: 164.

The Romans tried to renegotiate with the Visigothic king Athaulf when they proved unable to supply the Goths as promised in a previous treaty. "The barbarian [Athaulf] pretended to agree and advanced to the city named Marseilles, which he hoped to capture from treachery. There he was wounded by a blow from the most noble Boniface and, barely escaping death, he retired to his own tent, leaving the city rejoicing and full of praise and acclaim for Boniface." καὶ ὁ βάρβαρος τὰ ὅμοια ὑπεκρίνετο, καὶ πρὸς Μασσιλίαν, πόλιν οὕτω καλουμένην, παραγενόμενος δόλῳ ταύτην λαβεῖν ἤλπιζεν. ἔνθα πληγεὶς Βονηφατίου τοῦ γενναιοτάτου βαλόντος, καὶ μόλις τὸν θάνατον διαφυγών, εἰς τὰς οἰκείας ὑπεχώρησε σκηνάς, τὴν πόλιν ἐν εὐθυμίᾳ λιπὼν καὶ δι' ἐπαίνων καὶ εὐφημίας ποιουμένην Βονηφάτιον.

This particular fragment is taken from Photius' *Bibliotheca* (see Blockley's ed. for commentary and references). While Photius is known for providing fair summaries of the main contents of the books he read, in this case it is impossible to conjecture what exactly transpired between the hope for "treachery" and the wounding of Athaulf by Boniface. Since it came to violence from the Roman side, we must assume some sort of Gothic provocation, whether a formal siege or a betrayal of their intentions to capture the city by treachery.

Bazas		413

Paulinus of Pella, *Eucharisticos* 291-405; Wolfram 1988: 164 and n. 290 (p. 441).

Paulinus gives an inside view of how complex the situation caused by civil war was around 413; furthermore, he shows the use of "barbarian" troops by all sides. Clearly it was as difficult for the Goths and Alans to figure out which was the "right" side as it was for many Romans, some of whom fought against central authorities themselves by raising troops of slaves.

Trier	*incensa*	413

GT 2.9; Scharf 1993; cf. *Trier 457.

Gregory has found this information in Renatus Profuturus Friderigus. Originally in the context of mopping up at the end of Jovinus' revolt, there is a report of how Honorius' generals "cruelly put to death" a large number of aristocratic supporters (*nobiles*) of the usurper in the Auvergne (*apud Arvernus*). Then follows: "The city of Trier was sacked and burnt by the Franks in a second attack." || Treverorum civitas a Francis direpta incensaque est secunda inruptione. ||

In light of Frankish behavior at *Arles 411, it should be apparent that Franks involved in a Roman civil war was nothing out of the ordinary; the context of this notice makes it virtually certain that this was the case. Although Jovinus did have some Frankish client support, at least to begin with (Scharf), it would appear that some or all of the Franks decided to throw in their lot with the legitimist regime of Honorius and were only doing the same to a rebel capital as Honorius' generals were doing to rebellious nobles further south. The first destruction is not attested, but must either have been the result of the civil war or of the Rhine crossing in 406.

Theodosiopolis	πολιορκοῦντας	421/2

Theodoret of Cyrrhus, *HE*, 5. 36 [ed. 5.37].

"In the former war, too, these same Persians, when besieging the emperor's eponymous city, were providentially rendered ridiculous. For after Vararanes had beset the aforesaid city for more than thirty days with all his forces, and had brought up many helepoles, and employed innumerable engines, and built up lofty towers outside the wall, resistance was offered, and the assault of the attacking engines repelled, by the bishop Eunomius alone. Our men had refused to fight against the foe, and were shrinking from bringing aid to the besieged, when the bishop, by opposing himself to them, preserved the city from being taken. When one of the barbarian chieftains ventured on his wonted blasphemy, and with words like those of Rabshakeh and Sennacherib, madly threatened to burn the temple of God, the holy bishop could not endure his furious wrath, but himself commanded a stonethrowing machine (*lithobolon organon*), which went by the name of the Apostle Thomas, to be set up upon the battlements, and a mighty stone to be adjusted to it. Then, in the name of the Lord who had been blasphemed, he gave the word to let go, down crashed the stone on that impious chief and hit him on his wicked mouth, and crushed in his face, and broke his head in pieces, and sprinkled his brains upon the ground. When the commander of the army who had hoped to take the city saw what was done, he confessed himself beaten and withdrew, and in his alarm made peace. Thus the universal sovereign protects the faithful emperor, for he clearly acknowledges whose slave he is, and performs fitting service to his Master."

Καὶ ἐν τῷ προτέρῳ πολέμῳ τούτους αὐτοὺς τὴν ἐπώνυμον τοῦ βασιλέως πολιορκοῦντας πόλιν καταγελάστους ἀνέφηνε. πλείους γὰρ ἢ τριάκοντα ἡμέρας πανσυδὶ Γορορἀνου τὴν προειρημένην κυκλώσαντος πόλιν, καὶ πολλὰς μὲν ἑλεπόλεις προσενεγκόντος, μηχαναῖς δὲ χρησαμένου μυρίαις, καὶ πύργους ἔξωθεν ὑψηλοὺς ἀντεγείραντος, μόνος ἀντέσχεν ὁ θεῖος ἀρχιερεὺς (Εὐνόμιος δὲ τούτῳ ὄνομα ἦν) καὶ τῶν προσφερομένων μηχανῶν τὴν ῥύμην διέλυσε. καὶ τῶν στρατηγῶν τῶν ἡμετέρων τὴν πρὸς τοὺς πολεμίους ἀπειρηκότων μάχην καὶ τοῖς πολιορκουμένοις ἐπαρκεῖν οὐ τολμώντων, οὗτος ἀντιπαραταττόμενος ἀπόρθητον τὴν πόλιν ἐφύλαξεν. ἑνὸς δὲ τῶν ὑπὸ τῶν βαρβάρων τελούντων βασιλέων τὴν συνήθη βλασφημίαν τετολμηκότος καὶ τὰ Ῥαψάκου καὶ Σενναχηρεὶμ φθεγξαμένου καὶ μανικῶς ἀπειλήσαντος τὸν θεῖον πυρπολήσειν νεών, οὐκ ἐνεγκὼν τὴν λύτταν ὁ θεῖος ἐκεῖνος ἀνὴρ τὸ λιθοβόλον ὄργανον παρὰ τὴν ἔπαλξιν τεθῆναι κελεύσας, ὃ τοῦ ἀποστόλου Θωμᾶ ἐπώνυμον ἦν, καὶ λίθον μέγαν ἐπιτεθῆναι παρεγγυήσας, ἐν τῷ ὀνόματι τοῦ βλασφημηθέντος ἀφεῖναι προσέταξεν. ὁ δὲ κατευθὺ τοῦ δυσσεβοῦς βασιλέως ἐκείνου κατενεχθεὶς καὶ τῷ στόματι πελάσας τῷ μυσαρῷ, τό τε πρόσωπον διέφθειρε καὶ τὴν κεφαλὴν συνέτριψεν ἅπασαν καὶ τὸν ἐγκέφαλον διέρανε τῇ γῇ. τοῦτο θεασάμενος ὁ τὴν στρατιὰν ἀγείρας καὶ τὴν πόλιν αἱρήσειν ἐλπίσας ᾤχετο, τὴν ἧτταν διὰ τῶν πραγμάτων ὁμολογήσας, καὶ δείσας τὴν εἰρήνην ἐσπείσατο. οὕτως ὁ τῶν ὅλων παμβασιλεὺς τοῦ πιστοτάτου κήδεται βασιλέως. καὶ γὰρ δὴ καὶ οὗτος τὴν δουλείαν ὁμολογεῖ καὶ τὴν ἁρμόττουσαν τῷ δεσπότῃ θεραπείαν προσφέρει.

ARLES   *multa vi oppugnatum est*   425
Prosper a. 425 (1290); Wolfram 1998: 175.
   "Arles, noble city of Gaul, was assailed by the Goths with great violence,

until, threatened by Aëtius, they withdrew not without losses." || Arelas nobile oppidum Galliarum a Gothis multa vi oppugnatum est, donec inminente Aetio non inpuniti abscederent. ||

?ARLES   [haud procul de Arelate ... Gothorum manus extinguitur]   430
Hyd. s.a. 430 (92); Wolfram 1998: 175.

"Not far from Arles some bands of Goths were annihilated by Count Aëtius, and Anaolsus their leader was captured." || Per Aetium comitem haud procul de Arelate quaedam Gothorum manus extinguitur Anaolso optimate eorum capto. ||

NARBONNE   *obsidione et fame; obsideri coepta*   436f
Prosper a. 436 (1324); Hyd. a. 436 (107), a. 437 (110); *PLRE* 2 s.vv. Litorius, Aetius; Wolfram 1988: 176.

Prosper: "The Goths confounded the peace agreements and seized many towns in the vicinity of their settlements, attacking the city of Narbonne most of all (*Narbonensi oppido maxime infesti*). When it had suffered for some time from siege and hunger (*cum diu obsidione et fame laboraret*), the city was saved from both dangers by Count Litorius. For he put the enemy to flight and filled the city with grain, having each of his troopers bring along two measures of wheat."

Hydatius is more succinct but reports that in 436: "Siege of Narbonne begun by Goths." || Narbona obsideri coepta per Gothos. || The siege lasted until 437: "Narbonne freed from siege." || Narbona obsidione liberatur Aetio duce et magistro militum. ||

The Goths suffered a crushing defeat by Aëtius in 438 (Hyd. 112 a. 438), but rebounded to defeat Aëtius' subordinate, Litorius, at *Toulouse in 439.

NOVIODUNUM   ἐπὶ συνθήκαις ἡ πολιορκία ἐλύετο   c. 437
Priscus fr. 5.

Valips, leader of the Rubi (probably Rugi), had rebelled "against the eastern Romans, seized *Noviodunum, a city which lies on the river, slew some of its citizens, and having collected all the wealth in the city, prepared, together with those who had chosen to revolt with him, to overrun Thrace and Illyria. When a force sent by the Emperor was upon him and he was besieged within the city, he kept the besiegers at bay from the circuit wall (τοὺς πολιορκοῦντας ἐκ τῶν περιβόλων ἠμύνετο) for as long as he and those with him could hold out. But when they were worn out by the toil of continually fighting the great number of Romans, they checked the clouds of enemy missiles by placing the children of their prisoners on the ramparts. For the soldiers loved the Roman children and hurled neither missiles nor javelins against those on the wall. In this way Valips gained a breathing space, and the siege was ended on terms (ἐπὶ συνθήκαις ἡ πολιορκία ἐλύετο)."

Blockley assigns this fragment to the early part of Priscus' history, to any date between 434/5 and the fall of *Naissus 442. Since the revolt had been underway some time, and probably began before the Hun invasion (see *Viminiacum 441), I would suggest sometime around 437.

?TOULOUSE		439

Prosper a. 439; Salvian, *de gub. Dei* 7.39-43; *PLRE* 2 s.v. Litorius; Heather 2005: 303; Wolfram 1988: 176.

Litorius with Hun auxiliaries engaged the Goths near Toulouse after receiving positive responses from the pagan *haruspices*, but was captured after inflicting heavy casualties. In Salvian, who only alludes to the main protagonists and does not even name Toulouse, the proximity to the city is emphasized. He also contrasts the paganism of Litorius with the appropriately Christian prayers and humility of the barbarian leader, who even sent bishops to ask for peace (further on this, see Wolfram). It seems that on the basis of Salvian, Heather treats this conflict as a siege. If this is the case, Litorius must have been captured during a Gothic sally or a botched storm, but this is conjectural.

VIMINACIUM; OTHER FORTS AND CITIES	εἷλον; ἐκάκωσαν	441

Priscus fr. 6.1; Thompson 1948: 87ff; Maenchen-Helfen 1973: 116.

"The Huns invaded and took the Roman *phrouria* north of the Danube by surprise during market season; Viminacium and other cities and forts along the river taken by storm." ...καὶ περαιωθέντες τὸν "Ιστρον πόλεις καὶ φρούρια πλεῖστα ἐπὶ τῷ ποταμῷ ἐκάκωσαν. ἐν οἷς καὶ τὸ Βιμινάκιον εἷλον...

NAISSUS	ἐπολιόρκουν	442

Priscus fr. 6.2; Thompson 1945, 1948: 92; Maenchen-Helfen 1973: 116; Blockley 1972, 1981: 168 n.48. See chapter 1.1.2 for a full discussion of the events of the siege.

Thompson rejected Priscus' description as lifted off Thucydides, and arranged the fragments and chronology (placing the siege in 443 along with the siege of *Ratiaria, here listed in 447) such that the military operations described made no strategic sense, lending further support to his conclusion that the Huns were incompetent in siege warfare. Maenchen-Helfen effectively corrected the chronology, dating the attack on Naissus to 442 and that on *Ratiaria to 447. Blockley further demonstrated that the linguistic parallels between Thucydides and Priscus were very slight, the correlation between events described even less (1972), and that Maenchen-Helfen's chronology further supports Priscus' description of events (1981). See also *Singidunum and *Sirmium which were captured in the same campaign.

Due to the significance of the siege, a full translation of Priscus' account along with important terms and passages in the original is reproduced here:

"The Scythians were besieging (ἐπολιόρκουν) Naissus, a city of the Illyrians on the river Danuba. They say that it was founded by Constantine, who also built the city at Byzantium named after himself. Since the barbarians were destined to take this populous and also well-fortified city, they made progress with every attempt. Since the citizens did not dare to come out to battle, the Scythians, to make crossing easy for their forces, bridged the river from the southern side at the point where it flowed past the city and brought their machines up to the circuit wall.

"First, because their access was easy, they brought up beams mounted on wheels (δοκοὺς ἐπὶ τροχῶν κειμένας), upon which men stood who shot across at the defenders on the ramparts. At the other end of the beams [towers] stood men who pushed the wheels with their feet and propelled the machines wherever they were needed, so that one could shoot successfully through the openings made in the screens. In order that the men on the beam should fight in safety, they were sheltered by screens woven from willow covered with rawhide and leather (λύγοις διαπλόκοις ἐκαλύπταντο δέρρεις καὶ διφθέρας ἐχούσαις) to protect them against other missiles and whatever fire darts might be shot at them."

"When in this manner a large number of machines had been brought up to the wall with the result that the defenders on the battlements gave in because of the clouds of missiles and evacuated their positions, the so-called 'rams' were brought up also." πολλῶν δὲ τῷ τρόπῳ τούτῳ ἐπιτειχισθέντων ὀργάνων τῇ πόλει, ὥστε διὰ πλῆθος βελῶν ἐνδοῦναι καὶ ὑποχωρῆσαι τοὺς ἐπὶ τῶν ἐπάλξεων, προσήγοντο καὶ οἱ καλούμενοι κριοί.

"This is a very large machine. A beam is suspended by slack chains from timbers which incline together, and it is provided with a sharp metal point and, for the safety of those working it, screens like those described. With short ropes attached to the rear men vigorously swing the beam away from the target of the blow and then release it, so that by its force all the part of the wall facing it is smashed away." μεγίστη δέ ἄρα καὶ ἥδε ἡ μηχανή· δοκὸς ἐκ ξύλων πρὸς ἄλληλα νευόντων χαλαραῖς ἀπηωρημένη ἁλύσεσιν, ἐπιδορατίδα καὶ προκαλύμματα ὃν εἴρηται τρόπον ἔχουσα, ἀσφαλείας ἕνεκα τῶν ἐργαζομένων. καλῳδίοις γὰρ ἐκ τῆς ὄπισθεν κεραίας εἷλκον βιαίως ἄνδρες αὐτὴν ἐς τὸ ἐναντίον τοῦ δεξομένου τὴν πληγήν, καὶ μετὰ ταῦτα ἠφίεσαν, ὥστε τῇ ῥύμῃ πᾶν τὸ ἐμπῖπτον τοῦ τείχους ἀφανίζεσθαι μέρος.

From the walls the defenders tumbled down wagon-sized boulders which they had prepared for this purpose when the machines were brought up to the circuit. Some they crushed together with the men working them, but they could not hold out against the great number of machines. Then the enemy brought up scaling ladders (κλίμακας), so that in some places the wall was breached by the rams and elsewhere those on the battlements were overcome by the number of machines. The barbarians entered through the part of the circuit wall broken by the blows of the rams and also over the scaling ladders which were set against that part of the wall which was not crumbling, and the city was taken."

SINGIDUNUM   *exciderunt*   442

Marcellinus Comes s.a. 441 (3); Thompson 1948: 89; Maenchen-Helfen 1973: 116

Maenchen-Helfen believes that Singidunum fell after Naissus, with *Sirmium in 442, not 441 as Marcellinus, who nevertheless seems to place the cities in order:

"The kings of the Huns with many thousands of their men invaded Illyricum: they destroyed Naissus, Singidunum and other cities and many towns in Illyricum." || Hunnorum reges numerosis suorum cum milibus Illyricum irruerunt: Naisum, Singidunum aliasque civitates oppidaque Illyrici plurima exciderunt. ||

SIRMIUM   ἐπολιορκεῖτο   442

Priscus fr. 11.2.332-37, 368ff; Thompson 1948: 89; Blockley 1983(II): 168 n. 48; chapter 1.

Priscus refers obliquely to the siege of Sirmium; first in the context of a diplomatic row between Attila and the Romans over some valuable vessels that had been intended by the bishop for ransoming the citizens "at the time when Sirmium ... was besieged by the Scythians (κατὰ δὲ τὸν χρόνον, ἐν ᾧ ὑπὸ Σκυθῶν ... ἐπολιορκεῖτο τὸ Σίρμιον)," but misappropriated by a West Roman official "after the capture of the city (μετὰ τὸν τῆς πόλεως ἀνδραποδισμόν)." Later Priscus describes his encounter with one of the

Sirmian captives who had made a career for himself in Hun service, for which see discussion in ch. 7.1.3.

RATIARIA	προσέβαλε	447

Priscus 9.1, 4; Maenchen-Helfen 1973: 118.

In the first fragment, we learn that during peace negotiations, the Romans refused to hand over refugees to the Huns. "When the views of the Romans were reported to him, Attila reacted in anger and ravaged Roman territory, destroying some forts and attacking Ratiaria, a very large and populous city." ὡς δὲ τῷ Ἀττήλα τὰ δεδογμένα Ῥωμαϊκὴν ἐδῄου γῆν, καὶ φρούριά τινα καθελὼν τῇ Ῥατιαρίᾳ προσέβαλε μεγίστῃ καὶ πολυανθρώπῳ.

In the second fragment (from Theophanes), we only hear that Ratiaria had been sacked with a long string of other Balkan cities. Maenchen-Helfen redated this fragment (normally placed in 442) to 447.

METZ	*succenderunt*	451

LHF 5; Hyd. 150; GT 2.6; Halsall 1995: 228ff; Bachrach 2003.

There are two sources for this event, but *LHF* may have taken his account (*Eo tempore Chuni Renum transierunt, Mettis succenderunt...*) from Gregory of Tours' bald sentence that the Huns moved on from Metz to *Orleans, without any more information as to what they were up to in northern Gaul (see Gerberding 1987). Hydatius is thus the only direct source for the destruction of the first of "several cities destroyed" (*plurimae civitates effractae*) during the Hunnic invasion which led to the battle of the Catalaunian plains.

TRIER	*distruunt*	451

LHF 5; cf. *Metz above.

After attacking *Metz, the Huns "destroyed Trier." || ...Treveris distruunt... ||

For this, we only have the testimony of *LHF* which is very unreliable; cf. *Trier 456 where the *LHF* described an event that possibly never took place by recycling a passage from Gregory of Tours, or at least conflated two events more than a generation apart.

ORLEANS	*expugnare*	451

GT 2.7; Jord. *Get.* 194-96; Sid. Ap. *Ep.* 8.15; *Vita Aniani*; Thompson 1948: 153f.

Gregory has, with (or because of) some hagiographic elaborations, probably preserved the most trustworthy version of the assault on Orleans, which followed the Hunnic assaults on *Metz and *Trier (above): "Attila the King of the Huns marched forward from Metz and ravaged a great number of other cities in Gaul. He came to Orleans and did all he could to capture it by launching a fierce assault with his battering-rams. [The prayers of bishop Anianus for relief were finally successful.] The walls were already rocking under the shock of the battering-rams and about to collapse when Aëtius arrived, and with him Theodoric, the King of the Goths, and his son Thorismund. They hastened forward to the city and drove off the enemy and forced them to retreat." || Attela vero Chunorum rex a Mittense urbe egrediens, cum multas Galliarum civitates oppraemeret, Aurilianis adgreditur eamque maximo arietum inpulsu nititur expugnare. [...] Interea iam trementibus ab impetus arietum muris iamque ruituris, ecce! Aetius et Theudor Gothorum rex ac Thorismodus, filius eius, cum exercitibus suis ad civitatem adcurrunt adversumque hostem eieciunt repelluntque. ||

Already in Sidonius there is a reference to Anianus' intervention. The story is fully developed in the *Vita Aniani*, where Anianius' activities and miraculous interventions receive even more attention, but it also provides further information on how the Huns tried to capture the city.

Jordanes has an alternative report: "But before we set forth the order of the battle [of Chalons] itself, it seems needful to relate what had already happened in the course of the campaign, for it was not only a famous struggle but one that was complicated and confused. Well then, Sangiban, king of the Alani, smitten with fear of what might come to pass, had promised to surrender to Attila, and to give into his keeping Aureliani (*in eius iura transducere*), a city of Gaul wherein he dwelt. When Theodorid and Aëtius learned of this, they cast up great earthworks around that city before Attila's arrival (*magnis aggeribus eandem urbem ante adventum Attilae struunt*) and kept watch over the suspected Sangiban, placing him with his tribe in the midst of their auxiliaries. Then Attila, king of the Huns, was taken aback by this event and lost confidence in his own troops, so that he feared to begin the conflict. While he was meditating on flight—a greater calamity than death itself—he decided to inquire into the future through soothsayers. So, as was their custom, they examined the entrails of cattle and certain streaks in bones that had been scraped, and foretold disaster to the Huns. Yet as a slight consolation they prophesied that the chief commander of the foe they were to meet should fall and mar by his death the rest of the victory and the triumph. Now Attila deemed the death of Aëtius a thing to be desired even at the cost of his own life, for Aëtius stood in the way of his plans. So although he was disturbed by this prophecy, yet inasmuch as he was a man who sought counsel of omens in all warfare, he began the battle with anxious heart at about the ninth hour of the day, in order that the impending darkness might come to his aid if the outcome should be disastrous."

The two main versions are mutually exclusive, as the miracle stories has Aëtius and the Goths arrive while the storm was under way, and Jordanes has Aëtius and the Goths in the city, preventing a handover to the Huns by supervising the Alan garrison and organizing extra earthworks to protect the city. If the stories are to be reconciled, there are two possible explanations. 1) The story of Ananius was embellished by telescoping and inverting events in order to promote his miraculous intervention. 2) Jordanes' translation and epitome of Priscus has left out important connecting materials in a very complex sequence of events in 451, or more likely, conflated two conflated totally different events. The latter is most likely, and the *PLRE* 2 entry on Thorismodus, the Visigothic king (451-53), argues that in Jordanes, "the story is twisted into a victory for Thorismod over Attila in defence of the Alani," whereas it was actually a Visigothic invasion of Alan-held territory around 453 that ended in a siege of *Orleans (entry below). Thus, Jord. *Get.* 194 and possibly the first part of 195 belong to 453, while Attila's activities in 195f, such as the divination of entrails, are probably correct for the Hunnic siege in 451.

AQUILEIA	*obsidens, invadunt*	452

Jord. *Get.* 222 (= Priscus fr. 22.1); Proc. 3.4.29-35 (= Priscus fr. 22.2); Maenchen-Helfen 1973: 136f.

After his failure before *Orleans (451) and defeat at Chalons, Attila withdrew to Pannonia and invaded Italy the next year, arriving first at Aquileia. "Although he pressed the siege there long and hard, he made no progress at all, since from within the bravest of the Roman soldiers opposed him." || ibique cum diu multumque obsidens nihil paenitus praevaleret, fortissimis intrinsecus Romanorum militibus resistentibus ... ||

About to give up the siege, Atilla received a sign (from some storks that left the city with their young) that encouraged the Huns. Subsequently "... they built machines and brought up all kinds of artillery and quickly entered the city, which they despoiled, smashed asunder and devastated so thoroughly they left hardly a trace of it to be seen." || ... qui machinis constructis omniaque genera tormentorum adhibita, nec mora et invadunt civitatem, spoliant, dividunt vastantque crudeliter, ita ut vix eius vestigia ut appareat relinquerunt. ||

Procopius also tells the story of the storks, but has actually removed Priscus' description of how the city was stormed with engines as preserved by Jordanes. Instead, he claims that the city wall suddenly collapsed "for no reason" at the spot where the storks had nested.

MILAN   *devastant*   452

Jord. *Get.* 223 (= Priscus fr. 22.1); Maenchen-Helfen 1973: 138f; *Pavia (below).

"Milan, the metropolis of Liguria and once an imperial city, they destroyed the in the same manner and condemned Ticinum [Pavia] to a similar fate." || Mediolanum quoque Liguriae metropolim et quondam regiam urbem pari tenore devastant nec non et Ticinum aequale sorte deiciunt ... ||

Maenchen-Helfen argues that Aëtius may have deliberately abandoned the cities in the Po valley, forcing the Huns to engage in time-consuming sieges and limiting their range of movement: their armies were so slowed down by booty that they were unable to raid widely as customary. Furthermore, their supply situation soon became untenable, and finally the Huns were struck by the epidemics that so often plagued invaders of Northern Italy. The combination of logistical problems, disease, and the armies of Aëtius and Marcian approaching convinced Attila to withdraw.

PAVIA   *deiciunt*   452

See *Milan above.

ARLES   452/3

Sid. Ap. *Ep.* 7.12.3; *PLRE* 2 s.vv. Aëtius 7 (at p. 28), Ferreolus; Wolfram 1988: 178.

Sidonius wrote to his friend Tonantius Ferreolus in order to praise him for his defense of his province against the Huns and the Visigoths, and especially his cooperation with Aëtius, whose military skill he exceeded by convincing the Goths to withdraw peacefully from Arles "with a banquet."

ORLEANS   c. 453

See *Orleans 451 for sources and discussion; also Wolfram 1988: 178.

The Visigoths under Thorismund were probably besieging the Alans at Orleans, and one of the parties threw up great embankments (*magnis aggeribus*) around the city. In Jordanes, this is associated with Attila's siege in 451, but it is possible that this belongs to the Visigothic assault in 453, as the Visigothic king is mentioned (with Aëtius) as the one responsible for setting them up; if the association is correct, the siegeworks were probably constructed by the besiegers in 453.

[COLOGNE]	*coeperunt*	c. 456

LHF 8; Gerberding 1987; cf. *Trier below and discussion in chapter 1.3.3.

"In those days the Franks captured the city Agrippina upon the Rhine and called it Cologne, as if it were inhabited by *coloni*. They killed a great number of people from the party of Aegidius; Aegidius himself fled and escaped." || In illis diebus coeperunt Franci Agripinam civitatem super Renum vocaveruntque eam Coloniam, quasi coloni inhabitarent in eam. Multo populo Romanorum a parte Egidii illic interfecerunt; ipse Egidius fugiens evasit. ||

This passage has created the idea of a Frankish invasion under Childeric, providing an embryonic Frankish kingdom on conquered Roman soil. However, as with many other aspects of the *LHF*, this is a passage very likely to be fabricated (perhaps inspired by GT 2.9 via Fredegar 3.3, describing an event in 388) to explain the presence of the Franks to the exclusion of the Romans, and provide a link to *Trier (below), which *LHF* has transferred by several decades in order to fit his scheme. A hint is provided by the pseudo-etymology given by the author; he seems unaware of the fact that the full name of the city was *Colonia Claudia Ara Agrippinensium*, or for short [*Colonia*] *Agrippina*. The author presumably saw the normal shortest version in older literature, *Agrippina*, and tried to explain the form *Colonia* as a Frankish invention.

TRIER	*coeperunt*	c. 456

Fredegar 3.7; *LHF* 8; Gerberding 1987; cf. *Cologne above and chapter 1.3.3.

Fredegar's version is a creative fusion with Gregory's report on *Trier 413; new material relating to events in 456 (datable to the reign of Avitus) is italicized:

"The city of Trier was captured and burned *at the instigation of one of the senators named Lucius. When Avitus was emperor, given over to luxury, and this Lucius, having the most beautiful wife of all, Avitus feigning that he was bedridden with bodily infirmity, ordered that all senatrices should attend to him. And when the wife of Lucius came, she was taken forcibly by Avitus, who rose from his bed the next morning and said to Lucius: "You have a beautiful hot bath, but are shaking from the cold. As Lucius suffered these things,* the city was plundered and burned by the Franks *at his instigation*.

LHF: "They therefore [i.e. because of the flight of Aegidius from *Cologne] came to the city of Trier upon the river Moselle, devastating those lands, and took the city by setting it on fire." || Venerunt itaque Treveris civitatem super Mosellam fluvium, vastantes terras illas, et ipsam succendentes coeperunt. ||

This could also be a fabrication, as the author of *LHF* lifted a passage from GT 2.9 via Fredegar (where he seems to have found information on *Cologne as well), which described how the Franks participated in a Roman civil war that occurred in 411. Here *Cologne and *Trier have been compressed into one event and placed as a prelude to Childeric's career. Salvian's four destructions (*excidia*; *de gub. Dei* 6.89) in the early 5th century may have given the author of *LHF* the idea, but only two destructions are attested before Salvian wrote c. 440, only one directly (*Trier 413), and he also employed biblical rhetoric in the style of Solomon's Proverbs (cf. *Prov.* 30.18f, 21-31), so neither the number nor the degree of destruction should be taken literally. However, as discussed in some detail in chapter 1.3.3, there is good circumstantial evidence that a Frankish attack on Trier did take place at this time, but that it was indeed instigated by a Roman opponent to Avitus, and thus not a case of Frankish expansion.

The rest of the passage from *LHF* continues in much the same vein; the complex fighting at Orléans and *Angers (463) is transformed into yet another Frankish invasion.

ARLES	*obsideretur*	458/9

GT *de virtutibus sancti Martini* 1.2; see also *PLRE* 2 s.v. Aegidius; MacGeorge 2002: 85-88; Wolfram 1988: 180.

The preserved stories in e.g. Gregory of Tours' *Miracles* are highly legendary in nature, but clearly speak of a formal siege in which Aegidius was penned in by his enemies (*cum obsideretur ab hostibus*, MGH p. 587.13). A fuller version (used by Gregory) is found in Paulinus of Perigueux, *de vita Sancti Martini*, with full text, translation, and discussion in MacGeorge. The date is only suggestive and Paulinus' verse very convoluted, but his description indicates a very hard-fought siege that was broken by a sortie.

?NARBONNE	*tradidit*	c. 461

Hyd. a. 461 (217); *PLRE* 2 s.v. Agrippinus; Wolfram 1988: 180.

Agrippinus had been sent to Gaul in order to restore order in 451 or 452, i.e. immediately after the Hunnic invasion. He was replaced by Aegidius around 456/7 and faced charges of colluding with the barbarians (which of the many groups in Gaul is not mentioned), but after the death of Majorian, he was restored to favor and fought against Aegidius, who was in revolt against the new emperor Severus after 461. "Agrippinus the Gaul count of the city the enemy of count Aegidius, *vir insignus*, handed over Narbonne to Theodoric in order to gain the support of the Goths." || Agrippinus Gallus comes civis Aegidio comiti viro insigni inimicus, ut Gothorum mereretur auxilia, Narbonam tradidit Theodorico. ||

In light of previous and subsequent Gothic behavior, it is impossible to rule out that some level of threat was involved in the handover.

CASTRUM CAINONENSE   *obsederetur*   c. 463
Priscus fr. 39.1; GT *Glor. Conf.* 22; *PLRE* 2 s.v. Aegidius; MacGeorge 2002: 100.

The background for the war between Aegidius and the Visigoths is given in Priscus, while Gregory provides the details (MGH p. 762) through the eyes of St. Maximus, said to be a disciple of St. Martin. He founded a monastery *ad castrum Cainonensim* near Tours. It was apparently under Visigothic control, or perhaps loyal to the emperor Severus, and therefore besieged by Aegidius (*ab Egidio obsederetur*). The population of the *pagus* had clearly taken refuge there and was blockaded inside the fortifications (*inclusus; infra castri munitionem conclusus erat*). Aegidius' troops broke off the only water supply of the besieged which was a well excavated on the side of the mountain (*hostis adversus effossum a latere montis puteum, quem obsessi ad usum habebant bibendi, obturant*). Through the fervent prayers of St. Maximus, God provided heavy rains that not only replenished the water supply, but also caused the besiegers to withdraw.

ANGERS   *obtinuit*   463
GT 2.18; MacGeorge 2002: 101f; *PLRE* 3 s.v. Odovacer.

"Then Adovacrius himself came to Angers, and King Childeric arrived the next day; the comes Paul [of Angers] being killed, he occupied (obtinuit) the city." A great fire destroyed the church on the same day.

The mysterious figure Adovacrius (whom *PLRE* identifies as Odovacer) took control of Angers in a contest where the parties are nearly impossible to reconstruct. By whom Paul was killed is anyone's guess, since the Goths were also involved in fighting at Orleans against Paul and his Roman and Frankish supporters.

SINGIDUNUM   *invadens ... suae subdedit dicioni*   c. 472
Jord. *Get.* 282; *PLRE* 2 s.v. Theodericus 7 (at p. 1078); Wolfram 1988.

The young prince of the Ostrogoths, Theoderic, at first defeated the

Sarmatians in battle and killed their king; shortly after he "invaded the city of Singidunum, which the Sarmatians themselves had seized, and did not return it to the Romans, but reduced it to his own sway." || Singidunum dehinc civitatem, quam ipsi Sarmatae occupassent, indvadens, non Romanis reddidit, sed suae subdedit dicioni. || The implication is that he was doing this on the Romans' behest and was expected to hand over the city.

PAMPLONA	*obtinuit*	472/3

Chr. Gall. 511 a. 472 (651); *PLRE* 2 s.v. Gauterit; Wolfram 1988: 185.

The Visigoths expanded into Spain: "Gauterit, a count of the Goths, occupied Spain in the region around Pamplona, Caesaraugusta and neighboring towns." || Gauterit comes Gothorum Hispanias per Pampilonem, Caesaraugustam et vicinas urbes obtinuit. ||

Wolfram dates this to 472, *PLRE* to 473.

SARAGOSSA; OTHER CITIES	*obtinuit*	472/3

See *Pamplona above

TARRAGONA	*obsessa, obtinuit*	472/3

Chr. Gall. 511 a. 472 (652) and *Pamplona above; *PLRE* 2 s.vv. Heldefredus, Vincentius 3; Wolfram 1988: 185.

"Heldefred with Vincent, duke of the Spains, besieged Tarragona and took the coastal cities." || Heldefredus quoque cum Vincentio Hispaniarum duce obsessa Terracona maritimas urbes obtinuit. ||

Vincent was killed on an expedition to Italy as Euric's *magister militum*.

ARLES	*capta*	473/6

Chr. Gall. 511 a. 476 (657); *PLRE* 2 s.v. Euricus.

"Arles was taken by Euric along with Marseilles and other strongpoints." || Arelate capta est ab Eurico cum Massilia et ceteris castellis. ||

Wolfram dates this to 472, while *PLRE* places it in 476 based on other sources.

MARSEILLES	*capta*	473/6

See *Arles above.

NAISSUS	*invadit*	473

Jord. *Get.* 285; Wolfram 1988: 267

Due to a shortage in supplies in Pannonia, the Goths split up in two groups; one went to the West and were assimilated by the Visigoths; the other under king Theodemer invaded Illyricum: "Theodemer, seeing prosperity everywhere awaiting him, invaded Naissus, the first city of Illyricum." || videns Thiudimir undique sibi prospera provenire, Naissum primam urbem invadit Illyrici ... ||

| ULPIANA | *in deditione accipiunt* | 473 |

Jord. *Get.* 285f; Wolfram 1988: 269.

Theodemer sent his son Theoderic against Ulpiana and Stobi, the surrender of which the Goths accepted (*in deditione accipiunt*); they also captured a number of other Illyrian cities.

| STOBI; OTHER CITIES | *in deditione accipiunt* | 473 |

See *Ulpiana above.

| CLERMONT | *e semirutis murorum aggeribus* | 474 |

Sid. Ap. *Ep.* 3.3; Whittaker 1994: 273f; Harries 1994: 222-29; Halsall 2003: 124; ch. 1.1.2.

Sometimes this letter (written to Sidonius Appolinaris' to his brother-in-law Ecdicius after the latter had relieved Clermont from a Visigothic siege) is taken as the 5th-century "smoking gun" for the minimalist interpretation of siege warfare, as it "demonstrates" the limited scope of military activity conducted by microscopic armies (19 men in Ecdicius' relief party) and, implicitly, Visigothic incompetence in halting even a miniscule relief force. Both Harries and Halsall take the number quite literally and create a scenario in which this small band sweeps through the Gothic siege lines to relieve the city. However, if the number in Ecdicius' party is to be taken literally, the same logic must apply to Sidonius' reference to "a few thousand Goths" (*aliquot milia Gothorum*) encamped during the siege. Even hairy, innumerate barbarians could tell that there was a significant discrepancy between them and the relieving force if indeed these numbers had been accurate. It is clear from Sidonius' own statements (3.3.9) that his letter was not an accurate report of events. It was rather a highly emotional panegyric to Ecdicius in a semi-poetic and rhetorical form, emphasizing his personal role in raising the siege. It is not organized chronologically, either, requiring further clarification before any analysis is possible. It begins with a description of Ecdicius' birth, childhood and education in, and attachment to, the city (3.3.1f), followed by allusions to the fighting (3.3.3) and a description of his triumphal entry after defeating the Visigothic rearguard (3.3.3-6), before returning to further allusions to the fighting that had occurred in the course of the relief operation the day(s) before and the clearing of the many corpses that lay in the fields (3.3.7ff). Due to its importance, a translation of the whole letter and the original of some key passages in the original are reproduced below.

Once the chronological framework is clear, it is easier to pick out allusions to normal events taking place during a siege, as well as to the institutional framework described in chapter 1.2. As Whittaker observed, Ecdicius' companions, or *amici*, during the triumphal entry, must have been his most immediate trusted followers, the equivalent of the *doryphoroi*, but the *hypaspistai*, or common soldiers, are simply ignored. He further shows that such an omission was commonplace in 5th and 6th century literature, citing several pertinent examples. The lowly soldiers are however alluded to: the force Ecdicius had raised was described as a *publici exercitus speciem*, "a type of public army." The fighting on the day before the triumphal entry had been hard, although only two or three of Ecdicius' men (i.e. not the *amici*) had fallen. In contrast, the Gothic dead lay scattered over the field, and they had at first been unable to assemble the corpses, far less throw a mound over them. Instead they resorted to the expedient of beheading their fallen, presumably to avoid the identification of prominent individuals before they were forced to evacuate their camp and withdraw to the surrounding hills. Ecdicius' men pressed the rout into daylight, defeating the rearguard left to cover the withdrawal (3.3.4). This forced the Goths to place their dead in suburban houses and set them alight as they fled. All this happened over the evening, night and morning before Ecdicius' triumphal entry with his 18 followers at midday, which took place at a leisured pace (3.3.5) across the very field where the Goths had encamped. His other troops were probably busy pursuing the Goths, scouting the surrounding territory, plundering the Gothic camp, or were simply not noticed by Sidonius. The elation of the inhabitants is set in stark contrast to the danger they had been exposed to, as they observed their hero "from the half-ruined line of the walls." This must allude to breaches that were being made just as the relieving army arrived.

Reproduction of Sid. Ap. *Ep*. book 3, letter 3: *Sidonius Appolinaris to his brother-in-law Ecdicius*, AD 474:

"1. There never was a time when my people of Clermont needed you so much as now; their affection for you is a ruling passion for more than one reason. First, because a man's native soil may rightly claim the chief place in his affection; secondly, because you were not only your countrymen's joy at birth, but the desire of their hearts while yet unborn. Perhaps of no other man in this age can the same be said; but the proof of the statement is that as your mother's time advanced, the citizens with one accord fell to checking every day as it went by.

2. I will not dwell on those common things which yet so deeply stir a man's heart, as that here was the grass on which as an infant you crawled, or that here were the first fields you trod, the first rivers you swam, the first woods through which you broke your way in the chase. I will not remind you that here you first played ball and cast the dice, here you first knew sport with hawk and hound, with horse and bow. I will forget that your schooldays brought us a veritable confluence of learners and the learned from all quarters, and that if our nobles were imbued with the love of eloquence and poetry, if they resolved to forsake the barbarous Celtic dialect, it was to your personality that they owed all.

3. Nothing so kindled their universal regard for you as this, that you first made Romans of them and never allowed them to relapse again. And how should the vision of you ever fade from any patriot's memory as we saw you in your glory upon that famous day, when a crowd of both sexes and every rank and age lined our half-ruined walls to watch you cross the space between us and the enemy? At midday, and right across the middle of the plain, you brought your little company of eighteen safe through some thousands of the Goths, a feat which posterity will surely deem incredible. || illud in te affectum principaliter universitatis accendit, quod quos olim Latinos fieri exigeras barbaros deinceps esse vetuisti. non enim potest umquam civicis pectoribus elabi, quem te quantumque nuper omnis aetas ordo sexus *e semirutis murorum aggeribus* conspicabantur, cum interiectis aequoribus in adversum perambulatis et vix duodeviginti equitum sodalitate comitatus aliquot milia Gothorum non minus die quam campo medio, quod difficile sit posteritas creditura, transisti. ||

4. At the sight of you, nay, at the very rumor of your name, those seasoned troops were smitten with stupefaction; their captains were so amazed that they never stopped to note how great their own numbers were and yours how small. They drew off their whole force to the brow of a steep hill; they had been besiegers before, but when you appeared they dared not even deploy for action. You cut down some of their bravest, whom gallantry alone had led to defend the rear. You never lost a man in that sharp engagement, and found yourself sole master of an absolutely exposed plain with no more soldiers to back you than you often have guests at your own table. || ad nominis tui rumorem personaeque conspectum *exercitum exercitatissumum* stupor obruit ita, ut prae admiratione nescirent duces par-

ties inimicae, quam se multi quamque te pauci comitarentur. subducta est tota protinus acies in supercilium collis abrupti, quae cum prius applicata esset oppugnationi, te viso non est explicata congressui. interea tu caesis quibusque optimis, quos novissimos agmini non ignavia sed audacia fecerat, nullis tuorum certamine ex tanto desideratis solus planitie quam patentissima potiebare, cum tibi non daret tot pugna socios, quot solet mensa convivas. ||

5. Imagination may better conceive than words describe the procession that streamed out to you as you made your leisurely way towards the city, the greetings, the shouts of applause, the tears of heartfelt joy. One saw you receiving in the press a veritable ovation on this glad return; the courts of your spacious house were crammed with people. Some kissed away the dust of battle from your person, some took from the horses the bridles slimed with foam and blood, some inverted and ranged the sweat-drenched saddles; others undid the flexible cheek-pieces of the helmet you longed to remove, others set about unlacing your greaves. One saw folk counting the notches in swords blunted by much slaughter, or measuring with trembling fingers the holes made in cuirasses by cut or thrust.

6. Crowds danced with joy and hung upon your comrades; but naturally the full brunt of popular delight was borne by you. You were among unarmed men at last; but not all your arms would have availed to extricate you from them. There you stood, with a fine grace suffering the silliest congratulations; half torn to pieces by people madly rushing to salute you, but so loyally responsive to this popular devotion that those who took the greatest liberties seemed surest of your most generous acknowledgements.

7. And finally I shall say nothing of the service you performed in raising what was practically a public force from your private resources, and with little help from our magnates. I shall not tell of the chastisement you inflicted on the barbaric raiders, and the curb imposed upon an audacity which had begun to exceed all bounds; or of those surprise attacks which annihilated whole squadrons with the loss of only two or three men on your side. Such disasters did you inflict upon the enemy by these unexpected onsets, that they resorted to a most unworthy device to conceal their heavy losses. They decapitated all whom they could not bury in the short night-hours, and let the headless lie, forgetting in their desire to avoid the identification of their dead, that a trunk would betray their ruin just as well as a whole body. || taceo deinceps collegisse te privatis viribus publici exercitus speciem parvis extrinsecus maiorum opibus adiutum et infrenes hostium ante discursus castigates cohercuisse populatibus. taceo te aliquot

superventibus cuneos mactasse turmales e numero tuorum vix binis ternisve post proelium desiderates et tantum calamitatis adversae parti inopinatis certaminis inflictum, ut occulere caesorum numerositatem consilio deformiore meditarentur. siquidem quos humari nox succinct prohibuerat decervicatis liquere cadaveribus, tamquam minoris indicia foret quem nolles agnosci crinitum dimisisse truncatum. ||

8. When, with morning light, they saw their miserable artifice revealed in all its savagery, they turned at last to open obsequies; but their precipitation disguised the ruse no better than the ruse itself had concealed the slaughter. They did not even raise a temporary mound of earth over the remains; the dead were neither washed, shrouded, nor interred; but the imperfect rites they received befitted the manner of their death. Bodies were brought in from everywhere, piled on dripping wains; and since you never paused a moment in following up the rout, they had to be taken into houses which were then hurriedly set alight, till the fragments of blazing roofs, falling in upon them, formed their funeral pyres.

9. *But I run on beyond my proper limits; my aim in writing was not to reconstruct the whole story of your achievements, but to remind you of a few among them, to convince you how eagerly your friends here long to see you again*; there is only one remedy, at once quick and efficacious, for such fevered expectancy as theirs, and that is your prompt return. If, then, the entreaties of our people can persuade you, sound the retreat and start homeward at once. The intimacy of kings is dangerous; court it no more; the most distinguished of mankind have well compared it to a flame, which illuminates things at a short distance but consumes them if they come within its range. Farewell."

EDESSA	484

JS 16; *PLRE* 2 s.vv. Illus, Leontius ; chapter 2.3.1.

The Roman generals Illus and Leontius, who were in revolt against the emperor Zeno, were unable to hold their base at Antioch since the people there were afraid of a siege. They tried to shift their base to the east of the Euphrates, but when their envoy and 500 cavalry arrived to establish their authority and garrison the city, the Edessenes "rose up against him, and closed the gates of the city, *and guarded the wall after the fashion of war*, and did not let him enter."

Usurpers needed extensive political support in order to organize siege defense; battle is the last option.

5TH CENTURY

| PAPYRIUS | 4 years | [blockade; treason] | 484-8 |

See *Edessa 484 above; JS 17.

Deprived of a secure base, the rebels were forced to face battle, but lost and their troops were dispersed among the cities in the East. The rebels retired to a well supplied fort in Isauria, called Papyrius (JS 12), where they withstood a long blockade, since the geographical setting and good defenses made an assault impossible. They were finally betrayed by their own men and executed. The *PLRE* entries cited for *Edessa contain more source references.

| RAVENNA | 3 years | [intermittent blockade] | 489-93 |

Proc. 5.1.15, 24; Wolfram 1988: 281ff

The Goths under Theoderic establish blockade of Odovacar, who finally surrendered due to lack of supplies. The siege has been treated by virtually all standard works for on the period due to its political significance. Wolfram gives a brief but good summary of the main events and the complex fighting during this period.

| [PARIS] | *opsidionem ... perpessa est* | c. 490 |

Vita Genovefae c. 36-40; Daly 1994.

The question of a Frankish siege of Paris around 490 is fraught with difficulty. According to standard interpretations, Clovis besieged the city after defeating Syagrius (traditionally in 486) for an extended period of time. However, nothing else is known of such a siege. The explicit evidence from Genovefa's *Vita* is only: "Tempore igitur, quo opsidionem Parisius bis quinos, ut aiunt, annos a Francis perpessa est, pagum eiusdem urbis ita inaedia adflixerat, ut nonnulli fame interisse nuscantur." This prompted Genovefa to venture out in barges on the Seine to collect supplies (*ad comparandam annonam*) in nearby *oppida* (c. 36). Although she had to miraculously negotiate various obstacles, such as a large tree infested with little green monsters blocking the river, and cure many of the inhabitants of the places she traveled through, no mention is made of enemies, occupation, siegeworks, or any other sign of hostility. Her return was likewise laced with miracles, but none were directed against besieging Franks. She did however have a household (*obsequium*) that was responsible for baking bread and distributing to the hungry.

There is nothing to indicate a siege apart from the phrase *opsidionem Parisius ... a Francis perpessa est*. Firstly, the ten-year span (*bis quinos, ut aiunt, annos*) is completely unreasonable; no sieges of comparable length are reported during the period. Secondly, Daly points out that *obsidio* in ecclesiastical Latin sometimes means "occupation." This makes far more sense in the context. Genovefa had already advised Childeric not to execute captured enemies (c. 26), and after her death, Clovis erected a basilica in her honor (c. 56). From other sources, we know that Clovis had good relations with the clergy of northern Gaul, and his control of the region was inherited from or acquired in competition with his Roman colleagues in the northern Gallic regional command that had been outside imperial control since 461. Finally, the starvation afflicted the whole *pagus* (district), not only the *urbs* (walled city). The hagiographer seems to keep a clear distinction between these terms, and no raiders (who could cause similar effects, cf. chapter 5.1.1) are mentioned. The phrase must be taken thus: "At the time when the Franks had *occupied* Paris for ten years, as they say ..."

The supply expedition, then, was aimed at providing for the population in and around Paris during a famine. Genovefa mobilized resources from her household to acquire grain and prepare the bread in ovens afterwards. She clearly had a prominent position in Paris, as she had been advisor to Childeric and consorted with the elite in the surrounding communities. The date depends on when the Frankish occupation began. If the Roman military elite, who adopted Frankish identity in the 460s, is meant (cf. chapter 1.3.3), the date could be any time in the 470s. The accession of Clovis in 481 or shortly after is the latest probable date for the identification of a "Frankish" occupation; hence the event has here been dated to c. 490. Although it is not a case of siege warfare, Genovefa's ability to procure supplies and organize labor for construction projects (see chapter 1.2.5) is good evidence for how the elites in post-Roman Gaul *did* supply their cities and organized repairs in times of crisis. A siege would certainly be a suitable occasion to apply this ability.

SAINTES	*obtinuit*	496

Auctarium Havniense s.a. 496.

"Alaric in the twelfth year of his reign captured Saintes." [From the Franks, or perhaps local Gallo-Romans?] || Alaricus ann. XII regni sui Santones obtinuit. ||

BORDEAUX	*obtinuerunt*	498

Same as *Saintes (496), s.a. 498.

"In the 14th year of Alaric, the Franks captured Bordeaux, transferring it from Gothic power to their own possession, having captured Suarto, duke of the Goths." || Ann. XIIII Alarici Franki Burdigalam obtinuerunt et a potestate Gothorum in possessionem sui redigerunt capto Suatrio Gothorum duce. ||

Nisibis c. 498
JS 22, 24; discussion in chapter 2.3.1.

The Kadishaye, normally under the Shah's rule, revolted and began to besiege Nisibis, "and they fought against it for a considerable time." They were followed in revolt by the Tamuraye from the mountains and the Arabs, who raided villages, merchants and travelers in Persian territory (JS 22). Shah Kawad, displaced from the throne by his brother, returned to power with Hun allies in 498. Having defeated his domestic enemies, he then threatened the raiding tribes to join his upcoming campaign against Rome, including the Kadishaye, who abandoned their siege of Nisibis. The Arabs were enthusiastic, while the Armenians were compelled to join through fear (JS 24).

THE 6TH CENTURY

BURGUNDIAN CIVIL WAR   500
See *Dijon, *Avignon, *Vienne below.

?DIJON   [*cum omni instrumento belli ad castrum ... pervenerunt*]   500
GT 2.32; Mar. Av. a. 500; for context of this with *Avignon (500) and *Vienne (500), see Wood 1994:41ff; James 1988:85f; Wolfram 1988: 192; chapter 1.3 above.

According to Gregory, the Franks invaded Burgundy at the instigation of the Burgundian king Godigisel in order to oust his brother Gundobad, with whom he shared the kingdom, and then Godigisel pretended to come to his brother's aid: "The three kings each put his army in the field, Clovis marching against Gundobad and Godigisel. They arrived with all their military equipment at a fortified place called Dijon. When battle was joined on the River Ouche Godigisel went over to Clovis and their united forces crushed the army of Gundobad. Gundobad turned his back and fled when he saw the treachery of his brother, about whom he had no suspicion." || Moventesque simul hii tres exercitum, id est Chlodovechus contra Gundobadum et Godegiselo, cum omni instrumento belli ad castrum cui Divione nomen est pervenerunt. Confligentesque super Oscaram fluvium, Godigiselus Chlodovecho coniungetur, ac uterque exercitus Gundobadi populum adteret. At ille dolum fratres, quem non suspecabatur, advertens, terga dedit fugamque iniit ... ||

Marius of Avenches is slightly less detailed but concurs.

This conflict probably never reached the stage of a formal siege, but one appears to have been planned. Three features stand out: firstly, the fortification of Dijon is explicitly named as the goal of all the armies (*ad castrum ... pervenerunt*); secondly, these armies must also have carried substantial baggage trains, as it was worth mentioning specifically (*cum omni instrumento belli*); thirdly, Gundobad was betrayed in the field by his brother, thus preventing a defense if one was planned, and he then fled to *Avignon, which certainly was put under siege.

AVIGNON   *cum omni exercito circa murus urbis resedente*   500
See *Burgundian civil war, *Dijon (500), *Vienne (500).

GT 2.32: [continued from *Dijon] "He [Gundobad] made his way along the banks of the Rhône and took refuge in the city of Avignon." Godigisel "promised to hand over a part of his kingdom to Clovis" and went home to Vienne to celebrate a triumph. "Clovis called up more troops and set out in pursuit of Gundobad, planning to extract him from Avignon and kill him." Gundobad sent a close advisor, Aridius, who was to feign defection and then propitiate Clovis and prevent the destruction of the territory. Aridius was well received by Clovis, who had settled down for a formal siege. "Clovis remained encamped with his entire army around the city walls." || Denique Chlodovecho cum omni exercito circa murus urbis resedente ... || Aridius, however, began to persuade him to take a diplomatic solution that would be advantageous to Clovis as well as the "cities through which you propose to pass. What is the point of keeping all these troops under arms when your enemy is safe in a stronghold which is too well fortified for you to capture? You are destroying the fields, spoiling the meadows, cutting up the vineyards, ruining the olive-groves and ravaging the whole countryside, which is a very fruitful one. In doing this you are causing no harm whatsoever to Gundobad." Aridius suggested that Clovis instead demand tribute, which he did.

Since Clovis' goal was the death of Gundobad and the subjugation of his territory, Gundobad had no incentive whatsoever to surrender. In order to achieve this, Clovis settled down for a close blockade while systematically ravaging the countryside, even planning raids against the territories of other cities, a strategy clearly aimed at detaching Gundobad's supporters. Aridius' apparent defection was thus well received, but his arguments also made Clovis realize that his strategy was counter-productive: if he hoped to rule these territories, he was destroying the economic foundations of potential supporters throughout the region. In the event, the Franks would not be able to subjugate Burgundy until much later (see *Autun 534).

VIENNE	*obsidit*	500

GT 2.33; see *Burgundian civil war, *Dijon (500), *Avignon (500). Cf. *Naples (536).

Gundobad reneged on the promise of tribute to the Franks and proceeded to besiege (*obsidit*) his brother Godigisel in Vienne. Godigisel ordered the expulsion of the common people (*minoris populi*) due to lack of supplies, "but along with them was expelled an engineer (*artifex*) who was in charge of the aqueduct." The indignant engineer showed Gundobad's army how to break in via the aqueduct. "At their head marched a number of sappers with iron crowbars (*multis cum ferreis vectibus praecedentibus*).

There was a water-gate blocked by a great stone. Under the direction of the engineer they heaved this stone on one side with their crowbars, made their way into the city and attacked the defenders from the rear, who were still busy shooting their arrows from the wall. A trumpet-call was sounded from the centre of the city, the besiegers attacked the gateways, burst them open and rushed in. The townsfolk were caught between two fires and cut to pieces by two forces [...]. The Franks who had been with Godigisel banded together in one of the towers. Gundobad gave orders that none of them should be maltreated." The Franks were disarmed and sent to Alaric in Toulouse, but the senators, who probably defended their cities actively (see chapter 1.2.4f), and Burgundians were massacred.

Although no engines were reported in this brief anecdote, many features were typical of contemporary sieges: the expulsion of the non-essential surplus population, the presence of urban engineers who could contribute to the siege effort, specialists using tools in order to break through the wall (here the blocked aqueduct), continuous archery fire from the walls and towers, besiegers unable to approach the gates due to such defensive fire, and banding together in towers once the walls were overwhelmed. Gundobad was thus secure in his reign, and shortly after was allied to the Franks against the Visigoths, perhaps as compensation for discontinuing tribute. See *Vouillé campaign (507-8) for further discussion.

THEODOSIOPOLIS   *šrā 'al ... w-kabšāh*, ܚܪܒܗ... ܥܠ ܫܪܐ   502
JS 48; Zach. *HE* 7.3; Greatrex 1998: 79f; chapter 2.3.2.

Kawad invaded Roman territory and "encamped against" [i.e. besieged] Theodosiopolis, "and took it [in a few days]" (*wa-šrā 'al ... wa-kabšāh*, ܘܫܪܐ ܚܪܒܗ... ܥܠ; Chabot I.274.22f). According to Joshua, it was surrendered to the Persians by the governor Constantine, who rebelled for reasons of personal enmity towards the emperor and was made general in the Persian army. The city was then plundered, burnt, and a Persian garrison installed. This means that the destruction cannot have been overwhelming, although it seriously afflicted the countryside, as Kawad "laid waste all the villages in the region of the north, and the fugitives that were left he carried off captive." Clearly there was substantial resistance from the city and the surrounding region that was eventually made impossible by the defection of such a notable leader. According to Zachariah, Kawad "subdued the city, and he treated its inhabitants mercifully" while Constantine was taken prisoner.

| MARTYROPOLIS | ἐνδιδόντες | 502 |

Proc. *De Aed.* 3.2.4-7; Greatrex 1998: 81.

The city surrendered immediately when news of Kawad's approach was announced. The governor Theodore went out, presumably in a formal procession to greet Kawad with an *adventus*, since he was accompanied by the citizens, dressed in his robes of office, and presented the taxes for the last two years. The complete submission was rewarded as Kawad reappointed the existing administration, entrusting Theodore with "the tokens of the office," and Martyropolis with its province Sophanene was in effect regarded as a Persian province.

According to *De Aed.* 3.2.9-14, Anastasios was grateful to Theodore and the people of Sophanene after the war, understanding that they had wisely submitted when they lacked proper defenses. The situation was remedied by Justinian, who tripled the wall's thickness and added a ditch, since the wall alone was unable to resist engines as well as stop attempts to scale the wall.

| AMIDA | 97 days | *šrā 'al*, ܐܠ ܫܪܐ | 502f |

JS 50, 53; Zach. *HE* 7.3-4; Proc. 1.7.12-32; Greatrex 1998: 83-94; Luther 1997: 180-86; Lenski 2007; see also discussion in chapter 2.3.3.

The Persian army "encamped against" (*šrā 'al*, ܐܠ ܫܪܐ; Chabot I.276.8) the city of Amida on October 5. They constructed a mound to approach the top of the wall, assaulted with ladders, and broke down the walls with rams (mentioned by all sources). The citizens of Amida (and perhaps a small garrison) resisted by undermining the mound and using a large engine, the "Crusher" (Joshua), to destroy Persian rams and protective screens. The Persians stormed the city with ladders on 11 January, 503, after discovering that the monks guarding a section of the walls had fallen asleep. The storm resulted in systematic massacre, slavery, plunder, rape.

| CONSTANTINA/TELLA | *šrā 'lēh*, ܥܠܝܗ ܫܪܐ | 502f |

JS 51, 58; Proc. 2.13.8-15; Greatrex 1998: 87f, 101, 103; see discussion in chapters 2.3.4 and 6.1.2.

The prelude to the siege began in already in 502 with extensive raiding of the countryside that was intended to weaken the city before assaulting it; hence the dating to 502 rather than 503. The garrison chased the raiders off but fell into an ambush (JS 51). The formal investment when they "encamped against" the city (*šrā 'lēh*, ܥܠܝܗ ܫܪܐ; Chabot I.284.18) began in 503 (JS 58), but the raiding had in fact been too efficient: there was nothing for the Persians to forage on. Procopius reports how the bishop Bar-Hadad asked the Persians to spare the city, which was unprepared, lacking defenders and defensive engines. Joshua gives a fuller perspective. Bar-Hadad was important for maintaining the morale of the troops, perambulating the parapet, blessing the walls with holy water and giving those on guard the Eucharist, but in contrast to Procopius' assertion, this was a full-scale formal siege. The city was nearly betrayed by some of her Jewish defenders, who dug a tunnel from their section of the walls towards the Persians, but this was discovered by a Roman captive in the Persian camp who managed to tell the defenders. After the plot failed, a massacre of the Jews followed which was stopped by the commander Leontius and bishop Bar-Hadad. The Persians were approached by Bar-Hadad who negotiated their departure. They were convinced by lack of supplies, failure of the underwall plot, and fear of relieving armies, in addition to the bishop's holiness.

HARRAN   *la-mqarrābū 'am,* ܠܡܩܪܒܘ ܥܡ   502f
JS 52, 59; Greatrex 1998: 88, 104; see also chapter 2.3.4.

The Persians thoroughly raided the whole territory of Harran for captives, livestock and other property, before moving on to *Edessa (below). Just as at *Constantina, these raids were intended to break down morale and supplies for the next season (503). In the meantime, the Persians made an attempt at *Edessa, but failed and proceeded to Harran. After failed negotiations at *Edessa (below), the Persians dispersed their to various cities; one division went to attack Harran (*la-mqarrābū 'am ḥarran,* ܠܡܩܪܒܘ ܥܡ ܚܪܢ; Chabot I.287.15f). The garrison at Harran, however, sallied out and killed some of the Persians' Hun auxiliaries and captured the Hun chief. The Persians promised not to attack Harran if they returned the Hun chief alive. Despite the successful sally, the Harranites were afraid to fight the Persians and agreed to the terms offered, and in addition sent a gift of 1,500 rams "and other things," going to great lengths to avoid further conflict.

EDESSA   *šrā 'al,* ܫܪܐ ܥܠ ; *la-mqarrābū 'am,* ܠܡܩܪܒܘ ܥܡ   502f
JS 52, 59-63; Greatrex 1998: 88, 103-106; full discussion in chapter 2.3.5.

In the summer of 502, the Persians proceeded from *Harran (see above) to raid the villages around Edessa, taking 18,500 captives from both cities' *territoria*. As at *Harran, many of the citizens had gone out from the city into the villages to take part in the vintage, hence the great number of captives and vulnerability to raids. The raid prompted the Edessenes to make extensive repairs and preparations for the coming siege. The Persian army led by Kawad in person arrived in the summer of 503, first attempting to gain concessions or betray the Roman commanders during negotiations. The Romans were willing to pay large tribute, but not enough for the Persians. When negotiations failed and the opportunity for ambush never materialized, the Persians moved on to *Harran (see above), before they returned to encamp against (*šrā 'al*, ܫܪܐ ܥܠ; Chabot I.288.12) Edessa and resume negotiations; when these broke down, they approached to assault (*la-mqarrābū 'am*, ܠܡܩܪܒܘ ܥܡ; Chabot I.290.8). See Chapter 2 for further discussion of events.

THEODOSIOPOLIS   *nasbāh*, ܢܣܒܗ; *ḥrab*, ܚܪܒ   502
JS 52; Greatrex 1998: 89.

The Romans, unable to meet the Persians in open battle, retook Theodosiopolis by storm (*nasbāh*, ܢܣܒܗ; Chabot I.278.26), and massacred (*wa-ḥrab*, ܘܚܪܒ; Chabot I.278.25) the Persian garrison in the process.

NISIBIS   [*šrā 'al tḥūm ... lapay*, ܠܦܝ ܬܚܘܡܐ...ܗܘܐ ܥܠ ܫܪܐ]   503
JS 54f; Zachariah 7.5; Greatrex 1998: 96f.

"Areobindus went down and encamped on the border by Dara and 'Ammudin, towards the city of Nisibis; he had with him 12,000 men." A large Persian army of 20,000 men was defeated and fled into Nisibis, where many suffocated in the crush at the gates. However, another, larger Persian army arrived, forcing Areobindus and his army to abandon their camp and flee to Tella.

While the terminology is similar to that of a siege, it seems clear from the original that this was a field encampment, not a siege camp, although Greatrex argues for a siege: "he encamped at the border ... towards" (*wa-šrā 'al tḥūm ... lapay*, ܠܦܝ ܬܚܘܡܐ...ܗܘܐ ܥܠ ܘܫܪܐ; Chabot I.281.16f) Nisibis.

AMIDA   *šraw 'al Amid*, ܥܠ ܐܡܕ ܫܪܘ   503
JS 54-56; Greatrex 1998: 96-99; cf. *Amida 504.

The rest of the Roman army, 40,000 according to Joshua, besieged Amida (*šraw 'al Amid*, ܥܠ ܐܡܕ ܫܪܘ; Chabot I.281.19), and attempted to storm the city with three large siege towers reinforced with iron. However, the commanders failed to coordinate with Areobindus, and had to abandon the siege and burn the towers in order to pursue the Persian army that was now unchecked. The Romans left a force in the vicinity that ambushed a group of Persians who ventured out of the city to capture a flock of sheep.

This prompted the generals (and army) to return, hoping that a captured Persian officer would help negotiate a surrender, but when this failed, the Persian was impaled in view of the city. During winter the blockade fizzled out, only to be renewed in 504 (see below).

ASHPARIN   *ḥadrūy*, ܚܕܪܘܝ   503
JS 57; Greatrex 1998: 100f.

After a Roman army was surprised on the march and defeated by the Persians, most of the army fled to Samosata, but "... one of the Roman officers, whose name was Peter, fled to the fort of Ashparin; and when the Persians surrounded the fort (*w-kad ḥadrūy parsāyē l-ḥesnā*, ܘܟܕ ܚܕܪܘܝ ܦܪܣܝܐ ܠܚܣܢܐ ; Chabot I.283.25), the inhabitants were afraid of them (*dḥalū menhōn*, ܕܚܠܘ ܡܢܗܘܢ ; Chabot I.283.25f), and gave him up to them, and the Persians took him away prisoner. They slew the Roman soldiers who were with him, but the people of the castle they did not harm in any way."

BATNAN   *ašlem*, ܐܫܠܡ   503
JS 63; Greatrex 1998: 104, 106.

After giving up the siege of *Edessa (502f above), the Arabs went raiding across the Euphrates, while "... some few of the Persian cavalry went to Batnan, and because its wall was broken down, they could not resist them, but admitted them without fighting and surrendered (*ašlem*, ܐܫܠܡ ; Chabot I.291.20) the town to them."

KALLINIKOS   *saddar l-qūblāh kolleh ḥayleh*, ܣܕܪ ܠܩܘܒܠܗ ܟܠܗ ܚܝܠܗ   503
JS 64; Greatrex 1998: 107.

After *Edessa and the battles that led to *Ashparin, Kawad marched towards Kallinikos. He sent one of his generals ahead to besiege the city, but his army surprised and destroyed by the Romans, and the general taken captive to Kallinikos. "When Kawad arrived at the city, he drew up his whole force against it (ܣܕܪ ܠܩܘܒܠܗ ܟܠܗ ܚܝܠܗ ; Chabot I.292.3), threatening to raze it and to put all its inhabitants to the sword or carry them off as captives, if they did not give him up to him. The dux was afraid of the vast host of the Persians, and gave him up."

AMIDA   *šrā 'al*, ܫܪܐ ܥܠ   504f
JS 66, 69, 71-73, 75-77, 79-81; Greatrex 1998: 108-14; Lenski 2007; full discussion in chapter 2.3.6.

The Romans decided to resume the siege (šrā 'al, ܫܪܐ ܥܠ ; Chabot I.292.21) of Amida when they learned that the Persians were buying food from the local population as well as receiving arms and supplies from the Persian side. After cutting off supply lines, the Romans raided against the Persian reinforcements assembling at Nisibis from their base at Ras al-'Ain (69), while another division collected local craftsmen and farmers to begin undermining the wall (66). This failed to bring down the wall, and another mine into the city was discovered (71). Subsequently the Romans settled for a close blockade, during which fighting occasionally broke out, sometimes for trivial reasons, such as a fight over a donkey that strayed from the Roman camp towards the wall. The Romans began an assault "in a dense mass" resulting in 40 Roman fallen and 150 wounded; the number of Persians killed was around nine, but it was difficult to determine because they had built protective sheds on the parapet so they were hard to see (72). After the heavy losses, the Romans decided to only blockade the city at a safe distance, and avoid assaults altogether, since the Persian garrison would have to surrender as soon as the Persians were defeated in the field (73). In another incident, presumably better planned, the Romans successfully assaulted with archery, but the commander Gainas was hit by a Persian ballista bolt exactly where he had loosened his armor due to the heat (75). Afterwards no more fighting is reported, and a large part of the Roman army went off to raid Persian territory. The blockade was still efficient, as the Persians at first restricted rations to themselves and their Roman concubines, but let the men in the city starve (76). As supplies ran out, they completely abandoned their concubines who resorted to cannibalism while the Persians only sat on the wall. In contrast, the Romans were extremely well supplied thanks to special procurements of grain from Egypt (77). The Persians were unable to break the siege and sought negotiations, but these were inconclusive as the Romans were negotiating from a position of strength; furthermore, the Persians tried to smuggle in arms with food that had been allowed by the Romans (79-80). However, the Roman army began to drift away during a particularly cold winter, and this finally forced the local commanders to accept a negotiated peace, including the return of Amida (81).

Sirmium	*recepit*	504

Cassiodorus *Chr.* p. 160.

Cassiodorus only states "In this consulate by the valor of lord king Theoderic, the Bulgars were defeated and Italy retook Sirmium." || Hoc cons. virtute dn. regis Theoderici victis Vulgaribus Sirmium recepit Italia.||

Jordanes (300) is slightly more informative, although not about the siege; he does however say that a part of the Ostrogothic expeditionary force joined a Hunnic army to defeat the army of Illyria in 505. See also Marcellinus Comes s.a. 505 for the latter event. It is clear from these sources that in the absence of a number of "Goths" in Italy or Mesopotamia (see chapter 2.2.2 and 3.1.1), the East Roman government had recruited a large number of Bulgarians to replenish the Illyrian field army. As with the Goths, however, there is no reason to assume that this army was monoethnically "Turkic" as opposed to a previously "Germanic" force.

Vouillé campaign   507f
GT 2.37; for context and further literature, see chapter 1.3.4.

After the battle of Vouillé, "Clovis sent his own son Theuderic through Albi and the town of Rodez to Clermont-Ferrand. As he moved forward Theuderic subjected to his father's rule all the towns which lay between the two frontiers of the Goths and the Burgundes."

Following from the geography of the area and the order in which the cities are listed, we can deduce that the Franks pressed on from Vouillé straight to *Toulouse, where they were met by their Burgundian allies. They then split up into three columns: One under Clovis, which went on to *Bordeaux and *Angoulême, one under Theuderic to *Albi, *Rodez and *Clermont, and the last under Gundobad which reached *Narbonne. Clovis celebrated with a triumph at Tours, where he offered rich gifts to the cathedral of Saint Martin.

Toulouse   *incensa*   507
Chr. Gall. 511 a. 507 (689); cf. *Vouillé campaign (above).

"Toulouse was burned by the Franks and Burgundians ..." || Tolosa a Francis et Burgundionibus incensa ||

Gregory does not specifically mention the fall of Toulouse, only that Clovis received its treasures while wintering at *Bordeaux and before proceeding to *Angoulême.

Albi   507
See Vouillé campaign 507f.

Rodez   507
See Vouillé campaign 507-8

CLERMONT-FERRAND   507
See Vouillé campaign 507-8

NARBONNE [for Barcelona]   *capta*   507
Chr. Gall. 511 a. 507 (690); *PLRE* 2 s.v. Gesalicus; Thompson 1969: 8 n. 1.
 This is not mentioned by Gregory: "... and Barcelona was taken by Gundobad, king of the Burgundians." || ... et Barcinona a Gundefade rege Burgundionum capta. ||
 According to Thompson, this is a mistake for Narbonne, as per Isidore *HG* 37 who has the Burgundians seize Narbonne. It was probably confused by the invasion of the Visigothic claimant Gesalic in 513, who had Vandal and presumably Frankish and/or Burgundian support, as he had his base in Aquitaine before his invasion. He was defeated 12 miles from Barcelona by the Ostrogothic general Ibbas.

?BORDEAUX   *apud ... urbe hiemem agens*   507f
See *Vouillé campaign (above).
 After describing Theuderic's column (see Vouillé Campaign above), Gregory continues: "Clovis wintered in the town of Bordeaux. He removed all Alaric's treasure ..." || Clodovechus vero apud Burdigalinsi urbe hiemem agens, cunctos thesaurus Alarici a Tholosa auferens ... ||
 Depending on how to take *apud*, this could either have been a siege or a simple overwintering.

ANGOULÊME   *muri sponte corruerent; subiugavit*   508
See *Vouillé campaign (above).
 (Continued from *Bordeaux 507f) "... and went to Angoulême. There the Lord showed him such favor that the city walls collapsed of their own weight as he looked at them. Then, having driven out the Goths, he subjected the city to his rule." || Ecolisnam venit. Cui tantam Dominus gratiam tribuit, ut in eius contemplatione muri sponte corruerent. Tunc, exclusis Gothis, urbem suo dominio subiugavit. ||
 While this anecdote clearly reflects Joshua's miraculous capture of Jericho, there is reason to suspect that Clovis and his army had some part in the collapse, most likely as the result of an undermining operation. Cf. Procopius' treatment of the siege of *Aquileia (452), where we can demonstrate how he has removed references to siege engines that were present in his source Priscus, leaving a "spontaneous" collapse of the walls at the site where an omen had occurred.

ARLES   *obsidientibus*   508

Vita Caesarii 1.28-32; Cassiodorus *Chr.* and *Variae* 3.32, 36, 38, 40ff, 44, 4.36, 8.10; Klingshirn 1994: 22f; Thompson 1969: 8.

Arles was besieged by the Franks and Burgundians in alliance shortly after they had conquered most of Aquitaine from the Visigoths. We do not know whether the siege began in 507 or 508; most likely the latter. It was defended by a Visigothic (possibly Ostrogothic; this is not clear from the sources) garrison alongside the local population and relieved by an Ostrogothic army, who took control over the city and used it as a base to take control over Provence and march into Visigothic Spain.

Thompson dates the relief to 510, but Klingshirn places it in 508, a date secured by Cassiodorus' chronicle (s.a. 508) and the indiction cycle in Cassiodorus' letters. Theoderic remitted taxes for the Cottian Alps for the third indiction (509-10) to defray the cost of the army that passed through to Gaul (4.36). The army had passed through during early 508, so the administrative decisions must have been made in 509. Theoderic also decided to remit taxes due from Gaul in the fourth indiction (510-11) as a result of extensive damages and expenses during the war (3.32; 3.40 remission for Arles and damaged region only, requiring the rest of the province to pay for the Gothic expeditionary troops; 3.42 extended the remission to all of Ostrogothic Gaul). Presumably it took some time to survey the damages, report them to the king, receive a response in Gaul (3.40), and send new estimates that led to an extension of the tax remission (3.42). All this happened in time for the next tax season, so the Ostrogothic administration seems to have worked quite efficiently.

Cassiodorus refers to several other administrative measures taken in the course of the war; e.g. on the mobilization of the Ostrogothic army (1.24), others to ensure supplies by ship to Arles and the Ostrogothic garrisons (3.41, 3.44) and finally to encourage good relations between the Ostrogothic expeditionary forces and the locals (3.38, the garrison at Avignon was to behave *civiliter* towards the local Romans). The fighting is also documented in the *Variae*; letter 3.44 ordered the *possessores* (great landowners) to repair the walls and old towers at Arles, which were clearly damaged in the fighting (cf. the indirect evidence from Caesarius below). The fighting also extended to combat outside the walls: letter 8.10 is a later commendation of Tuluin, recounting his exploits as the general of the Ostrogothic forces at Arles and how he was wounded during a fight over the adjacent covered bridge across the Rhône.

In addition to the evidence from Cassiodorus, an oblique description of the siege is found in the highly apologetic *Vita Caesarii*. The hagiographers relate how the bishop Caesarius' endeavors to construct a convent outside the walls was rather rudely interrupted by the besieging barbarian Franks and Burgundians, who proceeded to tear down the roof beams and unfinished upper floors of the convent. While portrayed as an act of wanton barbarian destruction, the materials were probably useful for engines and siegeworks, as noted by Klingshirn. Cassiodorus' letter confirms that it was necessary to repair the walls immediately after the siege, so some damaged had been caused to the walls themselves. Caesarius was probably one of the *possessores* charged with rebuilding the walls by Theoderic, since he had formidable economic resources, controlled much of the population, and could organize large-scale construction.

Caesarius was accused of treason during the siege, as a fellow citizen (*concivis*; taken by Klingshirn to mean that he, as Caesarius, originated from Chalon, in Burgundian territory) apparently panicked in face of danger and defected to the besiegers. Caesarius was accused of treason by both the population and Ostrogothic garrison and arrested. However, the hagiographers deflected attention by calling his accusers "heretics" (i.e. the Arian Goths in the garrison) and "Jews" (representing the civilian segment of the defenders), and then accusing the Jews of treason instead. The Jews had allegedly attached a message to a stone hurled at the besiegers, inviting them to place their ladders against their section (... *nocte quadam unus ex caterva Iudaica de loco, ubi in muro vigilandi curam sorte susceperant, illigatam saxo epistolam, quasi inimicos percuteret, adversariis iecit, in qua nomen sectamque designans, ut in loco custodiae scalas nocte mitterent, invitavit* ...). This message was conveniently "found" in no-man's land outside of the city walls by a patrol from the city, but it is unclear how Caesarius escaped from his arrest.

Although it was probably a fraud to deflect attention, the anecdote is nevertheless revealing: the defenders were organized in social units (*caterva Iudaica*) that defended assigned sectors (*in muro vigilandi curam sorte susceperant ... in loco custodiae*), i.e. the equivalent of the *pedaturae* noted in chapters 1.2.5 and 2.4.2. Furthermore, the fighting involved a besieging army close to the city, regular exchanges of missiles (since patrols went out to pick up missiles and check for danger), patrolling of the no-man's land between the walls and the besiegers, and the imminent danger of scaling parties gaining control of a section of the wall. From this evidence we can conclude that the siege was fought in precisely the same manner as those in the East during the Anastasian war only five years earlier (see chapter 2.3 *passim*) or during the Justinianic wars in Italy a generation later (see chapter 3.1 *passim* and CO entries below for the years 535-55 *passim*).

How the siege ended (and the precise actions of the relieving force) is also uncertain; the *Vita* only states that the garrison consisted of "Goths" during the siege, and somewhat later, the "Goths" returned to the city with a number of captives. It is difficult to tell whether this was connected with Tuluin's battle on the bridge, or the result of a more extended campaign. The hagiographers' silence on the matter is in stark contrast to the aftermath, when Caesarius ransomed a large number of Burgundian prisoners, which were fed in his household, and he even received supplies from the Burgundian king to keep them fed. While framed as a pious endeavor, Caesarius' care for Burgundian prisoners and connections with the Burgundian court have clear political implications.

AUVERGNE CAMPAIGN; CLERMONT   *muros urbis evertere*   524
GT 3.11-13; GT *VP* 4, 5; *Formulae Arvernenses* 1; Wood 1994: 52ff; E. James 1988: 94; *PLRE* 3 s.v. Sigivaldus 1 (1149f).

According to Gregory (3.11), when Lothar and Childebert were planning an attack on Burgundy, Theuderic distracted his men from joining the expedition by sending his troops against Clermont instead, where they could punish its people for treason by ravaging their territory and take slaves. "When Theuderic arrived in Clermont with his army, he ravaged and destroyed the entire region (*devastat ac preterit* 3.12)." Arcadius, the instigator, fled, while his family was arrested and their goods confiscated. "King Theuderic entered the city of Clermont-Ferrand and quartered his troops in one of the suburbs." || Rex igitur Theudoricus ad urbem Arvernam usque accedens, in vici illius suburbana castra fixit. || His army was let loose on the whole region, but some were punished by St. Julian. The troops assaulted *Vollore and *Chastel-Marlhac from their base. (13:) "In the end Theuderic marched out of Clermont-Ferrand, but he left behind one of his relations called Sigivald to garrison the town." || Theudericus autem ab Arverno discendens, Sigisvuldum, parentem suum, in ea quasi pro custodia dereliquid. ||

Gregory is ambiguous as to whether there was a siege of Clermont in his *Histories*. Thorpe's translation makes it appear not, but the original states that Theuderic advanced *ad urbem* (i.e. did not necessarily enter it at first) and set up a suburban fortified camp (*suburbana castra fixit*) in one of the city's surrounding villages (*in vici illius*), which was typical procedure for sieges (see chapter 5). The city was clearly treated as hostile territory, with raiding, confiscations, the siege of other suburban fortifications, and a garrison of troops loyal to the king installed when the area was subdued. The destruction of the surrounding countryside was remembered in the Formulary of Auvergne, which refers to the loss of property documents (*cartolas nostras*) at a certain *villa ... per hostilitatem Francorum*, and Gregory refers to the great damage caused by the governor Sigivald and his troops, who were his personal retainers, before he fell out of favor (he migrated to Auvergne *cum omni familia sua*; see *PLRE* for references).

The smoking gun comes from Gregory's hagiographical writings, where he on at least two occasions gives further evidence on the nature of *Clermont. In the *Vita* of Quintianus, bishop of Clermont, we learn that Theuderic was breaking through the wall but was convinced to spare the population (but still captured *Vollore and killed the priest Proclus) in the city and surrounding countryside: "The blessed man [Quintianus] was assiduous in prayer, and he loved his people so much that when Theuderic came to besiege the town (*vallante cum exercitu urbem*), the holy man of God toured the walls all night singing psalms; and so that the Lord would promptly help the country and the people he prayed constantly, while fasting and keeping a vigil. Then King Theuderic, at the very moment when he thought that he would breach the walls of the town (*cum cogitaret etiam muros urbis evertere*), was softened by the mercy of the Lord and the prayers of this bishop whom he had though to send into exile." Theuderic was (miraculously) struck with panic and tried to flee down the road, before being advised by one of his own dukes not to attack the city due to its strong fortifications, churches and saintly protection. (*VP* 4.2)

In the *Vita* of the abbot Portianus, Theuderic was visited in camp during the siege by the saintly protagonist, who intended to ask for the release of captives. However, Portianus was intercepted by Sigivald, who insisted he join him for a cup of wine. Portianus refused but was nonetheless compelled, forcing him to (miraculously) break the cup he was offered and miraculously spill a serpent in the process. After the miracle he got what he asked for. Both stories throw light on the role of clerics in negotiating with the political actors and tempering the fate of cities and their territories. (*VP* 5.2)

There are several possible datings. James points out that there were several expeditions to Burgundy (524 and 534) which Gregory conflated, making it impossible to pinpoint Theuderic's Auvergne expedition exactly, since it occurred roughly at the same time as one of them, but before Theuderic's death c. 533. Thorpe dates it to 532, but Wood argues for an early date, 523/4. Perhaps it was simultaneous with Munderic's failed revolt that ended with the siege of *Vitry (524/33). Further possible dates are found in the entry on Sigivald in *PLRE*.

| Vollore | *expugnant, vastato* | 524/32 |

GT 3.13; see *Auvergne Campaign; Clermont 524/32.

While based at Clermont, Theuderic's "troops stormed the fortress of Vollore." || Lovolatrum autem castro hostis expugnant ... || Gregory blames the fall on the priest Proclus, who had mistreated Saint Quintinianus, one of Gregory's heroes. While he does not go into detail, Gregory indicates that the population "were lured into a false sense of security" when they heard that Theuderic's troops were preparing to withdraw. Then, "the fortress was destroyed" and the inhabitants led into slavery (*cumque vastato castello ducerentur captivi*).

| Chastel-Marlhac | *obsessi* | 524/32 |

GT 3.13; see *Auvergne Campaign; Clermont 524/32.

"The people of Chastel-Marlhac were besieged (*Tunc obsessi Meroliacensis castri*), but they retained their liberty, for they bribed the invaders not to take them captive. It was only because of their own stupidity that they had to pay anything at all." Gregory describes how well the site was fortified by nature, had constant water supply, and plenty land inside the walls (*munitio*) for growing crops. The people figured that they could sally out and capture the goods of the Frankish soldiers, but "about fifty of them were captured by the enemy. Their hands were tied behind their backs, they were dragged to a spot where their relatives could see them and the Franks drew their swords ready to kill them." The inhabitants ransomed the captives, which seems to have been the end of the siege.

| Vitry(-le-Brûlé) | *vallat, obsedit* | 524/33 |

GT 3.14; Wood 1994: 92; *PLRE* 2, s.v. Mundericus.

Munderic raised a rebellion (dated to 524/33 by *PLRE*) with his *amici*, claiming kingship with the support of a crowd of peasants (*rustica multitudo*). It is possible that this occurred at the same time as or shortly after the *Auvergne Campaign, but it is impossible to tell whether they were connected. King Theuderic raised an army to defeat him:

"When Munderic learned this, he decided that he was not strong enough to resist Theuderic. He took refuge with all his possessions inside the walls of the fortress of Vitry-le-Brûlé: and there he prepared to defend himself, with all those whom he had won over to his cause. The army which Theuderic had assembled surrounded the fortress and besieged it for seven days. Munderic and his men resisted stoutly. 'Let us stand firm,' said he, 'and fight to the death together, for we must never submit to our enemies.' The besieging troops hurled their javelins from their lines, but they made no progress." || Quod ille cognuscens et se non praevalens defensare, Victuriaci castri murus expetens com rebus omnibus, in eo se studuet commonare, his secum quos seduxerat adgregatis. Igitur commotus exercitus castrum vallat ac per septem dies obsedit. Mundericus autem repugnabat cum suis, dicens: 'Stemus fortes et usque ad mortem pariter demicemus et non subdamur inimicis'. Cumque exercitus a circuitu incontra iacula transmitteret nec aliquid praevaleret, nuntiaverunt haec regi. ||

The king sent Arigisel to negotiate Munderic's surrender. Arigisel pointed out to Munderic that he would run out of supplies and promised that he would return to the king's favor; however, this was a ploy to get him out of the fortification. Munderic killed Arigisel and several enemies when he perceived the betrayal, but along with his men (*cum suis*) "he was killed and his property passed to the King's treasury."

The king was in a hurry to end the siege due to grave political repercussions if a usurper was allowed to live too long, especially when other regions were proving volatile. The deceptively simple tactics of surrounding the fortification and bombarding it with javelins was one practiced by the Persians as well, but required some time and much skill to be effective—it was the job of the specialist Daylami mountain infantry. The *vallat* shows that Theuderic's army constructed siegeworks in order to enclose Munderic's *castrum*. Due to the brevity of the description we do not have any enlivening anecdotes on how the fighting over the walls was conducted, but it is quite possible that *ballistae* were used. Missiles (*iacula*) were discharged from the besiegers' lines (*a circuitu*). The verb *transmitteret* implies something more technical than a simple *iaceret*, while one of Theuderic's subjects, his younger contemporary bishop Nicetius of Trier, certainly had *ballistae* in his arsenal (see discussion in chs. 1.2.5 and 4.1.5). The defenders were in turn able to hold the royal Frankish troops away from the walls during a week of missile exchanges.

MARTYROPOLIS ἐγκαθεζόμενοι ἐπολιόρκουν, τῷ περιβόλῳ προσέβαλλον 531
Proc. 1.21.4-28; Greatrex 1998: 207-12.

The Persian army established a siege camp and assaulted the walls. The Roman garrison under generals Bessas and Bouzes lacked defensive siege engines and were low on supplies, the city was easily accessible "and could be captured very easily by a Persian siege," and the Roman relieving army was afraid to approach. However, the Persians withdrew after the death of Kawad and the Romans had convinced them that a Hunnic army recruited to assist the Persians had been bribed to join the Roman side.

Septimanian campaign; fortress of Dio   *obtinuit*   c. 532
GT 3.21; Thompson 1969: 13.

After the death of Clovis, the Visigoths had recaptured some territory, which the Frankish kings Theuderic and Lothar attempted to reconquer. Lothar's son "Gunthar advanced as far as Rodez then turned back." Theuderic's son Theudebert was more successful. He "went as far as Béziers. He captured the fortress Dio and sacked it (*Dehas castrum obtinuit atque in praedam deripuit*)" and went on to *Cabrières.

Cabrières (fortress)   *se subdant*   c. 532
GT 3.21f; *Septimanian campaign (532).

Theudebert threatened the fortress into submission (3.21): "Next he sent messengers to a fortress called Cabrières to say that, unless the inhabitants surrendered, the whole place would be burnt to the ground and they themselves made captive." || Deinde ad alium castrum nomen Caprariam legatus mittit, dicens, nisi se ille subdant, omne loco illud incendio concremandum, eosque qui ibidem resident captivandus. || He was invited into the fort by the wife of a prominent citizen, and thus the surrender was peaceful (3.22).

In addition to some support from local aristocrats, the fate of *Dio probably had some influence on the course of events at Cabrières.

Carthage   533
Proc. 4.1.1-12.

The Vandals cut the aqueduct and took control over the surrounding countryside, but did not raid their own country. In the meantime the Romans were busy repairing the fortifications. Procopius indicates that the Romans had no fear of Vandal poliorcetic capacity, but the Vandals were attacking the economic and logistical base of Carthage and tried to sow dissention among the ranks of the defenders. They targeted local Carthaginian citizens, Roman soldiers of the Arian faith, and disaffected allied troops, the Massagetae (i.e. Huns). Belisarius dealt with the former by impaling a prominent citizen outside the city walls in everyone's view, and the latter by promising them their share of the booty and a swift return to their homelands—their fear was to be stationed permanently in Africa. Belisarius began preparations for battle (represented as a set-piece oration by Procopius) as soon as the city walls were complete (4.1.12). The Vandal field army was routed shortly afterwards (4.3).

| LILYBAEUM | ἕξουσιν | 533 |

Proc. 4.5.11; Wolfram 1988: 338.

Belisarius sent soldiers to take a Vandal fort in Sicily which had recently been captured by the Goths; the Romans were repulsed and a minor diplomatic crisis ensued (4.5.12-25).

| MT. PAPUA | προσεδρεία | 533f |

Proc. 4.4.28-31; 4.6.1-7.17.

The Vandal king Gelimer took refuge after his defeat alongside Moorish allies on Mount Papua, where he was besieged over the winter by a division of Heruls. The Heruls attempted to storm the mountain, but were repulsed by the Moors with heavy losses, and Gelimer only gave up when food was scarce, disease (intestinal parasites) were beginning to spread among his relatives, and he was guaranteed an honorable surrender by Belisarius. Procopius also emphasized how the Vandals were accustomed to luxurious living and could not take living the simple life of the Moors.

| AUTUN | *obsidentes* | 534 |

GT 3.11; Mar. Av. s.a. 534; E. James 1988: 94; Wood 1994: 51-54.

Gregory seems to have been poorly informed about the conquest of Burgundy, as he probably conflated two different expeditions which occurred in 524 and 534 respectively; the context for the following quote is more appropriate for the situation in 524 (see *Auvergne campaign and Wood), but the events clearly occurred in 534: "But Lothar and Childebert were marching against Burgundy. They besieged Autun, forced Godomar to flee and occupied the whole of Burgundy." || Chlothacharius vero et Childeberthus in Burgundiam dirigunt, Agustidunumque obsedentes, cunctam, fugato Godomaro, Burgundiam occupaverunt. ||

Marius has an almost identical entry, but does not mention the siege of Autun: "In the year of these consuls, the kings of the Franks, Childebert, Chlothar, and Theudebert, took hold of Burgundy and, when they had put King Godomar to flight, divided his kingdom." || His conss. reges Francorum Childebertus, Chlotarius et Theudebertus Burgundiam obtinuerunt et fugato Godomaro rege regnum ipsius diviserunt. ||

| PALERMO | ὁμολογίᾳ ... παρέδοσαν | 535 |

Proc. 5.5.15f; Wolfram 1988: 339.

The Roman army (numbering 9,000 according to Wolfram; Proc. enumerates 7,500 at 5.5.2-4) and navy arrived from Catania and Syracuse, on the east coast of Sicily, which were captured without any resistance. Upon approaching the harbor of Palermo, which had a substantial Gothic garrison, Belisarius noticed that the masts overtopped the parapet, and ordered skiffs to be hoisted from which to fire upon the Gothic defenders. The Gothic garrison attempted to hold the wall but was overwhelmed by fire from above and immediately surrendered (ὁμολογίᾳ ... παρέδοσαν). Shortly afterwards, Belisarius made a triumphal entry into Syracuse.

| CARTHAGE | ἐς πολιορκίαν καθίστατο | 536 |

Proc. 4.15.1-10.

8,000 Roman mutineers marched on Carthage alongside the remaining 1,000 Vandals (who had either fled Constantinople or were simply not noticed when the Romans were assembling them, indicating how assimilated they were) and a throng of slaves (ἀφίκετο δέ οἱ καὶ δούλων πολύς τις ὅμιλος, 4.15.4), who may have been assembled to perform logistical tasks during the siege (cf. chapter 2). However, the besiegers fled when Belisarius arrived with about 100 of his own retainers from Italy; shortly after, the mutineers were defeated in battle (4.15.11-47), but they reassembled and the civil war in North Africa dragged on for some time (4.15.48-17.35).

| SALONA | ἀπαλλαγείς | 536 |

Proc. 5.5.2, 11; 5.7.1-10, 26-36; Wolfram 1988: 339; Amory 1997: 397ff. Cf. *Salona (537).

A Roman army led by Mundo captured Salona from the Goths after a field engagement in 535. A Gothic counterattack in early 536 was defeated by Mundo's force but both armies suffered so heavy losses that neither the Romans nor the Goths could control Salona. Shortly after, however, another Gothic expedition took the city, but the Romans prepared a major naval force at nearby Epidamnus. Gothic morale was low due to the news of the huge approaching fleet, poor state of walls and "exceedingly suspicious attitude" (5.7.31) of the population. The Goths having thus abandoned (ἀπαλλαγείς) the city in face of overwhelming force, the Romans immediately took over and repaired the walls.

| NAPLES | 20 days | κατὰ κράτος, τειχομαχοῦντα | 536 |

Proc. 5.8.-10.; Wolfram 1988: 541f. Cf. *Vienne (500).

Belisarius, supported by a fleet, based operations on a camp (στρατόπεδον) near the city and also took control of a suburban fort (φρούριον) by surrender. After failed negotiations with some of the city notables (τῶν τινας λογίμων), who were very divided among themselves, the Roman army "made many attempts on the circuit wall" (πολλάκις τε τοῦ περιβόλου ἀποπειρασάμενος ἀπεκρούσθη; 5.8.43) but suffered heavy losses, partly due to difficult terrain. The Romans also cut the aqueduct, but cisterns inside the city provided enough water. After a few days, Belisarius became anxious about reaching Rome before winter set in and Gothic mobilization began in earnest, and even gave orders to pack up. (5.8.1-9.10)

The Romans finally penetrated the walls by sending Isaurian troops to expand the aqueduct channel (which one of the Isaurians had examined on a hunch), allowing armored men to pass through the wall and deep into the city. The Romans attacked through the aqueduct at night and killed the guards in two of the towers before signaling those outside. Roman *tekhnitai* had prepared ladders in advance for the final storm, but miscalculated the height of the walls, making them too short. The situation was saved at the last minute by lashing ladders together two and two, and they stormed over the northern walls immediately afterwards, resulting in a collapse of the defense between fire from two directions. Other forces torched the eastern gates that had been abandoned. The city was thus taken "by force" (κατὰ κράτος). Naples was defended by a Gothic garrison and an urban militia that also included the Jewish population. The Jews had guaranteed supplies during the siege, and were the last to flee their towers, as they held an inaccessible part of the wall facing the sea and feared repercussions for their loyalty. (5.9.10-10.27)

The Roman soldiers killed indiscriminately, since many had "a brother or other relative slain in the fighting at the wall (τειχομαχοῦντα, 5.10.29)." Belisarius had to go round the whole city to stop the killing. Civilians taken as slaves were released, while the Gothic garrison of 800 men was taken prisoner and well treated. Presumably there were some casualties over 20 days of fighting, so it may originally have been somewhat larger, perhaps 1,000 men. The population banded together and lynched one of the Roman aristocrats that had advised resistance; another had died in a fit (perhaps suicide) but the mob impaled his corpse. (5.10.28-48)

ROME; TUSCAN CITIES	ἐθελούσιοι ... ἐδέχοντο	536(f)

Proc. 5.11.26; 5.14.1-14; 5.16.1-4.

Vittigis, the new king of the Ostrogoths, exhorted the Pope, Senate and people of Rome to remain loyal, "reminding them of the rule of Theoderic," and had them swear oaths; in addition, he stationed a garrison of 4,000 men to hold the city. However, the inhabitants of Rome, urged on by the

Pope, sent an envoy to Belisarius at Naples to invite the Roman army into the city. The Gothic garrison learnt of this only when Belisarius was approaching, and finding themselves unable to resist without support from the inhabitants, negotiated a withdrawal with the population. The Goths departed via the Flaminian gate as the Roman army marched in through the Asinarian on December 9, 536.

Smaller divisions were sent out to take other cities in Tuscany, specifically Narni, Perugia and Spoleto, which all submitted willingly (ἐθελούσιοι) around the turn of 536/7.

SALONA	κατὰ γῆν τε καὶ θάλασσαν ἐπολιόρκουν	537

Proc. 5.16.8-18, 6.28.2; Wolfram 1988: 341, 345. Cf. *Salona (536).

After losing the city in 536, the Goths dispatched an army and navy in 537. One division marched directly to Burnum, close to Salona, where it awaited the other division, which had recruited a large force of Sueve clients. The Roman garrison was reinforced by withdrawing troops from all the surrounding fortifications, and the defenses strengthened by preparing a moat around the circuit wall (περίβολος) as well as other, unspecified preparations. The Goths surrounded the city completely with a stockade (χαράκωμα) by land and fleet by sea. Roman naval attacks, presumably from the ships already present, defeated the Gothic fleet, but the Goths maintained a tight blockade on land. No information is given on when the Goths abandoned the siege, but one of the Gothic commanders, Uligisalus, was back in Italy to lead the garrison of Todi after Vittigis had abandoned Rome in 538 (6.11.1).

MILVIAN BRIDGE	καταπεπληγμένοι ... ἐκλιπόντες	537

Proc. 5.17.14-18.28.

Belisarius had fortified the Milvian Bridge just north of Rome in order to delay the Ostrogothic approach. It was the only bridge within several days' march—Belisarius estimated at least 20 days extra would be needed to find a crossing and reach Rome. The Goths began to prepare a storm; however, the garrison (φρουρά) of the tower (πύργος) panicked at the sight of the huge Gothic army (5.17.19). 22 cavalrymen deserted to the Goths, while the rest fled during the night, making their way south to Campania. Belisarius arrived unawares the next morning with a reconnaissance party of 1,000 cavalry only to be surprised by the Gothic vanguard on his side of the river, and barely fought his way back to the city.

| ROME | 1 year | πολιορκία ἑλεῖν ... οἰόμενοι | 537f |

Proc. 5.14.15ff, 5.18.34-24.21, 5.24.28-25.25, 5.27.1-6.4.20, 6.5.1-7.25, 6.9., 6.10.12-20; chapter 3.1.5 and *passim*.

The siege lasted from March 1, 537 to "around the spring equinox" of 538. The Ostrogothic field army is claimed by Procopius to number 150,000 men (perhaps paper total of Gothic armed forces, including navy); more likely 25-30,000 men. They established six large, well fortified siege camps (χαρακώματα) with ditches, high banks and palisades. These partially enclosed northern, western and eastern approaches, but did not establish a full blockade; that was attempted during the summer with another camp in the south but not completely successful. The garrison of the southern siege camp is given as 7,000 men, but apparently substantially less than what remained in the northern arc, as movement remained relatively free in and out of the city to the south, while most of the heavy fighting still occurred in the north. If they reassigned men evenly from the other camps (around 1,200 from each camp), the original six siege camps must have contained around three or four times that number, since the remaining forces in the northern sector could still face the Romans in multiple engagements. See also *Rimini II for Gothic numbers immediately after they abandoned the siege of Rome. Upon arrival, the Goths cut the aqueducts. Throughout the siege, there were intermittent assaults with archery and ladders, but a major assault with siege towers, rams, undermining took place on the 18th day of the siege. Otherwise there were sneak attacks, attempts at bribery and subversion, and 69 field engagements outside the walls against Roman sallies and raids. The brunt of the fighting occurred during the spring of 537, but when the Goths learned that the Romans were running low on supplies they decided to shift towards a blockade, establishing the southern siege camp and attempting to cut off sea-borne supplies by capturing *Portus.

Rome was held by the Roman expeditionary army that included military engineering specialists (τεχνῖται). About 5,000 regular troops at first, they were reinforced by 1,600 cavalry early in spring 537, and were supported by a citizen militia, sailors, army servants/followers that could double or triple their number. While some of these troops were sent to take nearby forts (e.g. Tibur and Albano) or as escorts for civilians, supplies and messengers between Naples and Rome, new reinforcements arrived via Naples during the summer and autumn (5,600 men listed at 6.5.1f, most of whom made it to Rome by land or sea). At most, Rome may have had in excess of 10,000 professional defenders with one or two thousand tied up in surrounding forts. The professionals handled siege engines, gates and fighting outside the walls. Civilians helped guard the walls and occasionally joined engagements outside, but could not be relied upon as they were too undisciplined. The Romans made a series of improvements to the defenses. The merlons on the battlement had been extended with flanking walls to prevent enfilading fire from the left. A deep trench (τάφρος) had been dug all around the city; archery was used extensively alongside various engines: ballistae on the towers, onagers on the parapets, trap doors at gates. Guard dogs outside the walls at night, and frequent sallies, especially at night, kept the Goths inside their camps. The capture of several fortifications along Gothic supply lines meant that the Goths began to starve late in 537.

Belisarius had ordered the citizens to bring in all their supplies from the countryside before the siege, and moved the old water-powered mills from the Janiculum onto boats in the Tiber when the aqueducts were cut. Further supplies were brought in by sea from Sicily to Portus. After the major assault of March 18 was repulsed, Belisarius had the citizens and soldiers send their women, children and servants to Naples in order to forestall starvation. When *Portus fell, the Romans had to rely on Ostia, which was less accessible. The morale of the Roman citizens was variable; they were indignant at Belisarius and the Emperor for having to endure a siege and at the Goths for raiding their properties, but for the most part fought enthusiastically alongside Roman troops, even insisting on participating in field battles outside the walls. At one point, they tried to open the doors of the temple of Janus. A continuous stream of deserters kept the Goths informed about Roman moves, but Belisarius was able to keep most high-ranking Romans in line or remove them (e.g. Pope Silverius) if they were suspected of treason. The removal of families to Naples was perhaps also intended to guarantee the good behavior of the remaining citizens.

The siege became increasingly difficult for the Goths to maintain as the Romans were able to bring in reinforcements and supplies during the summer and autumn of 537; soon, they had large enough forces to harass the Goths continuously and even raid Gothic supply lines, so that by winter, the besiegers began to lack supplies and had to abandon several forts, including Portus and Centumcellae (Civitavecchia). At the beginning of 538, a truce was arranged, but the Romans exploited Gothic supply problems by occupying the abandoned forts. During negotiations for a peaceful settlement, the Romans brought in even more supplies and sent an army ready to raid Picenum (which ended up occupying *Rimini I). As negotiations dragged on and the Romans kept expanding opportunistically, the

6TH CENTURY

Goths had enough and attempted a surprise assault in early March 538. When this failed and they received news of the Roman capture of *Rimini, they withdrew to protect Ravenna but were heavily defeated and lost many troops as they crossed the Milvian bridge.

Portus	ἀφύλακτον ... εἷλον	537

Proc. 5.26.3-27.1.

In order to tighten the blockade of *Rome, Vittigis ordered the capture of Portus, the main supply port for the Roman garrison, on March 21. Belisarius is criticized by Procopius for not having established a garrison; Proc. believed that 300 men would have been enough, but recognized that the defense of Rome required the limited manpower available. When the Goths took the city, they established a garrison of 1,000 men, but had to withdraw due to lack of supplies in the following winter.

Rimini (I)		538

6.7.26-34, 6.10.1-12.

During a truce with the Goths early 538, Belisarius sent out a force of 2,000 cavalry to Picenum who were to remain quiet as long as the truce lasted, but raid and capture as many forts as possible when it was broken. When this happened, they raided around Osimo, taking Gothic women and children captive and defeating a Gothic force that tried to stop them. Although Osimo and Urbinus were too well fortified (despite small garrisons) to capture, the Romans pressed on to Rimini, where Vittigis' wife, Matasuntha, enthusiastically received the Romans. In consequence, the Goths gave up the siege of *Rome.

Petra pertusa	ἀπεπειράσαντο τοῦ ... φρουρίου	538

Proc. 6.11.10-20.

On the way to *Rimini II, the Romans assaulted the fortified tunnel at Petra Pertusa, which was protected by a fort nestled below the cliff through which the tunnel ran. Unable to take the fort by frontal assault, the Romans climbed up the cliff above the fort and hurled stones and boulders onto the Gothic garrison. This convinced the Goths to surrender. Some of them were sent away and recruited into the Roman army; a few remained with their families, presumably serving alongside the Roman garrison that was installed there.

| Rimini (II) | ἐνστρατοπεδευσάμενοι ἐπολιόρκουν | 538 |

6.11.1-9, 6.11.21-12.25, 6.26.14-24, 6.27.12-24.

The Ostrogoths withdrew from Rome after the fall of *Rimini (I). Recognizing how exposed the cities of Picenum and Umbria were to Roman raids, they established large garrisons from 400 to 4,000 men in important cities and forts, altogether 9,800 men. The rest, presumably a similar number, marched on to retake Rimini from the Romans, but had to make long detours in order to avoid the Roman garrisons at Narni, Perugia and Spoleto (see *Rome and Tuscan cities 536). In the meantime, Belisarius sent troops (taking *Petra on the way) to Rimini, ordering them to replace the cavalry garrison at Rimini with infantry from Ancona, which had recently been captured, in order to save supplies and allow the cavalry to harass the Goths out in the open. However, the commander at Rimini, John, refused to do so and only sent away Belisarius' 800 *buccellarii* while retaining the 400 infantry from Ancona along with his remaining 1,200 cavalry (see *Rimini I). The division that had taken *Petra returned with the *buccellarii* to Belisarius.

The Goths set up camp and began the siege by constructing a tower that was to assault the wall at its most vulnerable; this time they propelled it by having men pushing from the inside and were able to approach the wall without incident. Since darkness fell before the assault could begin, they posted guards, expecting no difficult with the ditch (τάφρος). The Roman commander responded by exiting stealthily that night with a party of Isaurians with "pickaxes and various other tools of this kind (δικέλλας τε καὶ ἄλλα ἄττα τοιαῦτα ὄργανα)." They managed to extend the width and depth of the ditch and heaped the earth up against the city wall. Not only was the ditch widened, but the approach to the city wall beyond the ditch was now a steep embankment. The Gothic sentries, who were sleeping on duty, eventually discovered the excavations and attacked, but by then the Isaurians had completed the job and rushed back into the fort. When the Gothic assault began (after executing some guards), they compensated for the ditch by throwing in fascines (φάκελλοι), which had clearly been prepared in advance. This actually worked well: while the tower did sink somewhat as it was hauled into the ditch, it was nevertheless brought forward until it reached the steep embankment on the other side (6.12.12). By then, night was approaching again, so the Goths began to withdraw the tower so that is was beyond Roman reach during the night. As the Goths began to pull back the tower, the Romans sallied out with most of their force to prevent the withdrawal. Both sides suffered heavy losses during the fight.

While the Goths were able to pull the tower back to their camp, they decided to avoid direct assaults due to their losses, and the fact that the Roman garrison would soon run out of supplies (cf. *Amida 503f).

In the meantime, Narses had arrived in Picenum with an army of 7,000 (6.13.16ff; after attracting some of Belisarius' commanders, Narses' army counted 10,000; 6.28.6). After some discussion Belisarius and Narses decided to relieve the city, especially when they were informed that it would surrender in seven days. Belisarius divided the army into several divisions. One was detected by a Gothic patrol, and the Goths prepared to face them in battle, but were perplexed when an army arrived from another direction and a Roman fleet arrived as well. The Goths fled their camp, leaving their valuables and their sick behind.

Procopius does not comment explicitly on the tactics of the Gothic siege tower, but from his narrative we can deduce the following. Filling in the ditch with fascines took most of the first day after the ditch had been extended. Even though the fascines yielded to the weight of the tower, it did not prevent the Goths from moving the tower forward, only the steep embankment (which had only been recently created) in front of the city wall caused problems. While the tower had to be withdrawn for the night, the filling would obviously remain in place on the following day, so any assault could begin much earlier. Furthermore, when they reached the embankment below the wall, the Gothic sappers would be able to remove earth under cover of the tower. Then the tower could be brought right up to the walls without difficulty. There is nothing in Procopius' account to suggest that the Goths were particularly incompetent, apart from the sleeping guards. That is rather understandable after an arduous one-year siege at *Rome, a forced march across the Appennines, care for sick troops, and all the construction involved in collecting timber for the siege engines, materials for the fascines, and fortifying camp within a day or so of arriving.

MILAN; LIGUR. CITIES ἐνστρ. ἐπολιόρκουν, ἐς ἔδαφος καθεῖλον; *effracta* 538f
Proc. 6.7.35-38, 6.12.26-41, 6.21.; Mar.Av a. 538; *Ep. Austr.* 19 and 38.

When it became clear in the early spring of 538 that the Romans were able to hold on to *Rome, the Milanese under their bishop Datius asked for a few soldiers, promising the whole of Liguria (6.7.36): "For they declared that they were themselves able (αὐτοὶ γὰρ ἱκανοὶ ἰσχυρίζοντο εἶναι) without any trouble to detach from the Goths not only Milan, but the whole of Liguria also, and to recover them for the emperor." While *Rimini was under siege, Belisarius sent 1,000 men by sea to Genova. The troops brought the ships' boats on wagons to cross the Po, and defeated the Goths outside Pavia/Ticinum, almost taking the city in the process. However, they moved on to Milan and took Bergamo, Como and Novara at the same time. This meant that only 300 soldiers were left to hold Milan, while the rest were distributed in the other cities. At this point, the Goths arrived supported by 10,000 Burgundian "volunteers" secretly sent by the Frankish king Theudebert in order to circumvent his peace treaty with the Romans. The Romans held the city with active cooperation from the citizens, who helped guard the walls. Belisarius sent a relieving force, but it was unable/afraid to cross the Po, and further reinforcements were delayed by correspondence between Belisarius and Narses and the illness of one of the commanders who was to assemble boats for crossing the river. The Roman garrison despaired of help and tried to negotiate surrender, but the Goths would only guarantee the safety of the Romans, not the civilians. The Romans lost their nerve and surrendered. The whole male population was massacred, the women enslaved and given to the Burgundians as a reward, and the city itself was "razed to the ground" sometime around the turn of 538-9, after Belisarius had captured *Urbino in late December and marched on to *Orvieto. After the fall of Milan, the Goths took by surrender (ὁμολογίᾳ εἷλον) the other cities held by the Romans in Liguria.

Marius only mentions the end result: "In the year [John] was consul, Milan was stormed by the Goths and Burgundians (*Mediolanus a Gotis et Burgundionibus effracta est*), and there senators and priests along with other people were killed even in the holy places, so that the altars were stained with their blood." The *Epistulae Austrasiacae* 19 provides diplomatic evidence that Justinian sought the help of the Austrasians to hold Milan and Liguria. He asked for 3,000 men, but the Austrasians excused themselves and manipulated the situation for their own gain; see chapter 4.1.3 for a full discussion. Simultaneously, Theudebert had sent an envoy to the pope to inquire about a rather trivial issue of canon law (degree of consanguinity in marriage); in a duplicate reply to Caesarius of Arles (*Ep. Austr.* 38), the Pope instructed Caesarius to give Theudebert the standard explanation. The letter is accurately dated to 6 May 538, i.e. probably while Burgundian troops were mustering to cross the Pyrinees. Theudebert probably had his representatives appear to be in Italy on legitimate diplomatic business while investigating the military and political situation before an invasion.

6th century

Murray's translation of Marius is dubious, since neither source provides accurate evidence of what transpired when the Roman garrison surrendered. The original *effracta* does however sum up Procopius' description: it simply means "destroyed." Procopius provides ample evidence that the citizens were betrayed by the Roman garrison, but since they performed sentry duties we cannot assume they were completely defenseless; this had also been bishop Datius' assurance to Belisarius, who had few troops to spare. However, the professionals normally held the gates and could in effect hand over the city without the population discovering it before it was too late. Divided command in the Roman expeditionary army prevented the timely relief of the city; see *Orvieto below.

Todi	ξυμπολορκήσειν	538

Proc. 6.13.2ff.

Belisarius "sent some men to Tudera [Todi] and Clusium [Chiusi], with orders to make fortified camps (χαρακώματα) there, and he was intending to follow them and assist in besieging (ξυμπολορκήσειν) the barbarians at those places." However, the Goths surrendered as soon as they heard of the approaching army, and were all deported to Sicily and Naples.

Chiusi	ξυμπολιορκήσειν	538

See *Todi above.

Ancona	ἀποπειράσασθαι	538

Proc. 6.13.5-15.

The Gothic garrison at Osimo (4,000 men, 6.11.2) was reinforced by another army and sent to take Ancona. The Roman defenders (presumably very small after 400 of them were sent to reinforce *Rimini (II)) went out in formation to ward off the approaching Goths, but were startled by their huge numbers and fled; the inhabitants let them in but closed the gates at the Goths, who were in hot pursuit and killed many Romans. This left many Roman soldiers stranded up against the wall, but they were saved by climbing up ropes that the citizens threw down to them. The Goths nearly took the city by storm with ladders (κλίμαξιν), but were just barely warded off thanks to the bravery of two of the senior *doryphoroi* of Belisarius and Valerian.

Urbino	ἐστρατοπέδευσαν, προσεδρεία	538

Proc. 6.19.1-17.

Belisarius began to besiege Urbino, which had a Gothic garrison of 2,000 men (6.11.2), by establishing camps on two sides of the city, but Narses criticized his strategy and abandoned the siege with half (ἥμισυ) the army. Depending on how literally we should take "half," and how many troops had been assigned to other duties (e.g. *Orvieto), this must have left Belisarius with at the very most 10,000 men, but probably far less—a reasonable guesstimate would be 5,000 men (6.19.10, cf. *Rome 537f, *Rimini I and II 538). Although taunted by the Goths, the remaining troops persisted, having built a portable screen to cover their assault up the steep hill where the city was situated:

"Belisarius ... ordered the army to collect thick poles (ῥαβδοὺς παχείας) and to make of them a long colonnade (στοάν τε ἀπ' αὐτῶν ποιεῖσθαι μακράν). This device was destined to cover the men hidden inside as the moved forward close up to the gate at the particular point where the ground was level and carried on their operations against the wall. So they were engaged in this work." (6.19.6f) The screen is called both στοά and μηχανή (6.19.15f).

There remained enough troops to surround the hill completely for the assault. Just as the Romans were beginning the storm, the Goths ran out of water (after three days under siege) and decided to surrender as the Romans were making their way up the hill with their portable screen. They became subjects "on terms of complete equality with the Roman army (ἐπὶ τῇ ἴσῃ καὶ ὁμοίᾳ γεγενημένοι)." (6.19.17)

The city was captured "at about the winter solstice," i.e. at the end of December 538.

| CESENA | προσέβαλλόν τε καὶ τοῦ περιβόλου ἀπεπειρῶντο | 538/9 |

Proc. 6.19.18-21.

One of Narses' supporters, John, was sent from the new Roman base at Rimini with ladders (κλίμακας φέροντες) to take Cesena, but after a failed storm and heavy losses decided to move on (see *Imola below).

| IMOLA (FORUM CORNELII); EMILIA | ἐκ τοῦ αἰφνιδίου κατέλαβεν | 538/9 |

Proc. 6.19.22; see *Cesena above.

"And by a sudden move [John] succeeded in taking possession of an ancient city which is named Forocornelius; and since the barbarians [Ostrogoths] constantly retired before him and never came to an engagement, he recovered the whole of Aemilia for the emperor. Such was the course of these events."

Procopius is notably reticent about the achievements of Belisarius' rival Narses, and does not do justice to what was in fact achieved. Both

Belisarius and Narses were able to send reinforcements all the way through Emilia to the Po in order to attempt the relief of *Milan, while the Goths were unable to send reinforcements through the region to *Osimo (539) due to famine and lack of supplies.

| ORVIETO | πολιορκήσουσιν, λίμῳ παραστήσεσθαι | 538f |

Proc. 6.19.1, 6.20.1-14.

While besieging *Urbino, Belisarius had sent another army to besiege (πολιορκήσουσιν) Orvieto, which had a Gothic garrison of 1,000 men (6.11.1). The commander there asked for assistance from Belisarius after *Urbino, who decided to put off a prospective siege of *Osimo and instead install a garrison at Firmo who were to keep the Goths in check in Picenum. The hope was that the Goths in Orvieto would surrender immediately upon Belisarius' arrival. While hoping to be able to take Orvieto by storm (βιάζεσθαι) or stratagem (λάθρα ... ἐξ ἐπιβουλῆς), the former seemed impossible due to the strong position of the fort, so he decided to settle for a blockade and starve out the Goths, who were short on supplies and were soon reduced to eating "skins and hides."

Procopius never explains how the Goths surrendered, but refers to the fact early in his narrative (6.20.4). Since Procopius is reticent on the matter, it seems that Narses had a point when he criticized Belisarius for being too defensive and careful. The Romans controlled nearby Perugia, Narni and Spoleto, and already had Orvieto under siege. Belisarius' arrival seems to have had little impact on the siege already in effect (hence no further explanation for his activities there), it delayed the siege of *Osimo, and much time appears to have been wasted in relieving *Milan while Belisarius was corresponding with Narses about the deployment of the latter's junior officers, who refused to take orders directly from Belisarius. However, the siege of *Osimo (539) began while the whole region from Picenum to Tuscany was struck by a horrible famine (6.20.15-22) which prevented Gothic reinforcements from approaching from Ravenna. Thus, John's movement against *Cesena and *Imola above may have been designed to deliberately make Gothic reinforcements impossible by raiding the countryside. Thus, Belisarius' delay and shifting of troops may have been necessary as the Adriatic would effectively be closed in the winter and any army in Picenum would have difficulty getting supplies until spring anyway.

| Osimo | στρατοπεδεύεσθαι κύκλῳ, πολιορκίᾳ ... ἀκριβῆ | 539 |

Proc. 6.23.5-24.17, 6.26.1-27.24, 6.27.27-34; see also *Ancona 538.

Having received sole command, Belisarius marched from *Orvieto (538f) to Picenum in the spring of 539. Sending troops to besiege *Fiesole and another army to occupy Derthon near the Po in order to prevent Gothic reinforcements from breaking the sieges, Belisarius proceeded with 11,000 men to Osimo. The site was extremely well fortified and difficult to approach, situated on a hilltop. The Gothic strategy was to hold it and thus prevent a Roman march on Ravenna. It therefore had a very large garrison (at least 4,000; see *Ancona 538). Due to the strong fortification and large garrison, Belisarius decided on a close blockade (πολιορκίᾳ ... ἀκριβῆ) rather than try to storm the fortifications (προσβολὴν τῷ περιβόλῳ ποιήσασθαι, 6.23.16). The Romans set up an encircling camp according to *symmoria* (unit or division) in huts spread thinly around the base of the hill. Belisarius' private following, his *doryphoroi* and *hypaspistai*, were camped around him in a separate division to the east. The Goths attacked Belisarius' section while they were still making camp, but were repulsed. During the siege there were many engagements on the hillside, as the Goths had to venture outside the wall to collect grass for their horses, and the Romans constantly advanced to prevent this. The Goths tried different stratagems to defeat the Romans; at one point, they sent wagon wheels down the hillside against the advancing Romans, but missed their targets. Otherwise they used ambushes and noise diversions to confuse the Romans. Procopius, who was present at the siege, suggested using horns and bagpipes as separate signals for advance and retreat in order to prevent the Romans from being caught in ambushes, since the soldiers in camp could actually see the Goths emerging from ambushes but were too far away to warn their comrades far up on the hillside with shouts and other signals.

Eventually the Goths began to starve. There was already a famine in the region when the siege began, and a Gothic foraging party had been trapped outside by the arrival of Belisarius' army. The Goths managed to smuggle a letter to Vittigis in Ravenna, but he was unable to send an army for fear of the Roman army stationed at Derthon. Furthermore, the famine caused a complete lack of supplies on any possible Gothic route of march through Emilia and Picenum (see *Orvieto 538f), while the Romans could bring in supplies from Calabria and Sicily to the nearby port of Ancona. The Goths nevertheless persisted, and bribed a Roman soldier to carry messages to Ravenna. The Roman excused himself as having been ill and gone to a church to recover, and was thus able to carry several messages between

Ravenna and Osimo without arousing suspicion. The Gothic garrison held out far longer than expected due to this correspondence. The reason was discovered when Belisarius had a particularly strong Slav soldier hide in the bushes on the hillside and capture a high-ranking Goth as he went out to collect grass for his horses. The Roman traitor was revealed and subsequently burnt alive by his comrades.

Belisarius decided to destroy the Goths' main water supply, a cistern outside the circuit wall. The whole army advanced in close order with interlocking shields to cover "five Isaurians who were skilled in masonry (Ἰσαύρους πέντε τοὺς ἐς τὰς οἰκοδομίας ἐμπείρους)" who jumped in with their tools (ξύν τε πελέκεσι καὶ ἄλλοις ὀργάνοις, 6.27.5) while the Romans continued to approach the walls. The Goths "began to hurl stones and all sorts of missiles (λίθους τε καὶ τοξεύματα πάντα ἐπ' αὐτοὺς ἔβαλλον)" from the parapet to fight off Romans, but sallied out when they realized what the Isaurians were up to. A fierce battle ensued, which the Romans barely won after suffering heavier losses than the Goths, but they were encouraged by the presence and shouts of Belisarius. There were feats of particular bravery during the fighting: one of the *doryphoroi* took an arrow (βέλος) in the hand to save Belisarius, while seven Armenian soldiers were able to push back a large division of the Goths in difficult terrain. Despite the effort, it proved impossible to break up the cistern, so Belisarius had it poisoned with dead animals, poisonous herbs and lime before withdrawing, leaving only a single, inadequate well to the Goths. After the fall of *Fiesole, the starving Goths finally agreed to a surrender, handing half of their property over to the Roman troops as booty and "mingled with the emperor's army (τῷ βασιλέως στρατεύματι ἀνεμίγνυντο, 6.27.34)," i.e. the Goths became Roman subjects and were recruited into the army.

From the description of the battle for the cistern, it is clear that the Goths possessed formidable defensive capacities. Unfortunately, Procopius does not provide any details on engines used, but the barrage of λίθους τε καὶ τοξεύματα πάντα is described in similar terms as Roman defensive fire at *Rome (537f), which was conducted with onagers and *ballistrai* in addition to archers. It is therefore reasonable to infer that the Ostrogoths had the same defensive capabilities as the Romans in the early phase of the war.

FIESOLE   ἐξελεῖν, πολιορκοῦντες, ἐς λόγους, ὁμολογίᾳ παρέδοσαν   539
Proc. 6.23.1-5, 6.24.18-24, 6.26.1; 6.27.25ff; Bachrach 1970.

While preparing to attack *Osimo, Belisarius sent troops to besiege Fiesole: Cyprian and Justinus with their followers (ξὺν τοῖς ἑπομένοις), Isaurians, and 500 foot-soldiers (πεζοὺς) under Demetrius. Another division under Martinus and John with their followers (ξὺν τοῖς ἑπομένοις) and the rest of the army (καὶ στρατεύματι ἄλλῳ) under John the Glutton was sent to the Po region. The latter force camped at Derthon/Tortona to prevent Goths from advancing from Milan against either Fiesole or *Osimo.

To begin with, the Goths at Fiesole made frequent sallies, seeking battle rather than starve, but soon the Romans were able to prevent them from exiting at all. Vittigis ordered Uraïas with the Gothic army that had recently taken *Milan (538) to move south and assist Fiesole, but he stopped and camped opposite the Romans at Derthon. Both armies were unwilling to engage, but when both had encamped a few miles from each other, a Frankish force sent by Theudebert arrived. The Goths in Liguria thought they had come as allies and assisted them on the march, but the Franks revealed their true intentions when they were across the Po, sacrificing Gothic captives. They then defeated both the Goths and the Romans at Derthon on the same day. They soon withdrew due to dysentery (killing a third of the army) and lack supplies on account of their numbers (given as 100,000 at 6.25.2; the whole Frankish intermezzo is found at 6.25.1-26.1). This allowed the defeated Romans to reoccupy their positions and continue to protect those besieging Fiesole. The Goths at Fiesole, however, starving and in despair that no relief was on the way, negotiated a settlement and surrendered in return for safety, and the Romans brought them to *Osimo. Belisarius showed them to the defenders there, who surrendered shortly afterwards.

The Romans used a complex strategy of shielding the besieging forces who would otherwise be vulnerable to counterattacks from the region of Milan/Pavia or Ravenna. Another notable feature is the importance of "followers" of prominent generals. Bachrach has refuted Procopius' description of the Frankish army as a highly tendentious piece of classicizing ethnography. However, it is clear that the force was large due to its efficiency in defeating two armies in one day, and immediate difficulties with supplies and disease. If this was an accident, or the Franks only intended to attack one of the parties, it is indicative that the Franks could not tell the difference between Romans and Goths.

| RAVENNA | πολιορκεῖν, λίμῳ | 539f |

Proc. 6. 28. 1-6, 25ff.

After the surrender of *Fiesole and *Osimo, the Romans began to station themselves around Ravenna. The Romans, having full control of the Adriatic (θαλασσοκρατούντων), deployed troops on both sides of the Po and sent further divisions to Venetia and the Gothic *Alpine fortifications. Those on the Po seized Gothic grain boats from Liguria that were destined for Ravenna but had become stranded due to unusually low water levels. To increase pressure on the Goths, Belisarius bribed a citizen of Ravenna to burn the public warehouses.

With the failure to bring relief from Liguria due to the collapse of the last Gothic field army at the *Alpine forts, the Goths submitted after lengthy negotiations with the Franks and Romans (6.28.7-24, 6.29.1-34). Belisarius finally marched into Ravenna and took possession of the royal palace and treasury, but refused either solution acceptable to the Goths, who were most afraid of being sent to the East: either a transpaduvian Gothic kingdom (suggested by Justinian's ambassadors) or Belisarius becoming Western Emperor himself (suggested by the Goths). The last independent Goths had elected a new king but they also wanted Belisarius to take up the imperial title. Instead he dismissed the Goths at Ravenna to their homes, accepted the submissions of the last cispaduvian Gothic garrisons, and left for the eastern frontier in the spring of 540, as he was recalled to face the Persian invasion (6.29.32-6.30.30).

ALPINE FORTS	ἐπολιόρκουν, εἷλον	539

Proc. 6.28.28-35.

Some of the Gothic garrisons in the Cottian Alps decided to surrender when the Romans began blockading *Ravenna, and their commander received Roman troops. However, the last Gothic field army under Uraïas, 4,000 men, many from the same garrisons, refused to accept this and besieged the fort of the Romans and Gothic interlopers. More Roman troops soon arrived from the Po, capturing several other Alpine fortifications (τῶν ἐν ταῖς Ἄλπεσι φρουρίων τισίν) and enslaving the wives and children of the Gothic soldiers. Thereupon the Goths surrendered, and Uraïas was left without an army in Liguria.

ILLYRIAN FORTS	εἷλον	539

Proc. 2.4.4f; cf. *Kassandria, *Chersonese, *Thermopylai.

"... a mighty Hunnic army crossing the Danube River fell as a scourge upon all Europe ... they captured thirty-two forts in Illyricum ..." μέγα μὲν εὐθὺς στράτευμα Οὐννικόν, διαβάντες ποταμὸν Ἴστρον, ξυμπάσῃ Εὐρώπῃ ἐπέσκηψαν ... φρούρια μὲν δύο καὶ τριάκοντα ἐν Ἰλλυριοῖς εἷλον ...

| KASSANDRIA | κατεστρέψαντο βίᾳ, τειχομαχήσαντες | 539 |

Proc. 2.4.5; cf. *Illyrian forts, *Chersonese, *Thermopylai.

"... and they carried by storm the city of Kassandria ... never having fought against walls before." ... πόλιν δὲ τὴν Κασσάνδρειαν κατεστρέψαντο βίᾳ ... οὐ τειχομαχήσαντες πρότερον.

| CHERSONESE | τειχομαχήσαντες, βιασάμενοί | 539 |

Proc. 2.4.8(f); cf. *Illyrian forts, *Kassandria, *Thermopylai.

"This same people also assailed the wall of the Chersonesus, where they overpowered those who were defending themselves from the wall, and approaching through the surf and sea, scaled the fortifications of the so-called Black Gulf; thus they got within the long wall ..." οἳ δὴ καὶ ἐν Χερρονήσῳ τειχομαχήσαντες, βιασάμενοί τε τοὺς ἐκ τοῦ τείχους ἀμυνομένους καὶ διὰ τοῦ τῆς θαλάσσης ῥοθίου τὸν περίβολον ὑπερβάντες ὃς πρὸς κόλπῳ τῷ μέλανι καλουμένῳ ἐστίν, οὕτω τε ἐντὸς τῶν μακρῶν τειχῶν γεγενημένοι... whence they crossed over to Asia and plundered before returning.

| THERMOPYLAI | τειχομαχεῖν | 539 |

Proc. 2.5.10f; cf. *Illyrian forts, *Kassandria, *Chersonese.

"In another invasion they plundered Illyricum and Thessaly and attempted to storm the wall (τειχομαχεῖν μὲν ἐνεχείρησαν) at Thermopylai; and since the guards on the walls defended them most valiantly (τῶν δὲ ἐν τοῖς τείχεσι φρουρῶν καρτερώτατα ἀμυνομένων), they sought out the ways around and unexpectedly found the path which leads up the mountain which rises there. In this way they destroyed almost all the Greeks except the Peloponnesians, and then withdrew."

| SURA | στρατοπεδευσάμενος, τειχομαχήσων, κατὰ κράτος | 540 |

Proc. 2.5.11-26.

Avoiding the fort of Circesium and the city of Zenobia (Palmyra), the Persian army under Khusro stopped at Sura upon receiving an omen from the Shah's horse. "Khusro then made camp and led his army against the fortifications to assail the wall (ὁ δὲ στρατοπεδευσάμενος ἐπὶ τὸν περίβολον ὡς τειχομαχήσων τὸ στράτευμα ἐπῆγεν, 2.5.10)." The Roman garrison fought bravely from the parapets, but their commander was killed by an arrow. This broke the morale of the Romans, who sent their bishop the next day "to plead for them." He perfomed a ritual act of submission, taking "with him some of his attendants, who carried fowls and wine and clean loaves, and came before Khusro; there he threw himself on the ground, and with tears supplicated him to spare a pitiable population" and promised a "worthy" ransom. Khusro feigned agreement, but intended to make an example of the city as it had resisted him. The bishop was received joyfully by the *Sourenoi* when he apparently succeeded, but the Persian notables who accompanied him back to the city managed to jam the gate while the people were preoccupied celebrating their salvation. By this ruse the Persians could rush the gate and took the city "by force." The Persians sacked the city, plundering, killing, taking captives and finally setting the whole city on fire.

The inhabitants were afterwards ransomed by the bishop of *Sergiopolis (542) against a promise of future payments (2.5.28-33).

HIERAPOLIS	στρατοπεδευσάμενος, ὡμολόγησαν	540

Proc. 2.6.2-7, 21-24.

Most of the Roman army was at Hierapolis at the time of the Persian invasion. The commander Bouzes told the first men of Hierapolis (τοὺς Ἱεραπολιτῶν πρώτους) he would station the army in the hills around the city, leaving enough troops to hold the wall while the field army would harass the Persians. However, his army disappeared, only to show up again at *Edessa. When the Persians arrived and set up camp by the city, both sides agreed (ὡμολόγησαν) to a ransom of 2,000 pounds of silver; the Persians because of the strong fortifications and large garrison still in the city; the Hierapolitans because they were concerned that parts of the fortifications were difficult to hold, and they also "wished to preserve their land unplundered."

BEROEA	τειχομαχεῖν	540

Proc. 2.7.9-13, 19-37.

The Persian army arrived from *Hierapolis while Megas, the bishop of Beroea as well as Roman envoy, had gone from the Persian camp to *Antioch in order to arrange a peace treaty. At Beroea, the citizens agreed to pay twice the ransom of *Hierapolis since their circuit wall was deemed too weak, but they were unable to raise the full amount and fled to the acropolis. This led Khusro to order his troops to make a trial of the outer walls with ladders (τῷ τέχει κλίμακας ἐπιθέντας ἀποπειρᾶσθαι τῆς ἀνόδου ἐκέλευεν); they then opened the gate and admitted the whole army. The

Roman soldiers and civilians held the acropolis, but they had brought horses and other animals that consumed their finite water resources. Despite resisting Persian attempts at storming the acropolis, they were forced to surrender due to lack of water. Megas returned from Antioch and was allowed to negotiate their surrender. While the city had been destroyed, the civilians survived and most of the Roman troops defected to the Persians for lack of pay.

Antioch 540
Proc. 2.6.9-21, 25; 2.7.14-18, 2.8., 2.9.14-18; Haldon 1999: 100.

When the Persian army had crossed into Roman territory, Justinian sent his nephew Germanus with a small following to Antioch, where he oversaw the defenses along with the "architects of the public buildings" (τοῖς τῶν οἰκοδομιῶν ἀρχιτέκτοσι). They advised against constructing extra fortifications around or obstacles against a rock near the walls on Mount Sulpicius, since there was not enough time and any construction would alert the Persians to the vulnerability of the spot. The Antiocheians, fearing no reinforcements would arrive, held a council (βουλή) and sent Megas, bishop of *Beroea, to the Persian camp in order to secure a peace treaty in return for tribute. After exacting tribute from *Hierapolis, Khusro agreed to a tribute of ten centenaria of gold from the whole East. The proposal was turned down by Justinian's envoys "who forbade everybody to give money to the enemy, or to purchase the cities of the emperor." The bishop of Antioch was accused of wanting to hand over the city to the Persians and went (or was sent) off to Cilicia.

In June, after the surrender of *Beroea, the Persian army was approaching. Some citizens fled, others proposed to do the same, but were given hope by the arrival of 6,000 regular troops from Lebanon. Civilians also took part in the defense, variously called "the *demos* of the Antiocheians," "youths (νεανίας)" and circus factions. The hastily assembled Lebanese soldiers were concerned for the security of their home community, but morale was very high among local militias, who taunted the Shah and almost killed a Persian envoy from walls. The Romans defended themselves against Persian assaults with extensive archery from walls. While no engines are mentioned, they extended the parapet with scaffolds between the towers to provide extra firing platforms. However, when some of the scaffolds on Mount Sulpicius collapsed, many of the defenders thought that the walls had given in (to undermining or artillery) and began to flee. The Persians were then able to scale the wall. The demoralized soldiers fled, trampling women and children at the gates, but the civilian defenders remained. The "youths" sang victory songs to the emperor when they manage to temporarily drive back the Persians in street fighting, but were completely defeated by Persian reinforcements; massacre, rape and captivity ensued. The captives were resettled in Persian Mesopotamia.

Despite the reasonably long description in Procopius, there are few indications of how the actual wallfighting was conducted. There is clear evidence for archery on both sides, and the Persians used ladders; probably also heavy artillery. The fortifications at Antioch, built in traditional late Roman style, were well adapted for using artillery, and the scaffolding may have provided extra room for defensive siege engines—hence the collapse, which may have been caused by onager recoil. The low morale of the Lebanese troops is very understandable: there were only 20,000 soldiers in the whole of the eastern army, many of them had disappeared with Bouzes at *Hierapolis or were tied up in other cities in Syria and Armenia. This meant that their homes were extremely exposed to Arab raids, which were often encouraged by the Persians. Khusro celebrated his victory with sacrifices after bathing in the Mediterranean and at the grove of Daphne, a suburb of Antioch (2.11.1, 5f).

APAMEA	ἐστρατοπεδεύσαντο	540

Proc. 2.11.14-38.

As the Persian army approached Apamea, the inhabitants experienced a miracle of a flame above their bishop as he made a procession through the cathedral with their precious relic, a piece of the true cross. He subsequently negotiated the surrender of the city, as the citizens did not "marshal themselves on the wall." Khusro was admitted with "two hundred of the best Persians" after his army had set up camp (ἐστρατοπεδεύσαντο) beside the city wall. Once inside, he demanded all valuables in ransom, not only the 1,000 pounds of silver as originally agreed. He also presided over Hippodrome games, "being filled with a desire for popular applause," and cheered for the Greens (as opposed to Justinian, who supported the Blues). He also dispensed justice: when one of his men was accused of raping a Roman girl, he decreed that the perpetrator be impaled. While it was impossible to carry out the punishment openly due to protests from the Persian army, it was carried out secretly (but apparently publicized to the Romans) before the Persians marched on.

The miracle as retold by Procopius in effect gave a heavenly mandate to the bishop to ensure the safety of the city, despite Justinian's explicit orders *not* to pay any ransom. He could thus not be accused of betrayal, and at the same time protected the inhabitants who probably thought that resistance was futile after the fall and brutal treatment of *Antioch. In fact the Apameans could probably see the great crowd of Antiocheians captive in the Persian camp. Khusro's behavior is also interesting: he wanted to be seen as a rightful ruler of Roman territory and ostentatiously acted as such. While no garrison was left to hold the city, the formal submission, tribute, and performance of Roman rituals of power could be used to press claims to the city on a later occasion; cf. *Edessa below.

CHALKIS στρατοπεδευσάμενος, πολιορκίᾳ...αἱρήσειν, λύτρα διδόντες 540
Proc. 2.12.1-5.

Arriving from *Apamea at Chalkis, Khusro had his army encamp near the walls and threatened the city, "saying that he would take the city by siege (πολιορκίᾳ τὴν πόλιν αἱρήσειν), unless they should purchase their safety by giving ranson" and hand over the garrison and its commander. The Chalkidians (Χαλκιδεῦσι) paid ransom but swore that there were no soldiers present, hiding them inside the city.

EDESSA ἐξελεῖν 540
Proc. 2.12.6, 31-34.

While Khusro intended to take Edessa, his army lost its way several times, and when the Persians finally approached, Khusro's face was swollen with an infection in the jaw. Procopius effectively links this to Christ's particular concern for the city, as Khusro's approach is preceded by a digression on king Abgar (Augarius), who in the days of Augustus received a letter from Christ himself, containing a promise to protect his city. Instead of trying to besiege Edessa, the Persians demanded ransom, which the citizens paid in order to spare the countryside from being damaged. They had little fear of a siege of the city itself, as it had strong defenses and Bouzes was there, presumably with much of the army that had disappeared from *Hierapolis.

The presence of Bouzes is only revealed when the Edessenes, after paying their own ransom, offered to ransom the prisoners from Antioch by donating their personal wealth to the cathedral (2.13.1-7). The remarkable show of solidarity (prostitutes gave their jewelry, poor farmers their sheep) was hindered by Bouzes, who is alleged by Procopius on several occasions to have been corrupt. By this time the Persian campaign was winding down. Possibly the Persians were low on supplies and overburdened with plunder and captives whom they were anxious to ransom. Hence they were unwilling to pursue vigorous sieges and were fearful of relieving armies. For instance, on the way to Edessa, the Persians had constructed a bridge across the Euphrates but tore it down on the third day, leaving a part of the army stranded on the western bank (2.12.3-6). However, marching on from Edessa towards their own border, the Persians felt more secure. First they appropriated the ransom offered by the citizens of Constantina for the Antiocheians without releasing anyone, on the grounds the *Constantina had formally submitted to Kawad in 503 (2.13.8-15) and that any money they gave was the Shah's rightful property. They also found a pretext to refuse the ransom offered by the pagan citizens of Carrhae— apparently only Christians were obliged to pay the Persians tribute, in Khusro's view. Since they were no longer willing to accept ransom, the Persians by then must have felt confident they could bring all the captives to Persian territory. Finally, they took the time to besiege *Dara, the last Roman fortification before the border and in range of logistical support from Persian Mesopotamia.

| DARA | ἐς πολιορκίαν καθίστατο | 540 |

Proc. 2.13.16-29.

Dara had a double wall that provided extra security and an intervening space that was used to pasture animals in times of war. The Persian army overwhelmed the outer wall with missiles (πλήθει τε βελῶν βιασάμενος) on the western side, then approached the gates and set them on fire, but were afraid to follow up. Presumably defenders could use the height advantage of the inner walls and towers to shoot down at anyone who scaled the outer wall or entered through the gates. Instead the Persians decided to dig a tunnel from the ditch surrounding the city, which was outside of Roman view. However, they were betrayed by a soldier from their own camp (perhaps a soldier from Persia's Christian provinces or a Roman defector). He went out alone into the no-man's land between the Persian camp and the Roman walls, ostensibly to pick up missiles discharged by the Romans and taunt the defenders, but was able to reveal the Persian plan. The Romans frantically dug a counter-trench in the interval between the walls under direction of the city engineer, "Theodorus, a man learned in the science called mechanics (ἐπὶ σοφίᾳ καλουμένῃ μηχανικῇ, 2.13.26)." In straightforward Greek this would be a μηχανικός. This successfully intercepted the Persian tunnel. Khusro withdrew after negotiating a settlement and receiving silver. As the last Persian siege of campaign in 540, the Persians could rely on supplies from nearby Persian territory while sending booty across the border in order to pursue a siege.

Concerning the ditch whence the Persians dug their tunnel, Dewing's translation gives a clear impression that this was dug from the Persian fortified camp. However, the original is more ambiguous (ἀπὸ τῆς τάφρου ἀρξάμενοι), as Persian siegeworks are not mentioned explicitly. Furthermore, since the Persians had the ability to clear large stretches of wall with archery fire, they could probably begin digging directly in the city moat, which fits better with Procopius' description as being exceedingly deep and thus out of view of the defenders on the inner wall (or the few observers who ventured on to the stretches of outer wall which the Persians had not cleared).

ZERBOULE	3 days	ἐπολιόρκουν	540

Proc. 4.19.23-32.

The Romans assaulted a Moorish stronghold with archery, killing a great number of defenders along with their commanders. Unaware of this, the Romans began to withdraw on the last morning, but found that during the night the demoralized Moors had fled.

TOUMAR	ἐστρατοπεδεύσαντο, προσεδρεία	540

Proc. 4.20.1-22.

After winning at *Zerboule (above), the Romans moved on to the main Moorish camp at Toumar in the Aures mountains. The Romans were about to withdraw due to lack of water, but some brave Roman soldiers were able to defeat the guards of a difficult route up to the Moorish camp. This boosted Roman morale, and led to an unplanned assault that however took the Moors by surprise, and many of them were killed in the flight.

SARAGOSSA	49 days	*obsessa*	541

GT 3.29; *Chr. Caes.*; Isidore *HG* 41 (Wolf 1999: 98); Thompson 1969: 14f; chapter 4.1.1.

For the quotes from Gregory on the rituals performed on the walls by the Saragossans and Chr. Caes. on how long the expedition lasted, see chapter 4.1.1. Isidore adds: "During the reign of Theudis, the kings of the Frank came to Spain with innumerable forces and ravaged the province of Tarraconensis (*bello depopularent*). The Goths, under the general Theudegisel, closed off the passes (*obicibus... interclusis*) into Spain and laid low the army of the Franks, greatly amazed at their own victory. In response to their entreaties and to the offer of a large sum of *money*, the general provided the remaining troops a path of escape for the period of one day and one night. The miserable crowd of Franks who were unable to pass through within the allotted time were massacred by the swords of the Goths."

PETRA στρ. ἐς πολιορκίαν καθίστατο, τειχομαχεῖν, ὁμολογίᾳ παρέδοσαν   541
Proc. 2.17.3-28.

Khusro and his army came to Lazica in order to establish Persian hegemony over the region, traditionally a Roman client. They marched over mountainous, wooded terrain, so the Persians cut roads through the forest and used the timbers to make roads across "the rough places" (2.17.1). The Persian vanguard arrived at Petra to find an apparently abandoned city, but the Romans had hidden their forces within the gates. As the whole Persian army approached the seemingly unmanned wall with ladders (κλίμακας) and a ram (κριῷ τε τῇ μηχανῇ) to break through one of the gates, the Roman garrison stormed out "and slew great numbers of the enemy, and especially those stationed about the ram" (2.17.10). Khusro was furious, and after impaling either the responsible general or the officer directing the ramming team (τὸν ἄρχοντα ὃς δὴ ἐφειστήκει τοῖς τὸν κριὸν ἐργοῦσιν), moved the whole army to a camp closer to the city. The next stage of the siege began on the next day and was dominated by heavy archery exchanges:

Khusro "went completely around the fortifications, and since he suspected that they could not support a very strong attack, he decided to storm the wall (τειχομαχεῖν ἔγνω). And bringing up the whole army there, he opened the action, commanding all to shoot with their bows against the parapet (τοξεύειν ἅπαντας ἐπὶ τὰς ἐπάλξεις ἐκέλευε). The Romans, meanwhile, in defending themselves, made use of their engines of war and all their bows (ταῖς τε μηχαναῖς καὶ πᾶσιν ἐχρῶντο τοξεύμασι). At first, then, the Persians did the Romans little harm, although they were shooting their arrows thick and fast, while at the same time they suffered severely at the hands of the Romans, since they were being shot at from an elevation." (2.17.14f)

Unbeknownst to the Persians, the Roman commander was killed by an arrow, and they returned to their camp for the night while the Romans were utterly demoralized. The next day the Persians began undermining one of two vast towers that protected the only level approach to the city. When they fired the wooden props that replaced the stones they had taken out, the whole tower began to shake and the defenders managed to escape before it collapsed. As the Persians could now approach the wall on level ground, the Romans, "in terror, opened negotiations," surrendered the city in return for their lives, and "mingled (ἀνεμίγνυντο) with the Median army."

SISAURANON   στρατοπεδευσάμενοι ἐς πολιορκίαν καθίσταντο   541
Proc. 2.19.2-25.

Justinian had sent Belisarius with Gothic troops (also called Belisarius' ἑπόμενοι) to the east (2.14.8-13). Setting out from Dara, they first camped near Nisibis, hoping to lure out the Persian garrison and beat them in the field before commencing a siege. However, some of the Roman commanders allowed their soldiers to behave with indiscipline due to the heat and were hence defeated by a Persian surprise attack. The situation was only saved by a Gothic cavalry charge, but the Romans gave up on Nisibis (2.18.1-19.1) and instead began a siege of Sisauranon, which was held by a garrison of 800 Persian cavalry. At first, the Romans attempted to storm the fort, but were repulsed with heavy losses. Instead Belisarius decided on a blockade, and in order to shield his besieging force, he sent a raiding party of Arabs ("For the Saracens are by nature unable to storm a wall, but the cleverest of all men at plundering." Σαρακηνοὶ γὰρ τειχομαχεῖν μέν εἰσιν ἀδύνατοι φύσει, ἐς δὲ τὸ ληΐζεσθαι πάντων μάλιστα δεξιοί. 2.19.12) and 1,200 Roman regulars, most of which were Belisarius' own *hypaspistai*, across the Tigris. They received intelligence from a captured Persian that Sisauranon had almost no supplies, and Belisarius used this knowledge in negotiating a surrender with the Persian garrison, which was then shipped off to fight for the Romans in Italy.

The Romans ended their invasion due to serious illness in the Roman camp (the sick had to be transported back to Roman territory in carts) and the failure of the raiding party to return or give further information on their movements. Procopius alleges that the Arabs wanted to keep all the booty, so they fed the Romans who were with them with false information on Persian movements in order to make them take a different route back to Roman territory while the Arabs could made off with the booty (2.19.26-46)

VERONA   παραδοῦναι...χρήμασιν   541
Proc. 7.3.6-22; Wolfram 1988: 354.

A Roman army of 12,000 under eleven commanders assembled to defeat the resurgent Goths (7.3.4). A local notable arranged the betrayal of Verona through one of his clients, who guarded one of its gates. The Roman force that took over his gate were the Persians who had surrendered at

*Sisauranon, now under an Armenian commander. The Gothic garrison fled through another gate when they discovered the betrayal. After dawn, however, the Gothic garrison returned through the same gate they had fled and surprised the Romans inside Verona, since they observed that the Romans were poorly organized and most of the army was camped some distance away, preoccupied with a squabble over how to divide the booty. This led to a strange fight where the Romans on the walls defended themselves against an assault from the *inside*, and had to jump out to save themselves.

FLORENCE   ἐγκαθεζόμενοι ἀμφὶ τὸ τεῖχος ἐς πολιορκίαν καθίσταντο   c. 542
Proc 7.5.1-6.

After the success at *Verona (541), the newly elected Ostrogothic king Totila assembled 5,000 men (7.4.1) and defeated a large Roman army in battle (7.4.). He then sent some of his troops to besiege Florence (ἐγκαθεζόμενοι ἀμφὶ τὸ τεῖχος ἐς πολιορκιαν καθίσταντο), but the Goths withdrew to a nearby valley when scouts announced the arrival of a relieving army from Ravenna. There they again defeated the Romans in battle, leading to poor morale amongst Roman garrisons who stayed inside their fortifications. Totila in contrast persuaded most of his prisoners to join his cause. (7.5.7-19)

CESENA   τὰ φρούρια εἷλεν   542
Proc. 7.6.1.

"After this [the siege of *Florence] Totila took the fortresses of Cesena and Petra."

PETRA PERTUSA   542
See *Cesena above.

BENEVENTO   παρεστήσατο οὐδενὶ πόνῳ   542
Proc. 7.6.1; Wolfram 1988: 355.

From *Petra, Totila advanced through Tuscany to Campania and Samnium, "and with no trouble won the strong city of Beneventum, the walls of which he razed to the ground" so that it could not serve as a Roman base later.

NAPLES   πολιορκεῖν ἔγνω, ἐνστρατοπεδευσάμενος, ἀνάγκῃ τοῦ λιμοῦ   542f
Proc. 7.6.2-8.14; Av. Cameron 1985; Wolfram 1988: 355.

Naples had a garrison of 1,000 Romans and Isaurians and good hope of reinforcements. Having established camp near the city, Totila sent forces to take control of much of southern Italy (see also *Cumae below). This allowed him to control tax revenues, which made Roman troops elsewhere, already demoralized by their recent defeats, even more unwilling to mobilize due to lack of pay. Furthermore, two fleets with supplies and relieving armies were sent from the East but failed. One stopped in Epirus, while the other came to Sicily but was heavily defeated by the Goths in a naval battle. The fleet at Epirus finally arrived after a long delay, but a storm blew it ashore just beside the Gothic camp. After months of blockade, starvation and psychological warfare (such as forcing a captured Roman commander to persuade the inhabitants to surrender, and giving the city three months instead of one to wait for reinforcements), the besieged surrendered early in 543. Totila treated the civilians well, ensuring recovery from famine by gradual feeding and preventing rape and pillage. He also gave free passage and supplies to Roman soldiers, whom he had fraternize ("mix," ἀνεμίγνυσθαι) with the Gothic army befor they departed for Rome. Totila's final acts were to raze the walls and execute a Gothic soldier for raping a Roman girl.

The city avoided storm and was treated extremely leniently when captured, since Totila was actively trying to win over the Italian population and at least some of the Roman soldiers to his side, cf. the treatment of Roman soldiers at *Florence, senators' wives at *Cumae. According to Procopius, he also wanted to force the Romans to battle (having recently won in several field engagements) by depriving them of their bases, which the numerically inferior Romans had used so efficiently during their conquest.

CUMAE AND OTHER STRONGHOLDS	εἷλε	542

Proc. 7.6.3.

During the siege of *Naples, Totila "sent off a part of the army and captured the fortress of Cumae and certain other strongholds (Κύμην τε τὸ φρούριον καὶ ἄλλα ἄττα ὀχυρώματα εἷλε)," where he found much money and the senators' wives, whom he released in order to win favor among the Romans.

SERGIOPOLIS (RESAFA)	ἔς τε πολιορκίαν καθίστασθαι	542

Proc. 2.20.1-16.

Khusro, having returned from *Petra (541), began a new Persian offensive, which this time aimed at Palestine as Syria had been ravaged in 540. The bishop of Sergiopolis, learning that the Persians were on the way, went to the Persian camp in order to negotiate an agreement. However, Khusro detained and tortured him for failing to live up to the agreement made at *Sura (540), and demanded the city hand over all valuables. As this was being implemented in several rounds by Persian envoys, the citizens of Sergiopolis were warned by a Christian Arab in the Persian army that the Persians would exploit the situation to take control of the city. When the citizens prevented the Persian envoys from entering to take the last valuables, Khusro sent an army of 6,000 that was "to begin a siege and to make assaults upon the fortifications (ἔς τε πολιορκίαν καθίστασθαι καὶ προσβολὰς τῷ περιβόλῳ ποιήσασθαι, 2.20.12)." There was spirited resistance from the small garrison of 200 *stratiotai* supported by the *Sergioupolitai*, but soon the population became terrified and considered surrendering. The same Christian Arab, however, approached the walls at night, warning the citizens that the large Persian army was running out of water and would abandon the siege within two days. Thus the city was spared, but the bishop remained in Persian captivity.

KALLINIKOS	εἷλε, ἐς ἔδαφος καθεῖλεν	542

Proc. 2.21.30-33.

As the Persians moved on from *Sergiopolis, the Roman garrisons at first stayed in their fortifications. When Belisarius arrived, he restored morale and organized a field army from local resources and sent 1,000 men to hold the east bank of the Euphrates. This convinced Khusro to accept negotiations (2.20.17-21.20), and as soon as the troops across the Euphrates had been ordered to remain "quiet," the Persians again built a bridge across the Euphrates. Having crossed, they continued negotiations while the Roman army followed them. One condition that Khusro agreed to was to "treat the Romans as his friends" in return for a high-ranking Roman hostage (2.21.21-29). "But in the meantime, Khusro, disregarding the agreement, took (εἷλε) the city of Kallinikos which was entirely without defenders. For the Romans, seeing that the wall of this city was altogether unsound and easy to capture (περίβολον...σαθρόν τε καὶ εὐάλωτον), were tearing down portions of it in turn and restoring them with new construction." They had just torn down such a section when they learnt of the Persian advance, so the wealthy fled, leaving a "great number of farmers" in the city. The Persians enslaved them, razed the city and returned to their own territory. Note the contrast to Roman strategy in face of the Arabs in Appendix I.

| Anglon | ἐς χεῖρας τοῖς πολεμίοις ἦλθεν | 543 |

Proc. 2.24.10-25.35.

In response to the Persian invasions, the Romans made a counter-attack against Persarmenia. Assembling near the border, the Romans numbered 30,000 (2.24.16), but invaded in separate, uncoordinated detachments. The Persian army withdrew to a mountain-perched fort surrounded by a village called Anglon, where they blocked up the village entrances, dug trenches, set up ambushes in the houses and marshaled their army below. The Roman army had mostly reassembled by the time they were approaching the Persian fort, but were marching with poor discipline: at first they held together in strict formation without plundering any Persian territory, but when they found out that the Persian army had assembled at Anglon they spread out to plunder freely. They probably assumed that the Persians were hiding in the fortifications, but soon discovered the Persian army marshaled out in the open. The Roman commanders hastily reassembled their troops in battle formation and defeated part of the Persian army, but the Heruls who pursued them through the village were caught in the ambushes and heavily defeated. Then the Persians sallied out from the fortifications and completely routed the Romans, who panicked and fled, leaving all of their heavy equipment (arms, armor, draft animals) behind.

Due to this unmitigated disaster, Procopius is rightfully critical, but several aspects seem distorted. The division of the army and its staggered deployment may have had sound strategic reasons, such as logistical constraints and the need for mutually supportive columns. The transition from strict formation (to avoid ambushes) to one more suitable for pillaging the countryside was also sound based on the intelligence the Romans possessed. It may also have been that the soldiers, who were now "mixed in with the baggage train (2.25.14)" were actually moving to protect valuable siege equipment and supplies, since they expected to commence a siege and may have anticipated ambushes and sallies. Despite Procopius' criticism that the Romans were poorly organized and led, they managed to deploy to face the Persian army in battle and win the first clash before falling into ambushes.

| Edessa | προσεδρεία | 544 |

Proc. 2.26.-27.

The Persian army was accompanied by Hun mercenaries, whom they unleashed on sheep that pastured just outside the walls. When the shepherds defended their flocks (one *agroikos* showing great skill with his sling), both the Huns and the Romans sent reinforcements, leading to a major encounter. Both parties withdrew as the whole Persian army arrived and set up camp. When negotiations for a substantial ransom failed, the Persian *tekhnitai* began building a large mound to overtop the city wall from the 8th day of the siege, using timber, stone and earth. At first the Romans sent out their own Hunnic troops to stop the works, but afterwards the Persians strengthened their security. When the mound came closer to the wall, the Romans used slings and archery (σφενδόνας...τόξα), but the Persians responded with stretching "Cilician" mats, made of goat's hair cloth on a wooden frame, to cover the workers. Seeing this effective, the Romans reopened negotiations, but Khusro demanded all the wealth in the city to depart, and cited as justification the failure of Justinian to send envoys after his last campaign in 542 (cf. *Sergiopolis, *Kallinikos). The Romans also tried to build a counterstructure on top of the wall facing the mound, but this was abandoned since the mound was rising higher and presumably the Persians could shoot down on Roman workers. The Romans also dug a tunnel from within their walls under the mound, but when they were poised to begin undermining it, the Persians discovered the sounds made by the Roman sappers (τοὺς ὀρύσσοντας) and made their own countertunnels from each side of the mound. The Romans abandoned the attempt and filled in their own tunnel; instead they opened a section under their wall right in front of the mound, which they again began to undermine and filled with flammable materials soaked in oil, sulphur and bitumen. When the next round of negotiations failed and the Persians were set to storm the city, the Romans set fire to the mound and camouflaged the smoke by shooting fire arrows and incendiary projectiles all over the mound. The Persians finally realized what was happening and sent a large force to put out the fire, but the Romans, presumably attacking from the parapet, gained control of the mound which was then consumed by flames.

Despite this setback the Persians persisted, at first with a surprise night attack with ladders that was discovered by one of the farmers who was guarding the wall; next they filled in the moat to prepare an assault with engines. "And at every gate he stationed some of the commanders and a part of the army, encircling the whole wall in this way, and he brought up ladders and war-engines against it (κλίμακάς τε...καὶ μηχανὰς προσῆγεν,

2.27.29)." Among unspecified machines were great siege towers (πύργους καὶ τὰς ἄλλας μηχανάς). However, the Roman soldiers and civilians resisted bravely and repelled most of the Persian assaults, except for one section where the Persians breached the outer wall (προτείχισμα). While the soldiers and civilian men of military age held the walls, women, children and the elderly carried stones, boiled olive oil and otherwise supported those fighting on the wall. The farmers were particularly brave in the wallfighting, while the Persian breach was defeated by Roman regulars who were accompanied by some of the citizens (ξὺν...τῶν Ἐδεσσήνων τισιν). Two days later, the Persians made a final attempt to storm a section of the wall, but when this was repulsed by a sally, Khusro gave a written guarantee of safety to the population and withdrew after receiving money and burning his siegeworks (χαρακώματα).

Artillery is never specifically mentioned by Procopius, but its presence can be inferred from the narrative. The Romans carried stones on to the walls in large numbers and shot pots filled with incendiary materials at the mound. While the stones could be for the slingers, the pots probably required heavier ordinance. The Persians were able to approach the walls on several occasions and at one section broke through the outer wall; thus their "other machines" could encompass anything from siege sheds and rams to artillery. While the Huns were used actively by both sides, the Persians' Arab troops were kept in the rear during the main assaults in order to capture any fugitives.

LARIBUS	πολιορκούμενοι, προσεδρεία	544

Proc. 4.22.19f.

The Moors "are not at all practiced in the storming of walls (τειχομαχεῖν οὐδαμῇ ἤσκεται)," but caught the city unprepared with no provisions stored. Hence the Romans bought off the Moors with a large sum of gold.

HADRAMETUM (twice)	ἐξαπατηθέντες...τοὺς πολεμίους ἐδέχοντο	544

Proc. 4.23.1-17; 18-32.

The resurgent Moors captured some Roman troops, who were handed over to the Roman rebel Stotzas. They helped him and the Moors take Hadrametum by surprise, posing as victorious Romans bringing in a group of Moorish captives, but instead they took control of the gates and let in the whole Moorish army. Shortly after, however, a Roman army retook the city with the active help of the bishop and notables of the city; this prompted the Moors and rebel Stotzas to aggressive raiding against the Roman countryside.

6TH CENTURY

CARTHAGE [mobilization of troops] 544
Proc. 4.26.1f.

During the complex political and military maneuvers recorded in 4.25.-28., the "tyrant" Gontharis was at this point attempting to intimidate the loyalists under Areobindus as well as his own Moorish allies (who had marched against the city at his behest) by placing armored men with bows on the walls in great numbers. However, there was no fighting between any of the parties.

OTRANTO ἀποπειράσασθαι, ἐς πολιορκίαν καθίσασθαι 544
Proc. 7.9.22-10.12; Wolfram 1988: 355.

Totila sent troops to besiege Otranto while he himself advanced on central Italy, first to "make a trial" (ἀποπειράσασθαι, i.e. to storm), but when the resistance proved too strong, he ordered the Goths settled down for a siege (ἐς πολιορκίαν καθίστασθαι). Eventually the garrison was on the verge of surrendering. However, Belisarius was sent from the East despite Persian pressure, recruiting 4,000 fresh troops in Illyricum with cash bounties. While advancing overland to Salona, he had the starving garrison of Otranto replaced and new supplies brought in before the agreed date of surrender. The Goths did not guard the harbor but fled at the sight of the fleet. A Roman follow-up raid from Otranto was defeated and the Romans lost 170 fallen. However, Otranto was secure and the fleet with the emaciated garrison returned to Salona.

TIBUR προδοσίᾳ εἷλε 544
Proc. 7.10.19-23.

The local inhabitants (τῶν τινες οἰκητόρων) who were guarding the gates fell out with their Isaurian fellow-guards, and betrayed the city to the Goths, who were camping nearby. However, the Isaurians escaped before the Goths entered, but all males in the city were massacred. Rome was thus cut off from Tuscany.

FORTS AROUND BOLOGNA ὁμολογίᾳ ἑλών 544
Proc. 7.11.11-18.

Finally arriving at Ravenna, Belisarius sent some of the Illyrians who were stationed in Italy to secure Emilia, "and after taking some of the neighboring fortresses by surrender, remained inactive in Bologna." Due to lack of pay and news of a Hunnic invasion in Illyricum, the Illyrians turned homewards. This left the remaining Romans to face Totila's army that was heading for Bologna, but well prepared ambushes drove the Goths back.

Osimo	πολιορκουμένοις	544

Proc. 7.11.19-31.

As Osimo was besieged by the Goths, Belisarius sent 1,000 men in relief. They snuck past the Gothic camp at night, entered the city, and began to make plans of sallies, but it soon became apparent that the Goths were too numerous to confront openly, and the 1,000 extra men would cause the supplies to run out much faster. The reinforcements decided to escape at night but a deserter alerted the Goths who ambushed the Romans. They escaped but lost 200 men along with their baggage train and servants (θεραπεία).

Pesaro	ἀποπειρασάμενος...ἐξελεῖν οὐχ οἷός τε ἦν	544

Proc. 7.11.32-36.

The Romans rebuilt a fort, partly torn down by Totila, on the Adriatic between Rimini and Osimo. Prefabricated gates helped close the circuit wall. Totila "made an attempt on the town and tarried near it for some time, but since he was unable to capture it, he returned baffled to his camp at *Osimo."

Fermo	ἐνστρατοπεδευσάμενοι ἐς πολιορκίαν καθίσταντο, ὁμολογίᾳ εἷλεν	544f

Proc. 7.11.37-39, 7.12.12.

With all Roman garrisons confined to their fortresses, either by Belisarius' orders or by fear, the Goths were free to choose their targets. "They accordingly made camp in Picenum before Fermo and Ascoli and commenced a siege." The Romans were unable to provide any relief forces (7.12.1ff), so by spring 545 both Fermo and Ascoli surrendered.

Ascoli		544f

See *Fermo above.

Spoleto	ἐπολιόρκει, ἐς λόγους ἦλθεν	545

Proc. 7.12.12-16, 7.23.3.

From *Fermo and *Ascoli (544f), Totila advanced to Tuscany and "began the siege of Spoleto and Assisi." At Spoleto, the commander made an agreement to surrender after waiting a month for relief. Enmity between him and Belisarius may have contributed to the relatively rapid surrender. Totila tore down the city walls and fortified the amphitheater, probably due to demographic changes as the result of plague in combination with insecurity about the loyalties of the urban population.

| Assisi | ἐπολιόρκει, τὴν πόλιν...ἐνέδοσαν | 545 |

Proc. 7.12.12, 17f; *Spoleto above.

The inhabitants of Assisi despaired and surrendered to the Goths when most of the Roman garrison was wiped out during a sally.

| Perugia | ἐπολιόρκουν ἐς τὸ ἀκριβές | 545ff |

Proc. 7.12.18ff, 7.25.1ff, 24, 7.35.2.

Perugia refused to surrender to the Goths, even after the assassination of its commander in 545. Sometime later (when is not specified, but sometime after *Rome 545 began and well before *Rome 546f began), the Goths arrived to pursue a close siege of the city. After Totila's failure before *Rome (546f), he arrived himself and tightened the siege, making camp "hard by the circuit wall," and some time before Belisarius returned to Constantinople in 549, the city was taken by storm.

| Piacenza | ἐνστρατοπεδευσάμενοι ἐς πολιορκίαν καθίσταντο | 545 |

Proc. 7.13.8-11, 7.16.2f.

While the Goths where closing in on *Rome, they also established a close siege of Piacenza, "perceiving that the people in the city were in need of provisions." The city only surrendered after being reduced to cannibalism.

| Rome | ἐς πολιορκίαν καθίστατο | 545f |

Proc. 7.13.1-7, 7.15.-22.; cf. *Tibur 544 above.

When Gothic troops began appearing in the vicinity of Rome, a Roman sally to drive them off was ambushed after initial success. This destroyed the morale of the defenders. However, Totila banned raiding and thus ensured that the farmers in the surrounding countryside were still productive and provided the Goths with taxes and supplies. Furthermore, a Gothic fleet at Naples and in the Aeolian islands intercepted supply ships from Sicily. Belisarius now retreated towards Epidamnus to assemble reinforcements while Narses recruited Heruls for the forthcoming Italian campaign (7.13.12-21); some troops were sent to reinforce the Roman garrison at Portus and try to break the Gothic siege. The Romans made several attempts on the Gothic fortified camp (χαρακώματι), but Bessas, the commander of Rome, refused to send his troops in assistance. Thus, on the second raid, the Goths, warned by a deserter, set ambushes and nearly wiped out the Roman force. This allowed the Goths to intercept further supplies from Sicily, which had been organized by the Pope, since they could take control of the harbor without resistance from the remaining Roman troops that held Portus. The ships' crews misunderstood the warnings from Roman soldiers on the fortifications as rejoicing and sailed right into the trap.

In the spring of 546, the Romans were hard pressed and made overtures to Totila, but negotiations failed (7.16.). Famine was worsened by corruption, as the officers controlled the warehouses and sold supplies to the wealthy at exorbitant prices. After a spate of suicides, most of the starving inhabitants were allowed to escape, but many were captured by the Goths or perished otherwise (7.17.). Belisarius arrived at Portus with troops and supplies, but forwarned, Totila had the Goths build a fortified bridge across the Tiber (7.18.8ff). The Roman reinforcements were not strong enough to break through the Gothic blockade overland, so instead they built a large tower on two broad skiffs as well as two hundred other boats with wooden walls and arrow slits. These were loaded up with grain and soldiers, while the remaining troops guarded Portus and its approaches. The floating tower had a small boat on top filled with pitch, sulphur, resin (7.19.10). Advancing upriver, the fleet destroyed the first obstacle, a chain suspended across the river and heavily guarded. Then the Romans attacked the fortified bridge. After a heavy exchange of archery fire, the Romans burnt one of the towers—which must have been quite large, as 200 Goths perished in the flames—and broke the bridge. They also repulsed Gothic reinforcements from their fortified camps (ἐκ τῶν χαρακωμάτων). Belisarius had ordered a sally from Rome against the camp when the Goths had been drawn out, but since the officers at Rome were making money on the siege, they saw no reason to end it. Instead, some of the troops at Portus wanted part in the glory and assaulted the Gothic camp. However, when they were killed and the news reported to Belisarius, he turned back the fleet, believing that Portus had fallen (7.19.).

After this failure, some Isaurian troops betrayed Rome. Due to a breakdown in morale and discipline, most troops were unsupervised and lax in their guard duties. Thus the Goths managed to enter by night, climbing ropes thrown down the walls where the Isaurians were on duty. A general massacre was avoided, Totila forbade the Goths from raping any women, "although the Goths were extremely eager to have intercourse with them," and protected Boethius' widow from retaliation even though she had destroyed Theoderic's statue (7.20.). After settling affairs in the city, he had a third of its walls razed (7.22.7) and marched to southern Italy.

Spoleto 546
Proc. 7.23.3-7; cf. *Spoleto 545.

[The amphitheater now serving as the main fortification at] Spoleto was held by a garrison of Goths and ex-Roman soldiers. A Roman officer was able to convince some of the latter to return to Roman allegiance, so they betrayed the city and allowed a Roman army to enter.

Rome   ἐνστρατοπεδευσάμενοι, τειχομαχεῖν ἔγνω   546(f)
Proc. 7.24.

Belisarius retook Rome and had the "whole army" rebuild the broken sections of wall in only 25 days by using stakes to support the structure instead of mortar. Since there were now plenty supplies, the population returned. Totila advanced rapidly in order to storm before the gates were refitted—this had not been achieved for lack of artisans (τεχνιτῶν ἀπορίᾳ, 7.27.9). When the Goths arrived, very heavy fighting ensued over several days. In the first engagement, the Romans set their best forces to hold the open gates and with soldiers manning the walls they could shoot at the Goths from above. When the Goths finally withdrew, the Romans lay out large numbers of triboloi, four-pronged spikes, to cover the gateways in case of a surprise attack. The next day saw hard fighting at the walls, but the Goths were repulsed by Roman sally. A last attempt was defeated in a field battle; after this, the Goths withdrew to the fort of Tibur, which they rebuilt, while Belisarius could finally refit the gates and send the keys to Justinian.

Septem   *magna vi certaminis expugnarent*   c. 547
Isidore *HG* 42; Thompson 1969: 16.

"After this fortunate victory [see *Saragossa 541], the Goths undertook an ill-advised campaign across the straits of Cádiz. They crossed over to do battle with the soldiers who had assaulted the city of Ceuta and had expelled the Goths. The Goths initially assaulted the fortress with great power (*eundemque castrum magna vi certaminis expugnarent*). But on the following Sunday they put down their arms so as not to defile the sacred day with fighting. Seizing this opportunity, the opposing forces made a surprise attack and laid low the invading army, trapped as it was between land and sea, so that not a single man escaped death in the ensuing massacre."

ROSSANO   ἐστρατοπεδεύσατο...ἐγκαθεζόμενος ἐς πολιορκίαν καθίστατο   548
Proc. 7. 29.21, 7.30.5f, 9-24.

Learning that the Romans in the city were running low on supplies, the Goths established a siege at Rossano. When a fleet of reinforcements was driven back from the shore by Gothic cavalry, the Roman garrison of 300 Illyrian cavalry and 100 infantry surrendered on terms. Totila encouraged the Romans, in return for keeping their property, to join the Gothic army "on terms of complete equality; indeed this was the same procedure which he had regularly followed when the other strongholds were captured (7.30.21)." Only 80 men opted to return to Roman territory.

The Gothic besieging army must have been quite large and diverse. Those who opposed the fleet's landing were cavalry armed with spears or bows, just as Roman regulars, and many enough to completely discourage the landing of a "great fleet." Furthermore, Totila detached 2,000 men from his besieging force to go after a Roman army that was on its way to relieve unnamed cities that were under siege in Picenum.

MOUIKOURON   548(f)
Proc. 7. 35. 23ff.

Indulf, one of Belisarius' most experienced *doryphoroi*, deserted to the Goths, and was appointed commander of a large army and fleet that was sent against Dalmatia late in 548 or early 549. Exploiting his status as Roman commander, Indulf and his men were welcomed into Mouikouron, near Salona, but once inside, took control of the city and plundered it.

LAUREATE   548(f)
Proc. 7.35.26-29.

From *Mouikouron, the Gothic fleet raided Laureate. The Roman commander at Salona sent *dromones* against them, but the Roman fleet was defeated in the harbor and the Romans abandoned their ships.

ROME   ἐγκαθεζόμενος εἰς πολιορκίαν καθίστατο, τειχομαχεῖν   549f
Proc. 7.36.1-37.4.

"Totila now led his whole army against Rome, and establishing himself there entered upon a siege. But Belisarius had selected three thousand men noted for their valour and appointed them to garrison Rome, placing in command of them Diogenes, one of his own spearmen, a man of unusual discretion and an able warrior. Consequently a long time was consumed in the blockade (τῇ προσεδρείᾳ). For the besieged, on their part, showed themselves, thanks to their extraordinary valour, a match for the entire Gothic army, while Diogenes was ever keeping a strict watch that no one should approach the wall to damage it; furthermore, he sowed grain in all parts of the city inside the circuit-wall and so brought it about that they had not the least shortage of food. Many times indeed the barbarians attempted to storm the fortifications (τειχομαχεῖν) and make trial of the circuit-wall, but they were always repulsed, being driven back from the wall by the valour of the Romans. They did, however, capture Portus, and thereafter held Rome under close siege. Such was the course of these events." (7.36.1ff)

Justinan had failed to assign a replacement for Belisarius or send reinforcements. Hence, some unpaid Isaurian soldiers in the garrison, seeing how well other Isaurian deserters had done in the Gothic army, agreed with Totila to betray the city. The Goths arranged a diversion that drew most of the troops to another sector of the wall, but gained entry through the gate where the Isaurians were now on their own. Some of the garrison was massacred and others fled towards Centumcellae, but many were ambushed on the way. One last group refused to surrender, led by the Cilician Paulus, formerly in charge of Belisarius' household (ὃς τὰ μὲν πρῶτα ἐφειστήκει τῇ Βελισαρίου οἰκίᾳ), now in charge of a cavalry battalion (καταλόγου ἱππικοῦ ἄρχων) and co-commander with Diogenes, who was also an officer of Belisarius' household. His 400 men made a last desperate stand at Hadrian's mausoleum. Finding themselves without supplies, they decided to charge out rather than starve, but at the last minute received an offer from Totila to leave for the East without weapons and horses or keep it all and join the Gothic army (τῷ Γότθων στρατῷ ἀνεμίγνυντο). Most of the soldiers preferred the latter alternative rather than return as infantry (πεζοί), especially as they had not received payment for a long time. 300 other survivors received the same terms.

Afterwards, taunted by the Franks for his incompetence and lack of control, Totila decided to rebuild Rome and resettle the senators in his control there. Before setting out on his next campaign, he also presided over the hippodrome races (7.37.1-4).

Note how household commanders (*doryphoroi*) could be reassigned from household to regular units as a matter of course.

| CENTUMCELLAE | ἐνστρ. ἐς πολιορκίαν καθίσταντο, ἐνδοῦναι(?) | 549f |

Proc. 7.37.8-18, 739.25-28.

Totila besieged Centumcellae, offering terms of free passage or mingling with the Goths, but the soldiers had wives and children in the East, making it impossible for them to defect. Concerning a surrender, Totila agreed to wait until the garrison had received orders from the emperor. In the meantime he moved on towards the south and Sicily with a fleet of 400 warships as well as large ships captured from the Romans. By the end of the year or early 550, he returned to demand the surrender of the city, but the commander Diogenes (who had held command at *Rome 549) refused to surrender as he heard that a large Roman army was assembling to march on Italy.

Procopius never mentions the fall of Centumcellae to the Goths or the fate of Diogenes, but has Narses send soldiers to recapture the city in 552 (8.34.20), so it must have surrendered sometime in 550 or 551.

| REGGIO | τειχομαχοῦντας, ὁμολογίᾳ ἐνέδοσαν | 549 |

Proc. 7.37.19-23, 7.39.1f, 5.

The Goths wanted to secure Reggio before crossing to Sicily. They therefore tried to storm the city (τειχομαχοῦντας) in order to achieve this, but a successful Roman sally drove them from the walls. The Goths were too numerous to defeat, however, and established a blockade. There may have been another assault after *Taranto and *Rimini, unless Procopius is only repeating himself after a digression (compare 7.37.21f with 7.39.1f). Again the Goths settled down for a blockade while sending troops to *Messina. The garrison at Reggio surrendered when they ran out of supplies.

| TARANTO | παρεστήσατο οὐδενὶ πόνῳ | 549 |

Proc. 7.37.23.

While keeping *Reggio under blockade, Totila "sent an army against Taranto and took over the fortress there with no difficulty..." ἐς δὲ Ταραντηνοὺς στράτευμα πέμψας τὸ ἐκείνῃ φρούριον παρεστήσατο οὐδενὶ πόνῳ...

| RIMINI | προδοσίᾳ εἷλον | 549 |

Proc. 7.37.23.

"... likewise the Goths whom [Totila] had left in the land of Picenum also took the city of Rimini at that time [i.e. *Reggio and *Taranto], for it was betrayed to them." ...καὶ Γότθοι δὲ οὕσπερ ἐλίπετο ἐν Πικηνῶν τῇ χώρᾳ

6TH CENTURY

πόλιν Ἀρίμινον τηνικάδε προδοσίᾳ εἷλον. A Roman counterattack organized from Ravenna failed.

| MESSINA | προσέβαλε τείχει | 549 |

Proc. 7.39.2ff.

The Goths crossed over from *Reggio and "delivered an attack on the wall (προσέβαλε τείχει) of Messana." The Romans went out to face them in the open, but returned to the fortifications; the Goths were then free to plunder "practically the whole of Sicily."

| TOPEIROS; OTHERS | πολιορκίᾳ...εἷλον, τειχομαχήσαντες, κατὰ κράτος | 549 |

Proc. 7.38.

A Slav invasion of 3,000 men in two groups (1,200 and 1,800 men, respectively) defeated several armies in Thrace and Illyria, one of them even the (main?) cavalry divisions (τῶν δὲ ἱππικῶν καταλόγων) stationed at Tzurullum. In the process the Slavs plundered towns (χωρία) all over Thrace and Illyria "... and both armies captured many fortresses by siege (καὶ φρούρια πολλὰ πολιορκίᾳ ἑκάτεροι εἷλον), though they neither had any previous experience in attacking city walls (οὔτε τειχομαχήσαντες πρότερον), nor had they dared to come down to the open plain, since these barbarians had never, in fact, even attempted to overrun the land of the Romans. Indeed it appears that they have never in all time crossed the Ister River with an army before the occasion which I have mentioned above." (7.38.7f)

The latter army reached Topeiros in Thrace, near the Aegean, where the garrison of *stratiōtai* was lured out and ambushed. "But the inhabitants (οἰκήτορες) of the city, deprived as they were of the support of the soldiers, found themselves in a very difficult situation, yet even so they warded off the assailants as well as the circumstances permitted. And at first they resisted successfully by heating oil and pitch till it was very hot and pouring it down on those who were attacking the wall, and the whole population joined in hurling stones (λίθων βολαῖς) upon them and thus came not very far from repelling the danger. But finally the barbarians overwhelmed them by a multitude of their missiles and forced them to abandon the battlements, whereupon they placed ladders against the fortifications and so captured the city by storm (ἔπειτα δὲ αὐτοὺς πλήθει βελῶν οἱ βάρβαροι βιασάμενοι ἐκλιπεῖν τε τὰς ἐπάλξεις ἠνάγκασαν καὶ κλίμακας τῷ περιβόλῳ ἐρείσαντες κατὰ κράτος τὴν πόλιν εἷλον)." (7.38.15ff)

Procopius clams that 15,000 men were killed, while the women and children were taken captive at Topeiros. The numbers seem unlikely, especially considering the size of the Slav army (1,800 men in the division that took Topeiros, probably less after several battles). Either the horror was amplified by the proximity to Constantinople, as it was only twelve days away (e.g. Procopius details how the Slavs killed captives by impaling, bludgeoning or burning); otherwise, the numbers may be a grand total for those afflicted by the whole invasion.

| PETRA | ἐνστρ. ...ἐς πολιορκίαν καθίσταντο, τειχομαχούντων | 549 |

Proc. 2.29.10-13, 2.29.33-30.20; Braund 1994: 298f.

As the Persian grip on Lazica became firmer, the Lazi who had submitted to the Persians (cf. *Petra 541) wanted to return to Roman allegiance. The Romans sent troops retake Petra and thus thwart a Persian attempt to establish a Black Sea navy there. (The Persians had even sent timber meant for shipbuilding, claiming it was to be used for artillery at Petra (μηχανὰς ἐν Πέτρας τῷ περιβόλῳ), but this was destroyed by lightning—2.29.1ff.)

The Roman army (7,000) with Tzani (1,000) and Laz allies besieged the Persian garrison of 1,500, who bravely resisted the Roman storm but became worn down by archery/artillery. Although they were well supplied, the Persians endured extremely heavy losses (90% dead or wounded—see below). The Romans could thus approach the wall and undermine a long section, but a large building just behind the wall covered the collapsed section and thus prevented a storm. Despite heavy losses, the Persians persisted, even when a small group of Romans managed to enter the city and proclaim the emperor triumphant. Driving them back, the Persian commander promised to surrender, but this bought enough time for a Persian relieving army to arrive. The Romans had to withdraw, while the Persians only found 150 men unhurt plus 350 wounded inside the city: the rest (1,000 men) were dead. The Persians reinforced the damaged wall with sandbags due to lack of materials and installed a garrison of 3,000.

This could have been another Roman disaster (cf. *Anglon 543), but the Persians needed supplies and their army was later ambushed by the Romans and Lazi, who chased them far into neighboring Iberia by the end of 549. The Romans finally captured *Petra in 550.

| ABASGIAN FORTRESS | ἡλίσκοντο | 549 |

Proc. 8. 9.10-30; Braund 1994: 300f.

In reaction to gradual Roman annexation of their lands through quartering troops and imposing new laws, the Abasgians revolted and sought Persian protection. Justinian sent an army to quell the revolt, but the

Abasgians withdrew to a mountain fortress beyond a narrow gorge called Trachea. They intended to hold the gorge, but the Roman army found a way around by using their fleet, so that they could assault the Abasgians from two sides. The Abasgians fled back to their fortress, but the Romans were so intertwined with the fugitives that they managed to enter the fortress before the guards could close the gates. As the Abasgians fought for their homes and families, they defended themselves from the rooftops inside the city, hurling stones and missiles from above. The Romans defeated them by setting fire to their houses. Only a few escaped; the rest were burnt or taken captive, and the fortification razed.

CÓRDOBA   *adversus... urbem proelium movens*   549/50
Isidore *HG* 42-3 [45 in MGH ed.]; Thompson 1969: 16, 321ff; Collins 2004: 46f.

Because of a rebellion, the new Visigothic king Agila moved "against the city of Córdoba in battle (*iste adversus Cordubensem urbem proelium movens*)." "At the outset of his struggle against the citizens of Córdoba," he desecrated the church of the martyr Acisclus. Isidore attributed his failure to divine intervention; after losing his son and part of his army, presumably in battle, he fled to Mérida, and soon faced rebellion from Athanigild and an East Roman invasion.

PETRA   ἐς πολιορκίαν καθίστατο   550
Proc. 8.11.11-12.28; Braund 1994: 301f.

The siege of Petra pitched 6,000 Romans against 2,300 Persian defenders (8.11.42). The Persians were extremely well supplied with arms (five sets of arms and armor as booty for each Roman, 8.12.17), foodstuffs (enough to last five years), and water supply through hidden aqueduct pipes. After the Persians had taken *Petra in 541, they had laid down three layers of pipes on top of each other. During the siege, the Romans only uncovered the top two and hence did not break the Persian water supply.

At first the Romans undermined a large section of the walls (ἀμφὶ τὸ τεῖχος διώρυσσον) that they partly destroyed during their last attempt the previous year (*Petra 549). The Persians, when repairing the damaged section, had used large timbers instead of foundation-stones to tie the whole structure together, so the undermined section of wall moved in one piece without collapsing or tipping over. Due to the steep slopes of the site, it was impossible for the Romans to bring up the traditional, wheel-mounted rams' sheds, for "this engine (μηχανή) cannot be brought up to a wall except on smooth and very flat ground (8.11.21)." Instead, some Sabir Huns in Roman service invented a lightweight version: a framework of thick wands covered with hides, "which the throng of neither Roman nor Persian engineers (τεχνιτῶν) had thought of." Three were made, equipped with rams' heads taken from "standard" Roman rams, carried by a team of forty Roman soldiers who also operated the ram, and supported by sappers using poles with grappling hooks to pull out damaged masonry. In response, the Persians set up a wooden tower (ξύλινον πύργον) from which they hurled pots of incendiary materials against the roofs of these devices, but the sappers removed them with their grappling poles.

The siege culminated in a Romans storm with archery (ἔβαλλον ἐς τὸ τεῖχος) and ladders against the sunken stretch of wall (κλίμακας πολλὰς ἐς τὸ πεπτωκὸς τοῦ τείχους προῆγε). The Roman general Bessas, who was obese and over 70 years old, made two charges up the ladders, and after falling down the first time had to be dragged away by the foot by his *doryphoroi*, who formed a testudo with their shields to protect him. On Bessas' second charge, which greatly inspired the Roman troops, the Persians asked for terms, but refused Bessas' conditions and struggled on. However, another section of the wall that had been undermined previously suddenly gave in, leading to a violent struggle as both sides rushed to the gap. A few brave Armenians climbed a precipice where the city was considered impregnable, and fought their way onto the parapet, thus opening the way for the Romans. At about the same time, the Persians in the wooden tower were incinerated by their own incendiary bombs as they frantically tried to set fire to the Roman rams. 500 Persians retreated to the acropolis while 730 were captured, of whom only 18 were unhurt. The rest, 1,070 men, had apparently fallen. Almost every Roman soldier had been injured as well due to the hard fighting on the walls. The next day, the Romans tried to convince the Persians in the acropolis to surrender, referring to Christian pity for the vanquished. However, the Persians refused to come out, even after the Romans began throwing incendiary materials (πῦρ τῇ ἀκροπόλει ἐμβέβληνται); they allowed themselves to be incinerated rather than surrender.

Archaeopolis	τειχομαχεῖν	550

Proc. 8.14.4-34, 41-44; Braund 1994: 302-06.

The Persians sent reinforcements towards *Petra, mostly cavalry and eight elephants, that marched through rough terrain, with conscript infantry (*paygān*) leveling the road for cavalry and elephants (cf. *Petra 541). The

Persians were reinforced by 12,000 Sabir Huns, but fearing their great numbers, only kept 4,000 for the campaign. The Roman army in the region counted 12,000, but only 3,000 of them were in Archaeopolis (apparently with some Lazi, as one of them was bribed to set fire to the granaries). The other 9,000 were in a fortified camp at Phasis. Too late to relieve *Petra, and unwilling to besiege Archaeopolis with such a large enemy army in the vicinity, the Persians marched towards them first. However, the Romans fled on boats with their supplies when the Persians approached, so the Persians set fire to their encampment (χαράκωμα) and turned back to Archaeopolis. (8.13.1-10, 21-30)

The Persian commander had the Sabir Huns build large numbers of rams of the same sort as they had for the Romans at *Petra, "because he was quite unable to bring up the customary engines (μηχανὰς μὲν τὰς συνειθισμένας) to the circuit-wall of Archaeopolis, lying as it did along the lower slopes of the hill; for he had heard what had been achieved by the Sabiri who were allies of the Romans at the wall of Petra not long before, and he sought by following the method discovered by them to reap the advantage of their experience (8.14.4)." Daylami infantry, who were experienced mountaineers, attacked from the most inaccessible side with javelins, while the Persians and Huns attacked with arrows, rams and elephants on more level ground. The Romans were almost driven off the parapet, but a sudden Roman sally had a devastating effect as it took the Persians and Huns by surprise at close quarters. They were only armed with bows to shoot at the parapet or were otherwise unarmed inside the rams. An elephant struck by panic further confused the Persian lines. This led to the complete rout of the Persian army. 4,000 Persian soldiers and three of their commanders fell; four standards were lost and sent to Constantinople. Furthermore, 20,000 horses died of exhaustion and starvation due to lack of good fodder on the long march.

The political situation was still sensitive, as the Persians could bribe a Laz inhabitant to set fire to the granaries (8.14.23-28; cf. *Petra 541, 549). Procopius is furthermore extremely critical of the general of the East, Bessas, who had retreated to Armenia and concerned himself with collecting revenue instead of securing the passes. However, Procopius immediately continues to relate how difficult it was to bring supplies to the various border forts (8.13.11-20), while the resounding Roman victory at Archaeopolis, the loss of horses among the Persians, and the great efforts they had to spend in order to secure their supplies afterwards (8.14.45-54) may indicate that it was better to allow the Persians to wear themselves out rather than confront fresh armies close to their home bases. Braund has a good survey of the archaeology of the fortress, which is very well preserved.

| ANCONA | ἐξελοῦσιν | 551 |

Proc. 8.23.

The Goths besieged Ancona land and sea with 47 ships for some time (perhaps from 550, as it began "much before" the Gothic raid on Corfu in 551, cf. 8.22.17ff, 30ff). Hard pressed and still awaiting the large army assembling in the Balkans, the local Roman commanders organized reinforcements which arrived from Salona (38 ships) and Ravenna (12 ships). In the ensuing naval battle, the Romans proved more competent, ramming single ships and overwhelming crowded ones with archery. Only eleven Gothic ships escaped while the rest were sunk or captured, and many Goths perished. As aresult, the Goths burnt their remaining ships and give up the siege. Ancona was thus relieved and resupplied while reinforcements returned to their bases.

| SICILIAN FORTS | πολιορκῶν | 551 |

Proc. 8.24.1ff; cf. *Messina (549).

Liberius, who commanded the Romans in Sicily, "had laid siege to (πολιορκῶν) those Goths who had been left in the fortresses (ἐν τοῖς ἐκείνῃ φρουρίοις) of the island" and consistently defeated their small numbers whenever they sallied.

| CARANALIS | ἐνστρατοπεδευσάμενοι ἐς πολιορκίαν καθίστασθαι | 551 |

Proc. 8. 24. 31-37.

A Gothic fleet invaded Sardinia and Corsica, subjecting them to Gothic rule. A Roman fleet from Carthage was sent to Sardinia, and was preparing to besiege the Goths at Caranalis with a blockade, "for they did not consider themselves able to storm the wall (τειχομαχεῖν γὰρ οὐκ ᾤοντο οἷοί τε εἶναι), since the Goths had a sufficient garrison there." However, the Goths made a successful sally and drove off the Romans, who returned to Carthage, intending to make a new expedition the next year better prepared.

| CROTONE | πικρότατα...πολιορκούμενοι | 551f |

Proc. 8.25.24-26.4.

The Goths were tightly blockading Crotone. The population and Roman garrison were starving and asked for help from Sicily but none was forthcoming, even though the Romans were on the verge of surrendering. Early in the spring, however, the garrison at Thermopylai sent to reinforce Crotone. The Goths fled when they saw the fleet approaching. Subsequently high-ranking Goths and their soldiers began to negotiate the surrender of Taranto and Otranto to the emperor, but at least *Taranto (552) decided to hold out.

| ?RIMINI | [bypassed] | 552 |

Proc. 8.28.

Narses' vast army, setting out from Ravenna, pushed towards Rimini. The bridge crossing the river below the fortifications had been broken down by the Goths, and in order to stop the Romans from progressing further, the garrison at Rimini sallied out against them. The Heruls in Roman service defeated the Goths and killed their commander. Narses decided against a siege; instead he rebuilt the bridge and marched on to defeat the Goths at Busta Gallorum.

| NARNI | ὁμολογίᾳ εἷλε | 552 |

Proc. 8.33.9.

After his victory at Busta Gallorum, Narses marched into Tuscany and "took Narni by surrender and left a garrison at Spoleto," where he had the walls rebuilt.

| PERUGIA | παρέδωκε | 552 |

Proc. 8.33.10ff.

The "Gothic" garrison at Perugia was in fact commanded by two Roman deserters, one who wished to surrender, the other fight on. This led to infighting between the two parties, but those wishing to return to the Roman side won and handed over the city.

| ROME | προσβολή | 552 |

Proc. 8.33.13-27

The small Gothic garrison at Rome stashed their valuables in Hadrian's mausoleum, which had been turned into a fortress, but tried to hold the whole city. The Romans were also too few to cover the whole wall, so they attacked at intervals with a large number of archers (τοξοτῶν...μέγα τι χρῆμα). The Goths repelled most attacks but were forced to leave large sections unmanned. This allowed other units of the Roman army to scale with ladders. Gothic resistance crumbled, and those at the fortress surrendered. Dagisthaeus, who lost *Petra (541), won back Rome by leading the scaling party, while Bessas, who lost *Rome (545), won back *Petra (550).

| Taranto | πανταχόθεν αὐτὸν περιβεβλημένων | 552 |

Proc. 8.34.1-15.

The Gothic commander of Taranto, who had negotiated surrender after the Romans had relieved *Crotone (551f), refused to do so when he heard of the election of Teïas. After deceiving the Romans at Otranto, taking some of their soldiers hostage, the Roman garrison marched out against Goths. The Goths killed their hostages and went out to face the Romans but were defeated, and unable to return to Taranto which was now surrounded by the Romans.

| Portus | πολιορκήσαντες ὁμολογίᾳ εἷλον | 552 |

Proc. 8.34.16.

After the events at *Taranto, "the Romans took Portus by surrender after besieging the place, and likewise a fortress in Tuscany which they call Nepi, as well as the stronghold of Petra Pertusa."

| Nepi | | 552 |

See *Portus above.

| Petra Pertusa | ἐξαιροῦντα | 552 |

Proc. 8.34.24; see also *Portus above.

It is unclear whether the Romans captured it before the troops were recalled to defend the forces besieging *Cumae and the battle of Mons Lactarius: "And when Narses learned this [i.e. of the Gothic advance on *Cumae], he...called back Valerian, who was just capturing Petra Pertusa (καὶ Βαλεριανὸν ἄρτι Πέτραν ἐξαιροῦντα τὴν Περτοῦσαν...μετεκάλει)."

| Centumcellae | πολιορκεῖν | 552f |

Proc. 8.34.20; *Florence 553.

At the same time as the siege of *Cumae began, Narses "sent another force with orders to besiege Centumcellae." It surrendered with other Tuscan cities in the summer of 553; see *Florence 553.

| Gothic fortified camp | | 552 |

Proc. 8.35.7-15.

The Goths under Teïas had arrived in Campania and camped below

Vesuvius, on the banks of a small, but deep stream. Their camp was next to the bridge, "and placing wooden towers upon it they had mounted various engines in them, among them those called *ballistrai* (μηχανάς τε ἄλλας καὶ τὰς βαλλίστρας)" that shot down at anyone who approached. The Romans faced them on the other side of the stream for two months, but because of the deep stream and Gothic towers, there was little fighting apart from archery exchanges. The Goths were well supplied by sea. The situation only changed when their whole fleet was betrayed by one of their own, and Roman fleets arrived from Sicily and the East. By then, Narses had the Romans construct towers of their own, "thus completely humbling the spirit of his opponents." As a result of this pressure, the Goths withdrew to Mons Lactarius, where they were safe, but lacked provisions; they were therefore compelled to join battle and lost.

Procopius is remarkably reticent of how Teïas had reached Campania, what happened during the two-month stand-off, and the significance of the Gothic fleet, which must have operated from Liguria, Tuscany, Corsica and Sardinia (cf. *Caranalis 551). Presumably Procopius wanted to avoid getting into the Frankish invasion, which he must have heard of by the time of the fall of *Cumae. As the Gothic fortified bridge at *Rome (549), the towers may have been very large structures, especially since they supported artillery. His indication of *ballistrai* and "other machines" for shooting at approaching enemies demonstrates that the Goths had the full range of contemporary defensive artillery (see chapter 5.2).

| CUMAE | Over a year | πολιορκήσοντας, ἐξεῖλον | 552f |

Proc. 8.34.20ff, 34; Agathias 1.8.2-10.9; 1.11.5; 1.20.1-7; *PRLE* 3: 912-28 s.v. Narses 1 for detailed dating.

The garrison was held by high-ranking Goths and contained the Gothic treasury. "Narses, then, wishing to capture (ἐξελεῖν) them, sent some men to Cumae to besiege (πολιορκήσοντας) the fortress" in the summer of 552. The standoff at the *Gothic fortified camp and the ensuing battle of Mons Lactarius came as a result of Teïas mobilizing to protect Cumae, and Narses responding to "check the march of his opponents to Campania, in order that the force besiegeing Cumae might be able without fear of molestation to capture it either by storm or by surrender (ἢ βίᾳ ἢ ὁμολογίᾳ, 8.34.22)." Narses called in all his forces, including those guarding Tuscany and those who besieged *Petra Pertusa. Procopius ends his whole work after the battle of Mons Lactarius with a brief reference to the capture of Cumae: "Thus the Romans captured (ἐξεῖλον) Cumae and all that remained, and the eighteenth year, as it closed, brought an end of this Gothic war, the history of which Procopius has written."

Agathias provides the full story where Procopius only has a sentence. Returning victorious from Mons Lactarius, the Romans organized a storm (although no details are given for the preparations; contrast *Onoguris 555): "Meanwhile Narses gave the word of command and instantly led his troops forward. Toiling painfully up the hill they approached the fort, and immediately began to hurl their javelins at those who could be seen manning the battlements. The bows twanged as volley upon volley of arrows was discharged, stones were hurled high into the air from slings and all the appropriate siege-engines were set in motion." Εὐθὺς δὴ οὖν ὁ Ναρσῆς ἐγκελευσάμενος προσῆγε τὸν στρατόν. οἱ δὲ ξὺν πολλῷ πόνῳ ἐς τὸ γεώλοφον ἀναβάντες καὶ τῷ φρουρίῳ πελάσαντες αὐτίκα τοῖς δορατίοις ἠκόντιζον τῶν ἐν ταῖς ἐπάλξεσι φαινομένων, καὶ τὰ τόξα ἐπήχουν θαμὰ τῶν βελῶν ἀναπεμπομένων σφενδόναι τε ἐφέροντο μετάρσιοι καὶ μηχανήματα ὁπόσα πρὸς τειχομαχίαν ἀνεῖται, ἅπαντα ἐκινεῖτο. (1.9.1)

The Gothic response was formidable: "Aligern and his men, who were massed along the stretches of wall between the towers, were not slow to reply with javelins, arrows, huge stones, logs, axes and anything that seemed to serve their purpose. They had their war-engines too, and used them in an all-out effort to beat off the attackers." οἱ δὲ ἀμφὶ τὸν Ἀλίγερνον ἀνὰ τὰ μεταπύργια τοῦ τείχους ξυνειλεγμένοι οὐ σχολαίτερον ἀντακοντίζοντές τε καὶ ἀντιτοξεύοντες διεμάχοντο, λίθους τε ἐκ χειρῶν μεγάλους ἐπαφιέντες καὶ φιτροὺς καὶ πελέκεις καὶ ὅπερ ἂν οὐ πόρρω τῆς χρείας ἐδόκει, μηχαναῖς τε ἐχρῶντο καὶ οἷ ἀμυντερίοις, καὶ οὐδὲν ὅ τι παρεῖτο. (1.9.2)

Due to the fierce Gothic resistance—especially their efficient archery (perhaps also *ballistae*; see below)—the Romans turned to undermining and firing the walls, excavating from a cave that they found in the bedrock under the fortifications. When the wall collapsed, the Romans charged, but debris and Gothic resistance stopped them. Afterwards the Romans decided to settle for a tight blockade while Narses marched on to Tuscany with most of the field army. The Romans "constructed a continuous line of earthworks (καὶ οἱ μὲν χαράκωμά τε περιεβάλλοντο)" to pen in the Goths and starve them into submission, and left a considerable force to enforce the blockade. This was now a year into the siege, which first began late in the summer of 552. Aligern surrendered and joined the Roman side with his troops after Narses had captured *Florence, *Centumcellae, *Volterra, *Luni, *Pisa and *Lucca. The Frankish intervention that began in the summer of 553 helped convince Aligern, since he believed the Franks were only interested in establishing their own rule.

The assault and wallfighting described by Agathias must have begun well after the battle of Mons Lactarius (October 30), the undermining later in the fall or winter of 552/3, and construction of earthworks with palisades conducted over the course of at least several weeks, before the whole army departed. The Gothic war-engines must have included artillery, though some other machines were useful for wallfighting (see chapter 5.2 *passim* and 5.3.1). While Agathias credits Aligern as having a bow-shot powerful enough to split rocks, this sounds more like the *ballistae* described by Procopius at *Rome (537f).

FLORENCE   ἐθελονταὶ...παρέδωσαν   553
Agathias 1.11.1-4, 6; cf. *Centumcellae 552f.

Narses, leaving a force to blockade *Cumae (552f), set out towards Tuscany in the summer of 553 and sent another force to the Po to head off the Frankish invaders. "In a lightning campaign against the cities Narses annexed most of them without encountering any resistance. The Florentines went out to meet him (Φλωρέντιοι μὲν γὰρ ὑπαντιάσαντες), and on receiving an undertaking that they would suffer no ill-treatment, voluntarily surrendered (ἐθελονταὶ...παρέδωσαν) their persons and their property. The inhabitants of Centumcellae (Κεντουκελλαῖοι) did likewise, as did those of Volterra (Βουλοτερραῖοι), Luni (Λουναῖοι) and Pisa (Πισαῖοι)."

After the Gothic losses at Busta Gallorum and Mons Lactarius, there were not enough independent-minded Gothic forces to garrison potentially loyal cities—only 1,000 men broke off to go to Pavia. This left the urban populations with little will to resist (except *Lucca, *Parma below). They would therefore organize a formal *adventus* to receive their new rulers well outside of the city and in the process negotiate a benevolent transfer of power.

VOLTERRA   553
See *Florence above.

LUNI   553
See *Florence above.

PISA   553
See *Florence above.

| LUCCA | three months | ἐς τειχομαχίαν παρασκευάζετο |

Agathias 1.12.; 1.18.4-8; Av. Cameron 1970: 43, 51; chapter 4.1.3.

At first the Romans allowed the Luccans to renege on their commitment to surrender (for which they had already given hostages), and await a Frankish relieving force within 30 days. When the appointed day passed and the Luccans still did not surrender, the Romans "began to prepare for storm." In the meantime, during the blockade phase, which began in September, they used psychological warfare, manipulating public opinion. The city was defended by a Frankish garrison and citizen militia (simply termed Λουκανοί). The citizens were divided in pro- and anti-Roman factions; the anti-Romans had received Frankish support as the Gothic resistance crumbled all over Italy. Narses exploited this divide by pretending to execute the Luccan hostages in sight of the walls, but instead killed convicted criminals, before releasing the apparently dead hostages, who strongly advocated the Roman cause. In December, the Romans made a full assault:

"Narses, chafing at the thought that the citizens of Lucca might still hold out for a very long time, if the siege were continued in its present half-hearted form, closed in relentlessly on the walls. Siege-engines were brought up and fire-brands were hurled at the towers, while the archers and slingers directed their fire at anybody appearing on the battlements between the towers. Part of the wall was breached and the city was faced with imminent disaster." Ναρσῆς δὴ οὖν τὸ λοιπὸν οὐκ ἀνεκτὸν εἶναι ἡγούμενος, εἰ μέλλοιεν οἱ Λουκανοὶ ἐπὶ πλεῖστον ἀντέχειν οὕτω πως ἀνειμένα πολιορκούμενοι, ἐπέλαζε τοῖς τείχεσιν ἀφειδῶς· καὶ αὐτίκα αἵ τε ἑλεπόλεις μηχαναὶ προσήγοντο καὶ ἀνὰ τὰς τύρσεις πυρφόρα ἐρρίπτοντο βέλη οἵ τε ἐς τὰ μεταπύργια φαινόμενοι ἐβάλλοντο λίθοις τε καὶ τοξεύμασιν· καὶ διετέμνετο ἔστιν οὗ ἡ τοῦ περιβόλου οἰκοδομία καί ἅπασα ἰδέα κακοῦ περιεστήκει τὴν πόλιν. (1.18.4)

At first the Luccans and Franks defended themselves from the walls, and sallied out only when the walls were breached. The pro-Roman side prevailed and the militia (i.e. Λουκανοί) betrayed the Frankish troops by withdrawing while they were fighting outside the walls. This gave the Romans victory; Lucca and her inhabitants were again under the emperor's rule.

Cameron believes that Agathias probably had "highly placed" informant in Roman army in Italy, as he dates the surrender to December, after 3 months, which is correct (n. 12. p. 43), while the other source, Agnellus, must be referring to the beginning of siege in September. Despite this, Cameron criticizes him for the giving so much space to the mock execution at Lucca as typical of his moralizing use of *exemplum*. The criticism is somewhat unfair, as it ignores the psychological component of siege warfare and frailty of diverging loyalties which could wreak havoc among defenders. For multiple examples and discussion, see chapter 6.

PARMA   πρὸς τῇ Πάρμᾳ στρατοπεδεύσασθαι   553
Ag. 1.14.1-18.3.

A Roman army established camp and began raiding in the vicinity of Parma, but after a Herul contingent had been ambushed by the Franks near the city, withdrew to nearby Faenza. Narses sent a small force of 200 cavalry to help organize a close blockade, as "he expected their forces to be ranged like a continuous fortification and bulwark around the city of Parma" (ὁ μὲν γὰρ ἀμφὶ Πάρμαν τὴν πόλιν ὥσπερ ἐν προβόλου καὶ ἐρύματος μοίρᾳ τετάχθαι τὰ ξὺν ἐκείνοις στρατεύματα χρῆναι ἡγεῖτο, 1.17.2). Frankish morale was boosted and Gothic cities that had submitted to the Romans went over to the Franks en masse. The Romans sent by Narses had to evade Frankish foragers, but were able to move around them by marching at night. Arriving at Faenza, they organized supplies from Ravenna and had the generals "encamp hard by Parma (πρὸς τῇ Πάρμᾳ στρατοπεδεύσασθαι, 1.18.2)," presumably in the manner Narses had required, and return to *Lucca. Agathias does not reveal any further action or the ultimate result of the siege.

CONZA   over winter   ἐς πολιορκίαν καθίστατο   554f
Ag. 2.13.1-2.14.7; Wolfram 1988: 361.

The Roman army under Narses "settled down to a regular siege" of 7,000 Goths, led by the Hun Ragnaris. The site was well-fortified and the blockade had to be maintained over winter. These forces had recently been helping the Franks under Butilinus and Leutharis, but now found themselves without support. Wolfram believes that they also included a large number of survivors from the battle of Casilinum. The Romans avoided a direct assault due to the strong natural defenses. The Goths made frequent but inconclusive sallies. During negotiations in the spring, Ragnaris attempted to assassinate Narses, but was killed himself. His men surrendered and were sent "to the Emperor in Constantinople."

This force was obviously well supplied, and only began negotiating in the spring of 555, even after the Frankish defeat in the autumn of 554. From Agathias' narrative, it is impossible to reconstruct the Gothic role in the Frankish invasion, nor indeed the strategy of the Franco-Gothic alliance. As Wolfram suggests, it is possible that those Franks that survived Butilinus' expedition were also at Conza, but this not clear from the text either.

Onoguris τειχομαχήσοντες

Agathias 3.5.6-3.8.2; Av. Cameron 1970: 45f.

After some inconclusive, but logistically demanding maneuvers the previous years, the Roman field army in Lazica (claimed to number 50,000 men by Agathias, but see discussion below) made extensive preparations for the siege of Onoguris, which the Persians had recently occupied: "So all the generals and their men, who had been encamped on the plain of Archaeopolis, began to get ready the "wicker roofs" as they are called and the ballistae [sic Frendo] and other such engines of war with the idea of taking the place by storm if necessary (τούς τε καλουμένους σπαλίωνας ἐπεσκεύαζον καὶ τὰ τῶν μεγάλων λίθων ἀκοντιστήρια καὶ ἄλλα ἄττα τοιάδε ὄργανα, ὡς, εἰ δεήσοι, τειχομαχήσοντες). The "wicker roof" is a construction of osiers woven together so as to form a roof which is carried down on either side so as to enclose whoever gets under it. Skins and hides are then placed in layers over it and the device is completely overlaid with them in order to afford greater protection and to be proof against missiles. Inside, men conceal themselves under it in safety lifting it without being seen and moving it to wherever they which. When it is brought up to a tower or wall as the case may be, then the men underneath dig up the adjacent ground and drawing up earth lay bare the foundations. After that they keep striking it with hammers and crow-bars (μοχλοῖς τε καὶ σφύραις) until they cause the structure to collapse. These then were the sort of preparations the Romans were making for the siege." (3.5.9ff)

As the siege was about to begin, the Romans intercepted a Persian messenger going to Onoguris to inform the garrison that reinforcements were on the way. After some debate, the Romans decided to only send a small force of 600 cavalry against these reinforcements, and begin the assault on the gates and fortifications with the engines. The Persians defended themselves vigorously (3.6.11), "... dashing about on the battlements, raining down missiles and securing themselves against the oncoming ones by suspending canvas mantlets (ὀθόνας γάρ τινας καὶ ἁπλοΐδας) to soften and absorb the blows." The Romans began ascending the walls and were tearing down the mantlets (τά τε παραπετάσματα ἐκεῖνα καθεῖλκον) when the Persian relief force, 3,000 men from the forts of Mucheiresis and Kotaïs, was first chased off by the Roman detachment. However, outnumbering the small Roman force, they reengaged and emerged victorious. Free to relieve the fortress, they then nearly caused a total rout among besieging Romans. A complete disaster was barely averted by the commander Buzes, who held the rearguard as most of the army fled, but the Romans had to abandon their camp at Archaeopolis with all their equipment and supplies.

Great care and effort went into the engines, as Agathias found it worthwhile to devote a paragraph to the preparations. They seem to have been used in a coordinated fashion; the artillery may have been used to destroy the Persian mantlets and otherwise cover the approach of the *spaliones*. Cameron argues that the description of preparations was a rhetorical device, but does not provide any indication of what might be wrong, such as a critique of particular engines or methods, which are in fact all well attested in contemporary siege warfare (see chapter 5). There are even signs of contemporary innovations: the τὰ τῶν μεγάλων λίθων ἀκοντιστήρια, "the hurlers of great stones," which Frendo translates as "ballistae," may in fact have been trebuchets (see chapters 5.2.2 and 8.2.1). Cameron does however have a point about numbers, which Agathias regularly mishandles for rhetorical purposes. The inflated 50,000 may be a simple multiple of 5,000 (or grand total for all forces *per Armeniam*?), as there were no more than 12,000 Roman troops in the region during the Persian siege of *Archaeopolis (550), and the Romans had committed 6,000 men to retaking *Petra (550).

CLERMONT	*infra murus tenebatur inclusus*	c. 555

GT 4.16; Mar. Av. a. 555 ; *PLRE* 3 s.v. Chramnus; Wood 1994: 59.

Marius provides the dates, but most of the detailed information comes from Gregory. Chramn, in revolt against his father Chlothar, attempted to take over territory that had been on his itinerary. At first he "subjected to his own rule" (*in sua dominatione redigit*) the Limousin: "Then (*tunc*; rendered as "at that time" by Thorpe) the people of Clermont-Ferrand were shut up inside their city walls (*populous infra murus tenebatur inclusus*) and were dying off like flies, for they were attacked by one epidemic after another." The only rational explanation to this phrase is a close blockade that caused (famine and thus) an epidemic. Chlothar sent two of his other sons to relieve the city, but arriving at Clermont they found that Chramn was in the Limousin. He had presumably abandoned the siege as his brothers approached, or left a force to conduct the siege on his behalf. Unfortunately, Gregory fails to explain what exactly transpired. The two brothers caught up with Chramn and sought to force him to battle, but it was interrupted by a storm and then false news planted by Chramn that their father was dead. When the brothers hurried back to Burgundy, Chramn marched in their wake to *Chalon (below).

CHALON-SUR-SAÔNE   *obsidens adquisivit*   c. 555
GT 4.16; Wood 1994: 59; *Clermont (above).

"He came to the city of Chalon-sur-Saône, which he captured after a siege." || ...usque civitatem Cavillonensim venit eamque obsidens adquisivit. ||

DIJON   *infra murus... non est permissus intrare*   c. 555
GT 4.16; *Clermont and *Chalon (above).

Chramn marched from *Chalon to Dijon, but despite his significant military capabilities, "he was not permitted to go inside the walls" (*Infra murus tamen Divionensis non est permissus intrare*). Instead, the clergy gave him communion in one of the extramural churches before he went to meet his uncle. Here Gregory focuses on the miraculous intervention of St. Tetricius of Langres, who opened the holy books to prophecy the future career of Chramn (the result was negative). This anecdote allows Gregory to ignore any military preparations undertaken to defend the fort, and thus deflect attention from the considerable military force that his colleagues possessed and kept Chramn from attempting an assault. However, Gregory has previously described it as heavily fortified, like a city, and the favorite residence of the bishops of Langres (GT 3.19).

PHASIS   τῷ περιβόλῳ προσέβαλον   556
Agathias 3.20.9-3.28.10; Av. Cameron 1970: 46ff; Braund 1994: 306 n. 181; de Montpéreux 1839: III, 61; cf. *Avignon 583.

The Romans defended Phasis, which was situated near the sea and estuary of the Phasis river and whose fortifications were "entirely constructed of wood (ξύλοις ἅπαν ἐσκευασμένον, 3.19.8)." The garrison included regular eastern units (τοῖς ἑῴοις τάγμασιν) under Valerian; Justin son of Germanus with his men (ὁ ἀμφ' αὐτὸν ὅμιλος); Martin with his men (αἱ Μαρτίνου δυνάμεις); Angilas with Moorish troops armed with shields and lances; Theodorus with heavy Tzanian infantry; Philomathius with Isaurian slingers and javelin-throwers; Gibrus with Lombards and Heruls. The Romans reinforced the partly decayed wooden fortifications with a massive rampart (χαράκωμα...καρτερώτατα) and moat (τάφρος) filled with water from the sea; this submerged a mass of stakes driven into the moat. The Romans also used naval forces: large merchant-ships rode at shore and in the Phasis river estuary with boats "securely suspended about the mastheads." Soldiers and warlike sailors manned these towers, armed with "bows and slings and had set up catapults loaded and ready for action" (τόξα φέροντες καὶ σφενδόνας μηχανάς τε ἐκηβόλους ἐνθέμενοι, πρὸς τὸ ἐνεργὸν ἤδη ἐσκευασμένας, 3.21.4). More ships were stationed upriver, while river patrol boats were to control the fords and protect the ships in the estuary. These measures made it impossible for the Persians to blockade the city or besiege from all sides as they were exposed from elevated fire from two sides (city walls and mast-mounted troops), so they had to attack from one general direction away from the river and the sea. The Romans had also burnt the countryside and buildings around Phasis, which meant that the Persians had to cut timber from the forests some distance away and haul it to their camp.

The Persian field army included cavalry, Daylami infantry, elephants, and "an army of porters (ἀχθοφόροις ἐκ πλείστου, 3.23.1)" and attendants (e.g. ὑπηρέται τῶν Περσῶν καὶ ἀχθοφόροι, 3.28.1). They crossed the river Phasis at night on a pontoon-bridge between Phasis and the fort of Nesos (3.20.1); this forced the Romans to withdraw to their fortifications, as they had not expected such a rapid crossing. They also captured some of the Roman boats and blockaded those upstream by building a timber barrier on boats; the construction was supported by elephants wading into the river. Already on the first day, the Persians began to approach the walls with volleys of arrows, gradually increasing the rate of fire from a distance in order to lure out the Romans, as the full volleys were forcing the defenders off the walls. A Roman sally was nearly cut off by the Daylami, but a reckless Roman charge with leveled spears convinced them to open their ranks and let the Romans return. As the Persians were free to approach the walls, the porters filled in the moat with wood, earth and stones, so that they could bring in the siege engines (τὰς ἑλεπόλεις μηχανάς) on the following day. This seems to have been accomplished by the end of the first day of fighting, although Agathias claims they used "a disproportionate amount of time" on this operation (3.23.2).

On the next day, one of the Roman commanders, Martin, spread a rumor of a Roman relief army, which raised Roman morale. As calculated, the rumor also spread to the Persians who sent out part of their army to face it (3.23.5-24.6). Part of the Roman army under Justin (well over 5,000, according to Agathias) slipped out of the city unnoticed in order to visit a nearby church. At the same time, the Persians sent troops in after the false relieving army, and the remaining Persians were occupied with deploying for a storm, both out of sight of the exiting Romans (3.24.7ff). During the final assault, the Persians attacked with intense barrages of archery. "Meanwhile others were bringing up siege-engines, hurling fire-brands or hacking at the wall with axes from under the cover of the "wicker roofs" as they are called." ἄλλοι δὲ τάς τε μηχανὰς ἐκίνουν καὶ βέλη πυρφορὰ ἠκόντιζον ἔς τε τοὺς καλουμένους σπαλίωνας ὑπεισδυόμενοι πελέκεσι κατὰ τοῦ τείχους ἐπεφέροντο... (3.25.2). Others were undermining the earthen foundations. The Romans on the walls used javelins against the masses of unprotected, advancing Persians. Huge stones were dropped on the *spaliones* [Frendo's "wicker roofs"] and smaller stones slung against troops, "shattering shields and helmets." Those stationed on the mast-bound boats use bows and ballistae with great effect at long range.

At this point, the Romans who had gone outside with Justin returned and attacked the Persians who were assaulting the walls facing the sea. Due to a simultaneous Roman sally, some of the Daylami were rushing between threats to assist their compatriots, but their highly irregular formation and rapid pace convinced the rest of the Persians that they were fleeing. This caused a general panic among the Persians, and as they fled the Daylami ran as well. In fierce fighting, Persian cavalry (held in reserve) and elephant-borne archers drove some of the Romans back, but a wounded elephant threw its riders and caused total havoc among the Persian lines. At this point, the rest of the Romans sallied out in formation and the Persians were decisively defeated. The Persian equipment was set on fire; this attracted the porters, who were off in the forest gathering timber. They rushed to share in the booty, as they had been instructed by their commander to come to Phasis when he caught the city and set it on fire. This caused even greater Persian losses, given at 10,000. The Romans only lost 200 dead, who received an "honorable burial." Despite the great losses, the Persians withdrew in good order, protected by Daylami infantry, to their fort at Mucheiresis. On the way, they meet up with the detachment sent against the nonexistent Roman army.

Both sides expended immense labor on engineering projects. The Roman troops seem to have done this work by itself (see chapter 2.2.1). The Persian works were even more extensive; they built a pontoon bridge, filled in the large Roman moat, and hauled wood over great distances, since the Romans torched everything nearby in order to strip the surroundings of usable materials. The Persian laborers were probably the *paygān*, the conscripted infantry that normally accompanied the Persian cavalry and performed a range of support functions (see chapter 7.1.2). The Roman forces included regular units and federates. Agathias distinguishes the private retinues of the generals Justin and Martin from the regular troops, although his vocabulary not quite clear on this point until Agathias 3.27.1, where he refers to one of Martin's *doryphoroi*.

Cameron is negative to several points in Agathias' portrayal of events. Firstly, she is generally critical of the extensive physical works, and receives some support from Braund on this, who notes that the area is dominated by "impenetrable wetlands." However, the terrain has changed much, even since de Montpéreux's description of the site in 1839, so it would be unwise to conclude anything about the conditions of the plains in the 6th century on the basis of present-day conditions. Cameron further criticizes the description of Martin's ruse, calling it "patently absurd." However, news traveled fast during sieges, even to the enemy camp, and the *Strategikon* actually recommends using false good news to boost morale and confuse enemies (*Strat.* 8.1.9, 11, 12). She is also critical of Justin's miraculous visit to the church, but aside from the claim of divine inspiration (which soon surfaced after sieges anyway; see chapter 6.1.3), this was actually a clever move, if feasible, and the Persians were convinced it was the Roman relieving army that was upon them. The problem is whether it was possible to perform such a ruse based on the Persian vantage point and local topography. It is difficult to assess the merits of Justin's maneuver since the terrain has changed too much over the centuries and the fortifications moved with the meandering river and displacement of the sea shore. The Persians withdrew completely to camp after the first day, which meant that they had to redeploy from one direction the next morning. Roman control of the sea and river estuary, which surrounded at least some of the fortifications, meant that the Persians could not post scouts at will (cf. *Edessa 502), and certainly not blockade the gates. Finally, Cameron claims there were too many coincidences in the Persian collapse (the sequence Daylami regrouping > Persian flight > Daylami flight > massacre of porters appears incredible to her), but this seems in fact to be one of the best descriptions of the effect of misunderstanding and the "fog of war" that we have in this period. In general, Agathias provides far too many unique details to dismiss as a literary construction, and even Cameron believes he must have had an informant who participated, suggesting the *doryphoros* of Martin who wounded the elephant that broke the Persian lines.

The numbers provided by Agathias cause some problems however. 5,000 cavalry marching out of only one or two gates without detection, even if the scenario above is accepted, seems logistically difficult, and I would suggest that a much smaller number moved out, perhaps at night or early dawn, in order to escape detection. The figure may rather be a global total for the Roman garrison, with additional men in the ships (cf. *Onoguris 555, *Petra 550, *Archaeopolis 550 for comparable numbers). This further depends on the size of the fortification, which has not survived and is thus difficult to determine. Considering their orderly withdrawal, the Persian losses also seem large; perhaps the 10,000 should be considered the total force of professional troops, with attendants and porters coming in addition. If taken literally, a great proportion of their losses would have been among the poorly armed and organized *paygān* rushing to secure booty but were surprised in the open. It was clearly a political disaster, as the Persians ultimately lost control of Lazica (the treaty of autumn 557 divided the territory according to *de facto* zones of occupation, Agathias 4.30.7ff; cf. the failed Misimian revolt that ended in the Roman capture of *Tzacher/Siderun 557), and in 557 the Persian commander who lost at Phasis, Nachoragan, was flayed and his skin suspended on a pole for display (Agathias 4.23.2f).

?RHEIMS	*cuncta predis atque incendio devastavit*	556/7

GT 4.17; cf. *Clermont (above) and *Rheims.

In alliance with his nephew Chramn, Childebert tried to conquer parts of his brother Chlothar's kingdom. "While Chlothar was fighting against the Saxons, King Childebert marched to the district of Rheims. He pushed on as far as the city (*usque Remus civitatem properans*), pillaging far and wide and burning as much as he could (*cuncta predis atque incendio devastavit*)." He too had received news of Chlothar's death in battle and tried to exploit the situation by pressuring magnates and cities to join his side. This never amounted to a formal siege, but was a preparation to later actions that never materialized, as he died soon afterwards. However, the same policy was continued by his son Chilperic only a few years later.

TZACHER/SIDERUN	ἐτειχομάχουν	557

Agathias 4.15.4-20.9.

The Roman army was on an expedition to subdue the Misimians after regaining the allegiance of the Laz. As the Persians withdrew in the autumn of 557, the Romans approached Misimian territory and sent an embassy with representatives from their neighbors, the Apsilians. However, the Misimians killed the embassy and ambushed a Roman detachment, but were defeated and fled. They were hoping for Persian assistance and had assembled at the fort of Tzacher, in Greek Sideroun (*of iron*), which had an extensive but inaccessible settlement next to it. Due to weak leadership, the Romans "camped at a greater distance from the enemy than is normal when one is conducting a siege (4.17.1)." The siege was properly conducted after the arrival of a new general. The Romans found a secret passage up to the village, surprised the guards at night, and commenced a general massacre of the people, including women and children, and torched their houses. However, the Misimians soldiers in the fort sallied out and caught the Romans by surprise, killing many and chasing the rest down the precipitous rocks.

Afterwards, the Romans gave up that approach and "decided to attack the fort at its most vulnerable point and at the same time to fill in the moat. Assembling therefore a number of sheds and penthouses they brought them up and proceeded to attack the wall from a safe position (οἰκίσκους τινὰς καὶ καλύβας πλησιαίτερον τεκτηνάμενοι ἐκ τοῦ ἀσφαλοῦς ἐτειχομάχουν). They employed siege-engines (μηχαναῖς τε χρώμενοι), bows and arrows and every other available means of making life difficult for the defenders. The barbarians were in dire straits but they still put up a stiff resistance. Some of them brought up a wicker-roof (σπαλίωνα) and advanced against the Roman siege-works (τὰ ἔρκη) with the idea of demolishing them. But before they drew near and took cover under it, a Slav called Suarunas hurled his spear at the one that was most visible and struck him with a mortal blow. As the man fell the wicker-roof toppled over revealing and leaving unprotected the men inside it." (4.20.3f)

The exposed Misimians were shot down; one nearly escaped but was killed at the threshold of the gate. At this point, the Misimians asked for peace, having lost 5,000 young men and more women and children, and were granted a return to their status as Roman subjects.

The μηχαναῖς which the Romans used were probably some form of artillery, as they are mentioned after the description of the various protective devices but in the same context as missile weapons. It is notable how Agathias condemned the massacre of women and children who were blameless, despite the alleged crimes of Misimian soldiers and political leaders, and he also claims that Christian pity and solidarity (they were also Christians) were important arguments for both sides to return to their former allegiance without further bloodshed.

<u>Rhizaion</u> (Roman camp)   προεπῄεσαν τῷ ἐρύματι   558
Agathias 5. 1.-2.; *NJust* I *praefatio*.

When the majority of the Tzani revolted and engaged in brigandage in Pontus, the Romans sent an army to regain their allegiance. Setting up a fortified camp (στρατοπεδευσάμενος...χαράκωμα τῷ στρατῷ περιβαλόμενος) at Rhizaion, they began to distribute gifts to those who were somewhat friendly and prepare military action against the rebels. The latter however assembled on a hill overlooking the camp, and began an assault on the fortifications (προεπήεσαν τῷ ἐρύματι) by showering the Romans with missiles. The Romans, storming out, responded with an uphill assault, "holding their shields tilted over their heads and stooping slightly (5.1.6)." They were repulsed with javelins and rolling boulders, but as the Tzani followed up with an advance against only one side of the Roman camp, the Roman commander ordered some of his troops to move around the Tzani rear while others held their ground on the fortifications. The successful execution of this maneuver led to the complete defeat of the Tzani, who lost 2,000 men; the Romans lost 40 during their failed uphill assault. This victory meant that the Tzani were completely subdued, entered into official registers, and required to pay annual tribute, a feat notable enough to be commemorated by Justinian in one of his Novels.

CHERSONESE	τῷ περιβόλῳ προσέβαλλον	559

Agathias 5.21.1-23.6.

The Kotrigur Huns, jealous of the good relations between the Romans and the Utigurs, raided the Balkans in order to receive respect (i.e. subsidies). They spread out in three detachments: one raided Thrace and the vicinity of Constantinople, causing a general panic (see chapter 6.1.3), another Greece, and the last tried to storm the walls protecting the Chersonese. "Meanwhile the other detachment of barbarians which was besieging the Chersonese attacked the wall repeatedly, bringing up ladders and siege-engines (κλίμακάς τε προσάγοντες καὶ τὰς μηχανὰς τὰς ἑλεπόλεις), but was beaten off each time by the resolute resistance of the Romans defending it (5.21.1)." Subsequently the Huns made 150 reed boats for 600 men to try to sail around the walls, but were completely defeated by a flotilla (21.6-22.9). A sally forced the surviving Huns to withdraw. The group before Constantinople withdrew when one of their raiding parties was defeated by Belisarius; the group heading for Greece failed at Thermopylai. Justinian made peace and bought the freedom of captives. Although the people of Constantinople were dissatisfied with this solution, he soon after stirred up a brutal war between the Hunnic groups (5.24.f).

VERONA	*capta*	561

Agnellus 79.

"And they [the Romans] fought against the Veronese citizens [and the last independent Goths] and Verona was captured by the soldiers on the

20th day of the month of July." || Et pugnaverunt contra Veronenses cives et capta est Verona a militibus XX die mensis Iulii. ||

RHEIMS; OTHER CITIES   *pervadit; abstulit*   562
GT 4.23; *Soissons (below)

Sigibert, king of Austrasia, was fighting off the Avars to the east, when his territory was invaded by his brother Chilperic, who had claimed the largest (perhaps only) share of the inheritance after Chlothar's death in 561, but lost to his three brothers. "His brother Chilperic attacked Rheims and captured a number of other cities which were Sigibert's by right of inheritance." || ...Chilpericus, frater eius, Remus pervadit et alias civitates, quae ad eum pertenebant, abstulit. ||

Consigned to a few *civitates*, his attack on Rheims was a continuation of Childebert's policies (who had been similarly constricted) against the Austrasian kingdom (cf. *Rheims 556/57).

SOISSONS; OTHER CITIES   *occupant; in sua dominatione restituit*   562
GT 4. 23; cf. *Rheims (above).

"When he came back as victor over the Huns [i.e. Avars], Sigibert occupied (*occupat*) the city of Soissons." This was the capital of Chilperic, held by his son Theudebert. After winning a battle against Chilperic, Sigibert "brought his own cities once more under his dominion" (*civitatis suas in sua dominatione restituit*).

ARLES   *capere cupiens, ingressique urbem, sacramenta...exegerunt*   c. 566
GT 4.30; *PLRE* 3 s.vv. Guntchramnus, Sigibertus I, Firminus 1, Audovarius.

Sigibert, the king of Austrasia, "wanted to take over Arles" (*Arelatensim urbem capere cupiens*), which had been held by the Goths earlier (3.23) [taken from Guntram, the king of Burgundy]. Sigibert sent two armies, one consisting of the "men of Clermont-Ferrand" (Gregory only has *Arvernus*, i.e. acc. pl.) under the count Firminus, the other under Audovarius. They converged from two different directions on Arles, and entered the city ([*i*]*ngressique urbem*) and extracted oaths (*sacramenta...exegerunt*). However, this was clearly the result of compulsion, as the Arelatensians soon betrayed the new garrison (see *Arles II 567).

Sigibert's motives may have been a recent defeat against the Avars that caused loss of prestige—he had been captured and forced to pay a large ransom (GT 4.29). The dating is difficult and hence inconsistent in the different entries in *PLRE*, but since it involved armed forces from areas that had probably not taken part in the Avar war in 566, the invasion could have been organized very soon afterwards.

| ?Tours | *ad verum dominium revocare* | 567 |

GT 4.45; Wood 1994: 89; *PLRE* 3 s.v. Mummolus.

Guntram and Sigibert sent Mummolus to retake cities (*qui has urbes ad verum dominium revocare deberet*) that had been invaded (*pervasissit*) by Chilperic in 567. Mummolus "came to Tours, drove out Clovis, the son of Chilperic, and made the people swear an oath of allegiance to King Sigibert." || Qui Toronus veniens, fugato exinde Chlodovecho, Chilperici filium, exacta populo ad partem regis Sigyberthi sacramenta... ||

The dating again appears ambiguous; *PLRE* dates the first stage of the war to 567 or shortly after, while Wood does not mention these particular events but seems to group Tours and *Poiters (below) with the other conflicts that occurred in 573-75. This particular event appears to have been political, but backed by military force, since Gregory cited it as evidence of Mummolus' military prowess (cf. *Poitiers below).

| Poitiers | *circumdatus, obruit...accedens* | 567 |

As *Tours (above).

After *Tours, Mummolus "marched on Poitiers (...*Pectavum accessit*). Two inhabitants of the city, Basilius and Sighar, collected a mob together (*collecta multitudine*) and prepared to resist (*resistere voluerunt*). Mummolus hemmed them in on all sides (*quos de diversis partibus circumdatus*), overpowered them (*oppressit*), conquered them (*obruit*) and killed them (*interimit*). And coming to Poitiers in this manner (*et sic Pectavum accedens*), he insisted on an oath of fealty."

These anecdotes are in effect a panegyric to Mummolus and his prowess as a soldier; the scarcity of details is typical of Gregory. We have no idea how Clovis was driven out of *Tours, but the language here indicates that the fight for Poitiers occurred on the city walls: At first Mummolus surrounded (*circumdatus*) the city, then began an assault with archery that forced them from the walls, or "overpowered," "conquered" and "killed" them, a process that was summed up as Mummolus "marching on" (*accedens*) the city. It seems that Gregory may have epitomized a more extensive source by stringing together all the action verbs. The nature of the "mob", *multitudine*, is uncertain, but see ch. 4.1.5.

| Avignon | *abstulit* | 567/9 |

As *Arles (566); *PLRE* 3 s.v. Celsus 2.

In response to Sigibert's occupation of *Arles, Guntram sent an army under the patrician Celsus. "Celsus came to Avignon and captured (*abstulit*) the city." Guntram restored it to his brother after Sigibert's army had been routed at *Arles (below).

| Arles | *vallans, inpugnare...coepit* | 567/9 |

As *Arles (566) and *Avignon (above).

After Celsus had taken Avignon: "Then he, too, marched on Arles, surrounded the place and began to assault Sigibert's army, which was shut up inside the walls." || Accedens autem Arelate et vallans eam, inpugnare exercitum Sigyberthi, qui infra murus contenebatur, coepit. ||

The bishop of Arles encouraged Sigibert's men to make a sally (*"Egredimini foris"*), arguing that they had to protect the surrounding territory and promising to keep the city's oaths (cf. *Avignon above). When Sigibert's men were defeated, however, they found the gates closed. "Their army was assailed by javelins from the rear and showered with rocks by the townsfolk." || Cumque exercitus a tergo iaculis foderetur operireturque lapidibus ab urbanis... ||

Sigibert's army fled to the Rhône, where many of them drowned; those who survived swam across on their shields but lost their horses and other equipment. Guntram avoided escalating the conflict by sending Sigibert's commanders home and returning *Avignon to him.

| Sirmium | ἠβούλετο πολιορκήσειν, μετὰ τὴν τειχομαχίαν | 568 |

Men. fr. 12.3-7; Pohl 1988: 58ff; Whitby 1988: 84, 87.

We only know the immediate prelude to and aftermath of the Avar siege of Sirmium. The Avars claimed all the territory previously held by the Gepids, including the formerly Roman city of Sirmium, but the Romans managed to recapture Sirmium in 567 as the Gepid state was dissolving after their crushing defeat against the Lombards (see chapter 3.3.1). The description of the siege itself is not preserved in surviving fragments. In fr. 12.4, we learn how "Baian, the leader of the Avars, was intent upon the siege of Sirmium (ἠβούλετο πολιορκήσειν);" fr. 12.5 begins "After the assault upon the walls (μετὰ τὴν τειχομαχίαν) Baian sent envoys to discuss peace." Most of the remaining evidence deals with the negotiations, in which the Romans at Sirmium absolutely refused to give any tribute to have the siege raised, and Justin refused any concessions. Due to the loss of prestige, Baian unleashed the Kotrigur Huns under his control on Dalmatia, while he returned to Pannonia to establish control of former Gepid lands.

Based on the surviving vocabulary, the siege comprised a failed Avar assault on the walls (τειχομαχία). The city was defended by the Roman army. Blockley's translation implies a militia of Sirmians ("some of the inhabitants," fr. 12.5), but the original only speaks of "those who are in Sirmium" (ἔνιοι δὲ τῶν ἐν Σιρμίῳ) who were keeping watch from the roof of the baths inside the city. Since the city had an archbishop, a large civilian presence (and hence militia) is likely, but impossible to prove conclusively. It is possible that fr. 12.3, taken by Blockley to describe the prelude to a field engagement before or after the siege (p. 267 n. 155), actually derives from the prelude to the Avar storm of the walls. Avar motives are discussed by Pohl and Whitby; most notable is the fact that even a small amount of tribute established precedence: if the Romans had given *any* tribute, this implied an Avar hegemony that could be enforced at some later point. Eastern cities faced this problem with Persian invasions; the Persian shah used previous tribute to justify new claims and conflicts (see sieges in 540, 542).

VINCENZA		*cepit*	569

HL 2.14.

"Then Alboin took (*cepit*) Vincenza and Verona and the remaining cities of Venetia, except Padua, Mont Selice and Mantua."

VERONA		*cepit*	569

See *Vincenza (above).

MILAN AND LIGURIAN CITIES		*ingressus est; cepit*	569

HL 2.25.

"Alboin then came into Liguria at the beginning of the third indiction on the third day before the nones of September, and entered (*ingressus est*) Mediolanum... Then he took (*cepit*) all the cities of Liguria except those which were situated upon the shores of the sea." Archbishop Honoratus fled Milan to Genova.

PAVIA	3 years	*obsidionem perferens*	569ff

HL 2.26.

"The city of Ticinum (Pavia) at this time held out bravely, withstanding a siege more than three years, while the army of the Langobards remained close at hand on the western side." || Ticinensis eo tempore civitas ultra tres annos obsidionem perferens, se fortiter continuit, Langobardorum exercitu non procul iuxta ea ab occidentali parte residente. ||

?Málaga   *Malcitanae urbis repulsis militibus*   570
J.Bicl. 12, s.a. 570 (Wolf 1999: 60); Collins 1994: 52; Thompson 1969: 60.

"King Leovigild laid waste the region of Bastetania and the city of Málaga, defeating their soldiers, and returned victorious to his throne." || Leovegildus rex loca Bastetaniae et Malcitanae urbis repulsis militibus vastat et victor solio reddit. ||

Since there is no mention of a siege or capture, this may have been only a field engagement against the Roman soldiers garrisoned a Málaga, who had gone out to meet the raid against Bastetania. The distance was considerable, so, either the raid must have been too much for the local garrisons to handle, or it had come within Málaga's territory. It also indicates that Málaga was a significant base for Roman troops.

Medina Sidonia   *proditione...nocte occupant*   571
J.Bicl. 15, s.a. 571 (Wolf 1999:61); Collins 1994: 52; Thompson 1969: 60.

"King Leovigild seized Sidonia, that strongest of cities, by night through the treachery of a certain Framidaneus. He executed its garrison and restored the city to the jurisdiction of the Goths." || Leovegildus rex Asidonam fortissimam civitatem proditione cuiusdam Framidanei nocte occupant et militibus interfectis memoratam urbem ad Gothorum revocat iura. ||

The betrayal must have taken place during a formal siege of the city, since taking over the walls and overpowering the garrison was rare in case of a completely unexpected betrayal (e.g. *Martyropolis 589), and required significant manpower (*Rome 549f).

Improvized fortif. at Embrun   *circumdatisque...cum exercitu*   c. 571
GT 4.42; *HL* 3.4; *PLRE* 3 s.v. Mummolus 2; Wood 1994: 167.

Paul is based on Gregory, hence the latter is followed here. Mummolus, the patrician of Burgundy, turned back a Lombard invasion by building fortifications in the forest at a place called Plan de Fazi near Embrun: "He surrounded the Longobards with his army, made a rampart of trees which he had felled, and attacked them along the woodland paths." || Circumdatisque Langobardis cum exercitu, factis etiam condicibus, per divia silvarum, inruit super eos... || Thorpe's translation is one possible interpretation; a better rendering might be "...made a rampart of felled trees throughout the inaccessible areas and attacked them..." thus emphasizing that the fortifications were laid out according to the difficult terrain.

The Lombards were utterly defeated and most of them killed or captured. In Mummolus' army were two brothers, Salonius and Sagittarius, who also happened to be bishops, and they personally took part in the battle. One can surmise that they were leading their military retinues and craftsmen who could help construct the fortifications.

CORDOBA; OTHERS   *nocte occupat, in...dominium revocat*   572
J.Bicl. 20, s.a. 572 (Wolf 1999:62); Collins 2004: 52f; Thompson 1969: 60f.

"King Leovigild seized by night the city of Córdoba, which had rebelled against the Goths a long time before. He restored many cities and fortresses to the dominion of the Goths, killing a multitude of common people." ||
Leovegildus rex Cordubam civitatem diu Gothis rebellem nocte occupat et caesis hostibus propriam facit multasque urbes et castella interfecta rusticorum multitudine in Gothorum dominium revocat. ||

Just as at *Medina Sidonia (571); Leovigild killed the defenders, but here they included "a multitude of the common people." Collins and Thompson believe they were 6th-century *bacaudae* or peasant rebels. I would rather argue that they had sought refuge in the various fortifications with their patrons and landlords, and were probably involved directly with the defenses. See further chapter 3.2.2.

?TOURS   *pervadit*   573
GT 4.47; otherwise as *Tours (567). See also *Poitiers, *Limoges, *Cahors below.

The civil war recommenced after Sigibert had Chilperic's son Clovis driven from Bordeaux to Angers. Chilperic "sent his elder son Theudebert to invade (*pervadit*) the cities of Tours, Poitiers, and others south of the Loire," defeating an army in battle at Poitiers, slaughtering locals, burning the district around Tours until the citizens submitted, before doing the same in "the Limousin, district of Cahors and other territories nearby, all of which were ravaged and sacked. He burned the churches, stole their holy vessels, killed the clergy, emptied the monasteries of monks, raped the nuns in their convents and caused devastation everywhere."

As with *Tours (567), the fighting may never have reached the stage of formal sieges, but were designed to put pressure on the political elites who would be forced to submit before losing their economic foundations. In all cases, we can expect that many had fled to the cities for refuge, since it was the surrounding countryside that was deliberately targeted, although Gregory only notices how the clergy was affected. The forcible nature of their capture was recognized by the subsequent diplomatic resolution.

POITIERS	*pervadit*		573

See *Tours above.

LIMOGES	*pervadit*		573

See *Tours above.

CAHORS	*pervadit*		573

See *Tours above.

THEBOTHON	10 days	ἀπόπειραν...ἐποιήσαντο φρουρίου	573

J.Epiph. 1.3; Th. Sim. 3.10.1-5; Whitby 1988: 254ff.

Justin II decided to provoke a war against the Persians and ordered a raid against them in 572. Due to the short period of preparations, it was of limited size, and gave the Persians time to prepare defenses around Nisibis for a renewal of the conflict over the winter (see Whitby for references to the later Syriac chronicles). While the Romans won a battle outside Nisibis next spring, they bypassed it for now (see *Nisibis below), instead "making an attempt at the fortress of Thebothon (ἀπόπειραν μὲν τοῦ Θηβηθῶν ἐποιήσαντο φρουρίου) where they spent ten days. Unable to seize (ἑλεῖν) it, they returned to the city of Dara while it was still spring and again invaded enemy land planning to besiege Nisibis with the approval of the emperor Justin." The strategy was probably to cut off Persian supply lines to *Nisibis by taking Thebothon to the east of it, while other forces were preparing for the siege in Roman territory.

NISIBIS	šrā 'al, ܠ ܪܝܫ; παραστήσασθαι		573

J.Eph. 6.2; Men. fr. 16.2; Th. Sim. 3.10.5; Evagrius 5.8f; Whitby 1988: 254-58; further discussed in chapter 2.4.2.

Menander and Theophylact only briefly refer to the siege of Nisibis. According to the later Syriac chronicles (see references in Whitby), the Persians managed to delay the Roman siege enough to make extensive preparations, such as felling trees outside the walls, expelling the Christian population, and bringing in supplies. The fullest account of the siege itself is found in John, who relates how the Roman commander Marcian marched out from Dara (see *Thebothon), and "laid siege to Nisibis, the frontier town and bulwark of Mesopotamia, and then in possession of the Persians. And having strongly invested it, and constructed round it a palisade, he commenced, with the aid of the skilful mechanicians whom he had brought with him, to erect more scientific works, consisting of lofty towers and strong covered approaches." *Textus* 278.14-19:

ܘܚܝܠܬ ܓܪܐ ܥܠ ܢܨܝܒܝܢ ܡܕܝܢܬܐ: ܗ̇ܝ ܕܒܗ ܗܘܐ ܐܚܝܕܐ ܠܗ ܦܪܣܝܐ: ܘܩܡ ܥܠܝܗ̇ ܚܝܠܬܢܐܝܬ ܕܢܟܒܫܝܗ̇: ܘܒܢܐ ܥܠܝܗ̇ ܟܠܩܘܡܛܐ ܚܕ ܚܕܪܐ. ܘܡܛܠ ܕܐܦ ܐܝܬ ܗܘܐ ܥܡܗ ܡ̈ܟܢܝܩܘ: ܐܩܝܡ ܥܠܝܗ̇ ܡ̈ܟܢܢܡܛܐ ܕܡ̈ܓܕܠܐ ܪ̈ܡܐ ܘܕܦܘܪ̈ܓܘܣܐ ܬܩܝܦ̈ܐ:

"Quam cum fortiter expugnasset ut eam expugnaret χαλκώματα contra eam circa aedificavit; et quoniam mechanicos [apparatus: μηχανικός] etiam secum habebat, machinamenta [apparatus: μηχανήματα] turrium altarum et πύργων validorum contra eam erexit…" (*Versio* 210.29-211.3)

Just as the city was about to fall, Marcian was replaced as commander: the Romans were confused and began to withdraw. Before they had completely abandoned their camp, however, the Persians sallied out and defeated the infantry rearguard and captured all of the Roman equipment before it could be destroyed or removed. Most of the Roman forces were driven to Mardin, where they remained until after the fall of *Dara, although there seems not to have been a siege or blockade. The Persians brought the Roman siege engines to *Dara and used them to capture the city.

<u>DARA 6 months *īteb ʿal, d-nekbšīh*; χειροῦται, αἱρεῖ… βίᾳ 573</u>

J.Eph. 6.5; Th. Sim. 3.11.2; J.Epiph. 5; Evagrius 5.10; Whitby 1988: 257f.

The siege lasted for six months and the fall of the city was a great shock to the Romans; although John of Epiphania and Theophylact (based on J.Epiph.) has good concise descriptions, the most detailed account is found in John of Ephesus. See Whitby for the chronology and military movements.

According to John of Ephesus, Khusro arrived at *Nisibis "and found the engines and machines (*versio* 218.5: machinamenta/μηχανήματα … ballistas/μαγγανόν; *textus* 287.18: ܡ̈ܟܢܢܡܛܐ…ܡ̈ܢܓܢܘܢ) which Marcian had erected still standing before it. And with these he forthwith commenced the siege of Dara (*īteb ʿal dārā*), having removed thither all Marcian's engines of war (*ūmānwātā*), and applied them to his own use, for which purpose he had brought all kinds of artificers (*ūmānwān*, artifices—better rendered as "skills") with him. His first act was to command the stone-cutters (*pāsūlē*) and others to make a cutting through a hill which lay on the east of the city outside the aqueduct, in order to divert the water; and when, as was said, they found the stone hard, they lit fires upon it, and cooled it when hot with vinegar, and so made it soft for working. He further set up against the city all the engines (*versio* 218.14 ballistas; *textus* 288.1 قنبلة [sic]) which Marcian had constructed against Nisibis, and invested it (*wa-hwā yāteb 'lēh*), and used every device of war for its capture (*d-nekbšīh*; "break into, trample underfoot" etc.) during a period of six months. Among his machines were two towers, which he erected, but the Romans devised a plan for setting them on fire, and were successful, and burnt them, although all egress from the city was impossible. On the side of the besieged the generals were John, the son of Timus Esthartus, a man of great warlike ability, and Sergius, the son of Shaphnai, and others. But Sergius, as they said, was struck by an arrow and died. After a time the Persian king, not finding the siege making progress, removed his tent and pitched it on a mountain on the northern side of the city, whence he could see everything that was done within. And there also he ordered a tower to be built on more elevated ground, opposite a great turret which rose higher than the rest, and which they called Hercules. And against this the besieged found all their efforts unavailing, while the besiegers were able to strengthen their tower, and bring it up close to the city. Sometime before this, when the king saw that his vast works had not terrified the inhabitants, he had given orders for a brick wall to be drawn all round the outer fortifications, that if they made a sally, they might be caught within it."

Khusro fell ill when no progress was made, and opened negotiations. The Roman negotiator did not convey the demand for ransom to the city because he thought it was impregnable. In fact, further Persian assaults only provoked Roman scorn. "But this over-confidence led them to neglect the maintenance of a proper force upon the wall, especially as the cold was now great and intense; and they even came down from the ramparts, and went to their houses to eat and drink. But when the Persians saw that the wall was no longer guarded by the Roman soldiers, and that the tower which they had built exceeded the height of the fortifications, they set their invention to work, and fastened planks together, until they reached the wall; and passing over, they occupied the whole of it on one side of the city, and then began to descend within."

The population and garrison were unable to flee as the gates were locked, so they had no choice but to fight the Persians. House-to house fighting continued for seven days, until the Persians fled back to the ramparts and proposed a truce. This was however a ruse to get access to the city again, and when they began to fraternize, they took the opportunity to begin plundering and taking captives. Many of the people were "put to the sword," while notables were drowned in the river and valuables piled up. Khusro berated the remaining notables for failing to give ransom, which would have been only a small amount of the plunder he had taken. The surviving population was deported and the Persians left a garrison to hold the city.

Theophylact presents a summary version of the events: "The Persian king came to Daras like a hurricane and assailed the township for six months, circumscribing the city with mounds and ramparts (λόφους τε καὶ χάρακας τῇ πόλει περιγραψάμενος). After diverting the town's water supply, constructing towers (πύργους) to oppose its towers, and bringing up siege engines (τὰς ἐλεπόλεις παραστησάμενος), he subdued (χειροῦται) the city, although it was exceedingly strong."

John of Epiphania has some lacunae that can be reconstructed from Theophylact's version, but he also contains alternative Greek terminology for the siege engines that Theophylact described as *helepoleis*: "making use of projectile launching machines (τοῖς ἐμβόλοις χρησάμενος μηχανήμασιν) against it, and because no external aid came for its inhabitants, he captured (αἱρεῖ) the city with the Medians violently mounting onto (βίᾳ τῶν Μήδων ἐπιβάντων) the city."

Evagrius adds some details and further clarifies colloquial usage. The Persian mound is built close to the wall. Upon it they "placed city-taking (*helepoleis*) engines, and especially *katapeltai* that throw from above, which are commonly called stonethrowers (*lithoboloi*)." ... ἐλεπόλεις μηχανὰς ἑστώσας, καὶ μάλιστα τοὺς καταπέλτας ἐξ ὑπερδεξίων ῥιπτοῦντας, οὓς λιθοβόλους ἡ συνήθεια καλεῖ.

Theophylact's *helepoleis* are problematic and John of Epiphania's *embolois* more ambiguous than the translation indicates, but from John of Ephesus we have a full range of engines with which to compare the terminology found in the Greek authors: Evagrius uses *helepoleis* generically, but notes that the most important category were *katapeltes* of a specific type ("throwing from high above"), which we can equate with trebuchets, thanks to John of Ephesus, and further link to *lithoboloi*, as they were more commonly called, and frequently mentioned in other texts around 600 (see chapters 5.2 and 8.2 for further discussion).

APAMEA   *ptaḥūn leh*, στρατοπεδευσάμενος, περιβαλόμενος χάρακα
J.Eph. 6.6; Th. Sim. 3.10.7ff; J.Epiph. 4; Evagrius 5.9; Whitby 1988: 258; chapter 6.

John of Ephesus has the fullest account: while Khusro was besieging Dara, "he sent a Marzban, named Adormahan, with a large body of troops, to besiege Apamea (the original only has "sent...to Apamea." *Textus* 292.14). On his march thither, Adormahun stormed (*kbaš*) numerous castles, which fell in his way, and razed and burnt them, together with several strong and well fortified towns, and at length arrived at Apamea. Now, upon a previous occasion, the Persian king, after capturing Antioch, had once before laid siege to Apamea, and pressed it so hard, that finally it capitulated; and the king in person entered the walls, and was a spectator of an equestrian entertainment in the Hippodrome; and because he then destroyed none of the buildings, nor set fire to any thing, they now felt equal certainty that the Marzban on the present occasion would do them no harm. In this confidence, therefore, the princes (*rawrbānē*) of the city and the bishop went out to meet him, and carried him a dress of honour. And he treacherously said to them, "Inasmuch as your city is now ours, open unto me the gates, that I may enter in and inspect it." And they trusting to him, and not expecting that he would do them any injury, opened (*ptaḥūn leh*) the gates and admitted him within the walls. But no sooner had he entered than he seized the gates, and began to lay hands on and bind men and women, and spoil the city." The number of captives (presumably from the whole campaign) was 273,000.

John of Epiphania and Theophylact relate a shorter version of the same story; both specify that the Persians set up a fortified camp (as if for siege: στρατοπεδευσάμενος, περιβαλόμενος χάρακα) close (οὐ πόρρωθεν, πλησίον) to the city, and add that Adormahan had 6,000 men when he entered Roman territory (John of Epiphania also mentions nomadic barbarians, but it is unclear whether they are included in the grand total). For more information on Persian movements and relative chronology of the campaign, see Whitby. The ritual and political aspect of siege warfare is very important here: the citizens were confident that previous lenient treatment (cf. *Apamea 540) meant that they would receive generous terms if they made a formal submission, performed as a traditional *adventus* ceremony for the new magistrate. The Persians betrayed the rules of war; their right to "inspect" their new possession was probably meant as a ritual prelude to the reappointment of Roman notables and magistrates as "Persian" officials responsible for tribute (that would of course be rejected at the first possible opportunity by the Romans; cf. *Martyropolis 502). Since it involved a Persian delegation actually entering the city, however, it was a convenient ruse to gain control over the gates and subdue the city completely.

Evagrius only adds that there was a significant Arab contingent in the Persian army.

| ?PARIS | *[flamma consumpsit, direpti]* | 574 |

GT 4.49; Wood 1994: 89.

In response to Chilperic's invasion of *Tours (and other cities, 573), Sigibert mobilized an army, including troops from the eastern tribes (possible alternatives were Saxons, Thuringians and Alamans). They ravaged the district of Paris, burning villages, taking loot and even slaves, against explicit orders. However, this may have been a convenient excuse for Sigibert—he could be seen opposing the depredations but at the same time used them for political gains. They were also pursuing battle but were disappointed when not allowed to fight (Sigibert calmed their tempers but later had many of them stoned, perhaps also in response to the unauthorized raiding). The reason was that Chilperic sued for peace and returned the cities taken, "stipulating that the inhabitants should not be punished, for Theudebert had annexed them forcibly, coercing them with the fire and sword." || ...depraecans, ut nullo caso culparentur earum habitatores, quos ille iniuste igne ferroque obpremens adquisierat. ||

| ARLES | *debellavit* | 574 |

GT 4.44 [similar in *HL* 3.8]; *PLRE* 3 s.vv. Amo, Rhodanus, Zaban.

A massive Lombard invasion in three columns attacked Burgundy. The duke Amo led his troops via Embrun and camped at the villa of Saint-Saturnain near Avignon "and pitched his tents there...Amo captured the province of Arles and all the towns in the region" (*ibique fixit tenturia...Et Amo quoque debellavit provinciam cum urbibus qui circumsitae sunt*) before raiding around Marseilles.

The estate where Amo encamped had been given to Mummolus from the king, presumably in order to defend the south-eastern passes, and hence an important logistical hub in the region. The move was probably calculated to exploit stored-up supplies and deprive the Franks of an important staging post for defense; the move also presupposed very good intelligence and moving fast to avoid detection and Frankish mobilization.

| AIX | *obsidionem paravit* | 574 |

As *Arles above

After capturing the region of *Arles, Amo "made plans to besiege (*obsidionem paravit*) Aix, but he was bought off with twenty-two pounds of silver and marched on." The sum may have been symbolic.

6th century

| VALENCE | *obsedebat* | 574 |

As *Arles above.

"Zaban passed through the town of Die, came to Valence, and camped there. Rodan reached the town of Grenoble and set up his headquarters there...Rodan and Zaban did the same [as Amo] in the districts which they had invaded." Mummolus "raised and army and attacked Rodan, who was besieging (*debellabat*) Grenoble." After defeating this division, Mummolus headed on to "Zaban, who was besieging (*obsedebat*) Valence." Again victorious, the remnants of the first two divisions withdrew over the pass of Susa, which was held by the imperial *magister militum* Sisinnius. The last division under Amo pillaged their way out of Gaul, but were forced by snows to abandon their loot in the pass.

The vocabulary is indicative: while the straight enumeration of Lombard activities would make the reader believe these were raids in force, but not necessarily assaulting city walls (except *Arles, where this is prepared), only the arrival of Mummolus allows us to see what the Lombards were up to: besieging/assaulting the cities from their camps.

| GRENOBLE | *debellabat* | 574 |

See *Valence above.

| NANO; ?TRENT | *se tradidit; depraedatus est* | c. 575 |

HL 3.9; *PLRE* 3 s.vv. Ragilo, Chramnichis, Eoin.

The *PLRE* dates the Frankish capture of Anagnia (Nano) and subsequent events to 574 or 575: "In these days upon the approach of the Franks the fortress of Anagnia, which is situated above Trent within the boundary of Italy, surrendered to them. For this reason the count of the Langobards from Lagaris (Lägerthal), Ragilo by name, came and plundered Agnanis."
|| His diebus advenientibus Francis, Anagnis castrum, quod super Tridentum in confinio Italiae positum est, se eisdem tradidit. Quam ob causam comes Langobardorum de lagare, Ragilo nomine, Anagnis veniens depraedatus est. ||

Ragilo was defeated in battle by Chramnichis, who went on to plunder and ravage Tridentum, but he in turn was defeated by the Lombard duke Eoin.

| TOURNAI | *obsederent* | 575 |

GT 4.51; Wood 1994: 89f.

Sigibert defeated most of Chilperic's forces and won over the rest, and his army was besieging (*obsederent*) Chilperic in Tournai while Chilperic's former supporters acclaimed Sigibert as king. However, Sigibert was assassinated and Chilperic able to break free and re-establish his position.

| Tours | *praedas egit* | 575f |

GT 5.1, 4; *PLRE* 3 s.v. Roccolenus.

Chilperic sent Roccolen "with the men (*homines*) of Maine" to plunder (*praedas egit*) Tours. The strong-arm techniques were recorded in detail in GT 5.4, where the Roccolen tried to force Gregory to expel Duke Guntram (for the murder of Theudebert) from his church or see "the city and its suburbs ... burnt to the ground." Roccolen was admitted to the city on Epiphany (Jan. 6) 576, so the campaign must have begun very late in 575.

| Soissons | *volebant sibi subdere civitatem* | 576 |

GT 5.3; *PLRE* 3 s.v. Chilpericus 1 (292-96, at 293).

Troops from Champagne marched against Soissons "and they wanted to capture the city" (*volebant sibi subdere civitatem*). Chilperic relieved the city after defeating the besiegers in pitched battle.

| Theodosiopolis | πολιορκία [abortive] | 576 |

Men. fr. 18.6; J.Eph. 6.8; Whitby 1988: 262ff.

The Persians crossed the border unexpectedly before August, the normal campaigning season, and advanced to Theodosiopolis. The local peasants submitted or fled with their livestock if they had time. Despite the surprise, the Persians were unable to commence a siege once they reached the city, although they intended to take the city (John and Menander), made camp to the south of the city, had brought siege equipment (μηχανήμασιν) with them, and came as far as to accompany his army, which was drawn up in formation before the city, and reconnoiter the walls in order to plan the assault (Menander). The Roman army, which had at first been taken by surprise by the early invasion and had been scattered to deal with other issues, was beginning to assemble to the north of the city. Khusro was dissuaded from attacking, either because he "realized that the city really was very well prepared for war (κατενόησε τὴν πόλιν εὖ μάλα καὶ ὡς ἀληθῶς ἐς τὰ πολέμια παρασκευασμένην)" and feared that the Roman army would soon arrive from their camp (Menander), or because it had actually drawn up for battle near the Persians encamped before the city (John).

6TH CENTURY

CAESAREA   *nekbōš*   576

J.Eph. 6.8; Whitby 1988: 265.

Khusro was unwilling to give up the campaign after failing to reach *Theodosiopolis and intended to attack an unspecified city, but his movements were cut off by the Roman army. He then went towards Caesarea in order to take it (*textus* 298.9: *da-n'ōl w-nekbōš qsāryā*). The Romans had however anticipated the Persian plans and came to Caesarea before them. After a standoff over several days, the Khusro did not give battle and withdrew.

SEBASTEIA   *armī w-awqdāh la-sbāsṭyā b-nūrā*   576

J. Eph. 6.8; Whitby 1988: 265.

After the failures before *Theodosiopolis and *Caesarea, Khusro tried to score some minor victory on the way back, apparently inspired by the Magian priests: "and at their instigation he wheeled round, and leaving Cappadocia, advanced to attack Sebastia (the original has "advanced to the vicinity of Sebastia;" *textus* 298.19]); for though terrified as well as his men at the Roman armies, yet from shame of being ridiculed for not having accomplished any of his plans, he attacked and burned Sebastia with fire (*textus* 298.22f: *armī w-awqdāh la-sbāsṭyā b-nūrā*). But he could take neither booty nor captives, because the whole land had fled from before him."

Since the Romans had been able to anticipate Persian movements (cf. *Caesarea), they had good time to evacuate vulnerable cities, so it is uncertain whether there even was a formal siege. John is quite explicit that there were no civilians or valuables around or in the city, perhaps not even a garrison. Khusro thus found an empty city to "lay low" without a fight.

MELITENE   *w-'al(w) la-mlīṭīnā w-armīw bāh nūrā*   576

J.Eph. 6.8; Whitby 1988: 265f.

After the burning of *Sebasteia, the Roman army managed to outmaneuver the Persians again. After some fighting, Khusro fled, leaving his camp with all his valuables behind (even his fire-temple that he used for worship while on campaign; more details are found in John, but see Whitby for discussion of the course of events in relation to other sources). Due to a subsequent failure in Roman coordination, the Persians took heart at the end of the campaign, defeated one Roman division and "were emboldened to attack (the text only has *w-'al(w) l-* "enter," *textus* 300.9) and set fire to (*w-armīw bāh nūrā*, ibid.) the city of Melitene."

As with *Sebasteia, the vocabulary of the original does not necessarily indicate fighting, and again Khusro probably burnt down an evacuated city. As the Persians were withdrawing, the Romans even sent ambassadors to berate Khusro for his actions, referring to the burning of Melitene as nothing more than "a piece of mischief." After another standoff between the Roman and Persian armies east of Melitene, the Persians withdrew hurriedly at night towards the Euphrates, but they were heavily defeated as they crossed the river the next day (see J.Eph. 6.9 and Whitby).

CITIES IN ORESPADA   *occupat*   577
J.Bicl. 47, s.a. 577 (Wolf 1999: 66); Collins 2004: 54f; Thompson 1969: 61.

"King Leovigild entered Orespeda and seized cities and fortresses in the same province, making it his own. Not long after, in the same place, the Goths suppressed a revolt of the common people and after that Orespeda was held in its entirety by the Goths." || Leovegildus Rex Orospedam ingreditur et civitates atque castella eiusdem provinciae occupat et suam provinciam facit. et non multo post inibi rustici rebellantes a Gothis opprimuntur et post haec integra a Gothis possidetur Orospeda. ||

Cf. the comments made concerning *Medina Sidonia (571) on the militarization of the *rustici*.

AMIDA   3 days   *w-ḥadrūh w-īteb(w) 'lēh*   578
J. Eph. 6.14, 27; Whitby 1988: 269.

John has two versions of this raid, the second one adding some details to the first. The Persians began a raid against Sophene in order to cause a diversion just as Maurice was assembling troops for the projected Roman invasion. They pillaged, took captives, murdered, destroyed and burnt, and arriving at Amida, besieged (*textus* 331.11 *w-ḥadrūh w-īteb(w) 'lēh*, ܘܚܕܪܘܗܝ ܘܐܬܒܘ ܥܠܝܗ) the city for three (thus the *Textus* and Payne Smith's translation; Brooks' Versio (251.26) has *triginta*) days, threatening to burn the suburbs unless they received a ransom. The Amidenes refused, since they believed that the Persians would burn the suburbs anyway (which they of course did). After 15 days of devastation, the Romans set out after them, leading to the invasion of Arzanene and the sieges of *Aphum and *Chlomaron.

APHUM AND OTHER FORTS   παρεστήσαντο, κατεσκάψαντο   578
Th. Sim. 3.15.14; Whitby 1988: 269.

"So the Romans invaded Arzanene and, since there was no resistance, they reduced (παρεστήσαντο) the very strong fort whose name was Aphumon, razed (κατεσκάψαντο) some other forts, and administered great slaughter to the Persian state."

CHLOMARON   πολιορκούντων...προσβολάς... ποιουμένων   578
Men. Fr. 23.7, J.Eph. 6.15 and 27; Th. Sim. 3.15.15; Whitby 1988:269f.

While John of Ephesus and Theophylact only refer to other events on this campaign (cf. *Aphum above), Menander has a good description: "When the Romans were besieging Chlomaron, making assaults on the circuit-wall, bringing up artillery to take the place and also secretly digging mines underground ("Ότι πολιορκούντων Ῥωμαίων τὸ Χλωμάρων καὶ προσβολὰς ἐν κύκλῳ ποιουμένων τάς τε ἑλεπόλεις μηχανὰς περιστησάντων, πρὸς δέ γε καὶ ὑπονόμους ἔνερθεν ὑπορυττόντων ἐς τὸ ἀφανές)," the Persian commander sent the Christian bishop of the city to negotiate with the Romans. He was to ask the Romans to withdraw in return for all the gold and silver in the city so they would not act "impiously towards God" by attacking and killing a predominantly Christian population. The Romans then tried and failed to bribe the Persian commander with lands and offices, before Persians tried to buy off the Romans with valuable liturgical vessels, which the Roman general (the later emperor Maurice) refused to accept. "For, he said, he had not come to plunder holy objects or to wage war on Christ, but with Christ's help to fight and to free those of his own faith from the Persians with their erroneous beliefs." Further negotiations with the bishop failed. So did apparently the siege, but the Romans continued their raiding in Mesopotamia.

John relates how much of the Christian population of Arzanene came out to meet the Romans with their "holy vessels and crosses and the gospel, asking of them a pledge for their lives, and saying, 'Have mercy upon us; for we are Christians like you, and ready to serve the Christian king.'" Those who chose Roman rule were resettled on Cyprus. Theophylact says they were taken as captives and numbered 100,000. Although the Romans could be brutal when raiding Persian Christian territories, here it appears that John's version is more correct, as it is more detailed and he had closer sources. The number may be correct, but it was more suitable to present it as a huge Roman victory (with a suitable number of "barbarian" captives) for propaganda purposes, which Theophylact does elsewhere (see Whitby for further sources; for a discussion of Theophylacts tendency towards propaganda as opposed to John, see Whitby's treatment of *Theodosiopolis and the other 576 conflicts).

| Singara | καταστρεψάμενος | 578 |

Th.Sim. 3.16.2; Whitby 1988: 270.

At the very end of his Arzanene campaign, Maurice "laid waste (καταστρεψάμενος) the fort of Singara" before returning home.

| Classis | *invadens, spoliatam...nudam reliquid* | c. 579 |

HL 3.13; *PLRE* 3 s.v. Faroaldus; cf. *Classis 584.

"Faroald, first duke of the Spoletans, invaded Classis with an army of Langobards and left the rich city stripped, plundered of all its wealth." || Hac etiam tempestate Faroald, primus Spolitanorum dux, cum Langobardorum exercitu Classem invadens, opulentam urbem spoliatam cunctis divitiis nudam reliquid. ||

This was obviously not only a raid on its *territorium*, since the Romans took the city back with the help of Droctulf in c. 584 (*HL* 3.19).

| Sirmium | | 579ff |

Men. fr. 25.1-2, 27.2-3; J.Eph. 6.24, 30ff; *HL* 4.20; Pohl 1988: 70-76; Whitby 1988: 87f.

The context and known events of the siege are well described by Pohl and need only a brief summary here. While accepting the usual subsidies from the Romans after the accession of Tiberius in 578, the Avars at around the same time (probably 579, as John says the siege lasted for two years and Whitby dates the surrender to 581) began building a bridge across the Sava downstream from Sirmium. John erroneously places the bridge across the Danube; he also alleges that another bridge was built, which may be correct, but it is uncertain exactly where and when—perhaps because the first one had become unsound; cf. Men. 27.3. To accomplish this, they used Roman engineers and construction workers (who only complied after two of them were beheaded) originally sent by Justin to help build baths and a palace (John 6.24; he refers to them as "MYKNYQW [pl.] w-banāyē," i.e. "μηχανικοί and builders," *textus* 326.8). The Avars and their subjects provided the required manpower. In order to shield their activities, they also constructed a river fleet with Lombard assistance (Paul, *HL*). When the Roman commander at Singidunum protested, the Khagan swore dire pagan and Christian oaths that he meant no harm to the Romans. Instead they claimed they were preparing a punitive expedition against the Slavs and wanted to march across Roman territory south of the Danube, and requested Roman help to recross and attack the Slavs from the south, as they had done as allies in 577. However, when their intentions were clear (they in fact announced through ambassadors that the choice was either to hand over Sirmium or fight), they used their fleet and superior numbers to keep the Romans from interfering with the large workforce engaged in bridge-building.

The Roman garrison in Sirmium appears to have been completely penned in and unable to sally due to the size of the threat, but Menander explicitly blames the incompetence of the local commander. Roman field armies were all occupied in the East, so an attempt at sending reinforcements was only possible by sending troops from the garrisons (διὰ φρουρᾶς, Men. 25.2.93) of Illyricum and Dalmatia. Tiberius also attempted various diplomatic delays and to buy Lombard and Turkish mercenaries, but was unsuccessful. When all these measures failed, Tiberius ordered the military commanders to surrender the city on the condition that the whole population should be allowed to leave (without any possessions). As the blockade had been in effect for two years, the civilian population had become completely emaciated, and when the Avars gave the starving population food (for which John praised them highly), many gorged themselves and died before they were evacuated. According to John (6.33), the Avars moved into the city but it was destroyed by fire the next year. While his account implies that they were incapable of civilized life, one may speculate that it was a deliberate Roman act of sabotage to make the city uninhabitable and thus easier to regain, either through capture or negotiations.

Uzès	*conclusus in civitate*	581

GT 6.7; *PLRE* 3 s.v. Iovinus 1.

As the result of a contested election for bishop, the disappointed candidate, Jovinus, attacked the incumbent Marcellus. "He was besieged in his own city (*conclusus in civitate*), where he defended himself valiantly. He was not strong enough to beat Jovinus, so he bought him off with bribes." Jovinus was the king's nominee, a former governor (*rector*) of Provence, and would be in possession of considerable military resources in the form of private estates and a network of royal and secular allies.

Périgueux, (?Agen), others	*pervadit, abstulit, subegit*	582

GT 6.12.

King Chilperic sent an army against king Guntram. Chilperic's army emerged victorious in battle before proceeding to occupy (*pervadit*) and extract oaths from Périgueux and moved (*pergit*) to Agen and capturing (*abstulit et dicionibus regis Chilperici subegit*) other cities.

BOURGES	*vallant, obsedebant*	583

GT 6.31.

The civil wars raged on with complex maneuvers and several battles. Much of the fighting was centered on Bourges. Chilperic had several armies converge on the city and surround it with siegeworks (*vallant*) while they ravaged the surrounding countryside. This army besieged (*obsedebant*) Bourges for a time but the kings settled their differences by negotiations; the besiegers went home laden with booty.

AVIGNON	*obsidente*	583

GT 6.1, 26; Mar. Av. a. 581 (2); Murray 2000: 108; *PLRE* 3 s.v. Mummolus; cf. *Phasis 556.

Mummolus broke with Guntram in 581 "and took refuge inside the walls of Avignon." Marius adds some details to his exile: "Mummolus the patrician took refuge in the border country of king Childebert, that is Avignon, taking with him his wife and children, a host of household servants, and much wealth." || Eo anno Mummolus patricius cum uxore et filiis et multitudine familiae ac divitiis multis in marca Childeberti regis, id est Avinione, confugit. ||

Gregory provides the most detailed account of the siege, which took place in 583, only after he had received the pretender Gundovald. "When Mummolus had first entered the town, he had realized that there was a section of the city boundary which was not protected by the River Rhône: he had a channel dug from the main stream, so that the entire circuit of the town should be protected by the river-bed. He had great pits dug deep into the bottom of the river at this spot, and then the water concealed this booby-trap as it flowed in." When Mummolus was about to be besieged, he tricked his opponents into entering the river during negotiations, where the enemy commander Guntram Boso was barely rescued from drowning. "Mummolus and Guntram Boso exchanged a few insults, and then each drew back. Guntram Boso laid siege to (*obsidente*) Avignon with the help of the troops provided by King Guntram." However, Childebert, the king he was supposed to be representing, had not given his permission to do this and ordered the siege to be raised.

The preparations made by Mummolus were remarkably similar to those used by the Romans at *Phasis (556) against the Persians. He had probably had expert craftsmen among his many household servants; one of them, a *faber*, was taken prisoner after the siege of *Convenae (585).

| SINGIDUNUM | ἐλάμβανε | 583 |

Th.Sim. 1.4.1ff; Pohl 1988: 77; Whitby 1988: 142f, 151.

The Avars attacked Singidunum by surprise at harvest, when most of the population was camping out in the fields (ἐν τοῖς ἀγροῖς αὐλιζομένοις). Whitby rejects Theophylact's note of Roman "peacetime indolence" leaving the city without military equipment as moralizing rhetoric; it is also clear from *Sirmium (579ff) that there were military commanders and hence garrisons in both Singidunum and Sirmium. There is no reason to assume that the garrison of the former was removed after the fall of the latter, and certainly not that it had been stripped of its defensive equipment. His depiction of the city as a farming community that could be surprised at harvest time should not be in doubt; the insecurity of the region meant that open villages had been abandoned in most of the northern Balkans, cf. Curta 2001. Despite the surprise, there was heavy fighting at gates and many Avars fell before they took the city. Whitby dates the Avar attack to 583, Pohl to 584. Whitby's position is to be preferred, as it corresponds well with the chronology of *Sirmium, which lasted two years, followed by just under two years of peace. The Romans must have regained control of the city (as well as *Augustae and *Viminacium below) with the peace treaty of 584, and had certainly reestablished a naval base there by 588, since they destroyed the Avaro-Slav flotilla that was being constructed upstream (see *Singidunum 588).

| AUGUSTAE | ἀνελών | 583 |

Th.Sim. 1.4.4; Pohl 1988: 77f; Whitby 1988: 142f; cf. *Singidunum 583.

"After destroying Augustae and Viminacium..." the Avars marched all the way to *Anchialos (below). Again, Whitby's dating is to be preferred (cf. *Singidunum above), but Pohl notes that the Avar route bypassed several more important cities without attempting to assault them, such as Bononia and Ratiaria. This indicates that the Avars still had limited capacity for sieges, and were attempting to take lesser forts by surprise. Reports of extensive ravaging (λητζεται) of the neighboring city territories (1.4.3) may also be connected with the Avars' need to forage continuously in order to stay in the field.

| VIMINACIUM | ἀνελών | 583 |

See *Augustae above.

ANCHIALOS	over winter	στρ. καὶ ... περιτέμνεται	583f

Th.Sim. 1.4.4ff; Pohl 1988: 78f; Whitby 1988: 142f.

After taking *Augustae and *Viminacium, the Avars marched across the Haemus mountains. They encamped by and blockaded (στρατοπεδεύεται καὶ ... περιτέμνεται) the Black Sea port of Anchialos. Staying over the winter, they raided the surrounding villages and used the baths outside the city. Pohl, who dates this to 584-5, one year later than Whitby, connects Theophylact's evidence with the last surviving notices in John of Ephesus (6.45-49) on the Avar capture of Anchialos and how the Khagan donned imperial robes that they found in the city. However, Whitby demonstrates that this was a separate event that occurred in 588 (see *Anchialos s.a.). Thus, Pohl's discussion of John's evidence that two Roman cities were required to pay tribute after the fall of Anchialos probably belongs to the later date. For the first siege, this means that the Avars may have only been raiding, reconnoitering, applying pressure on the Romans (who sent ambassadors to Anchialos only after three months), and testing the defenses of Anchialos and perhaps surrounding cities.

SEVILLE; OTHERS	*gravi obsidione concludit, pugnando ingreditur*	583f

J.Bicl. 55, 65ff, 69 (Wolf 1999: 68ff); Collins 2004: 56-60; Thompson 1969: 64-73.

Hermenegild, the son of Leovigild, rose in revolt in Seville. "He made other cities and fortresses rebel with him against his father, causing greater destruction in the province of Spain—to Goths and Romans alike—than any attack by external enemies." || ...et alias civitates atque castella secum contra patrem rebellare facit. quae causa provincia Hispaniae tam Gothis quam Romanis maioris exitii quam adversariorum infestatio fuit. || (J.Bicl. 55, s.a. 579)

"Leovigild raised an army to subdue his rebel son."|| Leovegildus rex exercitum ad expugnandum tyrannum filium colligit. || At first he directed his efforts against Merida (GT 4.18, see further Thompson), thus securing control of the surrounding territory. (J.Bicl. 65, s.a. 582)

"After assembling his army, King Leovigild surrounded the city of Seville and trapped his rebel son with a very tight siege. King Miro of the Suevi came to relieve Seville in support of Hermenegild and there he ended his days. His son Eboric succeeded him as king of the province of Galicia. Meanwhile King Leovigild afflicted the city first with hunger, then with the sword, and finally with a blockade of the Baetis river." || Leovegildus rex civitatem Hispalensem congregato exercitu obsidet et rebellem filium gravi obsidione concludit, in cuius solacium Miro Suevorum rex ad expugnandam Hispalim advenit ibique diem clausit extremum. cui Eboricus filius in provincia Gallaeciae in regnum succedit. interea Leovegildus rex supra dictam civitatem nunc fame, nunc ferro, nunc Baetis conclusione omnino conturbat. || (J.Bicl. 66, s.a. 583)

"Leovigild restored the walls of the ancient city of Italica, which proved a great misfortune for the people of Seville." || Leovegildus muros Italicae antiquae civitatis restaurat, quae res maximum impedimentum Hispalensi populo exhibuit. || (J.Bicl. 67, s.a. 584)

"King Leovigild entered Seville by force after his son Hermenegild had fled to imperial territory. Leovigild captured the cities and fortresses that his son had seized and not long after apprehended him in the city of Córdoba. He exiled Hermenegild to Valencia, depriving him of his rule." || Leovegildus rex filio Hermenegildo ad rem publicam commigrante Hispalim pugnando ingreditur, civitates et castella, quas filius occupaverat, cepit et non multo post memoratum filium in Cordubensi urbe comprehendit et regno privatum in exilium Valentiam mittit. || (J.Bicl. 69, s.a. 584)

J.Bicl. 66 is somewhat ambiguous, as it is unclear to whom Miro came in aid, but this is clarified by J.Bicl. s.a. 585, when Leovigild conquered the Sueve kingdom—hardly the act of a grateful ally. Furthermore, the following compressed description gives three distinct stages: First a blockade to try and starve out the besieged, then attempts at storming the city, before finally the river itself was diverted. The fortification of Italica helped cap off the blockade or prevented relieving armies from the other cities under Hermenegild's control from breaking the blockade. Thompson dates the fall of Seville to the summer of 583 after a year of siege and the capture of Hermanigild to February 584; he also discusses the numismatic evidence which confirms John's description.

AKBAS	armīw 'law, ܡܠܟ ܐܟܒܣ; ἐνεχείρει αἱρήσειν	583

J.Eph. 6.36; Th. Sim. 1.12.1-7; Whitby 1988: 272.

According to Theophylact, this was a Roman failure. The Persians were attempting to take (ἐνεχείρει αἱρήσειν) the fort of Aphum, while the Romans assaulted Akbas. The Persian garrison at Akbas called for help with signal fires and the Romans were defeated by a Persian relieving force that arrived from Aphum. However, John, who was much closer to the events, records that the fort was illegally built by the Persians on demilitarized territory. When the opportunity arose, the Romans "attacked it, and ... invested it on all sides, and commenced a blockade," (ܟܪܟܘܗܝ ܚܕܪܘܗܝ ܡܠܟ ܐܟܒܣ

586   CORPUS OBSIDIONUM

,ܘܡܠܟ ܚܝܘܬܐ ܥܠ ܕܝܢ ܗܘܐ ܫܠܝܡܘ ܘܐܬܐ ... ,ܡܨܕܘ) starving the Persian garrison into surrender. Although they had been granted their lives and liberty, many of the Persians died when they emerged from the fort because they gorged themselves on the available water. The Romans demolished the fort completely.

BRESCELLO	*expugnare, exuperantes*	c. 584

HL 3.18; *PLRE* 3 s.vv. Authari 2, Droctulfus 1; chapter 3.3.

Authari besieged Brescello (*Brexillum...expugnare adgressus est*), which was strongly defended by the renegade Lombard *dux* Droctulf, who was fighting alongside imperial soldiers (*In qua Droctulft...sociatus militibus, Langobardorum exercitui fortiter resistebat*). "The Langobards waged grievous wars against him and at length overcame him together with the soldiers he was aiding, and compelled him to withdraw to Ravenna. Brexillus was taken and its walls were leveled to the ground." || Adversus quem Langobardi gravia bella gesserunt, tandemque eum cum militibus quos iuvabat exuperantes, Ravennam cedere conpulerunt. Brexillus capta est, muri quoque eius solum ad usque destructi sunt. ||

Authari then made peace with the Roman exarch Smaragdus.

TOURS	*ad conpraehendendas civitates, subdi*	584

GT 7.12

Guntram sent his forces to capture several cities (*ad conpraehendendas civitates*). Tours was forced to submit (*subdi*) to Guntram when the "men of Bourges" began their ravaging by burning a suburban church. Their new allegiance was made to avoid the complete destruction of their surrounding countryside.

POITIERS	*ad conpraehendendas civitates, cuncta vastarent*	584

GT 7.13; *Tours above.

Shortly after *Tours (above), the men of Tours joined the men of Bourges to ravage the territory (*cuncta vastarent*) of Poitiers, which submitted to Guntram after expelling the garrison of king Childebert's men.

CLASSIS	*pepulerunt*	584/5

HL 3.19; *PLRE* 3 s.v. Droctulfus 1; cf. *Classis (579).

"With the support of this Droctulf, of whom we have spoken, the Ravennans' soldiers often fought against the Langobards, and after a fleet was built, they drove out with his aid the Langobards who were holding the city of Classis." || Huius sane Droctulft, de quo praemisimus, amminiculo saepe Ravennantium milites adversum Langobardos dimicarunt, extructaque classe, Langobardes, qui Classe urbem tenebant, hoc adiuvante pepulerunt. ||

POITIERS   *pars maxima regionis devastata*   585
GT 7.24.

Another expedition was sent by Guntram to pound Poitiers into submission and accept a royal garrison. "When the invading armies came close to the city itself and most of the neighbourhood had been ravaged (*cum exercitus proprius ad urbem accederet et iam pars maxima regionis devastata*), they sent representatives who were to declare their loyalty to King Guntram." The bishop had to bribe the newly installed garrison with church plate in order to escape accusations of disloyalty. The territory of Tours, though loyal to Guntram, was also ravaged by the army.

TOULOUSE   [*resistentibus et bellum parantibus*]   585
GT 7.27; Bachrach 1996; Goffart 1957; Wood 1994: 93f.

The pretender Gundovald arrived at Toulouse with a "vast army" which convinced the Toulousains to submit, even though they had prepared for resistance. || His ita resistentibus et bellum parantibus, adveniente Gundovaldo cum magno exercitu cum vidissent, quod sustenere non possint, susceperunt eum. ||

CONVENAE   585
GT 7.28, 32-43; see *Toulouse above; the siege is discussed in detail in chapter 4.1.4.

Guntram's army mobilized against Gundovald. The "poor men" of Tours passed disorderly through the territory of Poitiers due to recent scuffles, while the professionals from Tours deserted on the march (7.28). Gundovald's messengers were tortured under interrogation and thrown in chains, in spite of the normal inviolability of envoys. This indicated that the ensuing struggle would be extremely brutal. In order to secure his position, Guntram declared his legitimate nephew Childebert as his heir (7.32-33). Gundovald fled to Convenae with a large army and convinced the inhabitants of the town to prepare for a siege by Guntram's army, including carrying in supplies and movable goods into the fortifications. He also convinced the local troops to sally out against Guntram's approaching army, but used this as a trick to get them outside the fortification. Gundovald's supporters found enormous stores of grain and wine, enough to last for years (7.34).

Guntram's army set out; part of it crossed the Garonne but most remained behind with the slow-moving baggage train. The vanguard were to close off the city and raid the countryside (7.35). Gundovald was taunted by the besiegers when he tried to address them, but he persisted and defended his actions (7.36).

"The fifteenth day of the siege dawned. Leudegisel spent this time in preparing new machines with which to destroy the city. He constructed wagons, which he fitted with battering-rams and protected with wattle-work, old leather saddles and planks of wood, so that the troops could rush forward under cover to knock down the walls. The moment they came near, rocks were dropped on their heads, so that all who approached the city walls were killed. The defenders hurled down on them flaming barrels of pitch and fat, as well as boxes filled with stones. Once nightfall had brought the struggle to an end, the besiegers returned to their camp." (7.37) There is also a description of Chariulf's granaries, and how Bladast, one of the conspirators, took fright and set fire to the church in order to cover for his escape. "As soon as morning came, the besiegers launched a second assault. They prepared fascines from bundles of sticks, and tried to fill in the deep ravine which lies on the eastern side. This device did not harm anyone. Bishop Sagittarius walked round and round the ramparts fully-armed, and with his own hand kept tossing rocks on to the heads of the besiegers."

The besiegers opened secret negotiations with the supporters of Gundovald and agree to betray him. Gundovald was handed over, killed and mutilated. The conspirators opened the gates to the besieging army. "All the common people were put to the sword, and all the priests of the Lord God, with those who served them, were murdered where they stood at the church altars. ...there remained not one that pisseth against a wall..." (I Sam. 25.22; I Kings 16.11; GT 7.38). The main conspirators were executed and Mummolus' treasure at Avignon revealed (7.39f). A member of Mummolus' household, a carpenter who had been taken captive, receives some attention from Gregory only because of his massive size (7.41).

"Some time later a decree was issued by the judges that anyone who had shown unwillingness to join this military expedition should be fined." The count of Bourges sent representatives to the local church of St. Martin to claim the fine, but the clerics resisted, claiming "They are not in the habit of taking part in military maneuvers." A miracle helped the clerics avoid further problems (7.42; cf. discussion in ch. 6.2 above). The last of the conspirators, Desiderius, Waddo and Chariulf fled to various places (7.43).

| CARCASSONNE | *nullo resistente ingressi* | 585 |

GT 8.30; Thompson 1969: 75; *PLRE* 3 s.v. Terentiolus.

Guntram organized an army to conquer Septimania and proceed further in retaliation for the treatment of Ingund and Hermangild (Thompson; GT 8.28). Two columns were to attack against *Nîmes (below) and Carcassonne; the latter consisted of the men of many *civitates* who marched on Carcassonne, but the whole affair appears rather bizarre: after raiding their way to the city, "When they reached this town they found the gates left open for them by the citizens and they marched in without meeting resistance (*nullo resistente, ingressi*). A quarrel arose (*scandalo commoto*) between them and the Carcassonnais, so they marched out again. At this juncture Terentiolus, one-time Count of Limoges and leader of the Frankish army, was struck by a stone thrown from the walls and killed (*lapide de muro commoto percussus, occubuit*). The enemy revenged themselves on him by cutting off his head and taking it back into town. Thereupon the entire army was stricken with panic and the men made up their minds to return home." The army abandoned their plunder, and suffered from Gothic ambushes as well as an attack by the people of Toulouse.

What must have transpired is the following: Guntram's troops reached the city, with most of the army remaining outside, but a delegation entering to negotiate in order to establish the new political situation. This is when the "scandal," or serious disagreement, arose that led the leaders to march out. As they went out they were vulnerable to fire from the walls; the rest of the army panicked, but it is unlikely that they would have been in the city at the time—Terentiolus would have been one of the first to leave, and it would be unwise to kill the commander of an army that was inside one's city.

| NÎMES; UNNAMED CITIES | *inclusis* | 585 |

As *Carcassonne above; *PLRE* 3 s.v. Nicetius 3.

"Those who had attacked Nîmes ravaged the entire neighbourhood, burning the houses and the crops, cutting down the olive-groves and destroying the vineyards, but they were unable to harm the beleaguered townsfolk (*nihil inclusis nocere potentes*) and so marched on to other towns. These, too, were heavily fortified and well provided with food and other necessities, so the invaders laid waste the surrounding countryside, not being strong enough to break into the towns themselves."

Duke Niketius took part in the expedition and broke his promises when he plundered a town that had opened its gates for him. 5,000 said to have died on these expeditions. Guntram, angered by incompetence, arranged a council of bishops and nobles to improve matters.

| CABARET | pace... occupant; obtinuit | 585 |

J.Bicl. 75, s.a. 585; Wolf 1999: 71; GT 8.30.

"The Franks entered Gallia Narbonensis with their army, wanting to seize it. Leovigild sent his son Reccared to meet them. He drove back the army of the Franks and the province of Gallia Narbonensis was freed from their attacks. Reccared captured two fortresses, one peacefully, and one in battle, along with a great multitude of men. In a violent battle, King Reccared attacked and seized the fortress called Ugernum, which is located very securely on the edge of the Rhône river. He returned victorious to his father and his country."

|| Franci Galliam Narbonensem occupare cupientes cum exercitu ingressi. in quorum congressionem Leovegildus Reccaredum filium obviam mittens et Francorum est ab eo repulsus exercitus et provincia Galliae ab eorum est infestatione liberata. castra vero duo cum nimia hominum multitude unum pace, alium bello occupant. castrum vero qui Hodierno vocatur tutissimus valde in ripa Rhodani fluminis ponitur, quod Reccaredus rex fortissima pugna aggressus obtinuit et victor ad patrem patriamque redit. ||

Presumably the first fortress, taken peacefully, should be equated with Cabaret below, while Ugernum is the equivalent of Beaucaire. According to Gregory, the Visigoths responded with an invasion led by Reccared, son of Leovigild: "He has captured the castle (*castrum obtinuit*) of Cabaret, he has ravaged the greater part of the land round Toulouse and he has led off a number of captives. He has attacked the castle (*castrum inrupit*) of Beaucaire near Arles, made off with all the inhabitants and their property, and shut himself up inside the walls of Nîmes." Guntram responded by sending over 4,000 frontier guards plus a force under Duke Nicetius of Clermont to patrol the border. There were failed Spanish embassies in 586 despite further aggression against Frankish shipping (8.35) and another raid against Narbonne by Reccared (8.38).

| BEAUCAIRE/UGERNUM | bello occupant, fortissima pugna; inrupit | 585 |

As *Cabaret (above).

| 8 MYSIAN AND SCYTHIAN CITIES | εἷλον | 586 |

Th.Sim. 1.8.10f; Pohl 1988: 82-85; Whitby 1988: 145ff.

The Avars found cause to break the peace made after *Anchialos (583) with the Bookolabras-affair (see esp. Pohl, who dates this and the campaign to 585). "The Chagan's men ravaged (ἐλυμήναντο) all the environs of Scythia and Mysia [Pohl p. 85 has "'Skythen und Myser" bot er auf"], and captured (εἷλον) many cities, Ratiaria, Bononia, Aquis, Dorostolon, Zaldapa, Pannasa, Marcianopolis, and Tropaion. The enterprise provided him with considerable labour: for he did not reduce these cities without sweat and trouble, even though he was helped along by a strong following wind in the suddenness of the invasion..." Theophylact in addition blamed Roman "indolence" and lack of preparations, but as with *Singidunum (583), this seem to be a rhetorical claim. The impressive list is rightfully criticized by Pohl (who also argues that *Appiaria (below) probably belongs in this context): "Viele dieser Städte behielten weiterhin ihre Bedeutung; man könnte sich fragen, ob die imposante Liste nicht vielmehr diejenigen angibt, die das Heer des Khagans angriff, die aber nicht alle bezwungen wurden." For instance, the Roman army sent against the Avars in 587 used Marcianopolis, one of the cities apparently taken by the Avars, as a base for their operations (Th. Sim. 2.11.3).

Whitby reconstructs the route of invasion (since Theophylact jumbled the names in apparently random order) along the lower Danube plain to the vicinity of the Black Sea; the Avars wintered in the Dobrudja south of the Danube delta. Whitby also associates the Avar invasion with desperate attempts to recruit a field army recorded in the later Syriac (but omits J.Eph. 6.45-49, who provides some pertinent evidence) and other sources in the autumn of 586.

Cf. *DAI* 30 and chapter 7.2.3 on how Avars surprise several cities in Dalmatia by capturing a Roman patrol and using their uniforms and equipment to gain access to Roman cities without a fight.

APPIARIA	κατεβέβλητο	586

Th. Sim. 2.15.13-16.11; Pohl 1988: 85, 87f; Whitby 1986: 65f nn. 38f; *idem* 1988: 145, 150f.

At a point that does not make much sense in Theophylact's account (see discussions by Pohl, Whitby and below), the Avars attacked the city of Appiaria, having caught the local military engineer outside the city (for a discussion of his position and additional evidence from Theophanes, see chapter 2.2.2f; for the anecdote in general chapter 7.2.3; for the engine chapters 5 and 8). The Avars made Bousas an offer he could not refuse. In order to survive, "Next Busas taught the Avars to construct a sort of besieging machine, since they had as yet no knowledge of such implements, and he prepared the siege-engine for a long-range assault. Shortly afterwards the fort was overthrown, and Busas exacted punishment for inhumanity by giving the barbarians skilled instruction in the technology of siegecraft. For, as a result, the enemy subsequently reduced without difficulty a great many other Roman cities by using the invention as prototype." καὶ δῆτα ὁ Βουσᾶς τοὺς Ἀβάρους ἐδίδασκε συμπήγνυσθαι πολιορκητικόν τι μηχάνημα ἔτι τῶν τοιούτων ὀργάνων ἀμαθεστάτους ὑπάρχοντας, ἀκροβολίζειν τε παρεσκεύαζε τὴν ἐλέπολιν. καὶ μετ' οὐ πολὺ τὸ φρούριον κατεβέβλητο, καὶ δίκας ὁ Βουσᾶς ἀπανθρωπίας εἰσέπραττε δεινόν τι διδάξας τοὺς βαρβάρους πρὸς πολιορκίαν τεχνούργημα. ἐντεῦθεν γὰρ πλείστας λοιπὸν τῶν Ῥωμαϊκῶν πόλεων ἀμογητὶ τὸ πολέμιον παρεστήσατο τῷ ἀρχετύπῳ σοφίσματι προσχρησάμενον. (2.16.10f)

This anecdote is highly problematic according to both Pohl and Whitby, as it fits better with Avar invasions of other regions. I would suggest that it was (mis)placed by Theophylact in order to provide a prelude to the Avar assaults on *Beroe and other Thracian cities (which were however all Avar failures). Whitby suggests that the Avars took the city after returning from their assaults on *Beroe and other Thracian cities in 587. While possible, similarly complex scenarios are rejected by Pohl for good reasons. I therefore follow Pohl's suggestion that it took place as the same time as the *8 Mysian and Scythian cities above, but with Whitby's 586 dating of the latter.

Note the meaning of *akrobolizein*, to sling (from afar).

THESSALONICA [storm] 586/597
Miracula St. Demetrii 1.13-14 (117-65); Pohl 1988: 101-107; Whitby 1988: 115-21; Vryonis 1981; Chevedden 2000; Curta 2001: 97f; discussion in chapters 7.2 *passim* and 8 *passim*.

[13th miracle:] The Avars decided to attack Thessalonica after being rebuffed by Maurice; the Khagan assembled all his Slav subjects and added to them barbarians of other races, ordering them to march against Thessalonica (117). It was the greatest army seen ever (like a new army of Xerxes), about 100,000 men (ἑκατὸν χιλιάδας), drying up rivers and sources wherever they camped (οἷς ἂν στρατοπεδεύσαντες παρεκάθισαν), laying the earth completely waste. They moved so fast the citizens did not know of their arrival until a day in advance (118). They appeared on September 22, four days before they were expected; hence the guard was still not fully mobilized. They were saved by a miracle of St. Demetrius, as the Slavs mistook the fort (φρούριον) of the martyr Matrone for the city itself at their nocturnal arrival. They only recognized their mistake at first daylight, when they began to attack the city itself with ladders prepared in advance (κλίμακας άνορθώσαντες...αύτοῖς προκατεσκευασμέναι, 119).

The saint miraculously appeared on the unguarded wall in the form of a soldier (ἐν ὁπλίτου σχήματι), killing the first enemy scaling the wall with a ladder by thrusting his spear "between two merlons of the breastwork" (κατὰ τὸ μέσον τῶν δύο ἐπάλξεων) so that he fell down dead and took the others with him; his blood was left on the parapet (120). Against the skeptics, John argues that this miracle must be true since no soldier had come forth to claim a reward. In fact, only a few men of the city were on the walls then, and even they went home at daybreak to rest (οὐδὲ τῶν ἀνδρῶν τῆς πόλεως ἦσάν τινες...πλὴν λίαν ὀλίγων, 121).

By next morning, a huge crowd surrounded the city from coast to coast; the population could see only heads instead of trees (122). The crowd pillaged and ate all around the forts (φρούρια), suburbs (προάστεια), and fields (ἀγροὺς). They did not bother to set up a palisade (χάρακα) and embankment (πρόσχωμα), since their shields and bodies were enough ("for their ditch is the intertwining of shields placed upon each other, not allowing any way through, and their embankment the denseness of their bodies," 123). This implies that they were in a hurry due to their huge numbers, only setting up a shield wall to protect themselves—or forming up in serried ranks in something resembling a phalanx or testudo formation. John attributes the sight to the sins of his people; no-one had ever seen such a barbarian phalanx except a few who had been in the army or been assigned to military duty far from the city (124). Note that the tactics seem parallel to those used by the Arabs at e.g. *Dvin 640.

The barbarians arrived shortly after a plague in July, which had left the city depopulated and in sorrow (125). Secondly there was a huge number of barbarians, more than all the inhabitants of Macedonia, Thessaly and Achaea together (126). Thirdly, only the least useful of those who survived in the city were actually in it; most were in the suburbs and fields when the enemy came, since it was time for harvest (διὰ τὸ τρύγης εἶναι καιρόν, 127).

There is good evidence for those who would normally serve in urban defense based on the list of those who were absent: The chosen young men of the army and of those serving in the great *Praitorion* (πλειόνων δὲ καὶ αὐτῶν τῶν ἐπιλέκτων νεανιῶν τοῦ τε στρατιωτικοῦ καὶ τῶν ἐν τῷ μεγίστῳ στρατευομένων πραιτωρίῳ) were absent on state affairs in Hellas (128), perhaps helping other communities in need due to persistent raids in the 580s. In addition, the wealthy and influential had combat-hardened servants in their prime (δούλοις ἀκμάζουσι καὶ ἐμπειροπολέμοις ἐκαλλωπίζοντο) and those belonging to the eparchy of Illyricum had gone to Constantinople with many more friends and all their retinue (σὺν φίλοις πολλῷ πλείοσι καὶ τῇ θεραπείᾳ πάσῃ, 129). In effect says: "No I'm not exaggerating" (130; apparently many people accused him of that).

[14th miracle:] The archbishop found himself at the theatre and was spoken to by an actor (τραγῳδός) concerning the fate of his daughter (Thessalonica) and her children (the citizens) just a few days before the arrival of the barbarian horde (131-34). But all that happened later. Collection of prisoners, booty and supplies from the countryside dominated the first day of the siege, and the besiegers consumed everything edible during the first two days (137). In the evening they collected wood to make a "river of fire" around the whole city and let out fearsome cries (138; cf. *Strategikon* 11.4).

139: "Then we heard noises from all around through all night and the next day, as they were preparing *helepoleis* and iron rams and stone-throwers of enormous size, and the so-called siege-sheds, which, with the stone-throwers, were covered with dry hides, (but) they changed their minds again because they could not resist fire or pitch thrown at them and covered their engines with the blood-soaked hides of newly slaughtered cattle and camels. And thus having brought them closer to the wall, from the third day onwards they threw stones, or rather hills as it would appear from their size, and further their archers (shot) arrows that resembled winter snow, so that no-one of those (standing) on the wall could peep out without danger and see anything (that was happening) outside. But the *khelōnai* were advancing towards the wall, with innumerable crowbars and axes they ate away at the foundations; for there were more than a thousand of them, I believe."

ἐλεπόλεις καὶ κριοὺς σιδηροῦς καὶ πετροβόλους ὑπερμεγέθεις, καὶ τὰς καλουμένας χελώνας, ἅστινας σὺν τοῖς πετροβόλοις δέρρεσιν ἐπισκεπάσαντες ξηραῖς, μεταβουλευσάμενοι πάλιν διὰ τὸ μὴ ὑπὸ πυρὸς ἢ πίσσης καχλαζούσης ἀδικεῖσθαι, δέρρεις νεοσφαγῶν βοῶν καὶ καμήλων ἡμαγμένας ἔτι τοῖς ὀργάνοις ἐκείνοις ἐνήλωσαν. Καὶ οὕτω ταῦτα πλησίον τοῦ τείχους προσάγοντες, ἀπὸ τῆς τρίτης ἡμέρας καὶ ἐπέκεινα ἔβαλλον λίθοις, μᾶλλον δὲ βουνοῖς τῷ μεγέθει τυγχάνουσι, καὶ βέλεσι λοιπὸν οἱ τοξόται αὐτῶν νιφάδας μιμουμένοις χειμερινάς, ὡς μή τινα τῶν ἐν τῷ τείχει δύνασθαι κ'ἂν προκύψαι ἀκινδύνως καί τι τῶν ἔξω θεάσασθαι· ἀλλὰ καὶ ταῖς χελώναις τῷ ἔξω τείχει προσφύντες, μοχλοῖς καὶ ἀξίναις ἀμέτρως περιετίτρων αὐτοῦ τὰ θεμέλια· ἦσαν γὰρ αὐταὶ τῶν χιλίων πλείους οἶμαι τῷ ἀριθμῷ.

The Thessalonians recovered their courage after two days of great fear and began to laugh at and taunt their attackers, inviting them to use the public baths (143; perhaps this was a spoof on the Khagan's bathing on other occasions; later, at *Anchialos 588, the Khagan's wives are said to have used the baths in the city, although this may have been conflated with earlier events). More on divine intervention (144).

The author claims it would take too long to tell everything that happened in the north and west of the town, including bravery of the barbarians, use of siege engines (μηχανήματα), and the failed building of a platform in the water to attack the city from the sea; it collapsed thanks to divine intervention. (ὅτε καὶ τῇ θαλάσσῃ τὴν ξυλινὸν γῆν καὶ πλατεῖαν ἐπιθεῖναι κατεμηχανήσαντο. Note that this looks like skills related to bridge-building, cf. Menander and John of Ephesus on the Avar bridge built at *Sirmium 579ff. Perhaps it collapsed through lack of skill or sabotage if prisoners of war were involved.) The author limits himself to a few incidents to the east, where he himself was witness (who protests that he has no other goal than to prove that the salvation of the city came from God alone, 145). Again, the enemies used the first two days to gather supplies and "prepare many and fearsome machines of various sorts against the city" (κατὰ τῆς πόλεως πολλὰ καὶ φοβερὰ διάφορα ηὐτρέπισαν ὄργανα). From the third to the seventh day, they attacked the walls with machines, including *helepóleis*, *kríous*, *petrobólous* and ground-crawling pole-working devices of the tortoises (τῶν χελωνῶν τὰ χαμερπῆ ῥαβδουργήματα) [μαγγανικὰ are mentioned in the apparatus; ms V, 12th century from Athos]. An attempted assault against the Cassandreotic gate with a iron-faced ram (σιδηρομέτωπον κρίον) but it was driven back in terror by sight of a hook (ἁρπαγά τινα), which the hagiographer belittles, but were probably very efficient, after burning the ram and other machines (146). The tortoises were successful in undermining the outer wall (προτείχισμα), while impervious to artillery and archery from defenders, but a brave sally by a few defenders armed with lances, spears and bows drove the enemy back in terror—before contact, it seems, as the defenders only reached the outer wall but did not engage (147). The miracle of the jammed gate (148f).

Next day the enemies abandoned their χελῶνας with their crowbars (μόχλους) and pickaxes (δικέλλας), and turned to the *petroboloi* (150). The first clear description of a traction trebuchet (151) is now famous among military historians. The translation most commonly used is that by Vryonis, but see Chevedden's emendations that clarify the mechanics and the discussion in chapter 8.2 *passim*:

"These were tetragonal and rested on broader bases, tapering to narrower extremities. Attached to them were thick cylinders well clad in iron at the ends, and there were nailed to them timbers like beams from a large house. These timbers had the slings hung from the back side and from the front strong ropes, by which, pulling down and releasing the sling, they propel the stones up high and with a loud noise. And on being fired they sent up many great stones so that neither earth nor human constructions could bear the impacts. They also covered those tetragonal ballistrae with boards on three sides only, so that those inside firing them might not be wounded with arrows by those on the walls. And since one of these, with its boards, had been burned to a char by a flaming arrow, they returned, carrying away the machines. On the following day they again brought these ballistrae covered with freshly skinned hides and with the boards, and placing them closer to the walls, shooting, they hurled mountains and hills against us. For what else might one term these extremely large stones?"

It is clear that the Avaro-Slavs were not able to coordinate efforts well, using different engines uncoordinated; tortoises and rams were deployed on the third day, abandoning them and turning to the trebuchets afterwards; then failing to cover them in fresh hides, forcing them to withdraw and refit for the next day; this was very wasteful from a logistical perspective as well. The Thessalonians set up countermeasures by suspending mats to soften the impact of artillery (152).

The hagiographer emphasizes that God, not man, saved the city. From dawn to the seventh hour (ἀπὸ τοῦ αὐγάσαι ἕως ὥρας ἑβδόμης καθ' ἑκάστην ἡμέραν—note that this took place daily, not in Lemerle's translation), the barbarians threw huge rocks, but missed consistently, overshooting or throwing too short. It seems that a combination of inexperience, inconsistent weight of rocks, and too few engineers to guide them made it difficult for the barbarians to successfully operate the engines for hours at a time. In contrast, the stones thrown from the city walls fell straight into the trestle of the barbarian *petroboloi*, killing those inside (153). Only one barbarian stone hit the parapet, destroying it completely down to the walkway (ἕως τοῦ περιπάτου). More than fifty *petroboloi* had been assembled at the eastern wall (154). When the barbarians withdrew to camp [one day], the Thessalonians made a sally from a gate by the sea, killing bathing barbarians, who must have been quite sweaty from hauling stones and pulling trebuchet ropes all day (155).

The last miracle occurred when the barbarians rested on the seventh day of the siege, deciding on an all-out assault the next day: deserters told the Thessalonians, who became terrified, but suddenly, at the eighth hour, the barbarians fled to the mountains yelling, only descending three hours later at dusk. They fought and killed each other once they were back at camp (156ff). The last night was passed in silence, but in the morning a large crowd of deserters stood outside the city, while the camp was empty of the barbarian horde. The deserters were admitted and interrogated (159). They said they surrendered so as not to die of hunger and recognized the Thessalonians as victors, telling the story of the huge army they saw sallying out the day before (160). Their description of the leader of army matched St. Demetrius; followed by prayers (161f).

The city was empty of people (cf. miracle 13) so that God could prove his mercy (135). This was further manifest through the miraculous appearance of unknown fighters on the walls, who were not only seen by citizens, but also barbarians, many of whom sought refuge with the town's leaders (προσρυέντες τοῖς ἄρχουσι) during the last days, despairing of victory and describing how the Avar *hēgoúmenos* (ἡγούμενος) had ascertained how there were few defenders in the city due to the harvest season, and sent them off to take the city immediately (αὐθημερόν), but they became totally demoralized by the sight of the "defenders" (136). The Slavs seem to have had very fragile morale on long-lasting expeditions.

The citizens sent out cavalry (ἵππεις) to scout for threats; they reported that the enemy had escaped in fear and disorder, leaving clothing, equipment, animals and bodies (163). The author ends with a reply to critics (who asked "were not the Thessalonians brave and skilled?") and skeptics; doxology (164f). Prayers and answers to the skeptics also occur elsewhere (140-142; 142, "Perhaps you ask: Whence is this clear? Who saw God?"—the answer given was "from his works—nobody ever saw God himself"). While this could be a *topos*, the text seems to contain evidence of actual use in the liturgy and this might explain criticism from the congregation who were familiar with the events.

Except for Vryonis, who argues for 597 (which is theoretically possible based on the indiction cycle), most scholars now accept the date 586 for this event. The connection between the Avars and Slavs is unclear (cf. *Thessalonica 604, 615 and 618, when the Slavs attacking the city were independent or only supported by the Avars), as the association between Avar leadership and Slav action might be an inference on the part of the author.

COMACINA	*obsidebant*	c. 587

HL 3.27; *PLRE* 3 s.vv. Eoin, Francio 1.

While one division of Lombards under Eoin invaded and ravaged Roman-held Istria, another besieged (*obsidebant*) the Roman garrison at

Comacina, the island in lake Como, which surrendered (*insulam tradidit*) after 6 months. The commander Francio was released and went to Ravenna. The dating is difficult, but *PLRE* places Eoin's expedition in 586 or 587. Paul also claims that Francio held his office of *magister militum* for twenty years, presumably since Narses had been dismissed in 568; hence the compromise suggestion 587, which does not clash with the Frankish invasion the next year (*HL* 3.29).

Beroe	προσπίπτει	587

Th. Sim. 2.16.12; Pohl 1988: 85-89; Whitby 1988: 147-51.

Outmaneuvering the Romans that had been sent against them in the Dobrudja (see *8 Mysian and Scythian cities 586), the Avars broke into the Thracian plain and attacked (προσπίπτει) Beroe "at the cost of a very great waste of time and after encountering many labors," but failed "because the local inhabitants arrayed themselves in opposition more spiritedly." The Avars were bought off with a small sum and moved on to *Diocletianopolis.

Again Pohl dates this campaign one year earlier than Whitby (586 vs. 587), who however identifies synchronicities with eastern sources and events and is thus to be preferred.

Diocletianopolis	περικάθηται...κατὰ τὸ κάρτερον	587

Th. Sim. 2.17.1f; cf. *Beroe above.

Moving on from *Beroe, the Avar Khagan "vigorously besieged (περικάθηται...κατὰ τὸ κάρτερον) Diocletianopolis, but the city marshalled itself in opposition strongly and prevented him from attacking with confidence; for they stationed catapults and other defences on the walls (καταπέλτας γὰρ ἐν τοῖς τείχεσιν ἀνεστήσατο ἄλλα τε ἀμυντήρια), and it was impossible for the barbarians to approach and engage at close quarters." Again the Avars were forced to move on, this time to *Philippopolis.

Philippopolis	περιβὰς...ἐνήθλει λαβεῖν	587

Th. Sim. 2.17.2f; cf. *Beroe above.

After withdrawing from *Diocletianopolis, "he moved to Philippopolis, invested the city, and strove to take it (περιβὰς...ἐνήθλει λαβεῖν). The town's inhabitants (οἱ τοῦ ἄστεως) fought back most skillfully and inflicted many injuries from their ramparts and battlements, so that the Chagan willingly abandoned the fight, respecting their inviolability on account of their courage." The next goal was *Adrianople.

| ADRIANOPLE | ἀγωνιστικώτερον...προσέβαλλεν | 587 |

Th. Sim. 2.17.4; cf. *Beroe above; also *PLRE* 3 s.v. Droctulfus 1; Whitby 1986: 67 n. 43; cf. *HL* 3.18f.

This was the last target of the Avars' Thracian campaign. "In the morning he crossed the forests of the Astike, as it is called, came up against Adrianopolis, and attacked the town fiercely (ἀγωνιστικώτερόν τε τῷ ἄστει προσέβαλλεν), but the townsmen (οἱ τοῦ ἄστεως) bravely resisted."

After some criticism at Constantinople, the emperor appointed an Armenian and a Lombard general to lead the relief of the city. Particularly Drocton (or Droctulf), the Lombard general, performed well when he broke the siege, and in a subsequent battle used feigned flight to defeat the Avars (2.17.8-12).

| PERSIAN FORT IN ARZANENE | προσβάλλει | 587 |

Th. Sim. 2.18.1-6; Whitby 1988: 284.

Herakleios the elder renewed the Persian war by assaulting a fort, probably in Arzanene, in 587. For events in the East (background, information on previous battles and abortive sieges, e.g., an attempted Roman blockade of Chlomaron) and dating, see Whitby 276-86.

"And so at this particular time Herakleios made another into the Persian state, and trouble became endemic among the Medes. When he had arrived, he attacked (προσβάλλει) a certain very strong fort; this was situated upon a lofty rock. The under-general arranged his siege engines and machines (καὶ διεκόσμει ὁ ὑποστράτηγος τὰς ἑλεπόλεις τε καὶ τὰς μηχανάς). The Persians also devised various counter-stratagems against his schemes, and wove things like robes: after collecting hairs and intertwining the warp with the weft, they produced long tunics and packed these densely with chaff; after making them solid, they hung them upon the wall (ἐξῆρτων τοῦ τείχους) and on these they received the bombardments (τὰς προσβολὰς ὑπεδέχοντο), mitigating the hardness of the discharges through the softness of the countering preparation. Many of the missiles flew right over the fort, but others were also brought down on the stronghold itself. Herakleios admitted no respite in the bombardment, alternating those engaged in the work (τοὺς ἐφισταμένους τῷ ἔργῳ) day and night. For those who had recently participated in the labours received relief from the succeeding force, while fellow labourers in turn replaced those, and others again took over the toil from those. It was for this reason that those protecting the stronghold grew weak and their strength grew faint. In this very way the fort was captured and came into Roman possession; after its capture the general installed a garrison in it."

| BEÏUDAES | λίαν ἰσχυρῶς…περικάθηνται | 587 |

Th. Sim. 2.18.7-25; Whitby 1988: 284.

While Herakleios the Elder was occupied with the *Persian fort in Arzanene, another group rebuilt Matzaron. Receiving intelligence from local farmers on an inadequately garrisoned Persian fort called Beïudaes (in Tur Abdin near Dara; it had been held by the Romans earlier in the war), the Romans decided to attack it. It was well situated and the defenders confident, as a strong tower (or literally, "rock" with fortifications) guarded the only approach.

2.18.11-14: "So the Romans dismounted from their horses, bombarded [προσαράττουσι; "dashed against" would be better] the rock, and the overtures of the conflict were effected by discharges of arrows; those in the fort defended themselves now with stones, now with catapults (καταπέλται), and created a deluge as if from some unseen lofty vantage-point, banishing as it were the alien enemy by means of the heights. While the Roman force was occupied, some brave Romans defended themselves with linked shields and, gradually moving step by step and enhancing their boldness with supreme heroism, led the way for the following troops; they moved forwards without regard for the deluge from the rock, and dislodged the barbarians from the rock. The besieged abandoned their allied rock, retired into the fortress, and surrendered the entrance to the enemy. The Romans took possession of the fortifications on the rock and besieged the fort exceedingly strongly (λίαν ἰσχυρῶς τὸ φρούριον περικάθηνται). Those standing on the parapet were unable to scare off the opposition, since they could not endure the innumerable missiles, but forthwith were suddenly to be seen showing their backs instead of their faces."

The fort was taken thanks to the exploits of Sapeir, who ascended the walls several times by driving spikes in the wall (2.18.15-25): "While the besieged were unable to endure the sight of the missiles, that man Sapeir firmly gripped the parapet" but was pushed off. At the second attempt he almost made it:, "But the Persian foe effected a stratagem kindred to the other: Since the parapet had recently been weakened by the Roman bombardment (σαθρωθείσης ἔναγχος τῆς ἐπάλξεως ταῖς Ῥωμαϊκαῖς προσβολαῖς), he pushed over the hero along with it and let them fall downwards." In both instances his comrades caught him on their shields. The last time he "finally mounted the garland of the rampart" and killed the Persian, spurring his comrades to follow. Roman soldiers were raised with ropes by their comrades and rushed (ἀράττουσιν; Whitby translates as "bombarded") the gates.

This anecdote is difficult, since Theophylact has the Romans dismount from their horses and rush (Whitby's translation as "bombard" makes little sense, just as the soldiers later rushed against rather than "bombarded" the gates) to the assault with bows and arrows. Only the Persians are mentioned as using stones and catapults. However, it is clear from Sapeir's exploits that the Romans had managed to damage the breastwork so much that it could be pushed over. Probably Theophylact has compressed events for dramatic effect and provide a suitably individualistic contrast to Herakleios' teamwork at the *Persian fort in Arzanene above. A cavalry vanguard may have redeployed as infantry, rushing the outworks on the rock in testudo formation under the cover of archery, but then more substantial machines, perhaps the Persians' own machines that had been captured, could be set up against the main fortifications.

?CARCASSONNE		587

J.Bicl. 86, s.a. 587; GT 8.45; Thompson 1969: 92; *PLRE* 3 s.v. Desiderius 2.

"Desiderius, a general of the Franks, was attacked and defeated by the Gothic generals of King Reccared. He died on the battlefield, along with a multitude of Franks." Gregory has a similar version, but emphasizes the proximity to the city walls, so a siege might have begun or preparations underway. The reconstruction may be made thus: Desiderius arranged an expedition against Carcassonne, but was headed off by a Visigothic army/sally from the city. Victorious at first, he pursued them up to the city, but was caught by a sally from inside the walls and killed.

BEAUCAIRE	*desolantes, nullo resistenti*	587

GT 9.7; Thompson 1969: 93.

After the failure of an embassy sent in 586 after fighting in 585 (see *Beaucaire s.a.), the Visigoths made a new raid on the province of Arles and laying waste to (*desolantes*) Beaucaire, "and returned home with the property which they had seized, and with some of the inhabitants, too, for no one offered resistance (*nullo resistenti*)."

The Visigoths again sued for peace, but Guntram refused (9.16).

FORTIFIED ESTATE CHURCH, WOËVRE	*cum armis vallat*	587

GT 9.12; *PLRE* 3 s.vv. Vrsio, Bertefredus, Rauchingus.

Childebert sent an army to deal with the rebels Ursio and Berthefried, who had supported the failed revolt of Rauching (GT 9.9) and fortified themselves on a hill-top church near the estate of Ursio, which used to be a fortification but was now more fortified by nature. Childebert's army surrounded the building where the rebels had withdrawn, the basilica, with their weapons (*basilicam cum armis vallat*). Gregory goes into great detail how the soldiers of Childebert tried to fight their way in, were driven back by Ursio, who himself fell in a sally; subsequently, the troops blockaded the church and had to climb onto the roof and break their way through in order to kill Berthefried by dropping tiles on his head.

Singidunum	7 days	πολιορκοῦσι	588

Th. Sim. 6.2.9-4.5; Pohl 1988: 128-34; Whitby 1988: 151-55.

The Avars resumed hostilities, beginning preparations late in 587 for an invasion the following year (Pohl dates this to 592 based on premises that Whitby effectively rejects).

"Therefore the Chagan ordered the Sclavenes to construct large numbers of boats so that he could control the crossing of the Ister. The inhabitants of Singidunum (οἱ μὲν τῆς Σιγγηδόνος οἰκήτορες) ravaged the Sclavenes' labours by sudden attacks, and consigned to the flames their nautical enterprises. It was for this reason that the barbarians besieged (πολιορκοῦσι) Singidunum; the city reached the extremity of disaster and had feeble hopes of salvation. But on the seventh day the Chagan ordered the barbarians to abandon the siege and to come to him. When the barbarians became cognizant of this, they left the city carrying off two thousand gold darics, a gold-inlaid table, and clothing."

Subsequently the Avars organized another fleet at Sirmium (6.4.4f): "Therefore the Chagan moved five parasangs, camped at Sirmium, and organized hordes of Sclavenes in timber operations (πλήθη τε Σκλαυινῶν ξυλουργεῖν παρεσκεύαζεν), so that he could cross the river Saos, as it is called, by boat. And so he pressed on with the campaign, while they provided shipping in accordance with his order: for such ar the things which fear of appointed officers can accomplish. So, shortly after the barbarian had acquired skiffs ready for use, the barbarians crossed the adjacent river."

Whitby argues that the "inhabitants" were in fact soldiers, and most likely those that had been organized in 586-7 and defeated the Avars at *Adrianople (587), since there were few field troops available in Thrace to face the Avars in the mountain passes and at *Anchialos later in 588. He believes that the inhabitants made a deal to pay ransom that went against the interests of the state at large, but Theophylact implies that the decision to withdraw came first, and was made by the Khagan, who was not present. The Avars may have offered to receive a ransom instead of pressing the siege, which was begun only to pen in the Romans and allow the rest of the Avars to move their ship-building enterprise further upriver. They may even have completed the necessary skiffs before the siege was abandoned, not afterwards as Theophylact's narrative implies.

| ANCHIALOS | ἐξεπολιόρκησαν | 588 |

Th. Sim. 6.4.7-5.3; J. Eph. 6.45-49; Ev. 6.10; Pohl 1988:134; Whitby 1988: 155 (cf. *Anchialos 583).

Having crossed the river, the Avar advanced against the Roman-held passes, which the latter held through several engagements against a vastly superior Avar force before withdrawing. The Roman withdrawal allowed the Avars to march to Anchialos. Theophylact fails to mention the capture of the city, which is attested by John of Ephesus and Evagrius. The latter conflated all the Slav and Avar invasions throughout the Balkans in the late 580s in his brief notice: including *Singidunum [below] and Anchialos "and all Greece," the Avars "stormed and enslaved (ἐξεπολιόρκησαν καὶ ἠνδραποδίσαντο)" a mass of cities and forts. This is the occasion when the Khagan donned the imperial robes deposed at Anchialos and demanded taxes from Roman cities (John and cf. *Anchialos 583, where Pohl assigned the reports of John and Evagrius, thus ignoring the city in the 588 campaign). Maurice led a campaign in person to the city in 590 in order to restore imperial authority and raise morale in Thrace.

| DRIZIPERA | παραστήσασθαι | 588 |

Th. Sim. 6.5.4-7; Pohl 1988f: 134; Whitby 1996: 165 n. 27; *idem* 1988: 155.

Whitby shows that Theophylact has left out much in his account between *Anchialos and Drizipera. For instance, the Avars had outflanked the eastern end of Maurice's ditch that had been constructed to protect the Thracian plains, and thus skirted the Roman forces stationed further to the west.

"After the fifth day had passed, he transported his camp to Drizipera, and made an attempt to reduce the city; but since the citizens arrayed themselves bravely, on the seventh day the barbarian constructed siege-engines. (...πρὸς τὰ Δριζίπερα μετοχετεύει τὸν χάρακα τήν τε πόλιν ἐνεχείρει πως παραστήσασθαι. ἐπεὶ δὲ οἱ τοῦ ἄστεος ἐς τὸ καρτερὸν παρετάττοντο, τὰς ἑλεπόλεις ἑβδόμῃ ἡμέρᾳ ὁ βάρβαρος ἐτεκταίνετο.) So, violent uproar afflicted the city and, as their hopes for safety were tossed at sea, they resorted to a pretence of boldness: for, opening the city gates, they threatened to spurn the rampart and do battle with the barbarians on equal terms. And so, after effecting their deployment to the extent of orders and formation, they were stricken by cowardice and did not go out of the city. But the barbarians were prevented from attacking by some divine solicitude. For at midday they imagined they saw countless Roman divisions in close formation, moving out of the city and hurrying to the plain in eagerness to do battle and die in combat (καὶ θανατᾶν πρὸς παράταξιν). Then the Chagan was glad to flee precipitately; the opposition was an illusion, a bogyman of vision and a bewilderment of perception."

Both Pohl and Whitby believe that the Avar withdrawal was due to the approach of a Roman army, which has been suitably recast as a miraculous intervention by Theophylact or his source. A (probably pre-planned) coincidence between the sally and approach of the relieving army would have helped to cement the impression of divine intervention. While the relieving army diverted Avar attention from the city, the Romans were defeated in battle, withdrew, and ended up besieged at *Tzurullon (below). Note the bellicose rhetoric at 6.5.7, implying a religiously motivated willingness to die in battle.

TZURULLON   περικαθήμενον... γεννικῶς ἐπολιόρκει   588
Th. Sim. 6.5.8-16; Pohl 1988: 135; Whitby 1988: 155.

Having engaged the Avars (see *Drizipera above), the Romans took refuge first at Didymoteichon, later at Tzurullon. "But the barbarians invested (περικαθήμενον) the city and vigorously besieged (γεννικῶς ἐπολιόρκει)" the Roman force (6.5.10). Several sources suggest that Maurice managed to convince the Avars to withdraw through a combination of threats of a Turkish invasion of their homelands and tribute (see Whitby and Pohl for references). Afterwards, the Romans disbanded in Thrace for the winter, where the soldiers "found subsistence in the villages" (6.6.1).

MARTYROPOLIS (twice)   προδοσία; στρ. καὶ χάρακα...ἐνέβαλλεν   589
Th. Sim. 3.5.11-16; Evagrius 6.14; Whitby 1988: 289, 299f.

Theophylact: The Persians took Martyropolis by treachery: 400 Persians pretended to come over to the Roman side under a Roman traitor, Sittas, but once inside the fortifications, they took control. A Roman army arrived, encamped and "encircled the city with a rampart" (στρατοπεδεύεται καὶ χάρακα περὶ τὴν πόλιν ἐνέβαλλεν) in order to regain it, but Persian reinforcements soon arrived and the Romans lost the ensuing battle, although with heavy Persian losses. The Romans only regained the city during the Persian civil war in 590.

Evagrius has the same basic story, but adds that the Persians kept the young women and expelled the rest of the population. He also helps clarify the meaning of some siege terminology. According to him, the Romans lacked *helepoleis*. Due to the generic nature of the statement, this must simply mean siege engines of whatever type. This also explains heavy losses incurred by the Romans, which were exacerbated by the Persian height advantage at this particular site. Instead of a direct storm, the Romans dug mines (διώρυχας) that brought down a tower, but the Persians were able to close the breach and the Romans lost many during the exchange of fire.

CARCASSONNE   [battle prevents siege]   589

J.Bicl. 91, s.a. 589 (Wolf 1999: 74); GT 9.31f; Thompson 1969: 94.

"The army of the Franks, sent by King Guntram under the general Boso, came to Gallia Narbonensis and set up their camp next to the city of Carcassonne. Claudius, the commander of Lusitania, was ordered by King Reccared to intercept him and hastened to that place. When battle began, the Franks were put to flight, their camp was seized, and the army was slaughtered by the Goths." John attributed the victory to God's intervention, claiming they numbered 60,000 while the Gothic force was only 300, which he equates with Gideon's force against the Midianites. Gregory's estimates below are probably much more accurate.

Guntram sent another army to attack Septimania (589), consisting of the men of Saintes, Périgueux, Bordeaux, Agen and Toulouse, but this was nearly wiped out by a Gothic ambush after a feigned flight. Some of those with horses fled, but the Goths "pillaged their camp and took all the foot-soldiers prisoner. Nearly five thousand men died in this engagement. The enemy seized more than two thousand. Many were later freed and found their way home." (GT 9.31)

Guntram attributed the defeat to treason from his son Childebert, who may in fact have warned the Visigoths, since they appear to have been extremely well informed on Frankish movements (9.32).

AKBAS   πολιορκήσας; αἱρεῖ κατὰ κράτος   590

Evagrius 6.15; for context and literature see *Martyropolis 589.

The Persian fort at *Akbas that had been captured by the Romans and demolished in 583 had clearly been rebuilt. Note how Evagrius explicitly credits the catapults with breaking down the walls.

"Comentiolus, having commenced the siege of Martyropolis, left the greater part of his army there, while he himself made an excursion with a chosen body of troops to Akbas, a very strong fortress, situated on a precipice on the bank opposite to Martyropolis, and commanding a view of the whole of that city. Having employed every effort in the siege, and thrown down some portion of the wall by catapults (*katapeltōn rhipsas*), he takes the place by storming the breach. In consequence, the Persians thenceforward despaired of keeping possession of Martyropolis."

Ἐν τούτῳ δὲ Κομεντίολος τὴν Μαρτυρόπολιν περικαθήμενος τοὺς μὲν πολλοὺς αὐτοῦ καταλείπει, αὐτὸς δὲ σὺν καί τισιν ἀριστίνδην ἐκλεγεῖσιν ἐκτρέχει κατὰ τὸ Ὄκβας ὀχυρώτατον φρούριον, ἀντικρὺ Μαρτυροπόλεως ἐς τὴν ἀντιπέρας ὄχθην διακείμενον ἐπί τινος σκοπέλου ἀποτόμου, ὅθεν καὶ ἄποπτος ἡ πᾶσα καθειστήκει πόλις· καὶ πολιορκήσας πείρας τε οὐδὲν ἀνιεὶς καὶ τοῦ τείχους τινὰ διὰ καταπελτῶν ῥίψας, ὑπερκαταβὰς αἱρεῖ κατὰ κράτος τὸ φρούριον. Ὅθεν λοιπὸν καὶ ἐν ἀπογνώσει Πέρσαις τὰ Μαρτυροπόλεως καθειστήκει.

FRANKISH INVASION OF ITALY   *coepit, sacramenta exegit; diruerunt*   590
HL 3.31; GT 10.3; *Ep. Austr.* 40; *PLRE* 3 s.vv. Romanus 7, Olo, Henus; Wood 1994: 167f; Goubert 1956 (vol. 2.1): 187-202.

After several failed ventures that came to nothing during the 580s (see Wood for references to campaigns in 584, 585, 589 and 590; Goubert 2.1 in general), the Austrasian Franks and the Romans managed to coordinate a major effort that aimed at conquering most of northern Italy from the Lombards.

Gregory lambasts Childebert's army for plundering Frankish territory on the way south, but he only mentions one poorly behaved duke—the rest seem to be guilty by association, and the tone is clearly satirical. It consisted of the forces of 20 dukes that separated into two columns when it entered Italy. The western column was led by seven dukes and advanced towards Milan; they seem not to have begun a formal siege, but camped in the surrounding countryside and began raiding. One of the dukes, Olo, was killed by a javelin during the assault on the fort of Bellinzona that belonged to the territory of Milan (see *HL: cum ...accessisset, iaculo... sauciatus*). They awaited Roman reinforcements that had been announced but did not show up as agreed; the exarch Romanus made the same claim about the Frankish eastern column that encamped near Verona (see *Roman invasion of Italy below).

The eastern column of 13 dukes was reasonably successful as they captured and extracted oaths from several forts (five according to Gregory;

quinque castella coepit, quibus etiam sacramenta exegit), but they were hampered by disease and lack of supplies, and had to withdraw before they joined up with the Romans. Again Gregory is scathing in his criticism: "He captured five strong-points and extracted oaths of allegiance...There is little more to tell. For nearly three months the troops wandered about in Italy..." The tone of Gregory's description is unfair. He admits himself that the Franks regained control of the areas that had been held by Sigibert, which was probably as far as Austrasian ambitions realistically stretched (especially when the *Teilreiche* were at loggerheads and sabotaged each other; see chapter 4.1), and that the weather conditions were horrible, with intense heat that caused dysentery. Paul, who used Gregory's account and extended it with material from Secundus of Trent's local chronicle, adds that the forces heading for Verona demolished (*diruerunt*) 13 fortresses in the territory of Trent. "When all these fortified places were destroyed (*haec omnia castra cum diruta essent*) by the Franks, all the citizens were led away from them as captives." The Franks were unable to subdue, but took ransom from, the fort of Verruca (which was apparently held by 600 men, ransomed for one solidus each, or alternatively from 1 to 600 solidi for each man, depending on status).

It seems that the forts were demolished in the recognition that most of them could not be held because of the epidemic of dysentery, but once demolished and depopulated, they could not easily prevent later Frankish invasions. Furthermore, the extractions of oaths from the remaining population established precedent and reinforced older Frankish claims from the age of Sigibert and before; they thus provided a foundation for later Frankish policies, e.g. those followed by Dagobert (see chapter 4.2). They may also have had considerable local support. Mimulf, the Lombard duke responsible for part of the border defenses against the Franks, was executed for colluding with the invaders (*HL* 4.3).

On a strategic level, withdrawal before the Lombards were crushed prevented the Romans from establishing a common border with the Franks—their involvement in Frankish politics around 585 was profound, and probably not many Franks were eager to see Roman armies in the Alps again. The policy also produced short-term advantages, as Agilulf, the new Lombard king, sent the bishop of Trent to ransom captives from the Franks (*HL* 4.1).

ROMAN INVASION OF ITALY   *ingredi...pugnando et rumpendo muros*   590
As *Frankish invasion of Italy above; also Ep. Aust. 40; *PLRE* 3 s.v. Romanus.
    The exarch Romanus, who was expecting to meet up with the Austras-

ian armies around Milan and Verona (see *Frankish invasion), relates in his complaint to king Childebert that: "However, before your *duces* had entered the border of Italy, God, for the sake of his mercy and your prayers, made us enter Modena fighting, likewise also Altino and Mantua fighting and destroying the walls, so that the Frankish army should see, that with God's help we had entered [those cities]; hurrying so that the nefarious nation of the Lombards should not be able to unite itself against the Franks, and awaiting the *vir magnificus* Henus, who was encamped twenty miles from the city of Verona, [and] whom we thought necessary for making decisions, hoping from him that we would meet face to face and, whatever would be useful in order to destroy that nefarious nation, we would decide through common council."

|| Ante vero quam fines Italiae vestri duces ingrederentur, Deus pro sua pietate vestrisque orationibus et Motonnensem civitatem nos pugnando ingredi fecit pariter et Altinonam et Mantuanam civitatem pugnando et rumpendo muros, ut Francorum videret exercitus, Deo adiutore, sumus ingressi: festinantes, ne gente nefandissimae Langobardorum se contra Francorum adunare liceret, et Heno viro magnifico, in viginti milibus prope Veronensi civitate resedente, ad quem necessarium duximus sine mora diregere, sperantes ab eo, ut nos videremus in comminus et, quae essent utilia ad delendam gentem perfidam, disponeremus communi consilio. ||

Contrary to Romanus' expectations, the Franks had already concluded peace—they had achieved their objectives and were furthermore troubled by dysentery and were low on supplies.

BERGAMO	*se communivit*	c. 591

HL 4.3; *PLRE* 3 s.v. Gaidulfus.

"Gaidulf, indeed, the Bergamascan duke, rebelled in his city of Bergamo and fortified himself against the king, but giving hostages, he made peace with the king." || Gaidulfus vero Pergamensis dux in civitate sua Pergamo rebellans, contra regem se communivit; sed datis obsidibus pacem cum rege fecit. ||

COMACINA	*se...seclausit, ingressus...expulit*	c. 591

As *Bergamo (591).

Gaidulf rebelled and fortified himself (*se...seclausit*) at Comacina, but was driven out by Agilulf (*Comacinam insulam ingressus, homines Gaidulfi exinde expulit*), who confiscated the treasure left behind by the Romans.

Gaidulf escaped to Bergamo, where he made peace again. However, Gaidulf was executed when he rebelled for a third time (*HL* 4.13).

| TREVISO | *obsessus* | c. 591 |

HL 4.3; *PLRE* 3 s.v. Vlfari.

"Also duke Ulfari rebelled against king Ago at Treviso, and was besieged and captured (*obsessus captusque*) by him."

| ROMAN CAMPAIGN IN ITALY | *retenuit* | 592 |

HL 4.8; *PLRE* 3 s.v. Romanus

The exarch Romanus reoccupied (*retenuit*) seven named cities: Sutri, Bomarzo, Orte, Todi, Amelia, Luceoli, and *Perugia (see 593 below) that had been lost to the Lombards, as well as others unnamed.

| PERUGIA | *obsedit* | 593 |

HL 4.8; *PLRE* 3 s.v. Maurisio.

Agilulf moved with a huge army against Perugia, held by Maurisio, a Lombard duke in Roman service. Agilulf besieged (*obsedit*) the city and killed Maurisio, but no further details are given.

| SLAV WAGON LAAGER | εἰσεπήδησε | 593/4 |

Th. Sim. 7.2.1-10; Pohl 1998: 141f; Whitby 1988: 159f.

By 593, the Romans established peace with the Avars and were conducting punitive actions (or *Buschkrieg*) against the Slavs on the lower Danube marshes and tributaries, even north of the river. There must have been many similar encounters that went unrecorded, but Theophylact provides a good example: 600 Slavs, returning from a raid (καταπρονομεύσαντες, 7.2.2) against Zapalda, Aquis and Scopi with wagons full of plunder and captives in tow, were surprised by a Roman party of 1,000 who arrived from Marcianopolis (these troops spoke Latin and were hence recruited in the Balkans). The Slavs reacted by slaughtering all the adult males, made a wagon laager and defended themselves with javelins that proved effective against the Romans' horses. The Romans were reluctant to attack but dismounted and began a missile exchange with the Slavs before they broke through (εἰσεπήδησε, 7.2.8), killing most of the Slavs. Unfortunately, the remaining captives had been killed by the time the Romans were in control of the improvised fortification.

SINGIDUNUM καταβαλεῖν, καθελών 595
Th. Sim. 7.10.-11.; Pohl 1988:143-46; Whitby 1988:161.

The Roman army moved to the Danube and crossed the river, which the Avars took as a breach of treaty. After fruitless negotiations, the Avars began besieging Singidunum. The Avars were attempting to destroy the walls (τὰ τείχη καταβαλεῖν, 7.10.1; τὰ τείχη καθελών, 7.11.1) and had begun to deport the population, presumably those caught outside the walls (cf. *Singidunum 583). Upon the arrival of Roman reinforcements and a fleet of *dromones* at Constantiola, the parties opened negotiations again but the Avars broke them off in anger, since they regarded this as their rightful territory. Roman naval units then approached the city and began surrounding it by sailing up both the Sava and Drava rivers, at the confluence of which it is situated. The besieging Avars at first set up wagons as a fortification around the city, but overcome by fear of the population (probably including a substantial garrison) remaining in Singidunum, they fled. The Romans subsequently περιβάλλουσι (7.11.8) the walls and the Avars set out for other targets in Dalmatia (7.12.).

The wording in the translation is slightly ambiguous (in part due to the deliberately rhetorical preference for circumlocutions), since it may give the impression that the Avars were already in control of the city, destroying ("razing down") its walls and deporting the population in the city (thus Pohl). This is clearly not the case, since a significant population remained that was not in their control as prisoners and potential hostages; the Avars must have been assaulting with wall-breaking engines (either trebuchets, rams, or a combination) that required a substantial baggage train, and tried to use the wagons as an improvised field-fortification. When they gave up and withdrew, the Romans either repaired the wall, or merely "embraced," i.e. manned it, although here the original is ambiguous.

BONKEIS AND 40 FORTS παραστησάμενος...τοῖς μηχανήμασι, ἐξεπόρθησε 595
Th. Sim. 7.12.1; Pohl 1988:146f; Whitby 1988:161; Curta 2001 *passim*.

After the Avar reversal at *Singidunum, they turned their attention to the southwest: "Near these regions the country of Dalmatia is situated. Then, after several camps, the barbarian came to the place called Bonkeis and, when indeed he had reduced the city with his siege engines (καὶ δὴ παραστησάμενος τὴν πόλιν τοῖς μηχανήμασι), he sacked forty forts (τεσσαράκοντα ἐξεπόρθησε φρούρια)." The general Godwin was sent to follow the Avar expedition with 2,000 men, and managed to ambush part of the Avar expedition and recover the booty they carried (7.12.2-8). Curta assembles archaeological evidence that many forts on the Danube were miniscule with garrisons of only 20 men or so; this explains the very large number, which is probably correct in this context.

Tomi [στρατοπεδεύοντες] 597f
Th. Sim. 7.13.1-8; Pohl 1988: 152f; Whitby 1988: 162.

The lull in the war that followed after *Bonkeis (595) was interrupted by the Avars in 597, who had in the meantime focused their energy against the Franks. Invading suddenly across the Danube plain they encamped near the city of Tomi on the Black sea. A Roman army arrived and both armies encamped in the vicinity of the city ('Ρωμαῖοι τοιγαροῦν καὶ οἱ βάρβαροι εἰς τὰ περὶ Τομέαν τὴν πόλιν στρατοπεδεύοντες, 7.13.2). It is unclear whether there was a proper siege with siegeworks (Theophylact mentions that they never left their χάραξ, but here clearly this means a fortified camp); whether the Roman army ever entered the town (the text clearly implies it was encamped in the region, not inside the walls); and if they did not, who was starving: the Romans inside the city or in the camp? And finally, if not a blockade or formal siege, the Avar presence made it impossible to bring supplies into the city/Roman camp, but this is also strange, considering it was a coastal town close to several naval bases under Roman control. It was under blockade from the autumn of 597 until Easter (which fell on 30 March) 598, when the Avars opened negotiations and offered supplies to the starving Romans in return for a truce. Whitby argues that they did this because they had heard of the Roman army coming to relieve Tomi (see *Drizipera 598).

From the unclear nature of the action, it appears that this story is less than trustworthy or has been garbled out of recognition. The simplest solution is that the Roman army entered the city, were holed up, and sea-borne supplies were made impossible by storms, ice and other bad weather, while the supplies were consumed more rapidly by the extra garrison. Alternatively, the Avars did blockade the city tightly, but were themselves blockaded by the Romans, who however were less well supplied. This begs the question how the Avars supplied themselves over the winter. They had managed to do so at *Anchialos (583), where they ravaged the countryside freely throughout the winter, and in the Dobrudja after the campaign against the *8 Mysian and Scythian cities (586), but there they were close to the Danube and could be supplied by Slav allies and subjects, and perhaps the stores captured in the Roman forts. At Tomi, however, they had wagonloads of supplies, and such a great surplus that they could buy off one Roman army so that they only had to face another. If this is correct, the Avar ability to organize supplies had matured since the early 580s (cf. *Singidunum 595); alternatively, it may have been supplies sent from Roman-held territory that were allowed through the blockade.

Pohl notes the psychological value of such a move at the most important feast-day in Christendom: he was clearly aiming for a large measure of goodwill, which could help him in negotiations as well as dampen resistance in coming conflicts—his ambition was after all to conquer and rule a substantial Christian population, and during the truce the two sides fraternized and feasted together.

Drizipera	ἐπόρθησαν	598

Th. Sim. 7.14.11; Pohl 1988: 153f; Whitby 1988: 162f.

The Avars managed to outmaneuver the Roman army that had been sent to relieve *Tomi, who had tried but failed to trap the Avars on the Danube plain, but had themselves been outmaneuvered. The Avars in fact trapped the Romans in the Balkan mountain passes, and they had to fight their way south. The Avars exploited the opportunity to sack (ἐπόρθησαν) Drizipera in the wake of the retreating Romans. A plague amongst the Avars and a Roman show of force at the *Long Walls facilitated a settlement, and after negotiations, the Avars were bought off with a large but in return had to relinquish effective control over the Slavs north of the Danube, whom the Romans could now pursue (7.15.).

Long Walls	[διεφρούρησε]	598

Th. Sim. 7.15.7; Pohl 1988: 154f; Whitby 1988: 163.

There was panic at Constantinople when the Avars advanced all the way to *Drizipera: "But the emperor took with him the bodyguards, whom Romans designate *excubitores*, assembled the army, and garrisoned (διεφρούρησε) the long walls; he also had with him a very large portion of the factions at Byzantium." Although this never came to a siege, it was necessary to calm sentiments in the capital, and in fact it appears to have been a real mobilization, as the imperial guards, the regular army and the circus factions took up positions together, and the show of force convinced the Avars to accept a settlement (see *Drizipera above).

THE 7TH CENTURY

| CITIES ALONG THE SEINE | *ruptas* | 600 |

Fredegar 4.20.

"In the same year Kings Theudebert and Theuderic took the field against King Chlothar and brought him to battle on the banks of the river Orvanne, near Dormelles. Chlotar's army was here massacred, but he himself took flight with the remainder. Then they laid waste the towns and districts along the Seine that had gone over to Chlothar (*pagus et ciuitates ripa Sigona qui se ad Clothario tradiderant, depopulant et vastant*). The towns were razed (*Ciuitates ruptas*) and from them the army of Theudebert and Theuderic took a great number of prisoners." Chlotar was forced to accept peace and only a small stretch of territory in the northwest of Gaul.

| PADUA | *iniecto igni... ad solum usque destructa est* | 601 |

HL 4.23.

"Up to this time the city of Padua had rebelled against the Langobards, the soldiers resisting very bravely. But at last when fire was thrown into it, it was all consumed by the devouring flames and was razed to the ground by command of king Agilulf. The soldiers, however, who were in it were allowed to return to Ravenna." || Usque ad haec tempora Patavium civitas, fortissime militibus repugnantibus, Langobardis rebellavit. Sed tandem, iniecto igni, tota flammis vorantibus concremata est, et iussu regis Agilulfi ad solum usque destructa est. Milites tamen qui in ea fuerunt Ravennam remeare permissi sunt. ||

Paul has left out too much to be certain of what happened here. It might have been a surreptitious move, the Lombards bribing someone burn the city in the middle of the night or something similar. However, the phrase *iniecto igni* suggests something more obvious; hence incendiary devices thrown in by artillery during a siege.

| MONSELICE | *invaserunt* | c. 602 |

HL 4.25.

"... the Langobards assaulted (*invaserunt*) the fortress of Monselice..." south of Padua.

CONSTANTINOPLE [περιφρουρεῖν...προστάττει] 602
Th. Sim. 8.7.-8.; Pohl 159-62; Whitby 1988: 164-69.

Just as the Romans were penetrating deep into Avar and Slav territory and the Avar Empire was beginning to buckle (8.5.12-6.1), the revolt against Maurice began (8.6.; see further Whitby). He organized the factions for defense, inscribing 1,500 Greens and 900 Blues (8.7.10f). They were then ordered to guard the Theodosian Walls (περιφρουρεῖν...προστάττει, 8.8.2), and Comentiolus was to take command of the guards (8.8.7); however, due to lack of support in the city, Maurice gave up Constantinople and tried to flee to the East, but was captured and executed.

?SAGUNTUM [obtinuit] c. 603
Isidore *HG* 58 (Wolf 1999): 104; Thompson 1969: 158; Collins 2002: 73.

On the reign of Witteric, Isidore observed: "Though active in the art of war, he never won a victory. For though he often exerted himself in battle against the army of the Romans, he accomplished nothing of particular glory except that he captured through his generals some soldiers at Saguntum." ||...vir quidem strenuus in armorum arte, sed tamen expers victoriae. namque adversus militem Romanorum proelium saepe molitus nihil satis gloriae gessit praeter quod milites quosdam Sagontia per duces obtinuit. ||

Thompson believes that the city was captured as well as the soldiers, but Collins is skeptical and the phrasing is too obscure even to guess what might have transpired. It is possible that an *et* dropped out between *quosdam* and *Sagontia*; one manuscript does indeed have the form *sangonciam*, but the majority has a form in *–a*. Although brief, this chronicle entry does show that Roman defenses in Spain for the most part still held firm as late as the first decade of the seventh century, and that any invader would have to face fortified cities and the regular Roman soldiers garrisoning them, whether in siege or battle.

CREMONA *obsedit, cepit* 603
HL 4.28; *PLRE* 3 s.v. Agilulf.

Due to renewed hostility with the Romans, "... king Agilulf departed from Milan in the month of July, besieged (*obsedit*) the city of Cremona with the Slavs whom the Cagan, king of the Avars, had sent to his assistance (*in solacium miserat*), and captured it (*et cepit eam*) on the twelfth day before the calends of September (August 21) and razed it to the ground (*et ad solum usque destruxit*)."

MANTUA   *expugnavit, interruptis muris...ingressus est*   603
As *Cremona above.

"In like manner he also assaulted Mantua, and having broken through its walls with battering-rams he entered it on the ides (13th) of September, and granted the soldiers who were in it the privilege of returning to Ravenna." || Pari etiam modo expugnavit etiam Mantuam, et interruptis muris eius cum arietibus, dans veniam militibus qui in ea erant revertendi Ravennam, ingressusque est in ea die Iduum Septembrium. ||

VALDORIA   *se tradidit*   603
As *Cremona above.

"Then also the fortress which is called Valdoria surrendered to the Longobards; the soldiers indeed fled, setting fire to the town of Brescello." || Tunc etiam partibus Langobardorum se tradidit castrum quod Vulturina vocatur; milites vero Brexillum oppidum igni cremantes, fugierunt. ||

EDESSA   *nqaš mašryāteh ʿal,* ܢܩܫ ܡܫܪܝܬܗ ܥܠ   603f
Dionysius 13 (Palmer 1993: 120f); Sebeos 31 (I: 58, II: 197f); Howard-Johnston 2010: 201f; Hoyland 2011: 55 n. 49.

Narses, a general appointed by Maurice, rebelled against Phokas in 602, basing himself in Edessa and possibly proclaiming (someone believed to be) Theodosius, Maurice's son, emperor (unless Khusro did this at Ctesiphon before the invasion). The city was besieged by Phokas' general in 603 (*textus* 220.20 *wa-nqaš mašryāteh ʿal ūrhay,* ܢܩܫ ܡܫܪܝܬܗ ܥܠ ܐܘܪܗܝ "and he set up his camps against Edessa"), but the siege was broken by Khusro who used the opportunity to invade the Roman Empire, formally recognize Theodosius on Roman territory, and attack *Dara in his name. The Syriac sources emphasize that Phokas' troops were able to capture Narses without fighting, although they provide no details, and the vocabulary as well as the Persian invasion indicates a normal siege camp. While the Persians were so occupied, another Roman army arrived, possibly from the Balkans, and renewed the siege of Edessa, which fell in 604.

DARA   9 months   *nqaš ʿal,* ܢܩܫ ܥܠ   603f
Dionysius 14 (Palmer 1993: 122); Sebeos 31 (I: 58, II: 197f); Flusin 1992: 71-74.

The Persians besieged (*textus* 221.16 *nqaš ʿal,* ܢܩܫ ܥܠ) Dara for 9 months.

THESSALONICA   [Βάρβαροι περὶ τὸ τεῖχος!]   604
Mir. St. Dem. I. 12 (106-10); Lemerle 1981: 41ff, 69-73; Pohl 1988: 240f; Curta 2001: 92-95.

A surprise Slav raid was miraculously discovered when a fire broke out at the church of St. Demetrius during his feast day. Needing to evacuate the church quickly, the presiding official was inspired by saint Demetrius to cry out "The barbarians are at the walls, to arms!" (106). Everyone rushed home to arm themselves and then mounted the walls, only to observe "below, on the plain of the sanctuary of saint Matrone, a barbarian army, not very numerous—we estimated them at five thousand—but redoubtable since they were all elite soldiers, accustomed to war." When they discovered the Slavs, the Thessalonians sallied out with a great cry and threw themselves at the enemies. After a difficult battle that lasted until the evening, the Thessalonians chased them away.

The text describes it as a contest between "the barbarian phalanx and the army of the city" (τῆς τῶν βαρβάρων φάλαγγος καὶ τοῦ στρατοῦ τῆς πόλεως, 110), which was probably the garrison and privately employed professional soldiers that had been mostly absent at *Thessalonica (586), although Lemerle thinks this was solely a civilian militia. He points to the lack of any military officers in organizing the defense (a civilian official had sounded the alarm), as opposed to the officers mentioned in 586, and terms referring to the "whole population" and the like. However, this ignores important differences in literary setting: in 586, the miracle hinged on the *absence* of trained defenders, so it was necessary to enumerate them in order to magnify the miracle. Furthermore, in 604, the whole point of the mobilization was again saintly intervention, not military chain of command. Demetrius thus chose a somewhat unlikely candidate to convey his message. This literary device amplifies the miraculous nature of the events.

Due to the size of the Slav raiding party that was effectively repulsed by the Romans, the garrison was likely supported by private retainers, armed servants, and any militia (see further discussion in chapter 6.2.2 and *Thessalonica 586). Although it never amounted to a siege, this shows the city's response to a serious threat: first to man the walls, and if possible, sally out to fight them off in order to protect the surrounding countryside.

Curta argues that this event took place during the Slav the raids reported by other sources in the 580s, and should hence be placed *before* *Thessalonica 586. While his argument is cogent, it is nevertheless impossible to place with any precision and hence the traditional date has been retained. Indeed, based on Curta's own argument that these Slavs were independent of the Avars (as were those who attacked *Thessalonica in

615), and that the Danube frontier was not breached until sometime around 615-20, it can be taken as representative for Slav raiding activities in the decades around 600.

| ORLÉANS | *circumdans* | 604 |

Fredegar 4.24ff; *PLRE* 3 s.v. Bertoaldus 1.

Bertoald, the mayor of the palace, was subject to a plot by the formidable Brunichildis. He "was sent to inspect the royal domains in the cantons and the cities along the banks of the Seine up to the Channel." While stopping at a villa with his small retinue of 300 men to go hunting, an army sent by Chlothar "dared to storm through" (*persumpsit...peruadere*) Theuderic's kingdom in order to get at Bertoald, who fled to Orléans, where he was received by the bishop. Chlothar's general "then invested the town with his men (*cum exercito Aurilianes circumdans*) and called on Bertoald to come out and fight, to which Bertoald replied from the ramparts" that he would come out for single combat if the besiegers withdrew from the wall, but they declined the offer. Theuderic sent a relieving army; when it began to engage Chlothar's force, Bertoald sallied out from the city but was killed. Nevertheless, Theuderic's army won the battle, raised the siege, and celebrated a triumph at Paris.

| BAGNAREA | *invasae sunt* | 605 |

HL 4.32; *PLRE* 3 s.v. Agilulf.

"Cities of Tuscany too, that is Bagnarea and Orvieto were seized by the Lombards (*a Langobardis invasae sunt*)" before a peace treaty between the Romans and Lombards came into effect; it is uncertain how the cities were captured.

| ORVIETO | *invasae sunt* | 605 |

See *Bagnarea above.

| MARDIN | *kabšūy*, ܟܒܫܘܗܝ | c. 606ff |

Dionysius 14 (Palmer 1993: 122); Flusin 1992: 74; see also *Dara above.

The Persians besieged (*textus* 221.20 *kabšūy*, ܟܒܫܘܗܝ) Mardin, which held out for 2,5 years, falling in AG 919 (607-08, Dionysius) or AG 920 (608-09 Flusin, based on *Chr.* 724). Since the city fell during or before the summer (of 608 or 609), it must have begun in the winter of 605/6 or 606/7. A number of other cities fell at the same time or shortly after; see Flusin and *Edessa (609) for lists and further sources.

5 EGYPTIAN CITIES (civil war)   [took...by storm]   608
J.Nik. 105.1f; *PLRE* 3 s.v. Theophilus 4 (p. 1309); Olster 1993: 122.

"And there was a man named Theophilus, of the city of Merada in Egypt, the governor of five cities in the reign of Phocas. And the officers of the city and a large body of men revolted against him. (And) they attacked Theophilus and put him and his followers to the sword. And they took the five cities by storm, i.e. Kerteba, San, Basta, Balqa, and Sanhur."

Theophilus was a representative (and presumably military commander) of Phokas' regime; hence this was the first stage in Herakleios' revolt in Egypt (cf. Olster), instigated by local provincials before Herakleios' brother Niketas arrived in person and took *Alexandria (below).

ALEXANDRIA   608
J.Nik. 107.15-25, 45-48; Olster 1993: 123-26; cf. *5 Egyptian cities (above).

The revolt spread to the rest of Egypt when the rebel army invaded from Pentapolis and defeated the loyalist military commander of Alexandria in battle. At Alexandria, a clique of clergy, notables and the Blue faction openly defied the regime of Phokas, confiscated the property of supporters, and admitted Niketas. However, a counterattack by Bonosus, who brought much of the army of the East, defeated a forward rebel army under Bonakis. Olster notes that during these conflicts, military officers were "far more important than the demes" (i.e. circus factions), and that the conflicts were "decided by local family alliances and animosities" rather than religious or ethnic dividing lines.

"And the rest of the troops, seeing these things, fled and betook themselves to the city of Alexandria. And all the notables in Egypt mustered round Nicetas, the general of Herakleios, and assisted him because they detested Bonosus, and they informed Nicetas of all that he had done. And Nicetas got together a numerous army of regulars, barbarians, citizens of Alexandria, the Green Faction, sailors, archers, and a large supply of military stores. And they prepared to fight Bonosus in the environs of the city. And Bonosus thus reflected: "By what means can I get possession of the city and deal with Nicetas as I did with Bonakis." And he sent Paul of the city of Samnud with his ships into the canal of Alexandria in order to cooperate with him. But Paul was not able to approach the environs of the city; for they hurled stones at him, and the ships took to flight."

DEMQARUNI   [purposing to...breach]   609
J. Nik. 107.49; Olster 1993: 126.

"And Bonosus likewise came with his troops and took up a position at

Miphamonis, i.e. the new Shabra. Next he marched with all his forces to the city of Demqaruni, and was purposing to make a breach in the city on Sunday. Now these events took place in the seventh year of the reign of Phokas."

| ALEXANDRIA | [blockade > sortie] | 609 |

J. Nik. 108.1-15; Olster 1993: 126f.

Niketas, the brother of Herakleios, had taken control of Alexandria but now faced the arrival of the army of Phokas' general Bonosus. Niketas consulted with a pillar saint that advised him not to fight from the ramparts, but sally out and meet Bonosus in battle, saying:

""Thou shalt conquer Bonosus and overthrow the empire of Phocas, and Herakleios will become emperor this year." And Nicetas was guided by the prophecy of the aged man of God and said to the inhabitants of Alexandria: "Fight no longer from the top of the wall but open the gate of On and meet Bonosus in close encounter." And they hearkened to the words of Nicetas and put the troops in array and placed the catapults and engines for hurling stones near the gate. And when a captain of Bonosus' troops advanced, a man smote him before he drew near to the gate, with a huge stone, and crushed in his jaw, and he fell from his horse and died forthwith. And another likewise was crushed. And when the battle pressed sore upon them they began to flee. And Nicetas opened the second gate, which was close to the church of S. Mark the Evangelist, and he issued forth with his barbarian auxiliaries, and they went in pursuit of the fleeing troops and they put some of them to the sword. And the inhabitants of Alexandria smote them with stones and pursued them and struck them with arrows and wounded them with grievous wounds. And some that sought to hide themselves from the violence of the battle fell into the canal and perished there." (108.4-9)

Many of the fugitives were trapped by thorn hedges and canals; only Bonosus and a few soldiers escaped, while many of his officers and notables fell. Niketas immediately began reorganizing Egypt in Herakleios' cause, gathering valuables (taxes?) "from the river" and assembling soldiers from the cities. He was assisted by the Blue faction, and the naval wing of Bonosus' army "intended to desert Bonosus and go over to Nicetas."

| MANUF | [capture] | 609 |

John of Nikiu, 109.9ff, 16f; Olster 1993: 127.

"And after he crossed the river, Nicetas abandoned the pursuit and marched to the city (?) of Mareotis, and left considerable forces there to guard the route. And he marched likewise to the city of the upper Manūf. And when he drew near the city, the party of Bonosus who were there took to flight, and he captured the city, and Abrāis and his people were taken prisoners, and (the troops of Nicetas) burnt their houses and likewise the way (?) of the city. And Nicetas directed a combined and powerful attack on the city of Manūf and compelled it to open its gates. Then all the cities of Egypt sent in their submission to him."

During the civil war, the artisan guilds of Egypt were involved in fighting against the Blues. The guilds exploited the situation to attack regime loyalists; the Greens, in contrast, supported Herakleios' revolt (cf. *Alexandria 608). Niketas repressed the fighting and organized a settlement that satisfied the Egyptians. 109.16f

EDESSA AND MESOPOTAMIAN CITIES   [besieged > negotiations]   609
Sebeos 33 (I: 63, II: 202); Howard-Johnston 2010: 201 n. 22; Hoyland 2011: 68f.

"Taking the host of his troops, Khoream went to the territory of Asorestan; on reaching Syrian Mesopotamia, they besieged the city of Urha [i.e. Edessa], and attacked it. But the [Edessans], because of the multitude of the [Persian] troops and their victory in the engagements, and since they had no expectation of salvation from anywhere, parleyed for peace, and requested an oath that they would not destroy the city. Then, having opened the city gate, they submitted. Similarly Amida, and Tʻela, and Rashayenay, and all the cities of Syrian Mesopotamia willingly submitted and were preserved in peace and prosperity."

Dionysius 32 (Palmer 1993: 134) describes the Persian procedure for deporting the population of Edessa street by street a few years later, but it stopped after only two streets because of Herakleios' invasion of Persia.

CIVIDALE   *obsidione, expugnare, ingressi*   c. 610
HL 4.37.

During a major Avar invasion of Friuli, the Lombards were defeated in battle and withdrew to their fortifications, including Cividale. The Avars surrounded the city (*obsidione claudunt*) and reconnoitered (*perambularet*) the site to find the best place for a storm (*ut qua ex parte urbem facilius expugnare posset*), but they were let into the city by the machinations of Romilda, wife of the *dux* Gisulf who had fallen. The Avars betrayed their oaths, destroyed the city and took the Lombards captive to Pannonia, where they killed the men and distributed their women and children. Romilda was executed but her daughters escaped rape.

ANTIOCH kabšūh, ܟܒܫܘܗ̇ 610/11

Dionysius 23 (Palmer 1993: 127f, AG 922); Sebeos 33 (I: 63, II: 202); Flusin 1992: 78.

"On October 8 of the following year the Persians took (*textus* 226.14 *kabšūh*, ܟܒܫܘܗ̇) Antioch. In the same year a Roman army joined battle with the Persians in Syria and the Romans came off much the worse..."

Flusin follows the dating of Dionysius (i.e. *Chr. 1234*), placing it in 611; Howard-Johnston places it in 610.

APAMEA kabšūh, ܟܒܫܘܗ̇ 610/11

See *Antioch (above).

The Persian success continued after *Antioch: "On October 15 they took (*textus* 226.15 *kabšūh*, ܟܒܫܘܗ̇) Apamea..."

EMESA ešta'bdat, ܐܫܬܥܒܕܬ 610/11

See *Antioch and *Apamea (above); also *Chr. 640* (Palmer 1993: 17) and cf. *Edessa 609.

Immediately after *Apamea, according to Dionysius, the Persians "... marched on Phoenician Emesa, which surrendered on the strength of an amnesty." Palmer's translation is rather liberal; cf. *textus* 226.16f *w-yab(w) lāh melltā. w-ešta'bdat l-parsāyē*, ܘܝܗܒܘ ܠܗ̇ ܡܠܬܐ. ܘܐܫܬܥܒܕܬ ܠܦܪܣܝܐ ("and they gave her [Emesa] an amnesty, and it surrendered to the Persians").The *Chr. 640* provides evidence of a Persian program of deportation: "AG 922: The Persians entered Emesa, where they found many people of eastern origin and these they sent away from there, each to his own country."

THEODOSIOPOLIS [camped around] 610/11

Sebeos 33 (I:63f, II: 202).

After a decisive victory over the Romans, the Persians under Khoream "camped around the city of Karin (Theodosiopolis) and initiated military action against it. They were opposed from within for a while, and not insignificant was the slaughter caused by those outside. Then the caesar T'ēodos came forward, saying 'I am your king.' They then acquiesced and opened [the gate]. The chief men of the city came out and presented themselves to him. On returning they persuaded the city that he really was T'ēodos, son of Maurice. Then, having opened the gate, they submitted." Several other cities or forts were also taken at the same time.

CAESAREA	šrā... 'al, ܐܠ ܩܣܪܝܐ	611

See *Antioch (above); Flusin 1992: 81f; Sebeos 33 (I 64f, II:202).

Dionysius continues: "... and the Persian general Vahrām besieged (*textus* 226.20f *šrā... 'al,* ܐܠ ܩܣܪܝܐ) Cappadocian Caesarea—some of the inhabitants were killed, but most went back with Vahrām as captives." Sebeos claims that the Christians abandoned the city, "but the Jews went out to meet him and submitted."

The Persians were then besieged in turn by Herakleios for over a year, but when supplies became scarce, they set fire to the city, broke the siege, and went to Armenia (Sebeos 34, I: 66; II 203).

ROMAN CITY IN HISPANIA	*obsedit*	c. 611

Isidore *HG* 59 (Wolf 1999: 105); Thompson 1969: 160; Collins 2002: 75.

Isidore reports the following activities during the reign of Gundemar (610-12): "He devastated the Basques during one expedition and besieged the army of the Romans on another." || hic Wascones una expeditione vastavit, alia militem Romanum obsedit... ||

Thompson dates this firmly to 611 based on surrounding events.

NAIX	*ceptum*	612

Fredegar 4.38.

Theuderic's army invaded Theudebert's kingdom (Austrasia). Assembling at Langres and marching through Andelot, "it took the stronghold of Naix..." || Nasio castra ceptum... ||

TOUL	*cepit*	612

Continued from *Naix (above).

"... and advanced upon the city of Toul, which also fell to it." || ...Tollo civitate perrexit et cepit. || Afterwards, Theuderic's army defeated Theudebert's Austrasian relieving army in open battle.

?COLOGNE	[*perrexit*]	612

Continued from *Naix and *Toul (above).

After his victories at *Naix and *Toul, Theuderic pursued Theudebert to Zülpich, where they fought a massive battle. Theudebert lost and fled, while Theuderic marched to (*perrexit*) Cologne, where Theudebert's treasure was confiscated and he himself ritually humiliated when captured shortly after.

DAMASCUS   ša'bdāh, ܫܥܒܕܗ   613
Dionysius 23 (Palmer 1993; 127f and n. 287, AG 922); Sebeos 34 (II: 202); Flusin 1992: 79.

According to Dionysius: "In year 4 of Herakleios Shahrvarāz subjugated (*textus* 226.23 *ša'bdāh*, ܫܥܒܕܗ) Damascus to the Persians and the Damascenes received an amnesty in return for the payment of tribute (*textus* 226.24 *wa-nsab(w) dārmasūqāyē melltā d-nettlūn madātā*, ܘܢܣܒ(ܘ) ܕܪܡܣܘܩܝܐ ܡܠܠܬܐ ܕܢܬܠܘܢ ܡܕܐܬܐ)." Michael the Syrian adds, after the fall of Damascus, "... in the following year he conquered Galilee and the region of (the) Jordan."

JERUSALEM   *nqaš...'al, w-kabšāh b-ḥarbā*, ܢܩܫ...ܥܠ ܘܟܒܫܗ ܒܚܪܒܐ.   614
Strategius; Sebeos 34 (I: 68ff, II: 207f); Dionysius 24 (Palmer 1993: 128); cf. Hoyland 2011: 64f; Flusin 1992: 151-81.

Flusin has an exhaustive analysis of the sources, context and events of the siege; here the focus is on military developments. To the quotes by Sebeos and Dionysius below, Strategius adds that the Persians used catapults to [help] bring down the wall and also has more details on the massacre.

According to Sebeos, when the Persians had taken control over most of Palestine: "At first they [the inhabitants of Jerusalem] agreed and submitted. They offered to the general and the [Persian] princes splendid gifts. They requested reliable officers, whom they installed in their midst to guard the city. But after some months had passed, while all the mass of ordinary people were complaisant, the youths of the city killed the officers of the Persian king, and themselves rebelled against his authority. Then there was warfare between the inhabitants of the city of Jerusalem, Jewish and Christian. The larger number of the Christians had the upper hand and slew many of the Jews. The surviving Jews jumped from the walls and went to the Persian army. Then Khoream, that is Ĕrazmiozan, gathered his troops, went and camped around Jerusalem, and besieged it. He attacked it for 19 days. Having mined the foundations of the city from below, they brought down the wall." A horrific massacre followed, leaving 17,000 dead according to Sebeos, while 35,000 were captured and the city burnt.

Dionysius: "In year 6 of Herakleios = 27 of Chosroēs, Shahrvarāz battered at the walls of Jerusalem and took it by the sword (the original is somewhat less informative; see *textus* 226.26f *nqaš...'al, w-kabšāh b-ḥarbā*, ܢܩܫ...ܥܠ ܘܟܒܫܗ ܒܚܪܒܐ), slaughtering 90,000 Christians in it. The Jews in their hatred actually bought Christians at a low price for the privilege of killing them. As for Zechariah, the Chalcedonian bishop of Jerusalem, Shahrvarāz took him captive and sent him down to Khusro in Persia, together with the venerable Wood of the Cross and the gold and silver treasure. He also banished the Jews from Jerusalem. The following year Shahrvarāz invaded Egypt and, with much bloodshed, subjected it with Alexandria to the Persians." Palmer (1993: 128 n. 289) adds from Michael the Syrian: "The first year of Herakleios' reign, there was a solar eclipse lasting four hours. There was also such a drought that all the crops failed, not only the wheat, with the result that there was a serious food-shortage. In the same year a party of Arabs came up from Arabia into Syria, capturing people and booty, wasting many regions and committing numerous massacres and acts of arson without the slightest compunction."

The Arabs were probably sent or encouraged by the Persians.

ROMAN CITIES IN HISPANIA   *pugnando sibi subiecit*   c. 614f
Isidore *HG* 61; Fred. 4.33; Thompson 1969: 162; Collins 2002: 75; Wolf 1999: 106.

On the reign of Sisebut (612-21), Isidore records: "Sisebut was famous for his military example and his victories. [After defeating his northern enemies:] He had the good fortune to triumph twice over the Romans in person and to subject certain of their cities in battle (*pugnando sibi subiecit*). He was so merciful in the wake of victory that he ransomed many of the enemy, who had been reduced to slavery and distributed as booty by his army, using his own treasure for their redemption." The fighting was so significant that there were reactions to the campaign as far away as Francia, where Sisebut's mercy towards the conquered Roman troops was noted by Fredegar. Thompson suggests that the two campaigns should be dated to 614-15.

CHALCEDON   περιεκάθετο   615
Nik. 6; Sebeos 38 (I: 78f, II: 210-13); Flusin 1992: 83-93.

The Persians under Shahin traversed Anatolia in 615. "Having done these things, he proceeded with his whole army against the city of Chalcedon, which he invested for a long time (καὶ περιεκάθετο ταύτην χρόνον ἐπὶ συχνόν), and requested that the emperor should come and parley with him. Indeed, the emperor assented and crossed over to meet him, surrounded by the imperial bodyguard and his retinue." Herakleios offered considerable concessions, but the siege was broken up when the general Philippikos approached or entered Persian territory, forcing Shahin to take his army back to the east.

The prelude and aftermath are given differently in the various sources; the chronology and context is sorted out in Howard-Johnston's commentary to Sebeos, who conflates this episode with the siege of *Constantinople in 626.

THESSALONICA   προσβαλεῖν τῷ τείχει   c. 615
Miracula St. Demetrii 2.1 (179-94); Pohl 1988: 250ff.

During the episcopate of John, the Sklavines in enormous numbers rose; Drogoubites, Sagoudates, Blegezites, Baiounetes, Berzetes and others, who made boats out of a single tree trunk; with these they proceed to raid all Thessaly and most of the Aegean, ravaging and laying waste to a number of cities and provinces. Finally they decided to attack Thessalonica (179).

The Slavs established a "camp" of innumerable monoxyla along the shore, while the rest invested the city on land from the east, north and west; they brought with them their families and their baggage, intending to settle in the city when it was taken (180): Εἶτα δὲ καὶ ἐπὶ τούτοις ὁμογνώμες γενόμενοι, ἄσπερ κατεσκεύασαν ἐκ μονοδένδρων γλυπτὰς νῆας, ἀπείρους τὸν ἀριθμὸν ὑπαρχούσας, κατὰ τὸ πρὸς θάλασσαν κατεστρατοπέδευσαν μέρος· τὸ δὲ λοιπὸν ἀναρίθμητον πλῆθος διά τε ἀνατολῆς, ἄρκτου καὶ δύσεως δι' ὅλων τῶν μερῶν τὴν θεοφρούρητον ταύτην περιστοιχίσαι πόλιν, μεθ' ἑαυτῶν ἐπὶ ξηρᾶς ἔχοντες τὰς ἑαυτῶν γενεὰς μετὰ καὶ τῆς αὐτῶν ἀποσκευῆς, ὀφείλοντες ἐν τῇ πόλει μετὰ τὴν ἅλωσιν τούτους ἐγκαταστῆσαι.

The Thessalonians were terrified due to the reputation of the barbarians. There was also a lack of ships in the city and surrounding region to defend the entrance to the port, and Christians who had fled from them knew from experience (as prisoners) of the Slavs' *"impitoyable comportement à la guerre"* (181): [...] δειλίαν δὲ πλείω θέσθαι τοῖς πολίταις ἐκ τῶν ἀποφύγων χριστιανῶν, τῶν ἐν πείρᾳ τῆς αὐτῶν ἀνηλεοῦς παρατάξεως γεγενημένων αἰχμαλώτων. Καὶ ἦν τότε καὶ τῶν δειλῶν καὶ τῶν ἀνδρείων ἡ ψυχὴ μία, καὶ ἕκαστος πρὸ ὀφθαλμῶν τὸν πικρὸν τῆς αἰχμαλωσίας ἑώρα θάνατον, οὐκ ἐχόντων ἑτέρως τοῦ φυγεῖν, κατὰ τὸ θεῖον λόγιον τὸ φάσκον· " Ἐάν τις ὑμᾶς διώκῃ ἐκ τῆς πόλεως ταύτης, φεύγετε εἰς τὴν ἑτέραν" · διότι καθάπερ στεφάνη θανατηφόρος τὸ βάρβαρον ἅπαν Σκλαβίνων τὴν πόλιν περιετείχει.

Divine intervention helped the citizens as the Slavs intended to make a general assault: those on the seaside decided they had to build covers to avoid being bombarded from the walls, so headed off to gather materials in a nearby field. This gave courage [and time] to the inhabitants (182): Ἀλλ' ὁ μὴ βουλόμενος ἡμῶν τῶν ἁμαρτωλῶν τὸν θάνατον, ἀλλὰ τὴν ἐπιστροφὴν καὶ τὴν ζωήν, οὐδὲ ἐν τούτῳ τοῦ ὡς ἀληθῶς γνησίου αὐτοῦ δούλου, τοῦ κηδεμόνος ἡμῶν ἀναξίων, τοῦ ἀειμνήστου μάρτυρος Δημητρίου τῶν πρεσβειῶν παρήκουσεν, ἀλλὰ πρώτην καὶ τοιαύτῃ πολιορκίᾳ ἐπίσκεψιν τῶν θαυμάτων ἐποιήσατο. Σύνταξιν γὰρ τοῦ παντὸς τῶν Σκλαβίνων ἔθνους ποιησαμένων ὁμοθυμαδὸν καὶ αἴφνης προσβαλεῖν τῷ τείχει, οἱ ἐν ταῖς ναυσὶν ὄντες Σκλαβίνοι σκέψιν ταύτην ἐποιήσαντο ἐφ' ᾧ ταύτας ἐπάνωθεν σανίσι τε καὶ ταῖς λεγομέναις βύρσαις σκεπάσαι, ὅπως τῷ τείχει μελλούσας προσορμῆσαι, ἀπληγας τοὺς ἐλάτας ἐκ τῶν ἀπὸ τῶν τειχέων λίθους ἢ ὅπλα ἀκοντιζόντων κατ' αὐτῶν φυλάξειεν. (Thus the martyr intervened on behalf of the city). [...] ἀλλ' εἰς τόπον κολπώδη ὁρμίσεώς τινος ὑπαρχούσης, τὸ ἐπικληθὲν ἐκ τῶν ἀρχαίων Κελλάριον, ἐκεῖσε παραγενόμενοι ἐφ' ᾧ τὸ μελετηθὲν αὐτοῖς ἐκπληρῶσαι τῆς τέχνης ἔργον, κἀκεῖσε ἐπὶ τοῦτο τῶν βαρβάρων ἐνασχοληθέντων, μικρὸν θάρσους τοὺς τῆς πόλεως ἀναλαβεῖν ὡς βραχείας ἐνδόσεως αὐτοῖς γεγενημένης,

They proceeded to construct obstacles in the water in several lines, including a "boulevard" whence to conduct impending battle (183): καὶ κατασκευάσαι τινὰς ἐκ ξύλων βάσεις ἐν τῷ λιμένι, ἐν αἷς τὴν ἀπόθεσιν τῆς ἁλύσεως ἐποιήσαντο, καὶ μηροὺς δὲ ὡσαύτως ἐξ ἀναλύτων σιδήρων ἑαυτοὺς ἀμπέχοντας, χιοειδῶς τινας ὀξείας φέροντας ῥάβδους, ἑτέρας δὲ ἡλωτὰς σπαθοειδεῖς ἐκ ξύλων ἐξεστώσας, ἐνδότερον δὲ τούτοις τὰς ἐπὶ παρακομιδῇ ξυλῆς τυχούσας νῆας, ἅσπερ κυβαίας ἐκάλουν, ἀπ' ἀλλήλων δι' ἀγκυρῶν συνεχόμενας, κατὰ τὸ στόμιον τοῦ λιμένος προσηλωθείσας, δίοδον πρὸς τὴν μέλλουσαν παράταξιν ἐποιήσαντο.

They also dug a trap ditch and covered it with branches in an undefended section near the port. Subsequently they prepared engines and commended themselves to God and St. Demetrius (184): Τάφρον δὲ τότε πρὸς τῷ πανυμνήτῳ τεμένει τῆς ἀχράντου Θεοτόκου τῷ ὄντι πρὸς τῷ αὐτῷ λιμένι ἐποιήσαντο, ἀτειχίστου τοῦ τοιούτου καθεστῶτος τόπου, ὡς ἅπαντες ἐπίστανται· καὶ ἡ τῶν πουλπίτων διὰ γονατίων ἡλωτῶν μηχανὴ κατεσκεύαστο ἐν τῇ γῇ κρυφηδὸν ἀποτεθέντων καὶ ἐξ ὀλίγης ὕλης τινὸς σκεπασθέντων, ὅπως τῇ τῶν τοιούτων ὀργάνων ἀορασίᾳ οἱ τὴν ὁρμὴν τῆς ἐπιβάσεως ποιεῖσθαι μέλλοντες πολέμιοι ἐν αὐτοῖς ἐμπαρῶσι· καὶ ἐν τῷ ἐκεῖσε δὲ μώλῳ, καὶ αὐτῷ ἀτειχίστῳ τότε, διὰ σανίδων καὶ ξύλων τινῶν ὡς μέχρι στήθους τειχίσαι, καὶ τὰ λοιπὰ δὲ τῶν ἄλλων μαγγάνων ἀμυντήρια τὰ πρὸς παράταξιν ἤτοι ὄργανα κατεσκευάσθησαν. Καὶ λοιπὸν τὴν ἐλπίδα πᾶσαν εἰς θεὸν καὶ εἰς τὸν ὑπερασπιστὴν τῆς πόλεως Δημήτριον ἀναθέμενοι, τῆς ἑαυτῶν προθυμίας τὸ σπουδαῖον ἐπεδείκνυντο, παραθαρρύνοντες τοὺς ἀσθενεστέρους τὸν λογισμὸν ἐπὶ τῇ ἐλπιζομένῃ τοῦ πολέμου παρατάξει.

The last sentence is significant for understanding the morale in the city: "People then put all their hope in God and in the defender of the city, St. Demetrius, and they demonstrated their great zeal, encouraging with the hoped-for *parataxis* [i.e. St. Demetrius fighting alongside them] in the war those who were weaker in faith."

This went on for three days; meanwhile the Slavs reconnoitered from a distance to determine where they could attack (τόπους εὐαλώτους ἐφ' ἑκάστης ἡμέρας κατασκοπούντων). This happened at dawn the fourth day with a great cry; general assault with *petroboloi*, scaling ladders, attempt to fire gates, and hail of missiles like snow in winter (185): [...] ἅπαν τὸ βάρβαρον φῦλον ὁμοθυμαδὸν ἀνακράξαν ἐκ πάντοθεν τῷ τείχει τῆς πόλεως προσέβαλον, οἱ μὲν διὰ πετροβόλων κατεσκευασμένων λίθους ἀκοντίζοντες, ἄλλοι προσάγοντες κλίμακας πρὸς τῷ τείχει ἐκπορθεῖν ἐπειρῶντο, ἄλλοι ἐν ταῖς πύλαις πῦρ ἀποκομίζοντες, ἕτεροι βέλη καθάπερ νιφάδας χειμερινὰς τοῖς τείχεσιν ἀνέπεμπον. The Slav fleet attacked the port from two angles, one against a postern, while another against the traps of which they were ignorant (186).

The Thessalonians, instead of relying on stones and missiles, prayed for help, and St. Demetrius appears on the walls and on the sea, even being seen by the Jews ("The Children of the Hebrews"), throwing the Slavs into complete confusion, especially at sea. The Slavs in the confusion and crowding end up using their own weapons against each other in the rush for each to save himself. The Thessalonians exploited confusion to strike back, some against ships, others sallied through a postern. St. Demetrius fought with them. The sea was colored red with barbarian blood, while a divine wind (only the second hour, long before normal winds) blew in from the sea to scatter the Slav fleet further. Bodies were cast ashore by the sea wall, so the soldiers went out, cut off their heads, and mounted them on the wall for the enemies to see (λοιπὸν οἱ τοῦ παραλίου παντὸς ὁπλῖται ἐξελθόντες, τὰς τῶν δυσμενεστάτων κεφαλὰς ἀποτέμνοντες, διὰ τοῦ χερσαίου τείχους τοῖς βαρβάροις ὑπέδεικνον). The barbarians abandoned most of their machines and their booty (τὰ πλεῖστα τῶν μαγγάνων καὶ τῶν σκύλων καταλιπόντες) in great sorrow. (187-191)

The Thessalonians gave thanks to St. Demetrius in his church (192). In contrast to the appropriate Christian rituals, the Slav chief, Chatzon, learnt through divination that he would enter the city; in fact, he was taken prisoner during a sortie from a postern, and taken inside by the nobles who wanted to use him for political gain (a sensible move for negotiations, ransom, bargains etc), but their women dragged him through the city and stoned him (193). The author ends with the usual rebuff of critics (194).

THESSALONICA πολιορκία c. 618
Miracula St. Demetrii 2.2 (195-215); Pohl 1988: 242f; Curta 2001: 107f; further discussion in chapter 7.2.2.

The Slavs had experienced a disaster last time they tried to take the city, and their prisoners fled to Thessalonica, taking with them part of their booty (196). They went to the Avar Chagan with great presents to ask for help, claiming it would be easy to take the city, as all other cities and territories around were depopulated and under Slav control; also the city was full of refugees from all over the Danubian regions—Pannonia, Dacia, Dardania and others (197). The Chagan was convinced, mobilized all his subject peoples—Slavs, Bulgars, and others unnamed—and after two years marched on Thessalonica. In advance he sent elite cavalry to kill or take prisoner everyone outside the city, following with the main army and machines (διαφόρους κατασκευὰς μαγγάνων πολεμιστηρίων πρὸς πόρθησιν τῆς καθ' ἡμᾶς πατρίδος) himself (198).

Most Thessalonians were out in the fields harvesting (τοὺς ἐν ἀμητῷ πάντες), and were caught unawares by the cavalry raid that suddenly appeared from many directions, coordinated to arrive at the fifth hour, capturing and killing many, μεθ' ὧνπερ εὗρον ἀγελαίων πλείστων ζῴων καὶ λοιπῶν τῶν ἐπὶ ἐργασίᾳ τοῦ ἀμητοῦ σκευῶν (199). The Thessalonians realized that they were unprepared, but tried to encourage each other; however, refugees from Naissos and Serdica told of sieges they had experienced and how one stone could destroy the walls of the city. " Ἐκεῖθεν φυγόντες ἐνταῦθα ἤκομεν μεθ' ὑμῶν ἀπολέσθαι, μία γὰρ τούτων λίθου βολὴ τὸ τεῖχος κατεάξει." (200)

Bishop John took a lead in encouraging the citizens and preparing defenses: ...μὴ ῥαθυμεῖν παρήνει, ἀλλὰ προθύμως τῶν δεόντων ἀνθοπλίζεσθαι, διαβεβαιούμενος μηδὲν λυπηρὸν ἢ ὀκνηρὸν φέρειν, τῷ θεῷ δὲ μᾶλλον καὶ τῷ μάρτυρι τὰς ἐλπίδας ἐπιρρίπτειν. Τούτοις δὲ καὶ τοῖς τοιούτοις τὸ θάρσος διὰ τῆς πόλεως παρέχων, καὶ τῷ τείχει μετ' αὐτῶν ἐνδιατρίβων καὶ τῶν πρὸς ἀντιμαχείαν εὐτρεπιζόντων (201), and after a few days the Chagan arrived with his host, including Bulgarians and other peoples, and surround the walls completely, καθάπερ λαῖλαψ χειμερινὴ τῇ ἁγιοφυλάκτῳ ταύτῃ πόλει διὰ πάσης τῆς χερσαίας προσῆψε τῷ τείχει, καὶ περιστοιχίσας ἅπασαν τὴν πόλιν, and the innumerable crowd could barely be supported by the earth and dried up every stream (202).

Barbarian preparations and Thessalonian reactions (203): Τότε δὴ ἑωρακότες οἱ τῆς πόλεως τὸ ἀνείκαστον τῶν βαρβάρων πλῆθος ἅπαν σεσιδερωμένον, καὶ τὴν τῶν πετροβόλων ἐκ πάντοθεν οὐρανομήκη παράστασιν ὡς ὑπερβαίνειν τῷ ὕψει τὰς τῶν ἔσω τειχῶν ἐπάλξεις, ἄλλους δὲ τὰς καλουμένας ἐκ πλοκῶν καὶ βυρσῶν χελώνας, ἄλλους πρὸς ταῖς πύλαις κριοὺς ἐκ ξύλων μεγίστων καὶ τροχῶν ἐμπειροκυλίστων, ἑτέρους δὲ πύργους ὑπερμεγέθεις ξυλοκατασκευάστους ὑπερβαίνοντας τὸ ὕψος τοῦ τείχους κατασκευάσαντας, ἔχοντας κατεπάνω νεανίας σφριγῶντας καθοπλισμένους, ἑτέρους δὲ τοὺς καλουμένους ὄρπηκας ἐμπεπηγότας, ἄλλους ὑποτρόχους κλίμακας ἐπιφερομένους, ἑτέρους διαπύρους μαγγανείας ἐπινοοῦντας, so that people said that "Even if God saved us during earlier sieges, we don't think we'll be saved this time, for never has such a multitude of barbarians been seen to attack the city."

Bishop John encouraged the population: ... ἀλλὰ τὸν ἀθλοφόρον αἰτεῖσθαι καὶ τὰ νῦν συμμαχεῖν. He then had a dream where a large man appeared to him, telling him that if the whole population would cry "Kyrie Eleison" as one, the city would be spared. John despaired, since the population was scattered across the city walls and the barbarians made noise from outside, but prayed for a miracle (204). In his prayer, he asked to be helped the same way as David, whom God gave salvation from Goliath through a stone (λίθον), "so that we may destroy slinging (σφενδονίσαντες) the barbarian phalanxes who are scheming (κακοτέχνους) against us, with you fighting beside us (205)." John was praying for the city's engines to work well, in effect.

John kept praying and encouraged the citizens to array themselves against the barbarians; when the barbarians began bombarding the city "with hills and mountains" (Τῆς οὖν πολιορκίας γενομένης, καὶ τῶν πετροβόλων πάντοθεν ἀκοντιζόντων οὐχὶ πέτρας ἀλλ' ὄρη καὶ βουνούς), one of the citizen catapult operators (εἷς ἐν τῇ ἔνδον τῶν πολιτῶν πετραρέᾳ) was inspired by God to give the battle cry "In the name of God and Saint Demetrius" ("Ἐν τῷ ὀνόματι τοῦ θεοῦ καὶ τοῦ ἁγίου Δημητρίου"), perhaps pronounced thus in colloquial Greek: στὄνομα θεοῦ καὶ ἁγίου Δημητρίου. The phrase can be pronounced with a strong rhythmical cadence in Greek (cf. modern Greek football chants), and may actually be the battle cry used by catapult crews when operating their engines. The stone he shot off slammed into a much larger enemy rock (clearly as it was being fired), so that both fell down and killed the crew inside the enemy *petrobolos* as well as the *manganarios* operating it (206).

A violent earthquake at midday caused the whole population to cry "Kyrie Eleison" at the same time. The enemies arrived to take advantage of the quake, only to find the walls sound. The bishop realized that God and St. Demetrius were protecting the city (207). There was also a third miracle:

arrows shot by the enemies were stuck in the city wall but with iron tips pointing out. John explains how the fear of the earthquake made the city cry in unison, while the enemies are struck with terror and the Thessalonians are eager for combat (208). Miraculously, ships (σιτοφόρους ὁλκάδας) arrive every day with supplies (although regular supplies to a besieged city might not qualify as miraculous), and sailors experienced in operating machines came to help operate the siege engines (τοὺς δὲ τούτων ναυτικοὺς ὡς ἐμπειρομαγγάνους ταῖς πετραρέαις καὶ τῶν λοιπῶν ἐξυπηρετεῖν κατεσκευασμένων ὅπλων). The barbarians claimed that the Thessalonians sent out the ships at night and let the same ships return during the day. The ship owners, however, said they had been directed to the city by the miraculous intervention by an unknown *kankellarios* who turned out to be the saint, as he also procured favorable winds (209). In fact, nobody knew about the siege, not even the emperor; an eparch named Charias only learnt about it upon disembarking. He therefore went to church of saint to pray, and then armed himself and went onto the ramparts (προσκυνήσας ἐν τῷ ναῷ τοῦ σῳσιπόλιδος Δημητρίου, πρὸς τῷ τείχει καὶ αὐτὸς μετὰ πάντων ὁπλισάμενος ἄνεισιν, 210).

The enemy siege engines were rendered useless and ridiculous; the mechanism of a large tower was destroyed on approaching the city and its occupants were killed; tortoises by the wall were hauled into the air by hooks suspended from the walls, exposing the soldiers underneath to blows from the top of the rampart (211). Εἶτα δέ, τῶν ἐκ τῶν ἀντιβίων κατασκευασθέντων μαγγανικῶν ὅπλων λοιπὸν καταπτυσθέντων, καὶ ἀπράκτων διὰ τῶν αὐτῶν ἀντιπαρατάξεων, καὶ ἀνεπιτηδείων διὰ τῆς τοῦ ἀθλοφόρου συνεργίας ἀποδειχθέντων, ὡς πᾶσι δεδήλωται· καὶ γὰρ τοῦ παρ' αὐτῶν ξυλοπύργου, ὅντινα ἐδόκουν ὑπὲρ πάντα φοβερώτερον καὶ ἐπιτήδειον εἶναι, καθοπλίσαντές τε καὶ προσορμῆσαι τῷ τείχει πειρώμενοι, θείᾳ προνοίᾳ αὐτομάτως ἐν τῇ αὐτοῦ κινήσει ῥαγέντος τοῦ ἐν αὐτῷ τὰ ὄργανα ἰθύνοντος, καὶ τοὺς ἐν αὐτῷ ὁπλίτας ἀποθανεῖν· ἄλλους δὲ ἐν ταῖς χελῶσι προσάπτοντας τῷ τείχει, ἐκ τῶν ἐπάνω τῶν τειχῶν διὰ ξύλων ἐχόντων ξίφος ὑνιοειδὲς ἐκ τῶν ὕπερθεν χαλώντων καὶ πηγνύντων, ταύτας ἀνήγειρον, ὡς λοιπὸν τοὺς ἔνδοθεν γυμνοὺς μὲν ταῖς ἐκ τῶν ὁπλιτῶν τοῦ τείχους τιτρώσκεται βολαῖς· ὅθεν οἱ τὸ πρὶν ἔκφοβοι γενόμενοι πολῖται εἰς τέρψιν εἶχον καὶ γέλωτα τὰ τῶν ὑπεναντίων ἀμυντήρια.

The barbarians realized that their chances were small, and asked for gold to leave the city, but the Thessalonians did not accept, so hostilities were resumed. The Chagan was furious, burnt all the sanctuaries and buildings outside the city and threatened to call in more barbarian peoples

(212). The siege lasted 33 days; the Thessalonians agreed to give concessions to the barbarians in order to secure peace, and the barbarians went home (213). After concluding the peace, the barbarians came right up to the wall in order to sell prisoners and objects at a low price; they proclaim the various miracles (214). The city is saved from many dangers by saintly intercession; doxology (215).

ALEXANDRIA   ἀνὰ κράτος εἷλε   619
Nik. 6 (p. 44f); cf. Hoyland 2011: 65.

"Now for Chosroes, king of Persia, collected a numerous army and sent it against the Romans after appointing Saitos commander of the Persian forces. This man came up to Alexandria, which he took by main force, and he captured all of Egypt (ταύτην ἀνὰ κράτος εἷλε καὶ τὴν ὅλην Αἴγυπτον ἠνδραπόδιζε). He devastated the entire oriental part <of the empire>, taking many prisoners and killing many without pity."

ANCYRA   παρέλαβον... πολέμῳ; *nazala 'alā...wa-'ftataḥahā*   622
Theoph. 302 (AM 6111, MS 434); Agapius 458; Hoyland 2011: 66; Foss 1977; Flusin 1981.

"In the same year the Persians took by war Ancyra in Galatia." Τῷ δ' αὐτῷ ἔτει παρέλαβον οἱ Πέρσαι τὴν Ἄγκυραν Γαλατίας πολέμῳ.

"The Persian commander Shahrbaraz attacked (*ghazā li-*, غزا لـ) the Romans and besieged (*wa-nazala 'alā*, ونزل على) Ancyra and took it (*wa-'ftataḥahā*, وافتتحها), killing and enslaving all who were in it. And at the end of the year he also captured Rhodes and enslaved its people." Note the lack of equivalent in Agapius to the Greek *polemō*.

ROMAN CITIES IN HISPANIA   *perdomuit, proelio concerto obtinuit*   c. 624
Isidore HG 62; Thompson 1969: 168f; Collins 2002: 77; Wolf 1999: 106.

The last Roman cities fell during the reign of Suinthila (621-31): "Having risen to the position of general under King Sisebut, he captured (*perdomuit*) Roman fortresses and overcame the Ruccones. After he ascended to the summit of royal dignity, he waged war and obtained (*proelio concerto obtinuit*) all the remaining cities which the Roman army held in Spain." His abilities to organize war is accentuated by HG 63, which records how Suinthila defeated the Basques, who "build the city of Ologicus for the Goths with their own taxes and labour," a typical method of organizing the construction of fortifications, cf. chapter 1.2.5 and *passim*.

Thompson places the final expulsion of the Romans to 623-25, while Collins suggests that Cartagena was taken in 625; he also refers to archaeological evidence that it was demilitarized by the Goths, who appear to have demolished its walls when it was captured. This would prevent the Romans from using it as a safe base if they were ever to send a fleet to recapture their province.

CONSTANTINOPLE   *nqaš(w) 'al*; μηχανήματα τειχομάχα ἐτέκταινον   626
Nik. 13; Dionysius 33; *Chronicon Paschale* 717-25; Howard-Johnston 1995b, 2010: 45-48 and *passim*; Pohl 1988: 248-55; Hoyland 2011: 68.

The siege and its many sources has been studied a number of times; this allows for more focus on the terminology employed by selected sources. The version of Dionysius 33 (Palmer 1993: 135, AG 936) is mostly interesting for the Syriac terminology and the Syrians' understanding of the geography on the Bosphorus, as he has the Persians arrive at *Constantinople* [i.e. Chalcedon] "with a great force and an arsenal of military equipment and they laid siege to (*textus* 231.21f *etaw nqaš(w) 'al*, ܐܬܘ ܢܩܫܘ ܥܠ) the city of Constantinople from the west. For nine months the Persians maintained their guard on the City and brought the emperor Herakleios, who was within, under great pressure. But after that the Persians rebelled against their king and made peace with Herakleios. [...]"

When the Avars had plundered Thrace, and the Persians destroyed the Asian side (cf. Dionysius), Nikephoros relates: "Now the Avars constructed siege engines, namely, wooden towers and "tortoise shells"; but when these machines approached the walls, a divine force undid them and destroyed the Avar soldiers inside." "οἱ οὖν Ἄβαροι μηχανήματα τειχομάχα ἐτέκταινον· πύργοι δὲ ἦσαν ξύλινοι καὶ χελῶναι τὰ κατασκευάσματα. καὶ ἐπεὶ προσῄεσαν τῷ τείχει τὰ ὄργανα, θεία δύναμις ἐξαπιναίως ταῦτα διέλυσε καὶ τοὺς ἐν αὐτοῖς τῶν Ἀβάρων μαχητὰς διώλεσεν." The Avars also planned to use subject Slavs to attack by sea "in their hollowed-out canoes" (ἅμα τοῖς μονοξύλοις ἀκατίοις) on a pre-arranged fire signal, but the Romans discovered this, lured the Slavs into a trap, and killed them. "When the barbarians beheld this, they gave up the siege and returned home. As for the archpriest of the City and Emperor Constantine, they proceeded to the church of the Mother of God at Blachernai to offer unto God prayers of thanksgiving (εὐχαριστηρίους λιτὰς τῷ θεῷ); and straightaway they erected a wall to protect that sacred church (τεῖχος δὲ εὐθὺς δωμησάμενοι τοῦ ἱεροῦ ἐκείνου ναοῦ φρούριον κατέστησαν)."

The version of Chronicon Paschale is one of the best siege descriptions of the 7th century and thus deserves to be quoted extensively (for supplementary sources, see Howard-Johnston, Pohl and Hoyland):

"And so on the 29th of the month June of the present indiction 14, that is on the day of the Feast of the holy and glorious chief apostles, Peter and Paul, a vanguard of the God-abhorred Chagan arrived, about 30,000 (... κατέλαβε πρόκουρσον τοῦ θεομισήτου Χαγάνου, ὡς ἄχρι χιλιάδων τριάκοντα). He had spread the rumour by means of reports that he would capture the Long Wall and the area within it, and as a result, on the same day, which was a Lord's Day, the excellent cavalry who were present outside the city came inside the new Theodosian Wall of this imperial city (ὥστε τοὺς εὑρεθέντας ἔξωθεν τῆς πόλεως ἐφίππους γενναιοτάτους στρατιώτας κατὰ τὴν αὐτὴν ἡμέραν, κυριακὴν οὖσαν, ἔνδον γενέσθαι τοῦ νέου Θεοδοσιακοῦ τείχους ταύτης τῆς βασιλίδος πόλεως...). The same advance guard remained in the regions of Melantias, while a few of them made sallies at intervals as far as the wall, and prevented anyone from going out or collecting provisions for animals at all (...καὶ ἔμεινεν τὸ αὐτὸ πρόκουρσον ἐπὶ τὰ μέρη Μελαντιάδος, ὀλίγων ἐξ αὐτῶν ἐκτρεχόντων μέχρι τοῦ τείχους ἐκ διαλειμμάτων, καὶ μὴ συγχωρούντων τινὰ ἐξιέναι ἢ ὅλως ἀλόγων δαπάνας συλλέγειν)." Some of these soldiers went out with civilians to harvest and collect supplies at the last minute (...ἐξῆλθαν στρατιῶται μετὰ παλλικαρίων καὶ πολιτῶν...); when they encountered the Avars, there was a fight that led to losses on both sides. The Romans could have won the engagement but had to divert forces to protect the civilians. (Whitby 170f; ed. 717)

Athanasius from Adrianopolis, a patrician, was sent by the Chagan to negotiate for concessions from Constantinople; however, he was upbraided by the authorities for his submissive attitude, which he said stemmed from his original instructions; "thereafter he had not learnt that the defences had been strengthened thus and that an army was present here; however, he was ready to tell the Chagan without alteration the message given to him. Then, after the same most glorious Athanasius requested that he first wished to inspect the army that was in the city, a muster was held and about 12,000 or more cavalry resident in the city were present." ...λοιπὸν δὲ μηδὲ μεμαθηκέναι αὐτὸν οὕτω τὰ τοῦ τείχους κατησφαλίσθαι καὶ στρατὸν ἐνταῦθα παρεῖναι· πλὴν ἑτοίμως ἔχειν αὐτὸν τὴν αὐτῷ διδομένην ἀπόκρισιν λέγειν ἀπαραλλάκτος τῷ Χαγάνῳ. εἶτα ἐπιζητήσαντος τοῦ αὐτοῦ ἐνδοξοτάτου Ἀθανασίου πρότερον ἐθέλειν θεωρῆσαι τὸν ἐν τῇ πόλει ὄντα στρατόν, ἀρμαστατιῶνος γενομένης ηὑρέθησαν τῶν ἐνδημούντων ἐν τῇ πόλει καβαλλαρίων περὶ τὰς ιβ' καὶ πρὸς χιλιάδας. Subsequently the officials sent a message to the Chagan, warning him not to approach; however he demanded the city and all the inhabitants (Whitby 172; ed. 718).

"On the 29th of the month July the same God-abhorred Chagan reached the wall of the whole of his horde, and showed himself to those in the city. After one day, that is on the 31st of the same month July, he advanced, arrayed for battle, from the gate called Polyandrion as far as the gate of Pempton and beyond with particular vigour: for there he stationed the bulk of his horde, after stationing Slavs within view along the remaining part of the wall. And he remained from dawn until hour 11 fighting first with unarmored Slav infantry, and in the second rank with infantry in corslets. And towards the evening he stationed a few siege engines and mantelets from Brachialion as far as Brachialion." ...τῇ κθ' τοῦ ἰουλίου μηνὸς αὐτὸς ὁ θεομίσητος Χαγάνος κατέλαβε τὸ τεῖχος μετὰ ὅλου τοῦ ὄχλου αὐτοῦ, καὶ ἔδειξεν ἑαυτὸν τοῖς τῆς πόλεως. μετὰ μίαν ἡμέραν, τουτέστιν τῇ λα' τοῦ αὐτοῦ ἰουλίου μηνός, ἦλθεν παρατασσόμενος πόλεμον [καὶ ἔμεινεν ἀπὸ ἕωθεν ἕως ὥρας ια' πολεμῶν] ἀπὸ τῆς λεγομένης Πολυανδρίου πόρτας καὶ ἕως τοῦ Πέμπτου καὶ ἐπέκεινα σφοδροτέρως· ἐκεῖ γὰρ τὸν πολὺν αὐτοῦ παρέστησεν ὄχλον, στήσας εἰς ὄψιν κατὰ τὸ λοιπὸν μέρος τοῦ τείχους Σκλάβους. καὶ ἔμεινεν ἀπὸ ἕωθεν ἕως ὥρας ια' πολεμῶν, πρῶτον μὲν διὰ πεζῶν Σκλάβων γυμνῶν, κατὰ δὲ δευτέραν τάξιν διὰ πεζῶν ζαβάτων. καὶ περὶ ἑσπέραν ἔστησεν ὀλίγα μαγγανικὰ καὶ χελώνας ἀπὸ Βραχιαλίου καὶ ἕως Βραχιαλίου. (Whitby 173; ed. 719)

The *brachialon* was a stretch of wall that protruded into the sea on both the Golden Horn and the Sea of Marmara, making it impossible to flank the landwalls and assault the weaker sea walls from land. The meaning is thus "all along the land walls."

"And again on the following day he [the Avar Chagan] stationed a multitude of siege engines close to each other against that part which had been attacked by him, so that those in the city were compelled to station very many siege engines inside the wall. When the infantry battle was joined each day [presumably storming of the walls], through the efficacy of God, as a result of their superiority our men kept off the enemy at a distance. But he bound together his stone-throwers and covered them outside with hides; and in the section from the Polyandrion gate as far as the gate of St. Romanus he prepared to station 12 lofty siege towers, which were advanced almost as far as the outworks, and he covered them with hides. And as for the sailors who were present in the city even they came out to assist the citizens. And one of these sailors constructed a mast and hung a skiff on it, intending by means of it to burn the enemies siege-towers. Bonus the all-praiseworthy *magister* gave commendation to this sailor for having dismayed the enemy not inconsiderably." [...] καὶ πάλιν τῇ ἑξῆς ἔστησε πλῆθος μαγγανικῶν εἰς τὸ μέρος ἐκεῖνο τὸ πολεμηθὲν παρ' αὐτοῦ σύνεγγυς ἀλλήλων, ὡς ἀναγκασθῆναι τοὺς τῆς πόλεως πάμπολλα στῆσαι μαγγανικὰ ἔνδοθεν τοῦ τείχους τῆς μάχης καθ' ἑκάστην τῶν πεζῶν συγκροτουμένης, καὶ τῶν ἡμετέρων κατὰ θεοῦ δημιουργίαν ἐκ τοῦ περιγεγονότος ἀποσοβούντων μήκοθεν τοὺς ἐχθρούς. ἐκαλάμωσε δὲ τὰς πετραρίας αὐτοῦ καὶ ἔξωθεν ἐβύρσωσεν. παρασκεύασεν δὲ εἰς τὸ διάστημα τὸ ἀπὸ τῆς Πολυανδρίου πόρτας ἕως τῆς πόρτας τοῦ ἁγίου Ῥωμανοῦ στῆναι ιβ' πυργοκαστέλλους ὑψηλούς, φθάνοντας σχεδὸν ἕως τῶν προμαχεώνων, καὶ ἐβύρσωσεν αὐτούς. καὶ οἱ εὑρεθέντες δὲ ναῦται ἐν τῇ πόλει καὶ αὐτοὶ ἐξῆλθον εἰς συμμαχίαν τοῖς πολίταις· καὶ εἷς ἐξ αὐτῶν τῶν ναυτῶν ἐμηχανήσατο καταρτίαν καὶ ἐκρέμασεν εἰς αὐτὴν κάραβον, ὀφείλων δι' αὐτοῦ ἐμπρῆσαι τοὺς πυργοκαστέλλους τῶν ἐχθρῶν, ὅντινα ναύτην καταπλήξαντα τοὺς πολεμίους οὐ μετρίως συνεκρότησε Βόνος ὁ πανεύφημος μάγιστρος. (Whitby 174; ed. 719f)

The *magister* Bonus, standing on the wall, tried to dissuade "the enemy from drawing near to the wall" (μετὰ δὲ τὸ ἐγγίσαι τὸν ἐχθρὸν τῷ τείχει), i.e. assaulting. The Chagan in turn demanded the city and property but not the population. The Avars then attacked with ready prepared canoes (μονόξυλα) on the third day. The Roman "cutters" (σκαφοκάραβοι) were unable to prevent the launch due to shallow waters, but blocked them from approaching closer (Whitby 174f). During a new round of negotiations, the Avar Chagan revealed three Persians dressed in silk. "And he said, 'Look, the Persians have sent an embassy to me, and are ready to give me 3,000 men in alliance (ἐπρεσβεύσαν πρὸς ἐμέ, ἑτοίμως ἔχοντες δοῦναί μοι τρεῖς χιλιάδας εἰς συμμαχίαν)." The Romans should cross over to the Persians with a cloak and shirt, but leave the city and property to the Avars, a condition the Romans could hardly accept (Whitby 175: ed. 721); later on, the Romans captured the Persian envoys and beheaded them in sight of the Persian camp on the Asian side (Whitby 176f). While the Chronicon Paschale has a report on the prelude to the naval engagement that was subsequently fought against the Slav canoes (Whitby 177f; cf. Nikephoros above), there is a lacuna concerning preparations for a Roman surprise that is covered by Theod. Sync. 308.2-28. As a result of a Roman surprise, the Slavs were slaughtered in their canoes; at the same time, the Armenians sallied out (further Whitby 178 n. 473).

"Our men drove all the canoes onto land, and after this had happened, the accursed Chagan retired to his rampart, took away from the wall the siege engines which he had set beside it and the palisade which he had constructed: by night he burnt his palisade and the siege towers and the mantelets, after removing the hides, and retreated. Some people said that the Slavs, when they saw what had happened, withdrew and retreated, and for this reason the cursed Chagan was also forced to retreat and follow them." ἐξέβαλον δὲ ὅλα τὰ μονόξυλα εἰς τὴν γῆν οἱ ἡμέτεροι, καὶ μετὰ τὸ ταῦτα γενέσθαι ὑπέστρεψεν ὁ ἐπικατάρατος Χαγάνος εἰς τὸ φωσᾶτον αὐτοῦ, καὶ ἤγαγεν τὰ μαγγανικὰ ἀπὸ τοῦ τείχους ἃ ἦν παραστήσας καὶ τὴν σοῦδαν ἣν ἐποίησεν, καὶ ἤρξατο καταλύειν τοὺς πυργοκαστέλλους οὓς ἐποίησεν, καὶ τῇ νυκτὶ ἔκαυσεν τὸ σουδᾶτον αὐτοῦ καὶ τοὺς πυργοκαστέλλους, καὶ τὰς χελώνας ἀποβυρσώσας ἀνεχώρησεν. Τινὲς δὲ ἔλεγον ὅτι οἱ Σκλάβοι θεωρήσαντες τὸ γεγονὸς ἐπῆραν καὶ ἀνεχώρησαν, καὶ διὰ τοῦτο ἠναγκάσθη καὶ ὁ κατάρατος Χαγάνος ἀναχωρῆσαι καὶ ἀκολουθῆναι αὐτοῖς. (Whitby 178f ; ed. 724f)

The Chagan said he withdrew due to problems of supplies, and furthermore, that he saw a woman walking around alone on the city walls, a reference to the Virgin Mary (Καὶ τοῦτο δὲ ἔλεγεν ὁ ἄθεος Χαγάνος τῷ καιρῷ τοῦ πολέμου ὅτι ἐγὼ θεωρῶ γυναῖκα σεμνοφοροῦσαν περιτρέχουσαν εἰς τὸ τεῖχος μόνην οὖσαν, Whitby 180; ed. 725). "On the Friday a rearguard of cavalry remained in the vicinity of the wall, setting fire to many suburbs on the same day up till hour 7; and they withdrew. They burnt both the church of SS Cosmas and Damian at Blachernae and the church of St. Nicholas and all the surrounding areas." However, a miraculous intervention by Mary prevented the burning of her church.

Whitby (171 n. 459) rejects Stratos' idea that Thessalonica was besieged the same year; the logistical problems alone would explain the slow advance. Furthermore, the Khagan's excuse for withdrawing is supported by other evidence (cf. Whitby n. 477; cf. Malchus fr. 2, ll. 20f; *Strategikon* xi. 2. 66f; p. 364).

Edessa	alṣāh ba-šdāyā d-kēpē d-manganīqē	630

Dionysius 38ff (Palmer 1993: 138f); Agapius 466; Hoyland 2011: 79ff; Sebeos 41f (I: 94f, II: 226f, 238f); Flusin 1992: 286ff.

After the peace treaty with the Persians, "[...] Herakleios marched towards Syria and his brother Theoderic went ahead to eject the Persians from the cities as agreed in the earlier pact with Shahrvarāz and as confirmed by the recent treaty with Shīrōē." See further Dionysius 38; Flusin dates this to 628 or after; Howard-Johnston to 630. Sebeos gives essentially the same picture, but Dionysius 39 (Palmer 1993:139) has a quite detailed description of the siege itself:

"So Theoderic began to make the rounds of the Mesopotamian cities, informing the Persian garrisons of their duty to return to their country. In fact they had already been informed of the treaty in letters from Shahrvarāz and from Shīrōē. Close on his brother's heels the King advanced, establishing governors and Roman garrisons in the cities. When Theoderic reached

Edessa, however, the Persians there turned a deaf ear to his proclamation. Their reply was, 'We do not know Shīrōē and we will not surrender the city to the Romans'. The Jews of Edessa were standing there on the wall with the Persians. Partly out of hatred for the Christians, but also in order to ingratiate themselves with the Persians, they began to insult the Romans and Theoderic was obliged to hear their sarcastic taunts against himself. This provoked him to an all-out attack on the city, which he subjected with his catapults to a hail of rocks (*textus* 235.25ff *w-alṣāh ba-šdāyā d-kēpē d-manganīqē,* ܘܐܠܨܗ ܒܫܕܝܐ ܕܟܐܦܐ ܕܡܢܓܢܝܩܐ). The Persian resistance was crushed and they accepted an amnesty (*mellṭā*) to return to their country. A certain Jew called Joseph, expecting a pogrom [n. 320, literally "the destruction of his people"], scaled down the wall and sped off to find Herakleios in Tella. He was admitted to the Royal Presence, where he urged the king to forgive his fellow-Jews the insults to which they had subjected Theoderic and to send an envoy to restrain his brother from exacting vengeance. Meanwhile Theoderic had entered Edessa and taken over control. After expelling the Persians and sending them off home, he had sent his men to herd together all the Jews who had insulted him. He had already begun to kill them and to plunder their houses, when Joseph arrived with a letter from the King, by which he forbade his brother to harm them."

Agapius adds a few details and Arabic terminology on the artillery bombardment: "He set up catapults (*'arrādāt*), loaded them with stones and fired them at them. He fired some forty missiles at them and killed many of them. They were unable to do anything against him and asked for a guarantee of safety (*amān*). He granted it and they went out of the city and headed for Persia."

Creating a lasting political settlement was complicated. The Jews and leading Monophysites had benefitted from the Persian occupation, and were unwilling to compromise (Dionysius 40; Palmer 1993: 140). The Monophysite metropolitan of Edessa made a public display of his opinion by refusing communion with Herakleios, who was infuriated, threw out the Monophysites and gave the church to the Chalcedonians. Sebeos even connects the Jews escaping Edessa with the rise of Islam.

| WOGASTISBURG | *inmurauerant circumdantes* | c. 630 |

Fredegar 4.68.

The peoples living on the Avaro-Slav frontier had submitted to Dagobert, promising that he would dominate all the peoples up to the borders of the Roman Empire (Fred. 4.58). However, Samo's state, established around

623 (Fred. 4.48), while it did recognize client status, would not tolerate Frankish domination beyond certain limits. According to Fredegar, the Slavs killed and robbed Frankish merchants inside Samo's kingdom, and when an embassy arrived to demand reparations, Samo only promised to conduct an investigation. Samo did recognize that "The land we occupy is Dagobert's and we are his men on condition that he chooses to maintain friendly relations with us." However, the tactless Frankish ambassador was dismissed and Dagobert had a pretext for war.

His army was raised in three divisions (*trebus turmis falange*) from Austrasia (one of which must have come from Alamannia, cf. below) and was joined by Lombards who attacked from the south. "But everywhere the Slavs made preparation to resist." The Lombard and Alaman divisions were successful and took many captives. "Dagobert's Austrasians, on the other hand, invested the stronghold of Wogastisburg where many of the most resolute Wends had taken refuge, and were crushed in a three-day battle. And so they made home, leaving all their tents and equipment behind them in their flight." || Aostrasiae uero cum ad castro Wogastisburc ubi plurima manus forcium Venedorum inmurauerant circumdantes, triduo priliantes pluris ibidem de exercito Dagoberti gladio trucidantur et exinde fogacetur, omnes tinturius et res quas habuerant relinquentes, ad proprias sedebus reuertuntur."

The Austrasian defeat led to Slav raids against Thuringia, while the Sorbs, longtime Frankish clients, placed themselves under Samo's rule. Slav successes prompted Dagobert to set up his son as ruler of Austrasia and organize the kingdom to stop the raids and begin counterattacks that restored the status quo (Fred. 4.74f, 77).

JERUSALEM   šrā 'al, aqīm(w) 'lēh qrābā ḥasīnā; qātalahum, قاتلهم   634f
Nik. 18; Chr. 640 (AG 945); Sophronius; Dionysius 73ff; cf. Hoyland 2011: 93f, 114-17; Hill 1971: 59f; Donner 1981: 151f; Busse 1984, 1986; Hoyland 1997: 63ff and *passim*; Howard-Johnston 2010: 380f; Appendix I.

Nikephoros preserves an early Greek record of the very beginning of the Arab expansion: "At about this time the Saracens began to appear from Aithribos, as it is called (this being a country of Arabia the Blessed) and attempted to lay waste the neighboring villages." Ὑπὸ δὲ τὸν αὐτὸν καιρὸν ἐκ τοῦ Ἀιθρίβου λεγομένου Σαρακηνοὶ δ' ἐφαίνοντο (χώρα δὲ τοῦτο τῆς εὐδαίμονος Ἀραβίας) καὶ τὰ ἐκεῖσε χωρία προσπελάζοντα ληΐζεσθαι ἐπεχείρουν. Mango (184) conjectures it may have been the unsuccessful raid against Muʾta in 629 (same story in Theoph. 335.14ff), but a similar report with a much firmer chronological indication is found in *Chr. 640* (Palmer 1993: 18f): "AG 945, indiction VII: On Friday, 4 February [the date is correct for 634], at the ninth hour, there was a battle between the Romans and the Arabs of Muhammad in Palestine twelve 'miles' east of Gaza. The Romans fled, leaving behind the patrikios the son of YRDN (BRYRDN), whom the Arabs killed. Some 4,000 poor village people of Palestine were killed there, Christians, Jews and Samaritans. The Arabs ravaged the whole region."

The Christmas sermon of Sophronius delivered in 634 (App. I) confirms the widespread destruction and chaos that made it impossible for the citizens of Jerusalem to reach Bethlehem, only a few miles away. Dionysius 73ff (Palmer 1993: 160ff) informs us that just before the Arabs besieged (*textus* 254.14 *šrā ʿal*, ܫܪܐ ܥܠ) Jerusalem, there was an attempt to defeat them in the field. "The people in the city came out, formed ranks and fought a battle with the Arabs, then went back within the walls. This episode was followed by a violent Arab assault." (*textus* 254.15-18: ܘܢܦܩ ܥܡܐ ܕܒܓܘܗ ܘܣܕܪ ܩܪܒܐ ܥܡ ܛܝܝܐ. ܘܐܬܩܪܒܘ ܥܡܗܘܢ ܘܗܦܟܘ ܥܠܘ ܠܡܕܝܢܬܐ. ܘܛܝܝܐ ܐܬܟܪܘܗ ܠܡܕܝܢܬܐ. ܘܐܬܚܝܠܘ ܥܠܝܗ ܩܪܒܐ ܚܣܝܢܐ.)

Dionysius 76 also refers to a plague in Palestine that may have been the result of malnourishment following extensive raiding and no access to the countryside for over a year. The rest of his account is similar to that of the Islamic traditions: The population lost hope of relief and began negotiating; Umar arrived with more soldiers, met the "leaders of the city" including Sophronius, and finalized negotiations on behalf of all Palestine. "They [the leaders of Jerusalem] received an agreement and oaths (*textus* 255.8f *nsab(w) melltā w-mawmātā*, ܢܣܒܘ ܡܠܬܐ ܘܡܘܡܬܐ)," whereupon Umar entered Jerusalem.

Baladhuri presents the traditionally recounted story of the surrender, with 'Umar who arrived to receive the submission personally (213f; De-Goeje 138): "He made the terms of capitulation with the people of Jerusalem to take effect and gave them a written statement." However, another account he reports provides an interesting alternative: "After fighting with the inhabitants (*faqātalahum*, فقاتلهم), they agreed to pay something on what was within their fortified city (*bihi ḥiṣnuhum*, به حصنهم) and to deliver to the Muslims all that was outside. 'Umar came and concurred, after which he returned to al-Madīnah."

The implications of this report are far-reaching, as Jerusalem did not actually surrender—no garrison or governor was installed, but the city agreed to a tribute and hand over the proceeds from countryside to those who had been its actual masters for at least the last year or so (cf.

Sophronius' Chrismas sermon, for which see Hoyland 1997). Thus, Umar's formal entry was achieved later than the first submission. The traditional dates are 636, 637 or 638. However, in a thorough dissection of the Islamic sources, Busse argues that Umar's arrival in Syria was in the autumn of 636, after which a final treaty with Jerusalem was made. Furthermore, Busse has argued that a preliminary submission, conflated with Umar's visit in later traditions, must have taken place at least a year earlier, suggesting Palm Sunday, 2 April, 635, in order to allow the Christians access to their extramural holy sites which had been denied to them during Christmas 634 (and perhaps even Easter 634; see App. I). Thus a preliminary submission was agreed upon after losing access to the countryside for 6-12 months; a final surrender came about when it was clear that Roman armies would not be returning anytime soon, after the loss at Yarmuk in August 636. It is clear that the original terms were incompatible with the building of the first Islamic shrine on the Temple Mount, which began between 636 and 639 (Hoyland) and was one of the first things the conquerors did when they actually gained control of the city itself, late in 636.

PALMYRA   *taḥaṣṣanū*, خصّنوا   634/5

Baladhuri 171 (DeGoeje 111); Hill 1971: 65, 69, 79; Donner 1981: 121-24.

"Tadmur [Palmyra]'s inhabitants held out against him [Khālid] and took to their fortifications. At last they sought to surrender (*fa'mtana'a ahluhā wa-taḥaṣṣanū thumma ṭalabū al-'amān*, فامتنع اهلها وخصّنوا ثم طلبوا الامان) and he wrote them a statement guaranteeing their safety on condition that they be considered *dhimmah* people."

DAMASCUS   *etā nqaš 'al; nazala bi,* نزل ب ; *taḥaṣṣana ahl al-madīna*   634/5

Chr. 819 (AG 945); Dionysius 56, 63; cf. Hoyland 2011: 96-103; Baladhuri 172, 182, 186-90; Hill 1971: 60ff; Donner 1981: 124f, 131f, 136f; Howard-Johnston 2010: 211ff.

An early chronology is provided by the *Chr. 819* (Palmer 1993: 76): "AG 945: ...This year the Arabs entered Syria and took Damascus." Note the tendency to telescope chronology in *Chr. 819*—i.e. the Arabs entered Syria in 945 (AD 633-34), *then* (normally this would mean in a subsequent year) took Damascus.

Dionysius 56 (Palmer 1993: 149f): "Khālid b. al-Walīd then led the Arab army to besiege (*textus* 245.5: *etā nqaš 'al*, ܐܬܐ ܢܩܫ ܥܠ) Damascus. He himself lodged in a monastery at the East Gate, Abū 'Ubayda lodged at the Gabitha Gate and Yazīd at the Gate of the Apostle Thomas. For the gates of the city were shut against them. The Arabs outside surrounded the wall and launched a determined attack on the city. The Damascenes were in a bad way and in great fear of the Arabs, when a Roman auxiliary force of 50,000 arrived and fought its way with set determination into the city. The Arabs, undeterred, continued fighting for all they were worth to capture Damascus." The size of the relieving force is clearly exaggerated, and did not break the siege. Damascus surrendered when no further reinforcements arrived; one of the gates was stormed but a treaty was made nonetheless, negotiated by John the deacon, son of Sargūn (Dionysius 63, Palmer 1993: 154). Dionysius is probably influenced by the Islamic historiographic tradition (see Hoyland and Howard-Johnston), exemplified by Baladhuri, who presents a similar version (172, DeGoeje 112): "Khālid then directed Busr ibn-abi-Artāt al-'Āmiri of the Quraish and Habīb ibn-Maslamah-l-Fihri to the Ghūtah of Damascus where they attacked many villages."... "Khālid camped at the East gate (*wa-nazala* [ونزل] *Khālid bi'l-bāb aš-šarqī*) of Damascus; and according to others, at the Jābiyah gate. The bishop of Damascus offered him gifts and homage and said to Khālid, "keep this covenant for me." Khālid promised to do so. Then Khālid went... to Bosra."

According to fuller traditions found elsewhere in Baladhuri (182, DeGoeje 118), the Arabs fough a bloody battle at Marj as-Suffār in *AH* 14. Despite heavy Arab losses, they won and the Romans fled to Damascus and Jerusalem. The Muslims then marched on Damascus (186, DeGoeje 120f): "Al-Ghūtah and its churches the Muslims took by force. The inhabitants of Damascus betook themselves to the fortifications (*wa-taḥaṣṣana ahlu'l-madīnah*, وتحصّن اهل المدينة) and closed the gate of the city. Khālid ibn-al-Walīd at the head of some 5,000 men whom abu-'Ubaidah had put under his command, camped at al-Bāb aš-Šarqī [the east gate]." The tradition includes information on the stations of the other Arab commanders, indicating a well-organized siege with specifically assigned sectors and gates.

Baladhuri presents several alternatives for how the city was captured (186ff, DeGoeje 121ff). The one commonly cited is that Khālid negotiated a settlement with the bishop, who surrendered the city in return for poll-tax, at the same time as another division entered the city from the other side. One version tells how the defenders abandoned the gate to attend a Christian festival, so the Muslims procured a ladder from a nearby monastery to clamber over the wall. While this is an obvious literary invention, some details on the fighting and negotiations appear genuine: the same traditions relate how the bishop encouraged the defenders, walking around on the city wall, and also engaged in negotiations with the Arabs from the walls; and later, when the Arabs gained control of one gate, the defenders rushed against them, leaving the walls on the other side exposed to attack. A final tradition in fact seems authentic: A funeral procession left one of the gates at night (again possibly a literary device), and when the Arabs perceived this (whatever the actual reason), they rushed at those defending the opened gate and won after a bloody battle. When the bishop learned that the Arabs were about to enter the city (*qāraba dukhūla 'l-madīnah*, قارب دخول المدينة), he "hurried to Khālid and capitulated (*fa-ṣālaḥahu*, فصالحه)." This tradition claims that the siege lasted 4 months (190, DeGoeje 124), but other traditions provide other possibilities.

The variety of anecdotes (for further references and possible dates, see Donner, Hill, Hoyland and Howard-Johnston) reflect the work of legal scholars who sought to explain why this particular city apparently had a large Muslim population and many disputed holy places as compared with other Syrian, Mesopotamian and Egyptian cities in the 9th century. Baladhuri cites scholars who believed that this situation arose from a peculiar type of surrender, which awarded half of the buildings to the conquerors; however, Baladhuri himself (and others; see Hill) believes that immediately after the siege ended, many of the inhabitants fled from Damascus, leaving abandoned houses to be occupied by the Muslims (189, DeGoeje 123). Its role as the first monumental capital of the Islamic Empire certainly had a similar effect. The various traditions (and chaotic situation on the ground) may in fact stem from a situation where the Arabs overpowered a gate or a section of the wall, causing the remaining Roman defenders to give up resistance.

EMESA	*qātalahum ahluhā*, قاتلهم اهلها	c. 635

Dionysius 64 (Palmer 1993: 155); Baladhuri 200; Hill 1971: 80f and *passim*; Donner 1981: 149 and *passim*; cf. *Damascus above.

Dionysius: "The Arabs wanted to take captives and loot, but Abū 'Ubayda, at the command of king 'Umar, prevented them and made the people tributaries instead. From there they went to Baalbek, Palmyra and Emesa. The Emesenes shut the gates against them and went up on the wall above the al-Rastan Gate, outside which the Arabs were encamped, to parley with them. Their proposal to the Arabs was this: 'Go and engage the king of the Romans in battle. When you have defeated him we will be your subjects. If you do not we will not open our gates for you.' When the Arabs began to attack the city (*textus* 249.7 *šrīū* [sic]... *'lēh*, ܥܠܝܗ...ܫܪܝܘ) regardless, the

Emesenes expected reinforcements to come and rescue them, but none came. Then they lost their will to fight and sued for peace. They asked the Arabs for an amnesty ("word"), a pact and oaths (*textus* 249.10 *melltā waqyāmā w-mawmātā*); and they received, like the Damascenes, a written covenant granting them their own lives and possessions and churches and laws and requiring them to pay 110,000 denarii as the tribute of the city. So the Arabs gained control of Emesa. The emir who was put in charge of collecting the tribute from them was Habîb b. Maslama. As for the Palestinians and the inhabitants of the coastal settlements, they all congregated within the walls of Jerusalem."

Baladhuri 200 (DeGoeje 130): "When the [Arab armies] met in Ḥimṣ, the people of the city resisted them (*qātalahum ahluhā*, قاتلهم اهلها), but finally sought refuge in the city and asked for safety and capitulation." The implication is that this was a field engagement.

BOSRA; OTHER CITIES IN JORDAN (i. e. Arabia)   *kbaš*, ܟܒܫ   c. 635
Dionysius 57 (Palmer 1993: 150 and n. 359); cf. Hoyland 2011: 95f; Baladhuri 126, 173; Hill 1971: 81; Donner 1981: 129, 135, 141, 145f.

The Arabs under Khālid b. al-Walīd moved on from *Damascus [and *Emesa?] south to Jordan, al-Balqā' and Hawrān, as several Arabic traditions claim (see Donner) and in Dionysius 57, which follows this particular reconstruction, but adds information from an old Christian source (cf. *Chr. 640*): "While the Arab armies were besieging Damascus they received the news of Abū Bakr's death; he reigned for two and a half years. The following king was 'Umar b. al-Khaṭṭāb. He sent a detachment to that part of Arabia called al-Balqā' and it took (*textus* 245.21 *kbaš*, ܟܒܫ) Bostra and destroyed the rest of the villages and cities. He also sent Saʿd b. (Abī) Waqqāṣ against the Persians. On their way these Arabs went up into the Mardīn mountains and they killed there many monks and excellent ascetics, especially [those in] the great and famous abbey on the mountain above Rhesaina, which is called 'The Abbey of Bnōthō', i.e. of the eggs." Michael the Syrian gives as the reason for the massacres that the Arabs suspected the monks were Persian spies, and that the attack was timed with internal Persian discord.

Baladhuri (173, DeGoeje 113) emphasizes that the conquest of Bosra occurred after a field engagement: "They drew close to it and fought its patrician until he was driven with his armed men inside the town. (...) At last its people came to terms stipulating that their lives, property and children be safe, and agreeing to pay the poll-tax." He also places the surrender of Bosra and *Adhriʿāt (636) after *Damascus (above); the Muslims then dispersed from Bosra to subdue the Ḥawrān, and from there they went on to "Palestine and Jordan, invading what had not yet been reduced." In the same campaign they also took ʿAmmān and al-Balqāʾ (193, DeGoeje 126).

| TIBERIAS | *fa-fataḥa... ṣulḥ^{an} baʿda ḥiṣār ayyām* | c. 635 |

Baladhuri 179 (DeGoeje 116); Hill 1971: 68, 82; Donner 1981: 137f.

"Shuraḥbīl ibn-Hasanah took Tiberias by capitulation after a siege of some days (*fa-fataḥa šuraḥbīl bin ḥasanah ṭabariyyah ṣulḥ^{an} baʿda ḥiṣār ayyām,* ففتح شرحبيل بن حسنة طبريّة صلحًا بعد حصار ايّام). He guaranteed for the inhabitants the safety of their lives, possessions, children, churches and houses with the exception of what they should evacuate and desert, setting aside a special spot for a Muslim mosque. Later, in the caliphate of ʿUmar, the people of Tiberias violated the covenant and were joined by many Greeks and others. Abu-ʿUbaidah ordered ʿAmr ibn-al-ʿĀsi to attack them, so he marched against them at the head of 4000 men. ʿAmr took the city by capitulation, the terms being similar to those of Shuraḥbīl..."

Note that the townspeople "violated the covenant" upon being "joined by many Greeks and others"; it is possible that the towns in the interior had lost many of their potential defenders during the fierce battles at Ajnadayn, Pella, and Yarmuk (although the extent, date and even historicity of these are contested), had few supplies and little hope of reinforcement, being full of refugees and demoralized soldiers.

| CITIES OF JORDAN (i.e. N. Palestine) | *bi-ghayri qitāl,* بغير قتال | c. 635 |

Continued from *Tiberias (above).

"In addition to that, Shurahbīl took easy possession of all the cities Jordan with their fortifications, which, with no resistance (*bi-ghayri qitāl,* بغير قتال), capitulated on terms similar to those of Tiberias. Thus did he take possession of Baisān [Bethshean, Scythopolis], Sūsiyah, Afik, Jarash, Bait-Rās, Kadas, and al-Jaulān, and subdue the district of the Jordan and all its land."

| ADHRIʿĀT | | c. 636 |

Baladhuri 214f (DeGoeje 139); Hill 1971: 72f; see also *Jerusalem 634f and *Bosra 635.

While the city surrendered at around the same time as *Bosra (635), Baladhuri provides a report on ʿUmar's reaction to submitting cities on his way to Syria in c. 636, when he went to negotiate the final submission of *Jerusalem (surrendered 635) and arrange his formal entry into the city:

"As 'Umar was passing, he was met by the singers and tambourine players of the inhabitants of Adhri'āt with swords and myrtle. Seeing that, 'Umar shouted 'Keep still! Stop them!' But abū 'Ubaidah replied, 'This is their custom (or some other word like it), Commander of the Believers, and if thou shouldst stop them from doing it, they would take that as indicating thy intention to violate their covenant.' 'Well, then,' said 'Umar, 'let them go on.'"

ALEPPO	šra 'al, ܠܥ ܚܝܬ	c. 637

Dionysius 72 (Palmer 1993: 160); cf. Hoyland 2011: 118f.

The Arabs besieged (*textus* 254.3f *šra 'al*, ܠܥ ܚܝܬ) Aleppo and Qēnneshrīn (Chalkis) which surrendered upon receiving oaths (*mawmātā*) and a covenant (*qyāmā*); these are the essential ingredients in the "word" (*melltā*) normally given according to the siege anecdotes, cf. *Emesa 635. The surrender was followed by a reorganization of the conquered areas and appointment of emirs.

QENNESHRIN		c. 637

See *Aleppo above

ANTIOCH	κατέτρεχον	638/9

Nik. 20; Dionysius 59, 77f; cf. Hoyland 2011: 118f; Howard-Johnston 2010: 216.

Again Nikephoros preserves an old Greek note of the Arab conquest of Antioch: "After a short lapse of time the Saracens overran the region round Antioch." Οὐ πολὺς δὲ χρόνος ἐν μέσῳ, καὶ Σαρακηνοὶ τὰ περὶ τὴν Ἀντιόχειαν κατέτρεχον. Dionysius 59 (Palmer 1993: 152) reports that the Arabs exploited the festival of Symeon the Stylite at Antioch: "[...] they appeared there and took captive a large number of men and women and unnumerable boys and girls. The Christians who were left no longer knew what to believe. Some of them said, "Why does God allow this to happen?" But a discerning person will see that Justice permitted this because, instead of fasting, vigils and psalm-singing, the Christians used to yield to intemperance, drunkenness, dancing and other kinds of luxury and debauchery at the festivals of the martyrs, thus angering God. That is why, quite justly, he slapped and punished us, in order that we might improve our behaviour." A similar account is recorded in Dionysius 77f (Palmer 1993: 162f): 'Iyād b. Ghanm conquered the rest of Syria, and made a deal with commander of Mesopotamia, who was consequently dismissed by Herakleios [AG 949]; next year [AG 950?] Mu'āwiya was appointed by 'Umar as commander over all Syria. "He took Antioch by siege and plundered all the villages around, leading the people away as slaves. Then the Arabs sent a demand for the tribute of Mesopotamia."

| Castrum at Unstrut | castrum undique circumdat | 639 |

Fred. 4.87.

Radulf, the Frankish duke set over Thuringia to defeat the Wends after *Wogastisburg (630), was so successful that he rose in revolt after the death of Dagobert. The king of Austrasia, Sigibert, was only around eleven, but he personally led a vast army organized by his dukes and consisting of Austrasians and trans-Rhenan clients. After an initial royal victory beyond the Rhine, "Radulf put up a wooden fortification round his position on the rise of the bank of [imprecise; rather "a hill" or "a mountain above"] the Unstrut (*castrum lignis monitum in quodam montem super Vnestrude fluuio*), in Thuringia, and when he had assembled as large an army possible from everywhere, fortified himself with his wife and children within this fortification to withstand a siege (*in hunc castrum ad se definsandum stabilibit*). Sigebert arrived with his army and invested the fort (*castrum undique circumdat exercitus*) while Radulf sitting tight within prepared for a vigorous defense. But battle was joined imprudently, owing to the youthfulness of King Sigibert. Some were for fighting on that same day, others preferred to await the morrow; and they were unable to give unanimous counsel." It emerges that Radulf was in contact with several dukes who had promised not to join the attack, while "the men of Mainz turned traitor in this battle" (*Macanensis hoc prilio non fuerunt fedelis*). Thus several thousand were killed in the confused fighting, as several divisions from Aquitaine advanced against the gate but were deprived of support and surprised by a Thuringian sortie. This was not a regular field battle, as the Austrasians withdrew to their camp and negotiated for safe passage across the Rhine on the following day. However, the collusion between besieged and besiegers meant that there was no possibility to continue the investment.

| Tella; Mesopotamian cities | kabšāh, ܟܒܫܗ | 639 |

Dionysius 78; Agap. 477; *Chr. 819* (Palmer 1993: 76f); Hoyland 2011: 120f; Appendix II.

Dionysius: In AG 951, Ptolemy, the new Roman commander of Mesopotamia, refused to pay tribute so the Arabs invaded, crossing the Euphrates towards Edessa, which surrendered along with Harran. "The Edessenes had also received an assurance with regard to Ptolemy and his Romans, so they returned to their country. But when 'Iyāḍ b. Ghanm came to Tella the arrogant [although *mšaqqlē* can also mean high-spirited, i.e. highly motivated] Romans in the city did not deign to accept assurances, so they were obliged to fight. In a determined assault 'Iyāḍ overwhelmed the city (*textus* 256.29 *kabšāh*, ܟܒܫܗ) and killed the three hundred Romans who were there."

Agapius adds some details on the siege: " 'Iyāḍ departed from (Edessa) and came to Tella because it had not been taken by guarantee along with the rest of the cities of Mesopotamia. When he came up to it, the Romans who were in it defied him. He was angry and erected siege engines (*'arrādāt*), and he and they kept at it until he conquered the city and killed the Romans who were in it.

Chr. 819 (Palmer 1993: 76f): "AG 947: The Romans and the Arabs fought a battle by the river Yarmūk and the Romans were utterly defeated. Umar took all the cities of Mesopotamia. The first of their leaders to enter Edessa and Harran was Abū Badr; and the one who invaded Dara, Amida, Tella and Rhesaina was 'Iyāḍ." Note the same tendency to telescope as for *Damascus, 634/5 above, and ibn Khalid's *Anatolian invasion, 664 below.

DARA; MESOPOTAMIAN CITIES *nqaš...'al, w-aqreb(w) 'ammāh, kabšāh* 640
Chr. Zuq. (Palmer 1993: 57); Dionysius 78; Agapius 477; Hoyland 2011: 120f; see also *Dvin below and Appendix II.

The Zuqnin chronicler explicitly links the Arab expedition Mesopotamia under 'Iyāḍ b. Ghanm (the name is provided by *Chr. 819* in *Tella 639 and Dionysius below) with a further expedition to Armenia, which took *Dvin: "AG 952: The Arabs laid siege to Dara and attacked it (Chabot II.150.30-151.1 *nqaš ṭayyāyē 'al dārā w-aqreb(w) 'ammāh*, ܒܡ ܢܩܫ ܛܝܝܐ ܥܠ ܕܪܐ ܘܐܩܪܒ). A great many people were slain on both sides, but especially among the Arabs. In the end they made an agreement (*melltā*) and they conquered the city. From that moment onwards no human being was killed."

Dionysius 78 (AG 951): "Next he went to Dara, assaulted it (*textus* 259.31 *kabšāh*, ܟܒܫܗ) likewise, took it and killed every Roman in the city. But Rhesaina, Mardīn and Amida he took by amnesty and covenant and oaths (*textus* 257.2f *melltā wa-qyāmā w-mawmātā*). It was at Amida that 'Iyāḍ b. Ghanm died by violence and was buried." Palmer 1993: n. 399, Michael adds: "After subjecting all Mesopotamia, 'Iyāḍ b. Ghanm returned to Syria. 'Umar ordered all the countries of his kingdom to be registered for the poll-tax. The poll-tax was imposed on Christians in AG 951." While adding information on the Christian population, here too Dionysius follows the early Islamic tradition as he does not mention the continuation of the invasion to *Dvin.

Agapius: " 'Iyād conquered the cities of Mesopotamia by agreement (*amān*) except Dara, which he conquered by the sword and killed the Romans in it. He organized his governors over all the cities of Mesopotamia and returned to Mu'awiya."

Dvin	*nqaš 'al,* ܢܩܫ ܥܠ	640

Chr. Zuq. (Palmer 1993: 57); Sebeos 42 (I: 100f, II: 246f); Kaegi 1992: 193; Manandean 1948; further discussion in Appendix II.

After taking *Dara (above), the Zuqnin chronicler explains 'Iyād b. Ghanm's next move: "The same year they laid siege to (Chabot II.151.4 *nqaš 'al,* ܢܩܫ ܥܠ) Adavīn (Dwīn), and in this city a large number of people were killed, as many as 1200 of the Armenians."

Sebeos 42 (I: 100f; II) gives details on what transpired between the capture of *Dara and the assault on Dvin. The Arabs defeated the Armenians in battle and pursued the routed troops to Dvin, which was full of refugees from the district: "On the fifth day they attacked the city. It was delivered into their hands because they surrounded it with smoke. By means of the smoke and the shooting of arrows they pushed back the defenders of the wall. Having set up ladders, they mounted the wall, entered inside, and opened the city gate. The enemy rushed within and put the multitude of the city's population to the sword. Having plundered the city, they came out and camped in the same encampment. It was the 20th of the month of Trē, a Friday. After staying a few days, they left by the same route that they had come, leading away the host of their captives, 35,000 souls."

Kaegi only mentions route but is hesitant as to whether the Arabs reached Dvin and does not connect it with *Dara above. According to Howard-Johnston's commentary, the precise indication of weekday and date allows a secure dating to 640. See further in App. II for a full discussion of the chronology and connection with *Dara above.

Caesarea Maritima	*etkarrkāh; ḥāṣara...ḥattā fataḥahā*	c. 640

Chr. Zuq. (Palmer 1993: 57); *Chr. 819* (ibid: 77); Dionysius 83 (ibid: 165f, 178 n. 445); cf. Hoyland 2011: 423f; Baladhuri 216ff (DeGoeje 140f); Howard-Johnston 2010: 469; see also *Caesarea (Cappadocia) 645.

The Syriac chroniclers mostly have brief notices, although they seem to confuse Caesarea Maritima in Palestine with Caesarea in Cappadocia. The Zuqnin chronicler reports "AG 953: The Arabs captured Caesarea in Palestine." *Chr. 819* has: "AG 954: Caesarea of Cappadocia was taken." Dionysius (Palmer 1993: 147) similarly has: "And so the Arabs entered Caesarea." The notice is placed in the completely wrong context, and probably taken from an earlier short chronicle with an alternative date. Dionysius 83 indeed has a more extended version of the siege comparable with the Islamic tradition (Palmer 1993: 165f): "AG 950: Mu'āwiya besieged Caesarea with vigorous assaults, taking captives from the surrounding country and laying it waste. He sustained hostilities day and night for a long time until he conquered it by the sword. All those in the city, including the 7,000 Romans sent there to guard it, were put to death. The city was plundered of vast quantities of gold and silver and then abandoned to its grief. Those who settled there afterwards became tributaries of the Arabs." However, the original text of Dionysios probably had a much fuller description of the siege, partly preserved by Michael the Syrian (Palmer 1992: 178 n. 445): "At this time (c. AG 953/4) the Arabs destroyed Caesarea in Palestine. [...] Mu'āwiya surrounded it (Michael the Syrian 4.423.4 *etkarrakāh*, ܐܬܟܪܟܗ) by sea and land and kept it under attack by day and by night, from the beginning of December until the month of May. Yet they would not take the word for their lives. Though "seventy-two catapults bombarded it continually with a heap of rocks" (Michael the Syrian 4.423.8 *wa-b-"72" mangānīqōs lā bāṭlīn waw men karyā d-kēpē*, ܘܒܫܒܥܝܢ ܘܬܪܝܢ ܡܢܓܢܝܩܘܣ ܠܐ ܒܛܠܝܢ ܘܘ ܡܢ ܟܐܦܐ, وحاقل ܕܢܐܦܐ), the wall was so solid that it did not crack. Finally the attackers made a breach, through which some entered, while others climbed onto the walls with ladders. For three days the fighting continued, before the ultimate Arab victory. Of the seven-thousand-strong Roman garrison some escaped on ships. Mu'āwiya took the treasures and obliged the population to pay tribute."

Baladhuri describes a hard-fought campaign to gain control of the coastal cities (179f, DeGoeje 116f): "... Abu-'Ubaidah directed 'Amr ibn-al-'Āsi to the sea-coasts of the province of the Jordan. There the Greeks [Rum] became too numerous for him being recruited by men from the district under Herakleios who was then at Constantinople." Mu'āwiyah distinguished himself in the following campaign, and Baladhuri emphasizes a long and hard-fought siege at Caesarea under Mu'āwiya's leadership (216ff; DeGoeje 140f); he "besieged Qaysāriyyah until he reduced it (*ḥāṣara ...ḥattā fataḥahā*, حاصر... حتّى فتحها), the city having been under siege for seven years" (217; DeGoeje 141). Mu'āwiya had begun to despair of capturing it, but he managed to take the city with the help of a Jew who pointed out a tunnel into the city. In addition to the great number of fallen Roman soldiers, perhaps as many as 7,000 men also mentioned in the Christian sources, 4,000 civilians were taken prisoner and resettled (218, DeGoeje

141): "They were then distributed among the orphans of the *Anṣār*, and some were used as clerks (*kuttāb*, كتاب) and manual laborers (*a'māl*, اعمال) for the Muslims."

BAHNASĀ (OXYRHYNCHUS)   [compelled the city to open its gates]   640f
J.Nik. 111.3, 5-9; *PLRE* 3 s.vv. Ioannes 246, 247, Theodosius 41; Howard-Johnston 2010: 186.

The Arabs invading Egypt in December 640 during the Nile's low season (see Howard-Johnston for date) outmaneuvered the Roman troops under the two Johns and Theodosius by marching out into the desert, supplying themselves by rustling sheep and goats. "And when they reached the city of Bahnasā, all the troops on the banks of the river came (to the succour) with John, but were unable on that occasion to reach Fayyūm. And the general Theodosius, hearing of the arrival of the Ishmaelites, proceeded from place to place in order to see what was likely to befall from these enemies. And these Ishmaelites came and slew without mercy the commander of the troops and all his companions. And forthwith they compelled the city to open its gates, and they put to the sword all that surrendered, and they spared none, whether old men, babe, or woman. And they proceeded against the general John. And he took all the horses: and they hid themselves in the enclosures and plantations lest their enemies should discover them. Then they arose by night and marched to the great river of Egypt, to Abūīt, in order to secure their safety. Now this matter was from God." [...] "And the chief of the faction who was with Jeremiah informed the Muslim troops of the Roman soldiers who were hidden. And so these took them prisoners and put them to death."

TENDUNIAS   [took possession of]   641
J.Nik. 112.10ff; Howard-Johnston 2010: 187.

"And the Muslim army took possession of the city of Tendunias; for its garrison had been destroyed, and there survived only 300 soldiers. And these fled and withdrew into the fortress and closed the gates. But when they saw the great slaughter that had taken place, they were seized with panic and fled by ship to Nakius in great grief and sorrow. And when Domentianus of the city of Fayyūm heard of these events, he set put by night without informing the inhabitants of (A)būīt that he was fleeing to escape the Muslim, and they proceeded to Nakius by ship. And when the Muslim learnt that Domentianus had fled, they marched joyously and seized the city of Fayyūm and (A)būīt, and they shed much blood there."

7TH CENTURY 651

BABYLON [besieged > evacuated] 641f
J.Nik. 117.1ff, 118f; Howard-Johnston 2010: 187f, 469.

"And 'Amr the chief of the Muslim forces encamped before the citadel of Babylon and besieged the troops that garrisoned it. Now the latter received his promise that they should not be put to the sword, and on their side undertook to deliver up to him all the munitions of war—now these were considerable. And thereupon he ordered them to evacuate the citadel. And they took a small quantity of gold and set out. And it was in this way that the citadel of Babylon in Egypt was taken on the second day after the (festival of the) Resurrection."

"Now the capture of the citadel of Babylon and of Nakius by the Muslim was a source of great grief to the Romans. And when 'Amr had brought to a close the operations of war he made his entry into the citadel of Babylon, and he mustered a large number of ships, great and small, and anchored them close to the fort where he was."

ROMAN FORTRESS NEAR ANTINOE [besieged] 642
J.Nik. 115.10ff; Howard-Johnston 2010: 188.

"And the inhabitants of the city (Antinoe) sought to concert measures with John their prefect with a view to attacking the Muslim; but he refused, and arose with haste with his troops, and, having collected all the imposts of the city, betook himself to Alexandria; for he knew that he could not resist the Muslim, and (he feared) lest he should meet with the same fate as the garrison of Fayyūm. Indeed, all the inhabitants of the province submitted to the Muslim, and paid them tribute. And they put to the sword all the Roman soldiers whom they encountered. And the Roman soldiers were in a fortress, and the Muslim besieged them, and captured their catapults, and demolished their towers, and dislodged them from the fortress. And they strengthened the fortress of Babylon, and they captured the city of Nakius and made themselves strong there."

KĪLŪNĀS [cast down the walls] 642
J.Nik. 118.11ff; *PLRE* 3 s.v. Theodorus 166.

"And the general Theodore, who was in command of the city, even the city of Kīlūnās, quitted (this) city and proceeded to Egypt, leaving Stephen with the troops to guard the city and contend with the Muslim. And there was a certain Jew with the Muslim, and he betook himself to the province of Egypt. And when with great toil and exertion they had cast down the walls of the city, they forthwith made themselves masters of it, and put to the sword thousands of its inhabitants and of the soldiers, and they gained an enormous booty, and took the women and children captive and divided them amongst themselves, and they made that city a desolation (lit. destitute). And shortly after the Muslim proceeded against the country (city?) of Cōprōs and put Stephen and his people to the sword."

| ALEXANDRIA | [assault > negotiations] | 642f |

J.Nik. 119.3f; Baladhuri 346f (DeGoeje 220f); Howard-Johnston 2010: 188, 469.

John presents the first, failed Arab assault on Alexandria thus: "And 'Amr sent a large force of Muslim against Alexandria, and they captured Kariun, which lies outside the city. And Theodore and his troops who were in that locality fled and withdrew into Alexandria. And the Muslim began to attack them but were not able to approach the walls of the city; for stones were hurled against them from the top of the walls, and they were driven far from the city." It happened amidst infighting amongst the Romans, which may have been fomented by officials and military officers who had been replaced (e.g. Domentiolus); this was also in the midst of a Roman civil war over the succession of Herakleios.

The city ultimately surrendered after lengthy negotiations, led by the Roman commanders and the Chalcedonian patriarch Cyrus, who was nearly stoned by the population for negotiating. Baladhuri also presents an image of internal discord among the Romans, but anachronistically equates the peace party with the Copts, implying that Muqawqis (i.e., Cyrus) was their representative, and goes on to relate how the city was only "reduced...by the sword" after three months of fierce resistance, although the population was spared of captivity.

| LOMBARD CONQUEST OF LIGURIAN CITIES | *cepit* | 643 |

HL 4.45.

"King Rothari then captured all the cities of the Roman which were situated upon the shore of the sea from the city of Luna (Luni) in Tuscany up to the boundaries of the Franks." || Igitur Rothari rex Romanorum civitates ab urbe Tusciae Lunensi universas quae in litore maris sitae sunt usque ad Francorum fines cepit. ||

| ODERZO | *expugnavit* | 643 |

HL 4.45.

"Also he captured and destroyed Oderzo..." || Otipergium...pari modo expugnavit et diruit. || He then defeated a large Roman force near Ravenna, killing 8,000.

| Artsap'k'; Nakhchewan; Khram | 643 |

Sebeos 44f (I: 109-12; II: 257ff).

Three Arab armies invaded Armenia in 643. They attacked several fortresses but one division failed in its attempts to take several of them, and suffered heavy losses against Artsap'k'; however, they manage to find an entrance and take the fortress by surprise. But the next day they were slaughtered to a man by the Armenian general T'ēodoros. Another Arab army attacked Iberia, Albania and Nakhchawan; failing to take the fortress of Nakhchawan they took the fortress Khram, where they carried out a massacre and took the survivors prisoner.

| Euchaïta | 644 |

Dionysius 84 (Palmer 1993: 166 and n. 407); cf. Hoyland 2011: 124f; Lilie 1996: 63; Howard-Johnston 2010: 219, 475.

Muʿāwiya led an expedition towards Euchaïta, "leaving a trail of destruction behind him. No one sounded the alarm. The Euchaitans were scattered over the countryside, harvesting the crops and working in the vineyards. They had seen the aggressors all right; but they were under the impression that they were Christian Arabs, from one or other tribe allied with the Romans. So they saw no reason to alter their dispositions, let alone to run away. The Arabs found the gates of the unhappy city open and the people sitting around without the slightest fear. The next moment they were entering it, plundering it, piling up great mounds of booty. They seized the women, the boys and the girls to take back home as slaves. Even the city-governor (*arkhōn*) was taken prisoner. Euchaita lay ravaged and deserted, while the Arabs returned, exulting, to their country." According to Michael the Syrian, "The Arabs enslaved the entire population, men and women, boys and girls. They committed a great orgy in this unfortunate city, fornicating wickedly inside the churches."

This may be connected to either of the raids that Lilie dates to 643 and 644; Howard-Johnston places it in 644.

| Tripoli | *ḥāṣarahum*, حاصرهم | c. 644 |

Baladhuri 194f (DeGoeje 127).

The conquest of Tripoli was achieved "at the beginning of the reign" of ʿUthmān, hence in 644 or soon after: "Sufyān erected on a plain a few miles from the city a fort which was called Ḥiṣn Sufyan [the fort of Sufyān], intercepted the recruits from the sea as well as from the land and laid siege to the city (*ḥāṣarahum*, حاصرهم)." The inhabitants were hard pressed and asked for ships from the emperor, which came and removed them and the

Arabs woke up one day to find the city empty. "Immediately he entered it and sent the news of the conquest to Muʿāwiya." The defense rotated between a summer and winter garrison; the latter was small since the sea was closed.

CAESAREA c. 645
Dionysios 86 (Palmer 1993: 167f n. 412); Lilie 1976: 63f.

Michael the Syrian provides evidence for an expedition against Caesarea sent in two columns by Muʿāwiya; the first headed for Armenia in October, under Habīb from Syria, forcing through snow by using oxen, taking the Armenians completely unawares. Muʿāwiya led the other column against Caesarea in Cappadocia and Amorion. He forced Caesarea to surrender, but Amorion refused, so he devastated the surrounding countryside, taking great quantities of booty.

Based on Arabic sources that refer to an attack on *Amorion, Lilie places this in 646, although a similar raid that reached Amorion also took place at about the same time.

AMORION c. 645
See *Caesarea above.

ALEXANDRIA (twice)   ghazat, غزت; wa-'ftataḥa, وافتتح   646f
Baladhuri 347ff (DeGoeje 221f); Agapius 479; cf. Hoyland 2011:130; Howard-Johnston 2010: 154 n. 55, 477 (214, 296 on Manuel).

Baladhuri gives several different versions, one only emphasizing capitulation, another that "ʿAmr reduced it and destroyed its wall." The first and most detailed version has the inhabitants appeal to Constans, who sent Manuel with a large fleet: "Manuwīl entered Alexandria and killed all the guard that was in it [...] Hearing the news, ʿAmr set out at the head of 15,000 men and found the Roman fighters doing mischief in the Egyptian villages next to Alexandria. The Muslims met them [...and defeated them in a hard-fought battle]; and nothing could divert or stop them before they reached Alexandria. Here they fortified themselves and set mangonels (majānīq). ʿAmr made a heavy assault, set the ballistae (ʿarrādāt) and destroyed the walls of the city (or rather: "and so its walls were overwhelmed," fa-'ukhidhat al-judurhā, i.e. by the majānīq). He pressed the fight so hard until he entered the city by assault (dakhalahā bi's-sayf ʿanwat^{an}), killed the fighters and carried away the children as captives. Some of its Roman inhabitants left to join the Romans somewhere else; and Allah's enemy, Manuwīl, was killed. 'Amr and the Muslims destroyed the wall of Alexandria in pursuance of a vow that 'Amr had made to that effect, in case he reduced the city."

Agapius apparently confirms the course of events in outline, but the other Common Source derivates only mention the revolt of Gregory in North Africa. It is therefore probable that Agapius has taken this from another source, but it seems quite specific in its geographic description and therefore somewhat different in nature from e.g. Baladhuri's account: "Gregory, the Roman patrician who was in Africa, rebelled. The Arabs attacked (*ghazat*, غزت) Alexandria, in which was Manuel, a patrician of the Romans. He and his men fled, taking to the sea, and they went to (the land of) the Romans. The Arabs conquered Alexandria and (*wa-ftataḥa*, وافتتح) Alexandria and destroyed its wall; they took control of it and of the coast between Alexandria and Pelusium (al-Farama). Then the Arabs raided Africa in this year and encountered there the Roman patrician Gregory. They defeated him and killed his men. Gregory made it to (the land of) the Romans and made peace."

CONSTANTIA	παρέλαβε	649

Theophanes 343f (AM 6140, MS 478); Dionysius 93-97 (Palmer 1993: 173ff); cf. Hoyland 2011: 131-34; Lilie 1976: 64; Howard-Johnston 2011: 219.

Dionysius provides most detail on the invasion of Cyprus prepared by Mu'āwiya, but no details on the siege of Constantia itself (Dion. 93, Palmer 1993: 173, AG 960); "and before long innumerable ships and many smaller boats, which had hearkened to the summons, were moored along the coast. The field-commander at Alexandria also received a letter from Mu'āwiya, bidding him to send ships bearing a numerous task-force from Egypt as reinforcements, which he promptly did." The size of combined fleet was altogether 1,700 ships (Dion. 94, 173f). Mu'āwiya intended to wait for the island to submit, but when this did not happen, in his mind (Dion. 95, 174) "a cruel doom of destruction took shape against the unfortunate population. Moreover, the Egyptian contingent put him under considerable pressure with their hostile recriminations and their angry insults because he had delayed and had held them back from an invasion of the island. At last he let the Alexandrinians have their head and ordered them suddenly to leap the fence, as it were, and invade." The horrors of the raid resulted in many being taken prisoner, while Mu'āwiya took up residence in the bishop's palace where he allegedly organized a harem of some sort. Dionysius states that this happened due to the sins of the Christians. When the campaign was over, the loot was divided between the armies and shipped off, causing immense human suffering (Dion. 96f, 174f): "What misery and lamentation were seen then! Fathers were separated from their children, daughters from their mother, brother from brother, some destined for Alexandria, others for Syria."

Theophanes is clearly abbreviated from the Common Source, but is only aware of one raid on Cyprus, and as the Zuqnin Chronicle (see next entry) links the invasion of Cyprus to an assault on *Arwad: "In this year Mauias invaded Cyprus by sea. He had 1,700 ships, and took (παρέλαβε) Constantia and the whole island, which he laid waste (ἐλυμήνατο). On hearing, however, that the *cubicularius* Kakorizos was moving against him with a great Roman force, he sailed away to Arados..."

Arwad	ἐπειρᾶτο παραλαβεῖν	649

Theophanes 344 (AM 6140, MS 478); *Chr. Zuq.* (Palmer 1993: 58); cf. Hoyland 2011: 134ff; Conrad 1992.

After completing his business on Cyprus, Mu'āwiya moved his fleet to the small fortified island city of Arwad, or Arados, just off the coast of Syria: "... and, after putting in his fleet, attempted to capture with the help of various engines the little town called Kastellos on that island (...καὶ καθορμίσας τὴν ἑαυτοῦ ναυστολίαν τὴν πολίχνην τῆς νήσου, τὸν Κάστελλον, ἐπειρᾶτο παραλαβεῖν παντοίαις μηχαναῖς χρώμενος). Meeting with no success, he sent to the inhabitants a certain bishop called Thomarichos to frighten them into abandoning the town, submitting to terms, and leaving the island (μηδὲν δὲ ἐξισχύσας πέμπει πρὸς αὐτοὺς ἐπίσκοπόν τινα, Θωμάριχον τοὔνομα, ἐκφοβῶν αὐτοὺς ἀφιέναι τὴν πόλιν καὶ ὑποσπόνδους εἶναι). When the bishop had come in to meet them, they held him inside and did not yield to Mauias. The siege of Arados having thus proved fruitless (καὶ ἀνονήτου γενομένης τῆς κατὰ Ἄραδον πολιορκίας), he returned to Damascus since winter had set in."

The Zuqnin chronicler conflates the two sieges of Cyprus and the two of Arwad: "AG 960: Mu'āwiya invaded Cyprus. In the very same year Aradus (arwād) was taken." For Dionysius' treatment, see *Arwad 650, when the city finally fell.

Lapethus	aqīm(w) 'leh mangānīqōs, ܐܩܝܡܘ ܠܗ ܡܢܓܢܝܩܘܣ	650

Dionysius 98 (Palmer 1993: 176f); cf. *Constantia (649).

There was a new invasion of Cyprus the next year; the Roman soldiers that had been sent to retake the island, and all those who had boats, fled upon their approach. The remaining population had to face the invading fleet: "As soon as the ships were ashore, the invaders filled all the mountains and plains, intent on plunder and slaves. They winkled the natives out of the cracks in the ground, like eggs abandoned in the nest. The general, Abū 'l-A'war, went down to Constantia and stayed there for forty days, enslaving the population and eating livestock head by head. At length, when they had had their way with the rest of the island, they all gathered at Lapethus [where some had fled]. For several days they tried the effect of promises (*mellē*, ܡܠܐ) of peace, but finding the Cypriots unreceptive to these, they began to bombard the city with catapults (*textus* 272.14f *aqīm(w) 'lēh managanīqōs*, ܡܢܓܢܝܩܘܣ ܥܠܝܗ ܐܩܝܡܘ) from all around. When the inhabitants saw that it was hopeless and that no help was on its way, they petitioned the general to proffer them his right hand in token of deliverance from death. He showed clemency readily and sent them the following instructions: 'The gold and silver and other assets which are in the city are mine. To you I give an amnesty and a solemn pact that those of you who so wish may go to Roman territory, and that those who wish to stay will neither be killed nor enslaved.' So the city was taken, its treasures were embarked on the ships with the rest of the booty and the Arabs sailed back to Syria in victory."

ARWAD   ἐπεστράτευσε... σφοδρῶς παραταξάμενος...λόγῳ παρέλαβεν   650
Theophanes 344 (AM 6141, MS 479); Dionysius 99 (Palmer 1993: 177f); cf. *Arwad (649).

Theophanes: "In this year Mauias set out against Arados with a great armament and took it by capitulation on condition that its inhabitants would dwell wherever they wished. He burned the town, destroyed its walls, and caused the island to be uninhabited to this day." Τούτῳ τῷ ἔτει ἐπεστράτευσε Μαυΐας κατὰ τῆς Ἀράδου σφοδρῶς παραταξάμενος, καὶ ταύτην λόγῳ παρέλαβεν εἰς τὸ κατοικεῖν αὐτοὺς ἔνθα βούλονται· τὴν δὲ πόλιν ἐνέπρησε καὶ ἀπετείχισεν, καὶ τὴν νῆσον ἀοίκητον κατέστησεν ἕως τοῦ νῦν.

Dionysius has the siege of Aradus last over two seasons, not as two separate assaults as indicated here. During the second season of the siege Muʿāwiya arrived better prepared and with more men to perform heavy assaults that ultimately led to the surrender of the city and deportation of its population.

SARAGOSSA   *circumseptus*   653
Taio Caesaraugustanus; Thompson 1969: 199f; Collins 2002: 84f.

The rebel Froia, supported by the Basques, had evidently taken control of the countryside around Saragossa, and was preventing egress from the city. Bishop Taio complained to his friend, bishop Quiricus of Barcelona, that the siege prevented him from getting any work done during the day, so he must have had some part directing the defenses; he only had time to make extracts from Gregory the Great's *Sentences* during the night. As Taio's letter shows, Froia lacked support from the local clergy and was defeated by king Reccesuinth, who had inherited the throne from his father Chindasuinth earlier that year.

Taio describes how Froia raised the revolt, using Vascones to ravage the countryside of *"Hiberia,"* wounding, killing and taking many captive and taking *"immensa spolia"* in addition desecrating churches (2). "When for this very reason the besieged circuit of the walls of the *urbs* of Caesaraugusta was containing us..." || Cum nos hujuscemodi causa Caesaraugustanae urbis circumseptus murorum ambitus contineret, adventumque supra taxati principis praestolaremur... || God intervened and protected the city, and the revolt was defeated (3).

?Tripoli 653

Chr. Zuq. (Palmer 1993: 58); Dionysius 101 (Palmer 1993: 179f and n. 450, AG 966); cf. Hoyland 2011: 139f, 141-44; Lilie 1976: 65, 67; Howard-Johnston 2010.

The chronology of this event is difficult to determine. The Zuqnin chronicler laconically has: "AG 963 [AD 651/2]: There was a battle between the Arabs and Romans at Tripolis."

Dionysius (partly reconstructed from Michael the Syrian) has a full version, and connects the events at Tripolis with the build-up for *Constantinople (654) and the Arab fleet establishing a series of bases in the *Aegean: "Then two dedicated men released the prisoners shut up in Tripolis, where the ships were moored. After killing the Arabs and the emir, they set fire to the whole fleet of ships. They themselves got away in a small boat and escaped to Roman territory."

Baladhuri 195 (DeGoeje 127f) notes a later event under 'Abd al-Malik (685-705) that seems similar: A Greek Patrician settled in Tripoli "with a large body of his men" in return for *kharāj*, but after two years he revolted, then fled "together with his followers to the land of the Greeks." He was later captured, either at sea or while reoccupying Tripoli: "I heard someone say that 'Abd-al-Malik sent someone who besieged him in Tripoli until he surrendered and was carried before 'Abd-al-Malik who killed and crucified him." Another report states "that Tripoli was conquered by Sufyān ibn-Mujīb, that its inhabitants violated the covenant in the days of 'Abd-al-Malik and that it was reduced by al-Walīd ibn-'Abd-al-Malik in his reign."

Tripoli clearly remained a center for (conscripted) craftsmen or former/ captive Roman soldiers even long after the 653 incident, which should therefore not be downplayed.

Aegean Islands	653

Dionysius 101; for full references, see *Tripolis above.

"Muʿāwiya, the emir of Syria and Damascus, equipped a great fleet to sail to Constantinople and lay siege to it. These preparations were made on the coast of Tripolis. Abū 'l-Aʿwar was appointed as admiral of the fleet and they sailed to Phoenicus on the Lycian coast, where they were met by the great fleet of Constans, the king of the Greeks {yawnōyē}, and his brother Theodosius." After a Roman defeat (presaged in a dream by Constans), the Arabs pressed the pursuit all the way to Rhodes, while 20,000 corpses pulled out of the water. Palmer adds a note from Michael: "AG 965: Abū 'l-Aʿwar and his army sailed to the island of Cos. By the treachery of the bishops (read 'bishop'?) there they took it, wasted and pillaged all its wealth, massacred the population, leading the survivors away as captives, and destroyed its citadel. Then he carried out a raid on Crete and another on Rhodes."

Lilie seems to place the event reported in the Syriac and Greek sources in 655, thus placing it before the naval battle at Phoenix according to the older dating scheme. See Appendix III for the problems with this dating and references in *Constantinople 654, as the islands were on the natural sea lanes towards Constantinople and thus essential for a naval expedition.

Constantinople	[ships...to attack the city]	654

Sebeos 48ff (I: 135-46, II: 270-77); Lilie 1976: 66f; O'Sullivan 2004; Cosentino 2008; Howard-Johnston 2010; Appendix III.

"Now when the king of Ismael saw the success of this victory and that the Persian kingdom had been destroyed, after three years of the peace treaty had fully passed he no longer wished to make peace with the king of the Greeks. But he commanded his troops to conduct war by sea and land in order to efface from the earth that kingdom as well, in the 12th year of the reign of Constans [652-3]." (48, I: 135)

"In the 11th year of Constans the treaty between Constans and Muawiya, the prince of Ismael, was broken. The king of Ismael ordered all his troops to assemble in the west and to wage war against the Roman Empire, so that they might take Constantinople and exterminate that kingdom as well." (49, I: 143)

After Mu'awiya had sent a message demanding submission, "All the troops who were in the east assembled: from Persia, Khuzhastan, from the region of India, Aruastan, and from the region of Egypt [they came] to Muawiya, the prince of the army who resided in Damascus. They prepared warships in Alexandria and in all the coastal cities. They filled the ships with arms and artillery [*mek'enayk'*]—300 great ships and with a thousand elite cavalry for each ship. He ordered 5,000 light ships to be built, and he put in them [only] a few men for the sake of speed, 100 men for each ship, so that they might rapidly dart to and fro over the waves of the sea around the very large ships. These he sent over the sea, while he himself took his troops with him and marched to Chalcedon. When he penetrated the whole land, all the inhabitants of the country submitted to him, those on the coast and in the mountains and on the plains. On the other hand, the host of the Roman army entered Constantinople to guard the city. The destroyer reached Chalcedon in the 13th year of Constans. He kept the many light ships ready at the seashore, so that when the very heavy ships might arrive at Chalcedon he could rapidly go to their support. And he had the letter of their king taken into the city of Constans." Constans received it, praying in sackcloth and ashes. "Behold the great ships arrived at Chalcedon from Alexandria with all the small ships and all their equipment. For they had stowed on board the ships mangonels [Grk. *magganon*, Arm. *manglion*], and machines [*mek'enay*] to throw fire, and machines to hurl stones, archers and slingers, so that when they reached the wall of the city they might easily descend onto the wall from the top of towers, and break into the city. He ordered the ships to be deployed in lines and to attack the city."

"When they were about two stades' distance from the dry land, then one could see the awesome power of the Lord. For the Lord looked down from heaven with the violence of a fierce wind, and there arose a storm, a great tempest, and the sea was stirred up from the depths below. Its waves piled up high like the summits of very high mountains, and the wind whirled around over them; it crashed and roared like the clouds, and there were gurglings from the depths. The towers collapsed, the machines were destroyed, the ships broke up, and the host of soldiers were drowned in the depths of the sea. The survivors were dispersed on planks over the waves of the sea. Cast hither and thither in the tossing of the waves, they perished; for the sea opened its mouth and swallowed them. There remained not a single one of them. On that day by his upraised arm God saved the city through the prayers of the pious king Constans. For six days the violence of the wind and the turbulence of the sea did not cease."

"When the Ismaelites saw the fearsome hand of the Lord, their hearts broke. Leaving Chalcedon by night, they went to their own land. The other army, which was quartered in Cappadocia, attacked the Greek army. But the Greeks defeated them, and it fled to Aruastan pillaging Fourth Armenia." The Arabs also try to subdue Iberia from their base at Dvin, but a fierce winter forced them back to Syria; Armenian princely politics. (50, I: 144ff)

Lilie is sceptical of this account, but for a recent analysis and vindication of this sequence of events, see Howard-Johnston's extensive commentary of Sebeos as well as his 2010 analysis of 7th-century evidence in context. O'Sullivan and Cosentino also support the historicity of this event, but propose slightly different reconstructions of events in 653-55.

DVIN	[sacked]	654/5

Sebeos 52 (I: 150, II: 279).

After the great disaster at *Constantinople (above), a Roman counteroffensive under Maurianos (winter 654-55) caused the Arabs to withdraw from eastern Anatolia across the Araxes. This gave the Romans the opportunity to sack Dvin, which must have had a very demoralized Arab garrison.

THEODOSIOPOLIS	[besieged; opened the gates]	655

Sebeos 52 (I: 150, II: 279).

After *Dvin (above), the Roman army went on to Nakhchawan, but were ambushed by Arabs during the spring; the army (at least its general Maurianos) fled to Iberia, while the Arabs advanced to Karin (Theodosiopolis): "Then the army of Ismael turned back from them, besieged the city of Karin, and attacked its [inhabitants]. The latter, unable to offer military resistance, opened the gates of the city and submitted." The Arabs plundered their wealth, "ravaged all the land of Armenia" and Caucasus, and took many hostages.

LYONS	*urbs fuisset...circumdata*	c. 662

Acta Aunemundi; Gerberding and Fouracre 1996: 166-79 (comm.), 179-92 (translation).

The bishop Aunemund of Lyons had fallen out of favor, possibly for conspiring with foreign peoples: "... three dukes had already been dispatched with orders to convey the blessed Bishop Aunemund under sure guard to the king, or, if he resisted, to leave him behind as a corpse. And with ever quickening step they approached Lyons." (c. 4)

"Now they hurried towards the aforesaid town to carry out what they had been ordered to do, God's servant [Aunemund] dreaded suffering and wanted to jump out of the way. Yet he turned back to the man of God [Waldebert] and said, 'It is better to undergo martyrdom as a result of this wickedly unfair charge than leave a bad example to others.' Thus he remained in the town which had been entrusted to his care, and so much did he exhaust himself in almsgiving, in fasting, in vigils and in continual prayer with his clergy, that, had he of late become guilty of anything through ignorance, lamenting here he erased it through penitence. Then the town was surrounded by troops as they came to kill him, and they kindled fires here and there at the crossroads and destroyed everything in the forts." The last sentence in original reads: || Cumque ad necem illius urbs fuisset cohortibus circumdata, passim incendia in compitis exurebant, atque in oppidis cuncta vastabant. ||

Upon these news Aunemund prayed, commended his soul to God, asked the citizens for forgiveness (c. 5), and bade them farewell before surrendering to the royal army (c. 6ff), which escorted him away (c. 9). "Thus the band of men hurried on into the diocese of Chalon and there, close by an estate of the town's church, they pitched their tents" (c. 10). || Igitur properante populi cohorte in Cabilonense territorim, inibi non procul a suæ ecclesiæ prædio tentoria collocaverunt. || That night he was killed by the royal troops, but his martyrdom demonstrated through various miracles (c. 11-15).

Aunemund was unable to marshal enough political support to hold out against his enemies, despite obvious attempts at bribing the population and his ostentatious display of sanctity. Hence his only named ally, the abbot Waldebert, advised surrender when the royal troops began burning the suburban settlements of the city. Unsurprisingly, Aunemund was murdered while Waldebert was spared, possibly as the result of a deal already made. The royal troops traveled regular routes from estate to estate; indeed, the same routes appear in the Carolingian period (cf. chapter 4.3.3).

THESSALONICA ἐν τοῖς τείχεσι παρατάξασθαι ἐπὶ πολιορκίᾳ καὶ ἁλώσει 662
Miracula St. Demetrii 2.4 (230-82); Lemerle II: 111-36; Curta 2001: 111f; Howard-Johnston 2010.

In his introduction to this miracle, the author protested that there was not enough papyrus in the Nile to tell the full story, but that he would give a few examples (269). If this *topos* contains a grain of truth, it demonstrates the extreme political and military complexity of an event completely unknown from any other sources, as the following description is one of the most detailed that is preserved for a late-7th century siege.

Perboundos, king of the Runchines, a Slavic tribe that lived near Thessalonica, was arrested by Roman authorities after being denounced in reports from the prefect to the emperor (231); the Runchine and Strymon Slavs, together with representatives from Thessalonica, went to the emperor, who however was busy with a war against the Arabs (cf. *Constantinople 670ff but see below). Perboundos was to be liberated after the war, so the envoys returned home as Perboundos was treated with respect (φορεσίαν καὶ πᾶσαν θεραπείαν, 232f). However, Perboundos was tricked by a *hermeneutes* (interpreter) into fleeing with him to his estate (ἐν τῷ αὐτοῦ προαστείῳ) in Thrace, thence back to secure lands (234). This was easily accomplished since Perboundos spoke Greek and dressed in Greek style (ὡς φορῶν ῥωμαῖον σχῆμα καὶ λαλῶν τῇ ἡμετέρᾳ διαλέκτῳ, 235). Having tortured and mutilated his keepers for dereliction of duty (236), the emperor warned Thessalonica to take measures of security and provisions (237). Perboundos was subsequently arrested again and executed; the Slavs began to stir in revenge (238-242).

Various tribes began to take prisoners systematically around Thessalonica, including on the sea with "carpentered" boats (διὰ τῶν ἐζευγμένων νηῶν, i.e. not log boats, Lemerle I: 200 n. 4). They attacked from different directions every day, reducing the Thessalonians to despair (243). A famine was caused by the corrupt administration of public granaries; officials exported grain instead of stocking it, even though they knew that the enemy was on the move (244). The famine intensified when raids began from two different directions; it was impossible both to sail in and out of the harbor and to tend the fields. The people were reduced to eating horses, asses and strange plants (245). Soon the besieged were in a very poor physical state, while those who ventured outside were captured or killed at once (246). Extramural churches became hiding places (the text calls them φρούρια, thus more like blockade camps, though it goes on to say they hid in them) for barbarian ambushes, swooping down on and killing anyone venturing outside. The same happened at sea; all those who went out fishing were captured by the *monoxyla* (253). The apparent frequency with which some of the citizens ventured out does however indicate that there is some reward for the risk. The famine was exacerbated by lack of water after the rains failed (247). Some even left their families and faith and went over to the barbarians, as they had no hope left (248). Some of these turncoats bought the freedom of Thessalonians who had been shipped off to interior for fear of their large number and proximity to Thessalonica (249). News of this stopped the general flight; otherwise the city would become completely deserted. A number of brave young men were ambushed by Slavs outside of the city (250; Lemerle believes the Slavs pretended to be on the Thessalonian side but betrayed the Romans. The interpretation of this passage is difficult and perhaps related to some politically embarrassing episode, cf. Lemerle I: 202 n. 9).

The emperor sent ten warships with provisions (δέκα ἐνόπλους καράβους μετὰ καὶ δαπανῶν), but could not spare more troops due to the war elsewhere. However, the crews exploited the desperate situation to charge exorbitant prices for only small amounts of food (251). While the hagiographer implies that this was a meager comfort, there could potentially be a considerable number of troops, sailors and supplies on ten ships, 1-2,000 men (cf. Pryor and Jeffreys 2006). The local authorities are afraid to be exposed (cf. 281), but the government may have had to allow private citizens to bring in supplies in order to procure them at all (cf. 254 below, where the fleet left to get more supplies when everything had been sold). The authorities ordered searches against those suspected of hoarding; the hagiographer laments the pitiful deaths of many for this reason (252). Again, however, this seems sensible to ensure rationing and prevent further speculation, as above, which may even have been nipped in the bud and provoked the searches in the first place. As with the youth who were ambushed by the Slavs, the context is probably far more complicated than the hagiographer would like to portray it, and different parties may have had different objectives (cf. *Thessalonica 682). The authorities and citizens decided to send all available boats with warships (βουλὴ τῶν κρατούντων καὶ τῶν πολιτῶν γίνεται ὥστε τὰ ὑπολειφθέντα σκεύη τε καὶ μονόξυλα μετὰ καὶ τῶν λεχθέντων δέκα καράβων) to territory of Belegezetes in Thessaly, who were at peace with the Thessalonians, in order to haul more grain; in the meantime, all unneeded hands were to stay on city walls (ἀχρήστου λαοῦ ἐν τοῖς τείχεσιν) while the flower of their age went out with the boats (254).

The king of the Drougoubites decided to exploit the situation and besiege the city (encouraged by other Slav princes), and made preparations: fire machines against the gates, engines with woven covers (p. 203 n. 12, "Probablement les tortues"), ladders, stonethrowers, other wooden engines, and other wooden machines which "we have never seen and cannot name." The attack began on 25 July, fifth indiction, from sea and land (255):

οἱ τῶν τοῦ ἔθνους τῶν Δρουγουβιτῶν ῥῆγες βουλῆς ταύτης γίνονται ὁμοθυμαδὸν ἐν τοῖς τείχεσι παρατάξασθαι ἐπὶ πολιορκίᾳ καὶ ἁλώσει τῆς πόλεως, τοῦ ἀδρανοῦς καὶ ὀλιγοστοῦ λαοῦ καταφρονήσαντες, ἄλλως τε δὲ καὶ διαβεβαιωθέντες παρὰ τινων τῶν αὐτοῦ Σκλαβίνων ἔθνους ἐκ παντὸς τρόπου πορθεῖν τὴν πόλιν. Ὅθεν λοιπὸν κατασκευάσαντες πυρφόρα κατὰ τῶν πυλῶν ὅπλα καί τινα ἐκ λυγοπλέκτων ὄργανα, κλίμακας οὐρανομήκεις, πετραρέας τε ὡσαύτως, ἑτέρας δὲ κατασκευὰς ξυλίνων μαγγανικῶν ἀπείρων, βέλη τε νεοκατασκεύαστα, καὶ ἁπλῶς εἰπεῖν ἅπερ οὐδεὶς τῆς καθ' ἡμᾶς γενεᾶς ἠπίστατο ἢ ἑώρακέ ποτε, ἀλλ' οὐδὲ τῶν πλείστων τὰς ἐπωνυμίας μέχρι τοῦ παρόντος ἐξειπεῖν ἠδυνήθημεν, καὶ οὕτως ἁπάντων Σκλαβίνων τοῦ Ῥυγχίνου ἔθνους μετὰ τῶν Σαγουδατῶν τῇ εἰκάδι πέμπτῃ τοῦ ἰουλίου μηνὸς ἰνδικτιῶνος πέμπτης τῇ πόλει προσέβαλον, οἱ μὲν διὰ τοῦ χερσαίου, οἱ δὲ διὰ τῆς θαλάττης μετὰ πλείστων ἀναριθμήτων πλωτήρων.

When the Thessalonians prayed for help (256), God provided a miracle: the Strymon Slavs made a detour three miles from the city, leaving the Runchines and Sagoudates to besiege the city alone (257). The Greek text appears to be of a liturgical nature; perhaps the hagiographer used church archives and found a specific liturgy made to save the city during this or a similar siege. The miraculous detour of the Strymon Slavs may in fact have been a deliberate strategy on their part to prevent overland relief from Thrace. At any rate, the remaining Slavs surround the city completely on land, launched their fleet, inspected the walls for places to attack and began assembling siege engines (258): Καὶ τῇ μὲν πρώτῃ ἡμέρᾳ ἀπὸ τοῦ δυτικοῦ βραχιονίου μέχρι τοῦ ἀνατολικοῦ πᾶσαν τὴν πόλιν κυκλωσάντων, καὶ τοὺς ἐμπειροπολέμους τοὺς τόπους ἅπαντας κατασκοπῆναι, ὅθεν εὐχερὲς αὐτοῖς ἐκ πολιορκίας τὴν πόλιν ἑλεῖν· ὡσαύτως δὲ καὶ οἱ ζευκτῶν Σκλαβίνοι τῶν παραλίων τὴν κατάσκεψιν ἐποιήσαντο, οἱ πάντες δι' ὅλων τῶν τειχῶν ἐπιφερόμενοι τὰ ἐπ' ἀπωλείᾳ αὐτῶν κατασκευασθέντα τῆς πορθήσεως ἀμυντήρια.

At this point, the Thessalonians lamented their fate, wailing over their impending deaths, and feared that the Belezegetes would kill those on the supply fleet when they heard of the siege, but despite their plans to do so (according to the hagiographer, but this seems unsubstantiated), this was prevented by St. Demetrius (259). At the evening of the first day, the saint also appeared on the rampart, chasing away Slavs from a postern using a mace (ῥάβδιον, 260). There was also another appearance of Demetrius stationing unknown troops of magnificent stature along the wall, lasting until dawn the second day. In contrast, the Thessalonians were in a miserable physical condition (261).

Despite such visions, the Slavs attacked with a cry that made the walls tremble with all their material, on land and sea (262): Καὶ τῆς ἡμέρας ἤδη λοιπὸν διαφοσκούσης, ἀναστὰν ἅπαν τὸ βάρβαρον ὁμοθυμαδὸν ἀνέκραξεν, ὡς σεισθῆναι τὴν γῆν ἅπασαν καὶ τὰ τείχη κλονηθῆναι. Καὶ αὐθωρὸν ἐν τῷ ἅμα πάντες τῷ τείχει μετὰ τῶν παρ' αὐτῶν κατασκευασθέντων ἀμυντηρίων ὅπλων τε καὶ μαγγάνων καὶ πυρός, οἱ μὲν διὰ τοῦ χερσαίου, οἱ δὲ διὰ τῶν ζευκτῶν ἐν τῇ παραλίᾳ πάσῃ, καθοπλισθέντες στοιχηδὸν οἱ τοξόται καὶ οἱ ἀσπιδιῶται καὶ οἱ ἀγριᾶνες καὶ οἱ ἀκοντισταὶ καὶ σφενδονισταὶ καὶ μαγγανάριοι καὶ οἱ εὐτολμότεροι ἅμα ταῖς κλίμαξι καὶ τῷ πυρὶ προσέρρηξαν τῷ τείχει.

Like snow in winter, the mass of missiles rained over the city, obscuring the day (263). The barbarians tried to burn a postern, only to find that the iron fittings miraculously remained to block the way. The barbarians left stupefied; many barbarians were also "miraculously" killed all along the walls (264), although hagiographers a generation earlier had no problem attributing this to the Roman artillerymen. After three days of failed storms against the easy approaches, with many chiefs killed or wounded and the remaining disputing among themselves, the Slavs withdrew (265). The Thessalonians took the materiel prepared by the enemies (266) while the Barbarians quarreled among themselves, blaming each other for thinking the city is full of old men and feeble women when they were fighting heavenly hosts instead. This is quoted as proof that St. Demetrius fought with the citizens (267), but fresh sailors and soldiers left behind by the fleet may have had a similar effect.

The supply party returned safe and sound with grain and dry vegetables, ensuring the survival of the city (268). The daily raids continued, however, and people were ambushed if they were imprudent (270). The Slavs even made naval raids, pillaged ships and took sailors from the islands, the Straits, Parion and the Proconnese, also the people at the toll station, returning with a great number of ships, πλοΐμων (277). The emperor ordered an attack on the Slavs of the Strymon (probably fixing the event to 662); the Slavs prepare to defend the approaches to their countries (τὰς κλεισούρας καὶ τοὺς ὀχυρωτέρους τόπους) and called in other barbarian kings (278). The army was victorious despite ambushes and the Slavs lost their best soldiers. The "people"—τινὰς εἰσδραμόντας [Lemerle I: 207 n. 18, "non pas, selon nous, des Sklavènes, mais des Grecs"—i.e. Greek population more or less under Slav control, but still loyal to Roman Empire] came to Thessalonica to encourage them to attack the Slav habitations and take their supplies; the Slavs fled in terror to the interior (279). The starved Thessalonians took huge amounts of food from Slav villages (280). Finally, grain was sent to Thessalonica before it was asked for; the local authorities had not made any requests because they did not want their export to become known. The emperor sent twelve times the necessary amount; the barbarians despaired and opened negotiations for peace (281). Thanks to God (282; also 266, 268).

This miracle also contains the story about a Slav military engineer who built a marvelous machine, but was unable to finish it due to insanity caused by St. Demetrius (271-276), is particularly revealing about Slav military organization at this time (see chapter 7.2 *passim* for further discussion).

The Slavs wanted new machines to attack the city, and there was competition over who could present the best project to their chiefs (271): ... Τῶν γὰρ προλεχθέντων Σκλαβίνων, ἐπὶ τῇ ἑαυτῶν ἀπωλείᾳ, τὴν τῶν ἀμυντηρίων ὅπλων τε καὶ μαγγάνων ἐπὶ παρατάξει τῆς πόλεως κακούργως ἐπινοούντων τε καὶ ἐργαζομένων, καὶ ἄλλος ἄλλας μηχανὰς ξένας ἐπινοῶν καὶ ἐφευρέσεις, ἄλλος ξιφῶν καὶ βελῶν νεοκατασκευάστους ποιήσεις, καὶ θάτερος θατέρῳ εὐδοκιμώτερος καὶ σπουδαιότερος σπεύδων τοῦ ἑτέρου δείκνυσθαι πρὸς παράθεσιν τῶν τῶν ἐθνῶν ἡγουμένων ἠγωνίζετο.

A particularly able and experienced Slav engineer asked for his king's permission and assistance to construct a wooden tower mounted on an ingenious device of cylinders (i.e., axles), to be covered with fresh hides, and consist of three stories—one for archers and slingers, one for the sappers, and the top for *petroboloi* and soldiers protected by a wooden breastwork (272): Ἐν δὲ οἷς τις ἐκ τῶν τοῦ αὐτοῦ Σκλαβίνων ἔθνους ὑπάρχων, καὶ τρόποις καὶ ἔργοις καὶ τῇ διανοίᾳ ἔμπειρος, τοιοῦτος πρὸς παράταξιν ἤτοι κατασκευὴν μαγγανικῶν τυγχάνων διὰ τῆς ἐνούσης αὐτῷ πολυπειρίας, τὸν ῥῆγα αὐτὸν ἠξίου παρασχεθῆναι αὐτῷ ἄδειαν καὶ τὴν αὐτῶν συνδρομήν, ἐφ' ᾧ κατασκευάσαι διὰ ξύλων εὐλήπτων πύργον ἔντεχνον ὑπὸ τροχοὺς καί τινας κυλίνδρους δι' εὐμηχάνου συνθέσεως· ἐνδύσεται τοῦτον ἐκ βύρσων νεοδάρτων, πετροβόλους ὕπερθεν ἔχειν φράσας, καθηλῶσαί τε ἐξ ἀμφιπλεύρου ξιφότευ[.] {in apparatu: post ξιφότευ desunt una vel duae litterae} εἴδη, ἐπάλξεις δὲ ἄνωθεν ἔνθα ὁπλίτας βαίνειν, τριώφορον δὲ τοῦτον καὶ τοξότας ἔχοντα καὶ σφενδονήτας· καὶ ἁπλῶς εἰπεῖν τοιοῦτον κατασκευάσαι ὄργανον δι' οὗ διϊσχυρίζετο τὴν πόλιν πάντως ἑλεῖν.

The Slav archons were astonished and incredulous at this, asking the engineer to sketch the design for them in the dust, which he did. Convinced, the chiefs put at his disposal a crowd of people craftsmen, lumberers, carpenters, ironworkers, as well as soldiers and fletchers (273): Τῶν δὲ λεχθέντων ἀρχόντων τῶν Σκλαβίνων ἐκπληττομένων τῇ διαθέσει οὗπερ ἔλεγε ξένου κατασκευάσματος, καὶ ἐν ἀπιστίᾳ τῶν λόγων γεγονότων, ᾔτουν μορφῶσαι ἐν τῇ γῇ τὴν τοῦ λεχθέντου ὀργάνου κατασκευήν. Μηδὲ ἐν τούτῳ μελλήσας ὁ τεχνίτης ὁ τὴν τοιαύτην κατασκευὴν ἐφευρών, τῇ γῇ δείκνυσι τὸν σκάριφον τοῦ ἔργου. Ὡς λοιπὸν πεισθέντας τῷ φοβερῷ τοῦ μέλλοντος γίνεσθαι, προθύμως παρέσχον νεανίας παμπόλλους, τοὺς μὲν κόπτοντας τὴν ὕλην πρὸς τὰ βάθρα, ἄλλους ἐμπείρους πελεκητὰς εὐφυεῖς, ἑτέρους τέκτονας σιδήρων εὐμηχάνων, ἄλλους ὁπλίτας καὶ βελοποιοὺς ἄνδρας· καὶ ἦν πολυπληθὴς συνδρομὴ τῶν ὑπουργούντων τῷ λεχθέντι μηχανήματι.

About to begin construction, Demetrius appeared and slapped the engineer. This made him lose his mind, and after some difficulty he ran off and lived like an animal in the mountains (274). "The said *manganarios*" (ὁ ῥηθεὶς μαγγανάριος) stayed there until the siege was raised (St. Demetrius made sure of that); he then returned to his senses and told everyone what happened. However, when the same happened again (more or less), he fled again (275). In his last appearance, St. Demetrius told him to go to Thessalonica and find him there; in the city, he recognized the saint and converted, telling everyone of his miracle (276).

Lemerle has found that the raiding, famine, siege and relief lasted for two years. Based on the indiction given (the fifth) for the siege itself, which only lasted three days, he suggests these dates: 647, 662, and 677. The former, he argues, is too early, leaving the latter two. Of these, Lemerle argues that the context of the Arab siege of *Constantinople in the 670s (traditionally 674-78) fits best, but he bases himself too strongly on the accuracy of Theophanes' dates. Curta also prefers 677, but does not provide any reasons. However, another possible context is Constans II's wars, which has only been considered Howard-Johnston, who argues that events in 647 provide the most convincing context with a Roman naval invasion of Egypt (*Alexandria 646) and other internal evidence (60 years since Avar raids in 586-7; but see *Thessalonica 618 and 682, which are explicitly connected by the hagiographer—cf. *Miracula St. Demetrii* 197 to 284—and do not, therefore, support Howard-Johnston's dating). However, Constans made two campaigns against the Slavs that was recorded by Theophanes before his *Italian expedition, which arrived in Italy in 663. The complex Roman-Slav relations reported here may then be aftereffects of a recent reorganization of client relations after his campaigns in the Balkans shortly before, i.e. 657/8 or 662.

Constantinople	*ballistrai*	663

Chr. Mar. [AG 974] (*textus* 72f; Palmer 1993: 32f).

A missing folio means that our knowledge of this expedition begins in mid-sentence; we have no idea of how it got to the vicinity of Constantinople and whether a fleet was involved: "... of the year, Yazīd b. Muʿāwiya went up again with a large army. While they were encamped in Thrace, the Arabs scattered for the purpose of plunder, leaving their hirelings and their sons to pasture the cattle and to snatch anything that should come their way. When those who were standing on the wall (saw) this, they went out and fell upon them and (killed) a great many young men and hirelings and some of the Arabs too. Then they snatched up the booty and went in (to the City). The next day, all the young men of the city grouped together, along with some of those who had come in to take refuge there and a few of the Romans and said, 'Let us make a sortie against them.' But Constantine told them, 'Do not make a sortie. It is not as if you had engaged in a battle and won. All you have done is a bit of common thieving.' But they refused to listen to him. Instead, a large number of people went out armed, carrying banners and streamers on high as is the Roman custom. As soon as they had gone out, all the gates were closed. The King had a tent erected on the wall, where he sat watching. The Saracens drew (them) after them, retreating a good long way away from the wall, so that they would not be able to escape quickly when put to flight. So they went out and squatted in tribal formation. When the others reached them, they leapt to their feet and cried out in the way of their language, 'God is great!' Immediately the others turned tail in flight, chased by the Saracens, who fell on them, killing and making captives right up to the point where they came within the range of the catapults (*ballistrai*) on the wall. In his fury with them Constantine was barely willing to open (the gates) for them. Many of them fell and others were wounded by arrows."

A possible solution to the context of this campaign is possibly found in the Greek evidence analyzed under *Constantinople 670ff, which several scholars now suspect is taken from an earlier event (654 or another raid in 668 or so) and misplaced in the framework of the Syriac Common Source. Here I propose that 663 is probably the most fitting context, as at least some of the evidence from 670ff seems to be describing similar events: most importantly the presence of (ibn) Khalid in support (i.e. on land; see 664 below) of a fleet, which encamps on the Marmara coast in a suburb of Constantinople, i.e. in Thrace as described above, under the command of Yazīd b. Muʿāwiya.

ITALIAN CAMPAIGN OF CONSTANS II 663
HL 5.6-10; *LP* 78.2.

Constans, intending to conquer Italy from the Lombards, "left Constantinople and taking his way along the coast, came to Athens, and from there, having crossed the sea, he landed at Tarentum." He visited a hermit when he landed at Tarentum; however, the hermit predicted that the venture would fail (5.6).

The dating is secured by the *Liber Pontificalis*: Constans arrived at *Benevento and Naples in the 6th indiction, i.e. 663, and visited Rome, presumably in the same year. He was killed at Syracuse in 669. On the way to Italy he possibly fought against the Slavs, visited Athens, and most likely tended to other affairs in Greece (cf. *Thessalonica 662). While each of the cities below may have taken some time to invest and may thus have been telescoped by Paul, for sake of convenience, they are all dated to 663, but it cannot be ruled out that some of the sieges perhaps occurred in 664 or even later. See further *Lucera, *Acerenza, *Benevento.

Lucera and other cities	*cepit, expugnatam diruit*	663

HL 5.7.

"Therefore after the emperor Constans, as we said, had come to Tarentum, he departed therefrom and invaded the territories of the Beneventines and took almost all the cities of the Langobards through which he passed. He also attacked bravely and took by storm Luceria, a rich city of Apulia, destroyed it and leveled it to the ground." || Igitur cum, ut diximus, Constans augustus Tarentum venisset, egressus exinde, Beneventoranum fines invasit omnesque pene per quas venerat Langobardorum civitates cepit. Luceriam quoque, opulentam Apuliae civitatem, expugnatam fortius invadens diruit, ad solum usque prostravit. ||

Acerenza	*capere minime potuit*	663

As *Lucera (above).

"Acerenza, however, he could not take on account of the highly fortified position of the place." || Agerentia sane propter munitissimam loci positionem capere minime potuit. ||

Benevento	*vehementer expugnabat*	c. 663

HL 5.7-9.

"Thereupon he surrounded Beneventum with all his army and began to reduce it energetically." || Deinde cum omni suo exercitu Beneventanum circumdedit et eam vehementer expugnare coepit. ||

The duke of Benevento, Romuald, sends for help from his father the king Grimuald. "When the king Grimuald heard this he straightaway started to go with an army to Beneventum to bring aid to his son." However, many Lombards deserted Grimuald on the way.

"Meanwhile the army of the emperor was assaulting Beneventum vigorously with machines of war and on the other hand Romuald with his Langobards was resisting bravely..." || Interim imperatoris exercitus Beneventum diversis machinis vehementer expugnabat, econtra Romuald cum Langobardis fortiter resistebat. || The Lombards sent out sallies with "young men" (*expeditis iuvenibus*) that assaulted the Roman siegeworks (*hostium castra inrumpens*), causing great damage. Grimuald sent a messenger to announce the arriving relief army, but the messenger was apprehended and interrogated by the "Greeks." Constans was alarmed by the news.

The messenger was threatened to lie to the besieged Beneventines about the relieving army, but revealed that Grimuald was nearby. "When he said this, his head was cut off by command of the emperor and thrown into the city by an instrument of war which they call a stone-thrower (*cum belli machina quam petraria vocant in urbem proiectum est*)." (*HL* 5.8)

Constans broke up the siege and went towards Naples, but was, according to Paul, defeated on the march by the duke of Capua (*HL* 5.9). A Roman division under Saburrus and his 20,000 men (the number may represent the total field army of Constans) was defeated at the hands of Romuald. Paul amplifies the scale of defeat by including an anecdote of a little Greek impaled during combat and raised over the battlefield (*HL* 5.10). The Roman commander Saburrus may in fact be the same as Saporios/Saborios, who revolted in the Anatolic region at about this time. See the discussion in Appendix III.

Anatolian campaign of ibn Khālid — 664
Chr. Mar. AG 975 (Palmer 1993: 33ff); fully discussed in Appendix III.

"In AG 975, the 22nd of Constans and 7th [Palmer corrects to 4th] of Muʿāwiya, (ʿAbd al-Raḥmān) b. Khālid, commander of the Arabs of Emesa, the capital of Phoenicia, went up with an army against Roman territory." See further *Skoutarion, *Amorion, *SYLWS, *Pessinos, *Kios, *Pergamon, *Smyrna below and *Amorion 666.

Skoutarion(?) — 664
Continued from *Anatolian campaign.

"He came and pitched camp by a lake called Scutarium ['*SQDRYN*]; and when he saw that a large number of people were dwelling in it, he wanted to take it. So he made rafts and boats and embarked a force on them and sent them towards the middle (of the lake). The lake-dwellers, seeing this, ran away and hid from them. When the Arabs got into the harbor, they disembarked and tied up the boats, then made off towards the interior to attack the people. At that moment the men who were in hiding got up and ran to the boats, cut off their moorings and rowed out into the deep water. Thus the Arabs were left on shore in the harbor, penned in by deep water and mud. The inhabitants then grouped together against them, surrounding them from all sides, fell upon them with slings, stones and arrows and killed them all. Their companions stood watching from the opposite shore, unable to come to their aid. The Arabs have not attacked that lake again up to the present day."

While this never amounted to a siege, the narrative is interesting since it shows Arab troops at work improvising rafts and boats, demonstrating their willingness to do whatever task necessary to achieve their objectives.

AMORION   *yab melltā*, ܝܒ ܡܠܬܐ   664
Continued from *Anatolian campaign.

"Ibn Khālid then set off from there and came to the city of Amorium and gave it the word. When they opened (their gates) to him he stationed an Arab garrison there and left that place." (*textus* 73.21ff:)

ܘܐܪܠ ܨܘܪܒܐ ܕܡܘܪܝܢ ܟܠܗܐ ܠܡܕܝܢܬܐ ܘܐܬܐ ܡܢ ܬܡܢ ܫܩܠ
ܗܟܢ ܡܢ ܘܥܡܕ ܒܗ̇ ܛܝܝܐ ܡܛܪܬܐ ܒܗ̇ ܘܐܩܝܡ.

The garrison remained at *Amorion until c. 666, when the Romans recaptured the fortress (cf. below and Appendix III).

SYLWS (SAGALASSOS?)   *kābšā*, ܟܒܫܐ   664
Continued from *Anatolian campaign; chapter 8 *passim*.

"He then came to the great fortress of SYLWS (possibly Sagalassos; cf. Palmer's suggestion for *Pergamon based on *pyrymws* below), because a master carpenter from Paphlagonia had played a trick on him. This man had said to him 'If you give me and my household your word (that our lives will be spared), I will make you a catapult (*manganīqā*) capable of taking this fortress.' Ibn Khālid gave him (his word) and gave orders for some long logs [*bld'*] to be brought; and so he made a catapult (*manganīqā*) such as they had never seen before. They went up and installed it opposite the gateway of the fortress."

"The men defending the fortress, trusting to its impregnability, let them get quite close. Ibn Khālid's men then drew back their catapult; a rock rose up in the air and hit the gate of the fortress. They then shot another rock and it fell a little short; then they shot a third rock, which fell shorter than the other two. The men above jeered and cried out, 'Pull your weight, Khālid's men, you are drawing badly.' They wasted no time in using their own catapult to propel a huge rock down onto Ibn Khālid's catapult from above, hitting it and wrecking it. In the process of rolling away, the boulder killed a large number of men." (*textus* 73.23-74.11:)

ܘܐܙܠ ܥܠ ܦܠܘܓܐ ܚܣܢܐ ܙܒܢܐ. ܕܐܝܬܝܗ̇ܝ܆ ܟܕ ܚܙܐ ܚ ܙܒܢܐ ܡܢ ܐܬܪܐ
ܗܘ ܕܦܩܠܘ ܠܚܣܢܐ ܘܐܡܪ ܠܗ. ܕܐܢ ܢܗܒ ܐܢܬ ܠܝ ܡܠܬܐ ܘܠܚܣܢܝ܆ ܐܢܐ ܐܚܪܒ
ܐܢܐ ܠܝ ܡܢܓܢܝܩܐ ܕܟܬܝܒܐ ܠܗ ܠܗܘܢܐ ܚܣܢܐ. ܘܗܘ ܝܗܒ ܠܗ. ܘܦܩܕ ܒܪ ܚܠܕ
ܘܐܬܬܘ ܩܠܝܐ ܐܘܪܚܐ. ܘܥܒܕ ܡܢܓܢܝܩܐ. ܕܠܐ ܝܕܥ ܐܟܣܬܗ. ܘܩܠܝܘ ܘܡܟܒܫܗ
ܘܫܘ ܠܗܘܟܠ ܦܘܪܓܐ ܕܚܣܢܐ.
ܘܡܪܘܡܗ ܕܚܣܢܐ. ܟܕ ܕܐܬܟܠܝܢ ܗܘܘ ܥܠ ܡܫܪܘܬܗ. ܫܒܩܘ ܠܩܠܝܢ ܕܢܦܪܚܘܢ
ܠܘܬ ܚܣܢܐ. ܘܟܕ ܐܬܬܠܝܘ ܕܒܝܬ ܟܠܝܕ ܒܡܢܓܢܝܩܐ ܕܝܠܗܘܢ. ܣܠܩܬ ܟܐܦܐ
ܘܢܩܫܬ ܠܘܬ ܬܪܥܗ ܕܚܣܢܐ. ܘܬܘܒ ܐܬܬܠܝܘ ܟܐܦܐ ܢܦܠܬ ܘܩܪܬ ܥܠܠ. ܘܬܘܒ
ܐܬܬܠܝܘ ܟܐܦܐ ܕܬܠܬ ܘܩܪܬ ܡܢ ܣܘܪܐ. ܘܗܘܘ ܕܠܠ ܡܗܠܠܬܬ ܡܚܘ ܟܕ
ܐܬܬܠܝܘ ܕܒܝܬ ܟܠܝܕ. ܒܝܫܐܝܬ ܓܝܪ ܡܬܬܠܝܢ ܐܢܬܘܢ. ܘܡܚܝܕܐ ܐܬܬܠܝܘ
ܗܢܘܢ ܕܡܢܓܢܝܩܐ ܡܢ ܠܥܠ ܟܐܦܐ ܪܒܬܐ. ܘܡܚܬܗ ܘܬܒܪܬܗ ܠܡܢܓܢܝܩܐ ܕܒܪ
ܟܠܝܕ. ܘܟܕ ܡܬܓܢܓܪܐ ܠܗ ܗܘܬ ܩܛܠܬ. ܘܐܝܟܢܐ ܩܛܠܬ ܓܒܪܐ ܣܓܝܐܐ ܡܠܬ.

The standard translation obscures the mechanics of the *manganīqā*, since it makes it appear like a classical *onager*, or one-arm catapult (see chapter 5.2.2) whose arm was drawn back and then released by a trigger mechanism. The key phrase "*w-kad ettelīw d-bēt kālīd b-manganīqā dīlhōn, selqat kēpā w-neqšat lwāt tar'eh d-ḥesnā*" should be translated as "and *when* the men of Khālid pulled at [the ropes of] the *manganīqā* that belonged to them, a stone went up and landed before the gate of the fort." The conjunction *kad* (when), omitted in Palmer's translation, thus indicates a causal connection between these two movements; the pulling at one end of the machine resulted in the stone being shot out. This is confirmed by the Roman troops jeering at the Arabs, who shout that their *pulling* is poor: "*ettelaw d-bēt kālīd, bīšā'it gēr mettlīn-tōn*" is in a slightly more literal rendering "pull men of Khālid, for you are pulling badly," referring to the pulling of the trebuchet ropes, which required a good deal of training (cf. chapter 8.1 *passim*). The large number of men killed when the Roman stone hit the Arab engine also indicates the presence of a substantial pulling crew required for the trebuchet.

PESSINOS *kbaš*, ܟܒܫ 664

Continued from *Anatolian campaign.

"Ibn Khālid went on from there and took the fortresses of Pessinos, Kios and Pergamon, and also took the city of Smyrna." *Textus* 74: 12f:

ܘܐܙܠ ܡܢ ܬܡܢ ܒܪ ܟܠܝܕ. ܘܟܒܫ ܠܚܣܢܐ ܦܣܝܢܘܣ. ܘܠܚܣܢܐ ܟܝܘܣ. ܘܠܚܣܢܐ
ܕܦܪܓܡܘܣ. ܘܐܦ ܠܡܕܝܢܬܐ ܙܡܘܪܢܐ.

Kios 664
See *Pessinos above.

Pergamon 664
See *Pessinos above.
Pergamon for *pyrymws* is the conjecture of Palmer.

Smyrna 664
See *Pessinos above.

Amorion 666
Theophanes 351 (AM 6159, AD 666/7, MS 490); Lilie 1976: 72ff; Howard-Johnston 2010: 490; Appendix III.

At about the same time as the revolt of Saporios, an Arab invasion (which may have been called in by those who raised the revolt), was said to have reached Chalcedon. "They also took Amorion in Phrygia and, after leaving there a guard of 5000 armed men, returned to Syria. When winter had fallen, the emperor sent the same *cubicularius* Andrew, and he reached Amorion at night when there was much snow. He and his men climbed on the wall with the help of planks and entered Amorion. They killed all the Arabs, all 5,000 of them, and not one was left." ...παρέλαβον δὲ καὶ τὸ Ἀμώριον τῆς Φρυγίας ε΄ χιλιάδας ἐνόπλων ἀνδρῶν ἀφέντες εἰς φυλακὴν αὐτοῦ, καὶ ἀνέκαμψαν εἰς Συρίαν. χειμῶνος δὲ γενομένου πέμπει ὁ βασιλεὺς τὸν αὐτὸν Ἀνδρέαν τὸν κουβικουλάριον, καὶ χιόνος πολλῆς οὔσης, ἐν νυκτὶ καταλαμβάνει· καὶ διὰ ξύλων ἀνέρχονται ἐπὶ τὸ τεῖχος καὶ εἰσέρχονται εἰς τὸ Ἀμώριον· καὶ πάντας κτείνουσι τοὺς Ἄραβας, τὰς ε΄ χιλιάδας, καὶ οὐχ ὑπελείφθη ἐξ αὐτῶν οὐδὲ εἷς.

Following the chronology of the Arab sources, Lilie and Howard-Johnston place this in 668/9. See App. III for an alternative argument.

?Nimis *inruentes* c. 670
HL 5.22.

The son of Lupus, Arnefrit, tried to regain the dukedom with Slav assistance, but "he was killed when the Friulians attacked him at the fortress of Nimis." || ...*aput Nemas castrum...inruentes super se Foroiulanis, extinctus est.* ||

The language is ambiguous, so it might have been a battle in the immediate vicinity of the *castrum*, but see *Cividale and the discussion of *Forlimpopoli (below).

| CIVIDALE | *voluerunt super ... castrum inruere* | c. 670 |

HL 5.23.

The Slavs were preparing to attack Cividale (Forum Julii) and camped nearby (*voluerunt super Foroiulanum castrum inruere; et venientes castrametati sunt ... non longe a Foroiuli*), but they were chased off by the arrival of duke Wechtari, who with a 25 followers allegedly defeated an army of 5,000.

Here Paul might deliberately be conflating the initial negotiations, when Wechtari first revealed himself with a few men, with the actual battle that followed. The number of Slavs is however realistic.

| ?FORLIMPOPOLI | *super eandem civitatem ... inopinate inruit* | c. 670 |

HL 5.27.

Grimuald surprised the population of Forlimpopoli (Forum Populi) on Easter Sunday (*super eandem civitatem ... inopinate inruit*), killing even the deacons who were baptizing infants.

As with *Nimis (above), it is difficult to deduce exactly what happened. It may have been a surprise raid on services held at extramural churches, but the phrasing *super ... inruit* is the same as *Cividale (above), which the Slavs certainly intended to assault.

| ?CONSTANTINOPLE | συμβολὴ πολέμου ἐκροτεῖτο | 670ff |

Theophanes 353f (AM 6164, 6165; MS 493f); Nik. 34; Lilie 1976: 75-82; Ostrogorsky 1969: 124; Haldon 2006; Howard-Johnston 2010: 231; Hoyland 2011: 166ff; Jankowiak 2013. Cf. Sebeos 45 and Olster 1995.

Theophanes has a full narrative drawn from a Greek source, most likely a brief chronicle by the patrician Trajan (also quoted in abbreviated form by Nikephoros). This Theophanes inserted into a quite different framework provided by the Syriac Common Source, which only mentions Arab naval raids against southern Anatolia (see Hoyland and Howard-Johnson). Indeed, both Howard-Johnston and Jankowiak argue that an actual siege of Constantinople at this time is a fiction produced by rather bold editing by Theophanes, who presents the following information:

"In this year [AM 6164] the deniers of Christ equipped a great fleet, and after they had sailed past Cilicia, Mouamed, son of Abdelas, wintered at Smyrna, while Kaisos wintered in Cilicia and Lycia. A plague occurred in Egypt. The emir Chale was also sent to assist them inasmuch as he was a competent and bold warrior. The aforesaid Constantine, on being informed of so great an expedition of God's enemies against Constantinople, built large biremes bearing cauldrons of fire and *dromones* equipped with siphons, and ordered them to be stationed at the Proclinesian harbour of Caesarius."

"In this year [6165] the aforesaid fleet of God's enemies set sail and came to anchor in the region of Thrace, b[e]tween the western point of the Hebdomon, that is Magnaura, as it is called, and the eastern promontory, named Kyklobion. Every day there was a military engagement (συμβολὴ πολέμου ἐκροτεῖτο) from morning until evening, between the *brachialion* of the Golden Gate and the Kyklobion, with thrust and counter-thrust. The enemy kept this up from the month of April until September. Then, turning back, they went to Kyzikos, which they captured (παραλαβόντες), and wintered there. And in the spring they set out and, in similar fashion, made war on sea against the Christians. After doing the same for seven years and being put to shame with the help of God and His Mother; having, furthermore, lost a multitude of warriors and had a great many wounded, they turned back with much sorrow. And as this fleet (which was to be sunk by God) put out to sea, it was overtaken by a wintry storm and the squalls of a hurricane in the area of Syllaion. It was dashed to pieces and perished entirely."

The naval expedition was apparently supposed to be coordinated with a land invasion, considering the following note in Theophanes: "Now Souphian, the younger son of Auph, joined battle with Florus, Petronas, and Cyprian, who were at the head of a Roman force, and 30,000 Arabs were killed." However, this probably happened later as the information, although in a more compressed form, with the following notice on Kallinikos, is from the Syriac Common Source and thus probably unconnected with reports of a naval expedition to Constantinople.

"At the time Kallinikos, an architect from Helioupolis in Syria, took refuge with the Romans and manufactured a naval fire with which he kindled the ships of the Arabs and burnt them with their crews (τότε Καλλίνικος ἀρχιτέκτων ἀπὸ Ἡλιουπόλεως Συρίας προσφυγὼν τοῖς Ῥωμαίοις πῦρ θαλάσσιον κατασκευάσας τὰ τῶν Ἀράβων σκάφη ἐνέπρησε καὶ σύμψυχα κατέκαυσεν). In this way the Romans came back in victory and acquired the naval fire."

Mango (comm. 193f) shows that Nikephoros' version is compressed from the same Greek source as used by Theophanes but without additional information. It deserves to be quoted in full for what it neglects: "After him [Constans II], his son Constantine was invested with the imperial office. *Immediately after his accession* the leader of the Saracens built many ships and sent them against Byzantium under the command of Chaleb (as he was called in his their language), a man most loyal and experienced in war. On his arrival, he put in at a seaside suburb of Byzantium called Hebdomon. When he became aware of his <presence>, Constantine, too, deployed a great fleet against him. Many naval battles were fought every day between the two sides, the engagements starting in the spring and continuing until autumn. When the bad weather came, the fleet of the Saracens crossed over to Kyzikos and wintered there, and in the beginning of spring it returned thence and continued in the same fashion the war on sea. So the war lasted seven years, and finally the Saracen fleet met with no success; on the contrary, they lost many fighting men. Badly injured and grievously defeated, they set out on their homeward journey. And when they came to the region of Syllaion, they were overtaken by violent winds and a tempest at sea which destroyed their entire armament." The Arabs then sued for peace, as did the Avars and other western peoples.

Mango follows Ostrogorsky's traditional dating of a series of naval engagements to 674-78, with references to further studies of the siege in his commentary (194). However, Lilie argues that the Arabs occupied Kyzikos in the Sea of Marmara in the winter of 671-72, with systematic buildups of naval bases along the southern coast of Anatolia via Rhodes to Smyrna then and in the following year. What happened next is unclear; the first wintering was probably a reconnaissance in force (perhaps with some fighting), since the last major expedition in 654 (see *Constantinople above) had gone so horribly wrong. The naval engagements would thus have recommenced later, when the sea-lanes along the coast of Anatolia were safe, in 673/674 (see MS and Hoyland for references to the other CS derivates). They lasted until defeated in 678 (hence the "seven years"), when the 30-year peace was concluded.

Although extremely important for military history in general and the survival of the East Roman (now Byzantine) state in particular, there are several problems with the famous story of Kallinikos: Theophanes has already mentioned Greek fire in his previous entry, which seems unconnected with the Kallinikos story from the Syriac Common Source. Its earlier use is furthermore attested by Sebeos 45, a notice dated to 649 in Howard-Johnston's commentary (I: 111f, II: 259f). We can also note the fact that Nikephoros, who relied on the same Greek source as Theophanes, does not mention Greek fire at all, which makes Theophanes' rendering even more complex. Furthermore, several of the important named, Arab commanders cannot easily be identified in the Arabic materials at that particular time, although Hoyland does provide suggestions for Mouamed and Kaisos, who may be equated with figures involved in raids just before and after 670 according to Arabic sources. However, if Khale[b] is indeed 'Abd al-Rahman ibn Khalid, as Hoyland also suggests, the events must have taken place prior to 667, when he was poisoned by enemies at court.

It is thus likely that significant elements of the siege story, including naval engagements outside the city walls, Greek fire, and the hurricane anecdote, have been transposed from another context in the original Greek source. The events leading up to and including *Constantinople 654 (Howard-Johnston), the mysterious Arab raid in Thrace right up to the walls of *Constantinople in 663, as well as the grand expedition of ibn Khalid in 664 (see Appendix III) all spring to mind. According to this interpretation, the successes of Constans II, who is explicitly lambasted in Theophanes (351; MS 490f) for his persecution of the orthodox (Chalcedonians), were transposed to the reign of his son Constantine IV because of his orthodoxy. This may indeed have been quite easily accomplished, as Constantine was junior emperor while Constans was away in Italy after 663.

Similarly, Kallinikos might have fled at a much earlier date; if indeed he was the inventor of naval fire, before 649 (cf. Sebeos); otherwise (e.g. if he helped refine the process), he may have taken part in the *Tripoli incident in 653, or fled at some other point during the Arab civil war; or at the latest in the 660s, when a number of craftsmen were moved by Arab authorities from inner Syria to coastal cities (see chapter 7.3). The Syriac Common Source does not provide any dates for this event, only a context of Arab naval raids against the Anatolian coast datable to the 670s, but such an advanced innovation must have needed some experimentation, refinement and development of infrastructure for such effective use (see Haldon for the infrastructure and technological competence needed to use Greek fire). We must therefore date the flight of Kallinikos (or just as likely, several anonymous models) to sometime in the late 650s or early 660s. None of this precludes the use of Greek fire during a possible siege or naval engagement in the 650s, 660s or 670s; in fact, its use on several occasions may have eased the transposition since events were so similar that additional material could be introduced seamlessly.

?NARBONNE   *aditum ... intercludere nisus est, subito ... ingrediens*   673
HW 7; Thompson 1969: 219ff; Collins 2002: 94.

The usurper Paul, supported by "Ranosindus, military head of the province of Tarraco, and Hildigisus, who held the rank of *gardingus*" prepared the revolt by mobilizing troops for an expedition against rebels in Gaul. "Argebadus, bishop of the see of Narbonne ... on finding out the truth from the most accurate report of certain men, attempted to block the road to the city (*aditum ... intercludere nisus est*) before the usurper. But this project too did not remain hidden from Paul. Before the prelate could carry out his plan, Paul suddenly entered Narbonne at great speed and with an army (*subito praepropero cursu Paulus cum exercitu Narbonensem urbem ingrediens*), frustrated promptly the plans made against him, and had the city gates closed by an appointed garrison of armed men (*sub delegato armatorum praesidio*)." Paul then publicly rebuked the bishop in front of the army.

?Barcelona	*in potestate ... adducitur*	673

HW 11; Thompson 1969: 221; Collins 2002: 94.

Wamba did not immediately rush to face the rebellion; he at first completed his week-long campaign against the Basques, raiding their territory until they submitted and provided hostages (*HW* 10)—this was probably necessary in order to prevent them from assisting Paul, who had called in their support. It may also have given Wamba time to mobilize troops on his route of march through Tarraconensis: "Barcelona, of all the cities involved in the rebellion, is first brought back into the king's authority ..." || Prima enim ex rebellione omnium civitatum Barcinona in potestate principis religiosi adducitur ... ||

Nothing further is reported about fighting; as with *Gerona (below), Paul may have instructed the local commanders to submit to whoever arrived first.

?Gerona	*subicitur*	673

HW 11; Thompson 1969: 221; Collins 2002: 94.

"... and after it Gerona is taken." || ...deinde Gerunda subicitur. ||

Paul had instructed the bishop Amator to submit to whoever of the contenders arrived first; most of the garrison troops were probably sent to Gaul in anticipation of Frankish reinforcements, and not expecting Wamba to attack before a full mobilization, which would have taken much longer to organize.

Collioure	*cepit atque perdomuit*	673

HW 11; Thompson 1969: 221f; Collins 2002: 94.

After *Gerona, Wamba's army split "into three divisions on the slopes of the Pyrinees, and so captured and subdued (*cepit atque perdomuit*) by an extraordinary victory the Pyrinean strongholds (*castra*) known as Colloure,

Ultrère, and Llivia, finding in these fortresses many objects of gold and silver that he handed over as booty to his teeming forces."

| ULTRÈRE | *cepit atque perdomuit* | 673 |

See *Collioure above.

| LLIVIA | *cepit atque perdomuit* | 673 |

See *Collioure above.

| CLAUSURAE | *inruptio facta est* | 673 |

As *Collioure above.

After capturing the *Collioure and the other Pyrenean strongholds, Wamba "... then sent out his troops before him, and led by two generals they took by storm (*inruptio facta est*) the bastion named Clausurae. There Ranosindus, Hildigisus, and a flock of other traitors, who had gathered to defend this fortress, were captured, and in this condition, with their hands tied behind their backs, were brought to the king."

| CERDANE | *inrupisse* | 673 |

HW 11; Thompson 1969: 221.

"Witimirus, however, one of the plotters, who was stationed in Cerdane and had entrenched himself there, seeing that our men were about to break in, fled forthwith (*nostros inrupisse persentiens, statim aufugiit*) and was able to reach Paul in Narbonne to bring him news of the great defeat. This report made the usurper very fearful. The pious king, however, having vanquished the armies of these fortresses, came down to the plains after crossing the Pyrenees and waited two days for his army to come together."

In this and the four preceding entries, Julian, the author of *HW*, has misunderstood the geography; it is tacitly corrected by Thompson, but the translator (Martínez Pizarro 2005) makes a rather larger issue out of it.

| NARBONNE | *ad expugnationem, ingrediuntur* | 673 |

HW 12; Thompson 1969: 222f, 266f; Martínez Pizarro 2005: 198 n. 74; chapter 3.2.3 above.

"When the multitude of troops gathered from various parts had assembled as one army, no delay was made, for he immediately sent out ahead of him a chosen band of warriors led by four generals to storm Narbonne, directing another army to carry out naval warfare there. Indeed, few days had passed since the rebel Paul fled form Narbonne like a slave, having found out with what success the king's party was advancing. Paul, claiming legal authority over the city, hedged it with an abundant garrison of traitors and gave military powers over it to his general Witimirus. He, mildly exhorted by our forces to hand over the city without shedding of blood, refused absolutely, and after having the gates of the city closed, cursed the pious king's army from the walls. He multiplies maledictions against the ruler himself, and tries to drive away the army by means of threats. The mass of our troops could not suffer this; tempers instantly became heated, and they aimed javelins at the rebels' faces. What use are many words? A savage battle is joined on either side, and the two parties confront each other by exchanging shafts. But when our men grew desperate, not only did they transfix with arrows the rebel soldiers fighting on the walls, but they hurled such showers of rocks into the city that from the clamor of voices and the noise of the stones the city itself would have appeared to be collapsing. From the fifth to the eighth hour of that same day, both sides struggled ferociously. Then the spirits of our warriors were ignited; they could not bear to have victory deferred, so they moved in closer in order to fight at the gates. Assisted by God, they set fire to the gates with victorious hand, leap over the walls, and enter the city as conquerors, compelling the rebels there to submit to them."

|| At ubi e diversis partibus collecta in unum exercituum multitude percrebuit, standi mora nulla fuit; sed statim per quatuor duces lectum numerum bellatorum ad expugnationem Narbonae ante faciem suam mittit, alium exercitum destinans, qui navali proelio bellaturus accederet. Et quidem iam erant parvi admodum dies, ex quo de Narbona rebellis Paulus serviliter fugiendo excesserat, comperto, quod tam feliciori proventu pars religiosi principis properaret. Quam civitatem Paulus ipse se iuris potestati adstipulans, multiplici perfidorum *praesidio* sepsit summamque proelii Wittimiro duci suo commisit. Quem quum nostrorum exercitus blanditer exortaret, ut civitatem sine sanguinis effusione contraderet, prorsus abnuit, obseratisque civitatis ipsius portis, e muro exercitum religiosi principis detestatur. In principem quoque ipsum maledicta congeminat minisque exercitum proturbare conatur. Quod nostrae partis multitudo non ferens, subita cordium accensione incanduit et telorum iactu perfidorum ora petivit. Quid multa? inmanis ab utrisque pugna conseritur, et vice sagittarum alternatim sibimet utraeque partes obsistunt. Sed ubi a nostris desperatum est, non solum in muro pugnantes seditiosos sagittis configunt, sed *tantos imbres lapidum intra urbem concutiunt*, ut clamore vocum et stridore petrarum civitas ipsa submergi aestimaretur. Unde *ab hora fere quinta diei*

usque ad horam ipsius diei octavam acriter ab utrisque pugnatum est. At ubi incalescunt nostrorum animi, victoriae dilationem ferre non potuerunt, sed ad portas propius pugnaturi accedunt. Tunc victoriosa per Deum manu portas incendunt, muris *insiliunt*, civitatem victores ingrediuntur, in qua sibimet seditiosos subiciunt. ||

Witimarus fled to a church, but was captured, taken away and flogged along with his troops.

Martínez Pizarro notes lack of late antique enceinte in the archaeological record. That does not necessarily mean much, as hardly *anything* Roman is found there.

| BÉZIERS | *subiugantur* | 673 |

HW 13.

Immediately after *Narbonne: "The cities of Béziers and Agde are quickly taken."

|| Deinde Beterris et Agate civitates illico subiugantur. ||

| AGDE | *subiugantur* | 673 |

See *Béziers above.

| MAGUELONE | *ad obsidionem* | 673 |

HW 13, following *Béziers and *Agde.

"In the city of Maguelone, however, Gumildus, the local bishop, watching the army that besieged him all around, noticed that the city was surrounded not only by those who had come over land to fight but also by those who had come over water to engage in naval combat; terrified by this form of warfare, he took a shortcut to Nîmes and joined there his associate Paul. Once the Spanish army learned that Gumildus had fled, it soon captured Maguelone by a victory not unlike the previous ones."

| NÎMES | *muros urbis petrarum ictibus petunt* | 673 |

HW 13-30; McCormick 1986: 297-327; de Jong 1999; Thompson 1969.

Four commanders (*duces*) were appointed to lead the vanguard against *Nîmes, which reached the city by a night march and began the assault at dawn of August 31. "When they are seen from the city, as they are few in number of warriors, the rebels make plans to intercept them with arms in the open fields. However, fearing devious stratagems, they choose to give battle from their walls (*de muris bellum conficere*), within their city, rather than expose themselves to the danger of unforeseeable chances outside, and to await the arrival of foreign armies sent to assist them. Yet, once the sun shone over the land, our forces join battle. The first phase of the struggle brings the ringing of trumpets and a shower of stones (*saxorum nimbo*). As soon as the trumpets are heard, our men gather from everywhere with a great roar of their voices and cast stones at the city walls (*muros urbis petrarum ictibus petunt*), which missiles, together with javelins and arrows, they aim at those standing on the walls (*misilibilibus* [sic] *quibusque constitutos per murum spiculis sagittisque propellunt*), just as they, resisting the onslaught, hurl missiles of all sorts back at our warriors (*cum tamen et illi in nostros ad resistendum multorum generum specula iacerent*). What shall I say? The fight grows more fierce on either side; the scales of battle lie even for both as they struggle, for indeed they are engaged in an equal contest. Neither our men nor theirs yield in the joined battle. So they fight all day under the double-edged blade of victory" (*HW* 13). A rebel commander taunted the besiegers and threatened them with the fate that would befall them when rebel reinforcements arrive, but only provoked a more fierce assault: "They draw close to the walls, fight more savagely than when they started, and take up the joined struggle with renewed ferocity" (*HW* 14).

Ending the fighting for the night, the besiegers sent word to Wamba calling for reinforcements; he sent 10,000 men in response under general Wandemirus. Marching at night, their arrival on September 1 strengthened the resolve of the besiegers, who feared that they would be unable to keep the enemy inside the city (*HW* 15). The rebels realized the scale of the reinforcements, and the rebel commander, Paul, surveyed their camp from "a prominent watchtower," recognizing the strategy of king Wamba. He argued to his followers that this was the whole army, but that Wamba had hidden his personal standards to give the impression that yet *another* army would soon approach (*HW* 16). The assault was renewed at the signal of trumpets. "But the enemy, finding their hopes of victory in walls rather than in courage, remain enclosed in the city and hurl shafts down from the walls (*intra urbem positi per murum specula iactant*), taking up once more the renewed struggle against our men." When the besiegers were gaining the upper hand, the morale of the rebels' foreign supporters began to waver (*HW* 17). The besiegers, on the other hand, were encouraged. "Burning with yet fiercer rage, until close to the fifth hour of the day they beat against the defenses of the city with continuous thrusts, cast showers of rocks with enormous noise, set fire to the gates, and break through the narrowest breaches of the wall." || Unde ferociori quam fuerant incensione commoti, usque in horam fere diei quintam cotinuis proeliorum ictibus moeniam civitatis inlidunt, imbres lapidum cum ingenti fragore dimittunt, subposito igne portas incendunt, murorum aditibus minutis inrumpunt. || With loyalists flooding the walls, the rebels withdrew to the amphitheater, "which is surrounded by a more massive wall and ancient defenses, and shut themselves in." The loyalists pursuing the rebels and (probably more preoccupied with) searching for plunder were ambushed and many were thus killed (*HW* 18).

The disastrous defeat led to infighting and despair among the rebels (*HW* 19f). They sent bishop Argebadus of Narbonne to implore for mercy. The bishop met Wamba outside the city, where Wamba assured him that the killing would stop, but did not give immunity against punishment. He ordered a halt in fighting until his arrival (*HW* 21f). Upon Wamba's triumphal entry (accompanied by angels, according to some observers), the rebels were pulled out of the amphitheater and Wamba gave prayers of thanksgiving for his victory (*HW* 23ff). On the same day (September 2), Wamba organized repairs of the breaches in the wall (*de reparatione inruptae urbis sollicitus statim murorum cava reformat*), replaced burnt gates (*incensas portas renovat*), and buried the dead (*insepultis tumulum praestat*). He also restored church vessels and other plunder to the locals and used treasury funds to repay them for damages (*HW* 26). The rebels were tried and sentenced to decalvation (either scalping or tonsuring) and ritually humiliated, and a Frankish army in the region of Béziers pursued, but apparently they had already given up and fled when they learned of the fate of the rebels (*HW* 27). Wamba then made a triumphal entry into Narbonne, discussed extensively with other ritual events by McCormick, and restored the status quo by disbanding the garrisons and driving away the Jews, before disbanding his whole army and returning to his capital for another triumph (*HW* 28ff).

AUTUN   *festinanter undique insistebant inrumpere civitatem*   c. 679
Passio Leudegarii; Gerberding and Fouracre 1996: 193-215 (comm.), 215-253 (translation).

The *Passio Leudegarii* has a scenario similar to that of *Lyons 662, but even more details on a rather hard-fought siege at *Autun, which may be connected to the repairs discussed in chapter 6.2.2. The text mentions the king's followers, *satellites* (ch. 11), as well as the importance of the king in raising an army (royal officers at various estates?) even at this late date (*multum colligerunt hostiliter populum*, c. 19). Ebroin's king was supported by bishops (who seem very eager to serve on military campaign) and nobles (c. 20).

The forces (*hoste*) were assembled in Burgundy(?) against Leudegar. He attempted to bribe the population, as his supporters were not confident of their support (c. 21): "his dependants and the clergy and his followers (*tam familiares quam clerici vel fideles*) were eager that he should carry off the treasures he had gathered together there and depart, so that when the enemy hear of this they might hold back from harrying and destroying the town, Leudegar would in no way agree to it." The commentary (p. 238 n. 176) emphasizes the outright attempt to purchase support from a reluctant population, as Leudegar doled out silver and treasure to both population and institutions.

Leudegar justified his actions and tried to bolster morale (c. 22): "'Let us therefore at one and the same time valiantly fortify our souls and the town's defences, lest either of the two kinds of enemy find a gap through which they could bring in danger.' He thus mobilized all the people of the town. They held a three-day fast, making the sign of the cross and carrying the relics of saints around the circuit of the walls, [n. 180 on possible shrinkage of city] stopping at the opening of each gate, where he bowed to the earth and tearfully beseeched the Lord that, if he should be called to suffering, God would not allow the people entrusted to him be led into captivity. And it has been proved that it came about thus." || Muniamus ergo virtutibus animam simul et civitatis custodiam, ne inveniant utrique hostes aditum, per quod inferre possunt periculum.' Commovens igitur universum urbis illius populum, cum triduano ieiunio, cum signo crucis et reliquias sanctorum murorum circumiens ambitum, per singulos etenim aditos portarum terrae adherens, Dominum praecabatur cum lacrimis, ut si illum vocabat ad passionem, plebem sibi creditam non permitterit captivari et ita praestatum est evenisse. || The description of both the ritual and stream of refugees militates against a shrunken perimeter, especially one only confined to the rather small *arx*. Even if part of the town center was uninhabited, the open spaces were useful to accommodate refugees, and the complex liturgical procession is more appropriate in the context of the circuit walls of a city.

The fighting is described in c. 23: "So, in fear of the enemy, the people from all around struggled to withdraw into the town, to close up the gateways with strong barricades and to fortify each position in turn. ... [He asks people for pardon in church] ... Not long after this the town was besieged by the enemy army, and that day until evening the people fought against it bravely. But then the forces of the enemy surrounded the town with a powerful blockade, and day and night they encircled it, howling like dogs.

The man of God considered the danger threatening his town and, calming the conflict before the walls, he went to his people and with these words implored them: 'I beg you, let us restrain ourselves from fighting these people. If they have come only to seek me, for my part I am prepared to grant their wish and quiet their fury. || Itaque cum ob metum hostium certatim populi undique se recipissent in urbe et meatus portarum forte obturassent serrate et super omnia stabilissint in ordine propugnacula, iussit vir Domini universos ingredi in ecclesia [...] Post haec nec diu vallatur civitas ab exercitu, eodemque die ab utraque populo fuit fortiter usque ad vespera demicata. Sed cum ab agmen hostium esset civitas obsidione valida circumdata et die noctuque vociferantes ut canes circuirent urbem, respiciens vir Domini civitatis inminere periculum, conpescuit omne supermurale conflictum et his verbis suum exortare adgressus est populum: 'Sinite, quaeso, contra hos pugnandum confligere. Si mei tantum causa huic isti advenerunt, de memet ipsum paratus sum eorum satisfacere voluntatem eorumque mitigare furorem. ||

Leudegar sent envoys to the besiegers, but enemy commander Diddo refused to withdraw until Leudegar submit. Leudegar refused and claimed he would rather die. "The enemy heard this, but they hurriedly pressed on with their attempt to break into the town on all sides, hurling darts and spreading fires." || Hostes vero, his auditis responsis, cum telorum iacula, cum incendia festinanter undique insistebant inrumpere civitatem. || (c. 24)

The language implies undermining and firing of the walls, and certainly indicates that there were exchanges of missiles and an attempt to storm the walls. This explains the phrase *supermurale conflictum*, which Gerberding and Fouracre found difficult to interpret. The royal forces clearly gained the upper hand in the fighting, because Leudegar subsequently gave himself up to be blinded and tortured. A new bishop was appointed, church treasure and citizens were despoiled as ransom, the booty was divided, but none were taken captive (cc. 25f).

The royal campaign continued against Lyons, but the bishop there, Desideratus, had better support from the population and allies, collecting a large force that prevented the royal troops from breaking into the city (*sed manu valida populi undique collecti urbem hanc maximam Deo praesule non permiserunt inrumpere*). In the meantime, Leudegar suffered martyrdom, and his brother, count of Paris, was also killed (cc. 27-35).

| Trent | *obsideret* | c. 680 |

HL 5.36.

Alahis, duke of Trent, raised a rebellion against king Perctarit. The king besieged (*obsideret*) him in Trent, but Alahis defeated Perctarit in a sally (*Alahis cum suis civitate egressus, regis castra protivit*) and forced him to flee.

| Bulgarian fortifications | προσεδρεύειν τῷ ὀχυρώματι | 680/81 |

Nik. 36; Beševliev 1981: 170-82.

The Bulgarian nation became a threat, "attempting to devastate by its incursions the neighboring places that were under Roman rule," so Constantine equipped a fleet, attacked the Bulgars with a "multitude of cavalry and ships", who fled to their fortifications (ὀχυρώματα) "and remained four days there." The Romans were unable to attack due to difficulty of terrain, so the Bulgars regained courage; Constantine went to Mesembria due to an attack of gout "after giving orders to the soldiers to keep on investing the fort and do whatever was necessary to oppose the nation." ... προστάξας τὰ τοῖς ἄρχουσι καὶ τοῖς λαοῖς προσεδρεύειν τῷ ὀχυρώματι καὶ ὅσα πρὸς ἄμυναν τοῦ ἔθνους κατεργάσασθαι.

However, the Roman troops fell victim to a rumor that Constantine had escaped, so they fled; the Bulgarians sallied out and pursued them, then crossed the Istros towards Varna to find that the country was good for settlement. "Furthermore, they subjugated the neighboring Slavonic tribes, some of which they directed to guard the area in the vicinity of the Avars and others to watch the Roman border. So, fortifying themselves and gaining in strength, they attempted to lay waste the villages and towns of Thrace." κρατοῦσι δὲ καὶ τῶν [ἐγγιζόντων] παρῳκημένων Σκλαβινῶν ἐθνῶν, καὶ οὓς μὲν τὰ πρὸς Ἀβάρους πλησιάζοντα φρουρεῖν, οὓς δὲ τὰ πρὸς Ῥωμαίους ἐγγίζοντα τηρεῖν ἐπιτάττουν. ἐν τούτους ὀχυρωθέντων καὶ αὐξηθέντων τὰ ἐπὶ Θράκης χωρία τε καὶ πολίσματα καταδῃοῦν ἐπεχείρουν. "Seeing this, the emperor was obliged to treat with them and pay them tribute."

See Theophanes 358.11-359.21 for further details. The expedition took place in 681, according to Mango; Beševliev has 680 and the Bulgar installation in Varna the next year.

| Thessalonica | δι' ἐμφυλίου πολέμου ἑλεῖν | c. 682 |

Miracula St. Demetrii 2.5 (283-306); Lemerle II: 137-62; Beševliev 1981: 159-72.

"We have previously spoken of the Sklavenes, namely Chatzon, and the Avars, how they ravaged almost the whole of Illyricum to wit the provinces of the two Pannonias, the two Dacias, Dardania, Mysia, Prevalitane, Rhodope, even Thrace and the region of the Long Walls besides Byzantium. The whole population was deported to the country near Pannonia, on the Danube, whose metropolis was once Sirmium, by the Chagan, who installed them as subjects (284)." Afterwards, they mixed (*mêlèrent*) with the Bulgarians, Avars, and other peoples, they had children with them, and became an innumerable people. But each child received from his father the traditions of his country and the *élan* of his race, according to Greek customs; and as the Hebrews in Egypt, through the orthodox faith and baptism, "and the people of the Christians increased in number;" and each speaking to the other of their ancestral homeland, they raised in their hearts the ardent desire to return (285). Sixty years later, most of them had become free, and the Chagan of the Avars, regarding them as a nation apart, set over them the Bulgar Kouber as chief (286). Kouber used their ardor to raise a revolt, including converts, and leave with arms and baggage (μετὰ καὶ τῆς αὐτῶν ἀποσκευῆς καὶ ὅπλων); the Chagan pursued them, but was defeated in five or six engagements (πολέμοις) before giving up (287). Victorious, Kouber crossed the Danube to "our" regions and settled with the people on the Keramasian plain. They wanted to return to their ancestral lands: Thessalonica, Constantinople, and the other cities of Thrace (288). However, Kouber advised them to remain as a block for negotiating with the emperor, so he could become their chief and Chagan himself. He sent an embassy to the emperor to ask permission to settle where they were and the Drougoubites to provide them with provisions (289).

Thus, the people of Kouber frequented the Slavs for provisions, learning they were not far from "our" city, so the Greeks, with their women and children, wanted to go there, whence the prefectural authority could send them on to Constantinople over sea (290). When Kouber understood this, he selected a particularly able archon fluent in "our language" (of Thessalonica, or Balkan Latin?), as well as Greek, Slavonic, and Bulgar. He was instructed to feign defection from Kouber, gain a position in the city, and then provoke a civil war that would allow Kouber to take the city (τὴν πόλιν δι' ἐμφυλίου πολέμου ἑλεῖν) and come and settle in it, and use it as base for expansion to the neighboring regions, the islands, Asia, and finally become emperor (291). The mole, named Mauros, arrived and gained support from the emperor, being appointed as head and *strategos* of all refugees from Kouber's camp (292). Anyone who knew of Mauros' intentions were killed and their children sold, so no-one revealed the plan (293). Mauros appointed subordinate officers (of hundred, fifty and ten men) who were his accomplices; his house was guarded night and day by soldiers supported by public funds. Their plan was to attack with combat-trained soldiers, burn important buildings, and make themselves masters of the city on the night of Holy Saturday when Thessalonica would be celebrating the resurrection (294).

The emperor by divine inspiration sends the strategos of the fleet, Sisinnios, in order to reinforce Mauros against Kouber with the soldiers of the Karabisians and encourage more to defect (295). Sisinnios discovered an abandoned church on the way, and celebrated the liturgy there (296). The army was ordered to celebrated the coming Friday; details of celebration and setting of camp with guards (297). St. Demetrius appeared to him twice, telling him to get moving, as the winds were favorable (298). The third appearance worked; Sisinnios gave orders to move despite the protestations of the sailors that the winds were contrary; instead, they feasted on the way (299).

The rapid passage of the fleet and its unexpected arrival ruined the plans of Mauros, who fell ill; Sisinnios, still ignorant, encouraged his recovery. Sisinnios gave the order that Mauros with all those under his authority along with the soldiers of the fleet should construct a camp (παραφοσσεύειν) west of the city, in order to aid those who wanted to escape from Kouber. This took several days (300ff). The emperor called Mauros and his men to Constantinople, where he was well received and given a command. However, Mauros was exposed by his own son, but the emperor decided not to punish him (304f). The miracle ends with thanks to God for his interventions and prayers (305f).

Beševliev dates these events to around 678, but Lemerle argues that they (possibly) took place late in the reign of Constantine IV, after Asparuch, said to have been Kouber's brother, led his Bulgars to the lower Danube in c. 680, suggesting the years 682-84. Mauros was found in Roman service later as well.

TARANTO	*expugnavit et cepit*	c. 685

HL 6.1.

"Romuald, duke of the Beneventines, after he had collected a great multitude of an army, attacked and captured Taranto and in like manner Brindisi and subjugated to his dominion all that very extensive region which surrounds them." || Romualdus Beneventanorum dux, congregata exercitus multitudine, Tarentum expugnavit et cepit, parique modo Brundisium et omnem illam quae in circuitu est latissimam regionem suae dicioni subiugavit. ||

690 CORPUS OBSIDIONUM

BRINDISI   *parique modo*   c. 685
See *Taranto above.

DAMASCUS   *šrā 'lēh*, ܫܪܐ ܠܗ; *melltā*, ܡܠܬܐ   c. 690
Dionysius 129ff (Palmer 1993: 200ff); cf. Hoyland 2011: 175-79, 182-85.

During the complex events of the second Arab civil war, 'Amr b. al-'Ās rebelled at Damascus. 'Abd al-Malik, who was campaigning in Mesopotamia, "returned to Damascus to besiege the city (*textus* 292.30 *šrā 'lēh*, ܫܪܐ ܠܗ) and launched a massive attack on it. 'Amr opened the gates on receiving 'Abd al-Malik's assurance (*melltā*) that he would not be harmed. But at a later stage 'Abd al-Malik did kill 'Amr by a trick."

MECCA   *aqīl 'lēh manganīqē*, ܡܓܢܝܩܐ ܠܗ ܐܩܝܠ   692
See *Damascus above for references and Robinson 2005a.

Subsequently, another rebel, ibn al-Zubayr (who had actually been recognized as Caliph in most Islamic provinces earlier in the civil war, cf. Robinson), was defeated by the combined armies of Hajjāj and Muhammad in the Hijaz. He sought refuge in the Ka'ba in Mecca: "Al-Hajjāj pursued them there and penned them up within the building, then used catapults (*textus* 293.21f *aqīl 'lēh manganīqē*, ܡܓܢܝܩܐ ܠܗ ܐܩܝܠ) to demolish the wall, thus enabling his men to rush in, take 'Abd Allāh b. al-Zubayr and kill him. They cut of his head and sent it to 'Abd al-Malik b. Marwān; then they rebuilt the sanctuary. After this 'Abd al-Malik made al-Hajjāj lord of al-Kūfa and of Yathrib, of Mecca and of all Iraq." In the aftermath of the war, Hajjāj exacted revenge in Iraq on both Arabs (i.e. Muslims) and Christians, killing and looting on a large scale, including impaling prominent Christian and Persian administrators at Nisibis and elsewhere, and burning Armenian leaders in a church, and murdering the administrator of Edessa. "Yet Christians still held office as scribes and leaders and administrators in the Arab territories."

?ANTIOCH   695
Chr. 819 (Palmer 1993: 78).

"AG 1006: The Roman armies went out to the vale of Antioch. They were met by Dīnār b. Dīnār who massacred them and only a few of them escaped, returning to the Roman Empire in disgrace. The same year the Romans rebelled against Justinian their king, cutting off his nose and banishing him. They brought Leontius out of prison and they made him their king."

Whether the objective of this expedition actually was Antioch is unclear; however, the Romans still had ambitions to recapture major cities such as *Carthage (697 below) and mobilized considerable resources for the purpose.

CARTHAGE (twice)   ἀνεσώσατο; παραλαμβάνει; *obsessam...cepit*   697f
Nik. 41; *HL* 6.10.

"At that time Carthage in Africa (which had previously been subject to the Romans) passed under the dominion of the Saracens who took it by war. On becoming aware of this, Leontios fitted out the entire Roman fleet (ἅπαντα τὰ Ῥωμαϊκὰ ἐξώπλισε πλόϊμα), to whose command he appointed the patrician John because of his experience in military matters and sent him against the Saracens at Carthage. Arriving there, he routed in battle the Saracens who were in Carthage and regained the city for the Romans (τοὺς μὲν ἐν αὐτῇ τῶν Σαρακηνῶν πολέμῳ ἐτροπώσατο, τὴν δὲ πόλιν Ῥωμαίοις ἀνεσώσατο). He also delivered from the dominion of that nation all the other towns that are there, left soldiers to guard them, and wintered in those parts (καὶ τἆλλα πάντα τὰ ἐκεῖσε πολίσματα τῆς τοῦ ἔθνους ἀπαλλάξας ἐξουσίας καὶ στρατὸν ὁπλίτην πρὸς φυλακὴν ἐν αὐτοῖς ἐγκαταλείψας αὐτοῦ διεχείμαζεν). When the king of the Saracens had heard of this, he mounted a stronger campaign against him, by which means he drove out John together with the Roman fleet and reoccupied (παραλαμβάνει) Carthage and all the towns round about."

HL is only interesting for its treatment of Arab-Byzantine warfare: "Then the race of the Saracens, unbelieving and hateful to God, proceeded from Egypt into Africa with a great multitude, took Carthage by siege and when it was taken, cruelly laid it waste and leveled it to the ground." || Tunc Sarracinorum gens infidelis et Deo inimica ex Aegypto in Africam cum nimia multitudine pergens, obsessam Cartaginem cepit captamque crudeliter depopulata est et ad solum usque prostravit. ||

CONSTANTINOPLE   δι' ἐπιβουλῆς παρέδωκαν   698
Theophanes 370f (AM 6190, MS 517); Nik. 41.

Theophanes: "Now, as Leontius was cleansing the Neorion harbour at Constantinople, a bubonic plague fell upon the City and, in the course of four months, killed a multitude of people. Apsimaros arrived with his fleet and anchored opposite the City at Sykai. For some of the people of the City did not wish to betray Leontios, but a betrayal was made through the single wall of the Blachernai by the provincial commanders who, under terrible oaths, had been entrusted over the altar table with the keys of the Land Walls: it was they who treacherously surrendered the City." τοῦ δὲ Λεοντίου ἐν Κωνσταντινουπόλει τὸν Νεωρήσιον λιμένα ἐκκαθαίροντος, ἡ τοῦ βουβῶνος λύμη ἐνέσκηψε τῇ πόλει, καὶ πλῆθος λαοῦ ἐν τέσσαρσι μησὶ διέφθειρε. καταλαμβάνει δὲ Ἀψίμαρος ἅμα τῷ συνόντι αὐτῷ στόλῳ, καὶ προσώρμησεν ἀντικρὺ τῆς πόλεως ἐν Συκαῖς. ἐπὶ χρόνον δέ τινα τῆς πόλεως παραδοῦναι Λεόντιον μὴ βουλομένης, προδοσία γέγονε διὰ τοῦ μονοτείχους Βλαχερνῶν καὶ ὑπὸ ἐξωτικῶν ἀρχόντων τῶν τὰς κλεῖς τοῦ χερσαίου τείχους μεθ' ὅρκου φρικτοῦ ἐκ τῆς ἁγίας τραπέζης ἐμπιστευθέντων· οὗτοι δι' ἐπιβουλῆς παρέδωκαν τὴν πόλιν.

According to Nikephoros, the fleet revolted when it reached Crete and appointed Apsimaros, a *droungarios*, emperor; it put in at Sykai opposite the City. Plague kills many in the City over four months. "For some time he joined battle with the inhabitants of the City, which he finally took by deceit after bribing the guards of the Blachernai walls and their officers." His army entered and confiscated "the monies of the citizens"; Leontios has his nose cut and is sent to a monastery.

THE 8TH CENTURY

| LODI | *expugnata...capta* | c. 701 |

HL 6.20.

King Aripert engaged his rival Rotharit who had begun a revolt: "... having first attacked and captured Lodi..." || ...expugnata primum et capta Laude... ||

| BERGAMO | *expugnans mox cepit* | c. 701 |

As *Lodi above.

After *Lodi, Aripert then "... besieged Bergamo, and storming it without any difficulty with battering rams and other machines of war, presently took it." || ...Bergamum obsedit, eamque cum arietibus et diversis belli machinis [id est cum manculis *add*. L 1] sine aliqua difficultate expugnans mox cepit... ||

In the process he also captured Rotharit, who was later killed.

| BENEVENTAN CAMPAIGN | *cepit* | c. 702 |

HL 6.27.

The Beneventan ruler Gisulf captured the cities Sora, Arpino and Arce "in the same way" (*pari modo cepit*) and raided Campania. Pope John VI (701-704) ransomed the captives.

| TARANTON | πολιορκήσας | 702 |

Theophanes 372 (AM 6193, MS 519); Dionysius 140 (Palmer 1993: 206 and n. 516, AG 1013); Lilie 1976: 114; Hoyland 2011: 104f.

Theophanes: "In this year Abdelas made an expedition against the Roman country. He besieged Taranton to no avail and returned home. He built up Mopsuestia and placed a guard therein." Τούτῳ τῷ ἔτει ἐπεστράτευσεν Ἀβδελᾶς Ῥωμανίαν, καὶ πολιορκήσας Τάραντον καὶ μηδὲν ἀνύσας ὑπέστρεψεν καὶ ᾠκοδόμησε τὴν Μοψουεστίαν καὶ ἔθετο ἐν αὐτῇ φύλακας.

Dionysius only provides the context and does not mention the siege, presumably because it was a failure, and may have had severe consequences for Roman client politics: "'Abd Allāh, the son of 'Abd al-Malik, carried out a lucrative raid on Roman territory and returned to Cilicia. He rebuilt Mopsuestia and garrisoned it with enough troops to protect it, before returning (to Damascus). The following year the Armenian leaders organized a revolt against the Arabs. Muhammad b. Marwān went up and crushed the Romans who had come to Armenia and he also killed many Armenians. Then Armenia reverted to Arab control." Michael the Syrian adds: "The Arabs rebuilt Mopsuestia, which they had occupied a short time before. They gave it very strong walls and other beautiful buildings and they put a garrison in it and made it their frontier with the Romans."

?Samosata	c. 705

Dionysius 142 (Palmer 1993: 208 n. 518); Hoyland 2012: 194.

Several *Common Source* derivates have a notice on a Roman raid sent by Tiberius Apsimar to Samosata, killing 5,000 Arabs and taking booty and 1,000 prisoners. The nature of the conflict is unknown and dating uncertain; possibly it took place before *Taranton 702.

Constantinople	προσεδρεύει δὲ τῷ τείχει; παραλαμβάνει	705

Nik. 42; cf. Hoyland 2011: 196-99.

Justinian II received help from the Bulgar khan Tervel to regain his crown: "He then armed his whole people and proceeded to the Imperial City along with <Justinian>. For three days <Justinian> encamped by the walls (προσεδρεύει δὲ τῷ τείχει) of Blachernai and demanded the inhabitants of the City to receive him as emperor; but they dismissed him with foul insults. However, he crept with a few men at night into the aqueduct of the City and in this way captured (παραλαμβάνει) Constantinople." He hung several representatives of the former regime.

Nikephoros adds some materials to Theoph. 374 (AM 6197, MS 522), but the differences are minor.

Anchialos	προσεδρεύουσιν	708

Theophanes 376 (AM 6200, MS 525); Nik. 43.

Theophanes: "In this year Justinian broke the peace between the Romans and the Bulgars and, ferrying the cavalry *themata* across to Thrace and fitting out a fleet, set out against Terbelis and his Bulgars. When he had reached Anchialos, he anchored his fleet in front of the fortress and commanded that the cavalry should encamp in the plains above, without guard or any suspicion. As the army scattered in the fields like sheep to collect hay, the Bulgarian spies saw from the mountains the senseless disposition of the Romans. Gathering together like wild beasts, they suddenly attacked and inflicted great losses on the Roman flock, taking many captives, horses, and arms in addition to those they killed. As for Justinian, he sought refuge in the fortress with the survivors and for three days kept the gates shut. On seeing the perseverance of the Bulgars, he was the first to cut the sinews of his horse and ordered the others to do the same. After setting up trophies on the walls, he embarked at night and stealthily sailed away, and so reached the City in shame."

Nikephoros: "Thereafter he broke the peace with the Bulgarians and, after conveying a great army to Thrace by land and by sea, proceeded to the city of Anchialos with a view to fighting them. While his men scattered heedlessly to gather hay, the Bulgarians fell suddenly upon them, killing many of them and taking many prisoners. Then for three days they besieged (προσεδρεύουσιν) Justinian, who had remained in the city of Anchialos. Embarking on a ship at night, he fled from there and returned to Byzantium."

Tyana	πολιορκοῦντες παρεχείμασαν	c. 708f

Theophanes 376f (AM 6201, MS 525f); Nik. 44; Dionysius 144 (Palmer 1993: 208, AG 1019); Lilie 1976: 116ff; Howard-Johnston 258, 260f, 511f; Hoyland 2011: 201.

Theophanes: "In this year Masalmas and Abas made an expedition against Tyana, incensed as they were on account of Maiouma's army that had been slain by Marianos; and, after laying siege to the town, they wintered there. The emperor sent against them two generals, namely Theodore Karteroukas and Theophylaktos Salibas with an army and a throng of peasant militia so as to fight and expel them. Rent by mutual dissensions, they made a disorderly attack and were routed; many thousands perished and many more were taken captive. The Arabs seized the camp equipment and the provisions and continued the siege until they had taken the city: for they had been short of food and were on the point of departing. On seeing this, the inhabitants of Tyana gave up hope. They accepted a promise of immunity and came out to the Arabs, leaving the city deserted until this very day. The Arabs did not keep their promise and drove some of them into the desert, keeping many others as slaves."

Τούτῳ τῷ ἔτει ἐπεστράτευσεν Μασαλμᾶς καὶ Ἄβας τὴν Τυάνον διὰ τὴν μανίαν τοῦ ἀποκτανθέντος στρατοῦ σὺν τῷ Μαϊουμᾶ ὑπὸ Μαριανοῦ, καὶ ταύτην πολιορκοῦντες παρεχείμασαν ἐκεῖ. καὶ ἀποστέλλει πρὸς αὐτοὺς ὁ βασιλεὺς δύο στρατηγοὺς, Θεόδωρον τὸν Καρτερούκαν καὶ Θεοφύλακτον τὸν Σαλιβᾶν μετὰ στρατοῦ καὶ γεωργικοῦ λαοῦ χωρικοβοηθείας πρὸς τὸ πολεμῆσαι καὶ ἐκδιῶξαι αὐτούς. αὐτοὶ δὲ εἰς ἔριν ἀλλήλων ἐλθόντες, καὶ ἀτάκτως συμβαλόντες αὐτοῖς τρέπονται, καὶ πολλαὶ χιλιάδες ἀπώλοντο, καὶ ἠχμαλωτεύθησαν πολλοί. λαβόντες δὲ τὸ τοῦλδον καὶ τὰ τούτων βρώματα παρεκάθισαν, ἕως οὗ παρέλαβον τὴν πόλιν. λειφθέντες γὰρ ἦσαν τὰ βρώματα, καὶ ἤμελλον ἀναχωρεῖν. οἱ δὲ τῆς πόλεως Τυάνων ταῦτα ἰδόντες καὶ ἀπογνόντες ἔλαβον λόγον τῆς ἑαυτῶν ἀπαθείας καὶ ἐξῆλθον πρὸς αὐτοὺς καταλιπόντες τὴν πόλιν ἔρημον ἕως τοῦ νῦν. οἱ δὲ τὸν λόγον μὴ φυλάξαντες τούτους εἰς τὴν ἔρημον ἐξώρισαν, καὶ πολλοὺς δούλους ἐκράτησαν.

Nikephoros: "At this juncture the king of the Saracens sent a large army under the command of Masalmas and Solymas (as they are called in their language) to besiege the city of Tyana. When they arrived there, they fought many encounters and threw down part of the walls by means of their siege engines, but were unable to achieve any further success and were intending to return home. Thereupon Justinian sent emissaries into the interior and, after collecting a numerous body of peasants and farmers, ordered them to proceed to Tyana and relieve the besieged. On seeing them unarmed, the Saracens rushed against them, and some of them were put to the sword, while others they captured. Thus emboldened, they pursued the siege of Tyana. Renouncing to do battle because of the lack of necessities and deprived of the emperor's help, <the inhabitants> surrendered to the enemy under treaty and departed to the country of the Saracens." Arabs proceed to raid all the way to Chrysopolis, massacre locals, and burn ferry boats.

Ἐν τούτοις ὄντων τῶν πραγμάτων ὁ τῶν Σαρακηνῶν βασιλεὺς λαὸν πλεῖστον ὁπλίτην ἐκπέμπει, ἡγεμόνας αὐτοῖς ἐπιστήσας Μασαλμᾶν καὶ Σολυμᾶν κατὰ τὴν αὐτῶν διάλεκτον καλουμένους, ὡς τὰ Τύανα τὴν πόλιν πολιορκήσοντας. οἱ δὲ ἐκεῖσε παραγενόμενοι, πολέμους πλείστους συνάψαντες, μέρος δὲ καὶ τοῦ τείχους ἐκ τῶν πρὸς τειχομαχίαν ὀργάνων καταβαλόντες καὶ πλέον οὐδὲν ἀνύσαι ἰσχύσαντες, ἀποχωρεῖν πρὸς τὰ οἰκεῖα ἐβούλοντο. ἐν ᾧ ἐκπέμπει Ἰουστινιανὸς πρὸς τῇ μεσογείᾳ, καὶ πλεῖστον λαὸν ἀγροικόν τε καὶ γεωργικὸν ἀθροίσας πρὸς τὰ Τύανα ἀφικνεῖσθαι ἐκέλευσεν ὡς τοὺς πολιορκουμένους ἐπαμυνόμενος. τούτους ἀόπλους οἱ Σαρακηνοὶ θεασάμενοι ὁρμῶσι κατ' αὐτῶν, καὶ τοὺς μὲν ξίφει ἀνεῖλον τοὺς δὲ αἰχμαλώτους συνέλαβον. ἐντεῦθεν θαρραλεώτερον διατεθέντες τῆς προσεδρίας Τυάνων εἴχοντο. οἱ δὲ ἀπορίᾳ δαπανημάτων τῶν πρὸς μάχην ἀπειπόντες τῆς τε παρὰ βασιλέως βοηθείας οὐκ εὐπορήσαντες, ὁμολογίᾳ ἑαυτοὺς τοῖς ἐχθροῖς παρέδοσαν καὶ πρὸς τὰ τῶν Σαρακηνῶν ἤθη ἀπῴχοντο.

Dionysius: "Maslama, the son of 'Abd al-Malik, the brother of Walīd, launched a raid into Roman territory and laid siege to Tyana. He persevered with the assault for nine months. Then the general Theophylact was sent at the head of a large force to relieve the city. But the Romans were defeated in battle by the Arabs and lost 40,000 men. So the Arabs penetrated the city and led the population into slavery in Syria."

This passage is very important and discussed extensively in chapter 2.4.3. This appears to be the normal Byzantine *modus operandi* (*contra*, as

8TH CENTURY

a peasant militia, see Howard-Johnston). The emperor recruits workers to rebuild the destroyed fortress walls in order to resist future sieges. However, for political reasons, the story has been compressed to leave out significant details; for instance, the Arabs actually seem to have withdrawn, but returned when they discovered the works going on in order to exploit the opportunity. The Taurus can be crossed in less than a day by a good rider at this point, so the Arabs may have begun a withdrawal but returned immediately when the opportunity presented itself. Justinian apparently recruited workers in a similar manner for his Chersonese fleet below; cf. also chapter 2.4.3 for the repairs of the Aqueduct of Valens.

CHERSON [preparations] 710
Theoph. 377 (AM 6203, MS 527); Nik. 45.

Theophanes: Justinian's first expedition against Cherson took place in the autumn of 710: "He fitted out a great fleet of every kind of ship—*dromones*, triremes, transports, fishing boats, and even *chelandia*—from the contributions raised by the senators, artisans, ordinary people, and all the officials that lived in the city." ...ἐξοπλίσας στόλον πολύν, μνησθεὶς τῆς κατ' αὐτοῦ γενομένης ἐπιβουλῆς ὑπό τε Χερσωνιτῶν καὶ Βοσφοριανῶν καὶ τῶν λοιπῶν κλιμάτων, πᾶσαν ναῦν δρομώνων τε καὶ τριηρῶν καὶ σκαφῶν μυριαγωγῶν καὶ ἁλιάδων καὶ ἕως χελανδίων, ἀπὸ διανομῆς τῶν οἰκούντων τὴν πόλιν συγκλητικῶν τε καὶ ἐργαστηριακῶν καὶ δημοτῶν καὶ παντὸς ὀφφικίου.

He had given orders to kill everyone in the city, but was furious that the children were spared. 73,000 were killed when the returning fleet was destroyed by a storm. Justinian was overjoyed, however, and sent another fleet. After failing, the remaining soldiers were sent captive to the Khazars.

Nikephoros does not have as many interesting details as Theophanes, but gives the impression that civilians also took part in the campaign: "Still remembering the denunciation made against him by the people of Cherson to Apsimaros, Justinian collected a large fleet of different ships and embarked in them as many as a hundred thousand men recruited from the army registers as well as among farmers and artisans and from the senate and the population of the City." ναῦς πολὺ πλείστας καὶ διαφόρους συναγείρας, ἐμβιβάσας παρ' αὐτοῦ ἄχρι εἰς ἑκατὸν χιλιάδας ἀριθμὸν ἀνδρῶν, [εἰδότας] ἔκ τε τῶν στρατιωτικῶν καταλόγων, ἔτι δὲ καὶ γεωργικοῦ καὶ τῶν βαναυσικῶν τεχνῶν τῶν τε ἐκ τῆς συγκλήτου βουλῆς καὶ τοῦ τῆς πόλεως δήμου...

Mango observes that as compared with Theophanes, Nikephoros probably misunderstood his source. In light of the argument in chapter 2.4, however, some of the above actually may have been craftsmen and laborers sent on the expedition as logistical support.

CHERSON πρὸς καστρομαχίαν 711

Theophanes 379 (AM 6203, MS 528); Nik. 45; cf. Hoyland 202-05.

The second expedition to Cherson was even more elaborate: "After which, he fitted out another fleet and dispatched the patrician Mauros, called Bessos, to whom he gave a battering ram and every other kind of siege engine, with instructions to destroy the walls of Cherson and the entire town, and not to leave a single soul alive there; furthermore, to inform him of his actions by means of frequent dispatches. This man, then, crossed the sea and threw down the battering ram the tower called Kentenaresios as well as the adjoining tower called the Wild Boar; but as the Khazars arrived on the scene, a truce was made." [...] εἶθ᾽ οὕτω τε πλώϊμον ἕτερον κατασκευάσας, ἀποστέλλει Μαῦρον τὸν πατρίκιον, τὸν Βέσσον, παραδεδωκὼς αὐτῷ πρὸς καστρομαχίαν κριόν, μαγγανικά τε καὶ πᾶσαν ἑλέπολιν, ἐντειλάμενος αὐτῷ, τὰ μὲν τείχη Χερσῶνος ἐδαφίσαι καὶ πᾶσαν τὴν πόλιν, μηδεμίαν δὲ ψυχὴν ἐξ αὐτῆς ζωογονῆσαι, πυκνοτέρως δὲ δι᾽ ἀναφορῶν τὰ αὐτῷ πεπραγμένα δηλοῦν. τούτου δὲ περάσαντος καὶ διὰ τοῦ κριοῦ τὸν λεγόμενον Κεντηναρήσιον πύργον καταβαλόντος, ἅμα δὲ καὶ τὸν πλησίον αὐτοῦ Σύαγρον καλούμενον, Χαζάρων δε καταλαβόντων, ἐγένετο ἀνοχὴ τοῦ πολέμου.

Justinian sent another fleet under Mauros and killed them all. Mauros began a siege but was stopped by Khazar sortie; Mauros and his troops joined the Chersonese to proclaim Bardas-Philippikos emperor. Justinian was suspicious and received 3,000 Bulgar mercenaries from Tervel, before marching north to check on affairs. The fleet arrived in Constantinople while Justinian was away, Philippikos was made emperor, and Justinian's army abandoned him after receiving assurances from the new regime.

The translation by MS is inaccurate concerning the *manganika*, which are trebuchets (cf. chapter 8.2.2). The phrase should thus read "he gave him a ram, trebuchets and every siege engine (*helepolis*; possibly another type of trebuchet, cf. chapter 8.2.1) for besieging a fortress." The ram was clearly very efficient; see further chapter 5.2 *passim* for a discussion of wallfighting and how these various engines would have been used in conjunction.

KAMAKHON; CILICIAN FORTS προεδόθη; ὑπὸ λόγον παρέλαβεν 711

Theophanes 377 (AM 6203, MS 527).

"In this year Outhman made an expedition against Cilicia and took many forts by capitulation. Kamakhon and the surrounding country were betrayed to the Arabs." Τούτῳ τῷ ἔτει ἐπεστράτευσεν Οὐθμὰν τὴν Κιλικίαν καὶ πολλὰ κάστρα ὑπὸ λόγον παρέλαβεν. προεδόθη αὐτοῖς καὶ τὸ Κάμαχον σὺν τοῖς παρακειμένοις τόποις.

TURANDA; OTHER FORTS   šrā 'al; aḥreb(w)-ennēn   712
Chr. 819 (Palmer 1993: 79); see also *Misthia and *Amasia below.

"AG 1021: Muhammad b. Marwān was deposed from (the governorship) of northern Mesopotamia {al-jazīra} and replaced by Maslama b. 'Abd al-Malik, who gathered an army and invaded the Roman Empire. He besieged (*textus* 14.28 šrā 'al, ܫܪܐ ܥܠ) the fortress of Tūranda [Gr. Taranton] and the cities of MWSY' and MWSTY' and he destroyed them (*textus* 14.29 aḥreb(w)-ennēn, ܐܘܪ ܐܢܝܢ) and led all who were in them away as captives."

Palmer suggests for the other two forts "Phrygian Mosyna and Lydian Mostene." However, they are clearly related to the events reported in Greek sources for *Misthia and *Amasia, with which they fit well. The dating of this and the following entries is difficult.

MISTHIA; OTHER TOWNSHIPS   παρέλαβον   712
Theoph. 382 (AM 6204, MS 532); Nik. 47; cf. *Turanda (above).

Theophanes: "And likewise, the Arabs occupied Mistheia and other forts, and captured a great many families and cattle without number." ὁμοίως καὶ οἱ Ἄραβες τὴν Μίσθειαν παρέλαβον καὶ ἕτερα κάστρα, πλείστων φαμιλιῶν καὶ κτηνῶν ἀναριθμήτων ἅλωσιν ποιησάμενοι.

Nikephoros: "After this, the Saracen nation, too, raided the Roman country and, as they overran many regions, inflicted much damage on men and animals and took (συμπαραλαμβάνουσι) Misthia as well as other townships (πολίσματα)."

Perhaps these raids should be equated with those reported by *Chr. 819* for *Turanda above, since the consonantal skeleton of the Syriac is fairly close.

AMASIA; OTHER FORTS   παρέλαβε; kabšāh, ܟܒܫܗ   712
Dionysius 147 (Palmer 1993: 209; AH 89); Theophanes 382 (AM 6204, MS 532); cf. *Turanda (above) and Hoyland 2011: 205.

Dionysius: "Walīd sent word to his uncle Muhammad that his term of authority over Mesopotamia was at an end and he appointed his brother Maslama instead. Maslama's first action on coming to Mesopotamia was to commission a survey of the arable land and a census of vineyards, orchards, livestock and human beings. They hung leaden seals on each person's neck. Maslama also raided Roman territory and sacked (*textus* 299: 9 kabšāh, ܟܒܫܗ) the city of Amasya, returning at the head of an endless train of booty and slaves." Palmer n. 525, Michael adds: "AG. 1022: Maslama took

TYBRND', Gargarūm and Tūranda {TWND'} and many other (forts) in the Hexapolis on the Roman border. An Arab post was established at Tūranda {TWND'}. [...] al-'Abbās b. Walīd also led an invasion and took Antioch in Pisidia; he returned with many captives."

Theophanes: "Masalmas captured Amaseia and other fortified places and took many captives." Μασαλμᾶς δὲ τὴν Ἀμασίαν παρέλαβε σὺν ἄλλοις καστελλίοις καὶ πολλῇ αἰχμαλωσίᾳ.

Galatian forts 714

Dionysius 151 (Palmer 1993: 211 n. 533); cf. *Amasia 712.

(From Michael the Syrian rather than *Chr. 1234*:) "In this same year (AG 1026) Maslama invaded the country of Galatia and occupied the fortresses which were there; he returned with many captives."

Constantinople [preparations] 715

Theophanes 384 (AM 6206, MS 534); Nik. 49ff; *Parastaseis* 3.

Details on the preparations made by Anastasios against the Arab expedition are found in Theophanes: "<Then the emperor commanded> that each man should store provisions for himself up to a period of three years, and anyone not having the means to do so should leave the City. He appointed overseers and started building dromones, <fire-carrying> biremes <and great triremes>. He restored the sea walls and likewise the land walls, and set up on the towers catapults for darts and stones and other engines. Having fortified the City as much as he was able, he stored a great quantity of produce in the imperial depots and so made himself safe." [...] ὅτι ἕκαστος φροντιζέτω τὰς ἑαυτοῦ δαπάνας ἕως τοῦ τριετοῦς χρόνου, ὁ δὲ τοῦτο ἀπορῶν ἀποτρεχέτω τῆς πόλεως. ἔστησε δὲ ἐπείκτας καὶ ἤρξατο κτίζειν δρόμωνάς τε καὶ διήρεις <πυρσοφόρους καὶ μεγίστας τριήρεις> · καὶ τὰ παράλια δὲ ἀνεκαίνισε τείχε, ὡσαύτως δὲ καὶ τὰ χερσαῖα, στήσας τοξοβολίστρας καὶ τετραρέας εἰς τοὺς πύργους καὶ μαγγανικά· καὶ κατὰ τὸ δύνατον αὐτῷ τὴν πόλιν ὀχυρώσας, γεννήματά τε πλεῖστα εἰς τὰ βασιλικὰ ὅρια ἀπέθετο καὶ καθ' ἑαυτὸν ἠσφαλίσατο.

Nikephoros (49) adds some information on how intelligence was gathered, as well as alternative terminology: "Anastasios bestowed care on military affairs and appointed capable commanders to take charge of them. Having been informed that the king of the Saracens was intending to invade the Roman country, he sent to him the patrician Daniel, a native of the town Sinope, who was at the time prefect of the Imperial City, on the pretext of negotiating peace, but in reality to observe (διοπτευσόμενον) their preparations against the Romans. When this man had returned, he announced that the foreigners were about to launch a major attack on the Roman State with both horse and sea-borne armies. On hearing this, <the emperor commanded that> each inhabitant of the City could remain if he had provisions for a period of three years, but anyone who was not so provided should depart wherever he wished. He restored carefully the walls of the City and refurbished the military engines (τὰ δὲ τείχη τῆς πόλεως ἐπιμελέστερον καινίζει καὶ τὰ πολεμηστήρια ὄργανα διασκευάζει). He also stored a great quantity of provisions in the City and fortified (κατοχυρώσας) it by such other means as befitted a hostile attack."

Nikephoros (50): "When a rumor spread that the Saracen fleet had come from Alexandria to Phoenix [n. probably Rhodian rather than Lycian Phoenix] for the sake of cypress wood <suitable> for shipbuilding (ναυπηγησίμης ξυλῆς κυπαρισσίνης ἕνεκεν), he selected some fast vessels on which he embarked an army from the region called Opsikion and sent them to the island of Rhodes. He also arranged for other Roman ships to gather there." The fleet dispersed when the Opsikians revolted and forced Theodosios, a tax collector, to become emperor.

Nikephoros (52): The Arabs were able to inflict heavy damage on the Byzantines, causing "much slaughter, abduction, and the capture of cities. For this reason also the Saracens advanced on the Imperial City itself, sending forth by land an innumerable host of horse and foot <recruited> from the various peoples subject to them (ἐκ διαφόρων ἐθνῶν τῶν ὑπὸ χεῖρα), as well as a great fleet numbering as many as eighteen hundred ships under the command of a certain Masalmas (as he was called in their tongue)."

CONSTANTINOPLE [naval blockade] 715

Nik. 51 (p. 118f); cf. Hoyland 2011: 207f.

During a Byzantine civil war, the emperor Anastasios controlled Constantinople and Nikaia, but the rebels were lodged in Chrysopolis across the Bosphorus. "From there they launched every day an attack on the inhabitants of the City, and for six months this battle continued." When Anastasios moved his fleet to the Neorion harbor, the others moved to the Thracian side. "After treasonably subverting certain persons, they took the City through the gate of Blachernai (as it is called) and, while it was night, broke into the houses of the citizens and inflicted heavy losses on them." Anastasios abdicated, was tonsured, and sent to Thessalonica.

SARDIS   kbaš, ܟܒܫ   c. 716
Chr. 819 (Palmer 1993: 76); *Pergamon (below).

"AG 1027: Sulaymān mustered an army of soldiers and workmen (*textus* 15.13 *qalīgrē*, ܩܠܝܓܪܐ) and they set sail and encamped in Asia, capturing (*textus* 15.14 *kbaš*, ܟܒܫ) two cities, Sardis and Pergamum, as well as other fortresses. They killed many and took many captive. As for the Syrians who had been exiled there, he set them free in safety."

PERGAMON   πρὸς αὐτὴν πολιορκίας εἴχοντο   c. 716
Nik. 53; Varvounis 1998; Herrin 1992; cf. *Sardis (above).

"The Saracen army, after destroying numerous Roman towns (πλεῖστα τῶν 'Ρωμαίων καθελὼν πολίσματα καταλαμβάνει), reached the city called Pergamon and set about besieging it (καὶ ἤδη τῆς πρὸς αὐτὴν πολιορκίας εἴχοντο). They captured it for the following reason. By some devilish intention the inhabitants of the city took a pregnant girl who was about to give birth to her first child, cut her open and, having removed the infant from inside her, boiled it in a pot of water, in which the men who were preparing to fight the enemy dipped the sleeves of their right arms. For this reason they were overtaken by divine wrath: their hands became incapable of taking up weapons and, in the face of their inactivity, the enemy captured the city without resistance (ἀμαχητί)."Although some traces of paganism persisted (Herrin), this incident is clearly fabricated for ideological reasons after the council in Trullo (Varvounis).

CUMAE   *pervasum est; noctu superveniente*   717
HL 6.40; *LP* 91.7.

Paul describes how Cumae was first captured by the Beneventan Lombards (*Cumanum castrum a Langobardis Peneventanis pervasum est*), but soon after recaptured by the duke of Naples in a night operation, who managed to take some of the Lombards by surprise (*sed a duce Neapolitano noctu superveniente quidam ex Langobardis capti*).

The Liber Pontificalis fills in some details on how Pope Gregory gave orders to the duke of Naples: "Obeying his instructions they adopted a plan and entered the walls of that Castrum by force in the quiet of the night— that is to say, the duke John, with Thodimus the subdeacon and rector, and the army; they killed about 300 Lombards including their gastald, and they captured more than 500 and took them to Naples."

Constantinople 717f

Theophanes 395(f) (AM 6209, MS 545); Nikephoros 54f; *Chr. Zuq.* (Harrak 1999: 62-65); *Chr. 819* (Palmer 1993: 80); Dionysius 154-63 (Palmer 1993: 211-19); Brooks 1899; Tritton 1959; Christides 2011; Zampaki 2012.

Due to the complexity of the events and the very different treatments in different sources, summaries are appended of the major Greek and Syriac sources with extensive quotes in translation. Theophanes is given in full in Greek; otherwise only important technical terms are given for Nikephoros and Dionysius. In outline, most of the descriptions are compatible, only varying in degree of detail and emphasis, but the Syriac *Chronicle of 819* shows how even significant events such as these could be manipulated out of existence; furthermore, no Syriac source mentions the previous Arab attempts at *Constantinople in 654 and 670s. How such an extensive manipulation was carried out is difficult to fathom, but is important to note as it affects our interpretation of events in the 7th century (cf. App. III).

Theophanes: "Now Maslamas, after he had wintered in Asia, was awaiting Leo's promises. But when he had received nothing from Leo and realized that he had been tricked, he moved to Abydos, crossed over to Thrace with a considerable army, and advanced towards the Imperial City. He also wrote to the Caliph Souleiman that the latter should come with the fleet that had been fitted out in advance. After devastating the Thracian forts, Masalmas laid siege to the City on 15 August. [The Arabs] fenced the land walls all round by digging a wide trench and building above it a breast-high parapet of dry stone. On 1 September of the 1st indiction Christ's enemy Souleiman sailed up with his fleet and his emirs. He had enormous ships, military transports, and *dromones* to the number of 1,800. He put in between the Magnaura and the Kyklobion." Further on ships' maneuvers and use of Greek fire (396); the Arab planned to beach ships by the sea walls in order to scale the walls with steering paddles.

Μασαλμᾶς δὲ χειμάσας ἐν τῇ Ἀσίᾳ, ἐξεδέξατο τὰς τοῦ Λέοντος ὑποσχέσεις· μηδὲν δὲ παρὰ Λέοντος δεξάμενος, καὶ γνοὺς ὅτι ἐνεπαίχθη ὑπ' αὐτοῦ, ἐλθὼν εἰς τὴν Ἄβυδον ἀντεπέρασε λαὸν ἱκανὸν εἰς τὴν Θρᾴκην, καὶ ἐπὶ τὴν βασιλεύουσαν πόλιν ἀπεκίνησεν, γράψας καὶ πρὸς Σουλεϊμὰν τὸν πρωτοσύμβουλον καταλαβεῖν μετὰ τοῦ προετοιμασθέντος στόλου. τῇ δὲ ιε' τοῦ Αὐγούστου μηνὸς παρεκάθισεν τῇ πόλει ὁ Μασαλμᾶς λυμηνάμενος καὶ τὰ Θρακῷα κάστρα. περιχαρακώσαντες δὲ τὸ χερσαῖον τεῖχος ὤρυξαν φόσσαν μεγάλην, καὶ ἐπάνω αὐτῆς περιτείχισμα στηθαῖον διὰ ξηρολίθου ἐποίησαν. τῇ δὲ α' τοῦ Σεπτεμβρίου μηνὸς τῆς α' ἰνδικτιῶνος ἀνέλαβεν ὁ χριστομάχος Σουλεϊμὰν μετὰ τοῦ στόλου καὶ τῶν ἀμηραίων αὐτοῦ ἔχων παμμεγέθεις καὶ πολεμικὰς κατίνας καὶ δρόμωνας τὸν ἀριθμὸν ͵αω', καὶ προσώρμισεν ἀπὸ τῆς Μαγναύρας ἕως τοῦ Κυκλοβίου.

Nikephoros: "Setting out from there, the Saracens proceeded to the straits of Abydos and crossed over to Thrace. After taking many towns there, they reached the Imperial City. They threw a palisade around it and began the siege, which they pursued for thirteen months <with the help of> many kinds of engines they had brought along." A supply fleet under Soliman arrived; however, its rearguard, heavily laden with troops and arms, was pushed back by a contrary wind. Leo launched the Byzantine fleet, broke their line, and burnt 20 Arab ships. The remaining Arab ships moved up the Bosporos to Sosthenion (Istinye on the European side). After a severe winter and snow for a hundred days, the Arabs lost a great number of men, horses, camels, and other animals. An Arab relief fleet arrived in spring from Egypt under Sophiam; another came from Africa under Iezidos. Both landed on the Asian side. "Now the Egyptian sailors entered at night the skiffs that were on their ships, came to Byzantium, and acclaimed the emperor (οἱ δὲ ἐν αὐτοῖς εἰσπλέοντες Αἰγύπτιοι νυκτὸς τοὺς ἐνεστηκότας λέμβους τῶν νηῶν εἰσβάντες πρὸς τὸ Βυζάντιον ἦκον καὶ τὸν βασιλέα εὐφήμουν)." Leo sent out his fleet with Greek fire, burnt all the Arab vessels, and took all the provisions of food and arms as booty. The Arabs withdrew on August 15. Many of their ships were destroyed by storm, others were scattered among the islands all the way to Cyprus.

Chronicle of Zuqnin: "AG 1028: Maslama invaded the Roman Empire. When a great and innumerable army of Arabs gathered and surged forwards to invade Roman territory, all the regions of Asia and Cappadocia fled from them, as did the whole area from the sea and by the Black Mountain and Lebanon as far as Melitene and by the river Arsenias as far as Inner Armenia. All this territory had been graced by the habitations of a numerous population and thickly planted vineyards and gran and every kind of gorgeous tree; but since that time it has been deserted and these regions have not been resettled. When the (Roman) king saw that a great company had invaded (his empire) and that his own general, Leo, had made a pact with them, his heart beat faster and his hands began to shake." Theodosius III abdicated and entered a monastery.

"As for this Leo, he is a courageous, strong and warlike man; moreover, he is by origin a Syrian from these borderlands. For his strength he was made a general. This man used his wisdom to spare (his) territory from drinking the blood of human beings by making a pact with Maslama to get him into Constantinople without a battle. He (Maslama), putting faith in his promise, did not attack anyone nor take any captives; instead, he made his way resolutely towards Constantinople, invaded (Europe) and attacked the City. As for Leo, once he had entered it and had seen that the Romans' hands were shaking and that the King had abdicated, he encouraged the Romans, saying, 'Have no fear!' When they saw his courage and reflected how little confidence they could place in the man who had been their king, they took him and made him their king. As soon as he had clasped the imperial crown (upon his brow), he was invested with might and heroism. He reinforced the wall of the City itself and sent an army to cut off the routes by which the supplies were brought to the (Arab) army from Syria. He also demolished the ships which formed the pontoon-bridge (over the Bosphorus), cutting it, so that the Arabs and their whole army were caught as in a prison. At this point Maslama ordered his men to plant a vineyard. They suffered cruelly from a terrible hunger, such that no bread was to be seen in the whole of their encampment; they actually consumed their own pack-animals and their horses."

Negotiations between Leo and Maslama kept dragging out; "the Arabs waiting outside and the Romans inside, with no battles fought, for about three years [n. 183 rather three months]. The famine among the Arabs became so severe that they resorted to eating their sandals and even the flesh of dead men. They went so far as to attack each other, so that a man was afraid to walk alone." The death of Sulayman and ascension of Umar led to the withdrawal of the Arab army after Maslama toured the city.

The governor of Tyana wanted to ambush the emaciated Arabs, but the Roman troops ended up being ambushed themselves by the well-prepared Arabs, who apparently took them while marching and pitching camp. "When the Romans arrived and descended into the meadowlands, neither knowing nor guessing what the Arabs had done, they pitched camp and even sent out their animals to graze, as an army usually does. At this point the Arabs came up out of their ambushes and the crannies all around the meadow where they had lain hidden, obeying the signal on which they had agreed, and they descended upon them, surrounded them and slaughtered them with all the blades of their swords; and there were no survivors from that army, which had numbered about 60,000 men. Then the Arabs pillaged the dead and returned to their companions. A second army of Romans had gone out after the first; but when they heard what had happened to them, they turned back in fear. As for the Arabs, they took captives and plundered everything which they came across and so came out and arrived back in Syria."

Chr. 819: "AG 1028: Once again Sulaymān mustered his armies, at the Meadow of Dābiq, and he sent a great army with 'Ubayda as its general to the Roman Empire. They invaded Thrace and encamped in that region. 'Ubayda invaded the country of Bulgaria, but most of his army was destroyed by the Bulgars. Those who were left were oppressed by Leo, the sly king of the Romans, to the point of having to eat the flesh and the dung of their horses. In that same year the fortress of Antigon (?) was taken by Dawūd, Sulaymān's son. Sulaymān himself died at the Meadow of Dābiq in September and 'Umar b. 'Abd al-'Azīz b. Marwān reigned after him for two years and seven months, (a good man) and a king more merciful than [all those] who had preceded him." Palmer notes how the source of the chronicler (presumably provided from an Arab perspective) clearly manipulates Arab intentions, as it has no reference to Constantinople whatsoever, and thus covers up a massive failure.

Dionysius: "The following year [AG 1027] Sulaymān, the king of the Arabs, told Maslama to get ready for an expedition into the Roman Empire in order to besiege Constantinople. [n. 535, there is no description in Syriac chronicles of first Arab siege in 674-78, but cf. *Constantinople 670ff] He mustered an army of 200,000 and built 5,000 ships, which he filled with troops and provisions. As leader of these troops he appointed 'Umar b. Hubayra, who was to be answerable to Maslama. He collected furthermore 12,000 workmen (*textus* 300.30, *qūlgārē*, ܩܘܠܓܪܐ), 6,000 camels and 6,000 mules to bear provisions for the animals and the workmen (*textus* 301.1, *qūlgārē*, ܩܘܠܓܪܐ). The camels he loaded with weaponry and catapults (*textus* 301.2, *zaynā saggī'ā, manganīqē*, ܡܢܓܢܝܩܐ, ܙܝܢܐ ܣܓܝܐܐ). For this force he prepared supplies to last for many years; for Sulaymān had said, 'I shall not cease from the struggle with Constantinople until either I force my way into it, or I bring about the destruction of the entire dominions of the Arabs.' On his invasion, (Maslama) was joined by about 3,000 unemployed and unoccupied people, who belong to the class of Arabs without possession whom they call 'volunteers'. They were also joined by many Arab owners of capital, who had provided mounts for the troops on the basis of hire or sale, in the hopes of being recompensed from the booty to be got out of the Royal City." (Palmer 1993: 211f; Dionysius 152)

"Maslama ordered Sulaymān b. Mu'awwid and al-Bakhtarī to proceed by land and 'Umar b. Hubayra by sea. After an extended march to the city of Amorium, al-Bakhtarī and Sulaymān encountered there Leo, the general who, as we have told, had held out against Theodosius. This man made a covenant with the Arabs, whom he led to believe that he would help them to capture Constantinople. Maslama, who was still on the road, travelling behind them, was informed about this in written dispatches; he was delighted with Leo's promises and he promised him in return that he would not permit the Arab army to cause any damage to Leo's province. So when the Arab army arrived, Maslama gave orders that no one should do any harm in that region, not even by (the theft of) a loaf of bread. Leo, for his part, gave orders that a travelling market be loaded up for the Arab army; and the Romans bought and sold in good faith without fear. But Leo's whole concern was to appropriate the Roman kingdom for himself." (Palmer 1993: 212f; Dionysius 153)

"As soon as the Arabs had left Leo's territory, they began to do all sorts of mischief and to commit all kinds of outrage in Roman territory, burning down churches and houses, looting, shedding the blood of men and taking children captive. Many cities in the region of Asia fell to them that summer and they ruined them and took captives and looted, slaughtering the men and sending the children and women back as slaves to their own country. That winter the Arabs spent in Asia. And Maslama sent Sulaymān b. Mu'awwid with 12,000 men to lay siege to the city of Chalcedon, to cut off supplies from that approach to Constantinople and to lay waste and pillage Roman territory in general." Leo, with the aid of Arab cavalry (6,000 men) brought Amorium over to his side, dismissed the cavalry (paying them 12 denarii each), and marched on Constantinople, where he was received joyfully and took command in AG 1028. (Palmer 1993: 213ff; Dionysius 154ff)

"When Maslama heard that Leo had become king, he was overjoyed, supposing that he would thereby find an opportunity to fulfill his promise and deliver the city to him. And Leo, from the moment of his elevation to the throne, wrote constantly to Maslama, encouraging him with vain hopes. At the same time he was restoring and strengthening the city and gathering into it plenty of supplies. He was also having ships prepared for combat with the enemy. And he came to a financial agreement with the Bulgars, by which they agreed to help the City. In short, he took every possible precaution to ensure the City's impregnability." Maslama understood the deceit, and crossed in June to face a land scorched by Leo's troops. Most of the Arabs landed 6 'miles' from the City, but Maslama landed 10 'miles' further down with 4,000 horse and was ambushed at night by the Bulgars, barely escaping to the large camp. (Palmer 1993: 215; Dionysius 157f)

"Then the whole army moved up to the west side of the City and pitched camp near the wall, opposite the so-called Golden Gate. They dug a ditch in front of the camp, between it and the City, and another behind it, between it and the Bulgars; to the right and to the left of the camp was the sea, with a force of about 30,000 Arabs on board the ships. Maslama also instructed the Egyptian crews to stay at sea and to defend his ships from the ships of the Romans. A further force of 20,000 under the command of Sharāhīl b. 'Ubayda was sent out to guard the (landward) approaches of the camp against the Bulgars and the seaward approaches against the Roman ships. On the opposite coast they had to combat the Roman scouts who tried to draw them off and to prevent supplies from reaching the Arabs." (Palmer 1993: 215f; Dionysius 159)

The Arabs were defeated by the Bulgars, and their supplies cut off by scouts, rough sea, and fear of Bulgar raids. Maslama tried to maintain morale by claiming the City would soon surrender and fresh supplies were on their way. In the meantime, the Arabs ate dead animals, corpses and dung; finally they resorted to pitch from the boats, but when 'Umar acceded as Caliph he planned a rescue operation. Having obtained intelligence from the second expedition, 'Umar sent orders to withdraw. Palmer n. 549 adds that Michael claims that the Arabs heard of the death of Sulaymān from the Byzantines on the city walls. The Arabs then withdrew, but most of their ships were destroyed by the Romans and Greek fire, while the survivors were destroyed by a storm. 'Umar sent reinforcements overland to the land army, bringing "20,000 mules and some horses" as mounts to the unmounted Arabs, as well as gold. He also encouraged the relatives of those on expedition to go out and meet them. (Palmer 1993: 216-19; Dionysius 160-63)

The Syriac sources appear somewhat reluctant to describe the role of Syrian Christian engineers in any detail. While Dionysius has preserved much valuable information on the composition of the army and prominence of conscripted Syrian engineers, neither he nor the Zuqnin chronicler reveal any details about their involvement in the fighting, and the give the impression that much of the siege was dominated by negotiations. From Arabic souces (Brooks, Tritton), however, we do learn of fierce artillery duels, and independent Syriac and Greek traditions (shared by Theophanes and Nikephoros) confirm the scale of Arab siegeworks. All sources emphasize the dreadful logistical situation of the Arabs, most recently studied by Christides. Finally, Zampaki has a survey of further literature on this and other Arab sieges of Constantinople.

CONSTANTINOPLE (civil war) c. 718
Nik. 57; cf. Constantinople 715.

Anastasios tried to foment a revolt from Thessalonica, e.g. by sending letters to an old friend, the current commander of the Walls (ἄρχοντα τειχῶν) named Niketas Anthrax, who was to open the City for him. He therefore marched on the city, with help from Bulgars and bringing monoxyla from Thessalonica; however, Leo scared off the Bulgarians with a stern letter, so they surrendered the culprits to him to be publicly mutilated and tortured in various horrific fashions.

TOULOUSE	*obsidione cingens*	c. 720

Byzantine-Arab chronicle of 741, § 42 (Hoyland 1997: 626); Collins 1989; Bachrach 2001.

"Also he made Narbonnian Gaul his own through the leader of the army, Maslama by name, and harassed the people of Franks with frequent battles. With inconsistent valour the general of the army, already mentioned, reached as far as Toulouse, surrounded it with a siege and strove to overwhelm it with slings and other types of machines. Informed of this news, the peoples of the Franks gathered together under the leader of the same people, Eudes by name. So having assembled they reached Toulouse. At Toulouse each battle line of the armies clashed in a great struggle. They killed Samh (Zema), the leader of the army of the Saracens, together with a part of his army, and they pursued the rest of the army as it slipped away in flight."

|| Galliam quoque Narbonensem per ducem exercitus Mazlema nomine suam facit gentemque Francorum frequentibus bellis stimulat atque incongruenti virtute iam dictus dux exercitus Tolosam usque pervenit eamque obsidione cingens fundis et diversis generum machinis expugnare conavit. Francorum gentes tali de nuntio certae apud ducem ipsius gentis Eudonem nomine congregatur sicque collecti Tolosam usque perveniunt. apud Tolosam utraque excercitus acies gravi dimicatione confligunt. Zema ducem exercitus Sarracenorum cum parte exercitus sui occident, reliquum exercitum per fugam lapsum sequntur. ||

It is clear that Aquitainians had to deal with a tremendous Arab capacity for siege warfare (see chapter 7.3) and did so successfully at Toulouse, both by resisting siege engines and engaging with a relieving army. The Aquitanians, then, must have matched Arab besieging skills.

ANGERS	*obsedit*	722/3

Fred. Cont. 11.

Sometime between events datable to 721 and 724, "Prince Charles pursued Ragamfred, laid siege to the city of Angers, laid waste the neighborhood and then returned home with rich booty." || His ita euulsis Carlus

Princeps insecutus idem Ragamfredo Andegavis ciuitatem obsedit; vastata eadem regione, cum plurima spolia remeauit. ||

| CLASSIS (twice) | *invasit; reddita est* | c. 723 |

HL 6.44.

Faroald duke of Spoleto attacked (*invasit*) Classis, but it was restored to the people of Ravenna by Liutprand (*iussu regis...reddita est*).

| NARNI | *pervasa est* | c. 725 |

HL 6.48; LP 91.13.

"In these days the city of Narni was conquered by the Langobards." ||
His diebus Narnia civitas a Langobardis pervasa est. || (*HL*)
"Then the Lombards seized the Castrum of Narni." (*LP*)

| NIKAIA | κυκλώσας | 727 |

Theophanes 405f (AM 6218, MS 560); Nik. 61; cf. Hoyland 226ff.

"At the summer solstice of the same 10th indiction, after the unhappy defeat of our fellow-countrymen, a multitude of Saracens led by two emirs was drawn up against Nicaea in Bithynia: Amer with 15,000 scouts led the van and surrounded the town which he found unprepared, while Mauias followed with another 85,000 men. After a long siege and a partial destruction of the walls, they did not overpower the town thanks to the acceptable prayers addressed to God by the holy Fathers who are honoured there in a church (wherein their venerable images are set up to this very day and are honoured by those who believe as they did). A certain Constantine, however, who was the *strator* of Artabasdos, on seeing an image of the Theotokos that had been set up, picked up a stone and threw it at her. He broke the image and trampled upon it when it had fallen down. He then saw in a vision the Lady standing beside him and saying to him: 'See, what a brave thing you have done to me! Verily, upon your head have you done it.' The next day, when the Saracens attacked the walls and battle was joined, that wretched man rushed to the wall like the brave soldier he was and was struck by a stone discharged from a siege engine, and it broke his head and face, a just reward for his impiety. After collecting many captives and much booty, the Arabs withdrew."

κατὰ δὲ τὴν θερινὴν τροπὴν ταύτης τῆς ι' ἰνδικτιῶνος μετὰ τὴν τῶν ὁμοφύλων κακὴν νίκην, καὶ κατὰ τῆς Βιθυνῶν Νικαίας παρατάττεται τῶν Σαρακηνῶν δύο ἀμηραίων στῖφος ἄμερ ἐν χιλιάσι μονοζώνων δεκαπέντε προσδραμών, καὶ ἀπαρασκεύαστον κυκλώσας τὴν πόλιν, καὶ Μαυΐας ἐπακολουθῶν ἐν ἄλλαις ὀκτὼ ἥμισυ μυριάσιν, οἳ μετὰ πολιορκίαν πολλὴν καὶ καθαίρεσιν τῶν τειχῶν μερικὴν τῷ τῶν τιμωμένων ἁγίων πατέρων αὐτόθι τεμένει, ταύτης μὲν οὐ περιγεγόνασι διὰ τῶν εὐπροσδέκτων εὐχῶν πρὸς τὸν θεόν, ἔνθα καὶ σεβάσμιοι αὐτῶν χαρακτῆρες ἀνεστήλωντο μέχρι τοῦ νῦν ὑπὸ τῶν ὁμοφρόνων αὐτοῖς τιμώμενοι. Κωνσταντῖνος δέ τις στράτωρ τοῦ Ἀρταβάσδου ἰδὼν εἰκόνα τῆς θεοτόκου ἑστῶσαν, λαβὼν λίθον ἔρριψε κατ' αὐτῆς, καὶ συνέτριψεν αὐτήν, καὶ πεσοῦσαν κατεπάτησεν· καὶ θεωρεῖ ἐν ὁράματι τὴν δέσποιναν παρεστῶσαν αὐτῷ καὶ λέγουσαν· οἶδας ποῖον γενναῖον πρᾶγμα εἰργάσω εἰς ἐμέ; ὄντως κατὰ τῆς ἑαυτοῦ κεφαλῆς τοῦτο πεποίηκας. τῇ δὲ ἐπαύριον προσβαλόντων τῶν Σαρακηνῶν τῷ τείχει, καὶ πολέμου κροτηθέντος, δραμὼν εἰς το τεῖχος, ὡς γενναῖος στρατιώτης, ὁ ταλαίπωρος ἐκεῖνος βάλλεται ὑπὸ λίθου τοῦ ἐκ τοῦ μαγγανικοῦ πεμφθέντος, καὶ συνέτριψεν αὐτοῦ τὴν κεφαλὴν καὶ τὸ πρόσωπον, ἄξιον τῆς ἑαυτοῦ δυσσεβείας κομισάμενος ἀνταπόδομα. αἰχμαλωσίαν δὲ πλείστην καὶ λάφυρα συναγαγόντες ὑπέστρεψαν...

Nikephoros 61 gives a somewhat more minimalistic picture: "The following summer a numerous force of Saracen cavalry again overran the Roman State. Led by Saracens called Ameros and Mauias, they came to the chief city of Bithynia, namely, Nicaea. After besieging it for some time, they finally departed without having accomplished anything (ἐπί <τε> τινα χρόνον τῆς πολιορκίας ἐχόμενοι τέλος ἄπρακτοι ἀπεπέμποντο)."

Mango's commentary (p. 212) dates it to 727 and notes an inscription preserved on the walls referring to the siege, as well as parallels in *Chr.* 1234 and Mich. Syr. 501, who treat this as an Arab success (see Hoyland for details).

NEOCAESAREA *etkabšat*, ܐܬܟܒܫܬ 729

Chr. Zuq. s.a. 728-29 (Harrak 1999: 161); cf. Hoyland 2011: 225.

"Maslama conquered Neocaesarea. He took all its citizens into captivity and sold them into slavery like cattle, except for the Jews who surrendered the city. For they went out secretly to Maslama, made an agreement with him, and treacherously directed him to the entrance of the city. These he took into captivity and did not sell them. So they went with him." Harrak notes that the incident involving the Jews was perhaps confused with the siege of *Caesarea Maritima in 640, where similar events are reported to have taken place.

Textus 171.9-16:

ܐܬܟܒܫܬ ܢܐܘܩܣܪܝܐ ܡܢ ܡܣܠܡܐ. ܘܐܫܒܝ ܠܒܢܝܗ̇ ܟܠܗ ܒܫܒܝܐ. ܘܙܒܢܗ ܒܕܡܘܬ
ܡܫܝܐ ܠܒܥܝܪܘܬܐ. ܣܛܪ ܡܢ ܝܗ̈ܘܕܝܐ ܕܗܢܘܢ ܗܘܘ ܡܫܠܡܢܗ̇ ܕܝܠܗ̇ ܕܡܕܝܢܬܐ.
ܒܗ̇، ܕܢܦܩܘ ܒܛܘܫܝܐ ܠܘܬ ܡܣܠܡܐ. ܘܥܒܕܘ ܥܡܗ ܫܠܡܐ. ܘܐܦ ܬܪ̈ܥܘܗܝ، ܥܠ
ܡܥܠܢܐ ܕܝܠܗ̇ ܕܡܕܝܢܬܐ ܒܢܟܠܘܬܗܘܢ. ܠܗܠܝܢ ܫܒܐ ܘܠܐ ܙܒܢ. ܐܠܐ ܢܦܩܘ ܥܡܗ.

LOMBARD CAMPAIGN   C. 731
HL 6.49; *LP* 91.13.

"At this time king Liutprand besieged Ravenna and took Classis and destroyed it." ... "Also king Liutprand attacked Fregnano, Monteveglio, Busseto and San Giovanni in Persiceto, Bologna and the Pentapolis and Osimo, fortresses of Emilia. And in like manner he took possession of Sutri but after some days it was again restored to the Romans." || Eoque tempore rex Liutprandus Ravennam obsedit, Classem invasit atque destruxit... Rex quoque Liutprand castra Emiliae, Feronianum et Montebellum, Buxeta et Persiceta, Bononiam et Pentapolim Auximumque invasit. Pari quoque modo tunc et Sutrium pervasit. Sed post aliquod dies iterum Romanis redditum est. || (*HL*)

"Liutprand king of the Lombards in a general campaign proceeded to Ravenna and besieged it for some days, and seizing the Castrum of Classe they took many captives and removed untold wealth." (*LP*)

Some of the castra mentioned in *HL* above surrendered to the Lombards as a result of Roman infighting over the iconoclastic policies of Leo (*LP* 91.18). Sutri was held by the Lombards for 140 days (91.21).

| RAVENNA | *obsedit* | C. 731 |

See *Lombard campaign (above).

| CLASSIS | *invasit atque destruxit* | C. 731 |

See *Lombard campaign (above).

| FREGNANO AND OTHER CASTRA | *invasit* | C. 731 |

See *Lombard campaign (above).

| SUTRI | *pervasit* | C. 731 |

See *Lombard campaign (above).

| BORDEAUX | *suiugauit* | 735 |

Fred. Cont. 15.

Charles cut through Aquitaine, but it is unclear from which direction (from Burgundy or Neustrasia, as the preceding passage would imply the former, but as passage below shows he was in Frisia during the preceding year). He took many cities during this campaign, but only Bordeaux and Blaye are mentioned.

"Duke Eudo died at this time. Learning this, Prince Charles consulted with his chieftains, crossed the Loire and went to the city of Bordeaux and to the stronghold of Blaye on the Garonne. Then he proceeded to occupy the whole area, including the cities and the strongholds. Then he returned in peace, victorious through Christ his helper, Who is King of kings and Lord of lords. Amen." || In illis quippe diebus Eudo dux mortuus est. Haec audiens praefatus princeps Carlus inito consilio procerum suorum denuo Ligere fluuio transiit, usque Garonnam uel urbem Burdigalensem uel castro Blauia ueniens occupauit illamque regionem coepit hac subiugauit cum urbibus ac suburbana castrorum. Victor cum pace remeauit opitulante Christo rege regum et domino dominorum. Amen. ||

BLAYE *subiugauit* 735
See *Bordeaux above.

AVIGNON *obsidionem coaceruat; obsedunt* c. 736
Fred. Cont. 20; *Chron. Moiss.* 291f; Fouracre 2000: 96f; Collins 1989; Bachrach 2001.

The Arab invasion was connected to the revolt of local Frankish magnate Maurontus, presumably to preempt Charles' reorganization of Burgundy under his brother Childebrand (see *Lyons 738). The invasion force had already taken Arles (*Chr. Moiss.* p. 291, § 89, Anno 734, Arabs *Arelato civitate pace ingreditur*), probably in cooperation with Maurontus, and an Arab garrison was also installed in Avignon. The forces that were involved in the siege of Avignon were led by *dux* Childebrand who was supported by a siege train (*cum apparatu hostile*) and the rest of the *duces* and *comites* of that region (Burgundy); Charles Martel with remaining army from Frankia and surrounding territories fought against the Arab expeditionary force, which may have been supported by local Franks sent by Maurontus. This was a complex siege, with gradual encirclement with camps (*tentoria instruunt*) around the city from all sides (*undique*), establishment of continuous siegeworks (*muros circumdat*) and fortified camps (*castra ponit*) and heap up embankments (*obsidionem coaceruat*—possibly as countervallation, or, more likely, for placing rams and artillery), storm with machines and trebuchets (rather than rope ladders) against walls and buildings. See discussion of replacing the probably corrupt *funibus* with *fundis*, attested in *Chr. 741* (see *Toulouse c. 720 above), in ch. 4.3.3 and 8.2.3. The Franks stormed the city, set it on fire, and massacred captives. The Arab expansion into the Rhône valley was partly interrupted by Berber revolt which followed shortly after.

Fred. Cont. 20: "But our noble Duke Charles sent against them his illustrious brother Duke Childebrand, who proceeded to that region in warlike array [rather: with military equipment/munitions] with the remaining dukes and counts." || ...Ad contra uir egregius Carlus dux germanium sum uirum inlustrum Childebrando ducem cum reliquis ducibus et comitibus illis partibus cum apparatu hostile diriget. ||

"Childebrand lost no time in bivouacking in the approaches and surrounding the countryside of this excellently provisioned city and in setting about his preparation for the forthcoming engagement until the arrival of Charles, the great commander, who forthwith shut up the city [rather: built walls], erected siegeworks [rather: built fortifications/fortified camps] and tightened the blockade [rather: threw up a siege, or even threw up siege {-works} i.e. embankments for machines; see below]." || Quique praepropere ad eandem urbem peruenientes tentoria instruunt, undique ipsud oppidum et suburbana praeoccupant, munitissimam ciuitatem obsedunt, aciem instruunt, donec insecutus uir belligerator Carlus praedictam urbem adgreditur, muros circumdat, castra ponit, obsidionem coaceruat. ||

"Then as once before Jericho, the armies gave a great shout, the trumpets brayed and the men rushed in to the assault with battering rams [rather: machines] and rope ladders [the text says "ropes of ropes;" with emendation to *fundis* it makes much better sense: "rope slings," i.e. traction trebuchets] to get over walls and building [rather: they attacked above the walls and {into} the intramural buildings—this explains the super then the building being destroyed]; and they took that strong city and burned it with fire ant they took captive their enemies, smiting without mercy and destroying them and they recovered complete mastery of the city." || In modum Hiericho cum strepitu hostium et sonitum tubarum, cum machinis et restium funibus [emend to *fundis*] super muros et edium moenia inruunt, urbem munitissimam ingredientes succendunt, hostes inimicos suorum capiunt, interficientes trucidant atque posternent et in sua dicione efficaciter restituunt. ||

The chronology is difficult; Fouracre gives 736 for Childebrand's expedition to Avignon, while the year 737 comes from Wallace-Hadrill. The chronicle of 754 reports a huge Arab expedition with naval assets in the era 775 (AD 738), partly interrupted by a Berber revolt in North Africa. It is clear from *Chr. Moissacense* that an Arab army had occupied Avignon in 734 and terrorized the surrounding territory for four years, but provides no other dates. It also confirms Fredegar's version of events in outline. The Frankish collaborator Maurontus probably gave intelligence and support to the invading Arabs as they raided northwards.

8TH CENTURY

NARBONNE	*obsedit*	736/7

Continued from *Avignon above; Wallace-Hadrill's ed. 95 n 1.

"Victorious, therefore, Charles, the dauntless, mighty warrior, crossed the Rhône with his men and plunged into Gothic territory as far as the Narbonnaise. He invested its famous capital, Narbonne itself. He threw up lines on the banks of the Aude in which he installed offensive armament of the battering type. He then continued his lines in a wide sweep round the Saracen emir 'Abd ar-Rahman and his viziers, investing them so. He added carefully constructed works at intervals." A better translation of the latter would be: "He built fortifications from every direction/everywhere around."

|| Victor igitur atque bellator insignis Carlus intrepidus Rodanum fluuium cum exercitu suo transiit, Gotorum fines penetrauit, usque Narbonensem Galliam peraccessit, ipsam urbem celeberrimam atque metropolim eorum obsedit, super Adice fluuio munitionem in girum in modum arietum instruxit, regem Sarracinorum Athima cum satellitibus suis ibidem reclusit castraque metatus est undique. ||

The Arabs sent reinforcements from Spain, but Charles marched against them and defeated them "on the banks of the Berre, at the palace in the valley of Corbières." Wallace-Hadrill locates this in Aude and equates the palace with the one built by the Visigothic king Athaulf. This means that it was probably used by the Arabs as a supply post for armies headed into Gaul. The Franks then pursued the Arabs as they fled through lagoons with boats (*nauibus*) taking plenty booty and prisoners, and *regionem Goticam depopulant*; however, there is no mention of a capture of the city.

NÎMES	*funditus muros et moenia destruens*	736/7

See *Avignon and continued from *Narbonne (above).

"The famous cities of Nîmes, Agde and Béziers were burnt, and their walls and buildings he razed to the ground. Their suburbs and the strongholds of that area were destroyed." || Vrbes famosissimas Nemausum, Agatem hac Biterris, funditus muros et moenia destruens igne subposito concremauit, suburbana et castra illius regionis uastauit. ||

The phrase *subposito igno* may reflect undermining and firing, rather than the wanton burning of the city. Thus, very complex siege operations are compressed to barely a sentence. However, it is important to read this description carefully, as it was commissioned by Childebrand, Charles Martel's brother and commander of many of these operations. Charles' efforts were very impressive, but perhaps eased by garrisons being defeated in open battle if they had been called out as reinforcements. However, interestingly enough, there is nothing about Narbonne being captured in Fredegar.

Later events that year (Fred. Cont. 21) show Charles clearing Provence of rebels (and Arabs?) under Maurontus, who withdrew to strongholds near the coast or on islands. Charles returned to the villa of Verberie on the Oise [n. by Senlis], where he became sick. Note importance of villas on marching routes.

| Agde | 736/7 |

See *Nîmes (above).

| Béziers | 736/7 |

See *Nîmes (above).

| ?Lyons | *sua dicione rei publicae subiugauit* | 738 |

Fred. Cont. 18; Fouracre 2000: 93.

"Charles, that shrewdest of commanders, now went with his army into the land of Burgundy against the city of Lyons. He subjected to his rule the chief men and officials of that province and placed his judges over the whole region as far as Marseilles and Arles. Then he returned to the Frankish kingdom and the seat of his power with gifts and much treasure." || Idcirco sagacissimus uir Carlus dux commoto exercitu, partibus Burgundie dirigit Lugdunum Gallie urbem maiores natu atque praefectus eiusdem prouintie sua dicione rei publicae subiugauit, usque Marsiliensem urbem uel Arelatum suis iudicibus constituit, cum magnis thesauris et muneribus in Francorum regnum remeauit in sedem principatus sui. ||

The events led to the complete reorganization of Burgundy under Carolingian rule. This is when the Carolingian system (see chapter 4.3 above) was firmly established, although the process had been under way for some time. Wallace-Hadrill notes the imposition of counts (*iudices*) and abolishment of the traditional patricianate that had prevailed since early Merovingian times. Presumably these counts and the subjected *maiores natu atque praefectus eiusdem provintie* were obliged to provide military resources to Charles Martel of the type we see later under Charlemagne. The nature of the conflict is unclear (or if indeed there was one), but the large army clearly backed up Charles' occupation and following reforms with a credible threat.

| PALOZONIUM | kabšāh; ܟܒܫܗ | c. 740 |

Chr. Zuq. s.a. 733-34 (Harrak 1999: 161f); cf. Hoyland 2011: 231ff.

"Sulaymān (son of the Caliph Hishām) invaded the Roman land, conquered Palozonium and took all its people into captivity." (*Textus* 171.17-19: ܥܠ ܣܘܠܝܡܢ ܒܪܗ ܕܗܫܡ ܘܟܒܫܗ ܠܦܠܘܙܘܢܝܐ. ܘܫܒܐ ܠܟܠܗܘܢ ܥܡܘܪܝܗ̇.)

See Harrak 161 n. 6 on the confused dating, actors and sequence of events—this probably happened several years later, in about 741-42, and in the context of the Byzantine civil war, during the rebellion of Artabasdos who had stripped P. of defenders before taking control of Constantinople, where he was besieged by Leo. The next event, the siege of *Synnada, probably occurred in 739-40, cf. Theophanes AM 6231.

| SYNNADA | nqaš 'al; ܢܩܫ ܥܠ | c. 740 |

Chr. Zuq. s. a. 734-35 (Harrak 1999: 162); cf. *Palozonium above and *Fort in Asia 741/2.

"Mālik, son of Shabīb, the amīr of Melitene (Malatya), and 'Abd-Allah al-Baṭṭāl marched in and besieged Synnada (*textus* 172.11f: ܘܢܩܫ ܥܠ ܣܘܢܕܐ; note difference from *SYLWS 664: ܣܘܠܘܣ). While they were camped in a meadow in Synnada, numerous troops that could not be counted gathered against them so as to take vengeance on them for the blows the Arabs had inflicted on Palozonium the year before. As these Arabs, who were about fifty thousand strong, were camped without worry, the Romans suddenly surrounded them on all sides, and slaughtered them all with the blades of their swords. Only a few Arabs remained alive, as the day reclined to sunset, though they too were wounded by the swords, lances and bows. They fled in this condition and marched the whole night. Barely five thousand out of the fifty thousand strong who came, escaped. Even their leaders fell along with them in the battle. Such a disaster had never befallen the Arabs before."

Another case of 50,000 invading and only 5,000 returning is reported below at *Kamakhon 766. The numbers are suspicious, but may also be indicative of the relative size of forces. The expression *nqaš 'al* means encamp against, i.e. besiege.

| LAON | obsidentes | 741 |

ARF 741 (in *Annales Einhardi*).

After the death of Charles Martel, Grifo the brother of Carloman and Pepin raised a revolt. "... he at once occupied the city of Laon and declared war on his brothers. Carloman and Pepin quickly gathered an army, besieged Laon, and captured Grifo." || ...ut sine dilatione Laudunum civitatem occuparet ac bellum fratribus indiceret. Qui celeriter exercitu collecto Laudunum obsidentes fratrem in deditionem accipiunt ||

The brothers restored order in breakaway provinces and took Grifo into custody.

FORT IN "ASIA"	*fa-nazala 'alā*, فنزل على	741/2

Agapius 509; cf. *Palozonium and Synnada 740; possibly the same campaign.

"En l'an 12 d'Hicham, Soleïman-ibn-Hicham fit une incursion et assiégea une des forteresses (*fa-nazala 'alā ḥiṣn ḥuṣūn*, فنزل على حصن حصون) d'Asie. La peste se mit dans ses rangs, et beaucoup de soldats moururent; la famine sévit; les Grecs en massacrèrent un grand nombre; la plus grande partie de leurs chevaux perirent; un très grand nombre d'entre eux se réfugièrent auprès des Grecs et se firent chrétiens à cause du malheur qui leur était arrivé. (Après cela) Soleïman retourna en fuyant."

LOCHES	*funditus subuertunt; ceperunt*	742

Fred. Cont. 25; *ARF* 742.

Fredegar: "Meanwhile the Gascons of Aquitaine rose in rebellion under Duke Chunoald, son of the late Eudo. Thereupon the princely brothers Carloman and Pippin united their forces and crossed the Loire at the city of Orleans. Overwhelming the Romans they made for Bourges, the outskirts of which they set on fire; and as they pursued the fleeing Duke Chunoald they laid waste as they went. Their next objective, the stronghold of Loches, fell and was razed to the ground, the garrison being taken prisoner. Their victory was complete." || Interea, rebellantibus Wascones in regione Aquitaniae cum Chunualde duce filio Eudone quondam, Carlomannus atque Pippinus germani principes congregato exercito Liger alueum Aurilianis urbem transeunt, Romanos proterunt, usque Beturgas urbem accedunt, suburbana ipsius igne conburent, Chunualdo duce persequentes fugant cuncta uastantes, Lucca castrum dirigunt atque funditus subuertunt, custodes illius castri capiunt; etenim uictores existent. ||

ARF: "Carloman and Pepin, mayors of the palace, then led an army against Hunald, duke of the Aquitanians, and took the castle of Loches." || Quando Carlomannus et Pippinus maiores domus duxerunt exercitum contra Hunaldum ducem Aquitaniorum et ceperunt castrum, quod vocatur Luccas... ||

Note the use of *funditus*, "completely" overthrew or destroyed, implying wall-breaking technology, that the Aquitanians had "Romans" defending the territory of Orleans, and that the Franks systematically ravaged the territories through which they marched.

CONSTANTINOPLE   τείχει παρακαθίσαντος; παρέλαβεν   c. 742f
Theoph. 415, 417f, 419f (MS 575, 578, 580f); Nik. 64ff; Speck 1995; Treadgold 1992; cf. Hoyland 2011: 238f.

The revolt of Artabasdos against the legitimate heir Constantine V was at first very successful, as he brought Constantinople over to his side. His henchman in the city was the *magistros* Theophanes Monotes: "Straight away Monotes sent a message to the Thracian region, addressed to his son Nikephoros, who was *strategos* of Thrace, bidding him collect the army that was there so as to guard the City. After closing the gates of the walls and setting a watch, he apprehended Constantine's friends whom he scourged, tonsured and threw in gaol. After Artabasdos had entered the City with the Opsikian army, Constantine, too, arrived at Chrysopolis with the two *themata*, namely the Thrakesian and the Anatolic, but he failed to accomplish anything and so returned to winter at Amorion. Artabasdos, for his part, restored the holy icons throughout the City." Nikephoros has a shorter version of the same.

Over the next year there was brutal fighting in northwestern Anatolia from which Constantine emerged victorious (most detailed in Nikephoros, but the siege is best served by Theophanes): "In the month of September, indiction 12, Constantine came to the area of Chalcedon and crossed to Thrace, while Sisinnios, *strategos* of the Thrakesians, had crossed by way of Abydos and laid siege to the land walls (καὶ τῷ χερσαίῳ τείχει παρακαθίσαντος). Coming to the Charsian gate, Constantine proceeded as far as the Golden Gate showing himself to the populace and then withdrew and struck camp at St Mamas (καὶ ἐλθὼν ἐπὶ τὴν Χαρσίου πόρταν διέδραμεν ἕως τῆς Χρυσῆς πόρτης ἑαυτὸν τοῖς ὄχλοις ἐπιδεικνύων, καὶ πάλιν ὑπέστρεψεν, καὶ ἡπλίκευσεν εἰς τὸν ἅγιον Μάμαντα.). Those in the City began experiencing shortages of supplies: accordingly, Artabasdos dispatched the *a secretis* Athanasios and Artabasdos, his *domesticus*, to bring supplies by ship. The fleet of the Kibyraiots found these men beyond Abydos, arrested them and brought them to the emperor, who donated the grain to his own men and straight away blinded Athanasios and Artabasdos. After this, Artabasdos attempted to open the gates of the land walls and give battle to Constantine, but the men of Artabasdos were routed in the engagement and many were killed, including Monotes. Then Artabasdos constructed fire-bearing biremes and sent them to St Mamas against the fleet of the Kibyraiots, but when these had set out, the Kibyraiots sallied forth and chased them away." There followed a severe famine, with increasing costs for food.

"As the people were dying, Artabasdos was forced to let them leave the City, but they took note of their faces and some he prevented from leaving. For this reason some painted their faces and put on female dress, while others donned monastic costume and garments of hair and in this guise they were able to escape detection and leave." A relieving army turned back when it came to Chrysopolis, but it was nonetheless pursued and defeated. After either two or fourteen months, Constantine ordered an assault: "On 2 November he suddenly drew up his forces in the evening and took the City through the land walls." ... ἄφνω παραταξάμενος τῇ δείλῃ διὰ τοῦ χερσαίου τείχους τὴν πόλιν παρέλαβεν. Artabasdos fled but was defeated again and arrested in Anatolia; Constantine punished the conspirators and celebrated his victory. Nikephoros adds that during the siege, due to the famine refugees bribed the guards in order to be let out secretly; they were well received by Constantine.

MS dates the entry of Ardabasdos into Constantinople to July 5, 741. Treadgold and Speck have differing interpretations as the siege may have lasted anywhere from two months to over a year. The dates found in Theophanes are AM 6233 (AD 740/1) for the beginning of the revolt and AM 6235 (AD 742/3) for the beginning of siege (i.e. 742). The extensive famine and construction of ships by Artabasdos would indicate a very long timeframe.

Hohenseeburg	*coepit castrum...per placitum*	743

ARF 743.

"Carloman and Pepin then started a war against Odilo, duke of the Bavarians. That year Carloman advanced alone into Saxony. By treaty, he got possession of the castle called Hohenseeburg and made Theoderic the Saxon submit." || Tunc Carlomannus et Pippinus contra Odilonem ducem Baiovariorum inierunt pugnam, et Carlomannus per se in Saxoniam ambulabat in eodem anno et coepit castrum, quod dicitur Hoohseoburg, per placitum et Theodericum Saxonem placitando conquisivit. ||

Emesa and other cities		745

Theophanes 422 (AM 6237, MS 584); *Chr. Zuq.* s.a. 745-746 (Harrak 1999: 174ff); cf. Hoyland 2011: 245-64.

"In this year Souleiman gathered his armies and, after engaging Marouam once again, was defeated with the loss of 7,000 men and escaped, first to Palmyra and then to Persia. The inhabitants of Emesa, Helioupolis, and Damascus raised a rebellion and shut their gates against Marouam. The latter sent his son at the head of an army against Dahak and himself came to Emesa, which he captured after a siege of four months. Dahak, for his part, was marching from Persia with a great force. Marouam engaged him in Mesopotamia and, after killing many more of his companions, captured him and slew him."

"From 10 to 15 August there was a misty darkness. At that time Marouam, after victoriously taking Emesa, killed all the relatives and freedmen of Isam. He also demolished the walls (καθαιρεῖ δὲ καὶ τὰ τείχη) of Helioupolis, Damascus, and Jerusalem, put to death many powerful men, and maimed those remaining in the said cities."

From the Zuqnin Chronicler, we learn that both partisans in the Umayyad civil war apparently recruited villagers and urban populations who were still predominantly Christians. For instance, Marwān recruited craftsmen and laborers (*textus* 189.1f: ܘܕܚܠ ܣܠܝܘ ܡܥܠܠܐ ܟܪ̈ܟܐ ܠܐ ܘܥܡܠܐ ܕܚܩܠܐ ܘܒܢܝ̈ܐ) in the region around Mosul, while Ibrahim in desperation recruited a host of villagers armed with slings—apparently both issues noted by the chronicler since they were local Syrian Christians.

At the beginning of the conflict, Marwān moved out from "the territory of the Turks" (i.e. his province facing the "Gate of the Turks", Armenia), to take part in the civil war. "After Marwān marched out to the Jazīra, which surrendered to him, he appointed governors for it in all the cities, including Mosul. He gathered a great army and rushed (to the Jazīra [This emendation appears erroneous; probably the laborers and craftsmen were recruited *in* the Jazīra to be brought on campaign to the west, cf. the following]) along with labourers and craftsmen."

"Then he crossed over to the West against the partisans of 'Abbās. Yazīd (II), who killed Walīd (II), died six months after, and Ibrāhīm, his brother, replaced him. When the last-named learned that Marwān had crossed the Euphrates, accompanied by numerous troops, and that the Jazīra surrendered to him, he trembled before him: *They reeled and staggered like drunken men.* [Ps. 107.27] Ibrāhīm first sent against him Nu'aym son of Thābit—it was reported about him that he had seventy sons—along with numerous troops. When they faced each other an waged battle, the entire army of Thābit was massacred and he was put to flight before Marwān. When the partisans of Ibrāhīm realized that Marwān had vanquished them in the first battle, the trembled. So they massed troops so numerous that they could not be counted and even gathered villagers to fight with slings (*textus* 189.13f: ܘܚܒܪ̈ܘܗܝ ܕܩܪܒܐ ܥܡܗܘܢ ܘܟܕ ܡܢܗܘܢ ܪܕܝ ܠܐܝܠܝܢ ܕܚܙܩܝܢ ܗܘܘ ܒܩܠܥܐ ܣܘܪ̈ܝܝܐ)." Presumably these were Christian villagers, who were still accustomed to fighting with slings in certain conditions, as during the 6th century, cf. *Edessa 502f.

"Then the armies marched toward each other, and pitched camp facing one another at 'Ayn Garrā. After waging many battles against each other, countless victims falling from both sides, in the end Marwān vanquished them and put them to flight, although Ibrāhīm and his brothers, as well as Sulaymān the son of Hishām, fled. A battle like this had never been seen seen in the world, nor did blood overflow in any other place as it did there. More than five thousand villagers were also killed. After this victory, Marwān besieged Emesa, subdued it [and] destroyed its wall (*textus* 189.22f: ܚܕ. ܕܝܢ ܝܘܕܝܐ ܐܚܕ ܡܢܗ ܕܗܒܐ ܒܕܡܝ̈ ܟܠܗ ܚܡܫܡܐܐ [ܐ]ܠܦܝ̈ܢ ܕܝܢ̈ܪ̈ܐ). He exhumed Yazīd (II) from the grave and crucified him, head down, on a stake. He took gold from one Jew, worth four hundred thousand (dinars)."

The events of the siege (including the destruction of the wall) provide a suitable context for the craftsmen he recruited in Jazira, who were thus employed as sappers and engineers during the siege.

GERMANIKEIA	παρέλαβεν	c. 745

Theophanes 422 (AM 6237, MS 584); Nik. 67; cf. Hoyland 2011: 264.

During the Arab civil war (cf. *Emesa) "At this juncture Constantine invaded Syria and Doulichia [Dulûk, Doliche or Dülük] and captured Germanikeia, taking advantage of the internecine war among the Arabs. The Arabs who lived in those parts he sent off unarmed under a verbal assurance. He took along his maternal relatives and transferred them to Byzantium together with many Syrians—Monophysite heretics, most of whom have continued to live Thrace to this very day and crucify the Trinity in the Trisagion in the manner of Peter the Fuller." ... ἐν τούτοις Κωνσταντῖνος Γερμανίκειαν παρέλαβεν ἐπιστρατεύσας τὴν Συρίαν καὶ Δουλιχίαν ἄδειαν εὑρὼν διὰ τὴν τῶν Ἀράβων πρὸς ἀλλήλους μάχην. λόγῳ δὲ τοὺς ἐν αὐταῖς Ἄραβας ἐξαποστείλας ἀόπλους προσελάβετο καὶ τοὺς πρὸς μητρὸς συγγενεῖς καὶ ἐν Βυζαντίῳ μετῴκισεν, σὺν καὶ πολλοῖς Σύροις μονοφυσίταις αἱρετικοῖς, ὧν οἱ πλείους εἰς τὴν Θρᾴκην οἰκοῦντες μέχρι τοῦ νῦν καὶ ἐν τῷ τρισαγίῳ τὴν τριάδα σταυροῦντες κατὰ Πέτρον τὸν Γναφέα διήρκεσαν.

Nikephoros: "After making such disposition concerning his own rule, Constantine undertook a little later an expedition against the country of the Saracens, from whom he captured the city (ὧν εἷλε πόλιν) of Germanikeia in the region of Euphratasia."

MELITENE; THEODOSIOPOLIS παρέλαβεν 750

Theophanes 427 (AM 6243, MS 590); Nik. 70; *Chr. Zuq.* s.a. 749-50 (Harrak 1999: 189f); cf. Hoyland 2011: 289f.

Theophanes has: "In this year the new masters slew the greater part of the Christians, whom they treacherously arrested at Antipatris in Palestine, because of their being related to the previous rulers. In the same year Constantine occupied (παρέλαβεν) Theodosioupolis as well as Melitene and conquered the Armenians."

Nikephoros adds Theodosiopolis: "After these events Constantine crowned his son Leo emperor and straightaway marched against the Saracens. He came to the town of Melitene, which he took by siege (πολιορκίᾳ εἶλε) and carried off from there a great number of captives and much booty."

A slightly different dating is found in Michael the Syrian and *Chr. 1234*, who also add some technical details, such as the use of ramparts and breaching of walls (see Hoyland); the Zuqnin chronicler places this in 750 and writes: "The year one thousand and sixty-one: Constantine (V), the Roman Emperor, marched out with numerous forces, tore down Melitene, and razed it. He ousted its inhabitants, without killing anyone or taking away any of their possessions. He simply brought them out and sent them away. All of them went to the Jazīra. He then destroyed Melitene's wall down to its foundation and burned its houses, after which he took his army and returned to his land." (*Textus* 207.10-16: ܢܦܩ ܩܘܣܛܢܛܝܢܘܣ ܡܠܟܐ ܪ̈ܗܘܡܝܐ ܒܚܝܠܘܬܐ ܣܓܝܐܬܐ. ܘܣܚܦܗ ܘܐܪܡܝܗ ܠܡܠܝܛܝܢܐ. ܘܐܦܩ ܐܢܘܢ ܠܥܡܘܪ̈ܝܗ ܡܢܗ. ܟܕ ܠܐ ܩܛܠ ܠܐܢܫ ܘܐܦ ܠܐ ܫܒܩ ܡܕܡ ܡܢ ܕܝܠܗܘܢ. ܐܠܐ ܐܦܩ ܘܫܕܪ ܐܢܘܢ ܡܢܗ. ܘܢܦܩ ܟܠܗܘܢ ܠܓܙܝܪܬܐ. ܘܣܚܦ ܫܘܪܗ ܘܐܘܩܕ ܒ̈ܬܘܗ̇. ܘܢܣܒ ܚܝܠܗ ܘܗܦܟ ܠܐܪܥܗ.)

IHBURG *occisus est a Saxonibus in castro* 753

Fred. Cont. 35; *ARF* 753.

Fredegar explains how Pippin ravaged rebellious Saxony *cum magno apparatu veniens*. Emerging victorious, he imposed heavy tribute; the king returned and crossed the Rhine at Bonn (*ad Renum ad castro cuius nomen est Bonna ueniens*). The *ARF* adds that "Bishop Hildegard was killed by the Saxon in the castle called Ihburg (*Hildegarius episcopus occisus est a Saxonibus in castro, quod dicitur Iuberg*). In spite of this Pepin had the victory..."

The Saxon campaign shows the military function of the bishop Hildegard, who commanded a fort on the Saxon border. Since he was from Cologne one can assume that Charles crossed the Rhine there. It is uncertain at what point he was killed during the campaign, but the implication is that he was killed in the course of a Saxon siege of his fort. Pippin's *magno apparatu* implies something of a siege train, which makes sense considering the Saxon practice of building forts (see chapter 4.3 and 7.2.5).

| SUSA | *cum telis et machinis et multo apparatu...defendere* | 755 |

Fred. Cont. 37.

When ambassadors failed to convince the Lombards to withdraw [from *Rome, cf. 756], the Franks decided to invade in support of the Pope. "So at the end of the year the king summoned all his Franks to meet him on 1 March (the customary Frankish date) at the royal villa of Berny-Rivière; and there he took counsel with his great men. About the time when kings go forth to battle, he set out for Lombardy with Pope Stephen and all the peoples who dwelt in his kingdom and his Frankish troops. The host passed through the Lyonnais and Vienne and reached Saint-Jean-de-Maurienne. King Aistulf of the Lombards got word of this, summoned the entire Lombard army and made for the defile known as the valley of Susa. There with all his force he established his base, and prepared to defend himself with the weapons, machines of war and mass of stores that he had most wickedly got ready for resisting both the Empire and the Apostolic Roman See."

|| ...euoluto anno, praefatus rex ad K. Mar. omnes Francos, sicut mos Francorum est, Bernaco uilla publica ad se uenire praecepit. Initoque consilio cum proceribus suis eo tempore quo solent reges ad bella procedere, cum Stephano papa uel reliquas nationes qui in suo regno commorabantur, et Francorum agmina partibus Langobardie cum omni multitudine per Lugduno Gallie et Vienna pergentes usque Maurienna peruenerunt. Aistulfus rex Langobardorum haec audiens, commoto omni exercitu Langobardorum usque ad clusas quae cognominatur ualle Seusana ueniens, ibi cum omni exercitu suo castra metatus est et cum telis et machinis et multo apparatu quod nequiter contra rem publicam et sedem Romanam apostolicam admiserat nefarie nitebatur defendere. ||

The Franks had difficulty penetrating the Lombard defenses, but a small group broke through or encircled and then broke through. There was fierce resistance from Aistulf but he lost "almost the whole army that he had brought with him, the dukes as well as the counts and all the nobility of the Lombard people." || ...pene omnem exercitus suum quod secum duxerat, tam ducibus quam comitibus uel omnes maiores natu gentis Langobardorum... || Aistulf fled with a few men to *Pavia.

| Pavia; others | *castra metatus undique; castra ... diripuit* | 755 |

Fred. Cont. 37; see *Susa (above).

"Thus, with God's help, the noble King Pippin had the victory; and he advanced with his whole army and the numerous columns of the Franks to Pavia, where he pitched camp [a better translation would be: Where he set up siege camps from every direction]. Far and wide in all directions he ravaged and burnt the lands of Italy until he had devastated all that region, pulled down all the Lombard strongholds and captured and taken in charge a great treasure of gold and silver as well as a mass of equipment and all their tents." (Wallace-Hadrill notes: "Tentoria are also mantlets or siege-shelters but this fits the context less well.")

|| Igitur precelsus rex Pippinus patrata Deo adiuuante uictoria, cum omni exercitu uel multitudine agmina Francorum usque ad Ticinum accessit. Castra metatus est undique, omnia quae in giro fuit uastans partibus Italie maxime igne concremauit, totam regionem illam uastauit, castra Langobardorum omnia diripuit et multos thesauros tam auri quam argenti uel alia quam plura ornamenta et eorum tentoria omnia capuit et cepit. ||

Aistulf submitted and the pope was restored. *ARF* 755 adds that he was accompanied by "Fulrad and his companions", while *Annales Einhardi* adds "a not inconsiderable body of Franks" (*non minima Francorum manu*), some of whom participated in the defense of *Rome in 756.

| Rome | over 55 days | *obsidentes et ex omni parte circumdantes* | 756 |

Codex Carolinus 8-9.

"On that very first of January the whole army of the same Aistulf, king of the Lombards joined together against this Roman city from the region of Tuscany and encamped next to the gate of the apostolic saint Peter and the gate of saints Pancratius and Portuensis; the same Aistulf joined with other armies from another region and [they] fixed [their] tents next to the Salarian gate and other gates, and he addressed us, saying: "Open up the Salarian gate for me, and I will enter the city; and give me you pontiff, and I [will] have compassion for you; otherwise, [when] I overthrow the wall, and kill you with one sword, I will see who can save you from my hands." But even the all the Beneventans joining [the Lombards] against this Roman city encamped next to the gate of saint John the Baptist and also next to the gate of saint Paul the apostle and other gates of this Roman city."

|| ...in ipsis Ianuarium Kalendis cunctus eiusdem Haistulfi Langobardorum regis exercitus Tusciae partibus in hanc civitatem Romanam coniunx-

erunt et resederunt iuxta portam beati Petri apostolici atque portam sancti Pancratii et Portuensi; ipse vero Haistulfus cum aliis exercitibus coniunxit ex alia parte et sua fixit tentoria iuxta portam Salariam et ceteras portas et nobis direxit dicens: 'Aperite mihi portam Salariam, et ingrediar civitatem; et tradite mihi pontificem vestrum, et habeo in vobis compassionem; alioquin, muros subvertens, uno vos gladio interficiam et videam, quis vos eruere possit a manibus meis'. Sed et Beneventani omnes generaliter in hanc Romanam urbem coniungentes resederunt iuxta portam beati Iohannis baptiste seu et iuxta portam beati Pauli apostolici vel ceteras istius Romane civitatis portas. ||

"And with fire and sword they consumed far and wide everything outside the city; they destroyed all the estates burning them almost to the ground." || Et omnia extra urbem praedia longe lateque ferro et igne consumpserunt, domos omnes conburentes poene ad fundamenta destruxerunt. || The Lombards also burned churches, destroyed holy images, raped nuns, burned papal and noble estates, and took captive members from the *familiam beati Petri*.

"Besieging this afflicted Roman city for fifty-five days and surrounding it from every direction, in fierce battles day and night they assaulted with terrible fury; they arrayed against us on the walls incessantly with various engines and many inventions and did not lack men fighting against us, so that, may God avert it, the same sinful Aistulf should subject us to his power and kill everyone with the sword. For mocking us thus with great fury, they declared: 'Behold, you are surrounded by us and you shall not escape our hands; now let the Franks come and deliver you from our hands'."

|| Quinquaginta et quinque dies hanc adflictam Romanam civitatem obsidentes et ex omni parte circumdantes, prelia fortissimo die noctuque cum pessimo furore incessantes [add. Ep. 9: incessanter cum diversis machinis et adinventionibus plurimis] contra nos ad muros istius Romane urbis commiserunt et non deficiebant inpugnantes nos, ut suae potestati, quod avertat divinitas, subiciens omnes uno gladio isdem iniquus Haistulfus interimeret. Ita enim, cum magno furore exprobantes nos, adserebant: 'Ecce circumdati estis a nobis et non effugietis manus nostras; veniant nunc Franci et eruant vos de manibus nostris'. ||

The Embolum on the envoys (p. 498) promises eternal life for effacing all barbarian nations: "... victorious through the intercession of Saint Peter, you will destroy all barbarian nations and possess eternal life." || ...victor, intercedente beato Petro, super omnes barbaras nationes efficiaris et vitam aeternam possideas. || It also describes the direct participation of clergy in the fighting: "The aforesaid abbot Warnehar donned armor for the love of saint Peter and kept guard on the walls of this afflicted Roman city day and night, and, as a good athlete of Christ, fought alongside all his men for the defense and liberation of all of us." || Praefatus vero Warneharius abbas pro amore beati Petri, loricam se induens, per muros istius afflicte Romane civitatis die noctuque vigilavit et pro nostra omnium Romanorum defensione atque liberatione, ut bonus atleta Christi, totis suis viribus decertavit. ||

Ed. p. 4 n. 1 states that "Hanc epistolam [i.e. ep. 8] exemplum esse illius, quae ad Pippinum, Carolum, Carolomannum omnesque Francos data est (ep. 9), exposuit Oelsner [etc...]." If the editor's argument is correct, the information *cum machinis diversis* and so on is original, but has been removed from the *exemplum*, along with some other specific information on the structure of the Frankish body politic.

NARNI; OTHER CITIES   *abstulerunt; conprehenderunt*   756
Codex Carolinus 8-9.

In addition to the siege of *Rome and raiding its territory, the Lombards committed other crimes: "For they also tore away the city of Narni, which Your Christianity conceded to saint Peter, and he also took other cities of ours." || Nam et civitatem Narniensem, quam beato Petro tua christianitas concessit, abstulerunt seu et aliquas civitates nostras conprehenderunt. ||

PAVIA   *circa muros...utraque parte fixit tentoria*   756
Fred. Cont. 38; *ARF* 756.

While Aistulf again ravaged the region of Rome, burning around St. Peter. Pippin marched with "the entire Frankish army" through Burgundy, Chalon, and Geneva to St-Jean de Maurienne. The Franks burst through passes, raided around, "... and encamped round the walls of Pavia so that none should escape from the city" || ...et circa muros Ticini utraque parte fixit tentoria, ita ut nullus exinde euadere potuisset. ||

Pippin reacted quickly to the news of the Lombard siege of *Rome, but used a different marching route from his campaign the year before. His staging post was the same, however. His swift reaction may have caught the Lombards off guard, as their campaign against Rome took place during the winter, and possibly they thought they were safe from intervention across the passes. They appear not to have taken the elaborate defensive measures of the previous year. This also has a good example of Franks attacking in two major columns, the other being led by Tassilo of Bavaria.

ARF only has *Papiam obsedit, Haistulfum inclusit*. Aistulf again submitted, paying treasure and tribute.

?Ravenna	*conquisivit*	756

ARF 756.

In addition to conducting the siege of *Pavia, Pippin conquered (*conquisivit*) Ravenna with the Pentapolis and the whole exarchate and handed it over to St. Peter.

Saxon strongholds at Sythen	*per virtutem introivit*	758

ARF 758.

"King Pepin went into Saxony and took the strongholds of the Saxons at Sythen by storm. And he inflicted bloody defeats on the Saxon people. They then promised Pepin to obey all his order and to present as gifts at his assembly up to three hundred horses every year." || Pippinus rex in Saxoniam ibat, et firmitates Saxonum per virtutem introivit in loco, qui dicitur Sitnia, et multae strages factae sunt in populo Saxonum; et tunc polliciti sunt contra Pippinum omnes voluntates eius faciendum et honores in placito suo prasentandum usque in equos CCC per singulos annos. ||

Annales Einhardi (see *ARF*) clarifies the violent nature of the storm: || Pippinus rex cum exercitu Saxoniam adgressus est; et quamvis Saxonibus validissime resistentibus et munitiones suas tuentibus, pulsis proelio propugnatoribus per ipsum, quo patriam defendere conabantur, vallum intravit. Commissisque passim proeliis plurimam ex ipsis multitudinem cecidit coegitque, ut promitterent se omnem voluntatem illius esse facturos et annis singulis honoris causa ad generalem conventum equos CCC pro munere daturos. His ita compositis et more Saxonico, ut rata esse deberent, confirmatis in Galliam sese cum exercitu suo recepit.||

Bourbon	*in giro castra posuisset, subito...captus*	761

Fred. Cont. 42.

The war Aquitanian broke out in 760 when negotiations broke down and Pippin raided Aquitaine (Fred. Cont. 41; *ARF* 760). Waiofar responded with a raid in force against Carolingian forward bases, including the region of Autun and Chalon and the royal villa of Mailly. In fury, Pippin "summoned all Franks to assemble in arms at the Loire. He then set out with a large force to Troyes and from there went through Auxerre to the town of

Nevers, where he crossed the Loire and reached the stronghold of Bourbon in the district of Bourges. And he surrounded it with (fortified) camps; and then the Franks stormed it and set it on fire. Such of Waiofar's men as were found there he took prisoner. || ...iubet omnes Francos ut hostiliter placito instituto ad Ligerem uenissent. Commotoque exercito cum omne multitudine iterum usque ad Trecas accessit, inde per Autisioderum ad Neuernum urbem ueniens Ligeris fluuium transmeato ad castrum cuius nomen est Burbone in pago Bituriuo peruenit. Cumque in giro castra posuisset, subito a Francis captus atque succensus est et homines Waiofarii quos ibidem inuenit secum duxit. ||

This is also mentioned more briefly in *ARF* 761 (see *Chantelle below).

CLERMONT *bellando cepit* 761
Fred. Cont. 42.

After taking *Bourbon (above), Pippin "laid waste a large part of Aquitaine, advanced with his whole force to the town of Auvergne and took and burnt the fortress of Clermont. A great many men, women and children perished in the flames. Count Bladinus of Auvergne was taken prisoner and brought in chains to the king's presence. Many Gascons were taken and slain in that engagement. And now that the town was taken and the whole region devastated, King Pippin returned home a second time, his army, by God's help, unscathed, laden with much plunder." || Maximam partem Aquitanie uastans, usque urbem Aruernam cum omni exercitu ueniens, Claremonte castro captum atque succensum bellando cepit et multitudinem hominum, tam uirorum quam feminarum uel infantum plurimi, in ipso incendio cremauerunt. Bladino comite ipsius urbis Aruernico captum atque ligatum ad praesentia Regis adduxerunt, et multi Vascones in eo proelio capti atque interfecti sunt. Igitur rex Pippinus, urbem captam hac regionem illam totam uastatam, cum praeda uel spolia multa Deo auxiliante inlesum exercitum iterum remeauit ad propria. ||

This is also mentioned more briefly in *ARF* 761 (see *Chantelle below), but it has no mention of atrocities committed against civilians.

CHANTELLE; FORTS IN AUVERGNE *per pugnam/placitum coepit* 761
ARF 761; cf. *Bourbon and *Clermont (above).

Pippin "set out on a campaign into that region and captured many castles, the names of which are Bourbon, Chantelle, and Clermont. These he took in battle (*istas per pugnam coepit*) and in Auvergne he obtained by treaty many other castles, which submitted to his authority (*et in Alverno alia multa castella coepit per placitum, quae se subdiderunt in eius dominio*). He went as far as Limoges, devastating this province because of Duke Waifar's slights."

The *Annales Einhardi* also mentions that Charles (i.e. Charlemagne) came along on campaign.

BOURGES *circumsepsit, fractisque muris cepit urbem* 762
Fred. Cont. 43; *ARF* 762.

"It happened that the year after the taking of Auvergne and the devastation of all that region, King Pippin—it was the eleventh year of his reign—came with his whole Frankish army to Bourges, fortified his position there and plundered the countryside round about [came with the full strength of the Frankish people, built fortifications from every side, and devastated everything that lay around]. He then surrounded the town with a strong fortification, with siege-engines and all manner of weapons so that no one would have dared to leave or enter it, and he also built a rampart. Finally he took the city after many had been wounded and more slain, and the walls breached. He restored it to his rule by right of conquest, but of his goodness showed mercy to the men left there as garrison by Waiofar, and dismissed them to go off home. Count Chunibert and such Gascons as he found there had to swear fealty to him and remain in his company: their wives and children were told to set off on foot for Francia. He ordered the repair of the walls of Bourges and left the city in charge of counts of his own."|| Factum est autem ut, post quod Pippinus rex urbem Aruernam cepit hac regionem illam totam uastauit, sequente anno, id est anno XI regni ipsius cum uniuersa multitudine gentis Francorum Bitoricas uenit, castra metatusque est undique et omnia quae in giro fuit uastauit. Circumsepsit urbem munitionem fortissimam, ita ut nullus egredi ausus fuisset aut ingredi potuisset, cum machinis et omni genere armorum, circumdedit ea uallo. Multis uulneratis plurisque interfectis fractisque muris cepit urbem et restituit eam dicione sue iure proelii et homines illos quos Waiofarius ad defendendam ipsam ciuitatem dimiserat clementiam sue pietatis absoluit dimissisque reuersi sunt ad propria. Vniberto comte uel reliquos Vascones quos ibidem inuenit sacramentis datis secum adduxit, uxores eorum hac liberos in Frantia ambulare praecepit, muros ipsius Bitorice ciuitatis restaurare iubet, comites suos in ipsa ciuitate ad custodiendum dimisit. ||

The siegeworks were extensive: first Pippin *circumsepsit* (fenced in) the city with a *munitionem fortissimam*; this cut the city off completely. He then set up siege engines, and "surrounded *them* with a rampart." The pronoun *ea* must refer back to *arma*; hence we have a double set of fortifications set up around the city: one to contain the besieged, another to protect the besiegers and their equipment from relieving forces. Interesting also the appearance of Gascons as garrisons, with their wives and children; all of these were taken by Pippin, whereas Waiofar's men were sent back to him. Note also repairs of walls, which presupposed craftsmen and supplies, while several counts were placed in the town as garrison; this implies several hundred men at least.

ARF 762 only has a summary version: For a third time King Pepin launched a campaign into Aquitaine and he took the city of Bourges and the castle of Thouars (*coepit civitatem Bituricam et castrum, quod dicitur Toarcis*).

THOUARS   *mira celeritate captus*   762
Fred. Cont. 43; *ARF* 762; *Bourges (above).

"Then he and the entire Frankish host made for the stronghold of Thouars and encamped round about it. The stronghold was captured with extraordinary speed, and was afterwards burnt. He took back with him to Francia the Gascons whom he found there, together with the count himself. And so, Christ going before him, King Pippin came home again with great spoils." || Inde cum omni exercitu Francorum usque ad castro qui uocatur Toartius ueniens, cumque in giro castra posuisset, ipse castrus mira celeritate captus atque succensus est et Vascones quos ibidem inuenit una cum ipso comite secum duxit in Frantia. Pippinus rex Christo duce cum omni exercitu Francorum cum multa praeda et spolia iterum reuersus est ad sedem propriam. ||

?NARBONNE   *custodias... capere aut interficere*   763
Fred. Cont. 44.

Waiofar attempted to take Narbonne by ambushing the garrison: "He sent his cousin, Count Mantio, with other counts to Narbonne with the object of capturing and killing the garrison, sent by the king to hold Narbonne against the Saracens, either as they entered the city or on their way back home. It so happened that Counts Australdus and Galemanius were making for home with their following when Mantio fell upon them with a crowd of Gascons." || ...nam Mantione comite consubrino suo partibus Narbone cum reliquis comitibus transmisit, ut custodias quod praedictus rex Narbonam propter gentem Saracenorum ad custodiendum miserat aut ad intrandum aut quando iterum in patria reuertebant capere aut interficere eos potuissent. Factum est ut Australdus comis et Galemanius itemque comis cum pares eorum ad propria reuerterent. Sic Mantio una cum multitudine gente Wasconorum super eos inruit. ||

There was a stiff fight, but the Carolingian counts killed "Mantio and all his companions" (*cum uniuersos pares suos*); the Gascons turned tail leaving their horses; a drawn-out pursuit ensued while the Franks made off with horses and other booty. While this never came to a formal siege, the conflict illuminates the dynamic of border defenses. Waiofar had also sent other armies who were to raid the important Carolingian staging posts, but they all failed miserably, and Waiofar afterwards gave up on holding his fortified cities (Fred. Cont. 45; chapter 4.3.3). On both sides of the border then, counts (and even abbots) with their followings had great importance in warfare, performing garrison duty along with their troops (*pares*). It also demonstrates that the Gascon troops of Waiofar fought on foot, as they lost their horses when they fled from the battlefield; this implies an infantry battle with the horses gathered in camp. It may conceivably refer to remounts, but it appears unlikely that these would be so close to an ambush site.

KAMAKHON	παρεκάθισε	766

Theoph. 444 (AM 6261, MS 613); Tabari II 353; *Chr. Zuq.* 766-67 (Harrak 1999: 206-16; ed. 228-43); Baladhuri 288 (DeGoeje 184f).

This was a major operation under several Arab commanders against the Byzantine border fortification of Kamakhon. The date is agreed by most sources as 766. Theophanes and Tabari both mention it briefly: "All summer Abdelas besieged Kamachon with 80,000 men, but he did not achieve any success and returned in shame." ... παρεκάθισε δὲ Ἀβδελλᾶς τὸ Κάμαχον μετὰ ὀγδοήκοντα χιλιάδων ὅλον τὸ θέρος, καὶ μηδὲν ἀνύσας, ὑπέστρεψεν κατῃσχυμμένος. (Note the use of verb and the stark contrast to the full description in *Chronicle of Zuqnin* below.) Tabari 353 (29: 42), AH 149 (February 16, 766–February 5, 767) notes: "In this year al-'Abbās b. Muḥammad led the summer raid on the Byzantine lands. With him were al-Ḥasan b. Qaḥṭabah and Muḥammad b. al-Ash'ath, and Muḥammad b. al-Ash'ath died on the road."

Baladhuri is somewhat more informative: "Al-'Abbās and al-Ḥasan advanced to Malaṭya from which they took provisions, and then camped around Kamkh. Al-'Abbās ordered that mangonels (*manājanīq*; another ms has the alternative pl. *majānīq*; see apparatus in DeGoeje) be set upon the fort. The holders of the fort covered it with cypress wood to protect against the mangonel stones, and killed by the stones they hurled two hundred Muslims. The Muslims then set their mantlets (*dabbābāt*) and fought severely until they captured it." Not only is the capture claimed by Baladhuri patently false (cf. *Kamakhon 793), but he also fails to mention the involvement of a large number of Syrian Christians in vast logistic demands made by the siege, as well as the actual fighting.

By far the fullest testimony, and the best description of a siege in the 8th century, comes from the *Zuqnin* chronicler, who personally spoke with many of the involved. He observed many nations in the Abbasid army attacking Amida and other cities on the Tigris; peoples listed are Sindhis, Alans(?), Khazars, Medes, Persians, Kufans, Arabs, Khurasanians and Turks, many of whom were idolaters and pagans (206f). Malnutrition ensued when the invading army gorged itself on fruits and vines. "Then all of them fell victim to various sicknesses, especially dysentery and haemorrhoids, in such a manner that wherever they settled or fled, unburied corpses of people could be seen alongside roads and high places and valleys, discarded and being devoured by animals. All their beasts of burden died too, especially the camels that followed them into the land; out of fifty or sixty that accompanied a man, not even five or six—and sometimes not even one—came out." (207; note the casualty rate of over 90 %.)

"After the whole army marched in, they besieged a fortress called Qamh, that was located on the borders. But among the many craftsmen from all of the Jazira who marched with the army, some were left by 'Abbas to rebuild a fortress called "Ziad", while others marched with the army. 'Abbas brought Armenian wagons in which he pulled numerous cedar beams. He ordered carpenters to make mangonels with them, which he placed on a peak facing the fortress so that they might hurl stones with them inside the fortress. The Romans, who were inside the fortress, also mounted mangonels against them." (207f)

(Ed. 230.27-231.9)

ܘܟܕ ܥܠ ܚܝܠܐ ܟܠܗ: ܘܢܩܦܘ ܥܠ ܚܣܢܐ ܐܚܕܐ ܕܐܝܬ ܒܝܬ ܬܚܘ̈ܡܐ
ܕܡܬܩܪܐ ܩܡܚ: ܡܠܓ̈ܪܐ ܓܝܪ ܣܓܝ̈ܐܐ ܐܝܠܝܢ ܕܥܠܘ ܥܡܗܘܢ ܡܢ ܟܠܗ
ܓܙܪܬܐ: ܡܢܗܘܢ ܫܒܩ ܥܒܣ ܠܡܒܢܐ ܚܣܢܐ ܚܕ ܕܡܬܩܪܐ ܙܝܕ. ܘܡܢܗܘܢ
ܥܠܘ ܥܡܗܘܢ. ܘܫܪܪ ܗܘ ܥܒܣ. ܘܐܝܬܝ ܠܗ ܥܓ̈ܠܬܐ ܐܪܡܢܝܬܐ. ܘܓܪܫ
ܒܗܝܢ ܩܝ̈ܣܐ ܣܓܝ̈ܐܐ ܕܐܪ̈ܙܐ. ܘܦܩܕ ܠܢܓ̈ܪܐ ܘܥܒܕ ܡܢܗܘܢ ܡܘܢܕܢܝܩܢ. ܘܐܩܝܡ
ܥܠ ܪܝܫܐ ܕܠܘܩܒܠ ܚܣܢܐ. ܕܢܗܘܘܢ ܫܕܝܢ ܒܗܘܢ ܟܐܦܐ ܠܓܘ ܚܣܢܐ. ܘܐܦ ܗܢܘܢ
ܪ̈ܗܘܡܝܐ ܕܓܘ ܚܣܢܐ. ܐܩܝܡ ܐܦ ܗܢܘܢ ܡܘܢܕܢܝܩܢ ܠܘܩܒܠܗܘܢ.

Note the word for catapult at 231.6, 8, *mwndnyqn* (ܡܘܢܕܢܝܩܢ, [wrongly given as ܡܘܢܕܢܝܩܝܢ, *mwndnyqyn*, by Harrak 1999: 208 n. 1] presumably pronounced *mōndanīqān* with absolute feminine plural ending; perhaps Persian plural, as there are no syame dots and this appears to fit better with the grammar.).

"Nevertheless, the besieged Romans devised for themselves an invincible weapon and built for themselves an impregnable wall—I mean God their creator—saying: "There is no salvation outside the Lord. It is better for us to trust the Lord than to trust a man or a ruler. Verily, all the nations surrounded us but the name of the Lord our God will destroy them." (p. 208 n. 2, taken from Ps 3:2 and 107: 8-10. Perhaps this is taken from actual liturgy.)

Sergius, commander at Kamakhon, had a good reputation for treating Syrian captives well ("They themselves testified about the man before us and before everyone.") when captured by the Romans for crossing the border in search of madder due to unemployment and Arab exactions. Sergius would allow them to return home with provisions or stay in Roman territory. "Truly, my brothers, God has rewarded this man in that he saved him, together with all the people who were with him inside the fortress, from the hands of the Assyrians!"

The fortress was surrounded and a storm commenced. "As the Persians [i.e. Abbasids] were fighting with all means, all their tricks proved failing. They made mobile wooden houses [presumably Baladhuri's *dabbābāt*] with which to fill the ravine beside the city-wall with dirt and stones, but this trick failed. The Romans were shooting stones from inside, and because they were shooting them accurately, they killed numerous people outside and even destroyed the mangonels of the Persians. Because there was only one side of the fortress which could be scaled, the Romans brought up long and strong beams, tied up big round stones to their top ends, and placed them in the opening. When the Persians gathered to climb up, the Romans released these beams and they swept all of them, driving them down before them and tearing them into small pieces." (209)

(Ed. 233.4-11)

ܟܕ ܕܝܢ ܗܢܘ̈ ܦܘܪ̈ܣܝܐ ܒܟܠ ܦܘܪ̈ܣܝܢ ܡܬܟܬܫܝܢ ܗܘܘ ܟܠܗܝܢ ܢܟܠ̈ܝܗܘܢ
ܒܛܠ̈ܢܝܬܐ ܡܫܬܟܚܢ ܗܘ̈ܝ. ܥܒܕܘ ܕܝܢ ܒ̈ܬܐ ܕܩܝܣܐ ܕܡܗܠܟܝܢ. ܐܝܟܢܐ ܕܢܡܠܘܢ
ܥܦܪܐ ܘܟܐ̈ܦܐ ܠܢܚܠܐ ܗ̇ܝ ܕܠܘܬ ܥܠ ܫܘܪܐ. ܐܦ ܗܕܐ ܒܛܠܢܝܬܐ ܗܘܬ.
ܪ̈ܗܘܡܝܐ ܕܝܢ ܐܝܠܝܢ ܕܫܕܝܢ ܗܘܘ ܟܐ̈ܦܐ ܡܢ ܠܓܘ: ܟܕ ܕܬܪܝܨܐܝܬ ܫܕܝܢ ܗܘܘ.
ܠܥܡܐ ܣܓܝܐܐ ܕܠܒܪ ܡܩܛܠܝܢ ܗܘܘ. ܘܐܦ ܠܡܢܓܢܝܩܐ ܕܦܪ̈ܣܝܐ ܡܬܒܪܝܢ ܗܘܘ.

The Persians attempted a sneak attack when the guards were asleep, and as they began to ascend, they cried "Allahu Akbar," but the guards woke up in time and were able to use similar tactics to destroy them again.

Two Persian commanders set out from their (unspecified) fort to raid the Romans; they crossed uninhabited mountains without supplies—indicating the great logistical problems experienced by the summer raiders—in order to reach Caesarea and beyond. They surprised the inhabitants of the villages, taking captives, cattle and valuables (209f). The raiders returned towards Syria. When they thought they had cleared the danger, they allowed themselves to relax and pitch camp in a large meadow. However, a Roman commander with almost 12,000 men came upon them by chance. When the commander heard of the Persian army, "he sent others—almost three hundred armed cavalrymen—in order to verify the matter, to see whether it was true or perhaps some apparition seen by those men through hallucination. When those who were dispatched went up and distinctly saw the other army, they informed the one who dispatched them, and he went up with four or five thousand men." The Persians were stunned by the sight (211).

Persian emissaries discovered that the Roman army was large and ready for battle, so they decided to negotiate, freeing the captives and giving back cattle and spoils hoping to be spared. "Nevertheless, the Romans did not abide by this, but quickly sent to notify the cities and the other military commanders of the matter. They dispatched a great army against them and divided it into four divisions. They attacked the Persians from the front and from the rear, and from this side and from the other side. Because it was still night, they gave to each other a signal: After they would come down all prepared, they would sound the trumpets and the whole crowd would shout at the same time *Kyrie eleison*! (ed. 237.10, 12: ܩܘܪܝܐܠܝܣܘܢ) When they went down and were ready, they sounded the trumpets and the crowd shouted like thunder: *Kyrie eleison*! The Persians heard their voices and trembled, becoming like the dead and the slain ones lying in graves." The Persian terror grew as they were unable to escape due to the Romans surrounding them completely. Furthermore, divine intervention confused Persians, from slave-keepers to prisoners, and in an hour they suffered a great defeat. "The Persians themselves, who escaped the battle while injured, testified before us under great oaths that they never saw or heard of so much blood in one single place as there; they said that the blood and the corpses rose up as high as a horse's belly on that meadow." (212)

Only 1,000 Persians under the commander Radād escaped to Melitene, the rest were killed or captured. The other raiding army under Mālik fled with 5,000 to Qalinqala (Theodosiopolis). These were the survivors of the army of 50,000 that had set out from Kamakhon. Back at Kamakhon, villagers had been recruited to bring supplies to the besieging army from Syria, but they heavy losses due to the deaths of their beasts of burden noted above. Consequently, the Persian besiegers began to starve. Unemployed people from "the Jazira, the West and even Inner Armenia" were attracted by the commanders to provide supplies for profit; many accepted the offer (213).

"Everyone brought what he could until everything became abundantly available. Traders, shopkeepers, textile dealers, and others like them sold wheat, barley, flour and products necessary for human life." Great amount of supplies were brought in; the opportunity for business prompted much greed among the traders. However, many villagers were killed by artillery bombardment from Romans (apparently among those who were recruited as laborers and set to work on siegeworks and man the trebuchets), so the Persian commander moved them to other tasks, assigning regular troops to man the *mwndanīq* instead: "Moreover, as the Persians used to attack the fortress, day and night, using many military devices, every day the villagers earned nothing but their own destruction, for many among them used to be killed daily by the Roman mangonels. But 'Abbās, because he was a compassionate man, granted a favour to these people who joined this labour. When he saw that many of them were killed by the stones of the mangonels which the Romans were shooting from inside, he gathered his military commanders and ordered that everyday someone from among them [i.e. the commanders of regular troops] should appoint men to shoot stones from mangonels, and that the villagers should be assigned to other works far away from the deadly danger." (214)

(Ed. 239.23-240)

ܘܟܕ ܕܝܢ ܒܦܘܪܣܐ ܡܫܚܠܦܐ ܕܩܪܒܐ. ܠܠܝܐ ܐܝܡܡ ܥܠ ܚܨܢܐ ܡܩܪܒܝܢ ܗܘܘ:
ܘܐܠܐ ܐܢ ܐܟܪܝܐ ܠܡܘܬܗܘܢ. ܟܠܝܘܡ ܠܐ ܡܬܬܓܪܝܢ ܗܘܘ. ܟܕ ܡܚܒܠܝܢ
ܕܪܗܘܡܝܐ: ܡܫܢܝܬܐ ܡܢܗܘܢ ܟܠܝܘܡ ܡܬܩܛܠܐ ܗܘܬ. ܥܒܣ ܕܝܢ ܟܕ ܪܚܡܢܐ
ܐܝܬܘܗܝ، ܡܪܚܡܢܐ: ܥܒܕ ܐܦ ܛܝܒܘܬܐ ܥܡ ܗܠܝܢ ܡܫܟܚܐ ܕܥܠܝܠܝܢ ܗܘܘ
ܒܦܠܚܢܘܬܐ: ܗ̇ܝ، ܕܥܒܕܐ ܗܘܬ ܡܢ ܠܓܘ ܕܪ̈ܗܘܡܝܐ. ܟܢܫ ܠܪ̈ܒܝ ܚܝܠܐ ܕܝܠܗ ܘܦܩܕ
ܠܗܘܢ ܕܟܠܝܘܡ ܢܩܝܡ ܚܕ ܡܢܗܘܢ ܠܓܒܪܐ ܕܢܫܕܘܢ ܒܟܐܦܐ ܕܡܘܢܕܢܝܩ. ܘܩܘܪ̈ܝܐ
ܢܗܘܘܢ ܠܥܒ̈ܕܬܐ ܐܚܪ̈ܢܝܬܐ ܕܪܚܝܩܢ ܡܢ ܡܘܬܐ. (ܘܟܕ ܚܙܐ ܠܗܘܢ ܬܡܢ ܐܘܟܝܐ
ܡܫܟܚܐ:)

The Persian commander vowed to stay for even ten years, but the Romans defied all threats. Persian morale was weakened by the onset of winter and knowledge of only 5-6000 returned from the raiding army of 50,000.

Finally, the Persians decided to evacuate, but the merchants were forced to leave behind their goods due to lack of beasts of burden (214f). The

Persians had to burn the supplies in order to prevent the Romans from benefiting, but still much was left behind. Other traders were simply despoiled without compensation by an army on the move to Qalinqala. "Moreover, when ʿAbbas left, he ordered all the agents working under his authority (ed. 241.21: ܥܡܪܐ ܕܚܘܬܗ ܟܠܗܘܢ; *ʿamalē* is Arabic), to take away from the laborers (ed. 241.20: ܡܠܐܟܐ) the wage which they had received when they used to come into (the camp), as well as (a fee) on the(ir) donkeys, and in this way they were dismissed. At this point ʿAbbas left and returned by the way he came, with disgrace, shame, and considerable and endless losses."

"The other forces came down to Amida and the Tigris. They left for the Persian territories, in want, hungry and weak. Not even half of them came out, chiefly because their beasts of burden and their slaves had escaped and entered the Roman territories."

The magnitude of the Arabs' humiliation is mirrored by the economic effects of their activities. The influx of new coin to pay for supplies and services made it easier to forge large quantities of money (215).

TOULOUSE; ALBI, GEVAUDAN   *cepit; in deditionem accepit*   767
ARF 767; Fred. Cont. 49.

Very early in 767, Pippin "continued his march through Aquitaine into Narbonne and conquered Toulouse as well as Albi and Gevaudan." || ... *postea perrexit iter peragens partibus Aquitaniae per Narbonam, Tolosam coepit, Albiensem similiter necnon et Gavuldanum...* || *AE* specifies that Pippin *in deditionem accepit* the latter two.

He then spent the rest of winter at Vienne; the annals emphasize that the army rested from its labor, indicating a rather hard campaign. Fredegar adds that Bourges was the most important base for the campaign itself, but says nothing of the cities captured.

ROCKS & CAVES; ALLY, TURENNE, PEYRUSSE   *conquisivit*   767
ARF 767.

In addition to the winter campaign against *Toulouse: "In August of the same year he marched for the second time into Aquitaine and came as far as Bourges. There he held an assembly in camp with all the Franks as was the custom. Continuing his march from here, he proceeded as far as the Garonne, captured many rocks and caves, and the castles of Ally, Turenne, and Peyrusse, and returned to Bourges. || *Et in eodem anno in mense Augusto iterum perrexit partibus Aquitaniae, Bituricam usque venit; ibi syn-*

odum fecit cum omnibus Francis solito more in campo. Et inde iter peragens usque ad Garonnam pervenit, [*AE* adds *castella multa*] multas roccas et speluncas conquisivit, castrum Scoraliam, Torinnam, Petrociam et reversus est Bituricam. ||

Again Fredegar says nothing of the captured fortifications, but relates how Remistanius revolted and raided "so efficiently that not a peasant dared work in the fields and vineyards" (Fred. Cont. 50).

The war ended with the death of Waiofar and capture of Remistanius in 768 and a mopping-up operation by Charlemagne in 769 (Fred. Cont. 51f; *ARF* 768, 769). While there were no further sieges, the Carolingian ability to bring in supplies to ravaged areas and build fortifications on enemy territory finally broke Aquitanian resistance.

?GERMANIKEIA 769

Theophanes 445f (AM 6262, MS 614).

"In this year Banakas invaded the Roman country and made many captives. The Romans overran Fourth Armenia and devastated it. Salech died and (the inhabitants of) Germanikeia were transferred to Palestine." ... καὶ Ῥωμαῖοι εἰσῆλθον εἰς τὴν τετάρτην Ἀρμενίαν καὶ ἐσκύλευσαν αὐτήν. τελευτᾷ δὲ Σαυλὰχ καὶ μετεποιήθη Γερμανίκεια εἰς Παλαιστίνην.

MS note: "Mich. Syr. ii. 526, AG 1080, reports that the inhabitants of Germanikeia (Mar'ash) on suspicion of being Roman spies were removed to Ramlah."

SYKE παρεκάθισεν 771

Theophanes 445(f) (AM 6263, MS 615); Lilie 1976: 170f.

"In this year Banakes invaded the Roman country and, after moving down from Isauria, laid siege to (παρεκάθισεν) the fort (κάστρον) of Syke. When the emperor had heard of this, he wrote to Michael, *strategos* of the *Anatolikoi*, Manes, *strategos* of the Buccellarii, and Bardas, *strategos* of the *Armeniakoi*. These men arrived and occupied the Arabs' exit, which was a very difficult mountain pass. Meanwhile the fleet of the Kibyraiots under their *strategos* the *spatharios* Petronas cast anchor in the harbour of the fort. On seeing this and losing all hope, Banakes encouraged and roused his men. He marched up to the cavalry *themata* and, with a great shout, routed them. He killed many of them and, after devastating all the surrounding country, returned home with much booty."

The Arabs resorted to battle only when completely hemmed in with no other choice left. Battle was only offered when all other attempts had failed. The event is only noted by Theophanes, as there are no clear parallels in Arabic sources, cf. those mentioned just above in different locations.

?Mopsuestia 772
Theoph. 446 (AM 6264, MS 616); Lilie 1976: 171.

"In this year Abdelas sent Moualabitos to Africa at the head of a numerous army. Afdal Badinar invaded the Roman country and took 500 prisoners, but the inhabitants of Mopsuestia (Μοψουεστεῖς) encountered them in battle and killed 1,000 Arabs. Abdelas went to Jerusalem for his fast and ordered that Christians and Jews should be marked on their hands. Many Christians fled to the Roman country by sea. Sergius Kourikos was apprehended outside Syke and Lacherbaphos, who was the representative of the local community, in Cyprus."

Note the efforts to control population for purposes of taxation, intelligence and providing labor, as well as their reaction to this treatment.

Beritzia c. 772
Theophanes 446f (AM 6265, MS 616f).

"In this year, in the month of May, indiction 12 [774 but the date is debated], Constantine dispatched a fleet of 2,000 *chelandia* against Bulgaria. He himself embarked in the red *chelandia* and set out with the intention of entering the river Danube, leaving the *strategoi* of the cavalry *themata* outside the mountain passes in the hope that they might penetrate into Bulgaria while the Bulgarians were occupied with him. When, however, he had gone as far as Varna, he took fright and was considering a retreat." The Bulgars however were also afraid and asked for peace, and a treaty was drawn up in writing. "The emperor returned to the City after leaving garrisons from all the *themata* in the forts he had built." καὶ ὑποστρέψας ὁ βασιλεὺς εἰσῆλθεν ἐν τῇ πόλει ταξάτους ἀφεὶς ἐκ πάντων τῶν θεμάτων καὶ εἰς κάστρα ἅπερ ἔκτισεν. [MS note this is a difficult construction, but translated according to Anastasius' rendering, omitting a kai: *taxatis derelictis ex omnibus thematibus in castris, quae condidit.*]

"In the month of October of the 11th indiction [the chronology is confused; cf. above] the emperor received a dispatch from his secret friends in Bulgaria to the effect that the lords of Bulgaria was sending an army of 12,000 and a number of boyars in order to capture Berzitia and transfer the inhabitants to Bulgaria (... πρὸς τὸ αἰχμαλωτίσαι τὴν Βερζιτίαν καὶ μεταστῆσαι αὐτοὺς εἰς Βουλγαρίαν). He gathered soldiers of the *themata* and the Thrakesians and joined the Optimati to the *tagmata* to a total of 80,000. He marched to a place called Lithosoria and, without sounding the bugles, fell upon the Bulgarians, whom he routed in a great victory. He returned with much booty and many captives and celebrated a triumph in the City, which he entered with due ceremony. He called this war a 'noble war' inasmuch as he had met with no resistance and there had been no slaughter or shedding of Christian blood."

Note the deliberate policy of population transfers, as Bulgaria needed economic and military expertise to compete with Byzantium. This anecdote also provides very clear evidence against the farmer/soldier theory: Most of these troops were very far from home, so thematic troops must be regarded as professionals who could be moved about according to military need. See further ch. 2.1.3.

Eresburg	*coepit*	772

ARF 772.

"The most gracious Lord King Charles then held an assembly at Worms. From Worms he first marched into Saxony. Capturing the castle of Eresburg (*Eresburgum castrum coepit*), he proceeded as far as the Irminsul, destroyed this idol and carried away the gold and silver which he found."

Eresburg; ?Büraburg	*destructum; pervenerunt usque ad castrum*	773

ARF 773.

"... the borderlands against the Saxons was exposed and not secured by any treaty. The Saxons, however, fell upon the neighboring Frankish lands with a large army and advanced as far as the castle of Büraburg. The inhabitants of the borderland were terrified when they saw this and retreated into the castle." || ...dimissa marca contra Saxones nulla omnino foederatione suscepta. Ipsi vero Saxones exierunt cum magno exercitu super confinia Francorum, pervenerunt usque ad castrum, quod nominatur Buriaburg; attamen ipsi confiniales de hac causa solliciti, cumque hoc cernerent, castello sunt ingressi. ||

The Saxons tried to burn church of Boniface at Fitzlar, but a miraculous intervention turned them to flight, while one Saxon was found dead in a compromising posture with wood and tinder inside the church. Due to the focus on the miracle history (divine assistance is only alluded to in *Annales Einhardi*), it is difficult to know whether the Saxons ever attempted a formal siege. However, they clearly captured Eresburg by storm during the same campaign, since it was destroyed and had to be rebuilt when Charlemagne took *Syburg in 775: *Annales Einhardi* 775 even says that "He fortified Eresburg, another *castrum* destroyed by the Saxons, and stationed a garrison of Franks in it." || Eresburgum aliud castrum a Saxonibus destructum munivit et in eo Francorum praesidium posuit. ||

<u>Pavia   *obsedit, coepit*   773f</u>
ARF 773, 774.

Under pressure from the Lombards, Pope Hadrian arrived via ship to Marseille to ask for help from Charlemagne, who deliberated with the Franks. After a decision was made, "the glorious king held a general assembly with the Franks as the city of Geneva. There the Lord King divided his army. He himself went on by way of Mount Cenis and he sent his uncle Bernard with his other vassals through the Great St. Bernard Pass. When the two armies united at the Cluses, Desiderius on his part moved to confront the Lord King Charles. Then the Lord King Charles with his Franks laid out his camp in the mountain valley and sent a detachment of his men through the mountains. When Desiderius realized what was going on, he withdrew from the Cluses. The Lord King Charles and his Franks entered Italy by the help of the Lord and the intercession of St. Peter, the blessed apostle, bringing his entire army through the valley without loss or disorder. He came as far as the city of Pavia, surrounded Desiderius, and besieged the city (*Papiam civitatem usque pervenit et Desiderio incluso ipsam civitatem obsedit*)."

The *Annales Einhardi* add that after Desiderius had fled to Pavia, Charlemagne "besieged [Desiderius] and because it was difficult, he spent the whole winter season laboring hard at the siege. || ...obsedit et in obpugnatione civitatis, quia difficilis erat, totum hiberni temporis spatium multa moliendo consumpsit. || Charles celebrated Christmas in his siege camp (*in sua castra*), and spent Easter in Rome. (*ARF* 773)

After Easter 774: "On his return from Rome the Lord King Charles came again to Pavia and captured the city of Desiderius (*ipsam civitatem coepit*), with his wife and daughter and the whole treasure of his palace besides. All the Lombards came from every city of Italy and submitted to the rule (*subdiderunt se in dominio*) of the glorious Lord King Charles and of the Franks. Adalgis, the son of King Desiderius, fled, put to sea, and escaped to Constantinople. After subduing Italy and setting it to rights, the glorious Lord King Charles left a Frankish garrison (*custodia*) in the city of Pavia and by God's help returned triumphantly (*cum magno triumpho*) to Francia with his wife and the rest of the Franks."

| SYBURG | *pugnando cepit* | 775 |

ARF 775.

"The pious and noble Lord King Charles held an assembly at the villa of Düren. From here he launched a campaign into Saxony and captured the castle of Syburg, restored the castle of Eresburg, and came as far as the Weser at Braunsberg. (...*Sigiburgum castrum coepit, Eresburgum reaedificavit*.) There the Saxons prepared for battle since they wished to defend the bank of the Weser. With the help of God and by their own vigorous efforts (*decertantibus*), the Franks put the Saxons to flight; the Franks occupied both banks of the river, and many Saxons were slain there."

After further campaigning, most of the Saxons submitted and gave hostages. The *Annales Einhardi* (see *ARF*) actually claim that while Charlemagne had wintered at Quierzy, he planned either to exterminate the Saxons or force them to submit and accept Christianity. The *Annales Einhardi* specify that the Syburg was indeed taken by storm at the first assault thanks to an overwhelming force. Eresburg had probably been destroyed by the Saxon invasion that occurred while Charlemagne was occupied in Italy: "Having crossed the Rhine as well with all the strength of the kingdom, he marched against Saxony and at the first assault, he took Syburg, where there was a garrison of Saxons, by storm." || ...Rheno quoque transmisso cum totis regni viribus Saxoniam petiit et primo statim impetus Sigiburgum castrum, in quo Saxonum praesidium erat, pugnando cepit. ||

| CIVIDALE, TREVISO, OTHER CITIES | *captas civitates* | 776 |

ARF 776.

Hruodgaud, a Lombard duke in Frankish service, had planned a rebellion that came to the notice of Charlemagne in 775, so he set out with his army before winter. In the spring: "The Lord King Charles entered Italy through Friuli. Hruodgaud was killed and the Lord King Charles celebrated Easter at the city of Treviso. He placed the cities he had captured (*captas civitates*) under the command of Franks, that is, Cividale, Treviso, and the other places which had revolted, and returned again to Francia, successful and victorious."

| ERESBURG | *sacramenta rupta* | 776 |

ARF 776.

The Saxons broke their oath, "and by tricks and false treaties (*sacramenta rupta*) prevailed on the Franks to give up the castle of Eresburg (*castrum Eresburgum*). With Eresburg thus deserted by the Franks, the

Saxons demolished the buildings and the walls (*muros et opera destruxerunt*)."

It seems that the Saxons had scared the Frankish garrison away somehow.

| SYBURG | *coeperunt pugnas et machinas praeparare* | 776 |

As *Eresburg (above).

"Passing on from Eresburg they wished to do the same thing to the castle of Syburg but made no headway since the Franks with the help of God put up a manly resistance (*viriliter repugnantibus*). When they failed to talk the guards (*custodes*) into surrender, as they had those in the other castle, they began to set up war machines to storm the castle. Since God willed it, the catapults which they had prepared did more damage to them than to those inside (*coeperunt pugnas et machinas praeparare, qualiter per virtutem potuissent illum capere; et Deo volente petrarias, quas preparaverunt, plus illis damnum fecerunt quam illis, qui infra castellum resedebant*). When the Saxons saw that their constructions were useless to them, they prepared fascines (*clidas*) to capture the fortress in one charge (*ad debellandum per virtutem ipsum castellum*). But God's power, as is only just, overcame theirs. One day, while they prepared for battle against the Christians in the castle, God's glory was made manifest over the castle church in the sight of a great number outside as well as inside, many of whom are still with us." The miracle consisted of two shields hovering above the church. The Saxons fled terrified, killing each other in stampede, and the Franks pursued the Saxons to the River Lippe.

There is no reason to assume that the *petrarias* made by the Saxons were literally causing damage to themselves; rather the annalist is being elliptic: *both* sides had *petrarias*, and since the Saxons were worsted in the artillery duel the besiegers were indirectly causing greater damage to themselves.

Perhaps the hovering shields were a trick or a post-hoc miracle, as *Mir. St. Dem.* It is noteworthy that the Franks followed up with sally that totally defeated the Saxons, indicating the "rational" explanation for the Saxon panic and defeat. *Annales Einhardi*'s version is simpler, stating that *facta eruptione*, the Franks chased the Saxons off.

| SAXON FORTIFICATIONS | *subito introivit* | 776 |

ARF 776.

Charles called an assembly at Worms, "and after deliberation suddenly broke through the fortifications of the Saxons with God's help." || ...consilio facto cum Dei adiutorio sub celeritate et nimia festinatione Saxonum caesas seu firmitates subito introivit. || The terrified Saxons came to the source of Lippe to submit and promise to accept Christianity. "The Lord King Charles with the Franks rebuilt the castle of Eresburg and another castle on the River Lippe." || Et tunc domnus Carolus rex una cum Francis reaedificavit Eresburgum castrum denuo et alium castrum super Lippiam... || The Saxons came there to be baptized and give hostages. "When the above castles had been completed and Frankish garrisons installed to guard them, the Lord King Charles returned to Francia." || Et perfecta supradicta castella et disposita per Francos scaras resedentes et ipsa custodientes reversus est domnus Carolus rex in Franciam. ||

KASIN (underground city)   ἤνοιξε... ἀπὸ καπνοῦ   776
Theophanes 449 (AM 6268, MS 620).

"In this year Madi sent Abasbali at the head of a great force against the Roman country. With the help of smoke he opened the cave called Kasin and, after capturing the men who were in it, returned home." Τούτῳ τῷ ἔτει ἀποστείλας Μαδὶ τὸν Ἀβασβαλὶ κατὰ Ῥωμανίας μετὰ δυνάμεως πολλῆς, ἤνοιξε τὸ σπήλαιον τὸ ἐπιλεγόμενον Καῦσιν ἀπὸ καπνοῦ, καὶ λαβὼν τοὺς ἐν αὐτῷ αἰχμαλώτους ὑπέστρεψεν.

BARBĀD   776
Tabari II 476 (29: 187).

An expedition set out from Baṣra by sea towards Barbād, an otherwise unknown city in al-Hind (India), in AH 159 (October 31, 775–October 18, 776). The expedition numbered some 9,200 men (460; 29: 171f), so probably reached the city of Barbād in the autumn of 776, after the beginning of AH 160.

"In the year 160 ʿAbd al-Malik b. Shibāb al-Mismaʿī reached the city of Bārbad with those volunteers and others who had set out with him. They attacked the city the day after their arrival and besieged it for two days. They then prepared a mangonel (*manjanīq*) and attacked it with all their equipment. Then the people gathered together and spurred each other on with the Qurʾān and praising Allāh, and Allāh allowed them to take it by force. Their horsemen entered it from all sides and forced them to take refge in their strongholds. They lit fires with oil, and some of them were burned while others attacked the Muslims but Allāh killed them all. More than twenty Muslims were martyred and Allāh gave (the city) as booty (*fayʾ*) to them."

| PAMPLONA | *destructa* | 778 |

ARF 778.

Charlemagne marched to Spain via two routes, himself over the Pyrenees against Pamplona, which he destroyed after returning from Saragossa (*Pampilona destructa*; the *Annales Einhardi* specify that the intention is to prevent rebellion: *Cuius muros, ne rebellare posset, ad solum usque destruxit...*). The other army "came from the regions of Burgundy and Austria [Austrasia] and Bavaria and Provence and Septimania and the land of the Lombards." || ...*venientes de partibus Burgundiae et Austriae vel Baioariae seu Provinciae et Septimaniae et pars Langobardorum*. ||

This is the first time the Lombards participated in a major Carolingian expedition, which was otherwise represented by much of the empire. The famous battle at Roncevalles is not mentioned in *ARF* but described in *Annales Einhardi*. Palace officers were in charge of the troops that were protecting the supply (and perhaps siege) train that fell victim to the ambush. || *In hoc certamine plerique aulicorum, quos rex copiis praefecerat, interfecti sunt, direpta impedimenta...* ||

| GERMANIKEIA | ἐκύκλωσαν | 778 |

Theophanes 451f (AM 6270, MS 623).

"The emperor Leo mobilized the Roman army: 100,000 men invaded Syria under the command of Michael Lachanodrakon of the Thrakesians, the Armenian Artabasdos of the Anatolics, Tatzates of the Bucellarii, Karisterotzes of the Armeniacs, and Gregory, son of Mousoulakios, of the Opsikians; and they surrounded Germanikeia (καὶ ἐκύκλωσαν τὴν Γερμανίκειαν). Isbaali, Madi's uncle, was there, and they took all his camels and were about to take (παραλαμβάνειν) Germanikeia itself, had not Isbaali prevailed upon Lachanodrakon by means of gifts to draw away from the fortified town (καὶ ἀνεχώρησε τοῦ κάστρου); he went forth to devastate the countryside and, after capturing the heretical Syrian Jacobites, returned to the fort (καὶ ἐξῆλθεν εἰς πραῖδαν τῆς χώρας, καὶ αἰχμαλωτεύσας τοὺς αἱρετικοὺς Ἰακωβίτας Σύρους πάλιν ὑπέστρεψεν ἐν τῷ κάστρῳ). Thoumamas sent an army and a number of emirs from Dabekon and made war on the Romans. It is said that five emirs and 2,000 Arabs fell. They withdrew on a Friday, having come on a Sunday."

"The emperor distributed rewards at Sophianai. He sat on a throne together with his son and the *strategoi* were given a triumph for their victory. He conveyed the Syrian heretics to Thrace and settled them there (ἐπέρασεν δὲ καὶ τοὺς αἱρετικοὺς Σύρους ἐν τῇ Θράκῃ καὶ κατῴκισεν αὐτοὺς ἐκεῖ)."

BOCKHOLT *fugientes reliquerunt firmitates* 779
ARF 779.
Charlemagne began a new Saxon campaign. "The Rhine was crossed at Lippenham, and the Saxons wanted to put up resistance at Bocholt. With the help of God they did not prevail but fled, abandoning every one of their bulwarks. The way was open for the Franks, and they marched into the land of the Westphalians and conquered them all." || Ad Lippeham transitur Renus fluvius, et Saxones voluerunt resistere in loco, qui dicitur Bohhoz; auxiliante Domino non praevaluerunt, sed abinde fugientes reliquerunt omnes firmitates eorum. Et Francis aperta est via, et introeuntes in Westfalaos et conquiserunt eos omnes... ||
The trans-Weser Saxons submitted and gave hostages and oaths.

DORYLAION; AMORION καθεσθέντων; παρεκάθισαν 779
Theophanes 452 (AM 6271, MS 624); Lilie 1976: 171f.
"In this year Madi, the leader of the Arabs, waxed angry and sent Asan with a great force of Maurophoroi, Syrians, and Mesopotamians and they advanced as far as Dorylaion. The emperor ordered the *strategoi* not to fight an open war, but to make the forts secure by stationing garrisons of soldiers in them. He appointed high-ranking officers at each fort and instructed them to take each 3,000 chosen men and to follow the Arabs so as to prevent them from spreading out on pillaging raids, while burning in advance the horses' pasture and whatever other supplies were to be found. After the Arabs had remained fifteen days at Dorylaion, they ran short of necessities and their horses went hungry and many of them perished. Turning back, they besieged Amorion for one day, but finding it fortified and well-armed, they withdrew without achieving any success."
Τούτω τῷ ἔτει θυμωθεὶς ὁ τῶν Ἀράβων ἀρχηγὸς Μαδὶ πέμπει τὸν Ἄσαν μετὰ δυνάμεως πολλῆς Μαυροφόρων τε καὶ τῶν τῆς Συρίας καὶ τῆς Μεσοποταμίας, καὶ κατῆλθεν ἕως τοῦ Δορυλαίου. ὁ δὲ βασιλεὺς διετάξατο τοῖς στρατηγοῖς μὴ πολεμῆσαι αὐτοὺς δημόσιον πόλεμον, ἀλλ' ἀσφαλίσασθαι τὰ κάστρα καὶ λαὸν εἰσενεγκεῖν πρὸς παραφυλακὴν αὐτῶν, ἀπολύσας καὶ ἄρχοντας μεγάλους κατὰ κάστρον, αὐτοὺς δὲ ἐπᾶραι ἐπιλέκτους στρατιώτας ἀνὰ τριῶν χιλιάδων καὶ παρακολουθεῖν αὐτοῖς πρὸς τὸ μὴ σκορπίσαι κοῦρσα καὶ προκαίειν πυρὶ τάς τε νομὰς τῶν ἀλόγων, καὶ εἴ που εὑρίσκοιτο δαπάνη. καθεσθέντων δὲ αὐτῶν ἐν τῷ Δορυλαίῳ ἡμέρας ιε' καὶ λειφθέντων αὐτῶν τὰ πρὸς τὴν χρείαν, ἐπτώχευσαν τὰ ἄλογα αὐτῶν· καὶ πολλὴ ἄλωσις ἐγένετο εἰς αὐτά. καὶ ὑποστρέφοντες παρεκάθισαν τὸ Ἀμώριν ἡμέραν μίαν, καὶ ἰδόντες αὐτὸ ὠχυρωμένον καὶ πολλὴν ἐξόπλισιν ἔχον ὑπέστρεψαν μηδὲν ἀνύσαντες.

8TH CENTURY

This entry succinctly describes the guerilla tactics used by the Byzantines throughout the 9th and 10th centuries and codified in military manuals, e.g. *De velitatione* from the late 10th century; see also chapter 2.1.3.

SEMALOUOS	παρεκάθισε, παρέλαβεν...ὑπὸ λόγου	780

Theophanes 453 (AM 6272, MS 624f); Lilie 1976: 172f.

The Caliph Madi organized a campaign under his son Aaron (Hārūn) while he himself led the persecution of Christians, including destruction of churches and forced conversions. "Aaron, after invading the Armeniac *thema*, besieged all summer the fort Semalouos and in the month of September he took it by capitulation. He had previously sent Thoumamas to Asia with 50,000 men. A small raiding party of his was met by Michael Lachanodrakon, who gave battle and killed the brother of Thoumamas." ὁ δὲ Ἀαρὼν εἰσελθὼν εἰς τὸ τῶν Ἀρμενιακῶν θέμα παρεκάθισε τὸ Σημαλοῦος κάστρον ὅλον τὸ θέρος καὶ τῷ Σεπτεμβρίῳ μηνὶ παρέλαβεν αὐτὸ ὑπὸ λόγου. ἦν δὲ πέμψας τὸν Θουμάμαν ἐπὶ Ἀσίαν μετὰ ν΄ χιλιάδων, καὶ συναντήσας Μιχαὴλ ὁ Λαχανοδράκων κούρσῳ αὐτοῦ μικρῷ ἐπολέμησεν αὐτῷ καὶ ἔκτεινε τὸν ἀδελφὸν τοῦ Θουμάμα.

The Arabs thus used a clever stratagem, sending a raiding party deep into Anatolia before committing a force to attack the fort of Semalouos, which they could besiege in peace. According to Tabari II 499f (29: 215), AH 163 (September 17, 789–September 5, 780), the Arabs had brought artillery: "Hārūn traveled until he stopped in one of the Byzantine districts in which there was a castle called Samālū. He besieged it for thirty-eight days. He set up mangonels against it, so that God conquered it after ruining it and after thirst and hunger had afflicted its inhabitants and after killing and wounds among the Muslims. Its conquest was according to the conditions that they would not be killed or deported or split up. They were granted that and they came out, and he fulfilled [the conditions] for them. Hārūn returned safely with the Muslims, except those who had been killed or wounded there."

NAKOLEIA	πολιορκεῖν	782

Theophanes 456 (AM 6274, MS 628f); Lilie 1976: 173-76.

"While the Roman army was busy with these matters [the revolt of Elpidios in Sicily], Madi's son Aaron sallied forth with an enormous armed force composed of Maurophoroi and men from all of Syria, Mesopotamia and the desert and advanced as far as Chrysopolis after leaving Bounousos to besiege (πολιορκεῖν) Nakoleia and guard his rear. He also sent Bouniche to Asia with a force of 30,000. The latter gave battle to Lachanodrakon and the Thrakesian *thema* at a place called Darenos and, the Arabs being 30,000 strong, killed 15,000. The empress for her part sent the *domesticus* Antony at the head of the *tagmata*; he occupied (ἐκράτησε) Bane and blockaded (ἀπέκλεισεν) the Arabs. But Tatzatios, *strategos* of the Bucellarii, defected to the Arabs because of his hatred towards the eunuch Staurakios." Tatzatios advised Arabs to ask for peace while the Byzantine envoys were careless and taken hostage. At last peace was concluded as both sides were under compulsion. "After peace had been concluded they departed, abandoning also the fort of Nakoleia. Tatzatios took away his wife and all his possessions."

Tatzatios was thus the second high-ranking Byzantine to defect and be well received by the Arabs that year. For further details of the siege, MS refer to the *Miracula Sancti Michaelis*, describing a siege which is only available in manuscript. It seems to include the story of trebuchet fired against St. Michael's church, whereupon the operators and Arab notables were struck by God and only cured when they make amends. Lilie however notes the raid-like quality of the Arab expedition.

| SAXON FORTIFICATIONS | *cepit* | 785 |

ARF 785.

Charles established his base at Eresburg during winter (due to flooding), and brought in his wife and children for Easter. "While he was staying at Eresburg, he sent out many detachments and also went campaigning himself. He routed the Saxons who had rebelled, captured their castles, broke through their fortifications, and held the roads open until the right hour struck." || Et dum ibi resideret, multotiens scaras misit et per semetipsum iter peregit; Saxones, qui rebelles fuerunt, depraedavit et castra cepit et loca eorum munita intervenit et vias mundavit, ut dum tempus congruum venisset. ||

The *Annales Einhardi* (see *ARF*) specify how Eresburg was used as a base for the various expeditions. Note also the "Byzantine" strategy of devastating enemy lands before attacking their fortifications.

| BRETON FORTIFICATIONS | *conquisierunt* | 786 |

ARF 786.

"The Lord King Charles sent his army (*exercitum sum*) into Brittany under his emissary Audulf, the Seneschal (*misso suo Audulfo sinescalco*). There they conquered many Bretons with their castles and fortifications in swamps and in forests (...*et ibi multos Brittones conquisierunt una cum castellis et firmitates eorum locis palustribus seu et in caesis.*). As was said before, the Franks proved they could overcome many fortifications of the Bretons (*in multis firmitatibus Brittonum praevalerunt Franci*)." He made a victorious return with the Breton leaders (*capitaneos eorum*) captive.

CITY OF DRAGAWIT   *sustinere non valuit... obsides...dedit*   789
ARF 789.

"From Aachen a campaign was launched with the help of God into the land of the Slavs who are called Wilzi. On the advice of Franks and Saxons he crossed the Rhine at Cologne, advanced through Saxony, reached the River Elbe, and had two bridges constructed, on one of which he built fortifications of wood and earth at both ends. From there he advanced further and by the gift of God subjected the Slavs to his authority. Both Franks and Saxons were with him in his army. In addition, the Frisians joined him by ship, on the River Havel, along with some Franks. He also had with him the Slavs called Sorbs and the Obodrites, whose chieftain was Witzan."

|| Inde [i.e. Aachen] iter permotum partibus Sclaviniae, quorum vocabulum est Wilze, Domino adiuvante; et una cum consilio Francorum et Saxonum perrexit Renum ad Coloniam transiens per Saxoniam, usque ad Albiam fluvium venit ibique duos pontes construxit, quorum uno ex utroque capite castellum ex ligno et terra aedificavit. Exinde promotus in ante, Domino largiente supradictos Sclavos sub suo domninio conlocavit. Et fuerunt cum eo in eodem exercitu Franci, Saxones; Frisones autem navigio per Habola fluvium cum quibusdam Francis ad eum coniunxerunt. Fuerunt etiam Sclavi cum eo, quorum vocabula sunt Suurbi, nec non et Abotriti, quorum princeps fuit Witzan. ||

Annales Einhardi (see *ARF*) has this on the bridge: "when [Charlemagne] came to the Elbe, he set up a camp on the bank and covered the stream with two bridges, one of which was fortified with a wall on both bridgeheads and strengthened with a garrison." || ...cum ad Albiam pervenisset, castris in ripa positis amnem duobus pontibus iunxit, quorum unum ex utroque capite vallo munivit et inposito praesidio firmavit. ||

AE continues: "Entering the country of the Wilzi he ordered everything to be laid waste with fire and sword. But that tribe, although warlike and confident in its number, was not able to withstand the attack of the royal army for very long. Therefore, as soon as he came to the city of Dragawit, who stands above the other kinglets of the Wilzi in age and lineage,

Dragawit at once with all his people came forth from the city, gave the hostages he was ordered to provide, and promised by oath to keep faith with the king and the Franks. The other magnates and chieftains of the Slavs followed suit and submitted to the authority of the king."

|| Ipse fluvio transito, quo consituerat, exercitum duxit ingressusque Wiltzorum terram cuncta ferro et igni vastari iussit. Sed gens illa, quamvis bellicose et in sua numerositate confidens, impetum exercitus regii diu sustinere non valuit ac proinde, cum primum civitatem Dragawiti ventum est, - nam is ceteris Wiltzorum regulis et nobilitate generis et auctoritate senectutis longe praeminebat, - extemplo cum omnibus suis ad regem de civitate processit, obsides, qui imperabantur, dedit, fidem se regi ac Francis servaturum iureiurando promisit. Quem ceteri Sclavorum primores ac reguli secuti omnes se regis dicioni subdiderunt. ||

AVAR FORTIFICATIONS	*dereliquerunt, fuga lapsi*	791

ARF 791.

Although the first Avar campaign failed due to a horse pestilence, it included preparations for storming Avar fortifications: two columns marched down each side of the Danube, supported by a fleet, in order to capture the elaborate fortifications that were on either side. There is some disagreement between *ARF* and *Annales Einhardi* how the fortifications were taken. The *ARF* claims that they were simply abandoned: "where the aforesaid Avars had prepared strongholds... the Avars... abandoned their fortified sites, which are named above, and having left behind their strongholds and devices they turned to flight... || ... ubi iamdicti Avari firmitates habuerunt praeparatas... Avari...dereliquerunt eorum loca munita, quae supra nominata sunt, firmitatesque eorum vel machinationes dimiserunt fuga lapsi.... ||

The *Annales Einhardi* (see *ARF*) claim that capture of the fortifications took place after defeating the Avar garrisons: "Having defeated the Hun garrisons and destroyed their fortifications, one of which was upon the river Cambus, the other next to the city of Comageni on the mountain Cuneoberg, constructed with a very strong wall, the [Franks] devastated everything with iron and fire. || Pulsis igitur Hunorum praesidiis ac destructis munitionibus, quarum una super Cambum fluvium, altera iuxta Comagenos civitatem in monte Cuneoberg vallo firmissimo erat exstructa, ferro et igni cuncta vastantur. ||

The campaign was preceded by three-day religious rituals.

| MARKELLAI | ἔστη ἐν τοῖς ὀχυρώμασι | 792 |

Theophanes 467f (AM 6284, MS 643); Sophoulis 2012: 168f.

"In the month of July [Constantine VI] made an expedition against the Bulgarians and built up the fort of Markellai (καὶ ἔκτισε τὸ κάστρον Μαρκέλλων); and on 20 July Kardamos, the lord of Bulgaria, went forth with all his forces and stationed himself on the fortifications (καὶ ἔστη ἐν τοῖς ὀχυρώμασι). Breathing hotly and persuaded by false prophets that victory would be his, the emperor joined battle without plan or order and was severely beaten. He fled back to the City having lost many men, not only ordinary soldiers, but also persons invested with authority, among them the *magistros* Michael Lachanodrakon, the patrician Bardas, the *protospatharios* Stephen Chameas, the former *strategoi* Niketas and Theognostos, and many other men in imperial service as well as the false prophet, the astrologer Pankratios who had prophesied that he would win. The Bulgarians took the whole train [*to touldon*], namely money, horses, and the emperor's tent with all his equipment."

A fight involving one or both fortifications must be inferred from the context, given their prominent role in the rather superficial narrative; according to both MS and Sophoulis, the Bulgarian fortifications were probably their border strongholds facing Markellai.

| BENEVENTAN *CASTRUM* | *potiuntur* | 792 |

Astronomus 6.

Charlemagne ordered Louis to send the Aquitainian army to Italy in 792-3, where he joined his brother on campaign against Benevento: "... Als dieser schliesslich selber vom Awarenzug heimkehrte, erhielt Ludwig vom ihm den Auftrag, nach Aquitanien zurückzugehen und zur Unterstützung seines Bruders Pippin mit so viel Mannschaft wie möglich nach Italien aufzubrechen. Gehorsam kehrte er im Herbst nach Aquitanien zurück, ordnete hier alles, was zum Schutz des Reiches nötig war, und zog durch die rauhen und gewundenen Schluchten des Mont Cenis nach Italien, feierte das Weihnachtsfest in Ravenna und stiess zu seinem Bruder. Mit vereinten Kräften fielen sie in die Provinz Benevent ein, werwüstete alles, was an ihrem Weg lag, und eroberten eine Festung (*castro uno potiuntur*)."

This section also shows how Louis had trouble controlling the royal estates that was his base for his personal troops. He had settled on four estates, one to be used every year for wintering, providing enough for the needs of his household. Depending on estate size, he may have had a personal following of some hundred men; cf. chapter 4.3 above.

| KAMAKHON | παρέδωκαν | 793 |

Theophanes 468f (AM 6285, MS 644).

A revolt of the Armeniac theme was subdued when the Armeniac rebels were betrayed by Armenian troops during battle. The leaders of the revolt were killed, others were punished with fines and confiscations, and one thousand men were taken in chains to Constantinople, tattooed in the face and sent to "Sicily and the other islands." However, the Armenian troops who had helped defeating the revolt were frustrated by lack of reward and handed over (παρέδωκαν) Kamakhon to the Arabs; according to Baladhuri, on 29 July 793 (see MS n. 6).

| THEBASA | παρέλαβον...ὑπὸ λόγον | 793 |

Theophanes 469 (AM 6286, MS 645).

"In this year, in the month of October of the 2nd indiction [AD 793], the Arabs took the fort of Thebasa [in Lykaonia] by capitulation; for which reason they let its commanders depart home." Τούτῳ τῷ ἔτει μηνὶ Ὀκτωβρίῳ, ἰνδικτιῶνι β', παρέλαβον οἱ Ἄραβες τὸ κάστρον Θήβασαν ὑπὸ λόγον· διὸ καὶ τοὺς ἄρχοντας αὐτοῦ ἀπέλυσαν πορευθῆναι εἰς τὰ ἴδια.

Presumably the garrison had same reason to surrender as at *Kamakhon above.

| ?MARKELLAI | | 796 |

Theophanes 470 (AM 6288, MS 646); Sophoulis 2012: 170f.

"Now Kardamos, the lord of Bulgaria, declared to the emperor, 'Pay me tribute or else I will come as far as the Golden Gate and devastate Thrace.' The emperor sent him some horse excrement in a kerchief and said, 'Such tribute as befits you I have sent you. You are an old man and I do not want you to take the trouble of coming all the way here. Instead, I will go to Markellai and do you come out. Then let God decide.' The emperor, after sending orders to the Asiatic *themata*, gathered his army and advanced as far as Versinikia, while Kardamos went as far as the wooded area of Avroleva, but lost courage and remained in the forest. The emperor encouraged his men and marched to the treeless part of Avroleva and defied Kardamos for seventeen days. The latter, however, did not dare give battle and fled back home."

The role of the fortification at *Markellai this time is uncertain, but the situation recalls that of 792, and the Byzantine response clearly shows that they intended the campaign to revenge their loss.

?AMORION 796
As *Markellai above.

"In the same year the Arabs came as far as Amorion, but did not achieve any success and withdrew after taking captives in the surrounding country."

LÉRIDA *subegit illam atque subvertit* 797
Astronomus 10; see also *Huesca (below) for background.

„Während der König nach Rom zog und dort das kaiserliche Diadem empfing, begab sich König Ludwig wieder nach Toulouse und brach von da unverzüglich nach Spanien auf. Als er sich Barcelona näherte, kam ihm Zaddo, der bereits unterworfene Herzog der Stadt, entgegen, übergab ihm aber die Stadt nicht."

„Der König zog an ih vorbei und erschien überraschend vor Lérida, das er unterwarf und vernichtete. Nach dessen Zerstörung und nachdem die übrigen Städte verwüstet und verbrannt waren, rückte er bis Huesca vor. Die ertragreichen Felder um diese Stadt wurden vom Kriegsvolk abgemäht, verwüstet, verbrannt, und alles, was man ausserhalb der Stadt antraf, wurde durch das verzehrende Feuer vernichtet. Als dies ausgeführt war, kehrte er vor Anbruch des Winters nach Hause zurück." || ... Quam transgrediens rex et Hellerde superveniens, subegit illam atque subvertit. Qua diruta et ceteris municipiis vastatis et incensis, ad Hoscam usque processit. Cuius agros segetibus plenos manus militaris secuit, vastavit, incendit, et quemcumque extra civitatem sunt reperta, incendio depascente sunt consumpta. ||

HUESCA *ad obsidionem* 797ff
ARF 797, 799; also *Lérida (above).

"The city of Barcelona in Spain which had previously revolted against us was returned to us by its governor Zatun. He came to the palace in person and submitted with his city to the Lord King." The version in *Annales Einhardi* (see ARF) adds: "After taking over the city, the king sent his son Louis with an army into Spain to lay siege to Huesca (*ad obsidionem Oscae cum exercitu...misit*)." Huesca finally sent its keys in 799.

Louis ravaged the city's territory during the first season, but seems not to have begun a formal siege then, since the Astronomer specifies that he went home after the first raids.

THE EARLY 9TH CENTURY

BARCELONA	*capta est*	801

ARF 801; Astronomus 13.

"In the same summer the city of Barcelona in Spain was captured after a two-year siege. Its governor Zatun and many other Saracens were taken prisoner. [...] Zatun and Roselmus were brought before the emperor on the same day and condemned into exile." || Ipsa aestate capta est Barcinona civitas in Hispania iam biennio obsessa; Zatun praefectus eius et alii conplures Sarraceni conprehensi. [...] Zatun et Roselmus una die ad praesentiam imperatoris deducti et exilio dampnati sunt. ||

The Astronomer's description of the capture of Barcelona in 801 is the most detailed in the West at this time when it comes to strategy, social aspects and morale, and deserves to be quoted in full in translation:

"In the period following the conclusion of this matter it seemed to the king and his counsellors that a campaign should be mounted to capture Barcelona. The army was divided into three corps: the king kept one with him at Roussillon, where he remained; he ordered a second, under Rotstagnus, count of Gerona, to invest the city; and he sent the third to take up position on the far side of the city so that those besieging it should not have to face a surprise enemy attack. Meanwhile those besieged within the city sent an appeal for help to Cordova, and the *rex* of the Saracens in fact immediately dispatched an army to their aid. But when those he had sent reached Saragossa they were told about the army stationed in their path to intercept them; William commanded, Ademar bore the standard, and they had a powerful force with them. Hearing this they turned upon the Asturians and, taking them by surprise, inflicted a defeat upon them, though they themselves suffered a much heavier one. Once the Saracens had retreated, our men returned to their companions besieging the city and joined them in the investment. Surrounded, and with all entry and exit forbidden, the inhabitants suffered at such length that eventually they were compelled by the anguish of their hunger to take down even the oldest hides from their doors and make miserable food out of these. Others, however, preferring death to such a wretched existence, hurled themselves from the walls. Some, indeed, were kept alive by empty hope, believing that the Franks would be prevented from maintaining the siege of the city by the harshness of the winter. But shrewd men devised a plan which dashed this hope of theirs; for building-material was brought in from all quarters and a start made on the construction of huts, as if our men were going to remain there in winter-quarters. When the inhabitants of the city saw this, they abandoned hope. In an extreme of despair they handed over their prince, whose name was Hamur and whom they had set up in place of Sadun, a relative of his, and, once they had been granted the freedom to depart in safety, surrendered themselves and the city.

The surrender occurred in the following way. When our men saw that the city was exhausted by the long siege and thought that it must be taken or surrendered at any moment, they made the worthy and appropriate decision to summon the king, since the fall of so renowned a city, should it happen when he was present, would give the king a glorious name far and wide. This worthy suggestion met with the full approval of the king, who therefore came to join his army besieging the city and carried on with the investment, showing extreme tenacity, for six unremitting weeks. At length, laid low, the city yielded to the victor. The king sent in guards on the first day after it had been surrendered and thrown open to him but delayed his own entry until he had settled how he might dedicate to God's name, by fitting thanksgiving to Him, the victory which he had hoped for and received. On the following day, accordingly, with *sacerdotes* and clergy preceding him and his army, he entered the city-gate in solemn pomp, to the singing of hymns of praise, and proceeded to the church of the Holy and Most Victorious Cross to give thanks to God for the victory divinely bestowed upon him. Then, leaving count Bera and Gothic troops there as a garrison, he returned home for the winter. Hearing of the danger which seemed to threaten him from the Saracens, his father had sent his brother Charles to his assistance; but when Charles, marching swiftly to his brother's aid, arrived at Lyons he was met by a messenger from his royal brother who reported the city's fall and bade him trouble himself no further. So Charles left Lyons and went back to his father."

CHIETI; SUBURBAN FORTIFICATIONS	*capta et incensa*	801

ARF 801.

At the same time as the capture of *Barcelona (above): "In Italy the city of Chieti was also captured and burned, and its governor Roselmus taken prisoner; the castles belonging to this city surrendered." || Et in Italia Teate civitas similiter capta et incensa est eiusque praefactus Roselmus conprehensus; castella, quae ad ipsam civitatem pertinebant, in deditionem accepta sunt. ||

Umayyad castella		802/6

Astronomus 14.

On a new Spanish campaign, Louis systematically raided around Tortosa and destroyed its *castella*. Astronomus describes how the army intended to go on to Barcelona, but Louis kept getting diverted to Italy (see *Ortona, *Lucera below) and Saxony, and organizing coastal defense. These distractions largely explain his failure to expand further on the Spanish border.

Ortona	*in deditionem accepta*	802

ARF 802.

The Franks had been campaigning in Italy since *Chieti 801, and their activities there are poorly documented. We suddenly hear that: "The city of Ortona in Italy surrendered." || Ortona civitas in Italia in deditionem accepta... ||

Lucera (twice)	*frequenti obsidione fatigata...in deditionem venit*	802

ARF 802.

At the same time as *Ortona (above): "Also Lucera, worn out by prolonged siege, was forced to surrender, and a garrison of our people was installed. [...] Duke Grimoald of the Beneventans besieged Count Winigis of Spoleto in Lucera, where Winigis was in command. Worn out by ill-health Winigis was made to surrender, but was held in honorable captivity." || ...Luceria quoque frequenti obsidione fatigata et ipsa in deditionem venit, prasidiumque nostrorum in ea positum. [...] Grimoaldus Beneventanorum dux in Luceria Winigisum comitem Spoletii, qui praesidio praeerat, adversa valitudine fatigatum obsedit et in deditionem accepit captumque honorifice habuit. ||

Herakles, Thebasa and other forts	παρέλαβε	806

Theophanes 482 (AM 6298, MS 661f).

"In the same year Aaron, the leader of the Arabs, invaded the Roman country with a great force composed of Maurophoroi, Syrians, Palestinians, and Lybians, in all 300,000. Having come to Tyana, he built a house of his blasphemy. He captured after a siege the fort of Herakles, which was very strong, as well as Thebasa, Malakopea, Sideropalos, and Andrasos. (καὶ πολιορκήσας παρέλαβε τό τε Ἡρακλέως κάστρον ὀχυρώτατον πάνυ ὑπάρχον καὶ τὴν Θήβασαν καὶ τὴν Μαλακοπίαν καὶ τὴν Σιδηρόπαλον καὶ τὴν Ἄνδρασον.) He sent a raiding contingent of 60,000 which penetrated as far as Ancyra and withdrew after reconnoitering (ἱστορήσας) it. Seized by fright and perplexity, the emperor Nikephoros set out also in a state of despair, exhibiting the courage that comes from misfortune. After winning many trophies," Nikephoros asked for peace through various clerics. The settlement included a tribute of 30,000 nomismata and the capitation tax of 3 nomismata for Nikephoros himself and 3 for his son. Hārūn rejoiced over their subjugation.

"They also stipulated that the captured forts should not be rebuilt. (ἐστοίχησαν δὲ καὶ τὰ παραληφθέντα κάστρα μὴ κτισθῆναι.) When the Arabs had withdrawn, however, Nikephoros immediately rebuilt and fortified the same towns (...ἔκτισεν εὐθέως τὰ αὐτὰ κάστρα καὶ κατωχύρωσεν). On being informed of this, Aaron sent out a force and, once again, took (ἔλαβε) Thebasa. He also dispatched a fleet to Cyprus, destroyed the churches there, deported the Cypriots, and, by causing much devastation, violated the peace treaty." According to MS n. 12, the treaty applied "Only [to] Herakleia according to Tabari;" ibid. n. 14, Tabari reports the deportation of 16,000 Cypriots at this time.

| ?RHODIAN FORT | ἀπόρθητον | 807 |

Theophanes 483 (AM 6300, MS 663).

"In this year, in the month of September of the first indiction, [i.e. AD 807] Aaron, the leader of the Arabs, sent Choumeid at the head of a fleet against Rhodes. This man sailed straight to Rhodes and, on arriving there, carried out much devastation, but the fort that is there remained uncaptured (ἀπόρθητον). On his return journey he was manifestly worsted by the holy wonder-worker Nicholas. For when he had come to Myra and attempted to break his sacred tomb, he smashed instead another one that stood nearby. Thereupon a great disturbance of sea waves, thunder, and lightning fell upon the fleet so that several ships were broken up and the impious Choumeid himself acknowledged the saint's power and unexpectedly escaped the danger."

| SLAV FORTIFICATIONS | *expugnatis...et manu captis* | 808 |

ARF 808.

The Danish king Godofrid organized a campaign against the Slavs who were subject to Charlemagne: "Since he was informed that Godofrid, the king of the Danes, with his army had crossed over into the land of the Obodrites, he sent his son Charles with a strong host of Franks and Saxons to the Elbe, with orders to resist the mad king if he should attempt to attack the borders of Saxony. Godofrid set up quarters on the shore for some days and attacked and took a number of Slavic castles in hand-to-hand combat (*expugnatis etiam et manu captis aliquot Sclavorum castellis*). Then he withdrew, suffering severe casualties (*cum magno copiarum suarum detrimento reversus est*). He expelled Thrasco, duke of the Obodrites, who did not trust the loyalty of his countrymen, hanged on the gallows Godelaib, another duke, whom he had caught by treachery, and made two-thirds of the Obodrites tributary. But he lost the best and most battle-tested of his soldiers (*optimos tamen militum suorum et manu promtissimos amisit*). With them he lost Reginold, his brother's son, who was killed at the siege of a town along with a great number of Danish nobles (*qui in obpugnatione cuiusdam oppidi cum plurimis Danorum primoribus interfectus est*)."

"But Charles, the son of the emperor, built a bridge across the Elbe, and moved the army under his command as fast as he could across the river against the Linones and Smeldinging. These tribes had also defected to Godofrid. Charles laid waste their fields far and wide and after crossing the river again returned to Saxony with his army unimpaired." || Filius autem imperatoris Carlus Albiam ponte iunxit et exercitum, cui praeerat, in Linones et Smeldingos, qui et ipsi ad Godofridum regem defecerant, quanta potuit celeritate transposuit populatisque circumquaque eorum agris transito iterum flumine cum incolomi exercitu in Saxoniam se recepit. ||

The Carolingians still had overwhelming force when centrally mobilized armies set out, but they faced an increasing number of well-organized, ambitious and aggressive petty kingdoms around the Frankish borders with their own agendas. For instance, the annals for this year goes on to describe how Godofrid deported merchants to his kingdom in order to control tax income from trade, and "decided to fortify the border of his kingdom against Saxony with a rampart..." from sea to sea with only one gate. "After dividing the work among the leaders of his troops he returned home." || ...limitem regni sui, qui Saxoniam respicit, vallo munire constituit ... Diviso itaque opera inter duces copiarum domum reversus est. || Clearly Godofrid controlled his magnates tightly, probably through scale of his personal property and resources. The only way of competing was to assemble riches from elsewhere—i.e. "Viking" raids.

The Slavs were causing trouble for Charlemagne too. "After having two castles built on the River Elbe by his envoys and placing troops in them for the defense against the attacks of the Slavs, the emperor spent the winter at Aachen..." || Imperator vero aedificatis per legatos suos super Albim fluvium duobus castellis praesidioque in eis contra Sclavorum incursiones disposito Aquisgrani hiemavit... ||

| SERDICA | παρέλαβεν δόλῳ καὶ λόγῳ | 809 |

Theophanes 484f (AM 6301, MS 665f); Treadgold 1988: 157ff; Sophoulis 2012: 188ff.

"In the same year, while the army of the Strymon was receiving its pay, the Bulgarians fell upon it and seized 1,100 lbs. of gold. They slaughtered many men together with their *strategos* and officers. Many garrison commanders of the other *themata* were present (ἦσαν γὰρ καὶ τῶν λοιπῶν θεμάτων ταξάτοι ἄρχοντες οὐκ ὀλίγοι) and all of them perished there. The Bulgarians took the whole camp train (τὸ τοῦλδον ὅλον) and withdrew. Before Easter of the same year, Kroummos, the leader of the Bulgarians, drew up his forces against Serdica, which he took by deceitful captitulation and slaughtered 6,000 Roman soldiers, not counting the multitude of civilians (...παραταξάμενος κατὰ Σαρδικῆς, ταύτην παρέλαβεν δόλῳ καὶ λόγῳ, στρατεύματα Ῥωμαϊκὰ κατασφάξας χιλιάδας ͵ϛ', χωρὶς ἰδιωτικοῦ πλήθους). Nikephoros pretended to be going on campaign against him on Tuesday of the Saviour's Passion week, but did not achieve anything worthy of mention. When the officers who had escaped the massacre requested from him a promise of immunity, he refused to give it and so forced them to desert to the enemy, among them the *spatharios* Eumathios, an expert in engines (ἐν οἷς ἦν καὶ Εὐμάθιος ὁ σπαθάριος μηχανικῆς ἔμπειρος)."

Nikephoros tried to manufacture a propaganda victory by manipulating his troops: "Wishing to rebuild captured Serdica (οἰκοδομεῖν βουλόμενος), but fearing the opposition of the host, he suggested to the *strategoi* that they should persuade the rank-and-file to petition the emperor for the rebuilding." The soldiers did not accept it and nearly mutinied, but Nikephoros appeased them; he then had the ringleaders betrayed and punished. Nothing more is said of Serdica at this point. However, Sophoulis discusses evidence of continuous habitation of the city since it fell to the Avars in c. 615 or 618 (see *Thessalonica s.aa.), so it was essential to both sides for controlling the central Balkans. The city may therefore have been the objective of the ill-fated return march of Nikephoros from Pliska in 811, when his army was ambushed crossing the Haemus Mountains. This long period outside of Roman control may also explain why Nikephoros refused immunity to his troops, as any sub-Roman population's loyalty must have been ambivalent, at best, and probably influenced the actions of any troops stationed there for an extended period of time.

| TORTOSA | *protrivit muralibus* | 809 |

ARF 809; Astronomus 16.

"In the west the Lord King Louis entered Spain with his army and besieged the city of Tortosa on the River Ebro. When he had devoted some time to the siege and had seen that he could not take the city quickly, he gave up and returned to Aquitaine with his army unimpaired." || At in occiduis partibus domnus Hludowicus rex cum exercitu Hispaniam ingressus Dertosam civitatem in ripa Hiberi fluminis sitam obsedit; consumptoque in expugnatione illius aliquanto tempore, postquam eam tam cito capi non posse vidit, dimissa obsidione cum incolomi exercitu in Aquitaniam se recepit. || (*ARF*)

Astronomus goes into some detail on the long siege of Tortosa in 809, which resulted in formal submission but not an actual conquest: "Im nächsten Jahr beschloss König Ludwig, in eigener Person, in Begleitung von Heribert, Liutard und Isembard und mit einer starken fränkischen Streitmacht, erneut gegen Tortosa zu ziehen. Dort angelangt, bedrängte und schwächte er mit Hilfe von Rammböcken, Wurfmaschinen, Sturmdächern und anderen Belagerungsmaschinen die Stadt so sehr, dass ihre Bürger jede Hoffnung aufgaben und, da sie die Ihren vom Kriegsglück im Stich gelassen sahen, die Schlüssel der Stadt aushändigten. Diese überbrachte (Ludwig) selber nach seiner Rückkehr unter grossem Beifall dem Vater. Das Ereignis jagte den Sarazenen und Mauren grosse Angst vor solchen Taten ein, denn sie fürchteten, das gleiche Los könnte auch die übrigen Städte treffen. So zog der König vierzig Tage nach Beginn der Belagerung von der Stadt ab und begab sich in sein Reich zurück."

|| Porro anno huic proximo Hludouuicus rex per semet ipsum Tortosam repetere statuit, habens secum Heribertum, Liutardum, Isembardum validumque Frantiae supplementum. Quo perveniens, adeo illam arietibus, mangonibus, vineis et ceteris argumentis lacessavit et protrivit muralibus, ut cives illius a spe deciderent, infractosque suos adverso Marte cernentes, claves civitates traderent, quas ipsas rediens cum multo patri attulit favore. Quae res magnum Sarracenis et Mauris pro talibus gestis incussit metum, verentibus ne singulas par sors involveret. Reversus est igitur rex a civitate post XL dies inchoate obsidionis et in proprium se contulit regnum. ||

Hohbuoki	*captum*	810

ARF 810.

Various news arrived at the court of Charlemagne: Godofrid had been murdered by a retainer (*a quodam suo satellite*—paid by Charlemagne?), and "that the castle of Hohbuoki on the Elbe, with Odo, the emperor's envoy, and a garrison of East Saxons, had been captured by the Wilzi." || ... castellum vocabulo Hohbuoki Albiae flumini adpositum, in quo Odo lega-

tus imperatoris et orientalium Saxonum erat praesidium, a Wilzi captum.
|| The fort was restored by a Frankish army in the next year (for which see
ARF 811).

DEBELTOS; OTHER FORTS	ἑλόντος...πολιορκίᾳ	812

Theophanes 495f (AM 6304, MS 679); Treadgold 1988:180f; Sophoulis 2012:
222-27.

"On 7 June Michael set out against the Bulgarians and was accompanied by Procopia as far as Tzouroulon. The Bulgarian leader Kroummos had taken Debeltos by siege and transplanted its population, which had defected to him together with their bishop (τοῦ δὲ Βουλγάρων ἀρχηγοῦ Κρούμμου ἑλόντος τὴν Δεβελτὸν πολιορκίᾳ καὶ τοὺς ἐν αὐτῇ σὺν τῷ ἐπισκόπῳ μετοικίσαντος προσρυέντας αὐτῷ). Following this, because of the great perversity of the emperor's evil counselors, the army and, in particular, the contingents of the Opsikion and the Thrakesians, raised a sedition and uttered insults. Michael calmed them with gifts and admonitions and so reduced them to silence."

"Having been informed that the troops had rebelled for fear of war and had been disorderly on garrison duty, the Bulgarians extended their power over Thrace and Macedonia. At that time the Christians abandoned Anchialos and Beroia and fled, although no one was pursuing them; the same at Nikaia, the castle of Probaton, and a number of other forts as also at Philoppoupolis and Philippi. Seizing this opportunity, the immigrants who lived at the Strymon also fled and returned to their homes." οἱ δὲ Βούλγαροι τὰ τῆς στάσεως μαθόντες τῶν στρατευμάτων, καὶ ὅτι τὸν πόλεμον καὶ τὸν ταξιτιῶνα ἀτακτοῦσιν, πλέον κατίσχυσαν Θρᾴκης καὶ Μακεδονίας. τότε καὶ Ἀγχίαλον καὶ Βέροιαν ἀφέντες Χριστιανοὶ ἔφυγον, μηδενὸς διώκοντος, Νίκαιάν τε καὶ τὸ Προβάτου κάστρον καὶ ἄλλα τινὰ ὀχυρώματα, ὡσαύτως καὶ τὴν Φιλιππούπολιν καὶ Φιλίππους· καὶ οἱ τὸν Στρυμῶνα οἰκοῦντες μέτοικοι προφάσεως δραξάμενοι ἐν τοῖς ἰδίοις φεύγοντες ἐπανῆλθον.

MESEMBRIA	παρετάξατο...κατὰ; παρέλαβεν	812

Theophanes 497ff (AM 6305, MS 681ff); Treadgold 1988: 184f; Sophoulis 2012: 227-34.

Krum asked for peace, on quite reasonable terms considering the circumstances. His conditions included tribute, the drawing up of a border, the return of refugees, "and that those who traded in both countries should be certified by means of diplomas and seals." It is clear that Bulgaria was well accustomed to a sophisticated administrative system based on Byzantine practice, and Sophoulis argues that the goal was to reestablish and perhaps expand the privileges awarded nearly a century earlier. Krum applied pressure by threatening to array himself against Mesembria, but the emperor Michael refused to make a treaty.

"In the middle of October Kroummos arrayed himself against Mesembria with an equipment of machines and siege engines in which he had become expert through the fault of Nikephoros, the destroyer of the Christians. For there was an Arab, highly skilled in engineering, who had accepted baptism and whom Nikephoros enrolled in imperial service and established at Adrianople, but offered him no suitable assistance or reward; on the contrary, he diminished his pay and, when the latter complained, had him severely beaten. Thereupon, the Arab in despair defected to the Bulgarians and taught them the whole art of making engines. So Kroummos took up his position; and since, out of stupidity, no one offered him any resistance all through that month, he occupied Mesembria." μεσοῦντος δὲ τοῦ Ὀκτωβρίου μηνὸς παρετάξατο ὁ Κροῦμμος κατὰ Μεσημβρίας ἐν μηχανήμασι μαγγανικῶν καὶ ἑλεπόλεων, ἃ τῇ προφάσει Νικηφόρου τοῦ καταλύτου τῶν Χριστιανῶν, μεμάθηκεν. Ἄραβα γάρ τινα προσελθόντα τῷ βαπτίσματι πάνυ ἔμπειρον μηχανικῆς ὑπάρχοντα στρατεύσας ἐν Ἀδριανουπόλει κατέστησεν, μηδεμίαν κατ' ἀξίαν ἀντίληψιν εἰς αὐτὸν ἢ εὐεργεσίαν πεποιηκώς, ἀλλ' ἢ μᾶλλον καὶ τὴν ῥόγαν αὐτοῦ κολοβώσας, τὸν δὲ γογγύσαντα ἔτυψε σφοδρῶς· ἀπονοηθεὶς δὲ ἐπὶ τούτῳ προσέφυγε τοῖς Βουλγάροις καὶ ἐδίδαξεν αὐτοὺς πᾶσαν μαγγανικὴν τέχνην. ἐν τούτοις παραστησάμενος, μηδενὸς ἀντιταξαμένου διὰ πολλὴν σκαιότητα δι' ὅλου τοῦ μηνός, παρέλαβεν αὐτήν.

When a chance for peace passed (1 November), a comet appeared on 4 November. "And on the following day we received the disastrous news of the capture of Mesembria, which frightened everyone by prospect of greater ills. For they found it filled with all manner of goods that are necessary for human habitation and took possession of it along with Debeltos, wherein they found 36 brass siphons and a considerable quantity of liquid fire that is projected from them as well as an abundance of gold and silver." καὶ τῇ ἐπαύριον ἡ περὶ τῆς ἁλώσεως Μεσημβρίας ἦλθεν ἡμῖν ἐλεεινὴ φάσις πάντας πτοοῦσα διὰ μειζόνων κακῶν ἀπεκτοχέν. εὑρόντες γὰρ αὐτὴν οἱ ἐχθροὶ πεπλησμένην πάντων τῶν ὀφειλόντων πρὸς κατοίκησιν ἀνθρώπων παρεῖναι πραγμάτων, ταύτην ἐκράτησαν σὺν τῇ Δεβελτῷ, ἐν οἷς καὶ σίφωνας χαλκοῦς εὗρον λϛ΄, καὶ τοῦ δι' αὐτῶν ἐκπεμπομένου ὑγροῦ πυρὸς οὐκ ὀλίγον, χρυσοῦ τε καὶ ἀργύρου πλῆθος.

Thus, after a siege of just two weeks, the Bulgars managed to storm the city, apparently causing substantial damage to the walls, including a four-meter breach in the walls (for references to archeological reports, see Sophoulis). Apparently this took the Byzantines by surprise, as relief was still being organized, and demonstrated that Bulgarian siege abilities had to be taken seriously.

ADRIANOPLE; CONSTANTINOPLE (prep.)   παρακαθίσας... ἑλών   813
Theophanes 500-03 (MS 684ff); Scriptor Incertus 345, 347; Treadgold 1988: 201ff; Sophoulis 2012: 249-57, 261-64.

"After the fall of Mesembria the emperor renounced the prospect of peace with Kroummos. He made a levy from all the *themata* and ordered that they should cross to Thrace before the spring. As a result, everyone was annoyed, especially the Cappadocians and the Armeniacs." The emperor Michael set out in May with *tagmata* and decided to have troops billet in Thrace rather than attack Bulgaria. "The presence of such a throng of our fellow-countrymen who lacked necessary supplies and ruined the local inhabitants by rapine and invasion was more grievous than a barbarian attack. At the beginning of June Kroummos, the leader of the Bulgarians, fearful of the great numbers of the Christian army, set out at the head of his own troops. When he had encamped at Versinikia, about thirty miles from the imperial army, the patricial Leo, *strategos* of the Anatolics, and the patrician John Aplakes, *strategos* of Macedonia, were very eager to give battle, but were prevented from doing so by the emperor on account of his evil counsellors."

At the Bulgarian threat, iconoclasts at Constantinople rushed to the tomb of Constantine and spread the rumor of his rising; apparently soldiers who were encouraged by those interested in a change of regime. Indeed, when the Byzantines were worsted in battle on 22 June near Adrianople, Michael abdicated in favor of Leo. Crowned on the following day, "he ordered the City to be placed in a state of defence. He himself toured the walls by day and night, encouraging everyone and bidding them be hopeful that God would soon work a miracle through the intercession of the all-pure Theotokos and all the saints and not allow us to be altogether shamed because of the multitude of our sins." τὰ κατὰ τὴν πόλιν προστάττει φρουρηθῆναι, τὰ τείχη νυκτὸς καὶ ἡμέρας αὐτὸς δι' ἑαυτοῦ περιπολεύων καὶ πάντας διεγείρων εὐέλπιδάς τε παραινῶν εἶναι, ὡς τοῦ θεοῦ παραδοξοποιήσαντος τάχιστα διὰ πρεσβειῶν τῆς παναχράντου θεοτόκου καὶ πάντων τῶν ἁγίων, καὶ μὴ πάντη καταισχυνθῆναι παραχωροῦντος διὰ πλῆθος πταισμάτων ἡμῶν.

"Puffed up by his victory, Kroummos, the new Sennacherib, left his brother with his own force to besiege Adrianople (μετὰ τῆς ἰδίας δυνάμεως πολιορκεῖν τὴν Ἀδριανούπολιν) and, six days after Leo's assumption of the imperial office, arrived at the Imperial City with might and horses and made a tour outside the walls, from Blachernai to the Golden Gate, exhibiting his forces." Krum was ostentatious about his ambitions: he performed sacrifices in view of city and demanded that his spear be affixed to the Golden Gate. However, he gave up a possible siege upon seeing strength of defenses and opened negotiations, but was ambushed. Enraged, he raided the suburbs of Constantinople and returned to press siege of Adrianople, which he took, παρακαθίσας Ἀδριανουπόλει καὶ ταύτην ἑλών. Scriptor Incertus adds that this was achieved by setting up *manganika*, having achieved nothing with a blockade.

Bulgar ravaging in Thrace was extensive, and a number of other fortifications were taken in the course of this campaign (for which see Treadgold and Sophoulis). Clearly the turmoil caused by regime change allowed the Bulgars to exploit the situation. Hoping to gain as great an advantage as possible, Krum ostentatiously paraded his army in front of Constantinople with great pomp and ritual, but as Sophoulis argues, he had no realistic ambition of taking the city and gave up when he realized that Leo's new regime could not accept too humiliating concessions to the Bulgars. After Adrianople, however, Krum began extensive preparations in the winter of 813-14 for another attempt to be launched in the spring of 814. Krum's death and internal Bulgar turmoil prevented the threat from materializing, and Sophoulis argues that the preparations described by Scriptor Incertus were exaggerated. Indeed, the arsenal at Krum's disposal was portrayed as fearsome in the extreme, containing "giant *manganika*, both *triboloi* and *petroboloi*, and *khelonai* and tall ladders, *sphairai*, crowbars and pick-axes (*oryai* or *orygai*), rams and artillery (*belostaseis*), both fire- and stone-throwing, and scorpions (*skorpidia*) for shooting missiles, and slings..." Most of these devices can indeed be identified in contemporary sources: basic artillery of differing marks, siege sheds and tools for engineering and mining is certainly not out of the question (ch. 5.2, 8 *passim*). However, the *skorpidia* are somewhat problematic (perhaps bow-*ballistrai*), while some information seems to be repeated only using different terms—for instance, slings, subsequently specified as a type of machine, should surely sort as various *manganika*; so should fire- and stone-throwing devices. Perhaps some exaggeration should be ascribed to imperial propaganda, which also claimed credit for Krum's death as a result of wounds suffered during the ambush months earlier. The threat was surely taken seriously by Leo, who transferred troops from Anatolia and built a new wall to protect the Blachernai suburb.

BIBLIOGRAPHY

Page/chapter references to text in editions are in most cases given whenever the original is quoted in parts I and II. If not indicated below, further information may be found in standard translations (e.g., Murray 2000, King 1987, Wolf 1999) or the relevant volumes of *MGH*, *CSCO*, etc.

Abbreviations of Series, Reference and Collective Works

5th-C Gaul	John Drinkwater and Hugh Elton (eds.), *Fifth-Century Gaul: A Crisis of Identity?* Cambridge, 1992.
AFRBA	Stephen Mitchell (ed.), *Armies and Frontiers in Roman and Byzantine Anatolia*. BAR International Series 156. Oxford, 1983.
AHR	*The American Historical Review*
AT	*Antiquité Tardive*
BEINE 1	Averil Cameron and Lawrence Conrad (eds.), *The Byzantine and Early Islamic Near East I: Problems in the Literary Source Material*. Princeton, 1992.
BEINE 3	Averil Cameron (ed.), *The Byzantine and Early Islamic Near East III: States, Resources and Armies*. Princeton, 1995.
BEINE 6	John Haldon and Lawrence Conrad (eds.), *The Byzantine and Early Islamic Near East VI: Elites Old and New*. Princeton, 2004.
BMGS	*Byzantine and Modern Greek Studies*
Borders	Florin Curta (ed.), *Borders, Barriers, and Ethnogenesis: Frontiers in Late Antiquity and the Middle Ages*. Turnhout, 2005.
ByzMajP	Karsten Fledelius with Peter Schreiner (eds.), *Byzantium: Identity, Image, Influence. Major Papers, 19th International Congress of Byzantine Studies, University of Copenhagen, 18-24 August, 1996*. Copenhagen, 1996.
ByzSlav	*ByzantinoSlavica*
BZ	*Byzantinische Zeitschrift*
CAH 13	A. Cameron et al. (eds.), *The Cambridge Ancient History Volume 13. The Late Empire, A.D. 337-425*. Cambridge, 1998.
CAH 14	A. Cameron et al. (eds.), *The Cambridge Ancient History Volume 14. Late Antiquity: Empire and Successors, A.D. 425-600*. Cambridge, 2000.
CC	*Corpus Christianorum*
CFHB	*Corpus Fontium Historiae Byzantinae*
CHGRW	Philip Sabin, Hans van Wees and Michael Whitby (eds.), *The Cambridge History of Greek and Roman Warfare, Volume II: Rome from the Late Republic to the Late Empire*. Cambridge, 2000.
CHI	*The Cambridge History of Iran*
CMMA	David Nicolle (ed.) *A Companion to Medieval Arms and Armour*. Woodbridge, 2002.
Cple	Cyril Mango and Gilbert Dagron (eds.), *Constantinople and its Hinterland*. Aldershot, 1995.
CRA	Paul Erdckamp (ed.), *A Companion to the Roman Army*. Malden, 2007.
CSCO	*Corpus Scriptorum Christianorum Orientalium*
SS	*Scriptores Syri*
CSHB	*Corpus Scriptorum Historiae Byzantinae*

DRBE	Philip Freedman and David Kennedy (eds.), *The Defence of the Roman and Byzantine East*. BAR International Series 297 (ii). Oxford, 1986.
DOP	*Dumbarton Oaks Papers*
EHB 1-3	Angeliki E. Laiou (ed.), *The Economic History of Byzantium From the Seventh through the Fifteenth Century* (3 volumes). Washington, D.C., 2002.
EHR	*English Historical Review*
EncIr	*Encyclopaedia Iranica*, online edition: www.iranica.com.
EncMedWar	Clifford J. Rogers (ed.), *The Oxford Encyclopedia of Medieval Warfare and Military Technology* (3 volumes). Oxford 2010.
Ethnogenese	Herwig Wolfram und Walter Pohl (eds.), *Typen der Ethnogenese unter besonderer Berücksichtigung der Bayern. Teil 1*. Vienna, 1990.
FCIW	*The Formation of the Classical Islamic World*
FCIW 5	Fred Donner (ed.), *The Expansion of the Early Islamic State*. Aldershot.
FCIW 8	Michael Bonner (ed.), *Arab-Byzantine Relations in Early Islamic Times*. Aldershot, 2004.
FCIW 10	David Waines (ed.), *Patterns of Everyday Life*. Aldershot, 2002.
FCIW 11	Michael G. Morony (ed.), *Production and the Exploitation of Resources*. Aldershot, 2002.
FCIW 12	Idem (ed.), *Manufacturing and Labour*. Aldershot, 2003.
FCIW 18	Robert Hoyland (ed.), *Muslims and Others in Early Islamic Society*. Aldershot, 2004.
Handwerk	Herbert Jankuhn et al. (eds.) *Das Handwerk in vor- und frühgeschichtlicher Zeit. Teil I: Historische und rechtshistorische Beiträge und Untersuchungen zur Frühgeschichte der Gilde*. Göttingen, 1981.
HE	*Historia Ecclesiastica*
JMMH	*Journal of Medieval Military History*
JöB	*Jahrbuch der österreichischen Byzantinistik*
JRS	*Journal of Roman Studies*
JSAI	*Jerusalem Studies in Arabic and Islam*
KdGr	Helmut Beumann (ed.), *Karl der Grosse: Persönlichkeit und Geschichte*. Volume I of *Karl der Grosse: Lebenswerk und Nachleben* (ed. Wolfgang Braunfels). Düsseldorf, 1965.
LAA	*Late Antique Archaeology*.
LAA 2	William Bowden, Luke Lavan and Carlos Machado (eds.) *Recent Research on the Late Antique Countryside*. Leiden, 2004.
LAA 4	Luke Lavan, Enrico Zanini and Alexander Sarantis (eds.) *Technology in Transition, A.D. 300-650*. Leiden, 2007.
Landscapes	Neil Christie (ed.), *Landscapes of Change: Rural Evolution in Late Antiquity and the Early Middle Ages*. Aldershot, 2007.
Langobarden	Walter Pohl and Peter Erhardt (eds.) *Die Langobarden. Herrschaft und Identität*. Vienna, 2005.
MASS	Anne Nørgård Jørgensen and Birthe L. Clausen (eds.), *Military Aspects of Scandinavian Society in a European Perspective, AD 1-1300*. Copenhagen, 1997.
MCuS	Ivy A. Corfis and Michael Wolfe (eds.), *The Medieval City under Siege*. Woodbridge, 1995.
MGH	*Monumenta Germaniae Historica*
SSAA	*Auctores Antiquissimi*
SSRG	*Scriptores Rerum Germanicarum*
SSRM	*Scriptores Rerum Merovingicarum*
SSRLI	*Scriptores Rerum Langobardicarum et Italicarum Saec. VI-IX*
LL	*Leges*
Epp.	*Epistulae*

MS	Mango and Scott's translation of Theophanes Confessor (see below).
NCHI 1	Chase Robinson (ed.) *The New Cambridge History of Islam. Volume I: The Formation of the Islamic World Sixth to Eleventh Centuries.* Cambridge, 2010.
NCMH 1	Paul Fouracre (ed.), *The New Cambridge Medieval History Volume I, c. 500– c. 700.* Cambridge, 2005
NCMH 2	Rosamund McKitterick (ed.), *The New Cambridge Medieval History Volume II, c. 700–c. 900.* Cambridge, 1995.
Ostrogoths	Sam J. Barnish and Federidco Marazzi (eds.), *The Ostrogoths from the Migration Period to the Sixth Century: An Ethnographic Perspective.* Woodbridge, 2007.
PG	*Patrologia Graeca*
PL	*Patrologia Latina*
PLRE 2	J.R. Martindale (ed.), *The Prosopography of the Later Roman Empire Volume II, A.D. 395-527.* Cambridge, 1980.
PLRE 3	J.R. Martindale (ed.), *The Prosopography of the Later Roman Empire Volume III, A.D. 527-641.* Cambridge, 1992.
PO	*Patrologia Orientalis*
P&P	*Past and Present*
PRT 1-2	Joachim Henning (ed.), *Post-Roman Towns, Trade and Settlement in Europe and Byzantium. Volume 1: The Heirs of the Barbarian West. Volume 2: Byzantium, Pliska, and the Balkans.* Berlin and New York, 2007.
Provinces	Thomas F.X. Noble (ed.), *From Roman Provinces to Medieval Kingdoms.* New York and Abingdon, 2006.
RBAE	E. Dąbrowa, *The Roman and Byzantine Army in the East.* Krakow, 1994.
Rebirth	Richard Hodges and Brian Hobley, *The Rebirth of Towns in the West AD 700-1050.* CBA Research Report 68. London, 1988.
Saxons	Dennis Green and Frank Siegmund (eds.), *The Continental Saxons from the Migration Period to the Tenth Century: An Ethnographic Perspective.* Woodbridge, 2003.
Settimane	*Settimane di studio del centro italiano di studi sull'alto medioevo.* Spoleto.
Settimane 15	*Ordinamenti militari in Occidente nell'alto medioevo,* 2 vols. 1968.
Settimane 18	*Artigianato e tecnica nella società dell'alto medioevo occidentale,* 2 vols. 1971.
Spätantike	Theo and Rudolf Schieffer (eds.), *Von der Spätantike zum frühen Mittelalter: Kontinuitäten und Brüche, Konzeptionen und Befunde.* Ostfildern, 2009.
Städtewesen	*Studien zu den Anfängen des Europäischen Städtewesens.* Vorträge und Forschungen 4, ed. Theodor Mayer & Konstanzer Arbeitskreis für mittelalterliche Geschichte. Konstanz and Lindau, 1958.
TIB	*Tabula Imperii Byzantini.* Vienna.
TLAD	Andrew Poulter (ed.), *The Transition to Late Antiquity, on the Danube and Beyond.* Oxford, 2007.
TRW	*Transformation of the Roman World*
TRW 1	Walter Pohl (ed.), *Kingdoms of the Empire: The Integration of Barbarians in Late Antiquity.* Leiden, New York and Cologne, 1997.
TRW 2	Walter Pohl with Helmut Reimitz (eds.), *Strategies of Distinction: The Construction of Communities, 300-800.* Leiden, 1998.
TRW 3	Richard Hodges and William Bowden (eds.), *The Sixth Century: Production, Distribution and Demand.* Leiden, 1998.
TRW 9	Gian Pietro Brogiolo et al. (eds.), *Towns and their Territories between Late Antiquity and the Early Middle Ages.* Leiden, Boston and Cologne, 2000.
TRW 10	Walter Pohl et al. (eds.), *The Transformation of Frontiers from Late Antiquity to the Carolingians.* Leiden, Boston and Cologne, 2001.

TRW 11	Inger Lyse Hansen and Chris Wickham (eds.), *The Long Eighth Century*. Leiden, Boston and Cologne, 2000.
TRW 12	Richard Corradini et al. (eds.), *The Construction of Communities in the Early Middle Ages*. Leiden and Boston, 2003.
TRW 13	Hans-Werner Goetz, Jörg Jarnut and Walter Pohl (eds.), *Regna and Gentes: The Relationship between Late Antique and Early Medieval Peoples and Kingdoms in the Transformation of the Roman World*. Leiden, 2003.
Visigoths	Peter Heather (ed.), *The Visigoths from the Migration Period to the Seventh Century: An Ethnographic Perspective*. Woodbridge, 1999.
VSNA	Edward James (ed.), *Visigothic Spain: New Approaches*. Oxford, 1980.
WGT	Kathleen Mitchell and Ian Wood (eds.), *The World of Gregory of Tours*. Leiden, Boston and Cologne, 2002.

Sources

Acta Aunemundi. Acta Sanctorum, Sept. vii. Brussels 1760, 744-46. Trans. Gerberding and Fouracre 1996: 179-92.

Aeneas Tacticus. Loeb. Cambridge, Mass. and London, 1928.

Agapius: Agapius of Manbij, *Kitab al-'Unvan*. Ed. and trans. Alexandre A. Vasiliev. *PO* 8.3. Paris, 1912. Trans. Hoyland 2011.

Agathias: *Agathiae Myrinaei Historiarum Libri Quinque*. Ed. Rudolf Keydell. *CFHB* 2. Berlin, 1967. Trans. Joseph D. Frendo. *The Histories*. Berlin and New York, 1975.

Agnellus: *Liber Pontificalis Ecclesiae Ravennatis. MGH SSRI*.

Amm(ianus) Marc(ellinus), *History* (trans. John C. Rolfe). Loeb rev. ed. 1950.

ARF: Annales Regni Francorum. Eds. G.H. Pertz and Friedrich Kurze. *MGH SS in usum Scholarum* 6. Hannover, 1895. Trans. Bernhard Walter Stoltz in *Carolingian Chronicles*. Ann Arbor, 1970.

Astronomus, *Vita Hludowici Imperatoris*. Ed. and trans. Ernst Tremp. *MGH SSRG* 64. Hannover, 1995. See also King 1987.

Auctarium Havniense. MGH AA 9 *Chronica Minora* 1.

Baladhuri: Imam Ahmad ibn Yahya ibn Jabir al-Baladhuri, *Futuh al-Buldan. Liber expugnationis regionum*. Ed. M.J. de Goeje. Leiden, 1866. Trans. Philipp K. Hitti (Vol. 1) and Francis C. Murgotten (Vol. 2) as *The Origins of the Islamic State*. New York, 1916-27 (1968-69).

Brevium Exempla. MGH Capitularia Regum Francorum 1. Trans. Loyn and Percival, 1975: 98-105

Byzantine-Arab Chronicle. MGH AA 11 (*Chronica Minora* 2) and J. Gil (ed.) *Corpus scriptorum Muzarabicorum I*. Madrid 1973. Trans. Hoyland 1997: 611-30.

Capitulare Aquisgranense. MGH Capitularia Regum Francorum 1.

Capitulare de villis. MGH Capitularia Regum Francorum 1. Trans. Loyn and Percival 1975: 64-73.

Capitulare Episcoporum. MGH Capitularia Regum Francorum 1.

Cassiodorus *Variae: MGH SSAA* 12. Trans. S.J.B. Barnish. *Cassiodorus Selected Variae*. Liverpool, 1992. Trans. Thomas Hodgkin. *The Letters of Cassiodorus*. Oxford, 1886.

Cassiodorus *Chr.: Chronicon MGH SSAA* 11, *Chronica minora*, 2, 109-161.

Chr. 637. Trans. Palmer et al. 1993: 1-4.

Chr. 640/724. Trans. Palmer et al. 1993: 5-24.

Chr. 819. See *Chr. 1234*. Trans. Palmer et al. 1993: 75-84.

Chr. 1234: Anonymi auctoris chronicon ad annum christi 1234 pertinens, I-II. Ed. I.-B. Chabot, *CSCO SS*, Series Tertia, XIV-XV (*Textus*). Paris, 1916-1920. Latin translation of Vol. I by Chabot in XIV (*Versio*), Louvain 1937; French translation of Vol. II by Albert Abouna in *CSCO* 354/SS 154, Louvain 1974. Partial trans. Hoyland 2011.

Chr. Caes.: *Chronicon Caesaraugustanum*. *MGH SSAA* 11.2.
Chr. Gall. 452. *MGH SSAA* 9.1.Trans. Murray 2000: 77-85.
Chr. Gall. 511. *MGH SSAA* 9.1.Trans. Murray 2000: 98-108.
Chr. Maroniticum. Chronica Minora, CSCO SS ser. III vol. 4, ed. and Latin trans. Ignatius Guidi. Trans. Palmer et al. 1993.
Chr. Moiss(iacense). *MGH SS* 1.
Chr. Paschale, 2 vols. Ed. Ludwig Dindorf. *CSHB*. Bonn, 1832. Trans. Michael and Mary Whitby. *Chronicon Paschale 284-628 AD*. Liverpool, 1989.
Chr. Zuqnin: Incerti Auctoris Chronicon pseudo-Dionysianum vulgo dictum, I-II. *CSCO* 91, 104, *SS* 43, 53. Ed. I.-B. Chabot. Louvain, 1927-33. Trans. Amir Harrak, *The Chronicle of Zuqnin: Parts III and IV AD 488-775*. Toronto, 1999.
CJC: Corpus Iuris Civilis
  CJ: Codex Iustinianus. Ed. Paul Krueger. Berlin, 1915.
  NJust: Novellae. Eds. Rudolf Schoell and Wilhelm Kroll. Berlin, 1912
Codex Carolinus. *MGH Epp*. 1.
Conc. Aur. 541. *MGH Concilia* 1.
Constantine Porphyrogenitus. *Three Treatises on Imperial Military Expeditions*. *CFHB* 28. Intr., ed., trans. & comm. John F. Haldon. Vienna 1990.
CTh: Codex Theodosianus. Theodosiani libri XVI cum constitutionibus Sirmondianis et leges novellae ad Theodosianum pertinentes (2 vols.). Eds. Theodor Mommsen and Paul Meyer. Berlin: Weidemann, 1904-05. Trans. Clyde Pharr. *The Theodosian Code and Novels and the Sirmonidian Constitutions*. Princeton: Princeton University Press, 1952.
NVal: Novels of Valentinian
DAI: Constantine Porphyrogenitus. *De Administrando Imperio*. Ed. Gy. Moravcsik, trans. R.J.H. Jenkins. *CFHB* 1. Washington, D.C., 1967.
De obsidione toleranda: Denis F. Sullivan, "A Byzantine Instructional Manual on Siege Defense: The *De obsidione toleranda*. Introduction, English Translation and Annotations." John W. Nesbitt (ed.). *Byzantine Authors: Literary Activities and Preoccupations*. Leiden and Boston, 2003: 139-266.
Desiderius *Ep. Epistulae S. Desiderii Cadurcensis*. Ed. Dag Nordberg. Stockholm, 1961.
De velitatione: Ed. and trans. Gilbert Dagron and Haralambie Mihăescu. *Le traité sur la guerilla (De velitatione) de l'empereur Nicéphore Phocas (963-969)*. Paris, 1986.
Dionysius: see "Conventions" for Part II. Trans. Palmer et al. 1993: 111-221.
Edict of Paris. MGH Capitularia Regum Francorum 1. Trans. Murray 2000: 566ff.
Ep. Austr.: Epistulae Austrasiacae. MGH Epp. 3.
Evagrius, *HE*: Evagrius Scholasticus. *Historia Ecclesiastica–Kirchengeschichte*, 2 vols. Trans. Alfred Hübner (w/ed. by Bidez and Parmentier, 1898). *Fontes Christiani* 57/1-2. Turnhout, 2007. Also Michael Whitby (trans.), *The Ecclesiastical History of Evagrius Scholasticus*. Liverpool, 2001.
Formulae Arvernenses. MGH LL Formulae Merowingici et Carolini Aevi. Trans. Rio 2008.
Fredegar (continuatus): *The Fourth Book of the Chronicle of Fredegar with its Continuations*. Ed. and trans. J.M. Wallace-Hadrill. London, 1960.
Genesius: *Iosephi Genesii Regum Libri Quattuor*. Eds. A. Lesmueller-Werner and I. Thurn. *CFHB* 14. Berlin and New York, 1978.
GT: Gregory of Tours. *Decem Libri Historiarum*. Ed. Bruno Krusch and W. Lewison. *MGH SRM* 1, 1. Trans. Lewis Thorpe. *Gregory of Tours: The History of the Franks*. Harmondsworth, 1974.
GT *VP*: Gregory of Tours, *Liber Vitae Patrum, MGH SSRM* 2. *Life of the Fathers* (2nd ed.) trans. Edward James. Liverpool, 1991.
Historia Wambae: Historia Wambae regis auctore Iuliano episcopo toletano. MGH SSRM 5. 501-526. Trans. Martínez Pizarro 2005.

HL: Pauli Historia Langobardorum. Edd. L. Bethman and G. Waitz, *MGH SRLI.* Trans. William Dudley Foulke. *Paul the Deacon: History of the Lombards.* Philadelphia, 1907/2003
HW: see *Historia Wambae.*
Hyd(atius): *The* Chronicle *of Hydatius and the* Consularia Constantinopolitana*: Two Contemporary Accounts of the Final Years of the Roman Empire.* Ed. and trans. R.W. Burgess. Oxford, 1993. Also trans. Murray 2000.
Isidore, *Etymologies: The Etymologies of Isidore of Seville.* Trans. Stephen A. Barney et al. Cambridge, 2006.
Isidore, *HG: Isidori Iunioris episcopi Hispalensis historia Gothorum. MGH SSAA* 11 *(Chronica Minora* 2), 391-488. Trans. Wolf 1999.
J. Bicl.: John of Biclar, *Chronicle. MGH SSAA* 11 *(Chronica Minora* 2). Trans. Wolf 1999.
J. Eph. *HE: Iohannis Epheseni historiae ecclesiasticae pars tertia. CSCO* 105, 106 *SS* 54, 55. Ed. and Latin trans. E.W. Brooks. Louvain, 1935-1964. John of Ephesus *The Third Part of the Ecclesiastical History of John Bishop of Ephesus.* Trans. R. Payne Smith. Oxford, 1860.
J. Epiph.: John of Epiphania. *Fragmenta Historicorum Graecorum* 4. Ed. C. Müller. Paris, 1851: 272-26.
J. Nik.: *The Chronicle of John, Bishop of Nikiu.* Trans. Robert Henry Charles. Amsterdam, 1916.
Jordanes *Romana et Getica. MGH SSAA* 5.1. *Getica* trans. Charles Mierow, *The Gothic History of Jordanes: in English Version.* Cambridge, 1915.
JS: (Pseudo-)Joshua the Stylite, *Chronicle.* Ed. in *Chr. Zuqnin I.* Trans. William Wright, *The Chronicle of Joshua the Stylite: Composed in Syriac A.D. 507.* Cambridge, 1882. See also Frank R. Trombley and John W. Watt. *The Chronicle of Pseudo-Joshua the Stylite.* Liverpool, 2000; and Luther 1997.
Leges Langobardorum. MGH LL 4. Trans. Katherine Fischer Drew. *The Lombard Laws.* Philadelphia, 1973.
Lex Salica. MGH LL Nationum Germanicarum 4.1-2. Trans. Katherine Fischer Drew. *The Laws of the Salian Franks.* Philadelphia, 1991. See also translation by Rivers listed under *Lex Ripuaria.*
Lex Ripuaria. MGH LL Nationum Germanicarum 3.2. Trans. Theodore John Rivers. *Laws of the Salian and Ripuarian Franks.* New York, 1986.
LHF: Liber Historiae Francorum. MGH SSRM 2. Trans. Gerberding and Fouracre 1996.
Liutprand: see *Leges Langobardorum.*
LJ: Liber Judiciorum. MGH LL Nationum Germanicarum 1. Trans. S.P. Scott, *The Visigothic Code.* Philadelphia 1908.
LP: Liber Pontificalis. Texte, introduction et commentaire (3 volumes.). Edited by L. Duchesne. Paris: E. de Boccard, 1955. Trans. Davis 2000 and 2007.
Lupus of Ferrières: *The Letters of Lupus of Ferrières.* Trans. Graydon W. Regenos. The Hague, 1966.
Malalas: *Ioannis Malalae Chronographiae.* Ed. Johannes Thurn. CFHB 35. Berlin, 2000. Trans. Elizabeth Jeffreys, Michael Jeffreys and Roger Scott.*The Chronicle of John Malalas. Byzantina Australiensia* 4. Melbourne, 1986.
Marc(ellinus) Com(es): *Marcellinus V.C. Comitis Chronicon. MGH SSAA* 11 *(Chronica Minora* 2), 37-104.
Marculf: *MGH LL Formulae Merowingici et Carolini Aevi.* Trans. Rio 2008.
Mar(ius of) Av(enches): *The Chronicle of Marius of Avenches. Marii episcopi Aventicensis chronica. MGH SSAA* 11 *(Chronica Minora* 2). Trans. Murray: 2000: 100-08.
Menander: R.C. Blockley. *The History of Menander the Guardsman.* Liverpool, 1985.
Miracles of St. Artemios: Virgil S. Crisafulli and John W. Nesbitt. *The Miracles of St. Artemios: A Collection of Miracle Stories by an Anonymous Author of Seventh-Century Byzantium.* Leiden, 1997.

Miracula Martinis Abbatis Vertavensis. MGH SSRM 3.
Miracula sancti Demetrii: See Lemerle 1979-81.
Movses Dasxuranci. *The History of the Caucasian Albanians*. Trans. C.J.F. Dowsett. London, 1961.
Nik(ephoros): *Nikephoros Patriarch of Constantinople: Short History*. Ed. and trans. Cyril Mango. *CSHB* 13. Washington, D.C., 1990.
ND: Notitia Dignitatum. Ed. Otto Seeck. Berlin, 1876.
NVal: see *CTh*.
Olymp(iodorus). Ed. and trans. Blockely 1983: 152-209.
Parangelmata poliorcetika. Ed. and trans. Sullivan 2000.
Parastaseis: see Cameron and Herrin 1984.
Passio Leudegarii. MGH SSRM 5. Trans. Gerberding and Fouracre 1996: 215-53.
Paulinus of Pella. *Eucharisticos*. Loeb.
Peri strategias: Anonymi Peri strategias, The Anonymous Byzantine Treatise on Strategy. Ed. George T. Dennis. *Three Byzantine Military Treatises. Text, transl. and notes. CFHB* 25, Washington, D.C., 1985: 1 -136.
Priscus. Ed. and trans. Blockley 1983: 222-377.
Proc(opius). *Wars*. Loeb (ed. Haury, trans. Dewing).
Proc(opius), *De Aedeficiis*. Loeb (ed. Haury, trans. Dewing).
Prosper: *Prosper Tironis epitoma chronicon. MGH SSAA* 9 (*Chronica Minora* I). Trans. Murray 2000: 62-76.
Rothari: see *Leges Langobardorum*.
Salvian, *De gubernatione Dei. MGH SSAA* 1.1 (partial trans. in Murray 2000)
Scriptor Incertus. *Historia de Leone Bardae Armenii filio*. In *Leonis Grammatici Chronographia. CSHB*. Ed. I. Bekker. Bonn, 1842.
Sebeos. *The Armenian History attributed to Sebeos. Part I: Translation and Notes. Part II. Historical Commentary*. Trans. and comm. R.W. Thomson, James Howard-Johnston and Tim Greenwood. Liverpool, 1999.
Sid(onius) Ap(ollinaris). *Epistulae et Carmina. MGH SSAA* 8.
Strategikon: Das Strategikon des Maurikios. Ed. George T. Dennis, trans. Ernst Gammischeg. *CFHB* 17. Vienna, 1981. Trans. George T. Dennis. *Maurice's Strategikon: Handbook of Byzantine Military Technology*. Philadelphia, 1984.
Strategius: see Conybeare 1910.
Tabari: Abū Ja'far Muḥammad ibn Jarīr al-Tabarī, *Tā'rīkh al-rusul wa'l-muluk*. Ed. M.J. DeGoeje. Leiden 1879 (1964-65). *The History of al-Tabari. An Annotated Translation* (39 volumes). Albany, 1985-1999.
Taio Caesaraugustanus. *Sententiarum Libri V, PL* 80.
Theodoret *HE. Theodoret Kirchengeschichte*. Eds. Léon Parmentier and Felix Scheidweiler. Berlin, 1954. Trans. *Nicene and Post-Nicene Fathers*, Series 2, volume 3.
Th. Sim.: Theophylact Simocatta. *Historiae*. Eds. Carl de Boor and Peter Wirth. Stuttgart, 1887/1972. Trans. Michael and Mary Whitby. *The* History *of Theophylact Simocatta*. Oxford, 1986.
Theophanes: *Theophanis Chronographia*. Ed. Carl de Boor. Leipzig, 1883 (Hildesheim and New York, 1980). Trans. Cyril Mango and Roger Scott. *The Chronicle of Theophanes Confessor: Byzantine and Near Eastern History AD 284-813*. Oxford, 1997.
Vegetius. *Epitoma rei militaris*. Ed. C. Lang, Leipzig, 1969/ M.D. Reeve, Oxford, 2004. Trans N.P. Milner. *Vegetius: Epitome of Military Science*. Liverpool, 1993.
Venantius Fortunatus. *Carmina. MGH SSAA* 4.1.
Vita Aniani episcopi Aurelianensis. MGH SSRM 3: 104-117.
Vitae Caesarii episcopi Arelatensi libri duo. MGH SSRM 3: 433-501. Trans. William E. Klingshirn. *Caesarius of Arles: Life, Testament, Letters*. Liverpool, 1994
Vita Desiderii Cadurcae Urbis Episcopi. SSRM 4: 547-602.

Vita Genovefae virginis Parisiensis. MGH SSRM 3: 204-38.
VPE: Vitae Patrum Emeritensium. Trans. A.T. Fear. *Lives of the Visigothic Fathers.* Liverpool: Liverpool University Press, 1997.
Zachariah, *HE: Historia Ecclesiastica Zachariae Rhetori vulgo adscripta.* I: *CSCO* 83, *SS* 38 (*Textus*), *CSCO* 87, *SS* 41 (*Versio*). II: *CSCO* 84, *SS* 39 (*Textus*), *CSCO* 88, *SS* 42 (*Versio*). Ed. E.W. Brooks. Louvain: Peeters, 1924. Translations: F.J. Hamilton and E.W. Brooks, *The Syriac Chronicle known as that of Zachariah of Mitylene.* London, 1899. Ed. Geoffrey Greatrex, trans. Robert R. Phenix and Cornelia B. Horn, with contrib. by Sebastian P. Brock and Witold Witakowski. *The Chronicle of Pseudo-Zachariah Rhetor: Church and War in Late Antiquity.* Liverpool, 2011

Secondary literature

Abels, Richard and Stephen Morillo (2005) "A Lying Legacy? A Preliminary Discussion of Images of Antiquity and Altered Reality in Medieval Military History." *JMMH* 3, 1-13.
Ahrweiler, Hélène (1966) *Byzance et la mer: la marine de guerre la politique et les institutions maritimes de Byzance aux VIIe–XVe siècles.* Paris.
Amt, Emily (2002) "Besieging Bedford: Military Logistics in 1224." *JMMH* 1, 101-124.
Amory, Patrick (1997) *People and Identity in Ostrogothic Italy, 489-554.* Cambridge.
Anderson, Thomas (1995) "Roman military colonies in Gaul, Salian ethnogenesis and the forgotten meaning of *Pactus Legis Salicae* 59.5." *Early Medieval Europe* 4 (2), 129-144.
Anderson, Benedict (1983, rev. ed. 2006) *Imagined Communities.* London and New York.
Antonopoulos, Panagiotis (2005) "King Cunincpert and the Archangel Michael." *Langobarden* 383-86.
Astachova, Natalia V. (1995) "The circus parties and the term δῆμος/οι according to Procopius (De Bello Persico I.24)." *ByzSlav* 56 (1), 19-22.
Bachrach, Bernard (1970a) "Charles Martel, Mounted Shock Combat, the Stirrup, and Feudalism." *Studies in Medieval and Renaissance History* 7, 49-75.
____ (1970b) "Procopius, Agathias and the Frankish Military." *Speculum* 45 (3), 435-441.
____ (1972) *Merovingian Military Organization.* Minneapolis.
____ (1974) "Military Organization in Aquitaine under the Early Carolingians." *Speculum* 49 (1), 1-33. Reprinted in Bachrach 1993a.
____ (1975) "Who were the Ripariolilbriones?" *Gräzer Beiträge* 3, 15-19.
____ (1983) "Charlemagne's Cavalry: Myth and Reality." *Military Affairs* 87, 181-87.
____ (1993a) *Armies and Politics in the Early Medieval West.* Aldershot.
____ (1993b) "Grand Strategy in the Germanic Kingdoms: Recruitment of the Rank and File." Kazanski and Vallet, 55-63.
____ (1994) *The Anatomy of a Little War: A Diplomatic and Military History of the Gundovald Affair (568-586).* Boulder, San Francisco and Oxford.
____ (2001a) *Early Carolingian Warfare: Prelude to Empire.* Philadelphia.
____ (2001b) "Military lands in historical perspective." *The Haskins Society Journal* 9, 95-122.
____ (2002a) *Warfare and Military Organization in Pre-Crusade Europe.* Aldershot.
____ (2002b) "Gregory of Tours as a military historian." *WGT* 351-63
____ (2003) "Fifth Century Metz: Late Roman Christian Urbs or Ghost Town?" *AT* 10, 363-82.
____ (2007a) " "A Lying Legacy" Revisited: The Abels-Morillo Defense of Discontinuity." *JMMH* 5, 153-93.
____ (2007b) "Are they not like us? The Carolingian fisc in military perspective." Celia Chazelle and Felice Lifshitz, eds. *Paradigms and Methods in Early Medieval Studies.* Palgrave MacMillan, 119-33.
____ (2010) "The Fortification of Gaul and the Economy of the Third and Fourth Centuries." *Journal of Late Antiquity* 3 (1), 38-64.

Bachrach, David (2003) *Religion and the Conduct of War, c. 300–c. 1215*. Woodbridge.
―――― (2012) *Warfare in tenth-century Germany*. London.
Barbero, Alessandro (2004) *Charlemagne: Father of a Continent*. Berkeley and Los Angeles.
Barbero, Alessandro and Maria Isabel Loring (2005a) "The formation of the Sueve and Visigothic kingdoms in Spain." *NCMH* 1, 162-92.
―――― (2005b) "The Catholic Visigothic kingdom." *NCMH* 1, 346-70.
Barth, Fredrik, ed. (1969a) *Ethnic Groups and Boundaries: The Social Organization of Culture Difference*. Oslo.
―――― (1969b) "Introduction." Barth 1969a: 9-38.
―――― (1969c) "Pathan Identity and its Maintenance." Barth 1969a: 117-34.
Berger, Albrecht (1998), "Viranşehir (Mokisos), eine byzantinische Stadt in Kappadokien." *Istanbuler Mitteilungen* 48, 349-429 [w/ 62 plates on following pp].
Bertolini, Ottorino (1968) "Ordinamenti military e strutture sociali dei Longobardi in Italia." *Settimane* 15, 429-607.
Beševliev, Ves. (1981) *Die protobulgarische Periode der bulgarischen Geschichte*. Amsterdam.
Bivar, A.D.H. (1972) "Cavalry Equipment and Tactics on the Euphrates Frontier." *DOP* 26, 271-91.
Blockley (1972) "Dexippus and Priscus and the Thucydidean Accound of the Siege of Plataea." *Phoenix* 26 (1), 18-27.
Blockley, R.C. (1981-83) *The Fragmentary Classicising Historians of the Later Roman Empire: Eunapius, Olympiodorus, Priscus and Malchus. Vol. 1 [Commentary]; vol. 2 Text, Translation and Historiographical Notes*. Liverpool.
Bonnet, Max (1887) *Le Latin de Grégoire de Tours*. Paris.
Bosworth, C.E. (1963) *The Ghaznavids: Their Empire in Afghanistan and Eastern Iran 994-1040*. Edinburgh.
―――― (1966) "Military Organization under the Būyids of Persia and Iraq." *Oriens* 18/19 [1965/1966], 143-67.
―――― (1968) "The Armies of the Saffarids." *BSOAS* 31 (3), 534-554.
―――― (1996) *The Arabs, Byzantium and Iran: Studies in Early Islamic History and Culture*. Aldershot.
Bowlus, Charles (1978) "Warfare and Society in the Carolingian Ostmark." *Austrian History Yearbook* 14, 3-26.
―――― (1984) "Two Carolingian Campaigns Reconsidered." *Military Affairs* 48 (3), 121-25.
Böhme, Horst Wolfgang (2009) "Migrantenschicksale. Die Integration der Germanen in spätantiken Gallien." *Spätantike*, 35-60.
Börm, Hennig (2007) *Prokop und die Persier: Untersuchungen zu den römisch-sassanidischen Kontakten in der ausgehenden Spätantike*. Stuttgart.
Bradbury, Jim (1992) *The Medieval Siege*. Woodbridge.
Brandes, Wolfram (1989) *Die Städte Kleinasiens im 7. und 8. Jahrhundert*. Berlin.
―――― (1999) "Byzantine Cities in the Seventh and Eighth Centuries—Different Sources, Different Histories?" *TRW* 4, 25-57.
Brandes, Wolfram and John Haldon (2000) "Town, Tax and Transformation: State, Cities and their Hinterlands in the East Roman World, c. 500-800." *TRW* 9, 141-72.
Brather, Sebastian (2008) *Archäologie der westlichen Slawen (2nd ed.)*. Berlin and New York.
Braund, David (1994) *Georgia in Antiquity: A History of Colchis and Transcaucasian Iberia 550 BC–AD 562*. Oxford.
Brock, Sebastian (1982) "Syriac Views of Emergent Islam." G.H.A. Juynboll (ed.) *Studies on the First Century of Islamic Society*. Carbondale and Edwardsville. 9-21 and 199-203. Reprinted as article 19 in *FCIW* 5.
Brogiolo, Gian Pietro (1999) "Ideas of the Town in Italy during the Transition from Antiquity to the Middle Ages." *TRW* 4, 99-126.
―――― (2000) "Towns, Forts and the Countryside: Archaeological Models for Northern Italy in the Early Lombard Period (AD 568-650)." *TRW* 9, 299-323.

_____ (2007) "Dwellings and settlements in Gothic Italy." *Ostrogoths* 113-133.
Brooks, E.W. (1899) "The Campaign of 716-718, from Arabic Sources." *The Journal of Hellenic Studies* 19, 19-31.
Brown, Peter (1971) *The World of Late Antiquity AD 150-750*. New York and London.
Brown, T.S. (1984) *Gentlemen and Officers: Imperial Administration and Aristocratic Power in Byzantine Italy A.D. 554-800*. British School at Rome.
Browning, Robert (1975) *Byzantium and Bulgaria: a comparative study across the early medieval frontier*. London.
Brubaker, Leslie and John Haldon (2001) *Byzantium in the Iconoclast Era (ca 680-850): The Sources. An Annotated Survey*. Aldershot.
_____ (2011) *Byzantium in the Iconoclast Era c. 680-850: A History*. Cambridge.
Brühl, Carlrichard (1968) *Fodrum, Gistum, Servitium Regis*. 2 vols. Cologne and Graz.
_____ (1975-90) *Palatium und Civitas. Studien zur Profantopographie spätantiker Civitates vom 3. bis zum 13. Jahrhundert. Band I: Gallien. Band II: Belgica I, beide Germanien und Raetia II*. Cologne and Vienna.
Bulliet, Richard W. (1987) "Medieval Arabic Tarsh: A Forgotten Chapter in the History of Printing." *Journal of the American Oriental Society* 107 (3), 427-38.
Bullough, D.A. (1970) "*Europae Pater*: Charlemagne and his achievement in the light of recent scholarship." *EHR* 85 (334), 59-105.
Burns, Thomas (1984) *A History of the Ostrogoths*. Bloomington and Indianapolis.
Busse, Heribert (1984) " 'Omar b. al-Hattab in Jerusalem." *JSAI* 5, 73-119.
_____ (1986) " 'Omar's Image as the Conqueror of Jerusalem." *JSAI* 8, 149-68.
Cameron, Alan (1976) *Circus Factions: Blues and Greens at Rome and Byzantium*. Oxford.
Cameron, Averil (1968) "Agathias on the Early Merovingians." *Annali della scuola normale superiore di Pisa*, Serie II vol. 37, 95-140.
_____ (1970) *Agathias*. Oxford.
_____ (1985) *Procopius and the Sixth Century*. London and New York.
Cameron, Averil and Judith Herrin (eds.) (1984) *Constantinople in the Early Eighth Century: The* Parastaseis syntomoi chronikai. Leiden.
Chevedden, Paul E. (1998) "The Hybrid Trebuchet: The Halfway Step to the Counterweight Trebuchet." *On the Social Origins of Medieval Institutions. Essays in Honor of Joseph F. O'Callaghan*. Leiden, 179-222.
_____ (2000) "The Invention of the Counterweight Trebuchet: A Study in Cultural Diffusion" in *DOP* 54, 71-116.
Chevedden, Paul E., L. Eigenbrod, V. Foley, and W. Soedel (1995) "The Trebuchet: Recent Reconstructions and Computer Simulations Reveal the Operating Principles of the Most Powerful Weapon of Its Time." *Scientific American*, July, 66-71.
Chevedden, Paul E., Zvi Shiller, Samuel R. Gilbert, and Donald Kagay, (2000) "The Traction Trebuchet: A Triumph of Four Civilizations." *Viator* 31, 433-486.
Christides, Vassilis (2011) "The Second Arab Siege of Constantinople (717-18?): Logistics and Naval Power." D. Bumazhnov, E. Grypeou, T.B. Sailors and A. Toepel (eds.) *Bibel, Byzanz und christlicher Orient. Festschrift für Stephen Gerö zum 65. Geburtstag*. Orientalia Lovaniensia Analecta 187. Leuven, Paris and Walpole, MS.
Christie, Neil (1995) *The Lombards*. Malden.
_____ (2000) "Towns, Land and Power: German-Roman Survivals and Interactions in Fifth- and Sixth-Century Pannonia." *TRW* 9, 275-97.
_____ (2006) *From Constantine to Charlemagne: An Archaeology of Italy, AD 300-800*. Aldershot.
Chrysos, Evangelos (1996) "The Roman Political Identity in Late Antiquity and Early Byzantium." *ByzMajP*, 7-16.
Claude, Dietrich (1981) "Die Handwerker der Merowingerzeit nach erzählenden und urkundlichen Quellen." *Handwerk* 204-66.

Clover, Frank (1971) "Flavius Merobaudes: A Translation and Historical Commentary." *Transactions of the American Philosophical Society (New Series)* 61 (1), 1-78.
Coates, Simon (2000) "Venantius Fortunatus and the Image of Episcopal Authority in Late Antique and Early Merovingian Gaul." *EHR* 115 (464), 1109-37.
Collins, Roger (1989) *The Arab Conquest of Spain 710-797*. Oxford and Cambridge, Mass.
____ (1996) "The Carolingians and the Ottonians in an Anglophone World." *Journal of Medieval History* 22 (1), 97-114.
____ (1998) *Charlemagne*. Toronto and Buffalo.
____ (1999) *Early Medieval Europe 300-1000* (2nd ed.). Hampshire and New York.
____ (2004) *Visigothic Spain 409-711*. Malden and Oxford.
____ (2010) review of McKitterick (2008). *EME* 18 (3), 356-60.
Conrad, Lawrence (1990) "Theophanes and the Arabic Historical Tradition: Some Indications of Intercultural Transmission." *Byzantinische Forschungen* 15, 1-44.
____ (1992) "The Conquest of Arwād: A Source-critical study in the historiography of the early medieval Near East." *BEINE* 1, 317-401.
Contamine, Philippe (1984) *War in the Middle Ages*. Trans. Michael Jones. Oxford and Cambridge, Mass.
Conybeare, Frederick C. (1910) "Antiochus Strategos' Account of the Sack of Jerusalem in A.D. 614." *EHR* 25, 502-17.
Cook, David (2007) *Martyrdom in Islam*. Cambridge.
Cosentino, Salvatore (2004) "Il Re Teoderico come constuttore di flotte." *AT* 12, 347-56.
____ (2008) "Constans II and the Byzantine navy." *BZ* 100 (2), 577–603.
Crone, Patricia and Michael Cook (1977) *Hagarism: The Making of the Islamic World*. Cambridge.
Crone, Patricia (1980) *Slaves on Horses: The Evolution of the Islamic Polity*. Cambridge.
Cuomo, S. (2007) *Technology and Culture in Greek and Roman Antiquity*. Cambridge.
Curta, Florin (2001) *The Making of the Slavs: History and Archaeology of the Lower Danube Region, c. 500-700*. Cambridge.
____ (2006) *Southeastern Europe in the Middle Ages, 500-1250*. Cambridge.
Daly, William M. (1994) "Clovis: How Barbarian, How Pagan?" *Speculum* 69 (3), 619-664.
Davis, Raymond, trans. (rev. ed. 2000) *The Book of Pontiffs* (Liber Pontificalis): *The ancient biographies of the first ninety Roman bishops to AD 715*. Liverpool.
____ (2007) *The Lives of the Eighth-Century Popes*. Liverpool, rev. ed.
Delbrück, Hans (1920) *History of the Art of War*. University of Nebraska Press, reprint edition, 1990. Translated by Walter, J. Renfroe. 4 Volumes.
Delogu, Paolo (1988) "The rebirth of Rome in the 8th and 9th centuries." *Rebirth*, 32-42.
____ (1995) "Lombard and Carolingian Italy." *NCMH* 2.
Demandt, Alexander (1970) "Magister Militum." *Paulys Realencyclopädie der classischen Altertumswissenschaft*, Supplementband XII, cols. 553–790.
____ (2007) *Die Spätantike Romische Geschichte von Diocletian bis Justinian 284-565 n. Chr.* (2nd ed.). *Handbuch der Altertumswissenschaft III.6*. Munich.
Dennis, George T. (1998) "Byzantine Heavy Artillery: The Helepolis." *GRBS* 39, 99-115.
DeVries, Kelly. *Medieval Military Technology*. Broadview Press: Peterborough, 1992.
____ (1997) "Catapults are not Atomic Bombs: Towards a Redefinition of 'Effectiveness' in Premodern Military Technology." *War in History* 4, 454-70.
Devroey, Jean-Pierre (2003) *Économie rurale et société dans l'Europe franque (VIe-IXe siècles). Tome 1: Fondements matériels, échanges et lien social*. Paris.
____ (2006) *Puissants et misérables. Système social et monde paysan dans l'Europe des Francs*. Brussels.
Díaz, Pablo (2000) "City and Territory in Hispania in Late Antiquity." *TRW* 9, 3-35.
Diehl, Charles (1896) *L'Afrique Byzantine: histoire de la domination byzantine en Afrique, 533-709*. Paris.

Dignas, Beate and Engelbert Winter (2007) *Rome and Persia in Late Antiquity: Neighbors and Rivals*. Cambridge.
Dobat, Andres Siegfried (2008) *Werkzeuge aus kaiserzeitlichen Heeresausrüstungsopfern*. Jysk Arkæologisk Selskabs skrifter 61. Højbjerg.
Donner, Fred McGraw (1981) *The Early Islamic Conquests*. Princeton.
Downey, Glanville (1946) "Byzantine Architects: Their Training and Methods." *Byzantion* 18, 99-118.
Drecoll, Carsten (2007) *Die Liturgien im römischen Kaiserreich des 3. und 4. Jahrhunderts*. Historia Einzelschriften 116. Stuttgart.
Drinkwater, J.F. (1998) "The Usurpers Constantine III (407-411) and Jovinus (411-413)," *Britannia* 29, 269-98.
Dunn, Archibald (1994) "The transition from *polis* to *kastron* in the Balkans (III-VII cc.): general and regional perspectives." *BMGS* 18, 60-81.
Durliat, Jean (1990) *Les finances publiques de Diocletien aux Carolingiens (284-889)*. Sigmaringen.
____ (1993) "Armée et société vers 600. Le problème des soldes." Kazanski and Vallet, 31-38.
Dyson, Stephen L. (2003) *The Roman Countryside*. London.
Edquist, Charles and Olle Edqvist (1979) *Social Carriers of Techniques for Development: A Comparative Economic Systems Approach*. Stockholm: Swedish Agency for Research Cooperation with Developing Countries.
Elton, Hugh (1996) *Warfare in Roman Europe AD 350-425*. Oxford.
____ (2000) "The nature of the sixth-century Isaurians." Mitchell and Greatrex, 293-307.
Ewig, Eugen (1965) "Descriptio Franciae." *KdGr* 143-177.
____ (1976-1979) *Spätantikes und fränkisches Gallien (Band 1-2)*. Beihefte der *Francia* Band 3.1-2. Zürich and Munich.
Fahmy, Aly Mohamed (1966) *Muslim Sea-Power in the Eastern Mediterranean from the Seventh to the Tenth Century A.D.* Cairo.
Finster, Barbara (1982) "Craftsmen and Groups of Craftsmen in the First Centuries of Islam." *FCIW* 12, 198-210 (article 8).
Foss, Clive (1975) "The Persians in Asia Minor and the End of Antiquity." *EHR* 90 (357), 721-47.
____ (1977) "Late Antique and Byzantine Ancyra." *DOP* 31, 27-97.
____ (1997) "Syria in Transition, A.D. 550-750: An Archaeological Approach." *DOP* 51, 189-269.
____ (2003) "The Persians in the Roman near East (602-630)." *Journal of the Royal Asiatic Society (Third Series)* 13 (2), 149-70.
Foss, Clive and David Winfield (1986) *Byzantine Fortifications: An Introduction*. Pretoria.
Fouracre, Paul (1990) "Merovingian History and Merovingian Hagiography." *P&P* 127, 3-38.
____ (2000) *The Age of Charles Martel*. Harlow.
____ (2005) "Francia in the seventh century." *NCMH* 1, 371-96.
Fouracre, Paul and Richard A. Gerberding (1996) *Late Merovingian France: History and Historiography 640-720*. Manchester and New York.
Flusin, Bernard (1992) *Saint Anastase le Perse et l'histoire de la Palestine au début du VIIe siècle. I: Les Textes. II: Commentaire: Les moines de Jérusalem et l'invasion Perse*. Paris.
France, John (1979) "La guerre dans la France féodale a la fin du IXe et au Xe siècle." *Revue Belge d'Histoire Militaire* 23, 177-98.
____ (1985) "The military history of the Carolingian period." *Revue Belge d'Histoire Militaire* 26 (2), 81-99.
____ (1994) *Victory in the East: A Military History of the First Crusade*. Cambridge.
____ (1999) *Western Warfare in the Age of the Crusades, 1000-1300*. Ithaca.
____ (2002) "The Composition and Raising of the Armies of Charlemagne." *JMMH* 1, 61-82.

_____ (2005) "War and Sanctity: Saints' Lives as Sources for Early Medieval Warfare." *JMMH* 3, 14-22

Frantz-Murphy, Gladys (1991) "Conversion in early Islamic Egypt: the economic factor." Yusuf Raghib, ed. *Documents de l'Islam medieval: nouvelles perspectives de recherché.* Cairo, 11-17. Also published as article 14 in *FCIW* 18.

_____ (2007) "The Economics of State Formation in Early Islamic Egypt." Sijpesteijn 2007b: 101-14.

Friedman, Jonathan (1992) "The Past in the Future: History and The Politics of Identity." *American Anthropologist* 94, 837-56.

Frölich, Annette (2009) *Jernalderens Lægekunst: en nytolkning af arkæologiske fund fra danske offermoser.* Jysk Arkæologisk Selskabs skrifter 63. Højbjerg.

Ganshof (1964) *Feudalism*, 3rd ed. Trans. Philip Grierson. London.

_____ (1968) "L'armée sous les Carolingiens." *Settimane* 15, 109-30.

Garsoïan, Nina (1997) Chapters 2-8 in Hovannasian (1997).

Gascou, Jean (1976) "L'institution des bucellaires." *BYZ* 76, 143-56

Gauthier, Nancy (2000) "Le réseau de pouvoirs de l'évêque dans la Gaule du haut moyen-âge." *TRW* 9, 173-207.

_____ (2002) "From the ancient city to the medieval town: Continuity and change in the early middle ages." *WGT* 47-66.

Geertz, Clifford (1973) "Thick description: Toward an Interpretative Theory of Culture." *The Interpretation of Cultures: Selected Essays.* New York, 3-30

Gerberding, Richard (1987) *The Rise of the Carolingians and the Liber Historiae Francorum.* Oxford.

Gillmor, Carroll (1981) "The Introduction of the Traction Trebuchet into the Latin West." *Viator* 12, 1-8

_____ (2005) "The 791 Equine Epidemic and its Impact on Charlemagne's Army." *JMMH* 3, 23-45.

_____ (2008) "The *Brevium Exempla* as a Source for Carolingian Warhorses." *JMMH* 6, 32-57.

Goffart, Walter (1957) "Byzantine Policy in the West under Tiberius II and Maurice: The Pretenders Hermengild and Gundovald." *Traditio* 13, 73-118.

_____ (1963) "The Fredegar problem reconsidered." *Speculum* 38 (2), 206-41.

_____ (1972) "From Roman Taxation to Medieval Seigneurie: Three Notes (Part I)." *Speculum* 47 (2), 165-187. "(Part II)," *Speculum* 47 (3), 373-394.

_____ (1982) "Old and New in Merovingian Taxation." *Past and Present* 96, 3-21.

_____ (1988) *The Narrators of Barbarian History (A.D. 550-800): Jordanes, Gregory of Tours, Bede, and Paul the Deacon.* Princeton.

_____ (2006) *Barbarian Tides. The Migration Age and the Later Roman Empire.* Philadelphia.

_____ (2008) "Frankish military duty and the fate of Roman taxation." *EME* 16 (2), 166-90.

Goldberg, Eric (2006) *Struggle for Empire: Kingship and Conflict under Louis the German, 817–876.* Ithaca.

Goubert, Paul (1956-65) *Byzance avant l'Islam, tome second: Byzance et l'occident sous les successeurs de Justinien. I, Byzance et les francs. II, Rome, Byzance et Carthage.* Paris.

Graff, David (2002) *Medieval Chinese Warfare, 300-900.* London.

Greatrex, Geoffrey (1998) *Rome and Persia at War, 502-532.* Cambridge.

_____ (2000) "Roman identity in the sixth century." Mitchell and Greatrex, 267-92.

Greatrex, G. and M. Greatrex (1999) "The Hunnic Invasion of the East of 395 and the Fortress of Ziatha." *Byzantion* 69 (1), 65-75.

Greatrex, Geoffrey and Samuel Lieu (2002) *The Roman Eastern Frontier and the Persian Wars: A Narrative Sourcesbook, pt. 2, AD 363-630.* London.

Grosse, Robert (1920) *Römische Militärgeschichte von Gallienus bis zum Beginn der byzantinischen Themenverfassung.* Berlin.

Haldon, John (1979) *Recruitment and Conscription in the Byzantine Army c. 550-950. A Study on the Origins of the Stratiotika Ktemata*. Wien.
____ (1984) *Byzantine Praetorians: An Administrative, Institutional and Social Survey of the Opsikion and Tagmata, c. 580-900*. Bonn.
____ (1990) *Byzantium in the Seventh Century: The Transformation of a Culture*. Rev. ed. 1997. Cambridge.
____ (1993a) "Military Service, Military Lands and the Status of Soldiers: Current Problems and Interpretations." *DOP* 47, 1-67.
____ (1993b) "Administrative Continuities and Structural Transformations in East Roman Military Organization c. 580-640." Kazanzki and Vallet, 1-20.
____ (1995) "Seventh-Century Continuities: the *Ajnād* and the "Thematic Myth"." *BEINE* 3, 379-423.
____ (1997) "Military Service, Military Lands, and the Status of Soldiers: Current Problems and Interpretations," *DOP* 47, 1-67.
____ (1999) *Warfare, State and Society in the Byzantine World 565-1204*. London.
____ (2000) "Theory and Practice in Tenth-Century Military Administration: Chapters II, 44 and 45 of the *Book of Ceremonies*." *TM* 13, 201-352.
____ (2004) "The Fate of the Late Roman Senatorial Élite: Extinction or Transformation?" *BEINEI* 6, 179-234.
____ (2010) "The resources of Late Antiquity." *NCHI* 1, 19-71.
Haldon, John and Hugh Kennedy (1980) "The Arab-Byzantine frontier in the eighth and ninth centuries: military organization and society in the borderlands." *Zbornik Radova Visantološkog Instituta* 19, 79-116.
Halsall, Guy (1995) *Settlement and Social Organization: The Merovingian Region of Metz*. Cambridge.
____ (1999) "Review article: Movers and Shakers: the barbarians and the Fall of Rome." *EME* 8 (1), 131-145.
____ (2003) *Warfare and Society in the Barbarian West, 450-900*. Abingdon and New York.
____ (2007) *Barbarian Migrations and the Roman West, 376-568*. Cambridge.
____ (2008) review of Ward-Perkins 2005. *EME* 16 (3), 384–386
Hamerow, Helena (2002) *Early Medieval Settlements: The Archaeology of Rural Communities in Northwest Europe 400-900*. Oxford.
Hannestad, Knud (1960) "Les forces militaires d'après la guerre gothique de Procope." *Classica et Medievalia* 21, 136-183.
Hardt, Mathias (2001) "Hesse, Elbe, Saale and the frontiers of the Carolingian Empire." *TRW* 12, 219-32.
Harries, Jill (1994) *Sidonius Apollinaris and the Fall of Rome AD 407-485*. Oxford.
Hawting, G.R. (2000) *The First Dynasty of Islam: The Umayyad Caliphate AD 661-750*, 2nd ed. London and New York.
Haywood, John (1999) *Dark Age Naval Power: A Reassessment of Frankish and Anglo-Saxon Seafaring Activity*, rev. ed. Anglo-Saxon Books, Frithgarth.
Heather, Peter (1991) *Goths and Romans 332-489*. Oxford.
____ (1996) *The Goths*. Oxford.
____ (1998) "Goths and Huns, c. 320-425." *CAH* 13, 487-415.
____ (2002) "State, Lordship and Community in the West (c.AD 400-600)." *CAH* 14, 437-68.
____ (2001) "The late Roman art of client management: Imperial defence in the fourth century west." *TRW* 10, 15-68.
____ (2005) *The Fall of the Roman Empire. A New History*. London.
____ (2007) "Goths in the Roman Balkans, c. 300-500." *TLAD*, 163-90.
____ (2009) *Empires and Barbarians: The Fall of Rome and the Birth of Europe*. Oxford.
Henning, Joachim (2007) "Early European towns. The development of the economy in the Frankish realm between dynamism and deceleration AD 500-1100." *PRT* 1, 3-40.

Hedeager, Lotte (1992) *Iron-Age Societies: From Tribe to State in Northern Europe, 500 BC to AD 700*. Trans. John Hines. Oxford and Cambridge, Mass.
Herrin, Judith (1987) *The Formation of Christendom*. Princeton.
____ (1992) "'Femina Byzantina:" The Council in Trullo on Women." *DOP* 46, 97-105.
Hill, Donald R. (1971) *The Termination of Hostilities in the Early Arab Conquests A.D. 634-656*. London.
____ (1973) "Trebuchets." *Viator* 4, 99-114.
Hill, David (1988) "Unity and diversity—a framework for the study of European towns." *Rebirth*, 8-15.
Hofman, Hanns Hubert (1965) "Fossa Carolina. Versuch einer Zusammenschau." *KdGr* 436-53.
Honigmann, Ernst (1935) *Die Ostgrenze des byzantinischen Reiches von 363 bis 1071*. Corpus Bruxellense Historiae Byzantinae 3, Brussels.
Hovannisian, Richard G. (1997) *Armenian People from Ancient to Modern Times Vol. I. The Dynastic Periods: From Antiquity to the Fourteenth Century*. New York.
Howard-Johnston, James (1984) "Thema." *Maistor: Classical, Byzantine and Renaissance Studies for Robert Browning*, ed. Ann Moffatt. Canberra, 189-197.
____ (1995a) "The Two Great Powers in Late Antiquity: a Comparison." *BEINE* 3, 157-226.
____ (1995b) "The Siege of Constantinople in 626." *Cple*, 131-42.
____ (2006) *East Rome, Sasanian Persia and the End of Antiquity: Historiographical and Historical Studies*. Aldershot.
____ (2010) *Witnesses to a World Crisis: Historians and Histories of the Middle East in the Seventh Century*. Oxford.
Hoyland, Robert G. (1997) *Seeing Islam as Others Saw it: A Survey of Christian, Jewish and Zoroastrian Writings on Early Islam*. Princeton.
____ (2011) *Theophilus of Edessa's Chronicle and the Circulation of Historical Knowledge in Late Antiquity and Early Islam*. Liverpool.
Hunt, David (1998) "The Successors of Constantine." *CAH* 13, 1-43.
Huuri, Kalervo (1941) *Zur Geschichte des mittelalterlichen Geschützwesens aus orientalischen Quellen*. Helsinki.
Innes, Matthew (2000) *State and Society in the early Middle Ages: The Middle Rhine Valley, 400-1000*. Cambridge.
Isaac, Benjamin (1988) "The Meaning of the Terms *Limes* and *Limitanei*." *Journal of Roman Studies* 78, 125-147.
____ (1990) *The Limits of Empire: The Roman Army in the East*. Oxford.
____ (1995) "The Army in the Late Roman East: the Persian Wars and the Defence of the Byzantine Provinces." *BEINE* 3, 125-55.
Ivison, Eric A. (2000) "Urban Renewal and Imperial Revival in Byzantium." *Byzantinische Forschungen* 26, 1-46.
____ (2007) "*Amorium* in the Byzantine Dark Ages (seventh to ninth centuries)." *PRT* 2, 25-59.
James, Edward (1977) *The Merovingian Archaeology of South-West Gaul. Part i: Text. Part ii: Catalogues and Bibliography*. BAR Supplementary Series 25 (i-ii), Oxford.
____ (1988) *The Franks*. Oxford and Cambridge, Mass.
____ (2009) *Europe's Barbarians AD 200-600*. Harlow.
James, Simon (1988) "The *fabricae*: state arms factories of the Later Roman Empire." J.C. Coulston (ed.) *Military Equipment and the Identity of Roman Soldiers. Proceeding of the Fourth Roman Military Equipment Conference*. BAR International Series 394, 257-331.
____ (2011) "Stratagems, Combat, and "Chemical Warfare" in the Siege Mines of Dura-Europos." *American Journal of Archaeology*, 115 (1), 69-101.
Jankowiak, Marek (forthcoming) "Le premier siège arabe de Constantinople (674-678): un malentendu historiographique", *Travaux et Mémoires* 17.

Jarnut, Jörg (2003) "*Gens, rex* and *regnum* of the Lombards." *TRW* 13, 409-27.
Jiménez Garnica, Ana Maria (1999) "Settlement of the Visigoths in the Fifth Century." *Visigoths* 93-115.
Johnson, Stephen (1983) *Late Roman Fortifications*. London.
Jones, A.H.M. (1964) *The Later Roman Empire 284-602: A Social, Economic, and administrative Survey, vol. I-II*. Baltimore.
de Jong, Mayke (1999) "Adding insult to injury: Julian of Toledo and his *Historia Wambae*." *Visigoths* 373-90.
Judah, Jamal (1989) "The Economic Conditions of the *Mawālī* in Early Islamic Times." *FCIW* 12, 167-97 (article 7).
Kaegi, Walter E. (1968) "Initial Byzantine Reactions to the Arab Conquest." *Church History* 38 (2), 139-49.
_____ (1990) "Procopius the Military Historian." *Byzantinische Forschungen* 15, 53-85.
_____ (1992) *Byzantium and the Early Islamic Conquests*. Cambridge.
_____ (2003) *Heraclius: Emperor of Byzantium*. Cambridge.
Kaiser, Reinhold (2004) *Das römische Erbe und das Merowingerreich*, Enzyklopädie deutscher Geschichte, 26. R. München.
Kazanski, Michel and François Vallet, eds. (1993) *L'armée romaine et les barbares du IIIe au VIIe siècle*. Association Française d'Archéologie Mérovingienne et la Société des Amis du Musée des Antiquités Nationales.
Keesing, R. (1989) "Creating the Past: Custom and Identity in the Contemporary Pacific." *The Contemporary Pacific*, 1 (1-2), 19-42.
Kennedy, Hugh (2001) *The Armies of the Caliphs: Military and Society in the Early Islamic State*. London and New York.
_____ (2006) *The Byzantine and Early Islamic Near East*. Aldershot.
_____ (2007) *The Great Arab Conquests: How the Spread of Islam Changed the World We Live In*. Philadelphia.
Kern, Paul B. (1999) *Ancient Siege Warfare*. Bloomington and Indianapolis.
King, D.J. Cathcart (1982) "The Trebuchet and Other Siege Engines." *Chateau Gaillard: Etudes de Castellologie médiévale* IX-X, 457-469
King, P.D. (1972) *Law and Society in the Visigothic Kingdom*. Cambridge.
_____ (1987) *Charlemagne: Translated Sources*. Kendal.
Klingshirn, William E. (1994) *Caesarius of Arles: The Making of a Christian Community in Late Antique Gaul*. New York.
Kobyliński, Zbigniew (2005) "The Slavs." *NCMH* 1, 524-44.
Koder, Johannes (1984) *Der Lebensraum der Byzantiner. Historisch-geographischer Abriß ihres mittelalterlichen Staates im östlichen Mittelmeerraum*. Vienna, rev. ed. 2001.
_____ (1993) *Gemüse in Byzanz. Die Versorgung Konstantinopels mit Frischgemüse im Lichte der Geoponica*. Vienna.
Kretschmer, Marek (2006) *Rewriting Roman history in the Middle Ages: the "Historia Romana" and the Manuscript Bamberg, Hist. 3*. Leiden.
Kulikowski, Michael (2004) *Late Roman Spain and Its Cities*. Baltimore.
_____ (2007) *Rome's Gothic Wars from the Third Century to Alaric*. Cambridge.
Labrousse, Michel (1968) *Toulouse antique, des origines à l'établissement des Wisigoths*. Paris.
Laiou, Angeliki E. and Cécile Morrisson (2007) *The Byzantine Economy*. Cambridge.
Landau-Tasseron, Ella (1995) "Features of the Pre-Conquest Muslim Armies." *BEINE* 3.
Landels, J.G. (2000) *Engineering in the Ancient World*, rev. ed. Berkeley and Los Angeles.
Lebecq, Stéphane (2000) "The role of the monasteries in the systems of production and exchange of the Frankish world between the seventh and the beginning of the ninth centuries." *TRW* 11, 121-48.
Lee, A.D. (2007) *War in Late Antiquity: A Social History*. Malden and Oxford.

Lemerle, Paul (1979-81) *Les plus anciens recueils des miracles de saint Démétrius et la pénétration des slaves dans les Balkans. I Le texte. II Commentaire.* Paris.

Lenski, Noel (2002) *Failure of Empire: Valens and the Roman State in the Fourth Century A.D..* Berkeley.

____ ed. (2006) *The Cambridge Companion to the Age of Constantine.* Cambridge.

____ (2007) "Two Sieges of Amida (AD 359 and 502-503) and the Experience of Combat in the Late Roman Near East." A.S. Lewin and P. Pellegrini, eds. *The Late Roman Army in the Near East from Diocletian to the Arab Conquest: Proceedings of a colloquium held at Potenza, Acerenza and Matera, Italy (May 2005).* BAR International Series 1717, 219-36.

____ (2008) "Captivity, Slavery, and Cultural Exchange between Rome and the Germans from the First to the Seventh Century CE." C. Cameron, ed. *Invisible Citizens: Captives and Their Consequences.* Salt Lake City, 80-109.

____ (2009) "Schiavi armati e la formazione di eserciti privati nella tarda antichità." G.P. Urso (ed.) *Ordine e Disordine nel mondo greco e romano.* Pisa, 145-75.

Leriche, Pierre (1993) "Techniques de guerre sassanides et romaines à Doura-Europos." Kazanski and Vallet, 83-100.

Lewis, Archibald R. (1976) "The Dukes of Regnum Francorum, A.D. 550-751." *Speculum* 51 (3), 381-410.

Lewit, Tamara (1991) *Agricultural Production in the Roman Economy, A.D. 200-400.* BAR International Series no. 568, Oxford.

Liebeschuetz, J.H.W.G. (1992) *Barbarians and Bishops.* Oxford.

____ (1993) "The end of the Roman army in the western empire." Rich and Shipley (eds.) 265-76.

____ (2001) *The Decline and Fall of the Roman City.* Oxford.

____ (2003) "The refugees and evacuees in the age of migrations." *TRW* 12, 65-79.

Lightfoot, Chris (1998) "The Survival of Cities in Byzantine Anatolia: The Case of Amorium." *Byzantion* 68 (1), 56-71.

Lightfoot, Chris and Mücahide (2007) *Amorion: An Archaeological Guide.* Istanbul.

Lilie, Ralph-Johannes (1976) *Die byzantinische Reaktion auf die Ausbreitung der Araber.* Munich.

____ (1984) "Die zweihundertjährige Reform: Zu den Anfängen der Themenorganisation im 7. und 8. Jahrhundert." *ByzSlav* 45 (1), 27-39; *ByzSlav* 45 (2), 190-201.

____ (1993) "Wie dunkel sind die „dunkelen Jahrhunderte"? Zur Quellensituation in der mittelbyzantinischen Zeit und ihren Auswirkung auf die Forschung." *JöB* 43, 37-43.

____ (1995) "Araber und Themen. Zum Einfluß der arabischen Expansion auf die byzantinische Militärorganisation." *BEINE* 3, 425-60.

Loseby, Simon (1998a) "Gregory's cities: Urban functions in sixth-century Gaul." Ian Wood (ed.) *Franks and Alamanni in the Merovingian Period: An Ethnographic Perspective.* Woodbridge, 239-70.

____ (1998b) "Marseilles and the Pirenne thesis, I: Gregory of Tours, the Merovingian kings, and "un grand port"." *TRW* 3, 203-29.

____ (2000) "Marseille and the Pirenne thesis, II: "ville morte"." *TRW* 11, 167-93.

____ (2005) "The Mediterranean economy." *NCMH* 1, 605-38.

Lot, Ferdinand (1946) *L'art militaire et les armees au Moyen Age en Europe et dans le Proche Orient.* Paris.

Loyn, H.R. and J. Percival (1975) *The Reign of Charlemagne. Documents on Carolingian Government and Administration.* Documents of Medieval History 2. London.

Luther, Andreas (1997) *Die syrische Chronik des Josua Stylites.* Berlin and New York.

Lynn, John (1996) "The Evolution of Army Style in the Modern West, 800-2000." *International History Review* 18 (3), 505-45.

MacGeorge, Penny (2002) *Late Roman Warlords.* Oxford.

Macháček, Jiří (2007) "Early medieval centre in Pohansko near Břeclav/Lundeburg: *munitio, emporium* or *palatium* of the rulers of Moravia?" *PRT* 1, 473-98.
Madelung, Wilferd (1969) "The Assumption of the Title Shāhānshāh by the Būyids and 'The Reign of the Daylam (*Dawlat al-Daylam*)'." *JNES* 28 (2), 84-108; 28 (3), 168-83.
Maenchen-Helfen, J. Otto (1973) *The World of the Huns: Studies in their History and Culture.* Berkely.
Malingoudis, Phaidon (1994) "Zur Wehrverfassung der slavischen Stämme im 7. Jahrhundert: Das Zeugnis der Miracula Sancti Demetrii." *JöB* 275-81.
Manandean, Hacop (1948) "Les invasions arabes en Arménie (Notes chronologiques)." *Byzantion* 18, 163-95.
Marazzi, Federico (1998) "The destinies of the late antique Italies: Politico-economic developments of the sixth century." *TRW* 3, 119-159.
____ (2007) "The last Rome: from the end of the fifth to the end of the sixth century." *Ostrogoths* 279-302.
Marcus, R.A. (1997) *Gregory the Great and his world.* Cambridge.
Marsden, E.W. (1969-1971) *Greek and Roman Artillery. I: Historical Development. II: Technical Treatises.* Oxford.
Martínez Pizarro, Joaquín, trans. (2005) *The Story of Wamba: Julian of Toledo's* Historia Wambae Regis. Washington, D.C.
Maspero, Jean (1912) *Organisation militaire de l'Egypte byzantine.* Paris.
Mathisen, Ralph (1979a) "Sidonius on the Reign of Avitus: A Study in Political Prudence." *Transactions of the American Philological Association,* 109, 165-71.
____ (1979b) "Hilarius, Germanus, and Lupus: The Aristocratic Background to the Chelidonius Affair." *Phoenix,* 33 (2), 160-69.
____ (1993) *Roman Aristocrats in Barbarian Gaul: Strategies of Survival in an Age of Transition.* Austin.
Mayerson, Philip (1964) "The First Muslim Attacks on Southern Palestine (A.D. 633-34)." *Transactions and Proceedings of the American Philological Association* 95, 155-99.
McCormick, Michael (1986) *Eternal Victory: Triumphal Rulership in Late Antiquity, Byzantium and the Early Medieval West.* Cambridge.
____ (2001) *The Origins of the European Economy: Communications and Commerce A.D. 300-900.* Cambridge.
____ (2007) "Where Do Trading Towns Come From?" *PRT* 1, 41-68.
McGeer, Eric (1995) *Sowing the Dragon's Teeth: Byzantine Warfare in the Tenth Century.* Dumbarton Oaks Studies 33, Washington, D.C.
McKitterick, Rosamond (1989) *The Carolingians and the written word.* Cambridge.
____ (2004) *History and memory in the Carolingian world.* Cambridge.
____ (2008) *Charlemagne: The Formation of a European Identity.* Cambridge.
Menestò, Enrico (1999) *Il corridoio Bizantino e la via amerina in Umbria nell'alto medioevo.* Centro italiano di studi sull'alto medioevo, Spoleto.
Menis, Gian Carlo, ed. (1990) *I Longobardi.* Mondadori Electa.
Mitchell, John (2000) "Artistic Patronage and Cultural Strategies in Lombard Italy." *TRW* 9, 347-70.
Mitchell, Stephen and Geoffrey Greatrex, eds. (2000) *Ethnicity and Culture in Late Antiquity.* London.
de Montpéreux, Frédéric Dubois (1839) *Voyage Autour Du Caucase: Chez Les Tcherkesses Et Les Abkhases, En Colchide En Géorgie, En Arménie Et En Crimée; Avec Un Atlas Géographique, Pittoresque, Archéologique, Géologique, Etc.* Paris.
Moorhead, John (1992) *Theoderic in Italy.* Oxford.
Morillo, Stephen (1994) *Warfare under the Anglo-Norman Kings.* Woodbridge.
Morony, Michael G. (1984) *Iraq after the Muslim Conquest.* Princeton.
____ (2004) "Social Elites in Iraq and Iran: After the Conquest." *BEINE* 6, 275-84.

Murray, Alexander (1994) "Immunity, Nobility, and the Edict of Paris," *Speculum* 69 (1), 18-39.
____ (2000) *From Roman to Merovingian Gaul: A Reader*. Peterborough.
Needham, Joseph and Robin D.S. Yates (1994) *Science and Civilisation in China: Volume 5, Chemistry and Chemical Technology; Part 6, Military Technology: Missiles and Sieges*. Cambridge.
Nelson, Janet Nelson, Janet (1986) "The Church's Military Service in the Ninth Century: a Contemporary Comparative View?" J. Nelson, *Politics and Ritual in Early Medieval Europe*. London, 117-132.
____ (1992) *Charles the Bald*. London.
____ (1995) "Kingship and royal government." *NCMH* 2, 383-430.
Nevo, Yehuda and Judith Koren (2003) *Crossroads to Islam: The Origins of the Arab Religion and the Arab State*. Amherst.
Nicasie, Martinus (1998) *Twilight of Empire: the Roman Army from the Reign of Diocletian until the Battle of Adrianople*. Dutch monographs on ancient history and archaeology v. 19. Leiden.
Nickel, Helmut (2002) "The Mutual Influence of Europe and Asia in the Field of Arms and Armour." *CMAA*, 107-25.
Niewöhner, Philipp (2007) "Archäologie und die „Dunkelen Jahrhunderte" im byzantinischen Anatolien." *PRT* 2, 119-57.
Nixon, C.E.V. (1992) "Relations between Visigoths and Romans in fifth-century Gaul." *Fifth-century Gaul*, 64-74.
Noble, Thomas F.X. (1984) *The Republic of St. Peter: The Birth of the Papal States, 680-825*. Philadelphia.
Noth, Albrecht (1994) *The Early Arabic Historical Tradition: A Source-Critical Study* (2nd ed. w/ Lawrence I. Conrad). Trans. Michael Bonner. Princeton.
Noyé, Ghislaine (2007) "Social relations in southern Italy." *Ostrogoths* 183-98.
Näsman, Ulf (1998) "The Justinianic era of South Scandinavia: An archaeological view." *TRW* 3, 255-78.
____ (2000) "Exchange and politics: The eighth-early ninth century in Denmark." *TRW* 11, 35-68.
Nehlsen, Herrmann (1981) "Die rechtliche und soziale stellung der Handwerker in den germanischen Leges. Westgoten, Burgunder, Franken, Langobarden." *Handwerk* 267-83.
Olster, David (1993) *The Politics of Usurpation in the Seventh Century: Rhetoric and Revolution in Byzantium*. Amsterdam.
____ (1995) "Theodosius Grammaticus and the Arab Siege of 674-78." *ByzSlav* 56 (1), 23-28.
Oman, Charles (1924) *A History of the Art of War in the Middle Ages*. Vol. I: A.D. 378-1278; Vol. II: A.D. 1278-1485, 2nd ed. London
Ostrogorsky, George (1969) *A History of the Byzantine State*. Trans. Joan Hussey. New Brunswick.
O'Sullivan, Shaun (2004) "Sebeos' Account of an Arab Attack on Constantinople in 654." *BMGS* 28, 67-88.
Ousterhout, Robert (1999) *Master Builders of Byzantium*. Princeton.
Palme, Bernhard (2007) "The imperial presence: Government and army." Roger Bagnall (ed.) *Egypt in the Byzantine World, 300-700*. Cambridge, 244-70.
Palmer, Andrew, Sebastian Brock and Robert Hoyland (1993) *The Seventh Century in West-Syrian Chronicles*. Liverpool.
Papaconstantinou, Arietta (2006) "Historiography, Hagiography, and the Making of the Coptic "Church of the Martyrs" in Early Islamic Egypt." *DOP* 60, 65-86.
Peeters, Paul (1933) "Πασαγνάθης–Περσογενής." *Byzantion* 8, 405-23.
Pentcheva, Bissera V. (2002) "The supernatural protector of Constantinople: the Virgin and her icons in the tradition of the Avar siege." *BMGS* 26, 2-41.

Pérez Sánchez (1989) *El ejercito en la sociedad visigoda*. Salamanca.
Périn, Patrick (2002) "Settlements and cemeteries in Merovingian Gaul." *WGT* 67-98.
Petersen, Leif Inge Ree (2010a) "Scandinavia: Narrative." *EncMedWar* 3: 227-34.
____ (2010b) "Scandinavia: Historiography." *EncMedWar* 3: 234-38.
____ (in press) "The Historiography of Byzantine Warfare: New Paradigms and Perspectives." *La Conducción de la Guerra c. 950-c. 1350: Historiografía: Actas del 1 Symposium Internacional Cáceres, Noviembre 2008*, ed. Manuel Rojas Gabriel.
____ (forthcoming) "The Logistics of Siege Warfare on Byzantium's Eastern Frontier from the Syriac Sources, 500-800." Proceedings of the conference "Medieval Frontiers at War" held at Cáceres in November 2010, ed. Manuel Rojas Gabriel.
von Petrikovitz, Harald (1971) "Fortifications in the North-Western Roman Empire from the Third to the Fifth Centuries A.D.," *JRS* 61, 178-218.
____ (1981) "Die Spezialisierung des römischen Handwerks." *Handwerk* 63-127.
Pohl, Walter (1988) *Die Awaren. Ein Steppenvolk in Mitteleuropa 567-822 n. Chr.* Munich, rev. ed. 2002.
____ (1990) "Verlaufsformen der Ethnogenese—Awaren und Bulgaren." *Ethnogenese* 113-24.
____ (1997) "The Empire and the Lombards: Treaties and Negotiations in the Sixth Century." *TRW* 1, 75-133.
____ (2000) *Die Germanen*. München.
____ (2003) "A Non-Roman empire in Central Europe: the Avars." *TRW* 13, 571-95.
Poláček, Lumír (2007) "Ninth-century Mikulčice: the "market of the Moravians"?" *PRT* 1, 499-524.
Pourshariati, Parvaneh (2008) *Decline and Fall of the Sasanian Empire: the Sasanian-Parthian Confederacy and the Arab Conquest of Iran*. London.
Pringle, Denys (1981) *The Defence of Byzantine Africa from Justinian to the Arab Conqest*. 2 vols. BAR International Series 99 (i-ii), Oxford.
Pryor, John H. and Elizabeth M. Jeffreys (2006) *The Age of the ΔΡΟΜΩΝ. The Byzantine Navy ca. 500-1204*. Leiden and Boston.
Purton, Peter (2009) *A History of the Early Medieval Siege*. Woodbridge.
al-Qadi, Wadad (2008) "Population Census and Land Surveys under the Umayyads." *Der Islam* 83, 341-416.
____ (2013) "Non-Muslims in the Muslim Conquest Army in Early Islam," forthcoming in the proceedings of the conference on "Christians, Zoroastrians, and Jews in the Umayyad Period," ed. Fred Donner and Antoine Borrut. Chicago.
Rance, Philip (2005) "Narses and the Battle of Taginae (Busta Gallorum) 552: Procopius and Sixth-Century Warfare." *Historia: Zeitschrift für Alte Geschichte* 54 (4), 424-72.
____ (2007) "Battle." P.G.A. Sabin, H. van Wees and M. Whitby, eds. *The Cambridge History of Greek and Roman Warfare: Volume II: Rome from the Late Republic to the Late Empire*. Cambridge, 342-378.
Rapp, Claudia (2004) "Bishops in Late Antiquity: A New Social and Urban Élite?" *BEINE* 6, 149-78.
Ravegnani, Giorgio (1998) *Soldati di Bisanzio in età giustinianea*. Rome.
____ (2004) *I Bizantini e la Guerra: L'età di Giustiniano*. Rome.
Retamero, Felix (1999) "As coins go home: Towns, merchants, bishops and kings in Visigothic Hispania." *Visigoths* 271-305.
Retsö, Jan (2003) *The Arabs in Antiquity: Their History from the Assyrians to the Umayyads*. London.
Reuter, Timothy (1985) "Plunder and Tribute in the Carolingian Empire." *Transactions of the Royal Historical Society (Fifth Series)* 35, 75-94.
____ (1990) "The end of Carolingian military expansion." P. Godman and R. Collins (eds) *Charlemagne's heir. New perspectives on the reign of Louis the Pious (814–840)*. Oxford, 391-405

Rich, John and Graham Shipley, eds. (1993) *War and Society in the Roman World*. London and New York.
Rihll, Tracey (2007) *The Catapult: A History*. Westholme.
Rio, Alice (2008). *The Formularies of Angers and Marculf*. Liverpool: Liverpool University Press.
Ripoll Lopez, Gisela (1999) "The transformation and process of acculturation in late antique Hispania: select aspects from urban and rural archaeological documentation." Alberto Ferreiro (ed.) *The Visigoths. Studies in Culture and Society*. Leiden, 263-302.
____ (2001) "On the supposed frontier between the *regnum visigothorum* and Byzantine Hispania." *TRW* 12, 95-115.
____ (2003) "Changes in the topography of power: from *civitates* to *urbes regiae* in Hispania." *TRW* 12, 123-48.
Robinson, Chase F. (2000) *Empires and Elites after the Muslim Conquest: The Transformation of Northern Mesopotamia*. Cambridge.
____ (2004) "The Conquest of Khūzistān: A Historiographical Reassessment." *BSOAS* 67, 14-39. Reprinted as article 18 in *FCIW* 5.
____ (2005a) *'Abd al-Malik*. Oxford.
____ (2005b) "Neck-Sealing in Early Islam." *Journal of the Economic and Social History of the Orient* 48 (3), 401-44.
Rogers, Everett (2003) *Diffusion of Innovations. Fifth Edition*. New York.
Rogers, Randall (1992) *Latin siege warfare in the twelfth century*. Oxford.
Rohrbacher, David (2002) *The Historians of Late Antiquity*. London and New York.
Roth, Jonathan P. (1999) *The Logistics of the Roman Army at War* (264 B.C.-A.D. 235). Leiden.
Rouche, Michel (1979) *L'Aquitaine des Wisigoths aux Arabes: Naissance d'une region*. Paris.
Rubin, Zeev (1986) "Diplomacy and War in the Relations between Byzantium and the Sassanids in the Fifth Century AD." *DRBE*, 677-95.
____ (1995) "The Reforms of Khusro Anūshirwān." *BEINE* 3, 227-97.
Sarantis, Alexander (2011) "The Justinianic Herules: From Allied Barbarians to Roman Provincials." Florin Curta (ed.) *Neglected Barbarians*. Turnhout, 361-402.
Sarris, Peter (2004) "The Origins of the Manorial Economy: New Insights from Late Antiquity," *EHR*, 119 (481), 279-311.
____ (2006) *Economy and Society in the Age of Justinian*. Cambridge.
____ (2011) "Restless Peasants and Scornful Lords: Lay Hostility to Holy Men and the Church in Late Antiquity and the Early Middle Ages." P. Sarris, M. Dal Santo and P. Booth (eds.), *An Age of Saints? Power, Conflict and Dissent in Early Medieval Christianity*. Leiden. 1-10.
Sawyer, Peter and R.H. Hilton (1963) "Technical Determinism: The Stirrup and the Plough." *P&P* 24, 90–100.
Scharf, Ralf (1993) "Iovinus—Kaiser in Gallien." *Francia* 20 (1), 1-13.
____ (1999) "Ripari und Olibriones? Zwei Teilnehmer an der Schlacht auf den Katalaunischen Feldern." *Mitteilungen des Instituts für Österreichische Geschichtsforschung* 107, 1-11.
Schmauder, Michael (2003) "The Relationship between Frankish *Gens* and *Regnum*: A Proposal based on the Archaeological Evidence." *TRW* 13, 271-306.
Schmitt, Oliver (1994) "Die *Buccellarii*. Eine Studie zum Militärischen Gefolgschaftswesen in der Spätantike." *Tyche* 9, 147-24.
____ (2001) "Untersuchungen zur Organisation und zur militärischen Stärke oströmischer Herrschaft im vorderen Orient zwischen 628 und 633." *BZ* 94 (1), 197-229.
Shahid, Irfan (1994) "Heraclius and the Unfinished Themes of Oriens: Some Final Observations." *Byzantion* 64 (2), 352-76.
____ (2002) "The Thematization of Oriens: Final Observations." *Byzantion* 72 (1), 192-239.
Sijpesteijn, Petra, and Lennart Sundelin, eds. (2004) *Papyrology and the History of Early Islamic Egypt*. Leiden.

____ (2007a) "The Arab conquest of Egypt and the beginning of Muslim Rule." Roger Bagnall (ed.) *Egypt in the Byzantine World, 300-700*. Cambridge, 437-59.

____ ed. (2007b) *From al-Andalus to Khurasan: Documents from the Medieval Muslim World*. Leiden and Boston.

Sivan, Hagith (1987) "On Foederati, Hospitalitas, and the Settlement of the Goths in A.D. 418." *The American Journal of Philology* 108 (4), 759-772.

Smail, R.C. (1995) *Crusading Warfare, 1097-1193*. 2nd ed. with a bibliographical introduction by Christopher Marshall. Cambridge.

Sophoulis, Panos (2012) *Byzantium and Bulgaria, 775-831*. Leiden.

Southern, Pat and Karen Dixon (1996) *The Late Roman Army*. London.

Speck, Paul (1995) "Das letzte Jahr des Artabasdos." *JöB* 45, 37-52.

Springer, Matthias (1998) "*Riparii*—ribarier—Rheinfranken." Dieter Geuenich (ed.) *Die Franken und die Alemannen bis zur "Schlacht bei Zülpich" (496/97)*. Berlin.

Squatriti, Paolo (2002) "Digging Ditches in Early Medieval Europe." *Past and Present* 174 (2), 11-65.

____ (2004) "Offa's Dyke between Nature and Culture." *Environmental History* 9 (1), 37-56.

Stroheker, Karl Friedrich (1948) *Der senatorische Adel im spätantiken Gallien*. Tübingen.

Sullivan, Denis F. (2000) *Siegecraft: Two Tenth-Century Instructional Manuals by "Heron of Byzantium."* Washington D.C.

Tausend, Klaus (1985-86) "Hunnische Poliorketik." *Grazer Beiträge* 12-13, 265-81.

Teall, John L. (1965) "The Barbarians in Justinian's Armies." *Speculum* 40 (2), 294-322.

Thompson, E.A. (1945) "Priscus of Panium, Fragment I b." *The Classical Quarterly* 39 (3/4), 92ff.

____ (1948) *The Huns*. Revised with an afterword by Peter Heather (1996). Oxford.

____ (1952) "Peasant Revolts in Late Roman Gaul and Spain." *P&P* 2, 11-23.

____ (1958) "Early Germanic Warfare." *P&P* 14, 2-29.

____ (1969) *The Goths in Spain*. Oxford.

Thompson, William R. (2006), "A Test of a Theory of Co-evolution in War: Lengthening the Western Eurasian Military Trajectory." *IHR* 28 (3), 473-503.

Todd, Malcolm (1998) *The Early Germans*. Oxford.

Treadgold, Warren (1980) "Notes on the Numbers and Organization of the Ninth-Century Byzantine Army." *GRBS* 21, 269-88.

____ (1988) *The Byzantine Revival 780-842*. Stanford.

____ (1992) "The Missing Year in the revolt of Artavasdus." *JöB* 42, 87-93.

____ (1995) *Byzantium and Its Army 284-1081*. Stanford.

____ (1997) *A History of the Byzantine State and Society*. Stanford.

Tritton, A.S. (1959) "Siege of Constantinople, A.D. 714-16." *Bulletin of the School of Oriental and African Studies* 22, 350ff.

Trombley, Frank (1997) "War and society in rural Syria c. 502-613 A.D.: observations on the epigraphy." *BMGS* 21, 154-209

____ (2004) "Sawirus ibn al-Muqaffa' and the Christians of Umayyad Egypt: War and Society in Documentary Context." Seijpestein and Sundelin, 199-226.

____ (2007) "The Documentary Background to the History of the Patriarchs of Ps.-Sawirus ibn al-Muqaffa', ca. 750-969." Sijpesteijn 2007b: 131-58.

Vallejo Girvés, Margarita (1996) "The Treaties between Justinian and Athanigild and the Legality of Byzantine Possessions on the Iberian Peninsula." *Byzantion* 66 (1), 208-18.

____ (1999) "Byzantine Spain and the African Exarchate: An Administrative Perspective." *JöB* 49, 13-23.

van Dam, Raymond (2005) "Merovingian Gaul and the Frankish Conquests." *NCMH* 1, 193-231.

Varvounis, M.G. (1998) "Une pratique de magie byzantine et la prise de Pergame par les Arabes." *Byzantion* 68 (1), 148-56.

Verbruggen, Jean François (1954), *De krijgskunst in West-Europa in de Middeleeuwen, IXe totbegin XIVe eeuw*. Translated by Sumner Willard and R.W. Southern (2nd ed. 1997) as *The Art of Warfare in Western Europe During the Middle Ages: From the Eighth Century to 1340*. Woodbridge.
____ (1965) "L'armée et la stratégie de Charlemagne." *KdGr*, 420-436.
Verhulst, Adriaan (2000) "Roman cities, *emporia* and new towns (sixth-ninth centuries)." *TRW* 11, 105-20.
____ (2002) *The Carolingian Economy*. Cambridge.
Vryonis (Jr.), Speros (1981) "The Evolution of Slavic Society and the Slavic Invasions in Greece. The First Major Slavic Attack on Thessaloniki, A.D. 597." *Hesperia* 50 (4), 378-90.
Walmsley, Alan (2000) "Production, exchange and regional trade in the Islamic East Mediterranean: old structures, new systems?" *TRW* 11, 265-343.
____ (2007) *Early Islamic Syria: An Archaeological Assessment*. London.
Ward-Perkins, Bryan (2005) *The Fall of Rome and the End of Civilization*. Oxford.
____ (2006) review of Halsall 2005. *War in History* 13, 523f.
Weidemann, Margarete (1982) *Kulturgeschichte der Merowingerzeit nach den Werken Gregors von Tours*, 2 vols. Mainz.
____ (1986) *Das Testament des Bischofs Berthramn von le Mans Vom 27. März 616. Untersuchungen zu Besitz und Geschichte einer fränkischen Familie im 6. und 7. Jahrhundert*. Mainz.
Werner, Karl Ferdinand (1965) "Bedeutende Adelsfamilien im Reich Karls des Grossen. Ein personengeschichtlicher Beitrag zum Verhältnis von Königtum und Adel im frühen Mittelalter." *KdGr* 83-142.
____ (1968) "Heeresorganisation und Kriegführung im deutschen Königreich des 10. und 11. Jahrhunderts." *Settimane* 15, 791-843.
West, G.V.B. (1999) "Charlemagne's involvement in central and southern Italy: power and the limits of authority." *EME* 8 (3), 341-67.
Whitby, Michael (1988) *The Emperor Maurice and his Historian: Theophylact Simocatta on Persian and Balkan Warfare*. Oxford.
____ (1994) "The Persian King at War." *RBAE*, 227-63.
____ (1995) "Recruitment in Roman Armies from Justinian to Heraclius (*ca.* 565/615)." *BEINE* 3, 61-124.
____ (2007) "The Late Roman Army and the Defence of the Balkans." *TLAD*, 135-61.
Whittaker, C.R. (1993) "Landlords and warlords in the later Roman Empire." Rich and Shipley, 277-302.
____ (1994) *Frontiers of the Roman Empire: A Social and Economic Study*. Baltimore and London.
Whittow, Mark (1996) *The Making of Byzantium, 600-1025*. Berkeley and Los Angeles.
____ (2010) "The late Roman/early Byzantine Near East." *NCHI* 1, 72-97.
Wickham, Chris (2009) *The Inheritance of Rome: A History of Europe from 400 to 1000*. London.
Wightman, Edith Mary (1985) *Gallia Belgica*. Berkeley and Los Angeles.
Witakowski, Witold (1996) *Pseudo-Dionysius of Tell-Mahré: Chronicle, Part III*. Liverpool.
Wolf, Kenneth Baxter, trans. (1999) *Conquerors and chroniclers of early medieval Spain*. Liverpool.
Wolfram, Herwig (1988) *History of the Goths*. Trans. Thomas J. Dunlap. Berkeley, Los Angeles and London.
Wood, Ian (1990) "Ethnicity and the ethnogenesis of the Burgundians." *Ethnogenese* 53-69.
____ (1994) *The Merovingian Kingdoms 450-751*. London and New York.
____ (1995) "Defining the Franks: Frankish origins in early medieval historiography." S. Forde, L. Johnson and A. Murray (eds.) *Concepts of National Identity in the Middle Ages*. Leeds, 47-57.

_____ (1998a) "Barbarian Invasions and First Settlements." *CAH* 13, 516-37
_____ (1998b) "The frontiers of western Europe: Developments east of the Rhine in the sixth century." *TRW* 3, 231-53.
_____ (1999) "Social relations in the Visigothic kingdom from the fifth to the seventh century: The example of Mérida." *Visigoths* 191-208.
_____ (2000) "Before or after mission: Social relations across the middle and lower Rhine in the seventh and eighth centuries." *TRW* 11, 149-166.
_____ (2001) "The creation of the Carolingian frontier-system c. 800." *TRW* 12, 233-45.
_____ (2003) "*Gentes*, Kings and Kingdoms—The Emergence of States: The Kingdom of the Gibichungs." *TRW* 13, 243-69.
Zampaki, Theodora (2012) "The Mediterranean Muslim Navy and the Expeditions Dispatched against Constantinople." *Mediterranean Journal of Social Sciences* 3 (10), 11-20.
Zanini, Enrico (1998) *Le Italie bizantine: Territorio, insediamenti ed economia nella provincial bizantina d'Italia (VI-VIII secolo)*. Bari.
_____ (2007) "Technology and Ideas: Architects and Master-Builders in the Early Byzantine World." *LAA* 4, 381-406.
Zuckerman, Constantin (2005) "Learning from the Enemy and More: Studies in "Dark Centuries" Byzantium." *Millennium* 2, 79-135.
_____ (2007) "The Khazars and Byzantium—The First Encounter." Peter B. Golden, Haggai Ben-Shammai and András Róna-Tas (eds.). *The World of the Khazars: New Perspectives*. Leiden, 399-432.

INDEX OBSIDIONUM

All cross references to sieges are by name (if in same year), year in parenthesis (used alone for same location in different year), or both. Page numbers in bold are discussions in main text that add to entry in Corpus Obsidionum. Corpus Obsidionum entries can be found by year, and has only been indexed for cross-references that are in a different year/campaign/context and hence difficult to find.

Abasgian fortress (549)
  defending families 320
  rooftop fighting 306
Acerenza (663)
  besieged by Constans II 122, 190
  defensible site 306
Adhri'āt (c. 636)
  adventus for 'Umar at 324, 334
Adrianople (377) not in Corpus Obsidionum
  given up 40
Adrianople (378) not in Corpus Obsidionum
  and Gothic siege competence 39ff, 362, 366
Adrianople (587)
  siege raised 181, 294, 602
  surprised by Avars 270n40
Adrianople; Constantinople (813)
  Bulgar arsenal at, preparations for Constantinople 385
Aegean Islands (653)
  Arab fleets in 392, 395, 398, 442, 443, 445
  bishop betrays Cos 329n58
  rape attested 355n102
Agde (673) see Wamba in general index
Agde (736/7)
  besieged by Charles Martel 247
?Agen (582) see Périgueux
Aix (574)
  Lombards bought off 180n122, 258n11
Akbas (583)
  fate of surrendered Persians 354n100
Akbas (590)
  artillery terminology 411n13
Aleppo (c. 637) see Arab conquests in general index
Albi (507)
  taken by Clovis 91

Albi (767) see Toulouse
  taken by Pippin II 250
Alexandria (608)
  and Roman civil war 330, 331n62
  artillery at 275
  defenders 340
  role of fleet 297, 347n96
Alexandria (609) see also (608)
  artillery at 275, 292
Alexandria (619)
  Persian massacre 354
Alexandria (642f)
  and Roman civil war 397
  artillery at 275, 394
  refuses surrender 329
  role of fleet 294
Alexandria (twice) (646f)
  artillery at 419
  temporarily retaken by Romans 110, 294, 350
Ally (767) see Rocks and caves
  context of 250
Alpine forts (539)
  Ostrogothic garrisons 155, 320
Amasia and other forts (712)
  and Arab census for organizing labor 400n90
  population taken captive 352
Amida (359)
  and Roman siege competence 37, 38f, 64
  compared to Amida (502f)
  rituals at 348n97
Amida (502f) **126-131**
  artillery, possibly trebuchet 275, 419
  civilian defenders 340
  craftsmen 135, 138n121
  house to house fighting 306
  Khusro present 270, 318

790   INDEX OBSIDIONUM

mining 286
morale at 317, 318, 319, 322
mound and countermeasures 281, 284, 290
narrative evidence 297f
pedatura of monks 339n86
Persian retaliation 135, 350, 351, 353
ram and countermeasures 284, 289
stormed with ladders 268n21
Amida (503) **133-135**
and Ostrogothic siege warfare 164, 509
and Roman strategy, 257n2
craftsmen at 141, 143
garrison size 337
Roman siege towers 282
siege abandoned 263
stratagems 292
terror tactics 319
Amida (504f) see also (503)
ballista used at 273
blockade of 261
local supplies 310, 329
mining 286, 287
starvation, cannibalism at 263, 341
Persian atrocities 353
tactics at 269n22
unplanned assaults 266
women kept as concubines 341, 351
Amida (578)
raiding before 314
Amorion (c. 645) see Arabs in general index
Amorion (664) **446f, 449-52**
historiography 439
settlement 452
Amorion (666)
Arab garrison at 272, 337
retaken by surprise 265, 272, 447
Amorion (779) see Dorylaion; also Abbasids in general index
?Amorion (796) see Abbasids in general index
Amorion (838) not in Corpus Obsidionum **428**
evidence of destruction 356n108
Anatolian campaign of ibn Khālid (664)
see: Skoutarion(?), Amorion, *SYLWS* (Sagalassos?), Pessinos, Kios, Pergamon, Smyrna
Anchialos (583f) see Anchialos (588), Tomi (597f)
Anchialos (588)
Avars and Roman rituals 323, 595
Avars use Slav fleet 381

Anchialos (708)
and Roman client management 384
Ancona (538)
civilian defenders 340n88
ladders repulsed 270n42
Roman sally 291
size of Ostrogothic force 343
Ancona (551)
Ostrogothic fleet 156, 157
question of Ostrogothic elite 154n16
Roman fleet 294, 296
Ancyra (622)
massacre, captivity 354
terminology 355n106
Angers (463) see also Trier (c. 456)
and Franks 45, 331
Angers (722/3)
and Charles Martel 246
Angers (873) not in Corpus Obsidionum **429**
Angoulême (508)
collapse of walls 12, 92, 288
Anglon (543) see also Petra (549)
nature of fort 304
Persian ambush 264n28, 291
size of Roman force 342
Antinoe, Roman fortress near (642)
Arabs take Roman arsenal 277, 394
local factions at 397n80
Antioch (540)
archery and ladders 268
artillery at 420
captives from 327, 351
civilians fight in streets 306, 310
illoyal bishop 328
interior scaffolds 286, 290
preparations 305
retinue of Germanus 60
Roman garrison from Lebanon 336
Antioch (610/11) see Roman-Persian wars in general index
Antioch (638/9)
captives taken during festival 310, 355
responses to 326
?Antioch (695) see Arabs in general index
Apamea (540)
submits to Persians 315, 324
Apamea (573)
Persian siegeworks 259, 279
population deported 315, 351

INDEX OBSIDIONUM 791

Apamea (610/11) see Roman-Persian wars in general index
Aphum and other forts (578)
    Perso-Roman border warfare 300n2
Appiaria (586)
    capture of *manganarios* Bousas 119, 381
    dating 407
    terminology 411n12, 415
Aquileia (361) not in Corpus Obsidionum 37, 64
    siege methods 40n25, 47n57, 210n63
Aquileia (452) see also Angoulême (508)
    Hun siege methods 48, 274, 284
    narrative evidence 358
    Roman garrison 51, 336
Archaeopolis (550) 327n56
    engineers at 364
    Persian army 343
    rams 284, 291, 313, 371
    Roman army 555, 559
    strategic context 257fn8, 295
Arles (411)
    Frankish participation 462
    in Roman civil war 43, 64, 85
Arles (425)
    Visigoths at 44
?Arles (430) see (425)
Arles (452/3) see (425)
Arles (458/9) see (425)
Arles (473/6) see (425)
    Visigoths conquer 45
Arles (508)
    bishop Caesarius and 322, 323
    Jewish defenders 333, 339n86
    nature of 92, 296
    Franks and Burgundians halted 152f
    repairs after 159f
    sources for 13, 195n8, 358
    supply of 158
    workforce, prisoners at 73f
Arles (c. 566)
    betrayal at 332n66
    in Frankish civil war 199
Arles (567/9) see also Arles (c. 566)
    defenders 341
    siegeworks 208n56
    sortie 207n47, 292, 312
Arles (574)
    Lombard invasion 180
    Lombard camp 310f
Armenian forts (643) see Artsap'k' etc.

Artsap'k'; Nakhchewan; Khram (643)
    Arab siege capacity 391
    massacre at Khram 355n104
Arwad (649) 395n76
    and Arab expansion 277, 391, 442
    bishop as Arab negotiator 329
Arwad (650) see also Arwad (649)
    Roman resistance 311n28
    terms of surrender 355
Arzanene, Persian fort in (587)
    artillery at 122, 181n125, 277
    artillery terminology 411n12
    Persian countermeasures 289
«Asia», fort in (741/2)
    epidemic in Umayyad camp 263
Ascoli (544f)
    Ostrogothic strategy 293
Ashparin (503)
    hands over soldiers 341
    population spared 353
Augustae (583)
    taken by storm 265, 381
Autun (534) 485
    nature of 196
Autun (c. 679)
    as refuge 308, 309
    bishop Leudegar's construction projects at 231
    context of 233
    power of bishops 328n57
    procession at 322
Auvergne campaign; Clermont (524/32)
    see: Vollore, Chastel-Marlhac, possibly Vitry
    dating 501
    nature of (sieges in) 197, 288, 353
Avar fortifications (791)
Avignon (500)
    siege abandoned by Clovis 196
Avignon (567/9)
    in Frankish civil war 199
Avignon (583) see also Phasis 556
    Mummolus' retinue at 74n155, 223
    Rhône diverted 208, 221n115, 304
Avignon (c. 736)
    Frankish artillery 247, 261n22, 272n46, 277fn65, 419
    Frankish siegeworks 281
    Frankish rams 285n72
    Frankish battle cries 321
    storm, massacre 350

Babylon (641f)
    Roman factions in 397n80
    terms of surrender 355n105
Bagnarea (605) see Lombards in general index
Bahnasā (Oxyrhynchus) (640f)
    context of 397n80
    massacre at 355
Barbād (776)
    artillery at 277fn65, 416fn32
    battle cry 321
?Barcelona (673)
    Wamba's strategy 175
Barcelona (801)
    context of 252n211
Batnan (503)
    abandoned 305
    repairs at 135
Bazas (413)
    in Roman civil war 43
    role of clients 44, 331
Beaucaire/Ugernum (585)
    reinforcements after fall of 338
    taken by Visigoths 173, 199
Beaucaire (587)
    population taken captive 350
    Visigoths storm 199
Beïudaes (587)
    artillery at 277
    artillery terminology 411n13
    Roman testudo against 269
Beneventan campaign (c. 702)
    rams at 189
Beneventan castrum (792)
    artillery at 253fn212
Benevento (542)
    walls razed 305n20
Benevento (663)
    Lombard sortie 280, 292
    petraria at 190, 417
    relief of 122, 293, 330
    various engines 284
Bergamo (c. 591)
    Lombard revolt 330
Bergamo (c. 701)
    engines used at 189f, 284, 423
Beritzia (c. 772)
    Bulgars plan deportation 384
    Roman triumph after 348
Beroea (540)
    assaulted with ladders 268n22
    cavalry garrison 337n79
    Roman soldiers defect 352
    walls and acropolis 305
    water supply consumed 262
Beroe (587)
    and Appiaria (586) 592
    as Avar failure 381
    ransomed 349
Béziers (673) see Wamba in general index
Béziers (736/7)
    and Charles Martel 247
Blaye (735)
    and Charles Martel 247
    suburban *castra* 303
Bockholt (779)
    and Charlemagne 251
Bologna, forts around (544)
    Illyrian troops abandon 320
Bonkeis and 40 forts (595) see Avars in general index
Bourbon (761)
    garrison captured 248
    siegeworks at 281
Bordeaux (498)
    and Clovis' strategy 91
?Bordeaux (507f) see also (498); also Clovis in general index
Bordeaux (735)
    and Charles Martel 247
    suburban *castra* 303
Bourges (583)
    siegeworks 208n56, 279
Bourges (762)
    ravaging, siegeworks 248
    cause of casualties at 272
Bosra and other cities in Jordan (c. 635)
    in Arab strategy 433
Brescello (c. 584)
    Droctulf at 181
Breton fortifications (786) 251n206
Brindisi (c. 685)
Bulgarian fortifications (680/81) see Bulgars in general index
?Büraburg (773) see Eresburg
    Saxon wooden fortifications 251n205
Burgundian Civil War (500) 90f, 196, 296
    ?Dijon, Avignon, Vienne

Cabaret (585)
    Visigothic storm of 173, 199
    Frankish response to 200, 338
Cabrières (c. 532) 198
Caesarea (576)

INDEX OBSIDIONUM 793

Caesarea (611)
  adventus at 333
  massacre and captivity 354
Caesarea (c. 645) 450
Caesarea Maritima (c. 640)
  Arab tactics and strategy 432, 438
  artillery terminology 416
  from blockade to storm 261, 266, 311n28
  narrative reproduced at Neocaesarea (729) 711
  origins of artillery 394
  Roman garrison at 337
  scale of artillery 275, 391
  walls breached 285
Cahors (573) 207n48
Caranalis (551)
  blockade broken 292
  Ostrogothic fleet at 157n27, 549
Carcassonne (585)
  betrayal and ambush at 199
?Carcassonne (587)
  Frankish defeat near 199, 291
Carcassonne (589) 332n66
  Frankish losses at 200
  Visigothic tactics 294
Carthage (533)
  and Arianism 335
  local *tekhnitai* 116, 141
  terror tactics at 319
  Vandal strategy 258, 262, 328
  walls repaired 305
Carthage (536) 342
Carthage (544) 331
Carthage (twice) (697f)
  Roman expedition 122
Castrum Cainonense (c. 463)
  besieged by Aegidius 45
  extramural well broken 262
  public prayers at 322
  refugees from countryside at 308
  torrential downpour at 263n25
Centumcellae (549f)
  context of 357
  morale of garrison 318, 320
Centumcellae (552f)
  surrender after battles of Taginae and Mons Lactarius 317, 551
Cerdane (673) see Wamba in general index
Cesena (538/9) see Ostrogoths in general index
Cesena (542) see Ostrogoths in general index

Chalcedon (615) 295
Chalkis (540) (=Qenneshrin)
  refuses to surrender garrison 341
  threatened to ransom 259
Chalon-sur-Saône (c. 555) 207
Chantelle; forts in Auvergne (761) 248, 258n13
Cherson (710)
  and range of Byzantine armies 106n41
Cherson (711) see also (710)
  fleet-born siege train 122
  in Roman civil war 331, 384
  ram used at 284
Chersonese (539)
  and origins of Hunnic siege competence 370
Chersonese (559) 315n27
  artillery terminology 414n22
  Hunnic tactics at 269n37
  nature of *helepolis* 371, 411, 421
  repulsed by sortie and fleet 292
Chastel-Marlhac (524/32)
  failed sortie by garrison 292, 338
  in Auvergne campaign 197
  water supply at 262
Chieti and suburban fortification (801)
  in Carolingian expansion 252n212
Chiusi (538)
  surrender in face of siegeworks 278, 320
  Ostrogoths recruited into Roman army 352
Chlomaron (578)
  archery cover fire 228n74
  artillery at 276
  as border fort 300n2
Christianity of population 324, 335n73
Cividale (c. 610)
  Avars inspect site 257n4
  betrayal of 328
  massacre and captivity 351
Cividale (c. 670)
  nature of 189
  numbers at 295
Cividale, Treviso and other cities (776)
  Carolingian expansion 252n212
Classis (c. 579) see Lombards in general index
Classis (584/5)
  recaptured by Droctulf 181
  role of Roman fleet 296
Classis (twice) (c. 723)
  nature of 189

Classis (c. 731)
    nature of 189
Clausurae (673) 175
Clermont (474)
    nature of siege, forces 45, 66f, 294f
    Visigothic losses 349
Clermont (507) 91
Clermont (524) see Auvergne campaign
    familia as garrison 216, 217n94, 353
    nature of 197, 207, 278, 358
    rituals at 207n51
    siege camp at 288, 309
Clermont (c. 555)
    epidemic at 263
    stormed 207n46
Clermont (761)
    population killed in fire 248
Collioure (673) 175
    troops rewarded 329
Cologne (c. 456)
    historicity of 87
?Cologne (612)
    nature of 225
Comacina (c. 587)
    duration of 263
Comacina (c. 591)
    Lombard civil war 330
Constantina/Tella (502f) **130f**
    Jewish defenders 333, 348
    mine betrayed 286
    pedatura of (Jewish) defenders 140, 339n86
    rituals and role of bishop 322, 324
    territory raided 257n2, 309, 312
Constantia (649)
    organization of Arab fleets 395n76, 442
    rape attested 355n102
Constantinople (602)
    circus factions mobilized 339
    Maurice lacks popular support 330
Constantinople (626)
    aqueducts cut 261f
    artillery terminology 121, 275, 414, 417
    Avar display of forces 323
    Avar *khelōnai* 285
    Avar siege towers 282
    Avar siegeworks 280
    Avars set up trebuchets 276fn62
    Avars' western frontier during 387
    ballistrai at 120n71, 273, 412
    complexity of 296
    composition of Avar army 258n9, 343, 382
    consequences of failure 350, 376, 379
    icon in rituals 322
    preparations 309n27
    role of moat 287
    Roman fleet 347n96
    Roman garrison 337
    sailors build rig at 144, 283, 289
    Slav losses 330, 349
    thanksgiving after 348
Constantinople (654)
    Arab armies in Anatolia 323
    Arab naval expedition 266, 297, 391, 450
    artillery terminology 415
    complexity of 296, 395n76
    consequences of failure 350, 397, 441, 445
    fleet-born artillery 277
    fleet-born siege towers 283
    strategic context 441-44
Constantinople (663) **439-53**, 133n106
    Arab battle cry at 321
    ballistrai at 120n71, 121, 273, 275, 412
    objectives for 397n82
?Constantinople (670ff) **439-53**
    Arab naval blockade 264, 266, 297
    Mu'āwiya strengthens infrastructure 398
    Kallinikos and Greek fire 398f
Constantinople (698)
    in Roman civil war 330
Constantinople (705)
    in Roman civil war 330
    reprisals at 353
    taken via aqueduct 288
Constantinople (preparations) (715) see also (717f)
    artillery arsenal, terminology 120n71, 273, 412, 414n22, 418
    repairs at 144n143, 305
    3-year supply required 314
Constantinople (715)
    duration of 257n5
    Roman civil war 330, 331
Constantinople (717f) see also (715 preparations)
    and Bulgars 113, 384
    Arab fleet 297, 311
    Arab forces 343
    Arab siegeworks 281, 293, 401
    artillery, terminology 277n65, 401, 416

INDEX OBSIDIONUM 795

complexity of 257n6
Coptic crews defect 396
infrastructure behind 395, 401
Roman collusion 447
Syrian craftsmen 401
Constantinople (c. 718) see East Roman civil wars in general index
Constantinople (c. 742f)
  role of icons at 322
  in Roman civil war 330, 331
Constantinople (preparations) (813) see Adrianople
  and Krum's ambitions 385f
Convenae (585) =St.-Bertrand-de-Comminges 208ff
  artillery, probable use of 274f &n56, 424
  bishop Sagittarius at 270n43
  complexity of 296
  defensive responses 289
  ditch filled in 287
  Fredegar on 224, 297
  local loyalties 314, 327n56
  logistics 220, 310, 311, 313
  massacre 332n66
  Mummolus' *faber* 74n155, 223, 343, 582
  tactics: rams and infantry 269n27, 284
Conza (554f)
  composition of garrson 343
  failed sorties 292
  garrison recruited into Roman army 352
  Ostrogoths hold out 203n22, 263
Córdoba (549/50)
  and Visigothic expansion 166n59
Córdoba and other cities (572) see also Córdoba (549/50)
  magnates and *rustici* 171n85, 341
  night attack 265n29
Cremona (603)
  rams used at 189, 284
  Slav participation 382
Crotone (551f)
  effect on Ostrogoths 548
  raised by Roman fleet 294
  starvation at 263
Cumae and other strongholds (542) see Ostrogoths in general index
Cumae (552f)
  artillery 274f &n56, 421
  mining, debris caused by 288
  missile exchanges at 268
  Ostrogothic surrender 352
  rams 284

Romans prefer blockade 261
siegeworks at 279
Cumae (717)
  size of garrison 338
  terminology at 189

Damascus (613) see Persian-Roman wars in general index
Damascus (634/5)
  Arab conquests, course of 433
  house-to-house fighting 306
  nature of Arab blockade 261, 281, 432n5
  sources for 647
  surrender conditions 354
  surrender terminology 355n106
Damascus (c. 690)
  artillery at 277n65
  fortified cities in Arab civil war 336
  reprisals after 319
Dara (540)
  archery effect 271
  mining, counter-mining 141, 271, 286, 287
  ransom for citizens of Antioch attempted 327
  walls at 304n19
Dara (573)
  artillery terminology 411n12, 416
  (civilian) craftsmen at 143, 344
  duration 263
  Persian storm 266, 282
  Persians use Roman engines 122, 277, 344
  progression of Persian siege methods 320
  Roman commander killed at 271
  siegeworks at 271, 279, 281f
  ransom refused by 317f
Dara (603f) see Roman-Persian wars in general index
Dara and Mesopotamian cities (640) **434-38**
  Arab siege tactics 391, 432, 448
  Roman resistance 398
Debeltos and other forts (812)
  Bulgars capture Roman infrastructure 385
Demqaruni (609)
Dijon (500)
Dijon (c. 555)
  adventus at 322
  rituals at 207n51

Diocletianopolis (587)
    artillery terminology 411n12
    Avars repulsed by artillery 275
    Avar tactics 381
Dorylaion; Amorion (779) see Abbasids in general index
Dragawit, city of (789)
    and relations with Franks 387
    and Slav acculturation to siege warfare 251
    Charlemagne builds bridges at 307
    hostages given 349
Drizipera (588)
    Avar fortified camp and trebuchets 276fn62, 381
    miraculous relief 294
    religious motivation 321
    Slav laborers at 381
Drizipera (598) see Avars in general index
Dvin (640) **434-38**
    captivity 355n103
    smoke used at 297
    tactics 391, 432, 448
Dvin (654/5)
    recaptured, abandoned by Romans 444, 450
    similar tactics at *Thessalonica (586/97) 593

Edessa (484)
    preparations for siege 124f, 258n9
Edessa (502f) **131ff**
    civilian defenders 135
    negotiations at 258n11
    rituals at 348n97
    territory raided 257n2, 309
Edessa (540)
    buys off Persians 314, 351
Edessa (544)
    civilian defenders 340
    complexity of 296
    mound and countermeasures 281, 286, 290, 291, 364
    Persian *tekhnitai* 344, 364
    preparations 304
    siegeworks 279, 280
    tactics (rams, artillery, missiles, ladders) at 268nn21f, 274fnn52 &56, 284
    walls of 304n19
Edessa (603f)
    accepts Maurice's son 331

Edessa and Mesopotamian cities (609)
    cause of surrender 323
Edessa (630)
    artillery subdues city 122, 277, 416
    artillery terminology 419
    Jewish community, role of 334
    surrender terminology 355n106
5 Egyptian cities (608)
    in Roman civil war 331n62
Embrun, improvised fortifications at (c. 571)
    Lombards defeated by Mummolus 180n122
    wooden ramparts 306
Emesa (610/11)
    population deported 351
    terminology of surrender 355n106
Emesa (c. 635)
    Arab blockade 261
    context of 430-33
    Jewish community, role of 334
    surrender at storm 293
    surrender terms 329
Emesa and other cities (745)
    Arab civil war; walls razed 336
    reprisals after 353
    soldiers and craftsmen from Byzantine frontier 402
Eresburg (772)
    in Charlemagne's Saxon wars 251n205
Eresburg; Büraburg (773)
Eresburg (776) see also (772)
    artillery terminology 417
    cause of surrender 265
    rebuilt 303
Euchaïta (644)
    context of 450
    population surprised in fields 309, 392, 448
    rape attested 355

Fermo (544f)
    Ostrogothic strategy 293
Fiesole (539)
    captured garrison paraded at Osimo 320
    Frankish invasion 212fn75
    Franks sacrifice humans 325n54
    Ostrogothic sortie 291
    retinues at 58n97
    shielding army 201n28, 257n6
Florence (c. 542)
    besieged after battle 257n8, 317
    Romans defect to Ostrogoths 352

INDEX OBSIDIONUM 797

Florence (553)
surrender after Taginae, Mons Lactarius 317, 354
?Forlimpopoli (c. 670)
nature of 189
Fregnano and other castra (c. 731)
nature of 189

Galatian forts (714)
population captive 351
Germanikeia (c. 745)
and Roman-Arab border warfare 123
population deported to Thrace 352, 403
range of Roman armies 106n41
?Germanikeia (769)
population deported to Palestine 352
Germanikeia (778) see also (c. 745)
outcome of 349
?Gerona (673)
role of bishop 328n57
Gevaudan (767) see Toulouse
in Pippin's conquest of Aquitaine 250
Gothic fortified camp (552)
ballistrai on towers 280, 413n15
bridge 280n68
Grenoble (574)
in Lombard expedition 180n122

Hadrametum (twice) (544)
recaptured after Roman mutiny 331
Harran (502f) =Carrhae
avoids siege 349
population surprised in fields 309
sortie captures Hun chief 292
territory raided 130, 131, 257n2
Herakles, Thebasa and other forts (806)
Hierapolis (540)
city and territory ransomed 259n14, 314
Hispania, Roman city in (c. 611) see Visigoths in general index
Hispania, Roman cities in (c. 614f)
Sisebut ransoms Roman captives 352
taken by force 173
Hispania, Roman cities in (c. 624)
taken by force 173
Hohbuoki (810)
and Slav acculturation 251
Hohenseeburg (743)
taken by negotiations 248n199, 258n113, 388
Huesca (797ff)
in Carolingian expansion 252n211

Ihburg (753)
bishop Hildegar of Cologne killed at 234n159, 248n199, 388
Illyrian forts (539) see Huns in general index
Imola; Emilia (538/9)
Imola=Forum Cornelii
surprise capture 265
Italy, Frankish invasion of (590)
dux killed during storm 271
Frankish siege capabilities 205, 300, 190n153
logistics of 220
Italy, Roman invasion of (590) see also Frankish invasion
context of 182n129
Roman successes 205n39
siege tactics 284
Italy, Roman campaign in (592) see (590)
Italian campaign of Constans II (663)
see: Lucera and other cities, Acerenza, Benevento
and Lombards 181, 184, 423
and Thessalonica (662) 668
Bulgars in 107, 112
organization of 122
Sapiorios/Saburrus, identity of 446
strategic purpose 444, 445
sources for 442

Jerusalem (614)
Arab raids around 310, 389
civilian defenders 340
duration of 266
massacre at 334f
mining at 288
original peaceful submission 354
sources for 27
Jerusalem (634f) **430-33**
Arab blockade 261, 311n28
surrender terminology 355n106
villagers massacred 310, 334
Jordan, cities of (c. 635)
context of 430-33

Kallinikos (503) = al-Raqqa
presence of shah 317
Kallinikos (542)
break of truce 312
captives taken 351
Persians build bridge to surprise city 307
taken during repair of walls 305

Kamakhon; Cilician forts (711) see Arabs in
    general index
Kamakhon (766)
    Abbasid logistics 296
    Abbasids use local conscripted, paid
        labor 310, 329, 403, 404
    "Armenian wagons" for transportation
        276
    artillery at 277fn65
    artillery terminology 416
    battle cries 321
    consequences of 349, 350
    dysentery in Abbasid army 263
    fiercely defended cf. (793) 327f, 402f
    numbers at 717
    pagan sacrifice in Abbasid army 325
    rituals 322
    rams, siege sheds 285, 286
    Roman countermeasures 290
    shielding army 257n6
    skill of Roman artillerymen 121
    sources for 14, 359
Kamakhon (793)
    surrendered (cf. 766) 328, 402f, 733
Kasin (underground city) (776)
    smoke used against 297
Kassandria (539)
    taken by storm 370
Khram (643) see Artsap'k' etc.
    massacre at 355n104
Kīlūnās (642)
    local parties to 397n80
    massacre at 355n104
    walls "cast down" 285
Kios (664) 446

Laon (741)
    in Carolingian civil war 248, 331
Lapethus (650)
    artillery terminology 416
    Arab fleet and artillery 277, 448
    infrastructure behind 395n76, 442
    surrender after hard fighting 355
Laribus (544)
Laureate (548(f))
    defected *doryphoros* leads Ostrogoths
        329
Lérida (797)
    in Carolingian expansion 252n211
Ligurian cities, Lombard conquest of (643)
    nature of 189

Lilybaeum (533)
    Ostrogothic garrison 154, 155
Limoges (573)
    territory raided 207n48
Llivia (673)
    troops rewarded after 329
    Wamba's strategy 175
Loches (742)
    garrison taken captive 248, 350
    territory raided 308
Lodi (c. 701)
    nature of 189
    rams, machines at 423
Lombard campaign (c. 731)
    see: Ravenna, Classis, Fregnano and
        other castra, Sutri
    nature of 189
Long Walls (598)
    circus factions mobilized 339
Lucca (553)
    artillery, archery tactics 270n39, 421
    artillery, incendiary weapons 274n52,
        275
    Frankish garrison, relations with locals
        203, 213n76, 329n59, 340n88
    Frankish sortie betrayed 292
    from blockade to storm 257
    psychological warfare 319
    walls breached 284
Lucera and other cities (663)
    taken by Constans II 122, 190
Lucera (twice) (802) 252n212
Luni (553)
    surrender after Taginae, Mons Lactarius
        317
Lyons (c. 662)
    role of bishop Aunemund 233
?Lyons (738)
    submits to Charles Martel 247

Maguelone (673)
    Visigothic fleet 175
?Málaga (570)
    and Visigothic expansion 166
Mantua (603)
    rams used at 189, 284, 382
    Roman garrison allowed to escape 352
    Slavs in Lombard army 382
Manuf (609)
    in Roman civil war 331n62
Mardin (c. 606ff)
    duration of 263

INDEX OBSIDIONUM 799

Mardin (690s?) **399f**
  pedatura of monks 339n86
  trebuchets at 392
  artillery terminology 416
Markellai (792) see also (796)
  astrologer predicts Roman victory 325
  base for Roman attack 384
?Markellai (796) see (792); also Bulgars in general index
?Marseilles (413)
  Athaulf wounded in fighting 43
  celebration at 348
Marseilles (473/6)
  captured by Visigoths 45
Martyropolis (502) **125f**
  adventus for Persians 324, 334
  as precedent at e.g. *Apamea (573) 573
  forgiven by Anastasios for submission 314
Martyropolis (531) see Persian-Roman wars in general index
Martyropolis (twice) (589)
  artillery terminology 411
  betrayed 264
  Romans defeated in battle 294
  siegeworks 279
Mecca (692)
  and siege warfare in Arab civil wars 336
  artillery terminology 416
  reprisals after 319
  trebuchet used at Ka'ba 392, 407
Medina Sidonia (571)
  and Visigothic expansion 166n59
  treason at 265n29, 330
  rustici at 578
Melitene (576)
  burned 318, 350
  Khusro's fire temple left behind before 325
Melitene; Theodosiopolis (750)
  and range of Roman armies 106n41, 123
  population deported to Thrace 352, 403
Mesembria (812)
  artillery terminology 414n22
  defected Roman engineer 123, 385, 404, 405
  population, arsenal captured 385
Messina (549)
  Ostrogoths cross to Sicily 157n28, 291, 546
Metz (451)
  lack of archaeological evidence 48

Milan (452)
  artillery, other machines at 274, 284
  defense against Huns 48
Milan and Ligurian cities (538f)
  Burgundian troops at 198n14, 201n25, 212
  Franks called in by Romans 212
  massacre at 351
  role of civilians and bishop Datius 154, 158n35, 328
  Roman garrison at, betrays population 212, 317, 341
Milan and Ligurian cities (569)
  Lombard expansion 180n120
Milvian bridge (537)
  abandoned by Romans 323
  tower built on 305
Misthia and other townships (712)
  population deported 351f
Monselice (c. 602)
  terminology for 189
?Mopsuestia (772)
  garrison ambushes Arabs 291
Mouikouron (548(f)) see also Laureate
Mt. Papua (533) see Vandals in general index
8 Mysian and Scythian cities (586) see Avars in general index

Naissus (442) **46f**
  and classicizing *topoi* 24, 46, 366
  and Hunnic wars 48
  and dating of Noviodunum (c. 437) 465
  archers on siege towers 269n36, 282
  evidence of artillery 274fn56
  evidence of Roman engineers 368
  Hun engines and tactics 265f, 296
  Huns build bridge 280, 307
  rams used, crushed at 210n63, 283, 289
Naissus (473)
  Ostrogothic motives 331
Naix (612)
  captured by force 225
Nakoleia (782)
  artillery at 277fn65
Naples (536)
  aqueduct cut, used to gain entrance 115, 262, 268, 287
  civilian defenders 154
  Ostrogothic garrison 154, 155, 338
  pedatura of Jewish community 333, 339n86

reprisals halted by Belisarius 353
Roman army at 343
similarity to Vienne (500) 485
storm, duration of 257n5, 268
suburban *phrourion* 300
tekhnitai, construct siege ladders 116f, 270, 344n92
Naples (542f)
role of fleet 257n7
surrender, lenient terms of 354
Totila bans raiding 313
Totila bans rape 319, 354
Nano; ?Trent (c. 575) see Lombards in general index
Narbonne (413)
population caught in fields 309
stormed by Visigoths 43
Narbonne (436f)
duration of 262
supplies 294
Visigoths repulsed 44
Narbonne (c. 461)
in Roman civil war 331
received by Visigoths 45
Narbonne (507)
captured by Burgundians 92
relieved by Ostrogoths 153
?Narbonne (673)
role of bishop 328n57
Narbonne (673)
artillery, evidence for 277, 418
in civil war 175
storm, duration, tactics of 266, 268, 272n47
taunts at 318
Narbonne (736/7)
artillery, evidence for 277fn65
missile exchanges, evidence for 272n46
siegeworks, rams 247, 261n22, 285n72
?Narbonne (763)
garrison ambushed 249
Narni (552) see Narses in general index
Narni (c. 725)
nature of 189
Narni and other cities (756)
nature of 189
Neocaesarea (729)
betrayal by Jewish community as *topos* 334n72
walls partially destroyed 285
Nepi (552) see Narses in general index

Nikaia (727)
artillery at 277n65, 414n22
Nîmes and unnamed cities (585)
abandoned due to strong fortifications 199
strategy at 311
Nîmes (673)
artillery, evidence for 418
nature of storm 175, 272n47
size of army 175f
Nîmes (736/7)
nature of 247, 285n72
role of suburban areas 308n24
?Nimis (c. 670)
nature of 189
Nisibis (c. 498)
and internal Persian unrest 125
Nisibis (503)
size of Roman army 342
Nisibis (573) see also Dara
artillery terminology 416
engineers, laborers, siegeworks at 121f, 142f, 279, 287, 344
house-to-house fighting 306
lack of cohesion leads to rout at 327
massacre at 353
Roman engines taken by Persians 122, 277, 282, 349
Roman logistics at 309n27
Roman strategy at 307
Noviodunum (c.437)
and Roman military infrastructure 49
human shields 269
Roman concern for civilians 320
storming tactics 268n31

Oderzo (643)
nature of 189
Onoguris (555)
Agathias' description 25, 550
akontistēria at 420
artillery, offensive use of 275
complexity of 296
forces involved 342n90
Persian relieving army 295
spaliōnes 285, 287
Orespada, cities in (577)
in Visigothic expansion 166n59
peasants at 171n85
Orléans (451) see also (c. 453)
Hunnic rams at 47f, 283f
relieved 293
role of bishop Anianus 322, 326

INDEX OBSIDIONUM

Orléans (c. 453) see also (451)
   Visigoths in Roman service 44
   siegeworks 279
Orléans (604)
   garrison, relieving army coordinate 293
   retinue as defenders 338
   triumph after relief 348
Ortona (802)
   Carolingian expansion in Italy 252
Orvieto (538f)
   blockade decided 257n4
   Ostrogothic garrison 338
Orvieto (605) see Lombards in general index
Osimo (539)
   artillery, evidence of Ostrogothic 162, 274n55
   besiegers starve at 260n19
   extramural cistern 262, 264, 306
   fight over pasture 291
   Isaurian masons at 115, 344n92
   missile exchange 268n31
   Ostrogoths recruited into Roman army 352
   psychological warfare 320
   retinues at 58n97
   siegeworks 278, 279
   traitor executed 319
   wagon wheels as defensive weapons 289
   Slavs, Antae at 372
Osimo (544)
   failed relief force 295
Otranto (544)
   garrison emaciated by blockade, replaced 349
   Roman fleet relieves 294

Padua (601)
   artillery, evidence for 189f, 274n52
   garrison allowed to escape 352
Palermo (535)
   Ostrogothic garrison 154, 155
   Roman fleet takes 283, 296
   Roman force 343
Palmyra (634/5)
   surrender, terms of 355nn101 &106
Palozonium (c. 740)
   population taken captive 351
Pamplona (472/3)
   Visigothic expansion 45

Pamplona (778)
   Charlemagne razes walls of 305
Papyrius (484-8)
   revolt fails 125
   betrayed 258n10
Paris (c. 490)
   meaning of *obsedit* 246n196
   St. Genovefa organizes supplies 73, 219
?Paris (574)
   in Frankish civil war 207n48, 332n66
Paris (885f) not in Corpus Obsidionum 408, 429
Parma (553)
   blockade 261, 279
   control of labor, resources 313
   Frankish garrison 202, 213n76
Pavia (452)
   defense against Hun arsenal 48, 274
   walls breached 284
Pavia (569ff)
   and Lombard invasion 180, 188
   blockade of 261
Pavia and other castra (755)
   arsenal available at 190
   besieged by Pippin 248n200
   blockade of 261
   siegeworks at 281
Pavia (756) see also (756)
Pavia (773f)
   and Charlemagne 252n212
   blockade of 261, 263
Pergamon (664)
   context of 446f
Pergamon (c. 716)
   alleged human sacrifice 325
   Arab fleet, (Syrian) craftsmen 401
Périgueux, ?Agen and other cities (582)
   submits after battle 208n52
   oaths extracted 332n65
Perugia (545ff)
   stormed after blockade 263, 266
Perugia (552)
   divided loyalties at 329n59
Perugia (593)
   Lombard competence 186
   Lombards in Roman service 181
   nature of 189
Pesaro (544)
   walls razed 305
Pessinos (664)
   context of 446f

Petra (541)
  missile exchange 274fn56
  Roman commander killed 271
  rams, defense against 284
  storm tactics 268n32
  surrender after undermining 288
  shah present 317
Petra (549)
  date set for surrender 293
  emperor proclaimed triumphant during storm 320
  mining, effect of 288
  mining under archery cover 271
  Persian garrison, losses 337n80, 349f
  relieving army 295
  repairs at 349
  timber for fleets or artillery 276
Petra (550) see also (549)
  artillery, possibly trebuchets at 420
  concealed water pipes 262, 307
  incendiary weapons 274n52
  last stand in acropolis 306
  old mine collapses 288
  rams, sheds constructed by Sabir Huns 284, 306, 370
  retinues at 58n96
  Romans beg Persians to surrender 320
  size of forces 342
  storm tactics, Bessas charges up ladders 270, 317n42
  tekhhnitai and rams 119, 344n92
Petra Pertusa (538)
  Ostrogoths recruited into Roman army 352
Petra Pertusa (542) see Ostrogoths in general index
Petra Pertusa (552) see Narses in general index
Peyrusse (767) see Rocks and caves
  captured by Pippin 250
Phasis (556)
  ballistrai on ships 273, 283
  countryside stripped before siege 309n27
  division (*pedaturae*) of walls 178, 339
  earth-and-wood fortifications 306
  incendiary weapons 274n52
  Persian archery tactics 268
  Persian laborers (*paygān*) 287, 364
  Persian losses 348
  Persian *tekhnitai* 344

retinues at 58n97
Roman fallen buried 348
ruse of false relieving army 291f, 319
ships, sailors at 144, 296, 347n96
spaliōnes at 285
Slavs, Antae at 372
water-filled moat, parallels to Avignon (583) 208n57, 304, 582
Philippopolis (587)
  Avar failure 381
  hard resistance despite surprise 270n40
Piacenza (545)
  cannibalism at 263
Pisa (553)
  surrenders after Taginae, Mons Lactarius 317
Poitiers (567)
  means of pressure 332
  stormed 207n45
Poitiers (573) see also (567)
  territory raided 207n48
Poitiers (584) see also (567)
  royal garrison at 340
  submits after raiding 207n48, 314n34
Poitiers (585)
  bishop bribes garrison 332
Portus (537)
  context of 357
  Ostrogothic garrison 338
Portus (552) see Narses in general index

Ratiaria (447) see also Naissus (442)
  Hunnic invasions 48
  taken with *phrouria* 300
Ravenna (489-93)
  context of 151, 152
  ends in political compromise 258n11
Ravenna (539f)
  complexity of 257n6
  failure of political settlement 354
  supplies fail, sabotaged 158
  supplies, organization of 312f
  taunt of "Greeklings" 318
Ravenna (c. 731) 188
  nature of 189
?Ravenna (756)
  nature of 189
Reggio (549)
  from storm to blockade 257n2
?Rheims (556/7) see also (562)
  pressure on magnates 314n34, 328
  territory raided 207n48

Rheims, other cities (562) see also (556/7)
  nature of 206n43
Rhizaion (Roman camp) (558)
  Roman sortie 292
  testudo at 269n33
  Tzani losses 348
?Rhodian fort (807) see Abbasids in general index
Rimini I (538)
  consequences of 337n78
  Romans invited by Matasuntha 328
  taken by raiding cavalry army 312
Rimini II (538)
  ditch-excavating vs. ditch-filling 287
  exhaustion among besiegers 263
  Ostrogothic army, camp 155n19, 338, 349
  question of barbarian incompetence 152
  retinue at 58n97
  Roman garrison at 337
  Romans near surrender due to supplies 293
  siege tower, stopped in moat 115, 164, 282, 291
Rimini (549)
  divided loyalties in urban community 329
?Rimini (552)
  bridge broken near 307
  sortie against Narses' army 291
Rocks and caves [in Aquitaine]; Ally, Turenne, Peyrusse (767)
  captured by Pippin 250
Rodez (507)
  context of 91
Rome (408ff)
  Visigothic intentions 42
Rome; Tuscan cities (536(f))
  population invites Romans, Ostrogoths abandon 328
Rome (537f)
  aqueducts cut 306
  archery, ladder tactics 269, 270
  ballistrai, use, effect of 273
  civilians evacuated 314
  civilians furious at Ostrogoths 315
  civilians mobilized, distributed among troops 73n150
  deserters disrupt plans, sorties 292
  fleet at 347n96

  fortified siege camps, blockade 162n48, 261, 264, 278
  garrison at Milvian Bridge flees 323
  historiographical treatment of 24, 295f
  Isaurians reinforce fortifications of Ostia 115
  Janus, doors opened at temple of 325
  Jordanes on 208n56, 357f
  mining 288
  moat 304
  music, victory songs by soldiers 320
  Ostrogothic army 155, 343
  Procopius on artillery 420
  question of Ostrogothic siege competence 152, 162ff
  rams, siege towers, tactics 162ff, 268n32, 282, 284
  refugees as defenders 140n130
  role of engineers (*tekhnitai*) 116f, 344n92
  Roman forces, gradual reinforcements 295, 336, 372
  69 engagements outside walls 290
  sorties against rams 291
  tactics at Hadrian's mausoleum 289f
  tekhnitai and *ballistrai* 119
  terror effect of siege engines 282
  urban community, divided loyalties in 327
  "wolf" trap door 289, 290
  walls repaired, parapet modified 290, 305
Rome (545f)
  blockade 261, 264n27
  disagreement among commanders 293, 327
  Ostrogothic fortified bridge 280
  role of Roman fleet 257n7
  ship-born towers 283
  suicides due to conditions 317
  supply fleet fails 294
  Totila bans raiding 313
  treason after hard fighting 164
  walls razed 305n20, 354
Rome (546(f))
  emergency repairs by "whole army" 116
  sortie ends siege 292
  triboloi (spikes) protect unfinished gates 290
Rome (549f)
  intramural fields used for cultivation 308

804 INDEX OBSIDIONUM

Isaurians defect to Ostogoths 156n21, 330
last stand in Hadrian's mausoleum 305
retinues at 58
Roman forces 337
sortie as last resort 291
storms attempted during blockade 266
Totila rebuilds, resettles 354
treason after hard fighting 164, 567
Rome (552) see Narses in general index
Rome (754) not in CO; see (756)
Rome (756)
and significance of sieges to Lombard warfare 188
artillery, possible use of 277fn65
clerics at war 234n59
Frankish intervention against Lombard siege 183, 248
Lombard siege engines, camps at each gate 189, 190, 261, 279
Lombards raid papal estates 309
pope promises heavenly reward to defenders 322
Rossano (548)
Romans defect to Ostrogoths 156n21, 352

Sagalassos(?) (664) see SYLWS
?Saguntum (c. 603) see Visigoths in general index
Saintes (496)
Clovis' strategy 91
Salona (536)
Ostrogothic and Sueve forces 155
population unsupportive of Ostrogoths 154, 158n20, 328
state of walls, repaired by Romans 305
Salona (537) see also (536)
improvement of fortifications 304
Ostrogothic blockade 261, 264
Ostrogothic land, naval expedition 156
siegeworks 279
Sueve client troops 177
surrounding forts stripped to hold 336n77
?Samosata (c. 705) see Arab-Roman frontier in general index
Saragossa and other cities (472/3)
Visigothic expansion 45
Saragossa (541)
countryside raided 310
Frankish expedition against 198f

Franks defeated, massacred 348
rituals at 322
Visigothic defenses 166n57
Saragossa (653)
Basque raids 310
in Visigothic civil war 173n89
Sardis (c. 716)
as base for Constantinople (717f) 401
captives taken at 351f
Saxon fortifications (776)
corvée labor 389
earth-and-wood 251
subdued by Charlemagne 388
Saxon fortifications (785) see (776)
Sebasteia (576)
burnt, possibly abandoned 350
campaign encouraged by Magian priests 325
Seine, cities along the (600)
stormed after battle 264n28
walls breached 225
Semalouos (780)
artillery at 277fn65
artillery terminology 416fn32
surrenders after hard fighting 258nn11 &13
Septem (c. 547)
Visigoths surprised at mass 324
Septimanian campaign; fortress of Dio (c. 532)
see: Cabrières
renewed Frankish expansion 198
Serdica (809)
military engineer at, defects to Bulgars 123, 385
taken by treason 385
Sergiopolis (542)
storm due to lack of water 265
Sevilla and other cities (583f)
and Visigothic expansion 166n59
complexity of 296
from blockade to storm 266
in Hermenigild's revolt 173
Italica rebuilt for blockade 261, 301
Sicilian forts (551)
failed sorties 292
recaptured by Romans 157n28
Singara (578) see Roman-Persian wars in general index
Singidunum (442)
in Hun invasions 48

INDEX OBSIDIONUM

Singidunum (c. 472)
   captured after battle 257n8
   taken by Ostrogoths 150n4
Singidunum (583) see also (588)
   population in fields 309
   rhetorical claims 591
   taken by surprise storm 265, 381
Singidunum (588) see also (583, 595)
   Avar river fleet 188
   fleet burnt by garrison 291
   ransom paid to Avars 349, 381
   Slav boatbuilders 375, 381
Singidunum (595) see also (588)
   improved Avar logistics 612
   Italian shipwrights sent by Lombards 381
   relieved by Danube flotilla 294, 296f
   walls broken down 284
Sirmium (442)
   in Hun invasions
Sirmium (504)
   captured by Ostrogoths 152
Sirmium (568)
   token ransom denied 314f
Sirmium (579ff)
   Avars inherit Gepid claim to 378f
   Avar river fleet 188
   Avars use pagan, Christian oaths 324f
   context of 181n124
   ordered to surrender 263
   population emaciated 354
   perceived Roman weakness before 374
   Roman craftsmen in Avar service 143, 382
   Roman craftsmen forced to build bridge 221n114, 280, 307
   Slav boatbuilders 375
Sisauranon (541)
   different Roman approaches 258n11
   from storm to blockade 271
   Ostrogothic cavalry charge saves Roman army 152, 353
   Persian garrison defects, sent to Italy 352
   Procopius on Arab plundering, siege skills 260n17, 389
   Roman forces, composition of 353
   troops ill from heat 263
Skoutarion(?) (664)
   Arab logistics 448
   context of 446f
Slav wagon laager (593/4) see Slavs in general index

Slav fortifications (808)
   and acculturation to siege warfare 251n207
   and Danes 388
Smyrna (664)
   context of 446f
Soissons; other cities (562)
   in Frankish civil war 206n43
Soissons (576)
   relief army wins in battle 208n52, 293
Spoleto (545)
   time limit for surrender 293
Spoleto (546)
   amphitheater fortified 306
   internal divisions 329
Stobi and other cities (473)
   Ostrogothic expansion 150n4
   possible West Roman involvement 331
Sutri (c. 731)
   nature of 189
Sura (540)
   attacked after omen 325
   commander killed 271
   consequences of ransom conditions (see Sergiopolis 542) 328, 529
   forces involved 342
   poor state of walls induces negotiations 305
   taken on false pretense 350
   stormed to spread fear 265
Susa (755)
   Lombard arsenal at 190
   Pippin's campaigns 248n200
Syburg (775) see also (776)
   in Charlemagne's Saxon wars 251n205, 303, 740
Syburg (776) see also (775)
   artillery, use of 277fn65, 388
   from negotiations to storm 265
   Saxon tactics 388
Syke (771)
   Arabs escape trap 294
SYLWS (664) or Sagalassos
   artillery duel at 275, 318
   artillery terminology 416
   context of 446ff
   logistics of, problems constructing a trebuchet 276, 409
   master carpenter forced to build trebuchet 121, 144, 39
   rope-pulling crews 416

Synnada (c. 740)
  Arab besiegers surprised, defeated 294
Sythen, Saxon strongholds at (758)
  taken by storm 248n199

Taranto (549)
Taranto (552)
  holds out on election of Teïas 318
Taranto (c. 685)
Taranton (702) =Turanda
  Mopsuestia rebuilt on expedition to 401
  captives taken from 351
Tarragona (472/3)
  in Visigothic expansion 45
Tella and Mesopotamian cities (639)
  Tella=Constantina
  Arab tactics at 435
  artillery terminology 419
  artillery at 436
  size of Roman garrison 336
Tendunias (641)
  garrison tries to prevent famine 394
  local factions at 397n80
  size of Roman garrison 336
Thebasa (793) see Abbasids in general index
Thebasa (806) see Herakles; see also Abbasids in general index
Thebothon (573)
  in blockade of *Nisibis 307
Theodosiopolis (421/2)
  artillery 275
  artillery chant/command 321
  individual names for artillery piece 419
  machines used 296
Theodosiopolis (502) 125f
  betrayed by Roman official 328
Theodosiopolis II (502) see above
  Persian garrison massacred 353
Theodosiopolis (576)
  abandoned due to approaching relief 293
  Persians draw up troops before city 258n12
  population surprised in fields 309
Theodosiopolis (610/11)
  adventus arranged by local dignitaries 354
  alleged son of Maurice used by Persians 331
  from storm to negotiations 265
Theodosiopolis (655)
  context of 444, 450

Theodosiopolis (750) see Melitene
  and range of Roman armies 123
  population deported to Thrace 352, 503
Thermopylai (539)
  and Hunnic siege skills 370
Thessalonica (586/597)
  artillery battery, size of 275, 410
  artillery terminology 411, 417
  complexity of 296
  context, dating of 374f
  divisions among besiegers 330
  hides to cover engines 409
  population surprised in fields 309
  poor barbarian aim 277n63
  question of Avar leadership 95n1, 382
  raiding around 310
  rams, *khelōnai* 285
  retinues at 60, 339
  siegeworks 278
  skill of Roman artillerymen 121
  terror, morale of population 316, 323
  traction trebuchet, first clear description of 274, 407f
Thessalonica (604) see also (586)
  civilian defenders mobilized 340
  dating of 376
  siege, raiding prevented by sortie 291, 312
  suburban fort mistaken for 264n28, 300
Thessalonica (c. 615) see also (586, 618) 375
  artillery terminology 421
  Avars' trebuchets set up 276n62
  civilian participation 144
  covers against missiles 269
  failed Slav attack with *monoxyla*, heads on display 319
  Jewish defenders, *pedaturae* 333, 339n86
  liturgical elements in *Miracula St. Demetrii* 325
  missile exchanges 268
  refugees at 308f
  Serdica, continuous habitation at 759
  Slav chieftain executed by women 348
  Slavs flee, abandon booty, machines 349
  Slav initiative 382
  Slav invasions, tactics, development of 376
  thanksgiving service after 348
Thessalonica (c. 618) see also (586, 615) 343
  artillery, effective defense, duels 275
  artillery terminology 414

INDEX OBSIDIONUM 807

Avar leadership 376, 379, 382
battle cry 321
chant of trebuchet crews 321
citizens refuse ransom 349
civilian catapult crews, sailors 144
khelōnai overturned 289
manganarioi 120, 415, 417
raiding before siege 258n9
refugees at 308f
Roman fleet at 347
siege towers 282
trebuchet crews inside protective framework 417
Thessalonica (662) see also (586)
artillery, duels 275, 277
artillery terminology 414, 417
failure of water supply 262
manganarios to build siege tower 120, 283, 415
missile exchanges 268
population emaciated 349
raiding before siege 261, 311
Roman fleet 294
sailors 144
St. Demetrius fights on walls 322
Slav acculturation 329, 376f, 444n18
Slav craftsmen at 147, 344, 377
three days' storm 266
trebuchets planned on siege tower 417
Thessalonica (c. 682)
Kouber's scheme at 112, 383
local loyalties 327n56
role of ex-Romans, Bulgars 383
Thouars (762)
context of 249
Tiberias (c. 635)
blockade of 261
revolt after surrender 432n5
terms of surrender 355n101
Tibur (544)
civilian defenders 340
locals betray Roman garrison 330
Todi (538) see Chiusi
Topeiros; other towns and fortresses (549) 373
citizens overwhelmed by archery, ladders 272n45
garrison ambushed, civilians lead defense 312, 340
massacre, atrocities 353
Slav army, skills 342, 373, 380

Tomi (597f)
Avar fortified camp 280
Tortosa (809)
artillery terminology 415
Carolingian arsenal 252
Toul (612)
taken by force 225
Toulouse (c. 413)
stormed by Visigoths 43
?Toulouse (439)
Hun participation 366
pagan rituals 324
possible siege 44
Toulouse (507)
burnt, sacked by Clovis 12, 91, 350
Toulouse (585)
prepatations for siege 207
surrenders in face of huge army 323
Toulouse (c. 720)
artillery at 246, 277n65
artillery terminology 418
rams used at 285
Toulouse; Albi, Gevaudan (767)
context of 250
Toumar (540)
water supply problems 262
stormed 266
Tournai (575)
stormed 207n45
?Tours (567)
nature of 206n43
oaths forcibly extracted 332n65
?Tours (573) see also (575f, 584)
territory raided 207n48, 314n34
Tours (575f) see also (573, 584)
loyalties of urban community 327n56
surrenders after threat to countryside 207n44
Tours (584) see also (573, 575f)
submits after raiding 207n4
?Trent (c. 575) see Nano
Trent (c. 680)
Lombard fortified camp at 279
sortie breaks siege 293
Treviso (c. 591)
revolt of Lombard duke at 330
Treviso (776) see Cividale
in Carolingian expansion 252
Trier (413) see also (456)
in Roman civil war, role of Franks 84f
Trier (451)
center of artillery production 48

Trier (c. 456) see also (413)
    historicity of 85, 87
Tripoli (c. 644)
    Arabs build fort to blockade 307, 391
    lengthy resistance 311n28
Tripoli (653)
    context of 329n58
    Kallinikos, inventor of Greek fire, possible background to 678
    losses to military infrastructure at 392
    military infrastructure at 395n76
    Roman sabotage at 397, 442
Turanda and other forts (712) see Taranton
Turenne (767) see Rocks and caves
    context of 250
Tyana (c. 708f) **145f**
    artillery at 277n65, 296, 401
    defeat of relieving army 294
    population deported, driven into desert 352
Tzacher/Sideroun (557)
    archery covers siege engines 269, 287
    arsenal used at 274n54
    from blockade to storm 257n3, 266
    fortification and settlement 304
    Roman atrocities 353
    siegeworks 279f, 285
    Slavs in Roman army 372
Tzurullon (588)
    Avars abandon siege 381
    Roman troops billeted in villages 310
    strategic relief 295

Ulpiana (473)
    Ostrogothic expansion 150n4
    possible West Roman involvement 331
Ultrère (673)
    context of, fortifications at 175
    troops rewarded by Wamba 329
Umayyad castella (802/6)
    in Carolingian expansion 252n211
Unstrut, castrum at (639)
    disloyalty in Frankish army 229, 387
Urbino (538)
    archery covers engines 269n35
    engines, screens 285
    Ostrogoths recruited into Roman army 352
    siege camps 278
    surrenders after well dries up 261f

Uzès (581)
    contested episcopal election leads to 208n54, 332

Valdoria (603)
    Avars send Slavs to help take 382
    rams probably used 284, 382
Valence (411)
    stormed by Goths 43
Valence (574)
    Lombards defeated by Mummolus 180n122, 293
Verona (541)
    betrayed to Romans by local 330
    Ostrogoths recapture 265
    Roman army 342
Verona (561) see Ostrogoths; Franks in general index
Verona (569)
    context of 180n120
Vienne (500) **196f**
    aqueducts used to capture 262, 287f
    archery exchange at 268
    artifex, shows way through aqueducts 223, 330, 343
    Frankish participation 343
    in Burgundian civil war 90f
    parallel to Naples (536) 502
    population expelled to preserve supplies 314
    towers as last refuge 305n21
Viminacium; other forts and cities (441)
    captive Roman merchant interviewed by Priscus 367
    in Hunnic invasions 48
    taken by surprise 265, 367
Viminacium (583)
    and Avar siege skills 381
    taken by surprise 265
Vincenza (569)
    context of 180n120
Vitry(-le-Brûlé) (524/33)
    common people at 218
    missiles, possibly artillery 197, 268n31, 274n55
    siegeworks 279
Vollore (524/32)
    defenders lax in guard duties 317
    population taken captive 350
    stormed 197

Volterra (553)
   surrenders after Taginae, Mons Lactarius 317
Vouillé campaign (507f)
   see: Toulouse, Albi, Rodez, Clermont-Ferrand, Narbonne, ?Bordeaux, Angoulême, Arles
   and relationship with Burgundian civil war (500) 486
   sources for, nature of 12f, 195n8

Woëvre, fortified estate church (587)
   nature of fortification 207n49, 302
Wogastisburg (c. 630)
   consequences of 387, 646
   nature of fortifications, siege 228

Zerboule (540)
   archery effectiveness 271

GENERAL INDEX

The index is limited to the most important technical terms, actors, structures and themes. The principal discussion(s) of a theme is marked in bold. For ubiquitous themes (e.g., blockade) and actors (e.g., Belisarius, Romans, Arabs, Persians), only substantive discussions are indexed; further discussion and illuminating examples for peoples and polities may be found under each theme. Entries in the Corpus Obsidionum are not for the most part indexed by page, but by name and/or year, and can also be found in the Index Obsidionum with further references.

Abbasids
    as besiegers 14, 121, 263, 296, 350
    religious ritual in armies 325
    revolution 9n34, 17, 402, 405
    use Christian irregular troops 402
    use conscripted and paid laborers 310, 402, 404
    see also CO s. aa. 771, 779, 793, 779, 806, 807, 808
Aeneas Tacticus 31, 258f, 261
Aegidius 44f, 88f
Africa
    and Arabs 97, 396
    and Vandals 49, 53, 54n87, 65, 67, 70, 82, 153, 363
Agathias
    on artillery 25, 380, 411f, 420f
    on Franks 201-04, 213, 214n77
Alamans 52, 76, 79f, 82, 460
    under Franks 90, 202, 205, 213, 214n77, 215, 228, 387
Ammianus Marcellinus
    on siege of Amida 37, 38
    on siege of Adrianople 39f
Anatolia
    settlement in 97, 308f, 356
    soldiers from 99, 102, 103, 104, 108, 115, 339, 377
    invasions of 130, 174, 263, 291, 296, 308, 316, 323, 343, 350, 391, 398f, 439-53
'anwat^{an} 8, 355n106
Apion family 59, 397
Arabs xviiff, 6-9, **389-405**
    and trebuchet 407, 409, 416, 418f, 423, 424
    and Franks 194, 246f, 346, 429
    before Islam 118, 125f, 130ff, 136, 137, 142, 260, 262, 310, 330, 343, 364

adopt technology, infrastructure 15, 32, 111, 121, 144, 276, 277, 303f, 318, 360, 364, 427
    see also Islam; Abbasids
architect, *architectus, arkhitekton* 36n9, 141, 143f, 146, 159, 161, 399, 409, 540, 677
Arianism 43, 52, 191, 327, 328, 335, 361, 495, 500
Armenia
    and religion 330, 335
    between Byzantium and Arabs 391, 393, 399, 402f, 404, 415, 434-53
    between Rome and Persia 125 &n91, 130f, 364
    in (East) Roman Empire 99, 102, 104, 107, 114, 180n119, 271, 384
army, Roman
    comitatenses 40n26, 49f, 54, 99, 154
    foederati, foideratoi, federates 29, 44, 50f, 53, 56n91, 70, 100, 102, 117, 150, 156, 167, 197, 214, 362, 370, 426, 342
    limitanei 49f, 53, 54, 99, 119, 151n7, 154
    military fashion 86
    of the Principate 36, 68, 370, 406
    reforms of Diocletian and Constantine 36, 38
    reforms of Valentinian III 51, 64f, 66
    5th century-developments 49-56, 63-67, 84-89
    See also craftsmen; artillerymen; Ammianus; Procopius; retinue; *buccellarii*
arsenal
    Arab 385, 394f, 397, 399, 442
    Roman 38f, 42, 105, 111, 117f, 120, 123, 128f, 223, 367, 398, 425
    see also *fabrica; fabricenses*

artillery 267, 272-77
  Arabs at *Caesarea (640), *Toulouse
    (720), *Kamakhon (766) etc. 246,
    266, 286, 391
  Franks 74, 197, 211, 223, 247, 281, 285n72
  Huns 284, 366, 379
  Lombards 189f
  Ostrogoths at *Osimo (539), *fortified
    camp (552) 162, 280
  Persians at Dura-Europos (256) 35
  Roman 36-40, 47f, 71, 117-22, 127f,
    181n126, 203, 270n39, 281, 283, 284,
    290, 292, 372, 394
  Slavs at *Thessalonica (586, 615, 618,
    662) 283, 375n37, 379
  terminology and diffusion 406-24, 425,
    428
  torsion 32
  Visigoths in 637 (see also Wamba) 175,
    268
  see also arsenal; artillerymen
artillerymen 37, 38f, 140, 348, 382, 667
  ballistarii, ballistrarioi 36, 38, 117-20,
    123, 159, 197, 273
  manganarioi 119f, 122, 123, 140, 414f, 417
  organization 117-23
  skill 121f
Attila 47, 51, 82, 326, 467, 468-71
  after death/see Hun 124, 150, 180n119
Avars
  appropriation 17, 119f, 277, 280*
  conquest 95f, 100, 174, 177, 178, 181
  earthworks 237, 246, 251
  influence on western Slavs 228f
  poliorcetic skill 6, 121, 186, 188, 221n114,
    257n4, 258n9, 262, 263, 265,
    270n40, 272
Avitus 44, 51, 84f, 472f

bacaudae 43, 85, 568
Baladhuri and reliability of Islamic
  historiography 432-36, 445
Balkans
  military culture 18n57, 81, 86, 88n197,
    100, 149ff, 152n10, 177, 370n19
  settlement 308ff, 356
ballista, ballistra 39, 118, 122n76, 272ff,
    289, 372n27, 397n82
  effect 37n16, 121, 163, 269, 446
  hand-held 38n20
  Nicetius' 74, 223, 273n49, 346, 424
  terminology 412f, 415, 418, 420f,

ballistarii, ballist(r)arioi see artillerymen
Belisarius
  defeats Gothic siegecraft 162ff
  devises floating towers 280, 283
  estates, retinue broken up 61f
  involves civilians 328, 340
  other inventions 290
  retinue 57, 329, 343
  restored to favor 116
Bishops in politics, warfare: individual
  Abbo on Danish siege of Paris (884f)
    408, 429
  Anianus and prayers for *Orléans (451)
    47f, 322
  Aunemund and defense of *Lyons /622)
    233
  Bertram of le Mans, estates of 227
  Bar-Hadad and prayers, liturgy for
    *Edessa (502f) 131
  appeals to shah for *Constantina
    (502f) 324
  Caesarius of *Arles (508) 13f, 22, 73f, 92,
    322, 323, 333, 345f, 510
  Datius and logistics, defection of *Milan
    (538) 158n35, 328
  Desiderius of Cahors, building
    activities, craftsmen of 231
  Eunomius of *Theodosiopolis (421/2)
    blesses catapult 275, 321
  Eusebius and dating of *Thessalonica
    (586/97) 375n37
  Hilary of Arles' retinue 65f, 230
  Hildegar of Cologne, killed at *Ihburg
    (753) 248n199, 388
  Leudegar of *Autun (679), building,
    military activities of 231, 233, 308
  and liturgical procession 322
  Maximus and prayers for *Castrum
    Cainonense (463) 322
  Megas of *Beroea negotiates at
    *Antioch (540) 520
  Praejectus of of Auvergne, building
    activities of 231f
  Nicetius of Trier's fortified residence,
    ballista 74, 221, 223, 273n49, 499
  receives craftsmen from Rufus of
    Turin 223, 424
  Sagittarius and defense of *Convenae
    (585) 211
  see also *Embrun (571)
  Thomas of Amida leads construction of
    Dara 136-39
  b. and *urbani* at *Arles (567/9) 341

b. betrays Cos (*Aegean Islands, 653) 329
b. election ends in war at *Uzès (581) 208n54, 332
b. helps Arabs defend Mardin (c. 690) 399
b. of *Antioch (540) suspected of treason 328
b. of *Constantina (502f) stops massacre of Jews 130, 348
b. of *Poitiers (585) bribes royal garrison 332
b. of *Sergiopolis (542) tortured by Persians 328, 350f
b. used by Arabs to negotiate with *Arwad (649) 329
b. of Langres' fortified residence at *Dijon (555) 221, 302
Bishops in politics, warfare: structural
  Coptic bishops justify Arab conquest 396f
  fighting bishops in West 165, 328
  Frankish bishops' retinues, military obligations 218f, 222, 229-32, 234, 239, 243, 302, 308
    logistical, building skills 74, 231f, 250, 345f
  Justinian sets bishops to oversee defenses 61, 139
  influence on Franks 232, 254
  limit of bishops' role in East 147
  Monophysite bishops cultivated by Persians 334, 335n75
  reflected in sources 22, 27f, 232f, 325f
  Visigothic bishops' retinues, building skills 74, 171, 346
    in revolt against Wamba 678-84
  b. org. defenses in Mesopotamia, Syria 59, 142
  see also *Toulouse (439), *Sirmium (442; 568), *Clermont (524), *Sura (540), *Apamea (540, 573), *Hadrametum (544), *Milan (569), *Chlomaron (578), *Frankish invasion of Italy (590), *Orléans (604), *Jerusalem (624), *Thessalonica (586, 618), *Damascus (634/5), *Saragossa (653); *Debeltos (812)
blockade, definition of 31, 256-66, 267, 278ff
  see also storm

boukellarioi (middle Byzantine theme) 62, 738, 745, 748
buccellarii
  organization 56nn90f, 57ff, 170, 508
  legislation 67n131, 99
  in Visigothic military organization 170
  drawn from regular units 57, 62
  see also retinue, *obsequium*
Bulgars 315, 383-86
  and Slavs 371f, 377f
  construct ditches 280
  in Italy 107, 179, 184
  origins 112f, 179, 369f, 376, 382f, 425
  relations with Byantium 123, 281, 293, 301, 329, 369, 403, 404, 426f
Burgundians
  allied with Franks 13, 73, 90, 92, 93, 152f, 343
  as Roman federates 66, 82, 84
  Frankish conquest 196ff, 215
  settlement of 53, 81
  warfare, military organization 197f, 268, 296, 426
  See also *Burgundian civil war (500), *Vouillé campaign (507f).
  Frankish Burgundy see Frankish kingdoms

catapult see artillery
civic munificence 34
  decline of 63, 425
  responsibility transferred to bishops 139
clients, Roman
  acculturation of 49, 93
  Byzantine 94, 111-15, 177, 178, 369ff, 374f, 384, 426
  Franks, Ostrogoths reestablish 83, 90, 387f, 426
  in civil wars 79-82, 85ff
  client management 75ff, 78ff, 82f, 87
  see also Alamans; Arabs: before Islam; Burgundians; Franks to c. 500; Goths; Heruls; Hunnic clients; Lombards; Ostrogoths; Sueves; Visigoths
Clovis
  and Burgundy 196f
  division of kingdom 192
  military, fiscal organization under 214f
  origins of Merovingian kingdom 89-92
  see also *Burgundian civil war (500), *Vouillé campaign (507f)

co-evolution 14f, 361, 363
craftsmen, engineers 32, **343-47**
    East Roman 115-23, 128f, 135-147, 384
    Frankish 222f, 231f, 250f
    in Arab service 9, 27, 392-404
    in Avar service 380-83
    in Bulgar service 385f
    in Hun service 368f
    Lombard 186ff, 381f
    Ostrogothic 160f
    Slav 375-78
    Visigothic 169, 171
    See also army; artillerymen; estates; *fabricae, fabricenses*; logistics; *munera*; naval forces

dediticii 41, 53, 76
doryphoroi 57f, 60ff, 167n64, 270
    equivalent at *Clermont (474) 477
    parallels in Frankish kingdoms 216
    Roman d. defects to Ostrogoths 329
    see also *Ancona (538), *Osimo (539), *Rome (549f), *Phasis (556)
Dura-Europos, siege of (256) 34ff, 128, 297, 347, 364

Ecdicius (fl. 474)
    relieves *Clermont (474) 66f
    supplies city from estates 313
engineers see architects; artillerymen; craftsmen, engineers
estate
    evolution of (bipartite) 10, 49, 57ff, 62f, 229
    obligations to state 67-74
    scale, income 63, 228f
    see also *munera*; logistics
ethnicity, ethnogenesis see identity
Euric
    consolidates Visigothic kingdom 45
    legislation on retinues 67, 170n76, 185
    see also *Tarragona (472/3), *Arles (473/6)

fabrica
    artillery 38f, 74, 117
    decentralization of production 70f
    in Anastasian War 138n121 &123f
    organization under Justinian 117f
    survival of competence in Gaul 74, 223
    survival of competence in Italy 157f, 159, 160f

fabricenses at Adrianople 40, 326
foederati, foideratoi, federates
    see army
forts, fortifications **300-07**
    earth and wood 228f, 246, 306, 385, 387f
    repair, construction of
        late Roman 36, 48, 71f
        Byzantium 118, 132, 135-47
        Ostrogoths 92, 159f
        Visigoths (Toledo) 169-72, 173, 175
        Lombards 186ff
        Franks 221f, 231f, 249, 250f, 262
        Avars, Slavs, Bulgars 382ff
        Wends, Saxons, Danes 387ff
        Arabs 345, 400-04
    see also *munera*
Franks before c. 500
    origins (ethnogenesis) 84-90
    petty kingdoms 82, 89
    Trojan ancestry 84, 90
    warfare 29, 30
    see also Rhine: fate of Roman army on
Frankish kingdoms
    Austrasia and Italy 200-06
    established by Clovis 89
    Frankish Burgundy 196-200
    military organization
        Merovingian 211-23, 225-28
        Frankish church 229-32
        Carolingian 238-45, 250-54
    warfare, politics 206-11, 224f, 228f, 232f, 245-50
    wars against Visigoths 90ff
frontier
    culture 29, 30, 77, 93
    Roman influence on 86
    involution of 50, 78, 82

Goths
    north of Danube 39, 86
    settlement of 39, 41, 100-03, 362
    See also Ostrogoths; Visigoths
Gregory of Tours
    in Frankish military organization 217f, 230, 338
    methodological problems using 193ff, 199, 216, 218f, 288, 326
    on Frankish origins 84ff, 90

Heruls 83, 100, 177f, 180n119, 339, 370
    see also *Mt. Papua (533f), *Anglon (530), *Rome (545f), *Rimini (552), *Phasis (556)

GENERAL INDEX   815

hospitalitas debate 54f
hypaspistai 57, 61f, 343
    Western equivalents 216
    see also *Clermont (474), *Osimo (539),
        *Sisauranon (541)
Huns under Attila 46ff, 365-69
    collapse of empire 89, 95, 99ff, 150
    effect of invasions 356
    empire building 82
    means of appropriation 17, 75
Hunnic clients (Sabirs, Utigurs,
        Kotrigurs) 369ff
    and Byzantium 101, 177, 292, 316, 320,
        328, 337n78, 373
    and Persia 130, 343, 364
    and Avars 378, 380, 382
    Sabir rams 210, 284

identity, ethnicity, ethnogenesis 34, 75-80
    conflation of Roman and barbarian 50,
        88
    diacritic markers 19
    ethogenesis 29, 49, 50, 76
    Frankish 30, 84-90, 215
    lack of importance in siege
        warfare 326f
    Lombard 176-79
    loss of Roman 29, 34, 82, 83f, 90, 363,
        325
    Ostrogothic 150ff
    Roman 18f, 100-03
    Romanization 50, 76, 80, 86ff
Islam see Arabs
Islamization 17, 405

Jericho (biblical siege) 12, 92, 247, 288, 413
Jewish communities 333f
    betrayal, pogrom at *Constantina (502f)
        130, 348
    defending *Arles (508) 13
    defending *Naples (536) 154
    in Arab invasions 431
    in revolt at *Edessa (630) 122
    military obligations of 140, 339
    role at *Jerusalem (614) 354
    topos of betrayal 333
    see also *Caesarea (611), *Thessalonica
        (615), *Neocaesarea (729)
Jordanes
    on *Aquileia (452) 48
    on *Orléans (451, 453) 48, 326
    on *Rome (537f) 208n56, 357ff

Justinian
    and Franks 212f, 226f, 254
    and Hunnic groups 316, 371
    and Lombards 177f
    and Slavs 371-74
    conquests of 95, 153
    laws on retinues 60
    laws on arms production,
        ballistarii 117f
    laws on role of bishops, magnates in
        public works 139
    services substitute for taxes 69f, 71,
        139f, 226f
    struggle with magnates 61f

labor obligations
    see logistics; *munera*; *pedatura*
laeti 53, 214
logistics, supplies 310-14
    Arabs 144, 166, 394- 404
    Franks 219-23, 226ff, 231f, 237f, 250-54
    Lombards 186ff
    Ostrogoths 157-61
    requisitioning 60
    role of 40f, 98, 104, 294
    Visigoths 169, 170ff
    see also blockade; estates; *fabricae*;
        arsenals; *munera*; taxation, role of
Lombards 176-190
    and Avars 378, 380, 381
    and Bulgars 112
    and Franks 198, 200, 204ff, 228, 230,
        248, 252
    East Roman influences on 107, 147, 346,
        426
    military organization 30, 167, 168, 301,
        338, 345, 423
    origins as East Roman clients 95, 96,
        100, 102
    warfare 261, 272, 277, 294
    wars with East Rome 113, 122

Majorian 44, 66, 82, 88f
manganarioi see artillerymen
maximalist 3f, 235ff
military see army
military lands 5, 53-56, 105
    see also army: *limitanei, comitatenses*;
        laeti
militia 72f, 339ff
    Byzantine 60, 109n48, 185, 292, 336, 422
    nature of armies 50, 108, 145f, 218

Roman 66, 72
Western 154, 171, 184n137, 203
minimalist 1, 3ff, 24, 194f, 235ff
Mummolus
  defense of *Avignon (583) 208
  defense of *Convenae (585) 208-11
  defense of Frankish Burgundy 180, 198, 293, 306
  household, carpenter of 74, 217n93, 223
  in Frankish civil wars 206n43
munera 67-74, **425**
  among Western successors 157, 159ff, 170ff, 188, 221f, 252ff, 345f
  Arab adoption of 394-404, 435, 452
  extraordinaria, sordida 68ff, 342
  levied on craftsmen, *possessores* 69ff, 92f
  substitution of, exemption from 69f, 106, 226

narrative
  episodic 9, 11-14, 24, 48, 189f, 195, 208n57, 296
  schematic 11-14, 296
naval forces
  Arab 9, 32, 266, 277, 343, 345, 391, 392, 395-99, 401, 427, 441-45, 448, 449, 453
  Byzantine 39, 102, 104, 110f, 122f, 160, 273, 292, 311, 317, 444
  in sieges 257, 264, 283, 294, 296f, 311, 347
  Ostrogothic 156f, 158, 161, 279, 280
  Scandinavian 78
  Slav 291, 319, 375, 376, 377, 381, 382
  Visigothic 45, 175, 176
North Africa see Africa

obsequium see retinue
Obsequium (East Roman army) 104, 108
  See *Opsikion* 449, 451
Ostrogoths **149-64**, 425f
  and Franks 13f, 92, 93, 194, 198n14, 200-04, 212
  and Visigoths 165f
  military organization 53, 54f, 58n99, 83, 177, 180, 190, 220n112, 222, 301, 312f, 338, 343
  origins in Balkans military culture 95, 100-03, 147, 331
  warfare 273, 280, 284, 289f, 293, 305, 306f, 328, 354, 357ff

see also naval forces; Corpus Obsidionum s. aa. 535-554

pedatura, pedatoura
  and civilians 339ff
  Arab use of 399f
  Frankish use of 222, 253f
  in Byzantium 140, 410
  Ostrogothic use of 159ff
  Roman origins 71f
  under papacy at Rome 187
  Slav, Saxon use of 389
Persians **363ff**
Priscus
  on *Aquileia (452) 358f
  on embassy to Attila's court 367
  on *Naissus (442) 24, 46ff
  on *Orléans (451, 453) 326
  see also Procopius
Procopius
  ignorance of traction trebuchet 420f
  on Ostrogothic siege competence 162ff
  on *Rome (537f) 357ff
  on survival of West Roman units 42, 215
  topoi in 201

retinue, *obsequium* 28, 49, 55, **56-67**, 92, 339, 425, 426
  Carolingian 238-45, 346
  Lombard 183f, 185
  Merovingian 74, 215-19, 222f, 230ff, 338
  Persian, Armenian 364
  Visigothic 167-70, 346, 418
  See also *doryphoroi; hypaspistai; buccellarii*
Rhine
  fate of Roman army on 30, 51f, 74, 84, 214f
  putative Frankish invasion across 85-90
  Vandal, Sueve and Alan crossing 42, 80ff
Ritual 32, **320-26**
  adventus 333, 334
  see also *Florence (553)
  degradation 128f, 350, 353
  liturgy, processions 133, 199, 207
  triumph 89, 348
Roman-Persian wars
  Anastasian War (502-06) 96, 123-35, 139, 260, 296, 312

GENERAL INDEX   817

war of 527-32 (Iberian War) see
    *Martyropolis (531)
war of 540-61(Lazic War) see CO s. aa.
    540-57
war of 572-91see CO s. aa. 573-90
war of 603-29 see CO s. aa. 603-630

Sassanids see Persians
Scandinavia 76ff, 251, 389, 427
Sidonius Apollinaris
    on Franks 85f
    on military organization, *munera* 66f,
        73
    on siege methods 45
    see also *Clermont (474)
Siege engines, tactics 267-95
    archery, missiles 267-72
        effect, tactical use of 278, 342
        civilians at *Arles (508) 13
        Byzantines at *Rome (546) 116
        Ostrogothic response to at *Rome
            (537f) 162f
        Franks, Burgundians at *Vienne
            (500) 196
        Franks at *Convenae (585) 209ff
        Huns at *Naissus (442) 265f
        Slavs at *Topeiros (549),
            *Thessalonica (615) 373, 376
        Arabs at *Dvin (640), *Dara (640),
            *Caesarea (640) 391, 448
        Visigothic response to at Adrianople
            (378) 39f
        Visigoths at *Narbonne (673), *Nîmes
            (673) 175
    ladders 268-72
        Visigoths at Adrianople (378),
            *Narbonne (673) 40, 268
        in Sidonius Apollinaris 45
        Huns at *Naissus (442) 47, 265
        Kotrigurs at *Chersonese (559) 371
        construction of at *Naples (536) 116
        Persians at *Amida (502f) 127f
        Ostrogoths at *Rome (537f) 163, 289
        Frankish "rope ladder" at *Avignon
            (736) 247n197
        Romans at *Amorion (666) 451
        Arabs at *Caesarea (640), *Dara
            (640), *Dvin (640) 266, 391, 437,
            448
        Franks at *Arles (508) 333
        Slavs at *Topeiros (549),
            *Thessalonica (615, 618) 373, 376
        in course of siege 267, 276, 287, 340

    see also *Cesena (538/9), *Beroea
        (540), *Antioch (540), *Petra (541),
        *Edessa (544), *Petra (550), *Rome
        (552), *Thessalonica (586, 662),
        *Damascus (634/5),
        *Constantinople (813)
    mining 286ff
        tactical use, effect of 267, 278, 283,
            285, 289, 347
        Persians, Romans at Dura-Europos
            (256) 35, 297
        Franks (possibly) at *Angoulême
            (508) 92
        Romans at *Amida (502f), *Petra
            (549) 127, 271, 281
        Arabs at Amorion (838) 424
        in Isidore of Seville 174
        in Vegetius 267
        Persian manpower used for 364
    rams 283ff
        in course of siege 267, 269, 274, 288
        terminology for 274, 296, 411
        mentioned by Sidonius 45
        Romans at *Cherson (711) 122
        Persians at *Amida (502f) 126, 275
        Sabir rams at *Petra (550),
            *Archaeopolis (550) 370f, 380, 420
        Roman *tekhnitai* responsible for at
            *Petra (550) 119
        Ostrogoths at *Rome (537) 162, 163f
        Kotrigurs at *Chersonese (559) 371
        Huns at *Naissus (442), *Orléans
            (451) 47f, 265
        Slavs at *Thessalonica (586) 283, 375
        Lombards at several cities 189f, 382,
            423
        Franks resist rams at *Lucca
            (553) 203
        Franks at *Convenae (585) 209ff
        Franks at *Narbonne (736) 247
        countermeasures 289, 291, 340
        in Isidore of Seville 174
        in Vegetius 267
        see also *Petra (541), *Edessa (544),
            *Constantinople (preparations,
            813)
    tortoises, sheds 285f
        tactical use of 267, 287
        mentioned by Sidonius 45
        in Isidore of Seville 174
        in Vegetius 267
        parallels to *Convenae (585) 210

Persians at Phasis (556) 287
Romans at *Onoguris (555) 287
Slavs (and Avars) at *Thessalonica
  (586, 615, 618) 289, 375, 376, 411
Persians at *Theodosiopolis
  (421/22) 296
see also *Constantinople (626),
  *Kamakhon (766)
siege towers 281ff
  Huns at *Naissus (442) 47, 265, 368
  Ostrogoths at *Rome (537f), *Rimini
    (538) 115, 162ff, 284, 287, 291, 357
  Slavs at *Thessalonica (662) 120, 377,
    417
  construction of 122, 142, 344, 377
  Romans at *Amida (503) 134, 277
  Persians at *Theodosiopolis, *Dara
    (573) 266, 277, 296
  Vegetius on 267
  Avars at *Thessalonica (618),
    *Constantinople (626) 285, 376
  on ships—*Rome (545f); Arabs at
    *Constantinople (654) 280, 283,
    443
  Abbasids at Amorion (838) 428
  countermeasures against 290, 420
  reintroduced to West 429
  see also *Edessa (544)
siegeworks, camps 278-81
  tactical use of 257, 261, 267, 269, 287f
  Ostrogoths at *Salona (537), *Rome
    (537f) 162, 261
  in Isidore of Seville 174
  Franks on multiple occasions 208,
    248f, 261
  Lombards at *Rome (756) 261
  Frankish engineering related to
    203f, 251
  Persians at *Dara (573), *Apamea
    (573) 271
  Arabs during early conquest 261
  Visigoths at *Sevilla (583f) 261
  Avars (Slavs) at *Drizipera (588) 381
  mounds 281f
  see also artillery
specialists see architects; artillerymen;
    craftsmen, engineers
Slavs 371-78
  and Romans 95, 112, 114
  as Avar subjects, clients 330, 380, 382
  as Lombard auxiliaries 382
  military organization 120, 251, 310, 344f

under Bulgars 383f
Wends (western Slavs) 228f, 387
storm, definition of 31, 256-59, 264ff,
  268-77, 281-88
see also blockade
Sueves
  and Lombard origins 177, 179
  as Ostrogothic clients 155, 177
  in Spain 44, 53, 64, 164
  see also Rhine: crossing
ṣulḥ^{an} 8, 355n106
  see also 'anwat^{an}
Syriac Common Source 26
  and Constans II 440ff, 446f

taxation, role of
  apotheke, kommerkiarioi in
    Byzantium 105f
  Arab-Islamic 8f, 110, 352, 395f, 404, 435,
    437
  Franks 215f, 220, 221f, 223, 225ff, 230f,
    234, 238
  Lombards 183f
  remitted by Ostrogoths 13, 92f, 157, 160f
  remitted by Anastasios 136
  services substitute for 62, 69f, 139f, 193,
    226, 254f, 425f
  Valentinian III cancels immunities 70
  see also hospitalitas debate; munera;
    logistics
technology, diffusion of 16-19
  Arabs and appropriation 389-405, 442f,
    445, 448f
  client integration/acculturation 147
  conquest appropriation 143
  institutional framework 32
  models of appropriation 361-69, 405
  northern barbarians 369-89
  practitioners 17
  social carriers 16f
  traction trebuchet 419-24
Theodosian Walls 48, 72
  and munera 72, 140
  ballistrai mounted on 446
  enclose agricultural space 308
  Hodegetria displayed on 322
  rebuilt in 447 after earthquake 48
Theophilos of Edessa see Syriac Common
    Source
thick description 10-12, 29, 295-98
  and Franks 195f
  and Huns 46ff

thin description 10-12, 357ff
  and Visigoths 39-45
traction trebuchet see artillery

Urbanism 307-316
  and Arabs 303f, 335f, 390
  and Persians 365
  in the East 7, 300f, 356
  in Gaul 3, 30f, 197, 253, 302f
  in Italy 149, 301, 356
  in Spain 171, 301f
  in the West 3, 227, 300

Valentinian III's legislation
  abolishes immunities, extends *munera* 70ff, 82
  on deserters in private service 51, 64f
  on retinue of Hilary of Arles 65f
  permits civilians to carry arms 66f, 72f
Vandals
  and East Roman reconquest 95, 141
  and Roman response 51, 52, 65, 70, 367
  and Ostrogoths 153
  in Spain 64
  means of settlement 53, 54n87
  see also Africa; Rhine crossing; *Carthage (533, 536)
Visigoths 39-45, 164-76, 361ff
  influence on Franks 234f
  military organization in Toulouse 45, 67, 73, 90
  military organization in Toledo 166-170, 313, 346
  origins see Goths
  settlement in Aquitaine 55, 81
  wars with Franks 88, 90ff, 152f, 196-200
  see also Euric; Wamba; *Sevilla (583f)

Wamba
  673 campaign (see also CO s.a.) 175f, 229f, 329, 330f, 343, 423
  legislation on retinues 168ff, 171, 346, 418
  use of fleet 311, 347
  Visigothic siege warfare 272, 346, 318, 418